THE GLIM SYS

Release 4 Manual

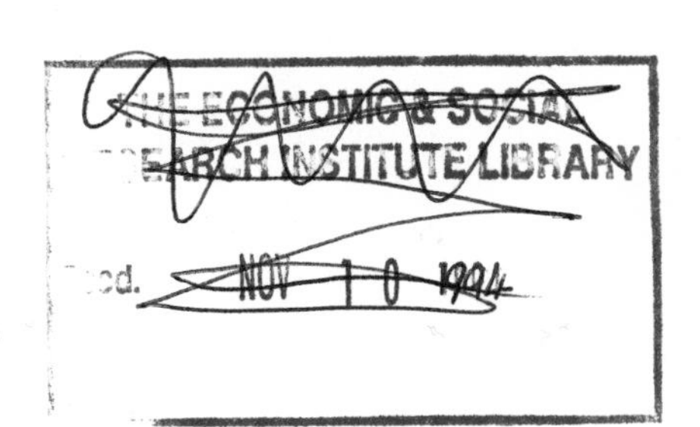

Edited by

BRIAN FRANCIS
University of Lancaster

MICK GREEN
University of Lancaster

CLIVE PAYNE
University of Oxford

Other Contributing Authors

TONY SWAN
Public Health Laboratory Service, London

ROBERT GILCHRIST
University of North London

MALCOLM BRADLEY
Halcrow-Fox Associates, London

MIKE CLARKE
Queen Mary and Westfield College, London

PETER GREEN
University of Bristol

ALLAN REESE
University of Hull

JOHN HINDE
University of Exeter

ANDREW STALEWSKI
Numerical Algorithms Group, Oxford

CARL O'BRIEN
University of London Computer Centre

CLARENDON PRESS • OXFORD

Oxford University Press, Walton Street, Oxford, OX2 6DP
Oxford New York Toronto
Delhi Bombay Calcutta Madras Karachi
Kuala Lumpur Singapore Hong Kong Tokyo
Nairobi Dar es Salaam Cape Town
Melbourne Auckland Madrid
and associated companies in
Berlin Ibadan

Oxford is a trade mark of Oxford University Press

Published in the United States
by Oxford University Press Inc., New York

First published 1993
Reprinted (with corrections) 1994

A catalogue record for this book is available from the British Library

Library of Congress Cataloging in Publishing Data
The GLIM system: generalized linear interactive modelling/editors-Brian Francis, Mick Green, Clive Payne; other contributing authors, Tony Swan . . . [et al.].
1. GLIM. 2. Linear models (Statistics)—Data processing.
I. Francis, B. (Brian) II. Green, Mick, Dr. III. Payne, Clive.
QA279.G64 1993 92–28937 519.5—dc20
ISBN 0 19 852231 2

Printed in Great Britain by
The Bath Press

Foreword

This manual describes how GLIM4 may be used for statistical analysis of data in its most general sense, including data manipulation and display, model fitting, and prediction. The statistical models which can be fitted in GLIM are contained in the Generalized Linear Modelling (GLM) scheme originally developed by Nelder and Wedderburn (1972). The manual has three parts.

1 The User Guide

This part introduces and illustrates all the facilities in GLIM4 in an informal way. Chapters have been arranged on a functional basis so that each describes the GLIM4 directives relevant to a particular type of activity involved in the statistical modelling of data. Chapter 1 gives an introduction to the basic ideas involved in fitting GLMs and provides an outline of the structure of the User Guide. It also contains a brief description of the differences between GLIM4 and the previous version, GLIM 3.77; full details of these are given in Appendices B and C. The heart of the GLIM4 system, the facilities for fitting and assessing standard models in the GLM scheme, are described in Chapter 8; the following chapter describes the fitting of more specialized models for advanced users. The intervening chapters describe the GLIM4 language and data structures and the facilities for defining, managing, and displaying data both as a preliminary and adjunct to the core activity of fitting and assessing models.

2 The Modelling Guide

Chapter 10 provides an introduction to Generalized Linear Models in the GLIM4 system. It contains descriptions of the standard models subsumed in the GLM scheme and covers extensions to related types of model. Chapter 11 covers the theoretical basis of the GLM scheme. Chapter 12 contains complete and self-contained examples of the application of GLIM4 to a wide range of models applied to a variety of types of data. Users new to the GLIM system are recommended to study relevant examples as an effective way of learning about the GLIM4 system. The datasets for all these examples can be obtained from NAG.

3 The Reference Guide

This contains a formal description of the syntax and semantics of the GLIM4 language, of the data structures it handles and of the directives provided. It thus constitutes a reference manual for the experienced user.

Other books describing Generalized Linear Models (and in some cases the use of earlier versions of GLIM) may be useful to the GLIM4 user. The Bibliography contains an annotated list of key sources on GLMs; it also lists other publications which illustrate the use of GLIM in particular types of application or which are aimed at particular substantive areas.

Manual conventions

The following conventions are used throughout the manual. References to other sections within a part are by section number, e.g [6.1.1] in the User Guide refers to Chapter 6, Section 1, subsection 1 on 'The LOOK Directive'. References to other parts contain the part name in addition.

Italics are used to denote instances of elements of the GLIM4 language. ***Bold italics*** are used for emphasis. **Bold** is used where words are being defined, or in the case of single letters in algebraic expressions, to denote vectors and matrices. `Mono` is used for GLIM statements and output.

The syntactical rules for the GLIM directives defined throughout the manual are set out in the Introduction to the Reference Guide.

Preface

GLIM4 is an extended version of the GLIM 3.77 system, a general-purpose program for fitting Generalized Linear Models together with associated facilities for managing, manipulating and displaying data. It is the result of a project sponsored by the GLIM Working Party of the Royal Statistical Society which has developed and supported all versions of the GLIM system. The contributors are all members of the GLIM Working Party. The new version, GLIM4, was developed at the Centre for Applied Statistics, University of Lancaster, UK under the direction of Brian Francis and Mick Green. It is distributed by the Numerical Algorithms Group.

Each copy of GLIM4 is distributed with the following documents:

(1) A Reference Card which gives abbreviated details of GLIM4 facilities.

(2) An Installers' Note which describes how to mount the software on the local computer system and gives technical details on how the program may be modified to extend the data space available and to support local graphical devices.

(3) A User Note which describes how the program is accessed on the local computer system, how data files are connected, the graphical devices supported, and so on. This is essential reading for any user of GLIM4.

The Numerical Algorithms Group (NAG) publishes a GLIM Newsletter, which includes contributions from users describing their applications of GLIM, and macros they find useful. All users are invited to communicate with each other through this Newsletter, and to submit articles for publication. Contributions to the Newsletter should be sent to the Editor at NAG. Many of the macros published in the Newsletter have been included in the GLIM4 macro library which is provided with the software. It is expected that this library will be revised and extended from time to time; new contributions or requests for updates should be sent to NAG.

For further information on the GLIM4 system contact:

NAG Ltd	NAG Inc.	NAG GmbH
Wilkinson House	1400 Opus Place, Suite 200	Schleissheimerstrasse 5
Jordan Hill Road	Downers Grove	W-8046 Garching bei München
Oxford	Illinois	Germany
OX2 8DR	IL 60515-5702	
United Kingdom	USA	
Tel: +44 865 511245	+1 708 971 2337	+49 89 3207395
Fax: +44 865 310139	+1 708 971 2706	+49 89 3207396
Telex: 83354 NAG UK G		

The authors wish to dedicate this manual to Mike Clarke, who died in early 1994, and whose contribution to the manual and to the numerical algorithms in GLIM was immense.

Acknowledgements

The GLIM4 software system is an extended and updated version of earlier versions of GLIM. This manual is a revised version of the GLIM 3.77 Manual, incorporating the revisions and new facilities available in GLIM4. Thus the new release draws heavily on the work done by earlier contributors to the development and documentation of the GLIM system. Particular mention should be made of Bob Baker, John Nelder, and Mike Clarke who were mainly responsible for the development of GLIM3 at the Rothamsted Experimental Station in the mid-1970s; contributing programmers also included John Beasley, V. Coulson, David Hill, Charlie Rodgers, S. Springer, and the late Robert Wedderburn. The next version, GLIM 3.77, was developed in the mid-1980s by John Nelder, Bob Baker, and Rodger White at the Rothamsted Experimental Station and Mick Green then at the Polytechnic of North London. The major contributors to the software for the new version are Brian Francis, Mick Green, and Malcolm Bradley from the University of Lancaster, and Mike Clarke of Queen Mary and Westfield College. Valuable contributions were also made by Peter Green, Allan Reese, Tony Scallan, and Mel Slater. There are many others who have converted and tested the various versions of GLIM on particular computer systems; all these also deserve our thanks.

A special debt is acknowledged to both John Nelder and Robert Baker for their leadership in the development of the GLIM system. The current manual draws heavily on the major contributions they made to previous versions of GLIM documentation. Our grateful thanks should also go to Doreen Hough, Moira Peelo, and Despo Panou for their help in preparing the manuscript.

Contents

I
The User Guide

1 Introduction to GLIM

Introduction

GLIM is an acronym for **Generalized Linear Interactive Modelling**. It is a program specially designed to facilitate the fitting of **generalized linear models** (**GLM**s). Here we give a brief introduction to the data types handled by GLIM and to the types of statistical model which can be fitted. An outline of the structure of the User Guide is provided together with a brief description of the new facilities in GLIM4.

1.1 Overview of GLIM facilities

GLIM is essentially designed for univariate statistical modelling, using the generalized linear model framework. We shall observe that the standard statistical models prove to be special cases of GLMs; these include the classical regression models, Analysis of Variance (ANOVA) and Analysis of Covariance (ANCOVA) models, a range of log-linear models for count data, and logit and probit models for the analysis of proportions (or quantal response data). Moreover, an even more extensive set of models can be fitted in GLIM by noting that they can be redefined as GLMs; a summary of some of these models is given in the Modelling Guide [10.6].

At the heart of GLIM are its **model specification** and **model fitting** facilities; these are complemented by a comprehensive tool kit for data input, manipulation, description, display, and presentation. The GLIM system is capable of a range of data descriptive tasks. For example, it can sort and display data, produce histograms, scatter plots and graphs; it can calculate and display descriptive measures such as the mean, variance, and range. Facilities are also provided to display and analyse tabular summaries of data. Thus multi-way tables of descriptive statistics such as counts, means, variances, and rates can be produced from a raw data matrix. Extensive facilities are provided to perform calculations on these data structures with a wide range of statistical and logical operators and functions.

The user gives instructions by using a simple but powerful interpretive language. This language allows blocks of instructions to be named as GLIM macros; the addition of branching and looping instructions which act on macros produces a programming language of considerable generality, well suited for describing (and communicating) more complex analytical techniques. The macro facility enables users to tailor GLIM to handle a very wide range of both standard and specialized applications; many of these specialist macros are made available to users through the macro library provided with the package, details of which are given in Appendix A.

The user has access to most of the data structures derived in model fitting, data summary, and calculations; access is also provided to system settings. This access, together with the macro facilities, is the key to the extensibility of the GLIM system.

GLIM is generally most useful when used in an exploratory manner, that is, as a tool for experimenting with the fitting of various models to a wide variety of types of data. But it can also be used effectively for simple summaries of data, including tabulation and the display of both data values and accompanying text. Text manipulation facilities are also provided.

1.2 GLIM data

GLIM is concerned with models in which the user seeks to explain the variation in a **response** (or *y*) variable in terms of certain **explanatory** variables.

1.2.1 Data types

Variables are classified into different **data types**. These include:

(i) **Continuous**: variables measured on an interval or ratio scale such as the income of a respondent in a survey, the yield of a crop in an agricultural experiment, or the time to recovery in a medical study.

(ii) **Count**: variables which represent a frequency of occurrence. Examples are the cell frequencies in a multi-way contingency table obtained from a social survey, the word frequency in a piece of text, or the number of bacterial colonies on an agar plate.

(iii) **Proportion**: these are handled in GLIM as a ratio of two counts with the numerator giving the number of 'successes' (r) out of a number of 'trials' (m), the denominator. Such data are alternatively known as 'quantal response' or 'bounded count'. Examples are the number of respondents voting for a particular political party out of the total number of respondents in a cell of a multi-way contingency table, and the number responding out of a total number of organisms exposed to a treatment in an assay. This type also includes variables where the numerator r is either 0 or 1 and the denominator m is 1, for example a respondent in a survey is either a member of a trade union (1 say) or not (0). Such data is often known as 'binary data'. Thus, with both types, values of the numerator r can be in the range $0, \ldots, m$.

(iv) **Categorical**: variables which are measured by assigning observations to one of a number of categories. These categories may be either unordered (for example, the variety names in a plant variety trial or the region of residence of a respondent in a social survey) or ordered (for example, the degree of agreement of a respondent to an attitude question in a social survey, or their age group). Such variables are called **factors** in GLIM. They can take only a finite set of values, called levels, and these values must be coded as 1, 2, 3, ...

1.2.2 The data matrix

GLIM expects to find the data for analysis in the form of a data matrix. This is a two-dimensional structure of a set of variables available for a number of observations. The observations refer to the entities for which the variables have been measured, for example, respondents in a social survey, plots in an agricultural experiment, patients in a medical study, agar plates in an assay, and constituencies in a set of election results. The variables may be any of the types described above. GLIM does not itself distinguish between types (i), (ii), and (iii), regarding them all as **variates**; however factors must be declared explicitly. Conventionally the observations define the rows of the matrix and the variables the columns. A data matrix is defined when we know the number of rows (observations) and the names and types (variate or factor) of the variables in each column.

The original data from an experiment or survey will often have undergone tabulation before an analysis by model fitting begins. A multi-way table is indexed by a set of factors with (in general) one cell for each combination of factor levels. Tables can be used as data for GLIM by turning them into data-matrix form. Each row of this data matrix then corresponds to a cell in the table and the columns contain the values of the factors defining the table, together with the continuous variate(s) containing the contents of each cell (for example, a count in a multi-way table, the yield of a crop in an agricultural experiment, or the numerator and denominator of a proportion in a multi-way table). The Modelling Guide [10] gives a detailed example of the conversion of a multi-way table into data matrix form. GLIM also has facilities for forming a table of counts from a raw data matrix input by the user; the resulting tabular structure can then be displayed and analysed directly.

1.3 Generalized linear models

GLIM is primarily a tool for fitting models subsumed in the Generalized Linear Modelling scheme originally developed by Nelder and Wedderburn (1972) and extended in McCullagh and Nelder (1989).

1.3.1 The GLM scheme

The modelling process may be thought of as one in which the observed values of the response or ***y*-variate**

$y_1\ y_2 \ldots y_n$ say

are matched by a set of **theoretical** values

$\mu_1\ \mu_2 \ldots \mu_n$ say.

For a good model, the μ's must have the following properties:

(i) they are all derived from a small number of basic quantities called **parameters**, and

(ii) the resulting set of μ's is close to the original y-values. Thus the μ's are highly patterned and, therefore, we hope, easier to understand and think about than the y's which will be 'rough' by comparison.

The model fitting process involves two basic decisions:

(i) the choice of the relation between the μ's and the underlying parameters of the model, and

(ii) the choice of a measure of discrepancy which defines how close a given set of μ's is to the y-values.

The first choice refers to the **systematic component** of the model, and the second is governed by assumptions we make about the **random component**. The latter is a statistical description of that part of the variation in the y-values which the systematic component does not account for.

In GLMs the systematic component is assumed to take the following two-stage form: for each row in the data matrix

(i) the **linear model** or **linear structure**, $\eta = \Sigma\beta_j x_j$, is a linear combination of explanatory variables x_j with parameters β_j, and

(ii) η is related to μ by the **link function** $\eta = g(\mu)$.

The random component in GLMs allows the y-variate to have a distribution with mean μ from the exponential family. This includes the Normal distribution (suitable for many continuous y-variates), the Poisson distribution (suitable for models of counts), the binomial distribution (suitable for models of proportions), plus the gamma, negative binomial, and inverse Gaussian distribution.

We have thus specified a GLM when we have defined:

(i) the terms to be included in the linear model,

(ii) the link function connecting the linear model to the theoretical values, and

(iii) the probability distribution of the random component.

GLIM then fits the model by choosing as estimates of the parameters β_j those values which minimize the discrepancy between the y-values and μ-values (called the deviance in GLIM) using maximum likelihood estimation. The actual form of the deviance depends on the mathematical definition of the probability distribution assumed.

The models are linear in the unknown parameters. However, the use of a link function between the linear model and the means of the response variable allows for a non-linear relation between the mean of the responses and the explanatory variables, whilst preserving the underlying linearity in the combination of the explanatory variables. The concept of differing random structure (other than the traditional Gaussian/Normal assumption) leads naturally to the fully generalized linear modelling concept.

1.3.2 Examples of GLMs

We can describe particular examples of GLMs by defining the types of the response and explanatory variables (the data side), and the link function and distribution (the model side)

Examples of some standard models thus contained in the GLM scheme are:

Model type	Response variable	Explanatory variables	Link function function	Distribution
Regression	Continuous	Continuous	Identity	Normal
ANOVA	Continuous	Factors	Identity	Normal
ANCOVA	Continuous	Mixed	Identity	Normal
Log-linear	Count	Factors	Log	Poisson
Logistic	Proportion	Mixed	Logit	Binomial

where 'mixed' refers to any combination of continuous or factor explanatory variables.

Many other combinations of these components can be constructed and GLIM provides a general tool for fitting the corresponding models. In particular it allows the explanatory variables to be arbitrary combinations of continuous and categorical variables in any context. In addition the user can specify their own probability distributions and link functions, and also intervene in the iterative process to solve the maximum likelihood equations. The Modelling Guide [10] gives a comprehensive treatment of the GLM scheme and the range of models subsumed within it; in particular the Examples chapter [12] gives complete and self-contained applications of GLIM to a wide variety of types of data and model.

1.4 Outline of the User Guide

The User Guide contains an informal description of all the GLIM facilities with each chapter arranged on a functional basis.

Chapter 2 describes the data structures which can be processed and the syntactical rules for the **directives** (the statements used to carry out operations) in the GLIM language.

Chapter 3 covers the definition of data structures and the facilities provided for inputting data values into these structures.

Chapter 4 deals with various utilities available in the system. Topics covered here include: the use of external files for input of data and GLIM statements for output of results, facilities for obtaining information about data structures and system settings, control over the GLIM environment, and on-line documentation.

Chapter 5 describes the facilities provided for generating, modifying, and manipulating data structures and their values.

Chapter 6 covers the directives provided for the display of values in data structures. These include listing of data values, histograms, plots, and high-resolution graphs. It also covers directives provided to display and store summary statistics in tabular form for display and/or analysis.

Chapter 7 describes the facilities for the specification and running of user procedures, called macros. These include looping, branching, argument passing, and user prompting.

Chapter 8 covers the specification, fitting, and assessment of models in the GLM scheme.

Chapter 9 deals with the specification and fitting of user-defined models. These allow the user to fit non-standard GLM models and modify the algorithmic process.

1.5 The new facilities of GLIM4

GLIM4 includes a range of new facilities not available in GLIM 3.77 or earlier versions. It incorporates a range of extensions to the facilities for model fitting and user control. We now briefly overview the new features; users new to GLIM should omit this section. A complete list of the new facilities is given in Appendix B while Appendix C sets out the incompatibilities between GLIM4 and earlier versions of GLIM.

The range of models covered has been extended to cover the inverse Gaussian distribution and the generalized power and inverse quadratic links. The model definition syntax has been extended to include orthogonal polynomials, user-defined model matrices, and model terms involving cross products of variates (previous versions of GLIM allowed these for factors, that is, categorical variables, but not for continuous variables). In applications where the values of certain explanatory variables are constant within subsets of observations a new facility for **indexing** in model formulae allows such data to be analysed efficiently without the need to replicate the values of these variables. A common example is when data points arise as repeated observations on each of a number of individuals, where some of the individuals' characteristics will be constant over the data points from each individual.

The new `ELIMINATE` directive offers considerable simplification of the analysis and computational savings when it is necessary to fit models which contain parameters whose estimates are not of direct interest. This is applicable in many useful GLIM applications such as the analysis of matched case-control studies, multinomial response models, and Cox proportional hazard models. Another example is models which require that fitted values be equal to observed values for some marginal subtable of the data, for example log-linear models where one variable is treated as a response variable. GLIM4 allows the user to specify a model term to be eliminated from subsequent fits (the term is implicitly still part of the model formula but its parameters are not explicitly estimated or displayed) with consequential savings in both storage and in computational time. These savings can be considerable over the equivalent procedures in which the term is explicitly included in the model.

GLIM4 provides users with the tools to access and manipulate system structures; for example, access is now provided to residuals, leverage values, deviance increments, and the weighted SSP matrix. The model matrix can be extracted and facilities for accessing subsets of the parameter estimates are provided.

Further useful facilities are provided by the new `PREDICT` directive. This allows calculation of the linear predictor, the variance of the linear predictor and the associated fitted values for specified values of the explanatory variables.

The range of functions in the language is considerably extended, to include the chi-squared, t, F, beta, binomial, and Poisson probabilities (given a deviate), and deviates (given a probability). The incomplete gamma function is provided, as are the log-gamma, digamma, and trigamma functions. The absolute value function (with the associated sign function) and cosine function are now available.

The algorithm at the heart of GLIM is a weighted least squares procedure. Earlier versions of the package used the Gauss–Jordan method to solve the weighted least squares equations. Gauss–Jordan is still available if required but the system now also provides the more stable and accurate Givens method. As an indication of the benefits of the Givens method, it has been found that in certain circumstances, Givens in single precision is as accurate as Gauss–Jordan in double precision.

GLIM's facility for user-defined models has been updated to allow users to use the standard GLIM specification together with their own personally defined specification (rather than requiring the specification of four user macros in GLIM 3.77). As a result the `OWN` facility is now redundant as GLIM4 uses the same fitting procedures for all fits, whether it be one of GLIM's default options or a customized method devised by the user. Hence the OWN directive has been removed. Thus, in GLIM4 the user can, for example, request a user-defined model with GLIM4's standard specification of the 'error' distribution while replacing the standard link function with the user's own specification. The fitting facility has also been extended to allow macros to be called: (i) before the first iteration, (ii) before the formation of the model triangle (the SSP matrix in the Gauss–Jordan case), (iii) after the formation of the model triangle but before inversion, or (iv) after inversion of the model triangle in each iteration. Moreover, any structure available to the user can be modified during the iterative fit. This, taken together with the greater selection of systems structures, provides a high degree of flexibility and allows the user to modify the iteratively re-weighted least squares algorithm to carry special actions for almost any type of statistical model, including a number of useful non-linear models.

GLIM4 has been designed to make macros more readable than they were in previous versions of GLIM; macros are now easier to develop, debug, and change. Keyword arguments can now be used in macros, and local structures can be defined which have context only within the body of the macro. A macro editor is now provided to edit existing macros or to create new ones. A simple macro debugger is now provided.

Plotting and histogram facilities have been enhanced; multiple plots with weights and user-defined scales are allowed, together with titles and labels for the axes. GLIM4 also provides facilities for generating high-resolution graphics with the new `GRAPH` and `GTEXT` directives.

A new structure, the list, which is a set of identifiers, is now provided. Lists allow the compact specification of sequences of identifiers in data input, calculations, and model fitting. They also provide a mechanism for passing an unrestricted number of arguments to macros.

GLIM4 also provides a summary of this manual through the new `MANUAL` directive which gives on-line information about the directives (including their syntax and examples of use), system structures, the local implementation, the macro library, and a glossary of GLIM terms.

2 The GLIM language

Introduction

This chapter outlines the basic components of the GLIM language and introduces the general syntax and mode of use of the package.

2.1 The character set

GLIM statements are composed of characters from the GLIM character set, which is listed in the Reference Guide [13.1]. Essentially, this consists of upper- and lower-case letters, digits, and a range of operators and brackets which will be introduced in the relevant chapters of the User Guide. Lower-case letters are treated as equivalent to the corresponding upper-case letters in statements when they are interpreted by the GLIM program, except in strings when the case is retained. The character '@' is used in fault messages to indicate an invalid input character.

There are also a number of special characters, whose representation may not be the same throughout the whole range of computers on which the package is implemented. The standard forms of these characters, used throughout this manual, are:

directive symbol	$	underline symbol	_
repetition symbol	:	item separator	,
system symbol	%	end-of-line symbol	!
substitution symbol	#	string symbol	'

The user may discover the local representation of these special characters from GLIM by using the statement

```
$environment I $
```

2.2 Numeric values

GLIM can handle two types of numeric values, that is, integers and real numbers. Some examples of permissible forms for numeric values are

2 -30 4.71 5. -.86 2000.5 0.000647

but they must not contain commas. To avoid many zeros in large or small numbers use can be made of the exponent notation. For example, one thousand can be written as

1000 or 1E3

The E means that the number on the left is multiplied by 10 to the power of the number on the right. The E can be in upper- or lower-case and can also be replaced by D or d, as in the following examples

–2.6e–10   2.6D–10   10d5

GLIM outputs very large or very small numbers in the exponent form.

2.3 Strings

Strings are used in a number of GLIM output directives and are specified by enclosing text by the string symbol `'` . The string can contain any GLIM character except the end-of-line symbol `!`. A single quote can be included by typing two adjacent string symbols. For example, the string

Bob's Income ($)

would be input as

```
'Bob''s Income ($)'
```

2.4 User structures

The general storage of data and text in GLIM is through user structures, which can be set up, modified, and deleted as required by the user.

2.4.1 User identifiers

User data structures are referred to by **user-defined identifiers**, which consist of a letter followed optionally by further letters and/or digits. An identifier may also contain an underline symbol, but not as the first character. Some examples are

```
Y    X2    YIELD    Mean_wt    A1xB2    y
```

but note that the identifiers `y` and `Y` are treated as equivalent.

Only the first eight characters of identifiers are stored, although longer identifiers are allowed. However, since for example

`GROUP_MEAN` and `GROUP_MEDIAN`

are treated as identical, it is strongly recommended that users confine identifiers to eight characters or less.

The storage of strings with user-defined identifiers is via macros, see [2.4.5] and [7] for further details.

2.4.2 User scalars

A **scalar** is a single number and **user scalars** may be defined by the NUMBER directive [3.1.1]. Scalars can be useful for storing such information as the number of items in a dataset, a conversion factor for a set of data, physical constants, or the measure of goodness of fit of a model. By default they are initially assigned zero values, but users may access them and assign values at any time in the GLIM program. Scalars may also replace real numbers and integer numbers where they are required in GLIM statements. This facility is very useful in writing general programs and macros.

2.4.3 User vectors

A **vector** is simply a set of numbers of some specified length. They are typically used for storing the data values under consideration, but can also be used to hold other quantities of interest such as the parameter estimates following a model fit, the values of an index for the cells of a table, and so on. In general the vector is referred to as a single composite data structure by its identifier; however, it is possible to access individual elements using the CALCULATE or PICK directives [5.4 and 5.5], see also [2.8] on indexing.

There are two kinds of vector: a variate, whose values are unrestricted, and a factor, whose values are restricted to the integers 1, 2, ... , k, where k is the number of levels of the factor. Factors are used for storing categorical data and indicating groupings of the observations. The distinction between variates and factors is particularly important in the specification and fitting of models [8].

2.4.4 Arrays

An **array** is a vector of numeric values with an additional dimensional structure. As with an ordinary vector it can be either a factor or a variate. Arrays are only really relevant in tabulation [6.10] and model fitting [8]; in other contexts the dimensionality structure is ignored and they are treated as ordinary vectors.

2.4.5 Macros

A macro is a piece of text made up of GLIM characters that is given a user-defined name. The contents of the macro may be viewed simply as a stored string which can be used in various contexts. Macros can be changed by using the macro editor, which is invoked by the EDMAC directive. The contents may be inserted into any GLIM statement by prefixing the macro name by the substitution symbol #. This facility can be very useful for building up complex options lists in producing high-resolution graphs, or storing large model formulae.

The text of macros may also consist of a set of GLIM statements to give a stored procedure which may be executed at a later stage. Macros may have up to nine formal arguments and their own local identifiers and so provide a powerful programming structure.

A full description of the macro facilities is given in [7].

2.4.6 Lists

A list is a data structure which enables the user to store a list of identifiers. This gives a very convenient shorthand method for referring to sets of identifiers. A full description of the formation and use of lists is given in [3.3].

2.5 System structures

System data structures are used by GLIM for storing certain information about the current GLIM run, such as system settings and useful information from the current model fit.

2.5.1 System identifiers

The identifiers for system structures are all prefixed by the special system symbol `%` to distinguish them from user-defined structures. The names are all predefined by the system, but unlike user identifiers, only the first four characters are significant, the first character being the `%` symbol.

2.5.2 Types of system structures

The system structures consist of scalars, vectors, and arrays; full lists and descriptions are given in the Reference Guide [14.2.2] and [14.3.2] and particular structures are described in more detail in the relevant chapters. In general the values stored in the system scalars are simply copies of the current system settings and changing the value of the scalar will not alter the setting. The system vectors and arrays are the key to the extensibility of the GLIM system and allow the user to access all of the information produced in the model fitting process; for example the vector `%fv` contains the fitted values for the model and the scalars `%dv` and `%df` contain the deviance and degrees of freedom respectively.

An additional set of 26 ordinary scalars `%a, %b, ..., %z` are available to the user (to retain compatibility with earlier versions of GLIM) and can be used as temporary storage of single numbers.

2.6 Directives

2.6.1 Syntax of directives

A GLIM program consists of a sequence of statements. In its simplest form a statement consists of a **directive name** followed by a set of **items**, for example

```
$data  4  X  Y  Z  $
```

A **directive name** is a word ('`data`' in the above example) prefixed by the special directive symbol `$`. No spaces may occur between the directive symbol and the accompanying word. The terminating `$` is optional, as the directive symbol at the beginning of the next directive also terminates the previous directive. However, it will be included throughout this manual as it ensures that the directive is executed immediately. Directives are predefined by the system and a full list is given in the Reference Guide [15]. Only the first four characters of a directive name are significant, the first character being `$`, so that

`$dat` and `$data`

are treated identically, and the first is an accepted abbreviation for the second. Some directive names may be further abbreviated for convenience, and the minimum number of characters required is indicated in the Reference Guide [15]. However, to avoid confusion throughout this manual, we will never use less than four characters for a directive name. When a directive name is the same as that in the previous statement it may be replaced by the repetition symbol : .

The *items* are associated with the directive type and their number and type vary with the directive. There may be no items (as in the `NEWJOB` directive), or a fixed number of items (as in `SLENGTH`) or a variable number of items (as in `DATA`). The simplest form of item is an identifier. Numeric values can also occur as items and these may either be entered directly or referred to by a scalar identifier. The `DATA` directive, for example, has one optional integer item at the beginning followed by at least one identifier. Throughout this manual square brackets are used in the syntax definition of directives to indicate optional items (see the Introduction to the Reference Guide for the formal rules of the GLIM syntax), so the `DATA` directive has the following syntax

`$data` [*integer*] *identifier* [*identifier*]*s*

For particular directives items may also consist of letters (as in `ENVIRONMENT`) or strings (as in `PAUSE`).

The directive can be thought of as the instruction and the items are the objects involved in the execution of that instruction. A number of directives can have their effect controlled through **option lists**, which occur as the first item after the directive name. They have the form

([*option*]*s*) where *option* is *name* = *setting*

For example

```
$graph (xlimit=0,40 title='Figure 1')  Y  X  $
```

produces a high-resolution graph of `Y` against `X`, with the x-range from 0 to 40 and entitled '`Figure 1`'. The *name* begins with a letter followed optionally by letters and/or digits, the first letter only of which is significant, so the above example can be abbreviated to

```
$graph (x=0,40 t='Figure 1')  Y  X  $
```

The type of the **setting** depends on the option and the options available depend on the directive. Any selection of available options may appear in any order; if an option is repeated its last instance will be used.

A few directives may also have **phrases** as items. These are descriptive items which either qualify identifiers or act as settings for the directive. For example in

```
$tabulate for A $
```

the phrase '`for`' specifies identifier `A` to produce a frequency distribution for the vector `A`, while in

```
$page on $
```

the phrase '`on`' switches on paging of the GLIM output. The form of the phrases available depends on the directive, but they are generally made up of names and as in option lists only the first letter of these is significant.

Operators are also a form of item although again their use is specific to particular directives. They occur most commonly in calculations with vectors and scalars using `CALCULATE` [5] and model fitting with `FIT` [8], but are also used with several other directives.

Finally, in some directives **separators** are used between items in lists. The basic rules are that the character '`,`' is used between items in lists and a space is used between separate lists or items not in lists. For example

```
$plot Y1,Y2 X1,X2 $
```

will produced two superimposed scatter plots, one of `Y1` against `X1` and the other of `Y2` against `X2`. For compatibility with earlier versions of GLIM some directives will allow other separators to be used (for example, `;` can be used to separate items in the `TABULATE` directive [6.10]).

2.6.2 Layout

The general layout rules for GLIM statements are very simple. Directive names, numeric values, identifiers, or the component parts of option lists and phrases may not contain spaces or straddle two lines; that is, a space or a new line acts as a separator between two items. Spaces and new lines may be inserted arbitrarily between statements or between items within statements. So a single statement may occupy more than one line, but also more than one statement may appear on a single line.

A set of statements ended by the statement `$newjob $` makes up a **job**. A **session** consists of several jobs and is itself terminated by `$stop`.

2.7 Functions

A **function** is a system-defined identifier consisting of the system symbol `%` followed by two or more letters. Functions are used to perform certain mathematical and logical operations in calculations with vectors and scalars [5]. The complete set of functions is listed in the Reference Guide [14.6].

2.8 Indexing

A powerful facility available for use in both calculations and model fitting is the ability to access indexed vectors. This provides a very simple mechanism for subsetting and replication of data values. The standard syntax for this is

vector_identifier (*identifier*)

where the identifier in brackets is the index and is a vector or scalar identifier with value(s) between 0 and the length of the vector to be indexed. This will provide a vector with the same length as the index identifier and values picked from the first vector according to the value of the index, with zero values for any zero values of the index. For example, if the identifier `I` is of length 4 with values 1, 2, 1, 3 and `J` is of length 3 with values 2, 0, 4, then the indexed vector `J(I)` would be of length 4 with values 2, 0, 2, 4 and similarly the indexed vector `I(J)` would be of length 3 with values 2, 0, 3. For further details see [5.4.8] and [8.2.4.5].

2.9 Files and subfiles

2.9.1 The file specifier

In a GLIM session it is possible to access both data and stored GLIM commands from a file; in addition text and graphical output may be directed to a file. Several directives are available for controlling this file handling. Files are made available to the GLIM program through a **file_specifier** which has the following syntax

[*channel*] [=] ['*filename*']

where *channel* is an integer for the channel number to be associated with the specified file. In the first occurrence of a particular file_specifier at least the channel number or the filename must be included, and if both are present the = is mandatory, for example

```
$input 20='example1.dat' $
```

If just a channel number is specified the system will prompt the user for the filename, while if only a filename is given as in

```
$input 'example1.dat' $
```

the file will automatically be allocated to the next free channel number. Details of the allocation of files to channel numbers can be found at any time by using the statement

```
$environment c $
```

and all files can be referred to either by their name or by their associated channel number. However, note that the full form of the file_specifier associating the file with the channel number should only be used once.

If the filename given exceeds the maximum length allowed in the particular implementation of GLIM, the filename will be truncated and a warning will be issued. Full details of file handling are given in [4].

2.9.2 Subfiles

When several GLIM programs, perhaps with accompanying data, are stored in a single file, it is convenient to have some way of naming different sections so that each can be retrieved from that file. These sections are called **subfiles** [4.2.5].

2.10 Comments

The full directive is `$comment` but this is usually abbreviated to `$c` as in

```
$c  This is a comment  $
```

Characters occurring between the directive name `$c` and the following directive symbol `$` are treated as comment and ignored by the program. Characters following an end_of_line symbol `!` and on the same line are also ignored, and so may be used to make an in-line comment.

2.11 Settings and internal switches

GLIM has a number of directives which can be used to control the system and switch certain facilities on or off. The standard syntax of the switches is

directive [*phrase*]

where phrase is `on` or `off`. If the phrase is omitted the status of the setting is reversed.

2.12 Mode of use

GLIM has been written to be usable in either of two modes, **batch** or **interactive**. Which modes are available to the user will depend on the computer system being used and the selection of an appropriate mode may depend upon the task being undertaken.

2.12.1 Interactive mode

The nature of statistical modelling is ideally suited to interactive working, and the GLIM system has been written with this in mind. In interactive mode the user simply types statements at a keyboard and obtains the output after each statement is executed and so can quickly adapt the analysis to take account of previous results. If the user makes a mistake in a statement a fault will be reported together with a comment to help correct it; the user may then retype the statement and execution will continue.

GLIM prompts for directives with the symbol `?`; however, if a statement is being input over more than one line then after the user types the 'return' key, GLIM will prompt with the abbreviated directive name preceding the `?`. If the 'return' key is typed in the middle of the input of a string, GLIM will prompt with `STRING?`. Since strings can include the directive symbol `$`, a new directive will not terminate the string but merely be included in the string, so care should be taken to terminate strings correctly.

2.12.2 Batch mode

In batch mode, a job is submitted to the computer, run without further intervention and the results either printed out or stored on file. If the user makes a mistake in the statements, execution of the program will be halted at that point with a warning but syntax checking will continue. Further runs will be required until all errors have been removed. The interactive mode has clearly a great advantage for error correction, although batch mode can be useful for performing an identical standard analysis on a number of datasets and for running analyses which take a large amount of time.

2.12.3 Session files

GLIM automatically stores a **transcript** of the input and output of the session on a file. This is particularly useful when working in interactive mode, where no record is otherwise available. What is to be written to the transcript can be controlled by the user at any stage through the `TRANSCRIPT` directive [4.3.1]. In interactive mode a **journal** of all of the statements issued by the user can also be kept, see [4.3.2]. This can be edited and replayed in batch or interactive mode.

On many systems the user can also set up an **initialization** file which can contain any valid GLIM directives and will be processed at the beginning of the session. This allows users to tailor the GLIM environment to their own requirements, see [4.2.1].

2.13 Machine dependencies

The GLIM program may differ between computer systems in respects other than simply the representation of the special characters. These include the amount of space available for data, the allowable range of values and accuracy for numbers, the channel numbers for input and output [4.1], the high-resolution graphics facilities [6.6], and the job control language required to gain access to the program. Information on space, graphics, channel numbers, and record lengths is available to the user through the `ENVIRONMENT` directive [4.5]; otherwise the user should consult the Users Note (see Preface) for the computer system being used.

2.14 A simple example

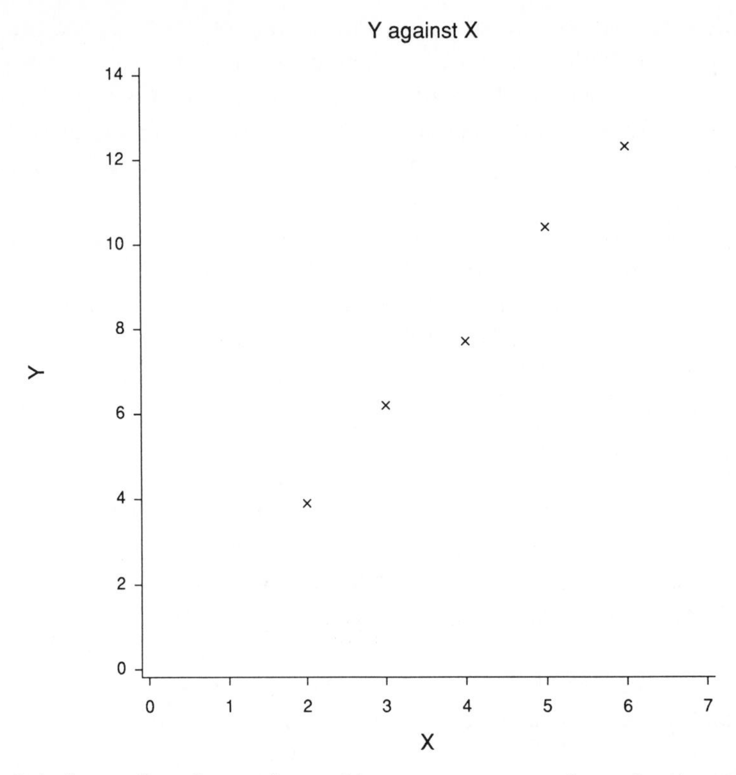

Figure 2.1 shows five data points with a strong suggestion of a linear trend. Suppose we want to plot this data, estimate the intercept and slope (model parameters) of a straight line fitted by least squares, and for each x-value calculate a fitted y-value from the parameters of the line. The following GLIM program will provide this information.

```
$c first define the data as consisting of 5
   observations and two variates X and Y
$slength 5 $data X Y $
$c read in the data point by point
$read
2     3.9
3     6.2
4     7.7
5     10.4
6     12.3
$c  declare Y to be the response variable
$yvariate Y $
$graph(title='Y against X' hlabel='X' vlabel='Y'
  x=0,7 y=0,14) Y X $
```

```
$fit X $      ! simple linear regression of Y on X
$c  print estimates of the parameters then the fitted
    values and residuals
$display e r $
$stop
```

An extensive set of more complex examples is given in the Modelling Guide [12].

3 Data structures: definition and input

Introduction

In Chapter 2, we saw that numeric data can be stored in different types of structure: scalars, vectors (consisting of variates and factors), and arrays. We now consider how these structures are defined and how values are given to these structures. We also introduce lists, which are structures in GLIM that are used to store collections of identifiers, and show how lists can be used and manipulated.

3.1 Declaring data structures

3.1.1 Declaring scalars

The `NUMBER` directive declares scalars, and may optionally assign values to the scalars. The syntax is

`$number` [*identifier* [= *value*]] *s*

For example

```
$number E=2.7183 $
```

declares a scalar `E` to take the value 2.7183, whereas

```
$number NITER  RT2PI = 2.507 $
```

declares two scalars, `NITER` and `RT2PI`, the first taking the default value of 0 and the second taking the value specified.

3.1.2 The standard length of vectors

The **standard length** for vectors is explicitly defined by the `SLENGTH` directive. The syntax is

`$slength` *integer* `$`

For example

```
$slength 60$
```

will define the standard length of vectors to be 60 elements; all vectors which are not explicitly defined as having a length by a directive will be set to be of standard length. So the directive

```
$variate  X Y $
```

following the above `SLENGTH` declaration, will define two variates with identifiers `X` and `Y` both of length 60.

The standard length is maintained in the system scalar `%sl`. If the standard length is not set, but the user has defined a *y*-variate as the response variate using the `YVARIATE` directive [8.1.1], then the default length for new vectors is taken as the number of units in the fit. In this case, the scalar `%sl` has the value 0, until the standard length is explicitly defined.

3.1.3 Variates

Variates can be declared with an explicit length by the `VARIATE` directive, which has the syntax

`$variate` [*length*] *identifiers*

For example

```
$variate 20  X  Y  Z
```

declares three variates `X, Y, Z`, of length 20.

The length may be omitted if a standard length has been previously defined with the `SLENGTH` directive. For example

```
$slength 30 $var X1 X2$
```

will define two vectors of length 30 called `X1` and `X2`.

In many directives, if a vector is undefined and otherwise unknown to GLIM, then it will be declared as a variate of standard length by default. So

```
$calculate A=4$
```

will define `A` to be a variate of standard length, with all elements of the variate taking the value 4. The individual descriptions of directives given in this manual will explain when this default action is taken.

3.1.4 Factors

In GLIM a factor is defined as a qualitative vector taking values from a finite set of alternatives called **levels**. The levels might be either qualitative, for example, different varieties, or quantitative, for example, age groups, and are taken from the set of positive integer values from 1 upwards.

Factors may arise naturally in the form of a variable with categorical values, for example, values which define the dimensions of a tabular structure. These may then be used, when declared as factors, as explanatory variables in model fitting.

Additionally, it is often required to group the data into several mutually exclusive subsets such that different models, or different plotting line styles or symbols may be specified for each subset. This grouping can be achieved in GLIM by creating factors.

The `FACTOR` directive is used to declare factors. The syntax is

`$factor` [*length*] [*identifier levels* [(*reference_level*)]]*s* `$`

For example

```
$factor GENDER 2 AGEGROUP 4$
```

declares two factors `GENDER` and `AGEGROUP` to be of standard length. For each factor, the identifier is followed by the number of levels of the factor, which must be positive. `GENDER` is defined as having two levels (and so `GENDER` should take only the values 1 and 2) and `AGEGROUP` has four levels and should take the values 1, 2, 3, or 4 when values are assigned to it.
Note however that factor values are not checked on assignment but only when the factor is used in model fitting, plotting, tabulation, and so on.

An explicit length for the declared factors can optionally be given:

```
$factor 30 GENDER 2 AGEGROUP 4$
```

would declare the above factors to have length 30.

It is also possible to specify a reference level for each declared factor. The reference level of a factor is used in model fitting and is that category to which all other categories are compared (see [8.2.6] for a detailed description of reference levels). If specified, then the reference level is enclosed in parentheses following the number of levels.

```
$factor  SUBJECT 12 (10)$
```

defines a factor called `SUBJECT` with 12 levels, with the tenth category as the reference level. The default reference level of a factor is 1, but if the reference level is specified, the value must be a valid level for the factor as defined by the current `FACTOR` statement.

The `GFACTOR` directive [3.2.8] may be used to assign values to a regular set of factors. Alternatively, the values can be read in directly with the `DATA` and `READ` directives [3.2] or calculated with the `CALCULATE` directive [5.4].

3.1.5 Arrays

Arrays in GLIM are simply vectors with dimensionality. Arrays can be variate arrays (taking any value), or factor arrays (taking only the integer values 1 up to the number of levels). Most of the time arrays are treated as vectors and the dimensionality is ignored. However, in model fitting [8.2] and in the TABULATE [6.10] , TPRINT [6.9], and PASS [4.10.4] directives the dimensionality is used.

An array is declared through the ARRAY directive, with syntax

$array *identifier* [*dimension_list*] $

The *identifier* can be an existing variate or factor, or can be a new identifier, in which case it is declared as a variate. The *dimension_list* consists of a list of positive integers separated by commas, each integer representing the length of a dimension. The length of the array is the product of all the integers in the dimension_list. For example

```
$array T 2,3,4$
```

defines T to be a $2 \times 3 \times 4$ array, of length 24. The length of the first dimension is 2, that of the second dimension is 3, and that of the third is 4.
Also

```
$array MAT1 5,20$
```

defines a two-dimensional array MAT1 with 5 rows and 20 columns. If the *dimension_list* is omitted, then the effect is to remove the dimensionality from the vector; a factor array will therefore be redeclared as a factor vector, and a variate array will be redeclared as a variate vector.

3.1.6 Redeclaration of variables

Any vector (a variate, factor, array) or scalar can be redeclared (for example, by changing its dimensionality or number of levels) or declared as a different type (for example, a factor can be redeclared as a variate). In most circumstances, this involves adding or removing attributes from the definition of the structure. Specific instances of redeclaration are:

- Any factor may be redeclared with a different number of levels and/or a different reference level.
- A variate may be redeclared as a factor of the same length, and vice versa.
- Arrays may be redeclared with different dimensions.
- A variate or factor may be redeclared as an array.
- An array may be redeclared as a variate or factor
- A user scalar may be redeclared as a variate of length 1, and vice versa.

Vectors are not allowed to be redeclared with a different length. If such redeclaration is required the vector should be deleted [4.6.1] and redeclared, but any existing values will be lost. Alternatively, the PICK directive [5.5] provides a method of reducing the length of a vector, and the ASSIGN directive [5.1] a method of increasing the length of a vector.

3.2 Data input

There are many ways in GLIM of assigning values to vectors. The `CALCULATE` [5.4] and `ASSIGN` [5.1] directives provide common methods of assigning values to a **single** vector or scalar. This section describes methods of reading a dataset into GLIM. Most datasets will consist of a collection of vectors, some of them variates and some of them factors, and all having the same length. This length is referred to as the **number of rows** of the dataset. Note that we reserve the term **number of units** to refer to the length of the *y*-variate [8.1.1], which may not be the same. The four directives `DATA`, `FORMAT`, `READ`, and `DINPUT` allow data from a dataset, in the form of real numbers, to be assigned to vectors by GLIM. The `DATA` and `FORMAT` statements may (optionally) be used to declare the structure of the data that is to be accessed by subsequent `READ` and `DINPUT` directives.

Factor values can be input directly but in many situations, particularly with tabular structures, it may be more convenient to generate their values directly in GLIM. This could be achieved by the `FACTOR` directive to define a vector as a factor, and using the `%gl` function [5.4.5] to generate the values. Alternatively, the `GFACTOR` directive [3.2.8] can be used which will both define vectors as factors and generate regular data values for these factors.

Note that GLIM does not support missing value codes. If a dataset has missing values, then the user should use a particular numeric value for the missing value (for example, 99, or -1). Observations containing missing values can then be omitted from model fitting, tabulation, plotting, and so on, by using weights.

3.2.1 Defining the data to be read

The `DATA` directive defines the identifiers of the vectors whose values will be read by subsequent `READ` statements (or `DINPUT` statements if the data only is stored in a separate file [3.2.7]). The syntax is

`$data` *[length] identifiers*

Each identifier must either be a variate or a factor. Thus, in reading the design and data in a one-replicate 2^3 experiment with factors `N, P, K` each at two levels and a variate `YIELD`, the `READ` or `DINPUT` statements might be preceded by

```
$slength 8
$factor  N  2  P  2  K  2 $
$data  N  P  K  YIELD $
```

The lengths of the vectors in the `DATA` statements may be declared directly, and independently of any `SLENGTH` statement. If the length is not declared, all identifiers in the `DATA` statement must be of equal length. If the identifiers are undefined or defined but with no length the current standard length is assumed.

The DATA statement sets up an internal list of identifiers, called the **data list**. This list stays defined after the data has been read, so further READ or DINPUT statements can be issued at a later stage, allowing different data values for the same list of identifiers to be read in. The current data list can be examined with the statement `$environment f$`.

3.2.2 Defining the format

Data can be read in either fixed or free format. In both cases the data matrix is read row by row, that is, the first value of each identifier in the data list, according to the ordering of the identifiers within the list, is read from the first row, and so on.

The format of the data is specified by the FORMAT directive with syntax

`$format` [*items*]

where the *items* can be null or the word `free` for free-format style, or can constitute a valid FORTRAN format statement for fixed-format style (except that the outer opening and closing brackets may be omitted, as these will be supplied by GLIM).

For fixed format the format statement can extend over as many lines as required. Format specifiers of the form F*w.d* should in general be used, as all data is stored in GLIM as real numbers. However, integer format specifiers of the form I*w* will be converted by GLIM to real formats, for example, I2 will be converted to F2.0. GLIM also recognizes E, G, X, and T formats, and the new line specifier / is allowed. After any such conversion the total number of non-space characters in the format statement must not exceed 238.

A FORMAT directive remains in force until replaced by another. If no FORMAT statement has been issued in the current job, free-format style is assumed. The current setting of the format can be obtained with the statement `$environment f$`.

3.2.3 Reading data in free format

If the data is to be read from the same source as the GLIM statements, that is, if both statements and data exist in a single file, or if they are to be input directly from the keyboard, the data values must be preceded by the READ directive, which has the form

```
$read
```

The READ statement must then be followed by the required number of real values (the number of rows × number of identifiers). The first real value is assigned to the first element of the first identifier in the data list, the second number to the first element of the second identifier in the data list, and so on, with values being assigned on a row by row basis. The values should be separated by one or more spaces, new lines, or tab characters, with or without a single comma.

Values do not have to be specified 'neatly'; GLIM does not expect a new row of the data matrix to start on a new input line, nor does GLIM expect the same number of data values on each input line. For example

```
$data 4 A B C $
$read
5 7 6
3 2 1
0 8 9
3 6 7
$
```

and

```
$data 4 A B C$
$read  5 7
6,3,2 1 0
                8 9
     3 6 7
$
```

are both valid, and both will assign the values 5, 3, 0, and 3 to A; 7, 2, 8, and 6 to B, and 6, 1, 9, and 7 to C.

Note that the width of the current input channel should be taken into account when data values are being specified. The width of the primary input channel cannot be adjusted, but it is possible to increase the width of other input channels to a maximum of 132 columns [4.2.2]. Users must use fixed format for datasets with records greater than 132 characters in length.

3.2.4 Reading data in fixed format

Where a fixed format has been specified, and the data is to be read from the same source as the GLIM statements, the data must be preceded by the READ directive with a terminating directive symbol:

```
$read  $
```

The data values are then read, starting on a new line, and possibly extending over many lines. When the first element of all the vectors in the data list have been read, a new row is started. Each new row will start reading from a fresh line of data, and the FORTRAN format will also be started afresh. After all the data has been read, the next GLIM directive must start on a new line. For example

```
$data 4 A B C$
$format (3f1.0) $
$read$
576
```

```
321
089
367
$c all data has been input.
```

The width of the input channel is ignored for the purposes of fixed-format reading of data.

3.2.5 Temporary data lists

An alternative and convenient way of reading in data is to use a temporary data list. This shortcut is allowed when no `DATA` statement has been issued or when no data list is in force. The identifiers of the vectors are given following the `READ` directive, and this sets up a temporary data list which is in force only while the data values are being read. The extended syntax of the `READ` directive for free format is

`$read` *identifiers*

and for fixed format is

`$read` *identifiers* `$`

So

```
$slength 4$
$data A B C $
$read
```

could be replaced by

```
$slength 4$
$read A B C
```

Note that the vector identifiers must all have the same length, either previously defined or implied by the `SLENGTH` directive. Note also that if a data list exists, the list of identifiers following the `READ` directive has a different meaning, given below [3.2.6].

3.2.6 Reading subsets of variables

A free-format dataset often contains more vectors than are needed to be read in to GLIM for a particular analysis. It is possible to select a subset of vectors from the data matrix which will contain data — this reduces the amount of storage needed for the data. The data list is used to define the complete structure of the dataset, by specifying the full list of vector identifiers. The list of identifiers following the `READ` directive is then used to specify those vectors whose values are to be stored. This will have the effect of ignoring certain vectors in the data set since the data values for the identifiers appearing in the data list but not the subset (read list) will be skipped over, and will not be stored. The syntax of the READ directive is identical to that given in [3.2.5].

For example

```
$data AGE DEPTH HEIGHT WIDTH GIRTH $
$read AGE HEIGHT WIDTH
```

will read the first, third, and fourth variables in the data set, ignoring the second and fifth.

Note that it is strictly not necessary to give different identifiers to variables which are being skipped over. So in the above example

```
$data AGE XXX HEIGHT WIDTH XXX $
$read AGE HEIGHT WIDTH
```

would also work.

3.2.7 Reading data from an external file

In many circumstances it may be advantageous to keep the data in a separate file. The DINPUT directive is equivalent to the READ directive, but instead expects the data and the GLIM statements to come from separate sources.

The DINPUT directive has a similar syntax to the READ directive:

$dinput *file_specifier* [*width*] [*identifier*] *s*

replaces the READ directive. As with the READ directive, the format of the data is specified by the FORMAT statement and the optional list of identifiers represents either a temporary data list [3.2.5] or a subset of the previously defined data list [3.2.6].

The *file_specifier* specifies the data input source, which may be a channel number, a file name, or both (see [4.2.2] on the assignment of external files to input channels). The width of this input source may optionally be specified to declare the input record length (up to a maximum of 132 characters). The default width is usually 80 characters, and the value for the local implementation may be found with the statement $environment c$. The input width is ignored for fixed-format read, where there is no maximum width and where the FORTRAN format determines the width of the input source.

Once the appropriate number of data items has been read, control is returned to the original input channel.

For example, if a dataset with 6 variables and 100 cases is stored in the file 'aphasia', the following directives will read and define the data:

```
$data 100 A B C D E F $
$dinput 'aphasia'$
```

or simply

```
$slength 100$
$dinput 'aphasia' A B C D E F $
```

3.2.8 Storing values in factors

As an alternative to reading the values of a set of factors, it is possible to generate them if they are in standard order (for example, in cross-classifying a multi-way contingency table). The `GFACTOR` directive gives a useful way of generating the values of one or more factors. The syntax is similar to the `FACTOR` directive, but will assign values to the identifiers as well as define them. Its simplest form is

`$gfactor` [*length*] [*vector integer*]*s*

Each *vector* becomes a factor with the number of levels given by the following *integer*. All vectors must be the same length, which will be their existing length or the standard length or the explicitly defined length if it is given.

In the simplest case, with only one vector, the length of the vector is divided by the number of levels, and each level is repeated that number of times. For example

```
$gfactor 6 A 2 $ ! generates  1  1  1  2  2  2
```

When more than one vector is generated, the factors are nested, with the right-most vector varying fastest. The factors are then said to be in **standard order**. For example

```
$gfactor 12 A 2 B 3 C 2 $
```

gives A: 1 1 1 1 1 1 2 2 2 2 2 2
B: 1 1 2 2 3 3 1 1 2 2 3 3
C: 1 2 1 2 1 2 1 2 1 2 1 2

A vector name may be substituted by a `*`, meaning a dummy factor (with the given number of levels), which will effectively be generated but not stored. For example

```
$gfactor 12  A 2  * 3  C 2 $
```

will give the same values for `A` and `C` as above.

The length of the factors must be an exact multiple of the product of the number of levels for all the factors. When the length is greater than the product, the last named factors have their values repeated, for example

```
$gfactor 16  A 2  B 4 $
```

gives A: 1 1 1 1 1 1 1 1 2 2 2 2 2 2 2 2
B : 1 1 2 2 3 3 4 4 1 1 2 2 3 3 4 4

```
$gfactor 24  G 2  H 3  I 4 $
```

results in

G: 1 1 1 1 1 1 1 1 1 1 1 1 2 2 2 2 2 2 2 2 2 2 2 2
H: 1 1 1 1 2 2 2 2 3 3 3 3 1 1 1 1 2 2 2 2 3 3 3 3
I: 1 2 3 4 1 2 3 4 1 2 3 4 1 2 3 4 1 2 3 4 1 2 3 4

A reference level may also be set for any factor, using similar syntax to the `FACTOR` directive. The full syntax is

`$gfactor` [*length*] [*vector integer* [(*reference_level*)]]*s*

Note that the `CALCULATE` function `%gl` [5.4.5] is also useful in generating the values of a single regular factor.

3.3 Lists

A list in GLIM is an ordered list of identifiers. In general, lists can hold identifiers of any type, except identifiers of type list. When reading data, the identifiers should be variates or factors [3.3.5]. Lists are particularly useful in reading data. The general characteristics of lists are discussed in [3.3.1] to [3.3.4], and their use in reading data is given in [3.3.5].

3.3.1 Defining lists

A list is defined by the `LIST` directive, with syntax:

`$list` [*length*] *list* = *identifier* [, *identifier*]*s*

For example

```
$list L=A,B,C $
```

will set up a list called `L`, with values the three identifiers `A, B,` and `C`. A list length can be specified:

```
$list 3 L = A,B,C $
```

and `;` may be used in place of `,` as a separator between list identifiers. The list of identifiers on the right-hand side must be equal to the specified list length.

If an existing list appears in the list of identifiers, then that list will be expanded; it is not possible to have a list identifier as an element in a list.

```
$list L1=X,Y,Z $
$list L2=A,B,C,L1 $
```

will set the list L2 to have six identifiers: A, B, C, X, Y, and Z.

An identifier can appear more than once in a list. A list may be null (of length zero), in which case it contains no identifiers; such a list might be generated by adding and removing elements from an existing list [3.3.4].

3.3.2 Generating identifiers

If no identifiers are given on the right-hand side of the LIST directive, and if no assignment symbol = is present, then an integer defining the list length must be present in the directive and a list of identifiers will then be generated automatically by GLIM. The LIST directive then has the syntax

$list *length list*

The identifier names of the list elements are constructed from the integers 1 up to the list length, prefixed by the name of the list. In addition, a list identifier *list* is created. For example

```
$list 10 X $
```

will generate the ten identifiers

X1, X2, X3, X4, X5, X6, X7, X8, X9, and X10

and define a list X containing these identifiers.

The identifiers may or may not exist already. If they do not exist, they will be created, and they can then be defined at a later stage. None of the generated identifiers should be a list. If the name of the list is too long and would generate non-unique identifiers, then the list name is truncated in the generation process. Thus

```
$list  3 MEASUREMENTS $
```

would generate identifiers

MEASURE1, MEASURE2, and MEASURE3

3.3.3 Accessing elements of lists

An element of a list may be accessed by using the list identifier followed by the list element number in square brackets. For example, in the list L2 defined earlier, L2[4] would refer to identifier `X`; `L2[1]` would refer to `A`. List elements may be used in place of identifiers in any directive in the GLIM system. Thus

```
$number L2[4] =100 $
```

would define `X` to be a scalar with value 100.

3.3.4 Adding and removing elements of lists

A list can be added to and elements can be removed from a list by using the operators + (or equivalently `,`) and `-`. If an identifier is added to a list, then that identifier becomes the last element in the list. If a list is added to an existing list, then the identifiers in the new list are added to the existing list. Thus

```
$list L = L + D $
```

will add the element `D` to the list defined above, and

```
$list L4 = L + L $
```

will create a list `L4` with eight elements, containing the identifiers in list `L (A, B,C,D)` repeated twice.

Identifiers may be removed from a list. If the identifier is present in the list, then the last occurrence of the identifier in the list is removed, and the gap is closed, with all identifiers to the right of the deleted identifier being moved one position to the left. The identifier to be removed cannot itself be a list. For example

```
$list L4=L4-B $
```

will produce a list with seven identifiers: `A,B,C,D,A,C,D` whereas

```
$list L4=L4-L $
```

is not allowed.
The `CALCULATE` function `%in` can be used to discover if an identifier is present in a list [5.4.5].

3.3.5 Using lists in reading data

List identifiers can be used in the `DATA`, `VARIATE`, `READ`, and `DINPUT` directives as a replacement for the lists of identifiers in these directives. The list should contain either undefined identifiers or identifiers which have been previously defined as variates or factors. For example

```
$slength 150 $
$list 20 X $
$factor X15 5 X17 3 $
$dinput 'votedata' X $
```

will read the first 150 rows of the 20 vectors named `X1, X2, ... ,X20` from the file '`votedata`' in free format, with `X15` and `X17` declared as factors, and the other vectors declared as variates.

3.3.6 Using lists in other directives

There are other directives in which lists are allowed; these include the `GRAPH` and `PLOT` directives for graphical output, the `LOOK`, `SORT`, `PICK`, `PRINT`, `TPRINT`, and `TABULATE` directives for data examination and manipulation, and the `FIT` and `TERMS` directives for model fitting. Lists may also be deleted together with their constituent identifiers through the `DELETE` and `TIDY` directives. See the relevant chapters in the User Guide for further information.

4 File handling and utilities

Introduction

GLIM has facilities for inputting statements and data from external files and similarly output can be directed to a file to provide hard copy output. These are accessed via **secondary input** and **secondary output channels.** Certain channels are automatically available to the user. One of these is for the **transcript** file which, by default, is a copy of everything that appears on the screen during an interactive session, including both input and output. Another is for the **journal** file which is a record of the GLIM statements input by the user during a session. This can be 'replayed' to reproduce a GLIM session. The user has control over what should be automatically stored in these files (see [4.3.1]). A third file which is available for input by the user is the **macro library.** This contains tried and tested macros which provide a valuable extension to the facilities available in GLIM [Appendix A]. There are also facilities for the user to save (to a **dump** file) the GLIM session in its current state and subsequently restart it at a later time, thus allowing an analysis to be spread over more than one session.

Directives are available to remove unwanted data structures thus releasing workspace. The `ENVIRONMENT` directive gives the user complete information about user-defined vectors, scalars, and macros, the use of workspace, and the settings of switches and system-dependent parameters.

Also on-line information is provided on the syntax and usage of GLIM commands and descriptions of technical terminology through the `MANUAL` directive.

The `PASS` directive allows the values of GLIM structures to be made accessible to FORTRAN subroutines written by the user and linked to the GLIM package. The user may thus extend the facilities of GLIM by providing efficient routines for non-standard operations on the data structures stored within the package.

File handling and other facilities in this section are unavoidably system-dependent, and the User Note should be checked for system-specific information.

4.1 Input and output channels

During a GLIM session further statements or data will be taken from the **current input channel** and any output will be sent to the **current output channel.** Unless otherwise directed GLIM will use the **primary input** and **primary output** channels. During an interactive session these will both be the terminal/computer with the keyboard for input and the screen for output. The user of a batch version may have to assign files to these channels in job control statements or specify them when invoking GLIM at the operating system level. The method of doing this is system dependent so that batch users may need to consult their local computer support staff or check with the User Note.

All other input and output channels are called **secondary** channels. These are identified internally in GLIM by integer numbers (following the FORTRAN convention); the particular values allowed vary with the implementation. The user can specify the filename to be assigned to a channel number during a GLIM job. Alternatively the user needs only to specify the filename and GLIM will select a suitable channel number. The channel numbers of currently open files are held in the system vector `%oc`. If the user simply specifies a channel number the program will prompt the user for a filename. On some systems this may be done before GLIM is run, whence the channel number may be used immediately within GLIM. The user can optionally specify the width (in columns) of a channel to be used and, for output channels, the height (in rows). These must not exceed those allowed by the system — ***the program is not protected against violation of this rule.***

Some channels are reserved by the system for use by certain directives, such as `DUMP`. Use of these channels for ordinary input/output may cause an error. The current settings for all input and output channels and those reserved for system use may be discovered with the statement `$environment c g $`. The User Note should be checked for system-specific information.

4.2 Secondary input

The main advantage of secondary input channels is that it allows the storage of datasets and/or pieces of GLIM program (including macros) that are wanted for more than one session. In particular, interactive users may want to set up the data definition and input section of a GLIM run by creating a file using a system text editor so that a permanent copy is kept. Then in a subsequent GLIM session the data entry stage can be achieved easily by use of secondary input from this file. This can be particularly useful if complex data transformations are necessary before analysis can begin or if the dataset is large. Another common use of secondary input files is to store a collection of the user's macros which can be input when required in an analysis. Note that `DINPUT` [3.2.7] can be used to take input from a file that contains only data.

4.2.1 The initialization file

At the start of each job within a session GLIM looks for an initialization file with a particular name in the user's filestore. The name is operating system dependent and can be discovered for the local system with the statement `$environment i $`. If this file exists then GLIM will input it automatically as a secondary input file. One use of this facility is to set some of the switches to non-default values, for example to attach a specific local graphics output device.

4.2.2 The `INPUT` directive

The directive to switch input to a secondary channel has syntax:

`$input` *file_specifier* [*width*]

The *file_specifier* must have at least one of channel number or filename, and the = sign is mandatory if both are given [2.9.1]. The *width* is an optional scalar with value between 30 and 132 giving the width of the input channel. If omitted the standard setting for the primary input channel is used.

Example Input is to be taken from a file `glimdata` which has records of length 100 characters.

```
$input 'glimdata' 100 $
```

Input will continue to be taken from this file until the statement

```
$return
```

is encountered; input then returns to the channel from which the `INPUT` statement was issued. Note that an `INPUT` statement must not be used instead of `$return` to resume input from the previous channel. If a later `INPUT` to the same file is given input is taken from the position in the file where the previous statement left off.

4.2.3 The `REINPUT` directive

If input has been taken from a file and at a subsequent time input is to be taken from the beginning of that file the user must use the `REINPUT` directive. This has the same syntax as `INPUT`.

4.2.4 The `REWIND` directive

The `REWIND` directive has syntax

`$rewind` *file_specifier*

and has the effect that all subsequent input from this channel will start from the beginning of the file.

4.2.5 Subfiles

If a user builds up a large collection of useful macros it is inconvenient to input the whole collection just to use one macro. Similarly when the user has many datasets that are accessed regularly it may not be convenient to have each in a separate file. Both these difficulties may be circumvented by the use of **subfiles**. An input file may be subdivided into sections, that is, subfiles, by beginning each section with a statement

`$subfile` *identifier*

where `$subfile` ***must be the first item on the line.*** (This restriction is imposed for the sake of efficiency in locating subfiles.) A subfile is ended by the next `RETURN` statement, which may be preceded by further SUBFILE statements. A file containing subfiles must end with the statement

```
$finish
```

which acts as an end-of-file marker. It must be the first item on its line.

Example

```
$sub  SF1
 block1
$sub  SF2
 block2
$return
$sub  SF3
 block3
$return
$finish
```

This illustrates a file with three subfiles, SF1 consisting of blocks 1 and 2 (because there is no $return at the end of block1), SF2 consisting of block 2, and SF3 consisting of block 3.

Subfiles may be selected for input by an extension of the INPUT and REINPUT directives of the form

$$\left\{ \begin{array}{c} \texttt{\$input} \\ \texttt{\$reinput} \end{array} \right\} \textit{file\_specifier } [\textit{width}] \textit{ subfiles}$$

The action is then to input one or more subfiles from the specified file. Other parts of the file will be ignored. Thus blocks 2 and 3 may be read from the file in the above example by the following statement

```
$input  'filename'  SF2  SF3 $
```

Note that the subfiles are searched for in the order given and thus it is more efficient to give them in the order that they occur in the file. REINPUT may be used in place of INPUT for efficiency when it is known that the subfiles will be found more rapidly by starting at the beginning rather than continuing from the current position in the file.

4.2.6 Nesting of input channels

Files being input on secondary input channels may themselves contain an INPUT statement. This will cause a temporary suspension of input from this file and input is then taken from the file specified through the latest INPUT statement. On encountering a RETURN input reverts to the original file at the point where it left off. Information on channels that are currently being used for input is stored internally in the program control stack [7.7.1] which can be examined with the statement `$environment p$.`

4.2.7 Opening and closing files

It is sometimes convenient to be able to open a file for input but not immediately start reading from the file. This is achieved using the `OPEN` directive with syntax

`$open` *file_specifier*

Similarly, it may be necessary to close files before the end of a session, for example because of limitations on the number of open files specified by the operating system. The directive is

`$close` *file_specifier*

4.2.8 Echoing of secondary input

Normally the user will not see what is being input from the secondary input channel. The user can specify that input from a secondary channel should be reproduced line by line on the screen as it is being input. This is referred to as **echoing** the input. This can be particularly useful when there are errors in the input file. The syntax for the `ECHO` directive is

`$echo` [*phrase*]

where *phrase* is `on` or `off.`

If echoing is `on` then all secondary input is reproduced on the primary output channel as it is read. If echoing is `off` secondary input is not echoed. If *phrase* is omitted the echoing state is reversed.

4.3 Secondary output

Secondary output to files has two main uses. Firstly it provides a means for the user to produce hard copy output of the results of an analysis and secondly numeric values may be output to form the input for further analyses at a later time or for input to other software. Two secondary output files are automatically produced in an interactive GLIM session, the **transcript** and the **journal** file. In the majority of cases a user will not need secondary output beyond these two files. However this facility is provided with the `OUTPUT` directive [4.3.3].

4.3.1 The transcript file

During an interactive session primary output is to the screen and thus would be lost at the end of a session. It is common that users would like a permanent hard copy of their analyses for perusal at their leisure. GLIM automatically makes a permanent copy of all output to the screen in the **transcript** file. The name of this file is system dependent and can be discovered with the statement `$environment c $`. The user can change what sort of output is saved by the `TRANSCRIPT` directive with syntax

`$transcript` [(*option*)] *names*

where *names* may be a subset, in any order of `ECHO, FAULT, HELP, INPUT, ORDINARY, VERIFY,` or `WARN` with only the first character of each name being significant.

Each *name* determines which input and output sources are written to the transcript file. The *names* stand for the following sources:

Transcript tag	*name*	Source	Value of *name*
`[i]`	`INPUT`	Primary input	1
`[v]`	`VERIFY`	Verification of macros	2
`[w]`	`WARN`	Warning messages	4
`[f]`	`FAULT`	Fault messages	8
`[h]`	`HELP`	Help messages	16
`[o]`	`ORDINARY`	Ordinary output	32
`[e]`	`ECHO`	Echoing of secondary input and output	64

The following option is available

`style` = *integer*

The `style` *option* determines the style of the transcript lines sent to the transcript file. If the *integer*, when rounded to the nearest integer, is zero or if the *option* is omitted, then the transcript tag precedes each transcript line and identifies the source of that line. If *integer* is negative then no such tag is produced.

If a *name* is used, then subsequent input or output from the corresponding source will be sent to the transcript file. If no *name* is supplied, then no input or output is sent to the transcript file. At the start of a job in interactive mode, the *names* `INPUT`, `WARN`, `FAULT`, `HELP`, and `ORDINARY` are assumed. This reproduces the user's screen in most installations.

Example

```
$transcript o      ! only ordinary output will be
                   ! sent to the transcript file
```

The meaning of `INPUT` and `ORDINARY` have changed from GLIM 3.77, and now refer to copying of the primary input and primary output respectively. Copying of all secondary input and output to the transcript file is now controlled by the *name* `ECHO`.

An alternative syntax allows the updating of a previous setting by the addition and/or removal of sources. If a *name* is preceded by a `+`, then the specified source is added to the contributing transcript sources. If a *name* is preceded by a `-`, then the specified source is deleted from the contributing transcript sources. Addition of a source which is already present, (or deletion of a source which is not present) is ignored. For example

```
$transcript  -H $
```

would stop subsequent help messages from being sent to the transcript file.

The third syntactic form allows an integer to be used in place of *names*. The integer may be formed by summing the required source values given in the table above. The system scalar `%tra` contains the sum of these source values for the current settings. For example

```
$transcript 10 $
```

would specify `VERIFY` and `FAULT` sources to be sent to the transcript file.

One main use for this is to allow the settings represented by `%tra` to be stored and recovered at a later stage. **Note that the *integer* may not be substituted by a *user scalar*** but only by an ordinary scalar [2.5.2].

Example

```
$calculate %s=%tra      !save current transcript settings
$use mymacro            !use a macro (which might change
                        !the transcript settings)
$transcript %s$         !recover the old settings
$transcript +e$         !and add echoing if not already
                        !present.
```

Note that the `VERIFY`, `HELP`, `WARN`, `ECHO`, and `OUTPUT` directives also act on the transcript settings.

4.3.2 The journal file

A user **journal** file contains a record of all input from the primary input channel during a GLIM session. The file may be replayed in a subsequent session to reproduce an analysis; errors in the interactive session may be edited out, using the system editor, before replay. The user journal file is also of use in the event of a system crash or communications failure. A journal file may be replayed by using the `INPUT` directive, renaming the file if necessary. The journal file cannot be closed by use of the `CLOSE` directive. No journal file is kept if the user is in batch mode.

The `JOURNAL` directive is used to control whether a user journal is stored for an interactive session. The syntax is

`$journal` [*phrase*]

where *phrase* is `on` or `off`.

If *phrase* is `on`, then journal information will be written to the journal file, the file being opened if necessary. If *phrase* is `off`, then the journal file is kept open but is not written to. If *phrase* is omitted the journal state is reversed.

The journal channel number and state of the journal switch may be found with the statement `$environment c $.` The system scalar `%jrc` also contains the journal channel number. The width of the journal channel at any point in a session is taken to be that of the primary input channel. The system scalar `%jou` indicates whether a user journal is in force.

The default state is implementation dependent (usually `on`) and the name of the default journal file is also implementation dependent.

Example

```
$journal on$      ! turns on the user journal file
```

4.3.3 The OUTPUT directive

The `OUTPUT` directive allows the user to direct all subsequent output from the package to a specified file until the next `OUTPUT` statement is given. The syntax of the `OUTPUT` directive is the same as that of the `INPUT` directive but with an extra optional setting for the height of the output channel. The syntax is

`$output` *file_specifier* [*width* [*height*]]

The *width* and *height* are optional scalars giving the width of the output channel and the height of a page on this channel. Note that the width and height values set here are used in directives such as `PLOT` to govern the layout of output so that a secondary output file can contain a plot that is too big for screen output. These settings can be displayed with the statement `$environment c $`

4.4 Controlling the amount of output

The following directives are provided to give the user control over the amount and detail of output from the program.

4.4.1 Setting the accuracy of numeric output

The accuracy of all numeric values output by GLIM can be controlled by the `ACCURACY` directive with syntax

`$accuracy` [*integer1* [*integer2*]]

Integer1 specifies the minimum number of significant digits (default 4) and *integer2* the maximum number of decimal places for all subsequent numeric values until another ACCURACY statement is issued. Note that the maximum number of decimal places takes precedence over the minimum number of significant digits, so that the number 0.01234 would be output as 0.012 after the following ACCURACY statement

```
$accuracy 4 3 $
```

4.4.2 Paging output in interactive mode

The amount of output from a directive may be more than can fit on the screen. In such a situation an important piece of information may have disappeared from the screen and, if the computer system does not allow the output to be 'scrolled back', is thus lost. Users may protect themselves against this occurrence by use of the PAGE directive with syntax

$page [*phrase*]

where *phrase* is on or off.

With paging on the output will pause after *n* lines of output, where *n* is the current primary output channel height [4.3.3]. The user is then prompted to press the 'return' key for output to continue. If *phrase* is omitted, the directive reverses the current pagination state.

4.4.3 Suppressing output from commands

It is sometimes useful to be able to switch off all output from the program until a particular task is complete, for example during cycles of an iterative procedure until it has converged. The OUTPUT directive with no items or which specifies channel zero, switches off all output until another OUTPUT statement is issued.

```
$output $
```

As this subsequent OUTPUT statement requires the user to provide an output *channel*, switching back to the primary output channel is best achieved by use of the system scalar %poc which holds this value, that is, with the statement

```
$output %poc $
```

4.4.4 Controlling the output of messages

When a statement cannot be successfully executed by GLIM, either because the statement was syntactically incorrect or because items specified in the statement are invalid or incompatible, the program outputs a **fault** message. The program may also output a **help** message. The user cannot stop the output of the fault message but does have control over the output of the extra help message. This is done through the BRIEF directive with syntax

```
$brief [phrase]
```

where *phrase* is `on` or `off`.

When brief is `on` the extra help message is suppressed. If brief is switched `on` and a fault had just occurred then its help message can be obtained by the statement:

```
$help $
```

If no fault had occurred the user will be referred to the `MANUAL` directive [4.11.1].

GLIM also outputs a **warning** message in those cases where the statement was successfully carried out but resulted in an unusual outcome (for example division by zero in a `CALCULATE` statement) which may cause the user future problems. The output of these messages can be controlled by

```
$warn [phrase]
```

where *phrase* is `on` or `off`. If warn is `off` the output of such warnings will be suppressed. If `phrase` is omitted the warning state will be reversed.

In normal operation, when a macro is invoked, the output will only be that specifically requested by statements in the macro, and the statements themselves will not be seen by the user. However, if a macro fails for some unexpected reason it may be useful for the user to see the statements from the macro as they are being executed. This is termed **macro verification** and can be controlled by

```
$verify [phrase]
```

where *phrase* is `on` or `off`.

With verify `on`, each line of the macro will be output before being executed. If *phrase* is omitted the verification state is reversed.

Note that these directives automatically change the transcript settings so that, by default, all screen output will be copied to the transcript file [4.3.1]. The current settings can be discovered with the statement `$environment C $` or by examining the values of the relevant system scalars; see the Reference Guide [14.2.2].

4.5 System information

The `ENVIRONMENT` directive gives the user access to information about the state of the program and its relation to the environment in which it is working. Note that system scalars also store the status of many of the settings described below (see the Reference Guide [14.2.2]). The syntax is:

```
$environment letters
```

The *letters* may be any subset, in any order, of:

```
A C D E F G H I M P R S U
```

with the following meanings:

Option A: Arguments

Option A gives information on **A**rguments to macros. For each macro, this provides a list of the argument identifiers defined for the macro and their currently set actual arguments. If a macro has no arguments, then no information is produced.

Option C: Channels

Option C gives information on **C**hannels. This gives a display of information on the primary input and output channels, channels used by the system for dump and macro library and secondary user files explicitly opened by the user, thus giving the user information on all open channels. The information supplied is the channel number and the filename of each secondary user file and the channel width and height where this is relevant.

Option D: Directory

Option D gives information on the **D**irectory of user-defined structures and system structures that currently have values. A column 'type' is displayed, giving the type of the structure. This will be one of: variate, factor, macro, scalar, list, array, or undefined. Other information provided is the length of the structure, the number of levels and reference category (if other than 1) for factors, and the size and shape of arrays. An example of the output is

```
Directory:   type        levels length  space
   %OC      variate                  6      3
   X        variate                 25      0
   F        factor        4(2)      30      1
   Q        variate                 12      3       3x4 array
   N        scalar                   1      1
   L        list                     4      2
   M        macro                   14     13
```

The length of a macro is the number of characters in the contents of the macro and does not include system space for the argument list. Information on the structures local to a macro are also displayed but only if that macro is currently being executed [7.7.1].

Option `E`: External

This gives details of **E**xternal subroutines accessible via the `PASS` facility [4.10.4]. These details are user-supplied (see the Reference Guide [17] for full details).

Option `F`: Format

Option `F` gives information on the `DATA` **F**ormat. It gives the current *data list* and the current `FORMAT` setting.

Option `G`: Graphics

The `G` option gives information on the **G**raphical facilities. The information displayed is a list of valid devices and the current default graphical device, the current settings of the `LAYOUT` directive, a table of available colours, line styles, and symbols, and a list of the current pen styles.

Option `H`: High-resolution

The `H` option produces a **H**igh-resolution display of the available colours, line styles, and symbols, and the current pen styles. The information is produced on the default graphical device (as shown by the `G` option to the `ENVIRONMENT` directive and set by the `SET` directive [4.7]), the device being opened if necessary.

Option `I`: Installation

This gives details of the **I**nstallation and system-specific information such as the representation of special characters [2.1], the largest integer, and the default names of the journal, transcript, and initialization files.

Option `M`: Maximum and minimum

Option `M` provides information on the **M**aximum and **M**inimum argument values to all GLIM functions available in `CALCULATE` which have a restricted domain.

Option `P`: Program control stack

This option lists the current contents of the **P**rogram control stack [7.7.1], indicating at each level the input channel number and record width or the macro identifier with an additional asterisk if the macro was invoked by a `WHILE` statement.

Option R: Random number generator

This lists the three current seed values for the standard psuedo-**R**andom number generator that is used with the CALCULATE directive.

Option S: System

The S option provides information on the space taken up by **S**ystem internals. These are GLIM data structures that are not accessible to the user. Space taken up by system constants is referred to as 'constants', and refers to space taken up by the ordinary and system scalars. 'User directory' refers to space taken up by the names of user structures and 'entity table' refers to space taken up by the directory. Neither may be deleted. 'Model fitting' refers to internal structures used in model fitting, and 'PCS' refers to space taken up by the program control stack.

Option U: Usage

U provides information on space **U**sage. The U option provides the columns 'used' and 'out of' when listing information on the usage of space within the program. The 'out of' value is the upper limit for that particular type of item (for example, the maximum number of user identifiers allowed by the system) and 'used' the number of such items currently in use.

4.6 Deletion of structures

Sometimes the program workspace becomes too full for the program to execute new statements. In this situation the user should remove any unwanted structures from the workspace before proceeding. This is achieved by use of the DELETE and TIDY directives.

4.6.1 The DELETE directive

$delete *identifiers*

where *identifier* is a *user_scalar, vector, macro*, or *list*.

This causes the vectors, macros, and lists specified to be deleted and the corresponding workspace to be made available for reuse. The identifiers themselves can be redefined and assigned new values. System vectors may be deleted to save space, but their deletion may affect certain aspects of GLIM such as display and extraction of results of a fit. Deletion of a user structure may also have side-effects, for example, if it is part of the specification of the current model. See the Reference Guide [15.14] for a full description of side-effects.

4.6.2 The TIDY directive

`$tidy` *identifiers*

where *identifier* is a *macro* or *list*.

The `TIDY` directive provides a mechanism for recovering workspace that cannot be achieved using the `DELETE` directive.

If *identifier* is a *macro* then the values of all the macro's local identifiers are deleted and the workspace recovered.

If *identifier* is a *list* then all the identifiers in the list are deleted and the list becomes null. Note that the list still exists but has no contents.

4.7 The SET directive

The standard operation of GLIM assumes certain default action settings. Some of these can be changed by the user through the `SET` directive.

4.7.1 Batch and interactive mode

The default mode of operation of GLIM is **interactive** with primary input from the keyboard and output to the screen. The alternative mode is **batch** in which primary input and output will usually be files. The user can change the mode by:

```
$set  { batch       }
      { interactive }
```

Note that many systems will set the mode automatically, according to whether the terminal is being used for primary input and primary output, and that there may also be constraints on changing the mode; see the User Note for details.

Defaults will be different for the two modes. In batch mode `PAGE`, `PAUSE`, `SUSPEND`, and the break-in facility [4.10.2] are not available and there will be no `TRANSCRIPT` or `JOURNAL` output.

4.7.2 Graphics devices

High-resolution graphics output will be assumed to be displayed on the default graphics device. The user can change this default with

`$set device =` *string*

where *string* is a valid graphics output device. The device names are implementation dependent; valid devices and the current default can be obtained with the statement `$environment G $`

4.7.3 Default model formula

The default model formula contains the constant term even when not explicitly included. This can be altered by

```
$set { noconstant }
     {  constant  }
```

If `noconstant` is specified then a model formula will only contain the constant term if explicitly included [8.2.4.1].

4.8 Dumping and restoring a GLIM job

When performing a large and complex analysis it may be necessary to complete it in several separate interactive sessions. It is then convenient if the current state of the session can be stored in a file from which subsequent sessions can be restarted. This is known as **dumping** and **restoring** the GLIM job. The directives for these operations are

`$dump` [*file_specifier*]

`$restore` [*file_specifier*]

If *file_specifier* is omitted GLIM uses the default channel and filename.

When a dump is performed the used part of the workspace is copied to the file specified by the directive. This includes not only the user's numeric variables but also any macros that have been input and the current status of all system structures, for example, the model formula with its current parameter estimates. When this file is restored it is copied back into the workspace and the session can be resumed from the point at which the dump was performed. Note that the file is ***not*** a text file and thus cannot be inspected using standard text editors. The file is in a format suitable for rapid reading and writing and thus `DUMP` can be used to produce a data file that can be input much more rapidly than a standard text file. This can be particularly valuable for very large datasets.

If more than one dump is performed in the same session the latest dump is appended to the dump file so that this file may contain several separate dumps. When restoring in this situation multiple `RESTORE` statements may be required to obtain the correct status. Each restore will read a dump, so that after, for example, three restores the third dump will have been read. Note that each restore overwrites any previous restored dump in the workspace. The `REWIND` directive, with syntax

`$rewind` [*file_specifier]*

may be used to rewind the dump file so that earlier dumps can be restored.

After a restore the program proceeds from the point at which the dump was instigated. In particular it will complete the reading of the input line that contained the `DUMP` statement. Thus if there were any more statements on this line they will be executed. This may be of use in some circumstances, but it is recommended that DUMP and RESTORE stand alone on a line. Note that GLIM4 dump files are not compatible with those of GLIM 3.77 [see Appendix B].

4.9 Terminating a GLIM job

A GLIM session is terminated by the statement

```
$stop
```

which causes all files to be closed, stops GLIM running and returns the user to the operating system. However if the user wishes to continue the GLIM session but start a completely new analysis the statement

$$\texttt{\$newjob [(close = } \left\{ \begin{array}{c} \texttt{no} \\ \texttt{yes} \end{array} \right\} \texttt{)]}$$

will reinitialize the system, delete all user structures and return GLIM to the status at the beginning of a session. Secondary files will be left open with input continuing from the last point of input unless the user specifies the option close=yes.

Note that all open graphics devices [6.6.1] are refreshed, and the LAYOUT setting [6.8] is reinitialized, but otherwise the graphics state is unaffected by a NEWJOB statement.

4.10 Additional facilities

The following features are system dependent; on some installations they may not be implemented and the form of operation may vary with different implementations. The user should check the User Note for details. The MANUAL directive also provides information on the local implementation ([4.11]).

4.10.1 The PAUSE directive

The PAUSE command will temporarily return the user to the operating system, without terminating the GLIM session, allowing the user to perform some operations at operating system level and then resume the GLIM session. When possible the PAUSE directive will allow an operating system command to be passed to the operating system without leaving GLIM (check the User Note for details). The syntax is

$pause [*string*]

where *string* is an operating system command.

For example, on a unix system:

```
$pause 'ls' $
```

or on a DOS system:

```
$pause 'dir' $
```

would list the files in the user's current directory and return to GLIM. If *string* is omitted the user will return to the operating system until the necessary operating system command is given to resume the GLIM program.

4.10.2 Break-in facility

It sometimes happens that a user realizes that a GLIM statement being executed will take a long time to complete, for example, printing a very large table. In such circumstances the user might wish to interrupt the execution in order to proceed without waiting for the statement to be completed. When implemented, the break-in (attention) facility allows the user to interrupt GLIM and abort the current GLIM command. The method of doing this is system-dependent, and the User Note should be checked for details, but it will often be done by typing 'control'-`C` (that is, holding down the 'control' key and pressing `C`). Repeated use of the break-in will ultimately terminate the GLIM session.

4.10.3 Command line parameters

Wherever the operating system allows, GLIM will be implemented to allow the user to assign files when invoking GLIM. The method of doing this is system-dependent, and the User Note should be checked for details, but it will include the assignment of the following files:

`pip`	primary input file
`pop`	primary output file
`sip`	secondary input file
`sop`	secondary output file
`dmp`	dump file
`log`	transcript file
`lib`	macro library file
`jou`	journal file

The three-letter keywords given will usually be used as option names when the files are assigned by setting options on the command line. Note that if the primary input is preassigned to a file, GLIM will usually operate in batch mode.

4.10.4 The `PASS` directive

In some circumstances the facilities of GLIM may not be adequate to perform certain non-standard operations efficiently. Thus users may wish to extend the GLIM facilities by providing their own FORTRAN subroutines, linked to the GLIM program, to carry out these operations. The PASS directive provides a mechanism for making data stored in GLIM accessible to such subroutines. Full details are given in the Reference Guide [17]. The syntax is:

$pass [(*option*)] { *keyword* / *integer* } [*vector_list* [*macro*]]

where *keyword* or *integer* is passed to the user's subroutine and is intended as a way of selecting a particular subroutine when the user has linked more than one to the GLIM program. The *vector_list* specifies the vectors (or scalars treated as vectors of length one) whose values are to be passed to the user's subroutine. Items in this list are separated by commas. The user may also specify a macro, in which case the contents of the macro will also be passed, characters being represented as integers (as defined by the FORTRAN ICHAR function). If any values are changed by the user's subroutine these become the new values within GLIM.

The *option*

```
array=fortran
```

specifies that the values of any GLIM arrays should be reordered to conform with the FORTRAN convention.

Information on what user-supplied subroutines have been linked to GLIM can be obtained with the statement $environment e$. Note that the User Note should be checked for any system-specific information.

4.11 Obtaining on-line information

While this manual provides the most comprehensive description of the usage of GLIM and definition of the directives, it is often sufficient to obtain a brief description of the syntax of a directive or meaning of some GLIM terminology. Such information is available on-line, that is, on the screen during a GLIM session, using the MANUAL directive.

4.11.1 The MANUAL Directive

The MANUAL directive has the syntax

$manual [*keyword* [*macro*]]

where *keyword* is one of *directive_name*, *system_identifier*, *topic* or *glossary_term*.

The MANUAL directive gives specific information as defined by the setting of the *keywords* given in the following table.

keyword	*Action*
directive_name	Meaning and usage of *directive_name*
system_identifier	Meaning and usage of *system_identifier*
glossary_term	Description of the meaning of *glossary_term* in the context of model fitting in GLIM
topics: one of	
DIRECTIVES	Lists all *directive_names*
GLOSSARY	Lists available *glossary_terms*
INFO	Gives this information
LOCAL	Gives information specific to the local GLIM implementation
LIBRARY	Gives information on available macros in the macro library
LIBRARY macro	Gives information on the usage of the macro from the macro library

The *glossary* is a collection of terms in common usage in the description of model fitting in GLIM and if *keyword* is a *glossary_term* then a description of the meaning of that term will be given.

Examples

```
$manual  extract        !  gives information on
                        !  the EXTRACT directive
$manual  %tp            !  information on the %tp
                        !  function
$manual  files          !  gives a description of
                        !  file handling
$manual  directives     !  lists all directives
$manual  library QPLOT  !  information on the
                        !  QPLOT macro from the
                        !  macro library
```

5 Calculations on data

Introduction

GLIM allows calculations to be performed using arrays, vectors, vector elements, scalars, and constants with the result of a calculation being stored or immediately printed out. Various facilities exist for specific statistical and modelling calculations to be performed, for example, GLIM will calculate summary statistics or model parameter estimates when appropriate directives are used. This chapter is concerned with arbitrary calculations and manipulations on data structures under the user's direct control. Some directives such as GFACTOR and TABULATE perform special-purpose calculations or data manipulations, and are described in the relevant chapters.

The most general data structure manipulation directive is CALCULATE which performs vector or vector/scalar arithmetic and includes a number of useful functions. ASSIGN is most useful when an arbitrary sequence of values is to be stored in a vector. The PICK directive may be used to extract values from vectors. There are two directives, GROUP and MAP, for recoding data, and the EDIT directive may be used to alter specific elements of vectors.

A value can be placed in a scalar or vector by means of the CALCULATE or ASSIGN directives. Values can also be copied from other data structures, in particular system structures set up by other parts of GLIM (such as %df, %dv) may be stored. CALCULATE is used to assign values obtained from the evaluation of arithmetical expressions.

Examples

```
$number BETA    $calculate  BETA = 12345.6789 $
$number COPY    $calculate  COPY = BETA $
$number LAST_DEV LAST_DF    $
$calculate  LAST_DEV = %dv  : LAST_DF = %df $
```

In all of the above examples ASSIGN could have been used instead of CALCULATE to the same effect. If the right-hand side of the assignment involves any arithmetic operations, then CALCULATE must be used. For example

```
$number  MEAN_SQ   $
$calculate  MEAN_SQ = %dv / %df $
$number  MS_REDN   $
$calculate MS_REDN =  (LAST_DV - %dv)/(LAST_DF - %df)$
$calculate  SIN_T = %sqrt( 1 - COS_T * COS_T ) $
$number  ROOT_2PI   $
$calculate  ROOT_2PI = %sqrt(2*%pi) $
```

The last two examples include references to `%sqrt`, which is the identifier of the GLIM function returning the square root. The last example includes the system scalar `%pi` which holds the mathematical constant π. GLIM provides a variety of functions which are described in full below [5.4.4].

The range of values and the accuracy with which any value can be stored is machine dependent. On most computers all data in GLIM is stored in single precision, usually giving approximately six decimal digits of accuracy. Internally, CALCULATE actually operates in double precision, so it is often more accurate to avoid storing intermediate results if possible. GLIM does not perform any integer arithmetic, although there are functions which return whole numbers as real values. There are other operators and functions which may be thought of as returning logical values, but these too are actually real values — with 0.0 representing 'false' and 1.0 representing 'true'.

5.1 The ASSIGN directive

Perhaps the easiest way of storing values in a GLIM data structure is by using the ASSIGN directive. This may be used when a complete set of values is to be stored in a vector, although it may also be used to store a value in a scalar. The syntax is

$assign *identifier* = *list_of_values*

The *identifier* is the name of the structure to receive the data values. The *list_of_values* may contain constants, scalars, or vectors, separated by commas or spaces, in any combination. The *identifier* may already exist or it may be new. If the *identifier* is already defined as a vector its length will be (re)defined by the number of values in the *list_of_values*. If the *identifier* does not already exist, ASSIGN creates it as a variate (of the appropriate length). Even when there is only one element in the *list_of_values*, if ASSIGN has to create the identifier, it will be a variate (of length 1) and not a scalar. If the resultant structure is required to be a scalar, it must be explicitly defined in a NUMBER directive [3.1.1]; in this case, it is easier to assign the value when it is defined.

Examples

```
$assign  RANG = 1 , 100 $        ! length 2
$number B=1 C=2   $assign  VALS = 1.5, B  C , 7 $
                                 ! length 4
$number E=8 F=9   $assign  VALS = E , VALS , 2 , F $
                                 ! length changed to 7
$assign  REG_RES = %df   %dv $
                                 ! length 2
$assign  %e = 2.71828175 $
```

We shall see later [5.4.5] how a vector may be created with regularly repeating elements. ASSIGN can also be used to create a vector of lagged values, for example:

```
$assign V = 1 2 3 4 $
 : UN_LAGED = V V V $
 : LAG_BY_2 = 3 4 V V 1 2 $
```

The *list_of_values* in the `ASSIGN` directive can also include abbreviated lists, for example `1 2 ... 10` indicating the first ten positive integers. The difference between the two values preceding the ellipses (`...`) is used to increment the values assigned until the last value is reached. The first and second values may be separated by a comma or space. The second value can be omitted, in which case an increment of 1.0 is assumed. No separator is required between the ellipses and the final value, but either a comma or any number of spaces are allowed.

Examples

```
$assign    T = 10 ... 17 $
             ! T has values 10,11,12,13,14,15,16,17
$assign INT = 0 30 ... 120 $
             ! INT will have the 5 values 0,30,60,90,120
$assign V = -2.5 -2 ... 1.9 $
             ! V : -2.5 -2 -1.5 -1 -0.5 0 0.5 1 1.5 1.9
$assign Y = 1950,1960,...2010 $
             ! Y : 1950,1960,1970,1980,1990,2000,2010
```

Abbreviated lists may be freely mixed with, and have priority over, other elements in the assignment list — scalars, constants, or vectors. But remember that an `ASSIGN` statement cannot include any arithmetical expressions.

In general, all identifiers on the right-hand side of an `ASSIGN` must have values. However, if an identifier appears on both the left and right, it need not already exist, and its original length will be taken as zero. Thus the statements

```
$delete Z    $assign Z = Z %df $
```

are valid, and will store `%df` in the only element of the variate `Z`.

This is useful in macros which create vectors with values calculated by other parts of GLIM. For example, when constructing a vector of the means of other vectors, the first value to be stored need not be treated as a special case. The creation of a vector of means may be achieved by a macro which repeatedly executes

```
$tabulate the %%i mean into %m $
$assign MEANS = MEANS , %m $
```

5.2 Recoding data with the `MAP` and `GROUP` directives

A common task in most practical analyses is to transform or recode the observed data before fitting a model. For example, it may be necessary to ‘group’ a continuous variable or collapse a number of levels of a factor. General transformations may be achieved in GLIM using the `CALCULATE` directive [5.4]. Recoding a vector, that is, transforming all values in a given range to a single new value, can be a cumbersome task with `CALCULATE`, and GLIM includes two directives which make this easier. Both directives work in the same general way, and have identical syntax:

```
{$group}      [result_vector = ] input_vector
{ $map }            [intervals interval_vector]
                    [values value_vector]
```

where *intervals* and *values* are keywords and only the first letter of the keyword is significant. `MAP` is used to form a variate of the recoded values, while `GROUP` produces a factor.

In order to recode a vector, its entire range must be split into contiguous intervals and a value associated with each interval. The intervals are specified by an *interval_vector* whose values are the interval end-points. Thus with an *interval_vector* with values x_1, x_2, ... , x_n the range of values is divided into the intervals shown below.

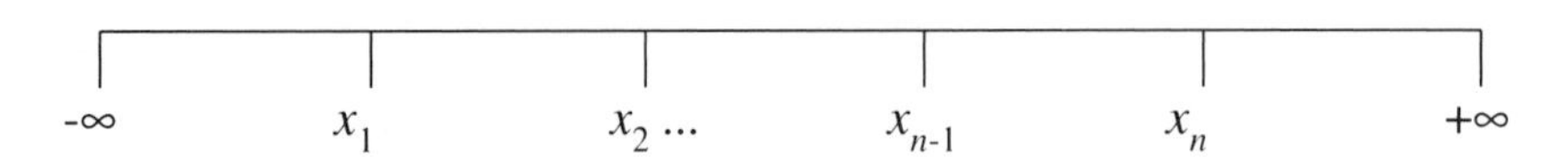

By convention the intervals are left-inclusive, that is, a value equal to an end-point value is included in the upper interval. The name of the *interval_vector* is usually preceded and followed by the symbol `*` which indicates additional left and right tail intervals extending to minus and plus infinity (∞). Thus the intervals phrase '`* X *`', where `X` has elements x_i $(i = 1, 2, \ldots, n)$ with $x_i < x_j$ for $i < j$, specifies the $(n+1)$ intervals

$$(-\infty,x_1)\,,\ \ [x_1,x_2)\,,\ \ [x_2,x_3)\,,\ \ldots,\ \ [x_{n-1},x_n)\,,\ \ [x_n,\infty)$$

where $[x$ specifies inclusion of the point and $x)$ specifies exclusion, so the second interval includes all values from x_1 up to (but not including) x_2.

If the `*`s are omitted from the phrase only the central $(n\text{-}1)$ intervals are specified and all values of the input vector must be in the interval $[x_1,x_n)$.

The values associated with each interval are specified as the elements of a *value_vector*, which must have the number of elements implied by the intervals phrase.

Examples

1. Assigning age groups: the statements

```
$assign INTS =    20 30  50 60 $
      :  VALS =  1  2  3   4  5 $
```

divide up the range as shown:

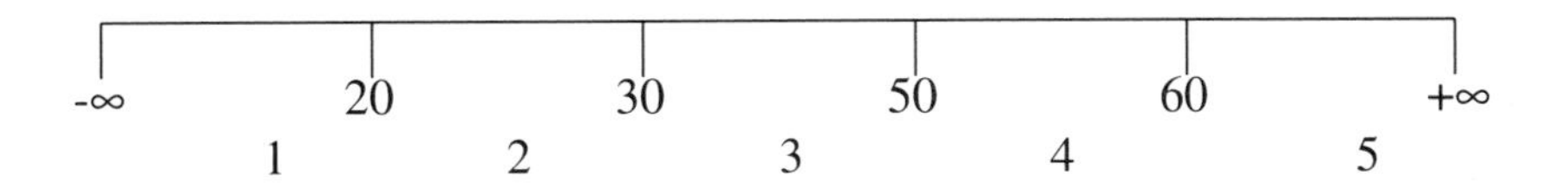

defining the intervals:

$(-\infty,20)$, $[20,30)$, $[30,50)$, $[50,60)$, $[60,+\infty)$

and

```
$group AGEGP = AGE intervals * INTS * values VALS
```

will apply the grouping, resulting in a five-level factor AGEGP.

2. A further example:

```
$assign X = 5 9 12 35 5 $
$assign INTS = 20 40 : MIDPTS = 10 30
     : REVLEVS = 2 1 $
```

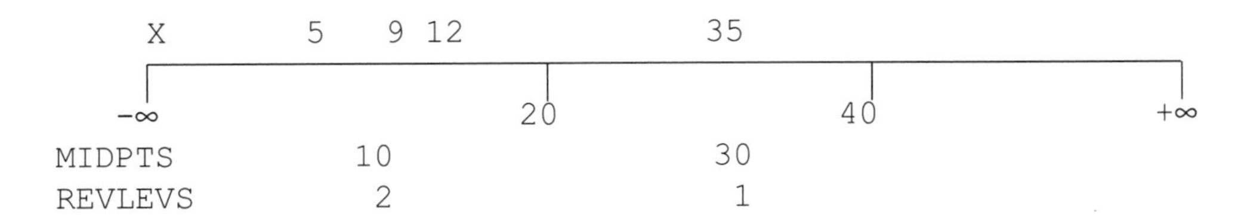


```
$map MX = X intervals * INTS values MIDPTS $
!    MX will have values    10   10   10   30   10
$group GX = X intervals * INTS values REVLEVS $
!    GX will be a 2-level factor with values 2 2 2 1 2
```

5.2.1 Default phrases in GROUP and MAP

The intervals and values phrases are optional and have generally useful default values. The default *interval_vector* (for both MAP and GROUP) is constructed from the distinct values of the *input_vector* and a right tail interval is included. For example, given an input vector with values x_1 , x_2 , ... , x_n;

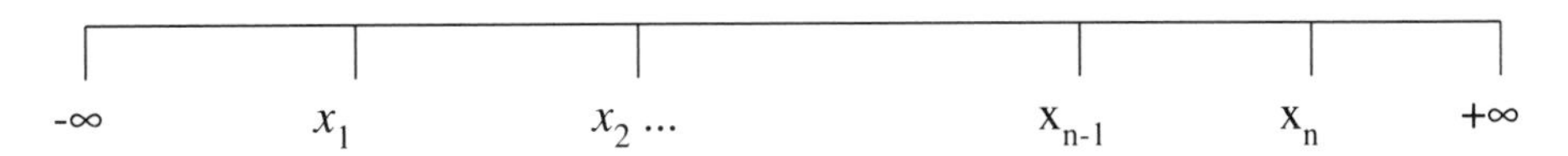

the default intervals are:

$[x_1,x_2)$, $[x_2,x_3)$, ... , $[x_{n-1},x_n)$, $[x_n,\infty)$

So `$group X $` is equivalent to `$group X intervals X * $` and `$map X $` is equivalent to `$map X intervals X * $`

When all the intervals have mid-points, MAP constructs the *value_vector* from the mid-point values. Intervals extending to infinity have no mid-point value so the mid-points cannot be used when the intervals phrase includes asterisks. If there is an upper-tail interval but no lower-tail interval, the default MAP *value_vector* has values equal to the lower end-points of the intervals. These may be represented graphically as:

(a) No infinite intervals:

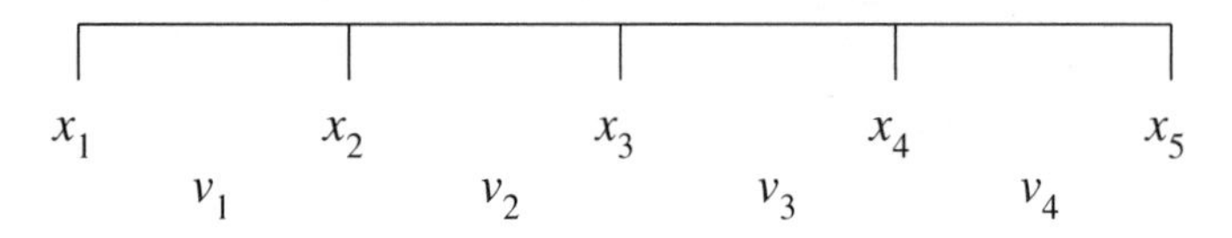

(b) An upper tail interval

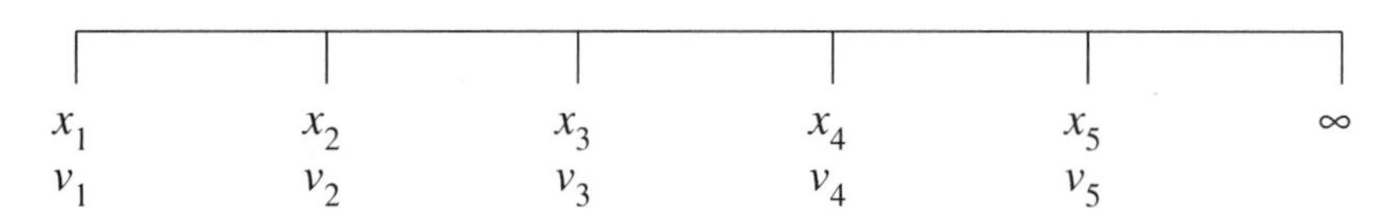

MAP cannot construct a default *value_vector* if there is a lower-tail interval.

Example

```
$assign X = 5 9 12 35 5   :   CUTS = 0 20 40 $
$map MX = X i CUTS $
  !   MX has values 10 10 10 30 10 (the mid points)
$map MX = X i CUTS * $
  !   MX has values 0 0 0 20 0 (the lower-end points)
```

The default *value_vector* for GROUP has values 1, 2, ... , m , where m is the number of intervals. Hence it creates factors with values most useful in GLIM, so it is often not necessary to specify a values phrase. For example

```
$assign CUTPTS = 0 5 15 45 65 $
$group AGE_GP = AGE intervals CUTPTS * $
```

will create AGE_GP as a five-level factor with levels 1, ... , 5 indicating the age groups 0–4 , 5–14 , 15–44 , 45–64, and 65+.

For an *input_vector* X with distinct values, `$group X $` is equivalent to assigning ranks; while `$map X$` leaves X unchanged.

5.3 The `EDIT` directive

As its name suggests, this directive is used for altering one or more values within existing vectors. The full syntax is

`$edit` [*integer1* [*integer2*]] *vectors list_of_values*

Here *integer1* and *integer2* denote the first and last indices of a contiguous set of elements within the *vectors*. The *list_of_values* may contain constants and scalars. These values will replace the indicated elements of the *vectors*. Note that `EDIT` applies exactly the same change(s) to each of the named *vectors*.

If only one element is to be changed, the second integer may be omitted. If both integers are omitted, *integer1* takes the value 1 and *integer2* the length of the first vector. The vectors in the list need not all be the same length but the range of elements chosen or implied must be valid for each vector. The number of values in the *list_of_values* must equal the number of elements implied by the integers.

Examples

```
$edit 13 A B 0.0 $
        ! sets the 13th element of A and B to zero
$edit 4  W  %a $
        ! changes the 4th element of W
$edit 2  4  W  10.9  8.1  7.6  $
        ! changes the 2nd, 3rd and 4th elements of W
```

Note the difference between the *list_of_values* in `ASSIGN` and `EDIT`. The `ASSIGN` list may contain vector names, scalar names, constant values, or abbreviated lists. The `EDIT` list can contain only constants and scalar identifiers. `EDIT` cannot be used to alter the value of a user scalar.

5.4 The `CALCULATE` directive

This directive takes the general form

`$calculate` [*identifier* =] *expression*

The *identifier* may be a scalar, vector, or vector element. The *expression* may contain any of these and also constants and function references linked by arithmetic or logical operators. All vectors in the expression must have the same length, which defines the length of the result. If the identifier does not already exist, it will be created as a variate with the resultant length. Note that, unlike `ASSIGN`, `CALCULATE` never adjusts the length of the resulting vector.

Although CALCULATE will store values in vectors declared as arrays, and arrays may be used on the right-hand side, it does ***not*** perform matrix arithmetic. It does provide vector arithmetic, where each operation is performed element-by-element.

Examples

```
$calculate  A = 1 $
! Unless A has been defined previously it is created
! with the standard length (as set by the last SLENGTH
! directive). Each element of A is given the value 1.

$calculate  X = Y $
! Assigns to each element of X the value of the
! corresponding element of Y.

$assign X = 1 2 ... 12 $
$calculate Y = 12*X  $look  X Y $
! will print the 12 times table
```

CALCULATE will not usually display the result of the operation. However, if there is no left hand side, the result will be output directly to the primary output channel. For example

```
$calculate  3/2 $
     ! will output 1.5
$calculate  %dv/%df $
     ! will output, but not save, the mean square
$calculate  X/Y $
     ! will output all values of the implied vector
     ! Z=X/Y
```

5.4.1 Expressions

Expressions are built up of operands (identifiers and constants), operators, functions, and round brackets. The functions are system identifiers [2.5.1] whose names start with %. They take one or more comma-separated arguments and return values derived from them. The arguments can themselves be expressions. When a function has a vector argument, it is evaluated for each element of the vector in strict sequence and a vector of values is returned; for example, %abs returns the absolute values of each element, not the norm of the vector. If a function has only one argument, the argument need not be enclosed in brackets, but omitting them may change the meaning because of the precedence rules. For example

%sqrt 9+7 gives the result 10, although
%sqrt(9+7) gives the result 4.

The simplest expression is a single number or identifier as used on the right-hand side of the first two examples in [5.4]. The arithmetic operators comprise

`+`   `-`   `*`   `/`   `**` (to the power of)   - (unary minus)

The logical and relational operators are described below [5.4.2]. Note that the operator - indicates both unary minus (to negate a value) and dyadic minus (to subtract one value from another). Simple operands may be combined with operators to form simple expressions, for example

```
X+Y   Z-1   -2/3.4   A**B
```

Sequences of simple expressions are allowed, for example

```
-A+B*C/D
```

If an invalid operand is given for the `/` or `**` operators (such as division by zero), or a function has an invalid argument, GLIM will issue the message

```
-- invalid function/operator argument(s)
```

It is not always possible to check that an operation will not cause arithmetic overflow, but some implementations of GLIM will trap such overflows, and print the message

```
-- floating point error(s)
```

Both of these messages are warnings; the calculation will proceed, assigning zero as the result of the invalid operation.

The interpretation (and hence the resultant value) of an expression depends on the precedences of the operators. In this respect GLIM follows the normal practice of arithmetic. Monadic operators, expressions in round brackets, and functions are evaluated first, then `**`, then `*` or dyadic `/`, then `+` and dyadic `-` .

The complete list of operators in their order of precedence is:

Highest	1	functions, monadic operators (/ (NOT) and -), brackets	
	2	**	
	3	* and dyadic /	
	4	+ and dyadic -	
	5	relational operators	== /= < > <= >=
	6	& (AND)	
	7	? (OR)	
Lowest	8	=	

It is always possible to use brackets to force a particular interpretation. Except when overruled by brackets, operations of equal precedence are performed from left to right.

Expressions in a CALCULATE statement can contain vectors, scalars, or a mixture of the two. Operations on vectors are carried out one element at a time in sequence. Any dyadic operation can be applied to two vectors of the same length, two scalars, or a vector and a scalar. In the last case the scalar is used repeatedly. The identifier on the left-hand side may be previously undefined; in this case it will become defined and given a length determined by the right-hand expression. Any other combination is non-conformable; the expression is faulted and no calculation is carried out.

As a simple example, if a set of Fahrenheit temperatures is stored in a vector named FAHRENHT, then the equivalent values in Celsius can be assigned to a variate CELSIUS which may be a new name or the name of an existing vector of the same length as FAHRENHT with

```
$calculate  CELSIUS = (FAHRENHT-32)*5/9 $
```

If a vector expression is assigned to a scalar identifier, the effect of the element-wise evaluation will be that only the last value will be stored after the statement has executed. For example

```
$calculate  %z = A $
! %z will contain the last element of A
```

Vectors of differing length cannot be mixed in an expression. The only ways of changing the length of a vector are to add more elements by using ASSIGN, to remove elements using PICK [5.5] or to DELETE the vector and redeclare it [4.6]. For calculation purposes it is possible to select or reorder the elements and hence alter the effective length by using an index vector (see [5.4.8]).

5.4.2 Beyond simple arithmetic

In addition to numeric values, GLIM expressions can contain operands whose value is either 'true' or 'false'. There is no distinct logical (or Boolean) type in GLIM; the value 0.0 is taken to be 'false' and any non-zero value is 'true' which is stored as 1.0. In this way logical quantities may be freely mixed with the ordinary arithmetic operations. For example:

```
$calculate HIGHER = A*(A>B) + B*(B>=A) $
```

sets each element of HIGHER to the maximum of the corresponding elements of A and B.

The relational operators comprise `==  /=  <  >  <=  >=`
and the logical operators are `&  ?  /`

Note that operators formed from more than one character must not have any spaces between the characters. The following table specifies the meaning of these operators:

`X == Y`	returns 1 when the corresponding elements of X and Y have the same value, or 0 if they do not. Note that a test of equality based upon real values will depend upon the machine accuracy for its result. For example, `%sqrt(8)==(2*%sqrt(2))` may or may not be returned as 'true'.
`X /= Y`	returns 1 when corresponding elements are unequal or 0 if not.
`X < Y`	returns 1 for each element of `X` which is less than the corresponding element in `Y` , 0 otherwise.
`X > Y`	returns 1 for each element of `X` which is greater than the corresponding element in `Y`, 0 otherwise.
`X <= Y`	returns 1 for each element of `X` which is not greater than the corresponding element in `Y`, 0 otherwise.
`X >= Y`	returns 1 for each element of `X` which is not less than the corresponding element in `Y`, 0 otherwise.
`X & Y`	returns 1 when corresponding elements are both non-zero, else 0.
`X ? Y`	returns 1 when either or both elements are non-zero, else 0.
`/X`	returns 0 for each element that is non-zero, or 1 if the element is zero (Not `X`).

The explanations above use `X` and `Y` as general operands. In practice the operands may be any system-defined or user-defined identifiers.

The relational operators are useful in forming subsets of data, weight vectors, and factors. For example:

```
$calculate  ELDERLY = AGE > 65 $
       ! creates a vector with values 0 or 1
       ! possibly for use as a weight variate
$calculate  ELDERLY = 1 + (AGE > 65) $
       ! generates values for a 2-level factor

$calculate  RETIRED =
   1 + ( SEX==1 & AGE>59 ? SEX==2 & AGE>64 ) $
$factor AGE_GP 6 $
$calculate  AGE_GP =
 1 + (AGE>4)  + (AGE>14) + (AGE>44)
   + (AGE>64) + (AGE>79)$
```

Provided that `AGE` has the standard length, the last two lines will create a factor with six levels (`AGE_GP`) from the continuous variate `AGE` (note that this can also be achieved directly using the `GROUP` directive).

5.4.3 Assignment in CALCULATE

The = sign was introduced at the start of this chapter with the ASSIGN and CALCULATE directives in a self-evident way to save a vector or the result of a calculation. In CALCULATE the = sign is, in fact, a dyadic operator with the restriction that the left-hand operand must be an identifier. The = sign can occur throughout an expression and the result of its operation is the value assigned. This has the following effects:

(a) More than one identifier can be initialized in a single statement, for example

```
$calculate  X = Y = Z = 1 $
```

(b) A number of different vectors may be formed in a single statement:

```
$calculate  X3 = X * (X2 = X*X) $
      ! will produce X2 = X**2 and X3 = X**3
```

(c) To keep track of a computation, a null addition may be used to force the result to be printed after the assignment, saving the user having to type an extra statement. Using this may make clearer the element-wise evaluation of vector expressions. For example

```
$calculate  0+Z = X * Y $
      ! will display and assign the
      ! element-by-element product of X and Y

$calculate  0+P = %np(Z) $
      ! assigns and prints the Normal probability
      ! of a given Z-score (using the system
      ! function %np)
```

5.4.4 Mathematical and statistical functions

The identifiers of the standard mathematical functions available in GLIM are given in the following table.

Function name	Meaning	Notes
%abs(x)	$\lvert x \rvert$	Absolute value
%sgn(x)	sign	Result is 1 for zero or positive x; -1 for negative x
%tr(x)	$[x]$	Integer part of x truncated towards 0
%sqrt(x)	square root	x must not be negative
%log(x)	natural logarithm	x must be positive
%exp(x)	e^x	Inverse function of %log
%sin(x)	sine x	x in radians
%cos(x)	cosine x	x in radians
%ang(x)	arcsine $(\sqrt{x})$	Angular transformation $0 \le x \le 1$; the returned value is in radians

The valid ranges for the sine, cosine, and exponential functions are limited by the type of computer in use. The User Note for the implementation of GLIM gives the details.

Note that GLIM measures angles in units of radians; values in degrees may be used (with the help of the system scalar `%pi`) by, for example

```
$calculate SIN_PHI = %sin( PHI_IN_DEG/180*%pi ) $
```

GLIM includes a number of functions which return the probability that a random deviate from a particular distribution is less than a given value x (the cumulative distribution function). There are corresponding functions which return the deviate associated with a given probability p. For all these functions the first argument is the deviate, or probability, and other arguments give parameters of the distribution.

Distribution	Lower-tail probability	Deviate function
Binomial(x,n,μ)	`%bip(k,n,mu)`	`%bid(p,n,mu)`
Beta(a,b)	`%btp(x,a,b)`	`%btd(p,a,b)`
Chi-squared(df)	`%chp(x,df)`	`%chd(p,df)`
F(df1,df2)	`%fp(x,df1,df2)`	`%fd(p,df1,df2)`
Gamma(a)	`%gp(x,a)`	`%gd(p,a)`
Normal(0,1)	`%np(x)`	`%nd(p)`
Poisson(μ)	`%pp(x,m)`	`%pd(p,m)`
Student's t (df)	`%tp(x,df)`	`%td(p,df)`

Example To output the p-value for a chi-squared statistic of 9.41 on 3 degrees of freedom, the following statement is used

```
$calculate 1-%chp(9.41,3) $
```

The following functions are included because of their usefulness in statistical modelling:

`%lga(x)`	Log gamma function
`%dig(x)`	Digamma function
`%trg(x)`	Trigamma function

The Reference Guide [14.6] gives a complete description of these functions.

5.4.5 Utility and attribute functions

GLIM has a function which assists the manipulation of dates and calculating the time intervals between them:

`%days(`*day* , *month*, *year*`)`

returns the number of days between the date(s) supplied and 13th September 1752.

The year should be specified as a four-digit number. For example `%days(5,7,1991)` returns 87222 indicating that July 5th 1991 is 87222 days after 13th September 1752. As with all functions, the arguments may be vectors. Thus from observed start and end dates stored in `STA_DAY`, `STA_MON`, `STA_YR`, `END_DAY`, `END_MON`, and `END_YR` (with years recorded as two-digit integers), the number of days between the start and end dates may be calculated by:

```
$calculate EXPOSED =%days(END_DAY,END_MON,1900+END_YR)
                  - %days(STA_DAY,STA_MON,1900+STA_YR) $
```

The function `%match(`*m* ,*y* `)` may be used to compare the contents of a macro with those of another macro, or a string. It returns a value of 0, 1, 2, or 3, indicating the degree of agreement between the strings. The `%match` function is described more fully in [7.2.2], but some examples of its results are given in the following table.

Macro `REPLY`	String/macro `TEST`	`%match(REPLY,TEST)`
`A`	`A`	3
`Yes`	`y`	2
`y`	`YES`	1
`n.      o.`	`N. O.`	3
`Maybe`	`y`	0

When working in GLIM, it is useful to be able to generate vectors whose values are sequences of the integers 1, 2, ... in some regular pattern. These may be used as index vectors in model fitting or `CALCULATE`, or as the values of a factor. Three functions are supplied which can be used to generate such values.

Name	Meaning	Values returned
`%cu(X)`	Accumulate	Cumulative sums of the elements of the argument `X`
`%gl(k,n)`	Generate levels	1, ..., *k* in groups of *n*
`%ind(X)`	Index	1, ..., *n* , where *n* is the length of `X`

The function `%cu` forms the cumulative element-by-element sum of its argument. For example, if `X` is a vector with values 1, 4, 3, 2, then `%cu(X)` is a vector with values 1, 5, 8, 10. The expression `%cu(1)` in an expression involving vectors will generate a vector with values 1, 2, 3, ..., the number of values being determined by the length of the vectors in the expression. Thus the statements

```
$number SUM    $calculate  0+SUM = %cu(X) $
```

cause successive subtotals of `X` to be printed, storing the final total in `SUM`.

The arguments (k and n) of the `%gl` function are usually positive integer expressions. If they are not integer valued scalars, they are rounded to the nearest integer. The `%gl` function generates values for a vector by assigning to it the integers 1 to k in blocks of n. It is particularly useful for assigning factor values in designed experiments, or for setting up multi-way contingency tables without reading in the values of classifying factors. For example

```
$calculate  PLOT=%gl(3,2) $
      ! would generate the values:
      ! 1  1  2  2  3  3  1  1  2  2  3  3  1   etc.
```

The number of values generated depends upon the length of the other vectors in the expression. In the simple expression above the number would be either the standard length (`PLOT` previously undeclared) or the length of `PLOT` (`PLOT` already declared).

The following example illustrates how the factors in a designed experiment may be created without reading their values

```
$units 12
$assign YIELD =  25.6 24.5 23.4 24.2
                      22.9 23.9 23.1 21.7
                      21.2 20.9 22.1 21.7 $
$calculate VARIETY = %gl(3,4)  : BLOCK = %gl(4,1) $
$factor VARIETY 3 BLOCK 4 $
```

This results in the following table of values for YIELD :

		BLOCK		
VARIETY	1	2	3	4
1	25.60	24.50	23.40	24.20
2	22.90	23.90	23.10	21.70
3	21.20	20.90	22.10	21.70

Note that this table can be printed in GLIM by the statement `$tprint YIELD VARIETY,BLOCK $`. See [6.9] for a full description of `TPRINT`.

Six functions are included to allow the user access to the attributes of an identifier:

Name	Returned value
`%dim(a,k)`	Length of kth dimension of array `a`
`%in(i,l)`	Position of identifier `i` in list `l`
`%len(x)`	Length of structure `x`
`%lev(f)`	Number of levels of a factor `f`
`%ref(f)`	Reference level of a factor `f`
`%typ(x)`	Type of structure `x`

The return value of %typ is a code number as follows: 4 for a scalar, 11 for a variate or array, 33 for a macro, 39 for a list.

The functions are useful in macros which need to check that their arguments are correctly set. For example, a macro may contain the following:

```
$calculate %a = %len(%1)/=%sl
$fault %a '%1 is the wrong length' $
```

If the argument of any of these functions is inappropriate, for example, requesting the reference level of a variate or the size of a non-existent dimension of an array, the function returns zero. Note that %typ returns the same value (11) for both variates and arrays. They must be distinguished by using the other functions. For example, to check that V is not an array, it must have %typ=11 and its length must equal the size of its first dimension. Thus to set %a to be 'true' (1.0) if V is not an array, the statement is

```
$calculate %a = %typ(V)/=11 ? %dim(V,1)/=%len(V) $
```

5.4.6 Pseudo-random numbers

The two functions %sr and %lr each return values that can be used as a 'random' values. %sr stands for Standard Random and %lr for Local Random. The values from %sr are generated by a mathematical operation and will always be the same series on any computer. The series of values will pass most of the tests for the output from a genuinely random process and the series will not repeat itself until 2^{35} (over 30 thousand million) values have been generated.

The operation generating %sr is relatively slow and on some computers %lr is more efficient. The process used by %lr depends on the implementation of GLIM that is being used, so in general it will not produce the same results on different machines. The characteristics of %lr are described in the User Note for the implementation.

The functions %sr and %lr each have one argument whose (rounded) integer value determines the range and type of the result. If the argument is negative or zero, the result value(s) will be uniformly distributed pseudo-random real numbers in the range [0, 1]. A positive integer argument n produces pseudo-random integers uniformly distributed in the range 0, 1, 2, ... , n.

The following statement assigns to X, pseudo-random deviates from a Normal distribution with mean zero and unit variance.

```
$calculate  X = %nd(%sr(0)) $
```

A pseudo-random sample (with replacement) of a vector X may be accessed by creating an index vector I in the following manner:

```
$calculate I = 1+%sr(%len(X)-1) $
```

The general use of index vectors in CALCULATE is described in [5.4.8].

The %sr function uses a multiplicative congruential generator, where each value x_i is derived from the previous value x_{i-1} by

$$x_i = 8404997x_{i-1} + 1 \quad (\text{modulo } 2^{35})$$

The process is started in each job by three special initial values (seeds) provided by the system which set x_0. These seeds may be examined by using the R option to the ENVIRONMENT directive. Users may set their own seed values for either generator at any time in a job, by using the directives SSEED or LSEED. For the standard generator the SSEED directive has the syntax

$sseed [*integer1* [*integer2* [*integer3*]]] $

The first two integers must lie in the range 0–4095 and the third in the range 0–2047. If the third, or second and third, or all the integers are omitted, the system default values will be used. Thus the statement `$sseed $` resets the generator to the same state as when GLIM started.

The system scalars %s1, %s2, and %s3 are initially set to the default seed values. They are continuously updated whenever %sr is used so that independent sequences may be generated in separate runs by recording the values of %s1, %s2, and %s3 at the end of a run, and using these values in a SSEED directive before values are generated in a following run.

5.4.7 Conditional and relational functions

There is one conditional function in GLIM. It takes the form:

%if (*expression1* , *expression2* , *expression3*)

The first argument, *expression1*, is a conditional expression, usually including a relational operator. If this expression is evaluated as 'true' (that is, non-zero), the result returned by the %if function will be the value of *expression2*; otherwise it will return the value of *expression3*. Note that when %if is executed, both *expression2* and *expression3* are evaluated, even though only one of them will be used. Thus,

```
%if( A<0 , 0 , %sqrt(A) )
```

will produce the same result as %sqrt(A) and both will generate a warning if A is less than zero. To avoid the warning use

```
%sqrt( %if( A<0, 0, A))
```

Examples

```
$calculate X = %if( X>0 , X , 0 ) $
      ! sets X to zero if it is negative
$number M   $calculate M = A $
      ! M will contain the last element of A
$calculate M = %if( A>M , A , M) $
      ! M becomes the largest element of A
$calculate NEW = %if(OLD==%lev(OLD) , OLD-1 , OLD) $
      ! combines the two highest levels of a factor
```

It is often shorter (and more efficient) to write an expression using logical values rather than use `%if`. For example

```
$calculate NEW = OLD - (OLD==%lev(OLD)) $
```

is exactly equivalent to the last example above. The expression

```
X*(A<B) + Y*(A>=B)
```

is, in various forms, very useful. It is exactly equivalent to

```
%if( A<B , X , Y )
```

The first example from [5.4.2] above

```
$calculate  HIGHER = A*(A>B)+B*(B<=A) $
```

could be written

```
$calculate  HIGHER = %if( B<=A , B , A) $
```

Similarly

```
$calculate  ELDERLY = AGE > 65 $
```

could be written

```
$calculate  ELDERLY = %if( AGE>65 , 1 , 0 ) $
```

and

```
$calculate RETIRED =
   SEX==1 & AGE>59 ? SEX==2 & AGE>64 $
```

could be written as

```
$calculate  RETIRED =
   %if( SEX==1 , %if(AGE>59,1,0) , %if(AGE>64,1,0) ) $
```

or

```
$calculate RETIRED = %if( SEX==1 , AGE>59 , AGE>64 ) $
```

In order to translate a date recorded as `DAY` and `MONTH` into the number of days since the start of the year, a very complex nested `%if` expression could be constructed. The following is much more straightforward

```
$calculate NDAY = DAY +31*(MONTH > 1) + 28*(MONTH > 2)
     + 31*(MONTH > 3) +...
```

Of course it is even easier using the `%days` function with

```
$calculate  NDAY =
     %days(DAY,MONTH,YEAR) -%days(31,12,YEAR-1) $
```

The following relational functions are retained for upward compatibility with earlier releases of GLIM. They all have equivalent relational operators, which are simpler to use.

Function	Operator
`%eq(X,Y)`	`X==Y`
`%ne(X,Y)`	`X/=Y`
`%lt(X,Y)`	`X < Y`
`%gt(X,Y)`	`X > Y`
`%ge(X,Y)`	`X>=Y`
`%le(X,Y)`	`X<=Y`

Note that although relational tests may be performed using operators (such as ==) or functions (such as `%eq`), the precedences of these differ and hence may produce different results. For example

```
%eq(2,2)+%eq(0,0)        ! yields 2
2==2+0==0                ! is interpreted as
                         ! (2==(2+0))==0  and so yields 0.
```

5.4.8 Indexed expressions

So far all expressions involving vectors have been evaluated for each element in sequence. Calculations may be required on a subset of elements only, or on a permutation of the elements. Any vector identifier in a `CALCULATE` directive may be followed by a vector expression in brackets. The length of the result is the length of the index (or subscript) expression and the elements selected are determined by the index values rounded to the nearest integer.

At its simplest, an index allows CALCULATE to assign to or from only a few elements of a vector. For example:

```
$calculate  X(2) = 1.5 $
        ! store a value in the second element of X
$calculate  X(2) = X(2)+1 $
        ! increment the second element of X
$calculate  %a = X(%len(X)-1) $
        ! save the penultimate element of X

$assign  SUB = 1 2 $
        ! set up an index vector and...
$calculate  X(SUB) = Y(SUB+2) $
        ! copy Y(3) to X(1) and Y(4) to X(2)
```

Note that it is not permitted to write a list of index values within the brackets in CALCULATE statements. In the last example X(1,2) would not be acceptable so ASSIGN must be used to set up the index vector.

The index values will normally lie between 1 and the length of the vector to which the suffix is attached. Values greater than the length or less than 0 will be faulted. In the last example, the only restrictions on the lengths of X and Y is that %len(X) ≥ 2 and %len(Y) ≥ 4. It is the index expressions which must be conformable. In the statement

```
$calculate X(SUBS) = Y(SUBS) + Z $
```

Z must have the same length as SUBS.

Index expressions can be very powerful; for example the following statements may be used to reverse the order of the elements of X:

```
$calculate %s = (%t=%len(X)) - 1 $
$assign REVERSE = %t %s ... 1 $
$calculate Y = X(REVERSE) $
```

Two vectors with the same length (A and B) may be merged, taking alternate values from each by the following statements:

```
$calculate %s = (%t=2*%len(A)) - 1 $
$variate %t COMBINED $
$assign I = 1 3 ... %s : J = 2 4 ... %t $
$calculate COMBINED(I) = A $
$calculate COMBINED(J) = B $
```

The use of index vectors and the `%gl` function can considerably reduce the amount of input required to create a dataset in GLIM. For example, consider the designed experiment described in McCullagh and Nelder (1989, page 366), which relates to the production of leaf springs for trucks. Here a factorial design with three replicates on the free height of leaf springs is applied to five factors, each with two levels, resulting in 48 observations of the response variate. The data is tabulated by McCullagh and Nelder as follows.

Factor levels					Replicates			Factor levels					Replicates		
					1	2	3						1	2	3
B	C	D	E	O	Free height			B	C	D	E	O	Free height		
-	-	-	-	-	7.78	7.78	7.81	-	-	-	-	+	7.50	7.25	7.12
+	-	-	+	-	8.15	8.18	7.88	+	-	-	+	+	7.88	7.88	7.44
-	+	-	+	-	7.50	7.56	7.50	-	+	-	+	+	7.50	7.56	7.50
+	+	-	-	-	7.59	7.56	7.75	+	+	-	-	+	7.63	7.75	7.56
-	-	+	+	-	7.94	8.00	7.88	-	-	+	+	+	7.32	7.44	7.44
+	-	+	-	-	7.69	8.09	8.06	+	-	+	-	+	7.56	7.69	7.62
-	+	+	-	-	7.56	7.62	7.44	-	+	+	-	+	7.18	7.18	7.25
+	+	+	+	-	7.56	7.81	7.69	+	+	+	+	+	7.81	7.50	7.59

Although the factor `E` is not in the standard order, it is repeated with a cycle length of eight for each replicate. The dataset may be constructed in GLIM by the following statements

```
$assign HEIGHT=
      7.78 7.78 7.81   8.15 8.18 7.88
      7.50 7.56 7.50   7.59 7.56 7.75
      7.94 8.00 7.88   7.69 8.09 8.06
      7.56 7.62 7.44   7.56 7.81 7.69
      7.50 7.25 7.12   7.88 7.88 7.44
      7.50 7.56 7.50   7.63 7.75 7.56
      7.32 7.44 7.44   7.56 7.69 7.62
      7.18 7.18 7.25   7.81 7.50 7.59 $
$factor 48 B 2 C 2 D 2 E 2 O 2 $
$calculate B = %gl(2,3) : C = %gl(2,6) : D = %gl(2,12)
   : O = %gl(2,24) $
$assign TEMP_E = 1 2 2 1 2 1 1 2 $
$calculate E = TEMP_E(%gl(8,3))    $delete TEMP_E $
```

The factors `B`, `C`, `D`, and `O` are generated by simple applications of `%gl`. For `E` the `%gl` function is used to generate an implied index vector so that values for `E` are extracted from a (short) temporary vector.

The special index value 0 (zero) can be used to produce expansion or contraction of vectors. If the index value 0 occurs for a vector providing values in an expression, a zero value is produced, and if it occurs for a vector being assigned to, no value is assigned. In this way the effective lengths of vectors may be altered to make conformable expressions.

For example, suppose it is required to extract the values of `X` corresponding to observations selected for another variate `Y` having a particular value in `%v`. We may not know in advance how many values will be found, but this can be calculated and used to declare the result vector.

```
$calculate  I = (Y==%v) $
      ! make a vector of 1s and 0s
 : %n = %cu(I) $
      ! count the 1s
$variate  %n  RES $
      ! declare vector to hold selected values
$calculate  J = I*%cu(I) $
      ! if I is zero then J is zero, otherwise J points
      !  to successive values from Y
 : RES(J) = X $
      ! non-zero values of J select elements from X
```

It is not necessary to form `J` explicitly, as the expression for it can be used directly; thus the last two statements can be collapsed into:

```
$calculate  RES(I*%cu(I)) = X $
```

This technique may be employed to form a normal plot of residuals from a regression when some observations have been weighted out. Provided that all the weights (in the system vector `%pw`) are zero or one, `%pw` may be used in an index expression as in the following example:

```
$calculate %n = %cu(%pw) $
      ! Count the number of points weighted in
$variate %n RESID NORMAL $
      ! and create the vectors to be plotted
$calculate RESID(%pw*%cu(%pw)) = %yv-%fv $
      ! %pw*%cu(%pw) is zero for %pw /= 1,
      !  otherwise the next index.
$sort RESID $
$calculate NORMAL = %nd((%cu(1)-0.5)/%n) $
$plot RESID NORMAL '*' $
```

In some circumstances is it necessary to 'subset' the data, for example to select the observations from all subjects in the age range 15–44. It is usually simpler to use the `PICK` directive [5.5] for this purpose, but subsetting is possible using indexing. For example, to extract the values from the variables `AGE`, `WEIGHT`, and `RESPONSE` for subjects whose age is between 15 and 44 the following could be used

```
$calculate I = (AGE>=15) & (AGE<=44) : %l = %cu(I) $
$variate %l SUBAGE SUBWGT SUBRESP $
$calculate NEW_IND = I*%cu(I) $
$calculate SUBAGE(NEW_IND) = AGE
  : SUBWGT(NEW_IND) = WEIGHT $
  : SUBRESP(NEW_IND) = RESPONSE $
```

As a practical example of the use of index vectors, consider a clinical trial where patient allocation to treatments is determined by a sequence of treatment category codes in random order linked to a list of identification or ID numbers. Frequently only a subset of ID numbers are allocated. Thus when the treatment allocation code is broken for analysis there is the problem of adding the appropriate treatment codes to the individuals in the dataset. For an array of 400 IDs and treatment codes and 126 subjects, the following statements (effectively a dictionary look-up sequence) do the job:

```
$data 400 ID_NUM TRT_CODE   $dinput 'dictionary' $
$units 126   $data TRIAL_ID AGE SEX RESPONSE
$dinput 'data' $
$calculate TREAT = TRT_CODE(TRIAL_ID) $
```

The following adds a new variable (`N_COMPS`: number of complications) to the data set, when values are only available for a subset of the individual units:

```
$units 126 $data TRIAL_ID AGE SEX RESPONSE
$dinput 'data' $
      !
$assign COMP_IDS = 7,11,15,301,27,292
      ! assign the ID numbers of the subset
  : N_COMPS = 1,2,1,1,2,1 $
      ! and the values of the new variable
      !
$variate 126 COMPLCNS   $calculate COMPLCNS=0 $
      ! Create the new variate
      !
$variate 400 J $
      ! Set up a dictionary variable J longer than the
      ! largest ID
```

```
$calculate J(TRIAL_ID)=%cu(1) $
      ! J is the position number of each
      ! trial id in the data set
$calculate COMPLCNS(J(COMP_IDS))=N_COMPS $
      ! Use J to get the appropriate
      ! positions in COMPLCNS to
      ! receive the values in the
      ! subset variable N_COMPS
```

5.5 The PICK directive

The previous section described how index vectors may be used in the CALCULATE directive to select subsets of vectors. For simple subsetting, it is easier to use the PICK directive whose syntax is

$pick [*target_vectors*] *source_vectors* *key_vector*

PICK selects those observations for which *key_vector*, a variate, has non-zero values, and puts the values of the vectors in the *source_vector* list into the equivalent vectors (that is, in the same position) in the *target_vector* list. Identifier names in these lists are separated by commas, and there must be the same number of identifiers in each list. If the *target_vector* list is omitted then the selected values are placed in the vectors in *source_vector* list whose length will be redefined as necessary; note that the original set of values will be lost.

For example

```
$calculate I = (AGE>=15) & (AGE<=44)
$pick SUBAGE AGE I $
```

will place the ages of all those between 15 and 44 in the new vector SUBAGE, while

```
$pick AGE I$
```

will place these observations in the original vector AGE, redefining its length as necessary.

```
$pick SUBWGT,SUBRESP  WEIGHT,RESPONSE I $
```

will create new vectors SUBWGT and SUBRESP containing the selected values of WEIGHT and RESPONSE respectively, for the same subset of observations.

The major advantage of using PICK is that GLIM automatically counts the number of non-zeros in the index vector, and creates the subset vector(s) with the correct length. In comparison, in the example using index vectors in [5.4.8], the length had to be calculated

(`%l=%cu(I)`) and the new vectors explicitly declared. Additionally, if only a subset of units is of interest, `PICK` can be used to reduce the length of variates, and this can result in space and time savings in subsequent analysis.

6 Data examination and display

Introduction

There are many facilities within GLIM for the examination and display of the contents of data structures. They can be used to explore the characteristics of a dataset before fitting a model, and to inspect the diagnostics of and present the output from a fit.

To output data values, use the directives `LOOK,PRINT,` or `TPRINT`. The `LOOK` directive processes only vectors or scalars and displays the output in parallel. The `PRINT` directive can output the contents of all types of identifier, and displays the output serially. The `TPRINT` directive will display a vector or an array as a table.

GLIM has facilities for graphical displays. The `HISTOGRAM` directive will produce a multiple histogram of a set of vectors and the `PLOT` directive will produce multiple scatter plots, both with optional point labelling. Both of these directives use character graphics and require no special graphics device or screen. Users with access to a graphics screen or other graphical output device can in addition produce high-quality line and point graphs by using the `GRAPH` directive. The style of the points and lines can be specified using the `GSTYLE` directive, text can be added to this graph through the `GTEXT` directive, and the position of the graph on the screen or page can be determined by the `LAYOUT` directive.

Elements within vectors can be reordered by the `SORT` directive. Tables of counts and of other summary measures of a vector cross-classified by a set of other vectors can be produced by the `TABULATE` directive. The resulting table can either be immediately output, or the table, and its cross-classifying vectors, can be stored as new GLIM structures.

6.1 Displaying values from vectors and scalars

The directives `LOOK` and `PRINT` can both be used to display the values of vectors and scalars. They differ in that `LOOK` outputs values in parallel, with one column for each identifier, whereas `PRINT` will print all the values of one identifier across the page before printing the values of the next.

For example, if `C` and `D` are vectors,

```
$look C D $
```

will produce two columns of values, with corresponding elements of `C` and `D` side by side, whereas

```
$print C D $
```

will output all of the values of `C` across the page, followed by all the values in `D`. We discuss the `PRINT` directive in detail in section [6.2].

6.1.1 The LOOK directive

The `LOOK` directive outputs the values of vectors or scalars as columns of real numbers. The syntax of the directive is either

`$look` [`(`*option_list*`)`] *scalar* [[`,`]*scalar*]`s`

which outputs the value of each *scalar*, or

`$look` [`(`*option_list*`)`] [*integer1* [*integer2*]] *vector* [[`,`] *vector*] *s*

which outputs the values of each *vector* in parallel. The *vectors* can be replaced by a *list*, in which case all elements of the list must be *vectors*.

In the simplest case, with *integer1* and *integer2* omitted, either all specified vectors should be the same length, or the first named must be the shortest. The range is then defined as the length of the first vector in the list.

If *integer1* and *integer2* are given, they specify an explicit range. These values denote the first and last elements to display. Omitting the second integer selects just one element from each vector.

The list of vectors can contain any number of identifiers but `LOOK` will display only as many columns as will fit on one line (the width of the current output channel can be set by the `OUTPUT` directive [4.3.3]) . The number of vectors that can be displayed will also be affected by the current accuracy setting [4.4.1] and the `STYLE` option. If too many vectors are specified, then the list of vectors is truncated and a warning is displayed:

```
-- list truncated
```

By default, the identifiers of the vectors are used as column headings, and the left-most column gives the element number of each row. The `STYLE` option in the option list can be used to change the format of this output. Values are always output as real numbers, with exponential notation being used if the number is too small or too big for the standard notation. Note that for vectors with large numbers of values, the `PAGE` directive can be used to paginate the output [4.4.2] , and the break-in facility [4.10.2] can be used to abandon the directive.

Examples

```
$number a=6 : b=3$
$look A B %sc %dv $       ! displays the values of the
                          ! scalars A,B, %sc and %dv
$look 1 10 C D $          ! displays the first ten elements
```

```
                              ! of vectors C and D
    $list ALL=C,D$
    $look ALL $               ! displays all elements of C and
                              !D (if C and D have the same
                              !length)
```

The only option in the option list is `STYLE`, which determines the appearance of the output.

`style =` *real*

If the value of *real* is zero or if the option is omitted, the vector identifiers are output at the head of each column and the left-most column gives the element number of each row in the output. If *real* is greater than zero, annotation on the output is increased, with cell borders added to the output. If *real* is less than zero, then the column headings and the column of element numbers are both supressed. This option is useful for writing vectors in parallel to an external file. The style setting does not affect the output of scalar values.

Example

```
    $output 'bj1.dat'$          ! redirect output to
                                ! the file bj1.dat
    $look(s=-1) C D %fv %rs$    ! write the vectors C,
                                ! D, %fv and %rs to the file
                                ! file without column
                                ! headings
    $output %poc$               ! restore output to the
                                ! primary output channel
```

6.2 General output: the PRINT directive

The `PRINT` directive provides a flexible method of outputting strings and values of identifiers, with a high degree of control available over the format of the output if this is required. The identifiers may include scalars, vectors, lists, or macros; it is also possible to output the names of identifiers. It may be used, for example, to output values with user-supplied labelling, to produce captions, or to assemble results into an analysis of variance table.

The syntax is

`$print [ (`*option_list*`) ]` *items* `$`

where an *item* will either be an identifier, or a string, or a **phrase** that controls the appearance or format of the output, or `;` or `/`. A phrase has the form

**integer*

or

**keyword real_list*

where the available keywords are `integer`, `real`, `margin`, `text`, `line`, and `name`. A keyword can be abbreviated to its first letter.

The types of the identifiers and other items can be freely mixed within one `PRINT` statement. A `PRINT` statement with no items will print a blank line.

6.2.1 The line buffer

`PRINT` uses a line buffer as temporary storage. Each item to be printed is added to the line buffer. When the line buffer is full, then the line is output to the current output channel and the buffer is emptied. Output is thus displayed line by line whenever the buffer is full or the terminating directive symbol is reached. If no terminating directive symbol is found, then the line buffer will not be emptied and the output will not be completed until the start of the next statement.

6.2.2 Printing scalars, vectors, and arrays

The simplest way to print a scalar or vector is to specify it as an item with no associated phrases. So

`$print` *scalars* `$`

outputs the values of the *scalars* in sequence across the line, and

`$print` *vectors* `$`

outputs the values of the *vectors*, writing them serially, element by element across the page. All the values of the first vector are output, then all the values of the second, and so on. The order and appearance will therefore be quite different from that obtained through using the `LOOK` directive.

As the values are output serially, scalars and vectors of different lengths can be mixed within the same `PRINT` statement.

An array is treated by the `PRINT` directive as a vector, with no account taken of its dimensionality. (Note that `TPRINT` [6.9] should be used to output an array in tabular form.)

By default, the output precision (the number of significant digits) and the number of decimal places of any value printed is controlled by the `ACCURACY` directive [4.1]. The default precision is four significant digits. By default, all values are output as real numbers in standard notation, not exponential notation.

Examples

```
$print %pi %nu$   ! prints the scalars %pi and %nu
                  ! on the same line
$print C D$       ! prints the values of C across
                  ! the page, followed by the
                  ! values of D
$list L=A,C,D$
$number A=2$
$print A L[2] $   ! prints the scalar A followed on
                  ! the same line by the values
                  ! of C
$number A=1E-15$
$print A$         ! prints 0.000
$number B=0$
$print B$         ! prints 0
```

Vectors may be indexed either by an integer, scalar, or vector. If a vector is indexed by an integer or scalar, then the specified element of the vector will be printed. If it is indexed by a vector, then the elements of the values vector specified by the index will be printed in the order given by the index. Each index vector is rounded to the nearest integer. An index of zero is ignored, and a negative index is faulted.

Examples

```
$variate 5 A$read A
20.2 22.7 24.3 27.4 29.9
$pr A(1)$                    ! prints 20.20
$assign B=1,0,2,1$
$pr A(B)$                    ! prints 20.20  24.30  20.20
```

The default output can be changed by the use of phrases.

**integer* is the accuracy phrase, and changes the current number of significant digits for the remainder of the statement. Examples of this phrase are `*5`, `*9`, and `*-1`. The *integer* corresponds to the first item of the `ACCURACY` directive, which sets the number of significant figures. As with the `ACCURACY` directive, any setting where *integer* >9 will be converted to 9. A negative *integer* causes values to be printed to the nearest integer, the decimal point remaining. Changes of accuracy within a `PRINT` directive do not affect the current setting of the `ACCURACY` directive, which is the value assumed at the start of each `PRINT` directive.

Example

```
$print *5 %a *2 %b %c $
      ! prints the value of %a with accuracy 5
      ! significant digits, then %b and %c with
```

```
        ! accuracy 2
$print %d $
        ! use the current accuracy to print %d (not
        ! necessarily 2)
$number a=1E-15$
$print  *9 A$
        ! will print A as 0.000000000, as exponential
        ! notation is not used by default
```

For greater control, the phrases INTEGER and REAL can be used.

The INTEGER phrase allows a value, scalar, or vector to be printed as an integer. Its full syntax is

$$\texttt{*integer} \left\{ \begin{array}{c} real \\ integer \\ scalar \\ vector \end{array} \right\} \quad real1 \; , \; real2 \; , \; real3$$

The three real values that may appear after the *real, integer, scalar*, or *vector* to be output are respectively: the field width; the minimum number of digits (extra zeros added are on the left if necessary); and a flag for justification within the field width (negative to print the value left-justified, positive or zero for right justification). If *real1*, *real2,* and *real3* are all omitted, the default is for the integer to be printed in the minimum field width possible. Each can be defaulted by specifying its value as zero. Trailing reals which have been omitted are set to the value zero.

The REAL phrase allows a value, scalar, or vector to be printed as a real. Its full syntax is

$$\texttt{*real} \left\{ \begin{array}{c} real \\ integer \\ scalar \\ vector \end{array} \right\} \quad real1 \; , \; real2 \; , \; real3 \; , \; real4$$

The four real values that may appear after the value, *scalar,* or *vector* to be output are respectively: the field width; the number of decimal places; a flag for justification within the field width; and a field to force exponential or standard notation. If *real1*, *real2*, *real3* and *real4* are all omitted but the REAL phrase is present, the default is for the integer to be printed in the minimum field width possible with GLIM making the choice of notation. Each real can be defaulted by specifying its value as zero. Trailing reals which have been omitted are set to the value zero.

Examples

```
$number X=123.456 : A = 1E-15$
$print *integer X        ! appears as 123
$print *integer X X      ! appears as 123123
$print *i X,8,5          ! appears as ___00123  (3 leading
                         ! spaces)
$print *i X,8,1,-1       ! appears as 123_____  (5 trailing
                         ! spaces)

$print *real X           ! appears as 123.5
$print *r X,8,2          ! appears as __123.46
$print *r X,10,2,-1,-1 ! appears as 1.23 e+02__
$print *r X,0,3          ! appears as 123.456
$print *r A              ! appears as 1.000 e-15
```

6.2.3 Printing text strings and macros

To print a string or macro, specify it as an item to the `PRINT` directive. Thus

`$print` *string* `$`

copies the string into the line buffer, and

`$print` *macro* `$`

outputs the text of the macro line by line with the first line joined onto any previous output from the directive in the line buffer.

Greater control over the format of the output of strings and macros can be obtained through the `LINE` and `TEXT` phrases.

The `LINE` phrase prints a string or the contents of a macro on a line by line basis taking account of the new lines in the string or macro. The `TEXT` phrase prints the string or macro ignoring new lines. The relevant syntax is

```
*line
```

or

```
*text
```

Examples

```
$print 'A directive starts with a $, like $print' $
$print 'the model''s deviance is ' *r %dv$
!
```

```
$print *t '
line 1 !
line 2 '$! produces    line 1 line 2
```

6.2.4 Printing names of identifiers

The NAME phrase allows the printing of the name of an identifier, rather than the values or contents of the identifier. The identifier can be a scalar, vector, macro, or list. Its main uses are where the identifier is a formal or keyword argument of a macro, or a system pointer, or a list element. The syntax is

`*name` *identifier* [*real1* [, *real2*]]

Real1 specifies the field width for the name, and *real2* specifies whether the name is left justified (*real2* negative) or right justified. The defaults are for the field width to be the number of characters in the name (*real1* = 0) and for the name to be right justified (*real2* = 0).

Example

```
$print 'The y-variate is ' *name %yv
```

6.2.5 Printing lists and list elements

`$print` *list_identifier* `$`

outputs the identifiers in the list, separated from each other by commas.

`$print` *list_identifier* [*scalar*] `$`

outputs the value(s) or contents of the element of the list specified by the scalar.

Example

```
$list ALL=C,D,T1,T2,S,PR,NE,CT,BW,N,PT !
$print ALL $
      ! prints C,D,T1,T2,S,PR,NE,CT,BW,N,PT
$print ALL[2] $          ! prints the values of D
$print *name ALL[2]      ! prints the identifier D
```

6.2.6 Specifying margins, new lines, and new pages

The character `;` forces a new line whenever it occurs as an item in a PRINT statement. The current line is printed, or a blank line if the buffer is empty.

The character `/` will skip to a new page if the output is being written to a file, otherwise it will have no effect.

The MARGIN phrase sets the left and right margins to the specified positions. The syntax is `*margin` [*real1* [, *real2* [, *real3*]]]

where *real1* specifies the left margin column and *real2* specifies the right margin column. *Real2* cannot be greater than the current output width set by the OUTPUT directive. *Real3* specifies the word-break size — the rounded value specifies that words of this length or less will not be split by the output routine.

Examples

```
$print ; $ !would print two blank lines.
$print : $ !would also print two blank lines.
$print *m 10,60 / 'Example output title';
 'The job number is ' *i %jn;
 'and the latest model has deviance ' *r %dv,8,0,0,1 $
```

6.2.7 PRINT options: saving output

Output from the PRINT directive goes by default to the current output channel. By using the STORE option, this output can instead be redirected and stored, saving the output as a macro. Thus the contents of macros may be defined dynamically during a GLIM run.

`$print (store=` *macro* `)` *items* `$`

If the STORE option is used, items are written to the *macro* line by line in the same format as would have been used if the items had been output to the current output channel. If the macro exists, then it must not be in use or on the stack [7.1]. If the macro exists and also appears as an item, then the macro contents are not changed until all items have been processed. See [7.9] for examples and further information.

6.3 Sorting

The SORT directive is followed by up to three items with syntax

`$sort` [(*option*)] *target_list* [$\left\{\begin{matrix} \textit{source\_list} \\ * \end{matrix}\right\}$ [*key_list*]]

where each *list* may be simply a single *vector,* or a *list identifier,* or a list of *vectors* separated by commas. In the simplest case the values of a vector X can be sorted into numerical order by

```
$sort  X
```

The default ordering is ascending, the option order=descending specifying descending numerical ordering. If it is required that the sorted values be stored in a different vector, Y say, the following form of the statement is used

```
$sort Y X
```

This leaves the values in X undisturbed. If the reordering is to be defined by the values of another vector, K say, the third item is specified

```
$sort Y X K
```

This reorders the values of X by a permutation that would sort K into numerical order and stores the result in Y. The vector K is referred to as the 'key'. This reordering can be applied to more than one vector simultaneously. For example

```
$sort Y,X,K Y,X,K K
```

The ordering defined by K is applied to each vector in the *source_list* (Y,X, and K in this case) and the values stored in the corresponding vector in the *target_list*. In the example above values of K would be sorted into ascending numerical order and the same ordering would be applied to Y and X. For example, the values of a complete dataset comprising several variables could simultaneously be put into chronological order using a 'date' variate as the *key vector.*

If some of the values of the key vector are equal, that is, 'tied', the resulting order of the sorted values for these tied values is arbitrary. Further key vectors can be specified to break these ties;

```
$sort Y X K1,K2
```

specifies that for tied values of K1 the order will be derived from the values of K2. This can be extended to any number of key vectors.

There are short forms for the syntax:

Short form	Equivalent form
`$sort X`	`$sort X X X`
`$sort Y X`	`$sort Y X X`
`$sort Y * K`	`$sort Y Y K`

6.3.1 Use of reals

The *source_list* and the *key_list* may contain *reals* rather than vector identifiers. These *reals* are replaced by vectors with values that are cyclically shifted versions of the integers 1, 2, ... , n where n is the length of the vectors in the command. If the *real* has rounded value m then the integers are shifted left m - 1 places for positive m and $|m|$ - 1 places right if m is negative. For example if $m = 2$ and $n = 5$ the values would be:

2 3 4 5 1

while for m = -2 they would be:

5 1 2 3 4

Reals as arguments have several uses. The current positions of the ranked elements of a vector can be obtained by

```
$sort  POS  1  X $
```

For example, `POS(1)` will now hold the position in `X` of its smallest element.

The ranks of the elements of a vector can be obtained in two steps by

```
$sort  RANK  1  X  :  RANK  1  RANK  $
```

Use of such *reals* in the *key_list* provides a mechanism for cyclically shifting the values in vectors in the *source_list* which can be useful in generating 'lagged' variables. The statement

```
$sort  XLAG  X  2  $
```

produces a lagged vector in XLAG, with `XLAG(2) = X(1), XLAG(3) = X(2),` and so on; but note that `XLAG(1) = X(`*n*`)`, so that for many applications it will be necessary to weight out the first unit in subsequent analyses.

Similarly, a vector of lead values can be obtained:

```
$sort  XLEAD  X  -3  $
```

would give `XLEAD(1) = X(3), XLEAD(2) = X(4)`, and so on.

6.4 Histograms

The `HISTOGRAM` directive draws a histogram using standard printing characters, no special graphics device being needed. The vector values are grouped into class intervals plotted on the vertical scale, and the frequencies are shown on the horizontal scale.

6.4.1 The HISTOGRAM directive

The `HISTOGRAM` directive has the full syntax:

`$histogram` [(*option list*)] [*vector* [/ *weight*]]*s* [*string* [*factor*]]

but this can be considerably simplified for basic applications. The simplest form of the directive is

```
$histogram vector $
```

Class intervals will be chosen to have 'nice' values. The histogram is scaled to fit the width of the output, with the number of intervals also being kept small where the total number of observations is small. The plot is annotated on the left with the lower and upper class intervals enclosed in brackets with a square bracket indicating inclusion of the end-point in the interval and a round bracket indicating exclusion. The count of the number of observations in the class is printed to the right of the class interval, and this is followed by the plotting symbols, with each plotting symbol representing one, two, or some other multiple of observations. The default plotting symbol used is the first letter of the vector identifier.

The plotting symbol can be substituted by a non-space *plot_character* by using the syntax:

```
$histogram vector 'plot_character'
```

Example

```
$var 200 C$
$calc C=%chd(%sr(0),1)$
      ! c has 200 approx. samples from the
      ! chi-squared(1) distribution.
$hist C '*'$
      ! gives the following output:

[ 0.000, 0.800) 124 **********************************
[ 0.800, 1.600)  34 ***********
[ 1.600, 2.400)  22 *******
[ 2.400, 3.200)   7 **
[ 3.200, 4.000)   5 **
[ 4.000, 4.800)   3 *
[ 4.800, 5.600)   1
[ 5.600, 6.400)   2 *
[ 6.400, 7.200)   1
[ 7.200, 8.000)   0
[ 8.000, 8.800)   0
[ 8.800, 9.600)   1
[ 9.600,10.400)   0
[10.400,11.200]   0
```

6.4.2 Multiple histograms

The `HISTOGRAM` directive can plot more than one vector at a time. The syntax is

```
$histogram vectors [ string ] $
```

which gives a joint histogram, with the full plot consisting of a number of interleaved subplots. By default, each subplot uses the first character of the identifier of the corresponding vector as its plotting symbol. Plotting characters can be specified as before, with the first non-space character of the string representing the first vector, the second character the second vector, and so on.

Example

```
$var 200 N$
$calc N = %nd(%sr(0))
$calc NSQ=N*N$
      ! get 200 approx samples from a Normal
      ! distribution and square them, then produce a
      ! joint histogram of C and NSQ.
$hist NSQ C$
      ! similar distribution!!
```

6.4.3 Grouped histograms

The values of a vector can be subdivided by a factor, which identifies separate groups within the dataset. The syntax can be extended to

`$histogram` *vector string factor* `$`

and this will produce a set of interleaved histograms as before, only with one for each level of the factor. Note that the string is now mandatory and must have k non-space characters, where k is the number of levels of the factor.

If more than one vector is specified, all of the specified vectors will be subdivided, and interleaving of both groups and vectors will occur. In this instance $n*k$ non-space characters must be present in the string, where n is the number of vectors.

Example

```
$histogram birthwt 'fm' sex $
      ! will produce a grouped histogram of birthwt
      ! sex=1 identified by f
      ! sex=2 identifiied by m
```

6.4.4 Weighted histograms

Each vector can have an associated weight vector of the same length. The values of the weight vector values can be zero or positive. For each class interval, the weight values corresponding to the contributing observations are summed, and the resulting weighted frequency is rounded before display.

Examples

```
$calculate W =%ind(Y)>4$        ! weight out the first 4
                                ! observations of Y before
$histogram Y/W                  ! producing a histogram.
                                !
$histogram C/B                  ! produce a histogram of the
                                ! totals of B for each
                                ! interval of C
```

6.4.5 **HISTOGRAM options**

The options to the HISTOGRAM directive give the user a greater degree of control of the appearance of the histogram. There are six options which can be used in the option list.

The STYLE option determines the appearance of the histogram, with syntax:

style = *real*

where if *real* is zero the default layout is produced, if *real* is greater than zero then plot borders are added which divide the class intervals, counts, and frequency plots, and if *real* is less than zero then all output is suppressed apart from the frequency plots.

The YLIMIT option specifies the lower (*real1*) and upper (*real2*) limits of the values for all vectors with syntax

ylimit = *real1* , *real2*

Only values of the vectors falling within or on these limits will be displayed, unless the TAILS option is also used.

The FLIMIT option is a synonym for YLIMIT.

The XLIMIT option, with syntax

xlimit = *real1* , *real2*

specifies the smallest (*real1*) and largest (*real2*) frequencies to be plotted as symbols. Counts outside this range are truncated on the plot but shown correctly as counts.

The size of the histogram can be determined by the COLS option, which specifies the maximum number of columns to be used in printing frequencies, and the ROWS option which specifies the number of rows and hence class intervals excluding tail rows. The syntax is

rows = *integer*
cols = *integer*

Finally, the TAILS option specifies whether an extra interval is to be added above or below the histogram; this option is only appropriate if the YLIMIT option is used. The syntax is

tails = *real1*, *real2*

If *y*-values above the upper *y*-limit exist and *real1* is positive, then an upper tail category will be produced; similarly if *y*-values below the lower *y*-limit exist and *real2* is positive, then a lower tail category will be produced.

Further details of the options available can be found in the Reference Guide [15.40].

Example

```
$hist (s=1 r=10 f=1,5) C $
```

produces

```
+--------------+------:------+
|[160.,240.)| 2   |CC     |
|[240.,320.)| 5+2 |CCCCC  |
|[320.,400.)| 3   |CCC    |
|[400.,480.)| 5+3 |CCCCC  |
|[480.,560.)| 2   |CC     |
|[560.,640.)| 3   |CCC    |
|[640.,720.)| 5+1 |CCCCC  |
|[720.,800.)| 0   |       |
|[800.,880.)| 0   |       |
|[880.,960.]| 1   |C      |
+--------------+------:------+
```

6.5 Scatter plots

Scatter plots are one of the most informative tools both prior to fitting a model and for checking whether it is adequate. Basic scatter plots are produced by the PLOT directive; high-quality scatter plots on a graphics screen or other specialist graphics output device are produced by the GRAPH directive which is discussed in [6.6]

6.5.1 The PLOT directive

The PLOT directive produces a scatter plot made up from printing characters; no special graphics device is required. The full syntax is

$plot [(*option list*)] *y_vector_list x_vector_list*
[*string* [*factor_list*]]

The syntax of the directive allows for a single *y*-vector to be plotted against a single *x*-vector, or several vectors to be plotted against a single vector, or multiple *y-x* plots to be superimposed on the same axes. Note that `PLOT` can represent data only by plotted points and not lines.
The simplest syntax of the directive is

`$plot` *y_vector x_vector*

The size of the plot will fit the screen or the default page size. The scales of the graph are derived from the smallest and largest values of each vector. Each point will be marked by the first letter of the identifier of the *y_vector*. If several points coincide to the resolution of the plot, then an integer showing the multiplicity is printed. Multiplicities greater then nine are printed as 9.

Example

```
$plot C S $
      ! produces a scatter plot with S plotted as the
      ! X_vector and C as the Y_vector.
      ! The plotting symbol is C. There are five incidences
      ! of two coincident points and one of three points
```

```
     |
     |                                                          C
     |
800. +
     |
     |                                      C
     |                                 C  C                        C
     |                                 C                      2
600. +                                    C
     |                                            C
     |
     |            C           C                             2
     |                                C   C                   2
400. +           C                     C    C
     |          C   C
     |     C
     |           C                        2    3
     |
200. +                             2
      +---------------+---------------+----------------+-------------
    400.             600.            800.            1000.
```

6.5.2 Multiple plots and plotting symbols

Multiple scatter plots are obtained by replacing the *y*-vector or the *x*-vector identifier by a list of identifiers. The list can either be a list of vectors, separated by commas or an identifier of type list, where the elements of the list are all vectors.

The allowed combinations for the lists are:

multiple *y*'s against a single *x*, with syntax

`$plot` *y_vector_list x_vector*

or a single *y* against multiple *x-vectors*

`$plot` *y_vector x_vector_list*

or pairs of *y-vectors* and *x-vectors*,

`$plot` *y_vector_list x_vector_list*

in which case there must be the same number of identifiers in each list, and the pairs are formed by taking the first element of each list, then the second, and so on.

In the first two cases, the lengths of all of the vectors must agree. In the third case, for each *y_vector-x_vector* pair to be plotted the lengths of the vectors must match, but each of the pairs may have a different length.

The first and third cases take the default plotting symbols from the first letter of the identifiers of the *y_vectors*; the second case instead takes the default plotting symbols from the first letter of the identifiers of the *x_vectors*.

Note that the `PLOT` directive in previous versions of GLIM allowed only a single *x_vector* and had a different syntax. The old syntax of

`$plot` *y_vectors x_vector*

is still accepted and is interpreted as though the *y_vectors* formed a *y_vector_list*.

User-specified plotting symbols can be used to identify different *y_vector-x_vector* plots and are specified via a string in a similar way to the `HISTOGRAM` directive. If there are *n* vector pairs, then the string must contain *n* non-space plotting characters, one for each pair. The syntax is

`$plot` *y_vector_list x_vector_list   string* `$`

Blanks and new lines within the string are ignored.

Examples

```
$plot C,%fv S $
      ! will superimpose the two plots C against S
      ! and %fv against S, using as plotting symbols C
      ! and %
```

```
$plot Y1 Y2 Y3 X$
      ! interpreted as $plot Y1,Y2,Y3 X $
$plot Y1,Y2 X1,X2 'ab' $
      ! plot Y1 against X1 with plotting symbol a and
      ! Y2 against X2 with plotting symbol b
```

6.5.3 Weighting and selection of points

Each *y_vector* in the *y_vector_list* can have associated with it a weight vector. The syntax for each appropriate *y_vector* is

y_vector /weight_vector.

If the *y_vector_list* is in fact a list identifier then the above syntax cannot be used, the syntax instead is

y_vector_list /weight_vector_list

in which case a *weight_vector* must be specified for each *y_vector.*

In any *Y, X* vector pair with a weight vector, if the weight for a particular point is zero, then the point will be omitted from the plot.

The points omitted are not used in defining the scale ranges. Any non-zero weight is treated as 1 and the point is plotted.

The system vector `%re` can also be used for storing weights. If `%re` has been assigned values by the user, these are treated as weights applied to all *y_vectors*. Weights in named vectors attached to individual *y_vectors* override the global effect of `%re`.

The lengths of weight vectors and `%re` if used, must be the same as their respective *y-vectors*.

Example

```
$calculate w=(%ind(y)>4)$     ! w omits the first 4 obsns.
$plot y/w,%fv/w x$            ! from the plot of y and the
                              ! fiitted values against x.
$calculate %re=w $
$plot y,%fv x$                ! An alternative method.
```

6.5.4 Labelling groups of points

It is possible to identify groups of points defined by a factor for each *y_vector-x_vector* pair, and to plot these subsets with different plotting symbols. A different factor can be specified for each *y_vector-x_vector* pair.

The list of factor names follows the string, with syntax

`$plot` *y_vector_list x_vector_list string factor_list*

In the simplest case, there will be a single *y_vector-x_vector* pair and a single factor

`$plot` *y_vector x_vector string factor*

where each level of the *factor* will be plotted with a different plotting symbol as specified in the *string*.

In more complicated uses, the *factor_list* will be a list of factors separated by commas. A factor is omitted by simply omitting it from the list, so the *factor_list*

```
,A,B
```

specifies no factor for the first pair, factor `A` for the second pair, and factor `B` for the third pair. The length of the *factor_list* must be equal to the number of *y_vector-x_vector* pairs. The *factor_list* can also be an identifier of type *list* with all of its elements of type *factor* (in which case factors need to be present for all *y_vector-x_vector* pairs, and none can be omitted).

If a factor has *k* levels, then different plotting symbols will be used for each of the *k* groups in the plot. The next *k* non-blank characters from the string will be used as plotting symbols. The total number of non-blank characters in the string must be at least as great as the sum of the number of levels for all the factors added to the number of omitted factors.

Example

```
      ! An analysis of covariance
$factor G 3$variate X$
$yvariate Y$
$fit G+X$
      ! linear regression with y-variate Y
      ! covariate X and factor G.
$plot y,%fv x 'abc %' G, $
      ! plots Y against X with factor G, identifying
      ! the different levels of G in the observed data
      ! with the plotting symbols a, b and c. Also
      ! plots %fv against X using % as the plotting
      ! symbol.
```

6.5.5 PLOT options

The `PLOT` directive has options to modify the size of the plot (`ROWS`, `COLS`), the range of values used to construct the axes (`YLIMIT`, `XLIMIT`), to add axis labelling and a title (`HLABEL`, `VLABEL`, `TITLE`), and to control the appearance of the plot (`STYLE`).

The `ROWS` option specifies the number of rows used for the *y*-axis of the plot. The *real* is rounded and must be positive and greater than 1. The `COLS` option specifies the number of columns used for the *x*-axis, and must be greater than 20. The syntax is

```
rows  = real
cols  = real
```

Note that the current output width and height, the second and third items in the `OUTPUT` directive [4.3.3], can also be used to control the size of the plot. If neither `ROWS` nor `COLS` is specified in the option list, the default size of the plot will be calculated from the current output width and height. The current values can be found with the statement `$environment c$.`

The `STYLE` option, with syntax

```
style  = real
```

chooses a general layout. The default (or `style = 0`) plot has y- and x- axes annotated with scale values. If the rounded *real* is positive, the whole plot is in ornate style, enclosed in a box with extra annotation. If the rounded real is negative, no annotation at all is produced.

These `YLIMIT` and `XLIMIT` options define ranges of values to construct the axes scales, with syntax

```
ylimit  = real1 , real2 [, real3 ]
xlimit  = real1 , real2 [, real3 ]
```

For each axis, the real values *real1* and *real2* are taken as the lower and upper end-points and the scale will be labelled with 'neat' values to span the interval. Points occurring outside the range specified by *real1* and *real2* are omitted from the plot. If *real3* is provided and the number of rows or columns is large enough, then it will be used as a step length, with axis annotation occurring after every increment of *real3*. The `YLIMIT` and `XLIMIT` options can be used either to increase the range so that the point cloud does not entirely fill the plot area and patterns can be better discerned, or to restrict the range of values so that outlying points are excluded.

The labelling options have syntax

$$\texttt{title} = \left\{ \begin{array}{c} \textit{string} \\ \textit{macro} \end{array} \right\}$$

$$\texttt{hlabel} = \left\{ \begin{array}{c} \textit{string} \\ \textit{macro} \end{array} \right\}$$

$$\texttt{vlabel} = \left\{ \begin{array}{c} \textit{string} \\ \textit{macro} \end{array} \right\}$$

The `TITLE` option allows the user to add a text title, which is centred above the plot.

The VLABEL option specifies a text label for the *y*- axis, which is printed above the *y*-axis and below any title.

The HLABEL option specifies a text label for the *x*- axis, which is centred and printed below the plot.

Each of these options can be provided as a text string or a macro. Each will reduce the size of the scatter plot unless the number of rows and columns is explicitly set.

Examples

```
$plot (style=1
       title='Cost vs size'
       vlabel='Size in MW'
       hlabel='Cost in millions of $'
       xlimit=0,1000) C S $
!  produces a labelled ornate plot
```

6.6 High-resolution graphs

The GRAPH and GTEXT directives differ from PLOT in that they expect to draw output on a high-resolution device such as a graphics terminal, a laser printer, or a pen-plotter. GRAPH can be used to produce scatter plots in a similar way to PLOT. It can also draw straight lines joining sets of points. GTEXT can be used to add further annotation to a graph. Two further directives — LAYOUT and GSTYLE — allow the user to position and scale the graph on the output device and to define pen styles.

The high-quality graphics facilities within GLIM are not intended to create presentation-quality graphics although many device drivers provide excellent print quality. GLIM does not support its own range of text fonts, nor can images be manipulated except by modifying statements and redrawing. Instead, GLIM will in general use the hardware capabilities of the device. Thus, for example, only four sizes of hardware characters are available on a Tektronix 4014, while full hardware character scaling is available on the Tektronix 4107 terminal.

6.6.1 Graphics devices

The GRAPH and GTEXT directives have the ability to generate output for a variety of devices. A small number of drivers for common devices are provided with all mainframe versions of GLIM. These include drivers for the Tektronix 4014, Tektronix 4107, Postscript output devices, and Hewlett Packard plotters.

Some devices are on-line, meaning they are directly connected to the program. An example is a terminal that can act in text or graphics mode. GLIM will then intermix text and graphics output, with graphics output normally being written to a separate screen. Other devices cannot be directly connected and are called off-line or file-based. GLIM writes the graphics output for these to a file and the user must take a separate action, probably outside GLIM, to send that file to the output device. A default filename for each device is used, but this name can be changed within a GLIM session if required.

Each graphics device in GLIM has a device_name which is a mnemonic of up to eight characters. A list of devices and associated device_names available on the local system may be obtained by issuing the statement

```
$environment g $
```

which produces output of the form

```
Graphical facilities:
  Available devices are:-

  Device description                   Name      Channel
  ------------------------------------ --------  -------
  Tektronix 4107/4109 terminal         T4107           6
  Tektronix 4014 terminal              T4014           6
  X11R3 X-windows device               X               6
  HP7550 plotter                       HP7550         97
  Monochrome A4 Postscript printer     POST           98
  Colour A4 Postscript printer         CPOST          99

       current setting T4107
```

For each device, a description of the device is followed by the device_name, the channel number used for graphics output by each device, and an indication of whether GLIM has used that device in the current session. Note that the inclusion of a device driver within the program may not imply that such hardware is available locally.

The GLIM graphics system has the concept of a default device, which is usually a graphics terminal or screen. The current default device is printed below the list of available devices. The user can choose another device to be used for subsequent graphics statements either by specifying a new device in the `DEVICE` option to the `GRAPH` or `GTEXT` directives, or via the `DEVICE` phrase to the `SET` directive [4.7]. In this context, the `SET` directive has the syntax:

`$set device='`*device_name*`'`

where *device_name* is an (up to) eight-character mnemonic for the graphics device required. The keyword `device` may be abbreviated to `d`. That device then becomes the current graphics device. Note that the name of the device must be given in full, and cannot be abbreviated.

In principle, there is no restriction to the number of devices open at the same time. For example, a user might want to produce two plots, both displayed on a Tektronix graphics screen but with the first copied to the Postscript printer, and the second to a plotter.

However, the number of actual devices which may be opened will be system dependent, and will depend on both the operating system in use and the implementation. The graph is drawn independently on each device, so the results may not be exact copies. GLIM will take advantage of the best resolution and range of features of each device.

6.6.2 The plotting area and current plotting region

For each device a **plotting area** is defined which is the largest square area on that device. The plotting area defines the area in which it is possible to produce graphical output. Associated with the plotting area is the **current plotting region**. The plotting region is initially defined to be the plotting area, but it can be redefined through the `LAYOUT` directive to be any region of the plotting area. Graphical output produced by the `GRAPH` or `GTEXT` directives will always be directed to the current graphics region.

6.6.3 Pen styles

Graphs in GLIM consist of one or more sets of *x-y* pairs consisting of a series of points. Each of these series of points is called a **polyline**, and each will be plotted in a certain style. The style chosen for the polyline will affect the symbol type, the type of the line joining the points, the line and symbol colour, and whether or not a symbol or a line is plotted at all.

For example, one style for a polyline might be a red dashed line joining the points with no symbol plotted; another style might be a yellow solid line with crosses marking each point.

— — — — — — — First style

—x—x—x—x—x—x—x—x—x— Second style

These styles are referred to as **pen styles** and each has a number (the **pen number**). There are 30 pen numbers available, of which the first 16 are predefined. The first nine pen styles are symbol only, and the later pen styles are line only with no symbols. The user may override these default pen styles and define new ones by using the `GSTYLE` directive [6.6.4]

A pen style consists of the following elements:

A line type: the line joins the points of the set in order: the first point to the second, the second to the third, and so on, up to the *n-1*th to the *n*th. The line type might be solid, dotted, dashed, or of different thicknesses, depending on the output device. If the line type index is 0 no line is drawn. If the line type is 1, a solid line is specified.

A symbol: the centre of the symbol marks the position of each point in the set of *x-y* pairs. The symbol set is defined within the program and is not hardware dependent. It consists of a cross, a triangle, an inverted triangle, a square, a dot, an asterisk, a plus symbol, and a diamond. The symbol index may be 0, in which case no symbol is plotted.

A colour: The colour represents both the colour of the line (if present) and the symbol (if present). The default colour is anti-background with index 1, which produces a black line on printers and plotters and light screens and a white line on dark screens. If colour is 0, no output is produced. On monochrome devices, all colour indices except 0 represent anti-background. On grey-scale devices, the colour indices represent different grey-scale intensities.

A list of available line types, symbol types, and colours and their indices, and a list of the currently set pen styles can be obtained with the statement `$environment g$`. A high-quality version of this produced on the current graphical output device can be obtained with `$environment h $`. This output may be produced on any output device by using the `SET` directive to change the current device before issuing the `ENVIRONMENT` directive.

6.6.4 The GSTYLE directive

The `GSTYLE` directive is used to define or redefine the pen style definition of any of the pen numbers used by the `GRAPH` directive. The directive has the syntax

`$gstyle` *integer1* [`colour` *integer*] [`linetype` *integer*] [`symbol` *integer*]

Integer1 must be in the range 1–30 and is the pen number to be defined. The keywords `COLOUR`, `LINETYPE`, and `SYMBOL` may occur in any order and at least one must be present. Only the first letter is significant. The integer following the keyword specifies the new index of the colour, line type, or symbol, as appropriate.The value 0 for `COLOUR` suppresses all plotting for that pen style; the value 0 for `LINETYPE` suppresses line plotting, and the value 0 for `SYMBOL` suppresses symbol plotting.

Each phrase acts independently of other phrases for that pen style, so for example setting a `COLOUR` will not affect the form of point or line type, and setting a `LINETYPE` will not in itself suppress any symbol set for marking points.

The `G` and `H` options to the `ENVIRONMENT` directive give a list of colour, line types, and symbols and their indices available on the current graphical output device.

6.6.5 The GRAPH directive

The `GRAPH` directive is used to produce scattergrams and line plots on a graphics device. The full syntax of the directive is

`$graph` [(*option list*)] *y_vector_list x_vector_list*
$\left\{ \begin{array}{c} \textit{integer\_list} \\ \textit{vector} \\ * \end{array} \right\}$ [*factor_list*]]

The *y_vector_list* , *x_vector_list* and *factor_lists* and the use and action of zero values in weight vectors and the system vector `%re` have the same form and effects as for the `PLOT` directive [6.5.2].

In its simplest form the syntax is

```
$graph y_vector x_vector $
```

which produces a scattergram of the *y*-vector against the *x*-vector on the current default device. The upper and lower limits of the *x* and *y* axes are calculated from the values of the *x_vector* and *y_vector*.

To produce a line plot, one method is to use pen style 10, whose default style is a solid line. Thus,

```
$graph y_vector x_vector 10$
```

will produce a line plot, joining (x_1, y_1) to (x_2, y_2), (x_2, y_2) to (x_3, y_3), and so on. However, if the default pen styles have been altered, an alternative method is to use the `GSTYLE` directive to redefine the style for pen 1.

Example

```
$graph y x$
        ! produces a scattergram of y against x
        ! in the current plotting region
        ! using pen style 1
        !
$gstyle 1 line 1 symb 0$
        ! redefine pen number 1 to be
        ! a solid line with no symbol
$graph y x$
        ! produces a line plot of y against x
        ! in the current plotting region
        ! using the redefined pen style 1
```

6.6.6 Multiple plots and plotting symbols

As with the `PLOT` directive, graphs with multiple *x-y* pairs are obtained by replacing the *y_vector* or the *x_vector* identifier by a list of identifiers. Again, the list can either be a list of vectors, separated by commas or an identifier of type *list*, where the elements of the list are all vectors.

Each *x-y* pair represents a polyline. By default, and starting with pen style 1, the `GRAPH` directive plots successive polylines with the style from the next pen number taken in order from the pen numbers 1, 2, ..., 30, 1, 2, ... and so on.

Alternatively, particular pen numbers can be associated with particular polylines by following the *x_vector_list* either with a list of integers or with a vector, where the elements of the vector are taken as the integers. If there are *n* polylines, then the *integer_list* must contain *n* integers, one for each polyline. The syntax is

$$\texttt{\$graph}\quad y\_vector\_list\ \ x\_vector\_list \quad \left\{ \begin{array}{c} integer\_list \\ vector \\ * \end{array} \right\}$$

If an asterisk is present in place of the *integer_list*, the default pen numbers are used.

Examples

```
$graph C,%fv S 1,10$    ! will graph C against S
                        ! using pen 1 and %fv
                        ! against S using pen 10
                        !
$graph Y1,Y2 X1,X2 $    ! graph Y1 against X1 with
                        ! pen  1 and Y2
                        ! against X2 with pen  2
$assign STY=2,4,6$
$graph Y X1,X2,X3 STY$ ! graph Y against X1 with
                        ! pen 2, Y against X2 with
                        ! pen 4 and Y against X3
                        ! with pen 6
```

The user should SORT [6.3] the points if necessary before drawing a graph which involves line plotting. On a screen device it may not be necessary to sort points that lie along a straight line, but the same plot intended for a pen-plotter might need to be sorted to avoid multiple passes over the same track. Sorting is also essential when plotting curved lines by a series of small straight-line segments.

Unlike PLOT, GRAPH does not recognize overlapping or coincident points. One way to distinguish coincident points is by adding a small random perturbation to the points to be plotted — this will need to be determined by trail and error.

Example

```
$calculate YP=Y+%sr(0)/10    ! Add a small perturbation
: XP=X+%sr(0) $              ! to Y and a larger
                             ! perturbation to X
$graph YP XP $
```

6.6.7 Factor levels and line plotting

Factors in a GRAPH directive are used in a similar way to factors in a PLOT directive [6.5.4]. Each factor level determines a separate polyline, which is defined as the subset of points in the *x-y* pair corresponding to that factor level. Each new polyline will cause a new pen style to be selected from the *integer_list*, or the next default pen style to be chosen. If the *integer_list* is given, there must be at least as many elements in the *integer_list* as the number of polylines generated by the directive. If the pen styles are to be defaulted, then the asterisk replacing the *integer_list* must be present.

Example

```
        ! An analysis of covariance
        !
$factor G 3$variate X$
$yvariate Y$
$fit G+X$
        ! linear regression with y-variate Y
        ! covariate X and factor G.
$graph y,%fv  x  1,2,3,10  G, $
        ! graphs Y against X with factor G, identifying
        ! the different levels of G in the observed data
        ! with the pen styles 1,2 and 3. Also plots %fv
        ! against X using pen style 10
$sort SY,SFV,SG,SX  Y %fv,G,X X$
$graph SY,SFV,SG 1,2,3,10  G, $
        ! the sorted version of the above graph.
```

6.6.8 Weighting

Weights can be attached to each point in an *x-y* pair either by a weight vector attached to the *y*-vector using a similar syntax to `PLOT` or through the user assigning values to the system vector `%re`.

The weights have the following effects:

1. A zero value for a weight excludes that point from the graph, and it is also not taken into account in deriving default scales. If a symbol is specified in the style, no symbol for that point is drawn. If a line type has been specified, the point is omitted from the polyline, and the previous point with non-zero weight is joined to the next point with non-zero weight, that is, no gap is left in the line.

2. A non-zero value for a weight has no effect on line drawing, but will cause the size of the symbol to be scaled by the value of the weight for that unit.

6.6.9 Output to more than one device

GLIM allows more than one device to be open at any one time, and a copy device may be specified in the `COPY` option in the option list of a `GRAPH` statement.

For example,

```
$graph (copy ='POST') D M$
```

will produce a graph of D against M using pen 1 on the default output device and will also reproduce the graph on the monochrome postscript device.

The graph is redrawn by GLIM, so the resolution of the copy in the above case will be better than the screen resolution. The output on the copy device will in general not be an exact copy of the default device, as the device may have different characteristics (for example, fewer colours, different line types, no scaleable characters). The copy device must not be the same as the default device.

A different copy device may be specified on each GRAPH command if required. GLIM will try to keep all devices open if it can, but on some implementations, old devices need to be closed.

6.6.10 Changing the default file name

The filename used by GLIM for file-based devices takes a standard default form for each device. The actual default name varies from implementation to implementation. It is possible on some implementations for users to specify their own filename for file-based devices. This is achieved by opening the channel for that device with an OPEN statement (with option STATUS=NEW) before the first use of that device.

Example

```
$open(status=new) 98='pscript.plt'
        ! (a different channel number might be used on
        ! the local system)
$graph(copy ='POST')D M$
        ! would change the default name of the
        ! postscript output file to pscript.plt
```

If the file for a particular device is open and has already received graphics output, the file should be closed using the CLOSE directive and reopened using the OPEN directive to direct subsequent output to a new file. All devices are left open at the end of a new job and are closed at the end of a session.

6.6.11 Options to the GRAPH directive

The GRAPH directive has options to specify the default device and copy device (DEVICE and COPY), control the range of values used to construct the axes (YLIMIT, XLIMIT), to add axis labelling and a title (HLABEL, VLABEL, TITLE), to control the appearance of the plot (STYLE, ANNOTATION), and to control actions before and after plotting (REFRESH, PAUSE).

Some options have the same syntax as for the PLOT directive and have similar actions. See Section [6.6.5] for YLIMIT, XLIMIT, TITLE, HLABEL, and VLABEL. Note however the special effect of a negative style with the REFRESH option described below.

DEVICE and COPY have syntax

```
device = string
copy = string
```

where string contains a valid *device_name*. The DEVICE option has a similar effect to the SET directive [4.7], [6.6.1]. It directs graphical output to the device specified, and makes that device the default device for further graphical output. The COPY option is described in [6.6.9].

The REFRESH option allows the user to draw multiple plots on the same page or screen. The syntax is

```
refresh = { yes }
          { no  }
```

The default, REFRESH=YES, causes the plotting area to be cleared at the start of each GRAPH directive; this means that the terminal screen will be cleared and a new page is started on a plotter or printer. If REFRESH=NO, then the plotting area is not cleared. If no LAYOUT directive has occurred since the previous graph, and if STYLE is negative, then superimposition on the previous graph is possible. Intervening LAYOUT directives between GRAPH statements using REFRESH=NO will cause graphs to be produced in different plotting regions on the same screen or page. An automatic refresh is carried out on all open devices following a NEWJOB statement.

The PAUSE option, with syntax

```
pause = { yes }
        { no  }
```

only has a relevance for screen-based devices in interactive mode. The default, PAUSE=YES, means that the program will output an audible indication and wait for the 'return' key to be pressed before continuing with the GLIM session. If PAUSE=NO, no such pausing will occur.

The ANNOTATION option with syntax

```
annotation = { yes }
             { no  }
```

allows the user to omit numeric labelling of the tick marks of the axes. This is helpful to avoid clutter where many scatter plots are drawn on the same page. If ANNOTATION=NO, the axes are calculated and drawn with tick marks, but the numeric labels are not displayed. The default, ANNOTATION=YES, is to display both the tick marks and numeric values.

Example

```
$graph (c='post' refresh=n ylim=-0.1,0.4
title='Residual Variance'
hlab='Sample size'
vlab='Parameter s.e.s with predicted confidence limits')
     y1,y2u,y2l,y3u,y3l  v1  10,11,11,14,14 $
      !
      ! produces the following graph
      !
```

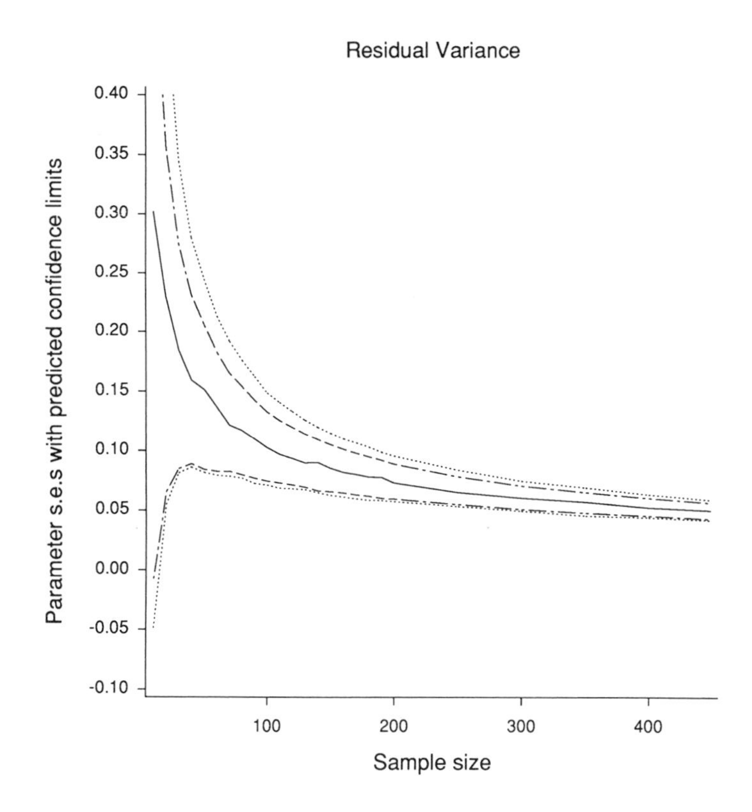

6.7 Adding further text to a graph

The GTEXT directive displays text on the current graphical output device. The text may either be superimposed on the current graph or may create a new graph consisting solely of text.

6.7.1 The GTEXT directive

The full syntax of the directive is

`$gtext [(`*option_list* `) ]` *y_item* [/*weight_item*] *x_item* *text_list*

The *y_item, weight_item,* and *x_item* can either be real values or vectors. The *text_list* can be one or more strings or macro names.

In its simplest form this is

`$gtext` *y_value x_value* *string* `$`

where *y_value* and *x_value* are reals. The effect is to superimpose on the current graph the text specified in the string centred vertically and to the right of the point (*y_value, x_value*), which is specified in user coordinates from the current graph.

This mode is the default mode, and may be specified explicitly by setting `MODE=PREVIOUS` in the option list. In this mode, the `GTEXT` directive must follow a successful `GRAPH` directive, and there must be no intervening `LAYOUT` directives or changes of device.

Example

```
$assign y=0,5,10,5,0,0$
       : x=1,2,3,4,5,1$
$graph y x 10$                  ! a line plot of y vs x
$gtext 10 3 'top of triangle'$  ! add some annotation.
```

An alternative coordinate system is available by setting the `MODE` option to `WINDOW` in the option list. The effect of this is to allow text to be drawn without the need for a previous graph from a `GRAPH` directive. The coordinate system in this mode is taken to be a unit square.

If the *y_values* and *x_values* are replaced by vectors of length *n*, then this specifies that there are *n* text strings to plot, and there must be as many *text_items* as values in the *y_items* and *x_items*. A vector of length *n* with the same value for each element may be replaced by a real value or scalar.

Example

```
$assign X=0,10 : Y=1,15 $
$graph Y X 10 $
$gtext Y X 'Start','Finish' $
```

If a *weight_item* is specified, then it specifies a multiplicative scaling which is applied to its associated text string. The size of the displayed text is increased or decreased in size. If no *weight_item* is specified, then a weight of 1.0 is assumed for all text.

6.7.2 GTEXT options

The `GTEXT` directive has options to specify the mode and coordinate system (`MODE`), the default device and copy device (`DEVICE` and `COPY`), to control the range of values used to construct the axes (YLIMIT, XLIMIT), to control the appearance of the plot (STYLE), to control the location and orientation of the text with respect to the plotting position (LOCATION, ORIENTATION), and to control actions before and after plotting (REFRESH, PAUSE). Many of these options are only available when mode is set to WINDOW.

The MODE option with syntax

$$\texttt{mode} = \left\{ \begin{array}{c} \texttt{previous} \\ \texttt{window} \end{array} \right\}$$

controls the coordinate system used. If MODE is PREVIOUS, the current plotting region defined by the last LAYOUT command and the current axes scaling defined by the last GRAPH command will be used. If MODE is WINDOW, the plotting region will be treated as a square with (0,0) at the lower left corner and (1,1) at the top right. The default MODE is PREVIOUS.

The ORIENTATION and LOCATION options control the relative position of the text in relation to the specified plotting point. The syntax is:

$$\texttt{location} = \left\{ \begin{array}{c} \texttt{left} \\ \texttt{right} \\ \texttt{centred} \\ \texttt{above} \\ \texttt{below} \end{array} \right\}$$

$$\texttt{orientation} = \left\{ \begin{array}{c} \texttt{horizontal} \\ \texttt{vertical} \end{array} \right\}$$

The default ORIENTATION for text is HORIZONTAL and text will be written left to right parallel to the x-axis. VERTICAL text will be written from the bottom to the top, parallel to the y-axis with the character text rotated ninety degrees anti-clockwise.

The default LOCATION is to start writing the text to the RIGHT of the plotting position from left to right. If LOCATION is LEFT, the text will end at the plotting position; if CENTRED, the text will be centred. If LOCATION is ABOVE or BELOW, the text will be centred, but above or below the plotting position respectively.

The options PAUSE and REFRESH have an identical function as for the GRAPH directive [6.6.11] and are also available under either mode. The default REFRESH setting is NO, and the default PAUSE setting is YES.

The following options are only available if MODE=WINDOW.

DEVICE and COPY have the same syntax and meaning as in the GRAPH directive.

The YLIMIT and XLIMIT options specify the coordinate system of the plotting region and have the following syntax

```
ylimit      = real1 , real2
xlimit      = real1 , real2
```

If the YLIMIT option is present, then *real1* specifies the lower bound and *real2* the upper bound of the coordinates for the *y*-axis. If the YLIMIT option is absent, then *real1* is taken as 0.0 and *real2* is taken as 1.0

The XLIMIT option has the same effect, but on the *x-axis*.

The STYLE option adds a border to the text. The syntax is

```
style = real
```

If the *real* value is positive, then a border is drawn around the plotting region. If *real* is zero or negative, or if the style option is not given, no such border is plotted.

Example

```
$gtext (mode=w location=centred )
      0.5  0.5 'the centre of the plotting region' $
```

6.8 The LAYOUT directive

The LAYOUT directive defines the current plotting region which will be occupied by subsequent graphics produced by the GRAPH and GTEXT directives. It will affect the position and size of graphs. LAYOUT can be used with the REFRESH=NO option to the GRAPH and GTEXT directives to produce multiple graphs on the plotting area.

The directive has the syntax:

```
$layout [real1 [,] real2   real3 [,] real4 ]
```

The plotting area is treated as a unit square, with the origin (0.0,0.0) as the lower left corner and (1.0,1.0) as the upper right.

If no *reals* are given, the whole plotting area will be used by the next GRAPH or GTEXT directive.

The four real values are read in pairs (the optional commas provide readability) and define the lower left corner (*real1*, *real2*) and upper right corner (*real3*, *real4*) of the plotting region to be used, in terms of (*x*, *y*) coordinates. Any scaling applies to the whole graph including symbols, titles, and axis annotation. However, if the plotting region defined by the LAYOUT directive is not square, the symbols and characters in any text will change size but not distort.

The current LAYOUT settings may be inspected with the statement $environment g$. The settings remain in force until changed by another LAYOUT directive or until a NEWJOB statement. The setting applies to all devices in use.

Example

```
$layout 0,0 0.5,0.5$              ! a graph of A against B
$graph A B$                       ! in the bottom left of the
                                  ! plotting area,
$layout 0,0.5 0.5,1.0$            ! a graph of C against D
$graph (refresh=no) C D$          ! in the top left, and
$layout 0.5,0.0 1.0,1.0$          ! a graph of E against F
$graph (refresh=no) E F$          ! on the right hand side
                                  !
$layout$                          ! Make the plotting region
                                  ! the whole plotting area.
```

6.9 Printing tables

The TPRINT directive enables the user to display one or more vectors in tabular form. The general syntax is

$tprint [(*option_list*)] *list1* *list2*

where *list1* specifies the vector(s) to be printed and *list2* defines the desired shape of the table. Each *list* can be a list identifier holding vectors only or a list of items separated by commas. The options available are STYLE which controls the form of the output and COLS which specifies the maximum width of the table.

6.9.1 Defining table structure

In the simplest form of *list2* the user specifies the number of levels for each dimension. For example, after defining a vector with 20 values by

```
$variate 20 X  $calculate X=10+%cu(1) $
```

it can be printed as a table with 4 rows and 5 columns as follows:

```
$accuracy 4 1 $
$tprint X 4,5 $

       1     2     3     4     5
  1 11.0  12.0  13.0  14.0  15.0
  2 16.0  17.0  18.0  19.0  20.0
  3 21.0  22.0  23.0  24.0  25.0
  4 26.0  27.0  28.0  29.0  30.0
```

If Y has already been defined as a 4×5 array *list2* may be omitted:

```
$array X 4,5 $
$tprint X $

        1     2     3     4     5
  1  11.0  12.0  13.0  14.0  15.0
  2  16.0  17.0  18.0  19.0  20.0
  3  21.0  22.0  23.0  24.0  25.0
  4  26.0  27.0  28.0  29.0  30.0
```

The table can have as many dimensions as required, as in the following four-dimensional table

```
$assign Y=11,12,...,26 $
$tprint Y  2,2,2,2 $

     1                          2
     1           2              1          2
     1     2     1     2        1     2    1     2
1  11.0  12.0  13.0  14.0     15.0  16.0  17.0  18.0
2  19.0  20.0  21.0  22.0     23.0  24.0  25.0  26.0
```

The user has control over the format of the table by use of / instead of a comma in the list. This separates the row dimensions from the column dimensions. For example

```
$tprint Y 2,2/2,2 $

         1           2
         1     2     1     2
  1 1  11.0  12.0  13.0  14.0
    2  15.0  16.0  17.0  18.0

  2 1  19.0  20.0  21.0  22.0
    2  23.0  24.0  25.0  26.0
```

Alternatively the shape of the table can be defined by factors which classify the values of the table. Thus

```
$gfactor 16 A 2 B 2 C 2 D 2$
```

will allow `Y` to be indexed in standard order by the four factors.

```
$tprint Y A,B/C,D $

    C    1           2
    D    1     2     1     2
  A B
  1 1  11.0  12.0  13.0  14.0
    2  15.0  16.0  17.0  18.0
```

```
2 1  19.0  20.0    21.0  22.0
  2  23.0  24.0    25.0  26.0
```

Use of factors allows the table to be printed with dimensions in a different order to that on input, for example:

```
$tprint Y  C,D/A,B $
```

```
   A    1             2
   B    1      2      1      2
 C D
 1 1  11.0  15.0    19.0  23.0
   2  12.0  16.0    20.0  24.0

 2 1  13.0  17.0    21.0  25.0
   2  14.0  18.0    22.0  26.0
```

The classifying vectors in *list2* can also be variates; these must then have values in numerical order. Note that the classifying vectors produced by TABULATE are always in numerical order. The levels are then labelled with the distinct values of the variate.

If the levels of factor A refer to years 1981 and 1991 then A could be replaced by a variate YEAR as follows;

```
$assign AVAL=1981,1991 $calculate YEAR=AVAL(A) $
$tprint Y C,D/YEAR,B $
```

```
 YEAR  1981.              1991.
    B    1       2          1       2
  C D
  1 1   11.0    15.0      19.0    23.0
    2   12.0    16.0      20.0    24.0

  2 1   13.0    17.0      21.0    25.0
    2   14.0    18.0      22.0    26.0
```

6.9.2 Parallel tables

More than one vector can be printed simultaneously, the identifiers in *list1* being separated by commas. For example, the vector Y2 could be printed in parallel with Y by

```
$calculate Y2=Y**2 $
$tprint Y,Y2 C,D/YEAR,B $
```

```
   YEAR  1981.              1991.
      B    1       2          1       2
C D
1 1  Y    11.0    15.0      19.0    23.0
    Y2   121.0   225.0     361.0   529.0
```

```
       2  Y    12.0   16.0    20.0   24.0
         Y2   144.0  256.0   400.0  576.0

     2 1  Y    13.0   17.0    21.0   25.0
         Y2   169.0  289.0   441.0  625.0

       2  Y    14.0   18.0    22.0   26.0
         Y2   196.0  324.0   484.0  676.0
```

6.9.3 Style options

There are three forms of output. The above examples are in the default format corresponding to option `style=0`. If the body of the table only is required, that is, without any indexing and labelling, `style=-1` should be used. The third form prints borders round the table and is invoked by using the option `style=1`.

Example

```
$tprint (style=1) Y,Y2 C,D/YEAR,B $
```

```
          +---------------------------------+
     YEAR |  1981.          |  1991.        |
        B |    1       2    |    1       2  |
  C D     |                 |               |
+---------+---------------------------------+
| 1 1  Y  |   11.0    15.0  |   19.0   23.0 |
|      Y2 |  121.0   225.0  |  361.0  529.0 |
---------------------------------------------
|   2  Y  |   12.0    16.0  |   20.0   24.0 |
|      Y2 |  144.0   256.0  |  400.0  576.0 |
+---------+---------------------------------+
+---------+---------------------------------+
| 2 1  Y  |   13.0    17.0  |   21.0   25.0 |
|      Y2 |  169.0   289.0  |  441.0  625.0 |
---------------------------------------------
|   2  Y  |   14.0    18.0  |   22.0   26.0 |
|      Y2 |  196.0   324.0  |  484.0  676.0 |
+---------+---------------------------------+
```

6.10 The TABULATE directive

This directive is used to produce tables of various statistics computed from raw data. The description that follows is intended to give the main uses of `TABULATE`, but cannot be fully comprehensive as the number of possible combinations of input and output parameter settings is very large. The syntax is

`$tabulate [(`*option_list*`)] [`*phrase*`]`*s*

The input and output parameters are specified in *phrases* preceded by the *phrasewords* `THE`, `FOR`, `WITH`, `COUNT`, `INTO`, `BY`, and `USING`. The *options* control the format of printed tables and are the same as for `TPRINT` [6.9].

6.10.1 The input phrases

The input phrases have *phrasewords* `THE`, `FOR`, `WITH`, and `COUNT` with the following syntax

`the` *vector statistic* [`,` *statistic*]*s*

defines the input *vector* from which summary statistics are to be computed and *statistic(s)* are keywords that specify which statistics are required.

statistic is one of:

```
mean
total
variance
deviation          (standard deviation)
smallest
largest
percentile    value
interpolate   value
weight
```

where *value* is a percentage between 0 and 100. Each *statistic* is fully defined in the Reference Guide [15.84].

`for` *vector* [`,` *vector*]*s*

specifies the vector(s) that define the classification of the table. If omitted the requested summary statistic(s) are calculated for all observations.

`with` *vector*

defines the weight vector whose values are used in the weighted form of the statistic. Alternatively the phrase

`count` *vector*

specifies a weight vector with values taken as frequency counts. This gives a different form of weighted statistic than that used via a `WITH` phrase. Specifying a weight vector can be used as a simple technique for omitting observations from the calculations. When neither a `WITH` nor a `COUNT` phrase is specified the weights for all observations are taken to be 1 (that is, every observation is included).

If only input phrases are given the resulting table will be output. However it is often useful to be able to store the resulting table for further manipulation or to form the basis of model fitting. This is achieved by specifying output phrases.

6.10.2 The output phrases

The output phrases are used to define what is to be stored and where. They are preceded by the *phrasewords* `INTO`, `BY`, and `USING` with the following meanings

`into` *vector* [`,` *vector*]*s*

specifies the *vector(s)* that are to be used to store the results of the *statistic(s)*, with one *vector* per *statistic*.

Although these vectors will be defined as arrays, it is often convenient to use the phrase

`by` *vector* [`,` *vector*]*s*

to request that GLIM generates the values of the vectors so that they classify the table in the manner that is required for subsequent model fitting. The values are generated in standard order (as in `GFACTOR`).

The phrase

`using` *vector*

requests that the output weights for the specified statistics are stored in *vector*. The output weight for a particular statistic reflects the amount of information that was used in its calculation and is intended to provide the basis for defining a formal weight variate in a GLIM analysis in which the results from `TABULATE` are defined as the *y*-variate. See the Reference Guide [15.84] for a formal definition.

6.10.3 Uses of TABULATE

The following examples use medical data on 113 patients including the variables:

`AGE`	age of patient
`SEX`	sex of patient
`SHTYP`	shock type (factor with 6 levels)
`SYSBP`	systolic blood pressure

`TABULATE` can be used to calculate various statistics.

```
$tabulate the AGE total $
          6173.
$tabulate the AGE mean $
          54.63
$tabulate the AGE variance $
          275.4
```

```
$tabulate the AGE deviation $ ! Standard Deviation
           16.60
$tabulate the AGE smallest $
           16.00
$tabulate the AGE largest $
           90.00
$tabulate the AGE percentile 90 $
           75.00
$tabulate the AGE fifty $ ! Median = Percentile 50
           56.00
$tabulate the AGE interpolate 90 $
           75.00
$tabulate the AGE weight$ ! the number of observations
           113.0
```

These values can be stored for future use, by adding an `INTO` phrase;

```
$number MEAN $tabulate the AGE mean into MEAN
$print MEAN $
   54.63
```

Find the mean age for each sex

```
$tabulate the AGE mean for SEX $

  SEX    1      2
 MEAN  56.24  52.87
```

The number of observations contributing to each mean can be accessed by inspecting the output weights specified in the `USING` option. The output means and weights can be stored for future use.

```
$tabulate the AGE mean for SEX into AMEAN using MWT $
$tprint AMEAN,MWT 2 $

           1      2
 AMEAN  56.24  52.87
   MWT  59.00  54.00
```

These output weights can also be obtained as a `WEIGHT` statistic;

```
$tabulate the AGE mean,weight for SEX $
```

```
     SEX      1      2
    MEAN  56.24  52.87
TOTAL_WT  59.00  54.00
```

The length of the resulting vectors can be stored using the `BY` phrase.

```
$tabulate the AGE variance for SEX into AVAR by %L $
$tprint AVAR %L $

         1      2
AVAR 253.2  299.0
```

Weights can be assigned to the observations using the `WITH` phrase. For example, the overall mean age can be found as a weighted mean of the means held in `AMEAN` using weights held in `MWT`.

```
$tabulate the AMEAN mean with MWT $
            54.63
```

Another use of input weights is to 'select out' observations. The mean age for males only could be obtained as follows.

```
$calculate SEL=(SEX==1) $
$tabulate the AGE mean with SEL $
            56.24
```

Since the default output weights are the frequencies of values in the `FOR` variable, frequency tables can be produced as follows.

```
$tabulate for SHTYP $

   SHTYP      1      2      3      4      5      6
TOTAL_WT  34.00  17.00  20.00  16.00  16.00  10.00
```

The classification vector(s) do not have to be factors. If a variate is used the distinct values contained in the vector are used to classify the dimension. For example, if `AGEGROUP` holds the recoded values of `AGE` by replacing the value by the class mid-point using classes 10–29, 30–49, ... , 90–109 a frequency distribution is produced by;

```
$assign LIM=10,30,50,70,90,110 $
$map  AGEGROUP=AGE  intervals LIM $
$tabulate for AGEGROUP using FAGE by MPNT $
$tprint FAGE MPNT $
```

```
MPNT    20.00    40.00    60.00    80.00   100.00
FAGE   12.000   22.000   59.000   19.000    1.000
```

The mean systolic blood pressure for each age group could be compared:

```
$tab the SYSBP mean for AGEGROUP into MSP $
$plot(cols=60) MSP MPNT $
```

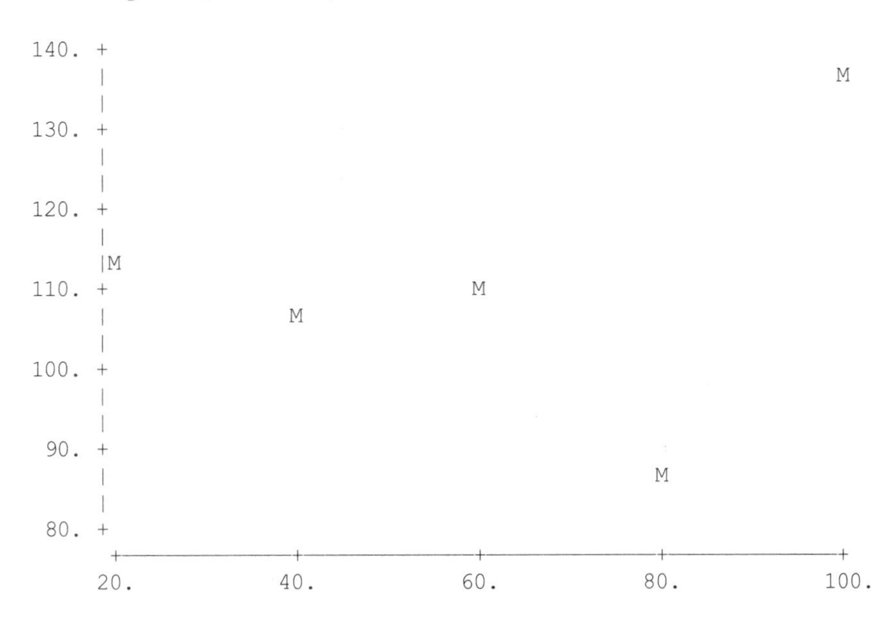

The variance of systolic blood pressure within each age group would be produced by

```
$tabulate the SYSBP variance for AGEGROUP into VSP
            using VWT $

- the table contains empty cell(s)
```

This warning is given if any of the cells do not contain enough values to calculate the statistic. In this case the last cell contains only one observation so that a variance cannot be calculated. Such cells can be identified by a zero output weight.

```
$tprint VSP,VWT MPNT $

MPNT    20.00    40.00    60.00    80.00   100.00
 VSP   1073.3    761.8    883.4    947.7      0.0
 VWT    11.00    21.00    58.00    18.00     0.00
```

Since the output weights for `VARIANCE` are the degrees of freedom the within-group variances can be pooled:

```
$tabulate the VSP mean with VWT $
            889.8
```

More than one statistic can be specified, so that the means and variances could have been obtained simultaneously using:

```
$tabulate the SYSBP mean,variance,weight for AGEGROUP
           by MPNT $

    MPNT   20.00   40.00   60.00   80.00  100.00
    MEAN  111.67  105.00  111.10   87.58  137.00
VARIANCE  1073.3   761.8   883.4   947.7     0.0
TOTAL_WT  12.000  22.000  59.000  19.000   1.000
```

Statistics can be calculated from frequency distributions by specifying the weights as a 'count'. Using the grouped AGE data frequency distribution:

```
$tabulate the MPNT mean,variance count FAGE $

    MEAN           55.58
VARIANCE           312.4
```

When input weights are specified PERCENTILE and INTERPOLATE will give different results. In PERCENTILE the weight is taken as a repeat count of the corresponding value, that is, the result would be the same as that for raw data with the value repeated the number of times given by the weight. In INTERPOLATE the weight is taken as a frequency of a class with mid-point given by the corresponding value so that the result is obtained by interpolating between the mid-points.

```
$tabulate the MPNT percentile 90,interpolate 90
           count FAGE $

P_90.00            80.00
I_90.00            79.59
```

Contingency tables are formed in similar fashion. In the following example a SEX by SHTYP table is produced. Any number of dimensions can be specified in the FOR phrase.

```
$tabulate for SEX,SHTYP $

SHTYP      1        2        3        4        5        6
  SEX
    1  21.000    9.000   10.000   10.000    5.000    4.000
    2  13.000    8.000   10.000    6.000   11.000    6.000
```

If the resulting table is to be analysed it is usually more useful to have factors that classify the contingency table. These are automatically generated if vector identifiers are used in the `BY` phrase.

```
$tabulate for SEX,SHTYP using TABLE by CSEX,CSHTYP $
$tprint TABLE CSEX,CSHTYP $
CSHTYP      1         2         3         4         5         6
  CSEX
     1  21.000    9.000  10.000  10.000    5.000    4.000
     2  13.000    8.000  10.000   6.000   11.000    6.000
```

An analysis could then be performed. For example

```
$yvariate TABLE $error P $fit CSEX+CSHTYP $

scaled deviance =  5.4570 at cycle 3
    residual df =  5

$tprint TABLE,%fv CSEX,CSHTYP $

     CSHTYP      1         2         3         4         5         6
CSEX
    1 TABLE  21.000    9.000  10.000  10.000    5.000    4.000
        %fv  17.752    8.876  10.442   8.354    8.354    5.221

    2 TABLE  13.000    8.000  10.000   6.000   11.000    6.000
        %fv  16.248    8.124   9.558   7.646    7.646    4.779
```

7 Macros

Introduction

Macros in GLIM have two functions:

(a) storing text;

(b) storing collections of GLIM statements. These statements may then be executed, possibly repeatedly, at appropriate points in the analysis.

In either case, the macro is stored in the workspace as text. It may be printed, edited, used as input to other directives (such as `GRAPH` or `PLOT`), and substituted as part of the command line. If it contains GLIM statements, it may also be executed, with optional argument passing. If the GLIM statements in the macro fail to achieve the desired effect, it may be traced and debugged interactively. A simple macro editor is provided to change the contents of macros.

Macros therefore provide the user with a powerful programming facility, allowing general routines to be written and stored in a file, which in turn allows the routine to be read into GLIM and used whenever it is needed.

7.1 Defining macros

The simplest way of defining a macro is by using the `MACRO` and `ENDMACRO` directives. The basic syntax is

`$macro` *identifier* *text* `$endmac`

A *space* or *new line* must follow the *identifier*

The *text* may consist of either:

(a) A set of one or more GLIM statements or parts of statements, possibly extending over many lines. The only GLIM directives not allowed in the macro are `$return`, `$finish`, and `$macro`.

or

(b) A text string of characters. Again the text may extend over many lines. The text is stored as it is input, with upper and lower case preserved.

Examples

```
$macro INFO
$print 'the standard length of vectors is ' %sl
$$endmac

$macro TITLE Recidivism in UK drink-driving $endmac
$graph(title=TITLE) y x$
```

The first macro `INFO` contains a single GLIM directive which prints out the current setting of the standard length of vectors with an informative message. The second macro `TITLE` contains text which is used as a title to a graph in a later directive.

Macros may be of any length, but compete for workspace with other GLIM structures. The length of a macro is defined to be the number of characters in the text; this will include new line characters and trailing spaces unless the end-of-record symbol is used to reduce the length of the macro [7.1.1]. The space occupied by a macro is related to the length of the macro with extra space needed for internal housekeeping. A list of macros currently defined in the workspace and their length and space used may be obtained with the statement `$environment d$`

Other methods of defining a macro are

(a) using the macro editor via the `EDMAC` directive [7.10];

(b) using the `store` option to the `PRINT` directive. [7.9];

(c) through user prompting with the `GET` directive [7.6].

7.1.1 Saving space when storing macros

As indicated above, GLIM will store, by default, all characters up to the end of each input record. When a macro consists of several lines, the total stored length of the macro may be large. The space taken may be reduced by indicating the end of each line of text with the end-of-record symbol `!`, which tells GLIM not to read any further characters from that record; all characters to the right of it are ignored. In particular, a blank line need only contain the single symbol `!`.

7.1.2 Commenting in macros

The end-of-record symbol provides a convenient method for commenting within macros stored in files. Text placed to the right of the end-of-record symbol is stored in the file, but is not stored as part of the macro text when it is read in. The macro will be more compact and will execute faster if this is done.

The alternative approach is to use the `COMMENT` directive [2.10] within a macro. In this case, both the directive and the text following it will be stored so increasing the length of the macro and slowing its execution. The advantage of the second method is that the comment is always available with the macro, and will, for example, be printed when the macro is printed.

7.1.3 Redefining macros

A macro may be redefined within a GLIM job at any time. The current contents of the macro are overwritten, in the course of which the length of the macro may change.

Example

```
$macro TITLE Sitka spruce analysis $endmac
```

overwrites the previous contents of TITLE.

A macro may be altered by using the macro editor [7.10]. The store option to the PRINT directive allows text to be appended to a macro [7.9].

7.1.4 Macros, functions, and system scalars

Various functions and system scalars are available to assist in the writing of general macros. They are most useful when used within macros but may be used at any time during a GLIM session. System scalars give information on the current state of model fitting (such as the current deviance) and information on the state of the system(such as the primary output channel number). Functions are available to find the length, type, number of levels, and number of dimensions of an identifier. A full list of functions and system scalars may be found in the Reference Guide [14.2, 14.6].

7.2 Examination of macros

7.2.1 Printing macros

A macro may be printed using the PRINT directive [6.2]. The PRINT *phrases* TEXT, LINE, and MARGIN may be used to control the appearance of the macro when it is output. Alternatively, the macro may be examined line by line by using the P command in the macro editor via the EDMAC directive [7.10].

7.2.2 Comparing macros

Through the %match function in the CALCULATE directive [5.4.5], GLIM provides basic checking facilities to test the contents of a macro with a character string or another macro. The first argument must be a *macro* identifier. The %match function compares the contents of this macro (the source) to target text supplied by the second argument to the function, and reports the level of the matching detected. The main use of this function is to help in processing macros input by the GET directive [7.6]. Further information on this function can be found in the Reference Guide [14.6.4].

Example

```
$mac S1  y $e
$calculate %a=%match(S1,'yes')$
```

Macro S1 contains text which is a substring of 'yes', so the scalar %a contains the value 1.

7.2.3 Deleting macros

A macro, like other structures, may be deleted from the workspace by the DELETE directive [4.6.1]. It is not possible to delete a macro which is currently being executed [7.7] or to which control will eventually return. The TIDY directive deletes the contents of all local identifiers of a macro [7.5] but not the macro itself.

7.3 Use of macros

7.3.1 Macro substitution

The substitution symbol # is used to substitute the contents of a macro within a GLIM statement.

Example

```
$macro MODEL A+B $e
$fit #MODEL $
$fit (#MODEL).C$
```

The macro MODEL contains text which may be interpreted as a model formula. The macro name preceded by the substitution symbol causes the contents of macro MODEL to be substituted, with the effect that the model A+B is fitted in the first fit, and the model (A+B).C in the second.

Example

```
$macro USAGE $environment u $endmac
#USAGE d$
```

The macro USAGE contains text which may be interpreted as the first part of a GLIM statement. The substitution of macro USAGE in the example will have the same effect as the statement $environment u d$

Macro substitution cannot occur ***within*** an item, such as an identifier or directive. The # symbol is not allowed to occur between $macro and its identifier, nor between $subfile and its identifier. Otherwise, it may be used anywhere. If substitution is included in a macro, that substitution will not take place until the macro is invoked.

7.3.2 Simple use

The statements in a macro may be executed sequentially by the statement

$use *macro*

Example

```
$number N=1 $
$macro RAISE $calculate N=2**N $print N$ $endmac
$use RAISE $use RAISE $use RAISE$
```

executes the macro RAISE three times, and results in the printing of the values 2, 4, and 16 and in the user scalar N having the value 16.

In this example, note that the repetition symbol : may not be used in place of the second and third instances of `$use`, as GLIM would take this to refer to the last statement in the macro (in this case `$print`).

For most applications it is important to ensure that a terminating directive symbol is placed ***after*** the last directive in the macro and ***before*** `$endmac`. This is to ensure that all statements are executed completely.

Example

```
$number N=1 $
$macro RAISE $ca N=2**N$print N $endmac

$use RAISE $
```

executes the macro RAISE once. This time, the terminating directive symbol has been omitted and the `$print` statement is therefore incomplete. The effect is that N is assigned the value 2, but no output is produced. Outside the macro, when GLIM encounters the next directive symbol, the `$print` statement will be completed and the value 2 will be printed.

The substitution symbol # may be used in place of the USE directive to execute a macro. The text of the macro is substituted, and the statements within the macro are executed sequentially. Note however that the use of the substitution symbol is less straightforward if macro arguments need to be specified [7.4]. Thus:

```
#RAISE
```

in the example above will have the same effect as `$use RAISE$`

Finally, care must be taken when the two methods of executing macros are mixed. Thus

```
$use M1 #M2
```

has the effect of executing M2 before M1 (because the USE directive does not have a terminating dollar).

7.3.3 Looping

The `WHILE` directive allows the statements in a macro to be executed repeatedly. It can be used to simulate loops and also to control iterative processes which continue until some criterion becomes sufficiently small. The syntax is

`$while` *scalar macro*

The value of the *scalar* is rounded to the nearest integer, and then examined. If the rounded value is initially zero, no action is taken. If the rounded value is initially non-zero, then the macro is executed and the scalar is re-examined (the macro will usually be written to contain statements to reduce the value of the scalar). If the rounded value is now zero, the directive terminates. If the rounded value is now non-zero, then the macro is executed again, and so on. The looping will continue indefinitely unless the macro itself eventually causes the scalar to attain the value zero.

Example

```
$number N=1 : TEST=1$
$macro RAISE
$ca N=2**N$print N $calculate TEST= (N<100)$
$endmac
$while TEST RAISE$
```

will cause `RAISE` to be executed four times. On the fourth iteration, `N` is 65536 and the scalar `TEST` will be set to zero by the logical test. The execution of the macro will then cease at the start of the next loop.

It is important to remember that macros which consist of sequences of statements should normally end with a directive symbol in addition to the one that starts `$endmac`. Failure to observe this simple rule can result in apparently inexplicable or unexpected results, as the following example illustrates:

```
$macro M $print %a $calculate %a=%a-1$endmac
$calculate %a=3$while %a M$
```

The macro `M` will be executed four times rather than the three required. The reason for this is that on the third iteration, as the `CALCULATE` statement has not been terminated with a directive symbol, the value of `%a` is still 1 rather than 0. The `CALCULATE` statement is not executed until a terminating `$` (the `$` of `$print`) is encountered on the fourth iteration. Insertion of a directive symbol before `$endmac` gives the required effect.

The break-in facility [4.10.2], if available, may be used to regain control if a macro is looping indefinitely.

7.3.4 Branching

The SWITCH directive allows branching to alternative macros, depending on the value of a scalar. The alternatives are to execute one of a set of macros or to do nothing. The syntax is

$switch *scalar macro1 macro2 ... macrok*

If the rounded value of the *scalar* (to the nearest integer) is *i*, and if *i* lies between 1 and *k* then the *i*th macro *macroi* is executed. If *i* lies outside the range 1 to *k*, then the directive does nothing (note that no fault message is generated).

Example A macro wants to check if the standard length has been set (available in the system scalar %sl) and to print a message if it is not positive. If the macro MESSAGE prints the message, the code to do this would be

```
$macro message !
$print 'Failure - standard length not set' $ !
$endmac
$calculate %a=(%sl<=0)$switch %a message$
```

It is possible to replace the list of macros by an identifier of type *list* :

$switch *scalar list*

In this case, the *i*th element in the list is checked to see if it is a macro, and if it is, it is executed. Thus, all elements of the *list* strictly do not have to be macros.

Example

```
$list MAC_LIST=MACRO1,MACRO2,MACRO3
$number A=3 $switch A  MAC_LIST$
```

will execute MACRO3. If A is not 1, 2, or 3, the SWITCH directive will do nothing.

A similar effect may be obtained by constructing a list of macros, and executing the *i*th element of the list directly.

Example

```
$list MAC_LIST=MACRO1,MACRO2,MACRO3
$number A=3 $use MAC_LIST[A]$
```

will also execute MACRO3. However, if A is not 1, 2, or 3, an error message will be generated.

7.3.5 Nesting of macros

Until now, only one macro has been executed at once. In general, however, it is useful for macros to call other macros. GLIM allows this, but there are some limitations related to the depth of nesting of macros, and discussion on this topic is postponed until [7.7].

7.4 Macro arguments

7.4.1 Basic concepts

All macros can have arguments, or parameters, which allow identifiers to be passed into the macro when invoked. This facility allows powerful, general macros to be written.

When the macro is written, the macro writer can refer to the first argument to the macro by using the special identifier `%1`, the second argument to the macro by using the special identifier `%2`, and so on. These special identifiers are known as **formal arguments.** Macros may have up to nine formal arguments. These names may be replaced by keywords through the use of **keyword arguments** [7.4.2] which provide an alternative method of referring to formal arguments. When the macro is invoked, **actual arguments** are substituted for the formal arguments. One way of assigning actual arguments is by the use of the `ARGUMENT` directive, with syntax

`$argument` *macro* [*argument*]*s*

The actual arguments may be macros, lists, scalars, vectors, subfiles, or functions. They may also be formal arguments or keyword arguments of other macros. They may not be integers, reals, constants, or expressions. Undefined identifiers may appear in an argument list, and their types will become defined when first used as actual arguments in the macro. Any occurrence of the formal argument `%1` will be replaced by the first actual argument in the list, `%2` will be replaced by the second actual argument, and so on.

The `ARGUMENT` directive does not cause the macro to be invoked, but sets the assignment of actual arguments to formal arguments for that macro. If a macro is invoked with the `USE` directive, the assignment can instead be made when the macro is invoked:

`$use` *macro* [*argument*]*s*

If a macro is invoked with the `$switch` or `$while` directives, then any actual arguments must be set using the `$argument` directive.

Example

```
$macro XMASDAYS
! prints the number of days to xmas
! three arguments :  %1 day, %2 month and %3 year
$number D1 D2 DAYS$ !
$calculate D1=%day(%1,%2,%3) !
```

```
  : D2=%day(25,12,%3) !
  : DAYS=%if( (D2>=D1),D2-D1,%day(25,12,%3+1)-D1) !
$print  *i DAYS ' days to xmas' $ !
$endmac
```

The above arguments can be set by the ARGUMENT directive

```
$calculate %d=12 : %m=4 : %y=1991
$argument XMASDAYS %d %m %y$
$use XMASDAYS $
```

or, when invoked, by the USE directive

```
$use XMASDAYS %d %m %y$
```

Actual arguments may also be assigned by user prompting [7.6].

7.4.2 Keyword arguments

Instead of using %1, %2, ..., %9 as formal arguments, GLIM provides a facility to give the formal arguments meaningful names. This aids readability of the macro. These names are called keyword arguments and are set up when the macro is defined. They are specified through the option list to the MACRO directive, with syntax

$macro (argument=*name1*, *name2*, ...) *identifier text* $endmac

Thus *name1* becomes a synonym for %1, *name2* for %2, and so on. Both keyword arguments and formal arguments may be mixed freely in a macro. The keyword argument names need to be different from one another and from any local identifiers [7.5] , but it is possible for a keyword argument to have the same name as another identifier in the directory. Within a macro, if a name is found which is present in both the keyword argument list and in the directory, then the keyword argument has priority.

Example The macro XMASDAYS above [7.4.1] may be rewritten to use keyword arguments:

```
$macro (arg=DAY,MONTH,YEAR) XMASDAYS
! prints the number of days to xmas
$number D1 D2 DAYS$ !
$calculate D1=%day(DAY,MONTH,YEAR) !
 : D2=%day(25,12,YEAR) !
 : DAYS=%if((D2>=D1),D2-D1,%day(25,12,YEAR+1)-D1)!
$print  *i DAYS ' days to xmas' $ !
$endmac
```

7.4.3 Setting and unsetting of arguments

Once an actual argument has been set for a macro, the setting remains in force until it is overridden by another assignment or the argument is unset explicitly. When a macro is redefined [7.1.3, 7.9] or edited [7.10], the argument assignments are lost.

Example

```
$calculate %d=12 : %m=4 : %y=1991
$use XMASDAYS %d %m %y$
$calculate %d=25 : %m=10
$use XMASDAYS$
```

The second invocation of XMASDAYS will work as the ordinary scalars %d, %m, and %y remain set as actual arguments to the formal arguments %1, %2, and %3.

Not all formal arguments need be assigned actual arguments. An asterisk may be used to indicate those arguments which are to remain unset. Trailing asterisks may be omitted.

Example

```
$use T1 * * A $
```

If no arguments have been set for macro T1, then this will set the third argument of T1 to A, but leave all other arguments unset.

After actual arguments have been assigned for a macro, a subset of them may be changed without repeating those that are not to be changed by using asterisks, this time to indicate no reassignment. Trailing asterisks may be omitted. Thus after

```
$argument TEST A B C$
```

the following reassignments are all equivalent

```
$argument TEST A D C$
$argument TEST * D *$
$argument TEST * D$
```

Arguments may be unset by using the minus sign - as an actual argument. This has the effect of unsetting that argument and all subsequent arguments. This is useful in situations where the macro takes a variable number of arguments and the system scalars described in the next section are being used. It is possible to unset the arguments to a macro while that macro is still being invoked.

7.4.4 Obtaining information on argument settings

The statement $environment a$ provides a list of all defined macros with their currently defined actual arguments and keyword arguments if used. Ten system scalars, with identifiers `%at` and `%a1`, `%a2`, ..., `%a9` , provide information accessible within a macro on the number of arguments set and whether a particular argument has been set or not. For $1 \le i \le 9$ the scalar `%a`*i* has the value 1 within a macro if the *i*th argument for that macro has been set, and the value 0 if it is unset. The system scalar `%at` is set to be the sum of the system scalars `%a1, %a2, ..., %a9` . (All ten system scalars have the value 0 if a macro is not being executed.)

Examples

```
$macro TEST $print %a1 %a2 %a3 %a4 %at $$endmac
$use TEST A B C$
```

will produce the following output:

```
1.000    1.000    1.000    0.0    3.000
```

```
$use TEST A - $
```

will produce the following output:

```
1.000    0.0    0.0    0.0    1.000
```

```
$use TEST * * * K$
```

will produce

```
1.000    0.0    0.0    1.000    2.000
```

7.4.5 The argument substitution mechanism

An actual argument is not substituted for a formal argument until the formal argument is reached in the execution of the macro. At that stage, the type, length, and and other attributes of the actual argument is checked and an error message is produced if inconsistencies are found. If an actual argument is itself a formal argument, say `%`*i*, then it will be replaced by the *i*th actual argument of the macro which invoked the current macro. If the actual argument for that macro is again a formal argument, then it will be referred back in a similar manner.

Example

```
$macro N1 $calculate %a=%1 $ $endmac
$macro N2 $use N1          $ $endmac
$macro N3 $use N2          $ $endmac
$argument N1 %2$ $argument N2 %c %1$
$use N3 %b$
```

The first argument of `N1` is the second formal argument of `N2`, which is the first formal argument of `N3`, which is `%b`. The effect of the statements above therefore reduces to the single statement `$calculate %a=%b$`

7.4.6 Extending the number of arguments

Some users may find that they need to use more than nine arguments when writing particular macros. If more than nine arguments are needed, then lists should be used. A list can be defined outside the macro, and passed into the macro as a single argument. Individual elements of the list may be obtained in the macro in the usual manner by following the formal argument by square brackets containing the element required; for example, `%2[5]` would refer to the fifth element of the second argument. The length of the list (that is, the 'number of arguments') may be obtained by using the `%len` function.

Example

```
$macro TEST $
$print 'the tenth element is ' *n %1[10] $ $endmac
$list L=A,B,C,D,E,F,G,H,I,J,K$
$use TEST L$
```

would produce

```
the tenth element is K
```

An alternative method which is slightly less flexible is to use the GET directive to prompt for new values of actual arguments for existing formal arguments, thus reassigning formal arguments dynamically within the macro [7.6].

7.4.7 Accessing the *i*th argument

A mechanism is provided in GLIM to enable the *i*th formal argument to be obtained, where *i* is a general scalar value rather than a known constant. Formal arguments of the form %*scalar* are allowed — in this special case *scalar* must be a system scalar or an ordinary scalar (for example, `%%d`) and not a user scalar. The scalar is rounded to the nearest integer value. If the resulting integer lies between 1 and 9 , then the *integer*th argument is accessed, otherwise an error is generated.

Example

```
$macro TEST $ca %%a = 1$ $endmac
$arg TEST A B C D E $
$calculate %a=3$use TEST$
```

would be equivalent to

```
$calculate C=1$
```

7.5 Local identifiers

Any macro may have a set of **local identifiers** associated with it when it is declared. These identifiers are local to the macro, that is, they are only available when the macro is currently being executed. Local identifiers provide a way of ensuring that the names of temporary identifiers used within a macro do not clash with names of identifiers set up by the user. Local identifiers are set up when the macro is defined via an option to the option list:

`$macro (local=`*name1*`,` *name2*`, ... )` *identifier text* `$endmac`

There is no limit to the number of local identifiers which may be defined for a macro. The local identifier names need to be different from one another and from any keyword argument identifiers [7.4.2], but it is possible for a local identifier to have the same name as another identifier in the directory. Within a macro, if a name is found which is both a local identifier and in the directory, then the local identifier has priority. The local identifier may have a different length, type, or contents to the identifier in the main directory. The statement `$environment d$` will list local identifiers for a macro if the directive is issued from within that macro.

Example The macro `XMASDAYS` introduced in [7.4] can be rewritten to use local identifiers; this removes the risk that identifiers `D1`, `D2`, and `DAYS` had been used for a different purpose outside the macro.

```
$macro (local=D1,D2,DAYS  arg=DAY,MONTH,YEAR)
XMASDAYS
! prints the number of days to xmas
$number D1 D2 DAYS$ !
$calc D1=%day(DAY,MONTH,YEAR) !
: D2=%day(25,12,YEAR) !
: DAYS=%if( (D2>=D1),D2-D1,%day(25,12,YEAR+1)-D1) !
$print  *i DAYS ' days to xmas' $ !
$endmac
```

A local identifier has no type, length, or values until these are explicitly declared during execution of the macro. Once declared, a local identifier keeps its type, length, and values. A local identifier therefore acts as any other GLIM identifier, but may only be accessed during execution of its macro. Deleting a local identifier causes it to lose its type, values, and length. For some applications the fact that a local identifier keeps its values will be of use. For other applications, the local identifiers will need to be redefined. This should be done by deleting the appropriate local identifiers via the `DELETE` directive at the start of the macro before they are (re)defined. Local identifiers may also instead be deleted at the end of a macro to recover their workspace. The `TIDY` directive provides a method of deleting all local identifiers of a macro — the statement `$tidy` *macro*`$` will delete all local identifiers of a macro but not the macro itself.

Example

```
$macro(loc=TVALS,TPROB)  tvalues
!  a simple macro for calculating t-values and
!  t-probabilities (two-tailed test) following a
!  Normal fit. It takes no account of aliased
! parameters.
!
! %pl is the length of the vector %pe
! %pe is the vector of parameter estimates
! %se are the standard errors of the estimates
!
$tidy tvalues $variate %pl TVALS TPROB$
$extract %pe %se$
$calculate TVALS=%pe/%se$
$calculate TPROB=%tp(TVALS, %df)$
$calculate TPROB=2*%if(TPROB>0.5,1-TPROB,TPROB)
$look(style=1) TVALS TPROB$
$endmac
```

7.6 User prompting in macros

The GET directive provides a method within macros of prompting the user for one of:

(a) a real number;

(b) a text string;

(c) an actual argument.

It enables macros to be written for the novice user who may not be comfortable with the conventional parameter passing mechanism of GLIM.

The basic syntax of the GET directive is

`$get (type =` *keyword* `prompt =` *string*`)` *identifier*

The *keyword* can be one of `scalar` or `macro` or `identifier`, and only the first letter is significant. If the `type` option is omitted altogether, `type=scalar` is assumed.

The `prompt` option allows the user to specify a prompt which will be output instead of the default prompt. A question mark is automatically added to the prompt string. The string may be substituted by a *macro* containing a text string.

If the type is `scalar`, then the user will be prompted for a scalar value, which will be stored in the *identifier* that will be created as a user scalar if it does not already exist.

If the type is `macro`, then the user will be prompted for a text string which will be stored in the *identifier* that will be created as a macro, or redefined if it exists already as a macro. The text string may be input in one of two ways:

(a) If the string does not start with the quote symbol `, then the string will be terminated by the end-of-line (normally when the 'return' key is pressed). This method is suitable for short strings.

(b) If the string starts with the quote symbol ` then the string is input in the usual way, and can continue over many lines. The string is terminated by a single quote symbol.

Note that the length of the prompt string will determine to some extent the maximum size of string which can be entered using the first method.

If `type` is `identifier`, then the *identifier* must be a formal argument or keyword argument. The user will be prompted for the name of an identifier, which will then be assigned as an actual argument to the formal argument, with any previous actual argument being replaced if one exists. This provides a means of dynamic assignment of arguments and of reassignment of arguments within a macro.

Example

```
$macro (local=CHOICE,MODELFRM arg=YVAR) NEWFIT
$delete CHOICE MODELFRM$
$get(prompt='model formula' type=macro) MODELFRM
$terms #MODELFRM$
$get(prompt='Y-variate name?' type=ident) YVAR
$print 'what type of model to do wish to fit?'
  :     ' 1= quasi-likelihood'
  :     ' 2= compound poisson-normal'
$get(prompt='type of model')CHOICE$
$argument METHOD1 YVAR : METHOD2 YVAR$
$switch CHOICE METHOD1 METHOD2$
$endmac
```

7.7 Nesting of macros and input channels

Each `USE`, (non-null) `WHILE`, and (non-null) `SWITCH` statement and macro substitution causes input to be read from the relevant macro. When reading from that macro is terminated, the program then reads the statements following the original `USE`, `WHILE`, or `SWITCH` statement or macro substitution.

A macro may itself contain `USE`, `WHILE`, or `SWITCH` statements or macro substitutions (which may refer to any macro, including itself) or statements which change the input channel (`INPUT`, `REINPUT`, or `DINPUT`). When such a statement is encountered in a macro, input is taken from the new source, and only when such input is terminated does the program process the statements that follow this original statement. Similarly, statements being read from secondary input channels may cause macros to be invoked, whereupon the input contained in that macro (and in any other macros called by that macro) is first executed before any further statements are read from the channel. See [7.7.4] for certain restrictions.

7.7.1 The program control stack

As outlined above, the directives `USE`, `WHILE`, `SWITCH`, `SUSPEND` (see [7.7.2]), `INPUT`, `DINPUT`, and `REINPUT` and the substitution symbol `#` all cause input to be read from a new source. Details of each new source are stored in the **program control stack** (or PCS). The first level of this stack corresponds to the primary input channel. When a new input source is encountered the stack is increased by a level, and details of that source are stored on the stack. When an input source is ended (for example, by a `RETURN` or `FINISH` directive or by encountering the end of a macro) the most recent level is released, the stack is decreased by a level, and reading continues from the input source at the new, lower level.

Example

```
$macro F
$data X1 X2 X3$
$dinput 12$
$fit #MODEL$
$use TVALUES$
$endmac
```

The sequence of operations on the PCS are:

- The statement `$use F$` is issued at the first level of the program control stack.
- Source is read from macro `F` at level 2 of the stack until the `$dinput 12$` statement is processed.
- Data values are then read from channel 12 at level 3 of the stack until sufficient numbers have been read, when control is returned to level 2 of the stack.
- Source then continues to be read from macro `F` at level 2 of the stack until the substitution symbol is reached in the `$fit #MODEL$` statement.
- The contents of macro `MODEL` are read at level 3 of the stack, then control is returned to level 2.

- Source then continues to be read from macro `F` at level 2 of the stack until the `$use tvalues$` statement is reached.

- The macro `TVALUES` is executed at level 3 of the stack, and control then returns to level 2.

- The rest of macro `F` is executed at level 2, and control returns to the primary input channel at level 1 of the PCS.

A macro may occur more than once in the stack; each time reading starts afresh from the beginning of the macro. Input channels (with the exception of the primary input channel through the `SUSPEND` [7.7.2], `GET`, and `DINPUT` directives) must not appear more than once in the stack at any one time.

If control is in a macro or a secondary input channel and an error occurs or the break-in facility is used, then control returns to the primary input channel and the level of the program control stack will be changed. See the Reference Guide [16] for further details on the error handling mechanism.

The current state of the program control stack can be inspected (with the statement `$environment p $`).

A `NEWJOB` statement clears the program control stack.

7.7.2 Interruption

The `SUSPEND` directive causes the execution of a macro or input from a secondary channel to be suspended and control is given to the primary input channel. The `SUSPEND` directive therefore increases the level of the program control stack by 1. Further statements can then be entered by the user (including invocation of other macros). A `RETURN` statement causes control to revert to the suspended macro or input channel at the statement following the `SUSPEND` directive.

One use for the `SUSPEND` directive is to place it at suitable points in an example file used for teaching. The file may then be read in to GLIM with echoing switched on, and will pause at predetermined intervals to allow students to examine the screen output, print vectors, and so on.

7.7.3 Changing the program control level

It is often useful to be able to leave a macro before the end, or to jump to the end of the macro. The directives `SKIP` and `EXIT` are provided for this purpose and, more generally, to provide a quick method for changing the program control stack level. The syntax is

`$skip` [*integer*]   and   `$exit` [*integer*]

If the *integer* is zero or is omitted, each directive has no effect. If the *integer* (say *n*) is positive and non-zero, then both directives will cause GLIM to return through *n*-1 levels of the program control stack, with the current source counting as the first level. The actions of SKIP and EXIT then differ. EXIT will exit unconditionally from the input source and return to the previous level, whereas SKIP will examine the new input source to see if it was invoked by the WHILE directive. If so, then the SKIP directive will test the WHILE scalar for further repetitions of the macro. Thus, for example, `$skip 1$` called from a macro means 'go to the end of the current macro', whereas `$exit 1$` means 'exit unconditionally from the current macro'.

When secondary input channels are skipped, the effect is the same as if a RETURN directive had been encountered on that channel. The file record pointer is ***not*** moved to the end of the file.

If *n* is greater than the number of levels of the PCS, then an error will be generated, and control will be returned to the primary input channel as described in [7.7.1]

Example

```
$macro (arg=VECTOR) SUM
$calc %a=%len(VECTOR) : %b=%type(VECTOR)
$calc %c=(%a>0)&((%b==11) ? (%b==17)) : %c=/%c
$exit %c$
$calc %d=%cu(VECTOR)$
$print 'the sum of the values in ' *n VECTOR ' is '
      %d $
$endmac
```

The macro checks the first argument to see if it has values and is of the correct type. If not, then %c has the value 1 and the macro is terminated. If all is correct, the sum of the values is found and a message is printed.

7.7.4 Terminating a macro prematurely

A common reason for wanting to terminate a macro prematurely is when an inconsistency in the actual arguments has been detected by the macro. In this case, it is useful to be able to print an informative message, then abandon the macro and all macro and secondary input sources and return to the primary input channel; in other words to duplicate the usual GLIM error handling mechanism. GLIM provides a method of doing this through the FAULT directive. The syntax is

`$fault` *integer* [*string*]

If *integer* is zero then the directive has no effect. If *integer* is non-zero, then GLIM will act as though a GLIM fault had been detected, with the current macro being abandoned and control returning to the primary input channel. The following fault message will be generated

```
** User generated error
```

followed, in the usual fashion, by the name of the macro and the program control stack level at which the user fault was generated. If the *string* is present, and the brief help setting is `OFF`, then the *string* provides a further diagnostic message which is printed on the following line. The *string* may be replaced by a macro identifier containing text. This allows more complicated messages to be generated, if desired, via the `store` option to the `PRINT` directive [7.9].

Example In the macro `SUM` above, instead of terminating the macro with the `$exit` directive, it is better to print an informative message. We can therefore replace the line

```
$exit %c$
```

with

```
$fault %c 'Argument is not a vector or has no values' $
```

7.8 Macro tracing and debugging

7.8.1 Macro tracing

GLIM refers to the process of tracing a macro, that is, echoing of the statements of an executing macro to the current output channel, as **verification**. Macros are executed by default with the verification state off. The directive `VERIFY` is used to change the verification state. The syntax is

`$verify [` *phrase* `]`

where *phrase* is `on` or `off`. If the phrase is omitted, the verification state is reversed. If macro verification is on, each line of the current macro is echoed to the current output channel as it is executed. If a transcript file is available, the verification output may additionally be copied to the transcript file if the `VERIFY` option to the `TRANSCRIPT` directive [4.3.1] is set. The statement `$environment c$` may be used to determine whether the `VERIFY` option to the `TRANSCRIPT` directive is set or unset. If macro verification is `off`, no such copying takes place. The scalar `%ver` contains the current verification state.

7.8.2 Macro debugging

It is possible to debug macros with the GLIM macro debugger. The macro debugger is a simple facility which returns control to the primary input channel during the execution of a macro. This allows the user to print values of vectors or scalars, to examine the directory, and to change vectors, declare new vectors, delete existing vectors, and so on.

By default, when a macro is executed, the debug state is off. The debug state is globally set and is controlled through the `DEBUG` directive. The syntax is

`$debug` [*phrase*]

where *phrase* is `on` or `off`. If the phrase is omitted, the debug state is reversed.

Once the debug state is `on`, GLIM will debug any macro being executed. The program will stop before every directive symbol to allow the user to issue GLIM directives. At this point the user has two choices.

(a) to continue to the next directive symbol, by issuing the `NEXT` directive, with syntax

```
$next
```

or

(b) to continue executing the macro with no further debugging, by switching the debugger off with

```
$debug off$
```

Note that the macro debugger will jump to the next directive symbol, not to the next directive. This means that the repetition symbol : is ignored by the debugger, and allows the macro writer some limited control as to which statements the debugger will stop at. The macro debugger will debug only macros; if secondary input channels exist on the program control stack, these will be read from without interruption; the debugger will next interrupt execution of GLIM
when the next directive symbol is found in a macro.

When the debugger stops the macro, an informative message is printed giving the next GLIM statements which are about to be executed:

`** DEBUG break on level` *k* `from macro` *macro* `before the directive`
GLIM statement(s)

where *k* indicates the level of the program control stack which is being debugged, and *macro* gives the name of the macro being debugged. The *GLIM statement(s)* gives the next GLIM statement or statements which are about to be executed; if `$endmac` appears here, then the debugger has reached the end of the current macro.

A debug prompt is then given:

<D>

At this point, most GLIM directives are allowed. In particular, the `PRINT, ASSIGN, CALCULATE,` and `ENVIRONMENT` directives are useful. However, directives such as `DINPUT, INPUT, WHILE, USE, SKIP, SWITCH, EXIT, SUSPEND, RETURN,` and `REINPUT` which change the level of the program control stack are not allowed. The `EDMAC` directive [7.10] is allowed but the macro being executed or any other macro on the program control stack cannot be edited. The system scalar `%cl,` which contains the current level of the program control stack, will contain the actual level of the macro being debugged plus one.

Local identifiers belonging to the currently invoked macro are available to the debugger and may be examined, with the statement `$environment d $` which gives names and lengths. The system scalars `%a1,...,` `%a9` and `%at` which indicate argument settings for the currently invoked macro are also available. Keyword and formal arguments are not available to the user in a debug session. Instead, the statement `$environment a $` may be used to find the actual argument associated with the formal argument.

Example A debugging session of the macro `SUM` defined above:

```
$debug on$
$assign A=1,2,3,4$
$use SUM A$

** DEBUG break on level 2 from macro SUM before the
directive
   $calc %a=%len(VECTOR) : %b=%type(VECTOR)

<D> $calc %type(A)
<D> $next

** DEBUG break on level 2 from macro SUM before the
directive
   $calc %c=(%a>0)&((%b==11) ? (%b=17)) : %c=1-%c

<D> $print %b$
<D> $debug off$
```

Debug output is treated as ordinary output and is copied to the transcript file unless the `o` transcript switch is off. However, directives issued while debugging are not copied to the journal file. The debugger may be used to debug macros defining user-defined fits by issuing the `$debug on$` statement before the model fitting is initiated. The debugger is not available in batch mode, when the debug setting is ignored.

7.9 Creating macros with the PRINT directive

The `store` option to the `PRINT` directive provides a mechanism for constructing macros dynamically within a GLIM session. Mixtures of text strings, integers, reals, and macros can be combined to create the contents of a new macro. The relevant syntax of the `PRINT` directive in this context is

`$print( store=` *macro* `)` *items*

Normally, the `PRINT` directive will display items on the current output channel. However, if the `store` option is present, then *macro* specifies the name of a macro to which instead all *items* will be written. If the macro does not exist, it will be created. Items are written to the macro line by line in the same format as that which would have been used if the items had been output to the current output channel. If the macro already exists, its contents will be replaced by text specified by *items*. The macro itself may be an *item*, in which case the macro text may be prefaced or appended to by other items. If a fault occurs during construction of the macro, then *macro* will be left unchanged. New-line characters are stored in the macro, but a terminating new-line character is not stored by default. Thus the `margin`, `text`, and `line` phrases can all be used and have an appropriate effect. If *macro* already exists, then it must not be present on the program control stack.

Examples Construction of a title for a graph:

```
$print(sto=TITLE)'Run '*i %a' with kappa = '*r %k$
$graph (t=TITLE) Y X$
```

One method of constructing identifier names:

```
$macro MODEL VAR1 + VAR2 $endmac
!
$macro ADD
$calculate %n=%n+1$
$print(store=MODEL) MODEL '+ VAR' *i %n $
$fit #MODEL$ca %m=(%n<=6)$
$endmac
!
$calculate %n=2 : %m=1 $while %m add$
```

7.10 The macro editor

The `EDMAC` directive gives the user access to a simple line editor for macros. The syntax is

`$edmac [ macro ]`

If the *macro* already exists, it will be loaded into the macro editor ready for editing. If the *macro* does not exist, it will be created. If no macro name is given, it is assumed that the macro editor will create the macro through the `S` or `T` editing commands.

7.10.1 How the editor works

The macro is loaded into the editor line by line, with the first line numbered 1 and subsequent line numbers numbered 2, 3, and so on. The last line of the macro is followed by the 'end-of-macro' line, which has its own line number and terminates the macro. If the macro has local identifiers or keyword arguments, then these may also be edited. The local identifiers and keyword arguments are decoded and placed in parentheses on the first line of the macro. After the macro has been edited, the user has the choice of reinterpreting this first line and redeclaring the local identifiers and keyword arguments (the usual choice) or treating this first line as text and part of the macro contents.

The macro editor has a concept of 'current line'. The current line may be changed by moving through the macro line by line or by jumping forwards or backwards. Many editing commands have a side-effect of changing the current line. When a line is added to the macro or deleted from the macro the line numbering changes to accommodate the change.

7.10.2 The editor command language

The macro editor has its own command language, and does not follow the syntactic rules which relate to the rest of GLIM. In particular:

(a) *new lines* may not be used to separate *items;*

(b) all macro editor commands start on a new line;

(c) a blank line is not ignored, but has a special meaning;

(d) *scalars* may not be used to substitute for *integers.*

Editing commands are of two types. Some commands are implemented simply by pressing the 'return' key, or typing an *integer* possibly prefixed by a + or - sign. These commands have no command word, and are used for moving around the file. Other commands consist of a command word, of which only the first character (known as the command character) is significant. These editing commands consist of a command word, optionally preceded and followed by *items* and *text*. Spaces may be inserted arbitrarily between *items*, and between the command word and its *items* and between the command word and *text*, but not between an *item* and *text* or within *text*. At least one space is required between a command word and *text*.

For example, the command

```
6I $gstyle 4 c 4$
```

is a valid editing command which will insert the text

```
$gstyle 4 c 4$
```

as a new line before line 6. It is valid to type this as

```
6     I          $gstyle 4 c 4$
```

or

```
6   INSERT $gstyle 4 c 4$
```

Most commands (except those relating to moving around the macro) have a common structure:

(1) Before the command word, optional *items* relate to line numbers of the macro and are either concerned with positioning or with ranges of the macro where the command is to be effective.

(2) After the command word, *items* refer to text changes or text additions.

We define the following syntactic structures:

line_number is either an *integer* or the character `*,` which refers to the 'end-of-macro' line.

line_number_separator is the character `,`

range is *line_number_1* [*line_number_separator* *line_number_2*] where *line_number_2* is not less than *line_number_1.*

text is *characters*

separator is any *character* in the GLIM character set but not *letter* or *digit* or *space* or *newline,* and is chosen to avoid characters appearing in the *texts*

item is a *line_number*, *line_number_separator* or *separator.*

Examples

`2` and `*` are valid *linenumbers*
`2` and `2,2` and `2,*` are valid *ranges*

`qwerty` is valid *text*
`/` and `'` are valid separators

7.10.3 The editor commands

7.10.3.1 Moving around the macro

The **move** command moves the current line to the line specified.

Syntax *linenumber*

The **relativemove** command moves the current line forward or backward in the macro the number of lines given by *linenumber.*

Syntax *+linenumber* or – *linenumber*

The **N or NEXT** command makes the **next** line the current line. The 'return' key or the command `N` or `NEXT` have the same effect.

Syntax `N` or *newline*

Examples

`5` Moves to line 5 of the macro and makes it the current line
`-2` then moves to line 3 of the macro and makes it the current line
`N` then moves to the next line (line 4) of the macro

7.10.3.2 Printing and displaying the macro

The **P or PRINT** command **prints** the lines specified by *range*. The current line becomes the line following the last line of the *range* . If *range* is omitted, it prints the next *k* lines, where *k* is the current output channel height.

Syntax [*range*] `P`

Example

`1,5 P` prints the first five lines of the macro

7.10.3.3 Line insertion, deletion and modification

The **I or INSERT** command **inserts** a line of text before the specified line. If *linenumber* is omitted, a line of text is inserted before the current line.

Syntax: [*linenumber*] `I` *text*

The **D or DELETE** command **deletes** the specified lines of the macro. If *range* is omitted, the current line is deleted.

Syntax [*range*] D

The **C or CHANGE** command **changes** the specified line, replacing it with the specified *text*. If *linenumber* is omitted, the current line is changed.

Syntax [*linenumber*] C *text*

Examples

`*I $endmac`	inserts the text '`$endmac`' after the last line of the macro
`1,3 D`	deletes the first three lines of the macro
`C $calculate a=a/2$`	changes the current line to the text '`$calculate a=a/2$`'

7.10.3.4 Finding and replacing text

The **F or FIND** command finds text in a macro. It searches the specified range of lines for the specified *text.* If *range* is omitted, the ***whole macro*** is searched. The first line which contains the specified *text* becomes the current line. If a question mark precedes the F command, then a query search mode is entered, with GLIM questioning the user to ask if the line containing the specified *text* is the one desired by the user. If not, GLIM proceeds to the next occurrence of *text.* Optional *separators* are provided to allow trailing spaces to be specified as part of the text.

Syntax [*range*] [?] F [*separator*] *text* [*separator*]

The **R or REPLACE** command finds text (*oldtext*) in a macro and replaces it with new text (*newtext*). It searches the specified range of lines for the specified *oldtext* and replaces it by *newtext.* If *range* is omitted, the ***whole macro*** is searched and replaced. If a question mark precedes the R command, then a query search and replace mode is entered, with GLIM questioning the user to ask if the specified *oldtext* found is to be replaced by *newtext*. GLIM then proceeds to the next occurrence of *oldtext.* and questions the user again, and so on. An optional *separator* is provided to allow trailing spaces to be specified as part of the replacement text.

Syntax [*range*] [?] R *separator oldtext separator newtext* [*separator*]

Examples

`F$fault`	will find the first line of the macro containing '`$fault`'
`?R/bd/%bd/`	will find all occurrences of `bd` and replace them with `%bd` if requested by the user.

7.10.3.5 Leaving the macro editor

The **`S` or `SAVE`** command **saves** the changes and exits from the macro editor. If *macro_identifier* is absent, the edited macro is saved under its default name and the editing session is ended. If *macro_identifier* is present, the edited macro is saved as a new macro with name *macro_ identifier* and the editing session is ended, leaving the original macro unchanged. If the first line of the macro to be saved starts with a left parenthesis, then the text following the left parenthesis is treated as an option list and is assumed to contain a local identifier and keyword argument list.

Syntax `S` [*macro_identifier*]

The **`T` or `TEXTSAVE`** command **textsaves** the changes and exits from the macro editor. If *macro_identifier* is absent, it saves the edited macro under its default name and ends the editing session. If *macro_ identifier* is present, it saves the edited macro as a new macro with name *macro_identifier* and ends the editing session leaving the original macro unchanged. All lines of the macro are saved as text, whether the first line starts with a left parenthesis or not.

Syntax `T` [*macro_identifier*]

The **`E` or `EXIT`** command **exits** from the macro editor and saves the edited macro under its default name. `E` is a synonym for `S`, but a new macro name must not be specified.

Syntax `E`

The **`Q` or `QUIT`** command **quits** from the macro editor without saving any changes made to the macro.

Syntax `Q`

7.11 The GLIM macro library

The GLIM macro library is a library of predefined macros supplied with the program which extend the facilities of GLIM. Further details may be found in Appendix A.

8 Model fitting and assessment

Introduction

The theory of Generalized Linear Models has been introduced in [1] and covered in more detail in the Modelling Guide [10]. However a brief summary is given here to help make this chapter relatively self-contained.

In simple terms models are algebraic structures to summarize the patterns of systematic variation seen in the mean or expected values of y (response/outcome /dependent) variables in terms of quantitative (variates or covariates) or qualitative (factors) explanatory variables.

The unknown constants or parameters in the models have to be estimated from data values, which include random variation, using some criterion to determine which are the best. GLIM uses the principle of maximum likelihood which makes an assumption that, given the probability distribution and form of model, the sample is the most likely to have occurred.

Thus the estimates are determined as those maximizing the likelihood function of the unknown parameters determined by the probability distribution. For the Normal distribution the estimation process is identical to classical least squares. For other distributions the process to obtain the parameter estimates involves solving non-linear equations iteratively, until some convergence criterion is met.

The directives available for model fitting and assessment can be classified into three types: those for model specification, for the control of the fitting process, and for the display and use of the results.

Specifying the model in terms of model elements

The relevant directives for this are

```
YVARIATE     TERMS   ERROR      LINK
WEIGHT       SCALE   OFFSET     SET
```

The models that GLIM fits are fully described in the Modelling Guide which also gives an extensive selection of illustrative examples. The elements of the modelling process are:

(a) a y-variate, the observed response, given as data, identified with the `YVARIATE` directive [8.1.1];

(b) a probability distribution for each element of the y-variate assigning it a vector of means μ, identified with the `ERROR` [8.1.2] directive;

(c) a model constructed from explanatory vectors, predicting η, specified in advance by the `TERMS` directive [8.1.7] or at the point of fitting the model by the `FIT` directive [8.2.1];

(d) a link function $\eta = g(\mu)$ connecting the linear predictor η with mean μ, of the probability distribution identified with the `LINK` directive [8.1.3];

(e) an optional weight variate for assigning prior weights to the values of the *y*-variate, identified with the `WEIGHT` directive [8.1.4].

Some error distributions have a scale parameter which needs to be estimated and the `SCALE` directive [8.1.5] allows this to be modified when necessary. The `OFFSET` directive [8.1.6] allows particular parameters in a model to be fixed in advance and not estimated. The `SET` directive [8.2.4.1] is used to specify whether the constant term is to be included in the model automatically (the default setting in GLIM specifies that it is to be included).

GLIM also provides facilities for fitting models where the user specifies all or part of these model elements, by providing macros which define the different elements. These are called OWN or user-defined models and the next chapter [9] gives a detailed description of them. However in this chapter we give the syntax for user-defined models where relevant.

Fitting the model

The relevant directives are

```
FIT          ELIMINATE     METHOD
ALIAS        CYCLE         RECYCLE
INITIAL      BASELINE      LOAD
```

The fitting is achieved by using the `FIT` directive to derive the linear model structure, possibly by reference to a model formula previously specified by the `TERMS` directive, and causing it to be fitted. The range of models is very wide indeed and a special algebra has been developed for the purpose of specifying them. This is based on the work of Wilkinson and Rogers (1973), but has progressed considerably beyond their original specification. Model specification is covered extensively in the discussion of the `FIT` directive [8.2].

The `ELIMINATE` directive [8.2.7] can be used to specify part of a model that is a necessary component, but is not of direct interest. The parameters represented by the term specified in the `ELIMINATE` directive are then implicitly fitted in each subsequent model fit, but those parameters are suppressed from the displays. It is most useful for simplifying the results obtained from fitting models with large numbers of nuisance parameters.

The `METHOD` directive [8.2.8] is used to select one of the two numerical methods, Givens and Gauss–Jordan, which can be used in the fitting process. The default, Givens, is usually the most accurate and appropriate, but in certain circumstances the Gauss–Jordan method will be necessary.

The `ALIAS` directive [8.2.6.1] is used to specify how GLIM should handle parameters which cannot be estimated because of the model formulation or limitations in the data. By default they will be included in the displayed estimates and relevant system vectors, but if required they can be suppressed.

The `CYCLE` and `RECYCLE` directives [8.2.9.1] are used to control the duration of the iterative process if the default settings prove insufficient.

The `INITIAL` directive [8.2.9.2] allows the specification of particular starting values for the fitted y-values used by the iterative fits if required.

The `BASELINE` directive [8.2.9.3] is useful in special cases where the 'deviance', measuring how well or badly the model can be made to fit the data, and its degrees of freedom, need to be measured from non-zero baselines.

The `LOAD` directive [8.2.9.4] allows the user to access the model matrices involved in the fitting process in various ways. For example, it can be used to modify the diagonal of the sums of squares and products (SSP) matrix suitably for a ridge regression analysis.

Displaying and saving the results of a fit

The relevant directives are

```
DISPLAY     PREDICT     EXTRACT
```

The `DISPLAY` directive has 15 options specified by a letter which allow immediate visual access to the details of the current model and the results from the previous fit. For example, `$display e $` displays the estimates of the parameters with their standard errors. A full list is given in [8.3.2].

The `EXTRACT` directive [8.4.2] gives access to those results which are not automatically available in a form that can be used in subsequent calculations.

The `PREDICT` directive [8.4.3] allows fitted values, that is, those predicted by the model, to be obtained for specified values of the explanatory variables.

8.1 The model specification directives

8.1.1 Declaring the response variate: the YVARIATE directive

The syntax is

`$yvariate` *identifier1* [`(`*identifier2* `)`]

Identifier1 must refer to a variate not a factor and the y-variate must be declared before a `FIT` statement can be used; the setting remains in force until redeclared. *Identifier2* is an optional index variate or factor identifying which elements of *identifier1* are to be used in the fit and in which order. The directive also (re)sets the number of units (`%nu`) for subsequent fits. This number will be the length of *identifier2* if present; otherwise it will be the length of *identifier1*. It will also set the standard length (`%sl`) for vectors if it has not been set previously (with `SLENGTH` [3.1.2]).

For binomial data, where the data can be represented as a proportion r/n (that is, r with a particular outcome out of n trials) the y variate is r and the variate n is specified as an item to the `ERROR` directive [8.1.2].

8.1.2 Declaring the probability distribution: the ERROR directive

The syntax is

`$error` *name* [*identifier1* [(*identifier2*)]]

The *name* can be `NORMAL`, `POISSON`, `BINOMIAL`, `INVERSE_GAUSSIAN`, `GAMMA`, or `OWN` and only the first character of the *name* is necessary, the remaining characters will be ignored. This directive has the default setting `N`(ormal). Identifiers are only required for the `OWN` and `BINOMIAL` cases. For `OWN` probability distributions *identifier1* must be a macro — see [9] for user-defined error structures. For the binomial distribution *identifier1* gives the denominator vector n in the proportion r/n; all values of n with non-zero weights must be positive; see `WEIGHT` [8.1.4]. *Identifier2* is needed if the binomial denominators are provided as an indexed variate and it identifies the vector holding the index values for *identifier1*.

8.1.3 Declaring the link function: the LINK directive

The syntax is

`$link` *letter* [$\left\{ \begin{array}{c} \textit{real1} \quad [\textit{real2}\ ] \\ \textit{macro} \end{array} \right\}$]

This defines the relation between the linear predictor, η, from the linear model and μ, the assumed mean of the y-variate (except in the binomial case where μ is the mean probability: the mean y-variate divided by the binomial denominator).

Possibilities for *letter* are given in the table below.

Letter	Name	Link function
I	identity	$\eta = \mu$
L	log	$\eta = \log(\mu)$
G	logit	$\eta = \log(\mu/(1 - \mu))$
R	reciprocal	$\eta = 1/\mu$
P	probit	$\eta = \Phi^{-1}(\mu)$
C	complementary log-log	$\eta = \log(-\log(1 - \mu))$
S	square root	$\eta = \sqrt{(\mu)}$
E	exponential	$\eta = (real2 + \mu)^{real1}$
Q	quadratic inverse	$\eta = \mu^{-2}$
O	own	as required

The *letter* is obligatory, but the *reals* are required only for the exponential link. The defaults are 0 for *real2* and 1 for *real1*. If both are omitted this is equivalent to the identity link.

In the case of the `OWN` link a *macro_identifier* must be supplied. The *macro* need not have been previously defined but, before a fit, must contain code which defines the fitted values μ in terms of the values of the linear predictor η and the value of the derivative of η with respect to μ. These are held in the system vectors `%fv`, `%eta`, and `%dr` respectively.

If no link has been explicitly declared the following default settings will be used

Error	Link
`N`	`I`
`P`	`L`
`B`	`G`
`G`	`R`
`I`	`Q`
`O`	No default

These are the canonical link functions which give sufficient statistics for their respective distributions. Note that with one exception — `OWN` — an `ERROR` statement resets the link to the default; a `LINK` statement should therefore follow (not necessarily immediately) the `ERROR` statement to which it refers. See [9] for further information on the `OWN` option for user-defined link functions.

8.1.4 Declaring prior weights: the `WEIGHT` directive

The weights given to each observation during fitting depend upon the distribution and link (see `ERROR` [8.1.2] and `LINK` [8.1.3]), but may be modified by multiplying by prior weights contained in a special weight variate, so that the variances of individual observations are divided by these prior weights.

The syntax is

`$weight` [*identifier1* [`(`*identifier2* `)`]]

Identifier1 must be a vector of weight values. For a `FIT` this must be the same length as the number of units in the fit (`%nu`) — see `YVARIATE` — or be associated with an index vector of length `%nu`. If an index vector is required it must be specified by *identifier2*. If *identifier2* is present the weights for each data point will be taken as the indexed values in the vector *identifier1*.

A weight-variate declaration remains in force until changed or cancelled. To cancel use a null statement, that is, `$weight $`. Non-declaration of a weight variate is equivalent to assigning all observations a prior weight of one. The values of a weight variate must be non-negative. Zero prior weights can be used as a technique for omitting observations from a fit although in some circumstances indexing or use of the `PICK` directive [5.5] may be more efficient for this purpose.

8.1.5 Fixing the scale parameter: the SCALE directive

The five standard distributions which GLIM allows to describe the random component fall into two groups. One group, comprising the Poisson and binomial, have mean–variance relationships which are known completely, while a Normal, inverse Gaussian, gamma distribution or user-defined OWN error implies that the relationship contains a scale parameter, ϕ, which is generally unknown. For the Normal distribution, we have $\mathrm{var}(y) = \phi$, that is, a constant with ($\phi = \sigma^2$) the usual variance. For the gamma distribution $\mathrm{var}(y) = \phi\ (\mu^2)$ with $\phi = k$. For the inverse Gaussian the scale parameter is the same as for the Normal, that is, $\phi = \sigma^2$, and $\mathrm{var}(y) = \phi\ (\mu^3)$.

In the Modelling Guide [11] it is pointed out that the variance–covariance matrix for the parameter estimates, β, is a function of ϕ so that a value for ϕ must be supplied in order to calculate this matrix and other values, such as standard errors, derived from it. GLIM offers several options through the use of the `SCALE` directive.

The syntax is

```
$scale [ real [ { deviance   } ]]
                { chisquared }
```

where *real* is a non-negative value and `deviance` and `chisquared` are keywords.

If *real* is present and is positive then its value will be assumed for ϕ. If *real* is zero or null then ϕ will be estimated by the mean deviance or by the mean chi-squared statistic of the current model according to which keyword is present. If no keyword is present `deviance` will be assumed.

Without a `SCALE` statement, a binomial or Poisson error implies that ϕ will be given the default value of 1, while a Normal, gamma, inverse Gaussian, or OWN error implies that the scale parameter is estimated using the current mean deviance. Any `ERROR` [8.1.2] statement resets the scale parameter to its default value.

8.1.6 Fixing parameter values - the OFFSET directive

The `OFFSET` directive allows an offset (see the Modelling Guide [11.4.7]) to be defined and used.

The syntax is

```
$offset [identifier1 [ ( identifier2 ) ]]
```

Identifier1 must be a vector of values to be used as offsets in the fitting process. For a `FIT` this must be the same length as the number of units in the fit (`%nu`) — see `YVARIATE` [8.1.1] — or be associated with an index vector of length `%nu`. If an index vector is required it must be specified by *identifier2*. The (indexed) values in the vector *identifier1* are taken to be a constant part of the linear predictor η. This situation arises when certain parameters in the model are, for some reason, known. This means that part of the model will be constant throughout the fitting process. The value of these constants, conventionally known as offsets, can be subtracted from the linear predictor η before the remainder is used in the consequently simplified estimation process.

An offset, once declared, remains in force for all subsequent fits until changed by another OFFSET directive or the *y*-variate is changed. A null OFFSET statement cancels any existing offset.

8.1.7 Specifying the model structure: the TERMS directive

The TERMS directive is used to specify a particular linear model structure without fitting it. The actual fitting is initiated by a subsequent FIT directive. A model specified by the TERMS directive becomes the current model and can be fitted by employing a FIT directive with one of the operators + - . * / by itself without any other model terms (see the FIT directive [8.2.1] for full details). Any model or model modification changes the current model specification and thus overrides the TERMS directive. Permanent storage of complex model formulae can be achieved by use of macros [7.3.1]. The current model can be inspected using the DISPLAY directive [8.3.2] with option L for the terms in the linear model, option M for the full model specification details, and option P for a full list of the parameters in the order that they will be fitted. The syntax is

$terms *model_formula*

The model formula algebra and details of how to construct appropriate *model_formulae* following Wilkinson and Rogers (1973) is covered extensively in [8.2].

8.1.8 The ordering of declarations

Because certain LINK [8.1.3] and SCALE [8.1.5] settings are the most frequently used for some ERROR settings the ERROR directive resets LINK and SCALE to appropriate default settings [8.1.3]. When the ERROR setting, either by specification or default, is not B, the binomial denominator becomes undefined.

Apart from this restriction, the directives ERROR, YVARIATE, WEIGHT, OFFSET, ALIAS [8.2.6.1], and CYCLE/RECYCLE [8.2.9.1] as required can appear in any order. If they are not specified then, respectively, Normal error is assumed, the *y*-variate is not defined, weight and offset are set to 1 and 0 respectively, intrinsic aliasing [8.2.6] is switched ON, and the CYCLE directive is assumed with its default values [8.2.9].

8.2 Fitting the model

After setting the response variate, the error distribution and link function, and possibly a prior-weight variate, and so on, it remains only to describe the structure of the linear model (also known as the linear structure) and initiate the fitting process. The TERMS directive described above can be used to specify a linear model in advance of fitting it. The FIT directive can be used both to define the linear model and also cause the numerical estimation procedure to be carried out. The ELIMINATE directive is used to specify a subset of parameters in the linear model whose estimation are not of direct interest. All three directives result in the formula that follows being stored internally as part of the currently defined statistical model.

8.2.1 The FIT directive

The syntax is

$fit *model_formula*

The model formula resembles an algebraic expression in appearance, having operands, operators and brackets. The operands are either factors (such as A, B) or vectors (such as X1, X2); the operators comprise the following set with the precedences indicated:

	**	.	/	*	,	+	-	-/	-*	
highest	1	2	3	4	4	4	4	4	4	lowest

For example, a model might take the form

```
A + B
A*B
1 + X
A + B*X
```

or
```
A*B/(C*(X1+X2)
```

The meanings of the operators, however, are different from those in arithmetic expressions and follow the model algebra specification of Wilkinson and Rogers (1973).

8.2.2 The syntax and meaning of model formulae

A *model_formula* is a set of **terms**, each term corresponding to a set of parameters in the underlying linear model. So, for example, a linear model that might be written in statistical theory as

$$\beta_0 + \beta^{X}x + \beta^{A}{}_{j} + \beta^{B}{}_{k} + \beta^{AB}{}_{jk}$$

would be expressed in the GLIM model formula notation as

```
1 + X + A + B + A.B
```

where there are five terms, namely 1, X, A, B, and A.B. The symbol '1' in this context denotes the constant term, X will have been declared as a GLIM variate which stands for the covariate X, while A and B will have been declared as GLIM factors to represent the classifications that give rise to the parameters $\beta^{A}_{j} + \beta^{B}_{k}$ of the statistical model. The interaction parameters β^{AB}_{jk} are represented by the **compound** term `A.B`. This is in fact one of the models necessary for a classical analysis of covariance. In this model the slopes of the linear relationship between the response variate and the covariate *X* have been assumed parallel in all the categories of the *A* by *B* classification. The presence of the interaction term `A.B` is allowing for the possibility that the effects of *A* after adjustment for the effects of *X* are not the same for each level of *B*.

By standard statistical theory we can also think of a term as standing for a set of columns in the model matrix that corresponds, for a given dataset, to that model. So the constant term `1` corresponds to a column of ones, the term `X` corresponds to a column vector whose components are the values of `X`. The term `A` corresponds to the block of ones and zeros that specifies the incidence of units within the classification levels represented by the factor `A`.

In full generality a **term** is a set of one or more distinct **model components**, where each component is one of the following types:

Type of model component	Syntax	Example
Constant	`1`	`1`
Variate	*variate*	`X`
Factor	*factor*	`A`
Matrix	*array*	`Q`
Indexed variate	*variate* (*index*)	`X(I)`
Indexed factor	*factor* (*index*)	`A(I)`
Indexed matrix	*array* (*index*)	`Q(I)`
Variate polynomial	*variate* < *integer* >	`X<3>`
Factor polynomial	*factor* < *integer*>	`A<N>`
Unit polynomial	< *integer* >	`<N>`

where the examples pre-suppose that `X` has been declared as a variate, `A` as a factor, `Q` as a two-dimensional array, and `N` as a scalar. The **index** `I` can be either a variate or a factor but its values have to satisfy certain compatibility conditions [8.2.4.5].

A full explanation of the meaning and use of these model components is given below. But before this explanation is given, it is necessary to complete the description of the model formula syntax.

Although conceptually (and internally) a model formula is a set of terms, and each term is a set of model components, at the user level a wider selection of operators is provided to enable formulae to be specified succinctly in a way that corresponds to some standard notions of experimental design.

The full set of model operators with their associated precedence order, as given earlier, is as follows:

**	.	/	*	,	+	-	-/	-*
1	2	3	4	4	4	4	4	4

As in ordinary algebra, round brackets can be used to enclose subformulae, however the meaning of the operators is very different from their meaning in ordinary algebra (and from their meaning in the `CALCULATE` directive [5]).

As indicated above, '+' and '.' are fundamental in the sense that every model formula, however specified, is reduced internally to a set of terms linked by '+' , each term being a set of model components separated by '.'. When internally stored formulae are displayed [8.3.2.8] they are displayed in this form. All operators other than '+' and '.' are translated and lost on input. Because formulae are sets of terms, and terms are sets of model components, the operators '+' and '.' have the property that `T+T` and `T.T` both reduce to `T`, whatever term `T` is. In their commutative, distributive, and associative properties '+' and '.' behave like their algebraic counterparts so that `A+B` and `B+A` represent the same model (whether it is stored as `A+B` or `B+A` depends on the order of input, see below), while both `A.(B.C)` and `(A.B).C` are the same as `A.B.C` and `A.(B+C)` expands to `A.B+A.C`

The remaining operators have their meaning defined in terms of '+' and '.'. The symbols '*' and '/' correspond respectively to the statistical notions of crossing and nesting. So, for example, `A*B` stands for `A+B+A.B` while `A/B` stands for `A+A.B`. The comma ',' is a synonym for '+' while the exponentiation operator '**' followed by an integer can be used to abbreviate the specification of formulae with many terms. For example `(A+B+C+D)**3` stands for `(A+B+C+D)*(A+B+C+D)*(A+B+C+D)` which expands to

```
A+B+C+D+A.B+A.C+A.D+B.C+B.D+C.D+A.B.C+A.B.D+A.C.D+B.C.D
```

The remaining three operators are used to remove terms from the model formula. The unary operator '-' deletes the term or bracketed subformula within its scope, so that, for example, `A*B-(A+B)` reduces to `A.B`. Because '-' denotes set deletion rather than algebraic subtraction the question of negative formulae does not arise; `B-(B+C)` is the null formula and `A+B-(B+C)` is the same as `A`.

More specialized forms of deletion are provided by the operators '-/' and '-*'. The operator '-*' removes the following term and all terms to which it is marginal, while '-/' retains the following term but removes terms to which it is marginal.

For example

`A*B*C-*A` is equivalent to `B+C+B.C`, all terms involving `A` being removed.

`A*B*C-/A` is equivalent to `A+B+C+B.C`, with the term A being retained.

Because the fitting procedure is sequential it is sometimes useful to know the order in which terms appear in the internally stored model formula. In general, terms are stored in order of their cardinality or, in statistical terminology, in the order of the interaction that they represent. So main effects appear first in the model, and hence will be fitted first, before terms representing interactions. Within this ordering, terms are stored in the order in which they appear in the input model formula. For example

```
$fit  B*C + A/B - C
```

results in the model being internally stored, and hence parameters being fitted and estimates output, in the order

```
B + A + B.C + A.B
```

Both alphabetical order, and the order in which the vectors were originally declared, are irrelevant in determining their ordering in the model and hence the order in which parameters are estimated. It is also worth noting that, at the stage the model formula is read in, no distinction is made between variates and factors, so some types of invalid model term are not detected on definition but only when actually fitted.

8.2.3 Incrementing the stored model formula

Both the `TERMS` and `FIT` directives cause the model formula that follows to be stored internally (this is the sole effect of `TERMS`, whereas `FIT` goes on to carry out the appropriate numerical estimation). Model formulae that begin with an operator are said to be **incremental**; those that start with a term, or a left-hand bracket, are **non-incremental**. Executing a `TERMS` or `FIT` statement in which the following formula is incremental causes the stored model formula to be updated as if it had been enclosed in brackets with the incremental formula written after it. So that

```
$terms 1+A $terms +B $
```

would result in the model `1+A+B` being stored as the current model formula without fitting it, while

```
$fit 1+A   $fit  +B $
```

results in a final model formula equivalent to

```
$fit  1+A+B $
```

Of course the sequence of output is different because in the incremental model fitting approach the results of fitting both `1+A` and `1+A+B` are output with a measure of how the deviance has changed and this is what would usually be required.

8.2.4 The meaning of the model components

This section covers the detailed meaning of the ten types of model component listed above together with some examples of their use.

8.2.4.1 The constant term and the SET *directive*

The constant term is used to represent a parameter in the statistical model corresponding to the mean of a reference subgroup of the data. In a simple linear regression it represents the intercept. It is represented in model formulae by the special term '1'; for example,

```
$fit 1+X $
```

In many models it is an important and useful term, but there are some circumstances in which it is not required. By default GLIM will automatically assume that it is required and include it in the fitted model even though it is not explicitly specified. However it can be excluded from the model with the '-' operator [8.2.2]. If the user wishes to have the constant term not included by default this can be achieved by using the NOCONSTANT phrase in the SET directive [4.7].

The relevant syntax here is

```
$set { constant   }
     { noconstant }
```

Only the first letter of each keyword is required so

```
$set n $
```

will change the default from constant, which means that a constant term will be added to each model automatically, to noconstant which means that if a constant term is required in a model it will have to be added explicitly.

```
$fit  1 $
```

therefore, corresponds to fitting the simple model of one constant parameter β_0. The corresponding model matrix is a single column vector of ones.

```
$fit 1+X $
```

corresponds to the model $\beta_0 + \beta^X x$, and a model matrix with two columns in which the first column is a vector of ones and the second column contains the values of the variate X. Note that the constant term '1' is considered to be a zero-order term and so $fit X+1 results in the internally stored, and fitted, formula being the same, namely 1+X. The constant term, where specified, is always fitted first.

The usual requirement is for the constant term to be present and so this is the default setting for non-incremental models. Thus

```
$fit X $
```

has the effect by default of fitting the model `1+X`. If, with this default in force, the constant has to be excluded then this has to be done by explicitly deleting it from the model, as for example in

```
$fit X-1 $
```

to fit a line through the origin.

Automatic inclusion of the constant only applies to non-incremental models but not in incremental models. Because of this models can be slightly confusing in some circumstances, as in the sequence

```
$fit X $fit . A $
```

which has the effect of first fitting `1+X` and then `A+A.X`, which has introduced `A` as well as `A.X`, but does not have the constant term.

Note that the null model is a legal model formula. With the above default switched off this can be obtained simply by

```
$fit $
```

or with the default in operation by

```
$fit 1-1 $
```

Null models are useful for calculating deviances for models with known parameters via the `OFFSET` directive.

8.2.4.2 Variates

Variates are used to represent quantitative variables, sometimes called covariates, but more properly known as **explanatory variables**. In an algebraic linear model they contribute a term $\beta^X x$ to the statistical model, where β^X is the parameter to be estimated. So, with the default for including the **constant**, that is, the intercept, in operation, we can carry out a simple linear regression of a *y*-variate `Y` against `X` by

```
$yvariate Y $
$fit  X $
```

corresponding to the model $\beta_0 + \beta^X x$. If, for example, the variate `X` had been assigned values by

```
$assign X= 1.1, 2.2, 3.3, 4.4, 5.5  $
```

then the corresponding model matrix would be

1	1.1
1	2.2
1	3.3
1	4.4
1	5.5

If we had also previously input V with

```
$assign V= 2, 3, 2, 2, 1 $
```

then, for example,

```
$fit X+X.V $
```

would result in the model $\beta_0 + \beta^{\mathrm{X}}x + \beta^{\mathrm{X.V}}xv$ being fitted, corresponding to the model matrix (assuming that the standard default for including the constant term was in force):

1	X	X.V
1	1.1	2.2
1	2.2	6.6
1	3.3	6.6
1	4.4	8.8
1	5.5	5.5

where the headings to the columns represent the GLIM representation of the parameter which is estimated by that column.

Note that because terms such as X.X are always reduced to X, quadratic terms in the statistical model have to be introduced by duplicating X in another, differently named, variate. However, this is not always necessary because polynomial regression is better carried out using orthogonal polynomials, for which a special syntax is provided below [8.2.4.6].

8.2.4.3 Factors

Factors are used to represent qualitative variables (such as classifications, treatments, and blocks). They contribute a **set** of parameters to an algebraic linear model, so that

```
$fit A $
```

(with the standard default) corresponds to the model $\beta_0 + \beta^{\mathrm{A}}_j$ and if A had been declared by

```
$factor A 4  $
$assign A = 2, 1, 4, 3, 2  $
```

the corresponding model matrix would be

1	A(1)	A(2)	A(3)	A(4)
1	0	1	0	0
1	1	0	0	0
1	0	0	0	1
1	0	0	1	0
1	0	1	0	0

where the headings to the columns represent the GLIM representation of the parameter which is estimated by that column.

Note that there is another default relating to aliased parameters which would usually result in a slightly different parametrization [8.2.6].
To specify interactions compound terms are used. Extending this example

```
$assign B=2,1,2,1,2 $factor B 2$
$fit A*B   $
```

results in the cross classification model represented by

$$\beta_0 + \beta^{A}_{j} + \beta^{B}_{k} + \beta^{A.B}_{jk}$$

being fitted, which for this set of data corresponds to the model matrix

β_0	β^{A}				β^{B}		$\beta^{A.B}$							
1	A(1)	A(2)	A(3)	A(4)	B(1)	B(2)	A(1).B(1)	A(2).B(1)	A(3).B(1)	A(4).B(1)	A(1).B(2)	A(2).B(2)	A(3).B(2)	A(4).B(2)
1	0	1	0	0	0	1	0	0	0	0	0	1	0	0
1	1	0	0	0	1	0	1	0	0	0	0	0	0	0
1	0	0	0	1	0	1	0	0	0	0	0	0	0	1
1	0	0	1	0	1	0	0	0	1	0	0	0	0	0
1	0	1	0	0	0	1	0	0	0	0	0	1	0	0

or, with the variate X defined as in Section 8.2.4.2 immediately above,

```
$fit A+A.X
```

corresponds to the model $\beta_0 + \beta^{A}_{j} + \beta^{A.X}_{j} x$ with the model matrix

β_0			β^{A}				$\beta^{A.X}$	
1	A(1)	A(2)	A(3)	A(4)	A(1).X	A(2).X	A(3).X	A(4).X
1	0	1	0	0	0	1.1	0	0
1	1	0	0	0	2.2	0	0	0
1	0	0	0	1	0	0	0	3.3
1	0	0	1	0	0	0	4.4	0
1	0	1	0	0	0	5.5	0	0

8.2.4.4 *Two-dimensional arrays*

Although the set of **model components** and **operators** provided as standard should enable nearly all commonly occurring models to be specified, it may occasionally happen that users want to think in terms of model matrices rather than **model formulae**. Two-dimensional arrays are provided to enable the model matrix to be specified directly by the user. For example, the model to be fitted might be `X1 + X2` given by the model matrix

1	X1	X2
1	2.2	3.3
1	4.4	5.5
1	6.6	7.7
1	8.8	9.9

Assuming that the constant term will be included by default, this could be fitted by

```
$assign Q= 2.2, 3.3,...,9.9 $
$array Q 4,2 $
$fit Q $
```

The model formula interpreter detects that `Q` has been declared as a two-dimensional array and checks that its dimensions are appropriate.

Example A dataset for a two-factor classification with `A` having four and `B` two levels with one observation per cell and with *Y* as the response variate is

	Y	A	B
1	8.0	1	1
2	4.0	1	2
3	0.5	2	1
4	0.4	2	2
5	4.1	3	1
6	6.2	3	2
7	3.0	4	1
8	8.1	4	2

The model can be fitted in the ordinary way by

```
$yvariate y $fit A + B $
```

Now `A + B` has the model matrix

1	A(2)	A(3)	A(4)	B(2)
1	0	0	0	0
1	0	0	0	1
1	1	0	0	0
1	1	0	0	1
1	0	1	0	0
1	0	1	0	1
1	0	0	1	0
1	0	0	1	1

This model can instead be fitted using an array, with the above matrix being assigned to a two-dimensional array and that array used to fit the model. This is done with the following statements

```
$assign M =      1,0,0,0,0,
                 1,0,0,0,1,
                 1,1,0,0,0,
                 1,1,0,0,1,
                 1,0,1,0,0,
                 1,0,1,0,1,
                 1,0,0,1,0,
                 1,0,0,1,1 $
$array M  8,5  $fit M $
```

Arrays can also be used to specify **contrasts** of one set of levels of a factor against another within a factor. Suppose that the first category of factor `A` was a control group and the other three categories represented increasing 'levels' of some fertilizer. Interesting contrasts might be the control category against the other three and those to test for a linear and quadratic trend, that is,

	Coefficients for levels of A
(1) 1 versus 2, 3, 4	`1, 0, 0, 0`
(2) `0*A(1), -1*A(2), 0*A(3), 1*A(4)`	
A linear trend over levels 2, 3 and 4	`0,-1, 0, 1`
(3) `0*A(1), -1*A(2), 2*A(3), -1*A(4)`	
A quadratic trend over levels 2, 3 and 4	`0,-1, 2,-1`

This is specified as follows:

```
$array C  4;3 $
$assign C = 1, 0, 0,
            0,-1,-1,
            0, 0, 2,
            0,-1,-1 $
```

Then, using the indexing facility described in Section [8.2.4.5]

```
$fit C(A) + B  $display e $
```

will provide the estimates equivalent to those contrasts adjusted for the effects of factor B.

Because the contrasts above are actually orthogonal (the sum of the cross products of the coefficients is zero) their estimates, in the full model estimating them all, may be interpreted directly. If contrasts are not orthogonal then they should be tested in different models and interpreted with considerable care.

Finally it should be noted that arrays can be combined with variates and factors to form compound terms.

8.2.4.5 Indexing

Indexing in model formulae is intended to provide a corresponding facility to the indexing available with the CALCULATE directive. It is a powerful facility which makes some commonly used types of analysis much easier to specify, and we give examples of such analyses later on.

In general the syntax is of the form

model_vector (*index_vector*)

where *model_vector* can be a variate, factor, or two-dimensional array and *index_vector* can be a variate or a factor. Without indexing, if V is a vector in the model formula with values v_i , i, = 1, ... m, then the ith row of the model matrix is constructed using v_i , and the length of V, namely m, is the same as the length of the currently declared y-variate.

With an index vector J, say, if V(J) appears in the model formula then the value used for the ith unit is v_j , where j is the value of the vector J for the ith unit. It is now the length of the index vector which must be the same as the length of the y-variate. The length of V need not be related to the y-variate but the values of the index J must not yield integers less than unity or greater than m, the length of V. If J contains other than integer values, then the nearest integer is taken. For example

```
$assign X=1.1, 2.2, 3.3, 4.4, 5.5$
$assign J= 5, 2, 4, 5, 2, 4, 1, 3, 1, 3$
$fit X(J) $
```

fits the model corresponding to the model matrix

1	5.5
1	2.2
1	4.4
1	5.5
1	2.2
1	4.4
1	1.1
1	3.3
1	1.1
1	3.3

The example shows an indexed variate, but the principle is exactly the same for factors and two-dimensional arrays, with a whole row rather than a single value being selected by the index in the latter case, of course.

The same variate may appear indexed and unindexed, or indexed by different vectors, in the same model formula. For example, a formula such as `X+X(I)+X(J)` is permitted. However, double indexing as in `$fit X(I(J))` and `$fit X(I)(J)` is not allowed.

A common requirement is for many terms in a formula to have the same index. To facilitate this, bracketed subformulae can be indexed. For example

```
(A*B)(I) + (X+Y)(J)
```

is an efficient way of specifying

```
A(I)+B(I)+A(I).B(I)+X(J)+Y(J)
```

However, the prohibitions above still apply, so `(A(I)+X)(J)`, which would lead to double indexing, is not allowed.

8.2.4.6 Variate polynomials

This type of model component is used when models involving powers of an explanatory variate are fitted. To fit directly a model of the form

$$\beta_0 + \beta_1 x + \beta_2 x^2 + ... + \beta_k x^k$$

is not only numerically unstable (Biometrika Tables Volume I, Pearson and Hartley, 1958), but also statistically inconvenient, because the β_i will be highly correlated and only β_k has an immediate interpretation. If it is non-significant then a lower-order polynomial will represent the data adequately, but with a completely different set of coefficients. A useful alternative is to fit **orthogonal polynomials** of degree one or two higher than that required, that is, a model of the form

$$\beta_0 + \beta_1 x_1 + \beta_2 x_2 + \dots + \beta_k x_k$$

where x_i is a polynomial of degree i and where the product of x_i and x_j summed over the data is zero for $i \neq j$. The polynomials used are discussed thoroughly in Biometrika Tables Volume I [Table 47, ibid]. These tables give the coefficients in integer form which is convenient, but not essential. Fitting the model with the x_k term removed, in non-iterative fits, will give the same estimates for $\beta_1, \dots, \beta_{k-1}$. In unweighted analyses, this means that the highest-order polynomial required to model the data can be identified from any fit of an orthogonal polynomial of a power higher than that required. In weighted analyses and iterative fits these polynomials are no longer strictly orthogonal, but are still useful for fitting high-order polynomials. For a variate X, a set of such polynomials can be specified with a model component of the form

```
X<K>
```

where K is either an integer or a scalar identifier. The model above would then be fitted by executing

```
$fit 1+X<K>  $
```

For example

```
$assign X=1,2,3,4  $fit 1+X<2> $
```

would result in the following model matrix being fitted.

1	X<1>	X<2>
1	0.6708	0.5
1	-0.2236	-0.5
1	0.2236	-0.5
1	0.6708	0.5

Note this is

1	-3	1
1	-1	-1
1	1	-1
1	3	1

with each column divided by the square root of the sum of their squares. This ***normalizes*** the coefficients so their squares sum to 1.

This provides a means of doing in a numerically stable way what the user might otherwise be tempted to do by fitting the model `(1+X)**K` but cannot because `X.X`, `X.X.X`, ..., are all reduced to X by the model algebra.

Note that the coefficients of powers of x in these polynomials, and hence its explicit mathematical form, depend on the data. There is a DISPLAY option O [8.3.2.10] to output the coefficients of the powers of x from which the explicit mathematical form may be deduced. However, within GLIM, values or plots of the fitted polynomial for selected x-values may easily be obtained by use of the PREDICT directive.

8.2.4.7 *Factor polynomials*

The analysis of some experiments or contingency tables, in particular, requires estimates of linear, quadratic, cubic, ... trends (often known as contrasts) over the levels of a factor. If `A` is a factor and `K` is an integer then the appropriate model component is `A<K>`. The statements

```
$factor A 4 $fit A<3>  $
```

will produce estimated coefficients labelled `A<1>`, `A<2>`, and `A<3>` where:

`A<1>` is the linear trend or contrast across the four levels of A;
`A<2>` is the quadratic contrast; and
`A<3>` is the cubic contrast.

The `A<3>` model component will generate the coding array for `A` as

-3	1	-1
-1	-1	3
1	-1	-3
3	1	1

This is then applied to the sequence of `A` values in the data to generate appropriate columns in the model matrix for the three contrast parameters. Thus data values as follows will produce

A	A<1>	A<2>	A<3>	
3	1	-1	-3	i.e. row 3 from above
1	-3	1	-1	row 1 from above
1	-3	1	-1	etc.
2	-1	-1	3	
4	3	1	1	

instead of the usual pattern of ones and zeros.

8.2.4.8 *Unit polynomials*

In many important circumstances *y*-values are obtained for equally spaced explanatory *x*-variates, for example, from a series of responses observed over time. In this case the explanatory variable could be considered as a factor with values from 1 to the number of observations. Without explicitly declaring and assigning values to such a factor the model component `<K>` can be used to fit polynomial contrasts up to order `K` over the full sequence of *y*-values, where `K` is an integer or scalar.

For example, the component <2> in a model to be fitted to eight values of a y-variate obtained at equal intervals along an x scale would generate the following columns in the model matrix

<1>	<2>
-7	7
-5	1
-3	-3
-1	-5
1	-5
3	-3
5	1
7	7

The parameter estimates are identified by <1>, <2>, ...

8.2.5 The mathematical meaning of the '+' and '.' operators

We have already seen that the '.' operator is used to specify interaction, and the '+' operator addition, of terms in the underlying algebraic linear model, and that all other operators are reduced to these. In terms of the model matrix the '.' operator corresponds to **direct product** (the element by element product of each column in the first matrix with each column in the second) and the '+' operator to **concatenation** (effectively the reverse of matrix partition) of blocks of columns.

More precisely, suppose, in the context of a particular dataset, that $[F]$ and $[G]$ stand for the model matrices corresponding to the GLIM model formulae F and G, and suppose that $[F]$ and $[G]$ have elements f_{ij}, $i = 1,...,l, j = 1...,m$ and g_{ik} , $i = 1,...,l, k=1,...,n$. Let

$$[E] = [F] \mid [G]$$

stand for the partitioned matrix with elements

$$e_{ij} = f_{ij}, \qquad j \leq m$$
$$e_{ij} = g_{i,j-m} \qquad j > m,$$

and let

$[E] = [F] \times [G]$ be defined by

$e_{ih} = f_{ij}\, g_{ik}$, where $h = j + n*(k - 1)$ for $j = 1,...,m$, $k = 1,...,n$, and $h = 1,...,mn$.

The meaning of '+' and '.' is then given by

$[F{+}G] = [F] \mid [G]$
$[F.G] = [F] \times [G]$

For example

[*F*]			[*G*]		[*F*+*G*]					[*F*.*G*]					
1	1	0	1	0	1	1	0	1	0	1	1	0	0	0	0
1	0	1	1	0	1	0	1	1	0	1	0	1	0	0	0
1	0	1	1	1	1	0	1	1	1	1	0	1	1	0	1
1	0	1	1	1	1	0	1	1	1	1	0	1	1	0	1

The actual order in which columns appear in the model matrix is determined by the order in which the corresponding terms appear in the internally stored model formula, which was specified above.

8.2.6 Parametrization and aliasing

It often happens that there is not enough information in the data to determine uniquely all the parameters specified by the GLIM model formula. In this situation the rank of the model matrix is less than the number of columns and one or more of the columns is a linear combination of the remainder. This always happens with the default coding for factors, if there is more than one in the model, or if the constant term is included. Standard statistical theory usually handles this by imposing additional constraints on the parameter estimates. In GLIM the numerical estimation procedure involves fitting terms sequentially according to their ordering in the internally stored model formula. If, during the numerical fitting process, a parameter is found to be linearly dependent on those already estimated then the corresponding estimate is set to zero (the mathematically correct value). Such parameters are said to be **extrinsically aliased**.

For example, we might (perhaps inadvertently) define three variates `U`, `V`, and `W` in such a way that `W` is the sum of `U` and `V`, and then try to carry out a multiple regression analysis with `U`, `V`, and `W` as explanatory variates.

```
$assign U= 1.1, 2.2, 3.3, 4.4    $
$assign V= 5.5, 6.6, 7.7, 8.8    $
$calculate W=U+V                 $
$fit U+V+W $display e            $
```

The parameter corresponding to `W` will be aliased because the terms will be fitted in the order `U`, `V`, `W`. If on the other hand we do

```
$fit W+U+V  $display e$
```

we find that it is V which turns out to be extrinsically aliased because the terms are fitted according to the order in which they are entered in the model.

Because arithmetic with real, as opposed to integer, numbers cannot be carried out with perfect precision on a computer, the test for extrinsic aliasing does not work very well if an attempt is made to test for exact collinearity. In fact the level at which extrinsic aliasing is detected is determined by an internally maintained tolerance level whose default setting is 10^{-10}, corresponding to signalling collinearity at a squared multiple correlation of $1\text{-}10^{-10}$. This constant is under control of the user via the CYCLE and RECYCLE [8.2.9.1] directives and experimenting with the setting may be necessary to force estimation of parameters associated with nearly collinear explanatory vectors, particularly if these have very large or small numerical values. Estimates obtained from such models should be viewed with caution.

For many linear models it is possible to determine mathematically that a certain number of parameters will be aliased whatever the data. Unless otherwise specified GLIM automatically selects a set of such parameters and omits them from the fitting process, thus saving time and space and enabling larger problems to be handled. Parameters omitted from the fitting process as a result of such prior analysis of the model are said to be **intrinsically aliased**. Again it is mathematically correct to take their parameter estimates as zero.

Suppose, for example, that we have two factors A and B, with three and two levels respectively, declared by

```
$assign A=1,1,2,2,3,3 : B=1,2,1,2,1,2 $factor A 3 B 2$
```

If, with the standard default for the constant in operation, we execute

```
$fit A  $
```

corresponding to fitting the model $\beta_0 + \beta^{A}_{i}$ then, given that β_0 is estimated, only two linearly independent combinations of the β^{A}_{i} will be estimable. The corresponding model matrix is

1	A(1)	A(2)	A(3)
1	1	0	0
1	1	0	0
1	0	1	0
1	0	1	0
1	0	0	1
1	0	0	1

It will be seen that the column corresponding to the constant term 1 is the sum of those corresponding to the three levels of the factor A. With the left-to-right sequential fitting algorithm used by GLIM, if there were no intrinsic aliasing then the parameter corresponding to the third level of `A` would be extrinsically aliased and its estimate would be zero. With intrinsic aliasing switched on, which is the default, it is possible (via the `FACTOR` directive) to specify a particular factor level, the **reference level**, as the one to be intrinsically aliased. So, for example,

```
$factor A 3  (2) $
```

specifies that `A` is a factor at three levels with reference level 2, and in the example above estimates would be obtained for β^{A}_{1} and β^{A}_{3}, but not for β^{A}_{2}. The actual model matrix generated by GLIM in this case would be

1	A(1)	A(3)
1	1	0
1	1	0
1	0	0
1	0	0
1	0	1
1	0	1

Parameter estimates corresponding to levels other than the reference level may be interpreted as relative to the reference level. If no reference level is specified then by default the reference level is taken to be level 1.

Parameters may become intrinsically aliased against terms other than the constant. For example, with the factors `A` and `B` defined above,

```
$fit A+B-1  $
```

gives the model matrix

A(1)	A(2)	A(3)	B(1)	B(2)
1	0	0	1	0
1	0	0	0	1
0	1	0	1	0
0	1	0	0	1
0	0	1	1	0
0	0	1	0	1

The second column for the `B` parameters is equal to the sum of the `A` parameter columns minus the first `B` parameter column. This results in all three parameters for `A` being estimable, but the parameter corresponding to the second level of `B` being intrinsically aliased. More generally, several of the parameters for the interaction term `A.B` would be intrinsically aliased if `A` and/or `B` were in the model.

8.2.6.1 The `ALIAS` *directive*

The syntax is:

`$alias` [*phrase*]

where *phrase* is `ON` or `OFF`. If the keyword is omitted the directive causes the current setting to be reversed.

Intrinsic aliasing can be switched on or off via the `ALIAS` directive, the default being to have it switched on. The aliasing tolerance parameter controlling extrinsic aliasing has no effect on intrinsic aliasing which is based on algebraic rather than numerical considerations.

Note that if the setting is `NOCONSTANT` and a constant is not included explicitly in the model the estimated effects of the first factor will be measured from the origin 0. The estimated effect of the reference category (by default level 1) of the first factor will take the place of the constant parameter '`1`'. The reference category effects for subsequent factors will be measured from those of the first level.

Terms are fitted sequentially, as are parameters within terms when a term containing factors implies more than one parameter. The order of fitting can be seen with the `M` and `P` options to the `DISPLAY` directive. When a new parameter is to be added to the model it may be found that there is no information about it in the data because it is aliased with parameters fitted previously. The aliased parameter is set to zero and thus excluded from the model.

See `DISPLAY` options `E` , `A`, and `U` [8.3.2] for display of various combinations of intrinsically aliased, extrinsically aliased, and unaliased parameters.

By default the model matrix formed during the fitting of a model does not contain columns for intrinsically aliased parameters. It is possible, although not usually desirable, to cause the model matrix to contain a column for all parameters, whether intrinsically aliased or not. When the `ALIAS` directive is used to switch the intrinsic aliasing off the parametrization will generally change. Factors that would have had their first parameter intrinsically aliased will generally be found to have their last parameter extrinsically aliased as a consequence of the sequential fitting procedure.

8.2.7 The `ELIMINATE` directive

It often happens that large groups of otherwise uninteresting nuisance parameters have to be included in a model, for example to force equality of observed and fitted marginal totals in a log-linear model analysis. If terms corresponding to these parameters are explicitly included in the model formula then not only are large numbers of unnecessary parameter estimates output, but much space and time is unnecessarily used up as well, perhaps precluding interesting analyses that would otherwise be possible.

If the nuisance parameters can be represented by a single term, (possibly compound or higher-order) as is the case with several useful standard analyses, then a series of models can be fitted relative to this term without incurring the time and space penalties.

The syntax is:

```
$eliminate [compound_term ] $
```

where the *compound_term* can be any model formula that reduces to a single term, or in which there is a single term to which all others are marginal. For example, A might be a factor with many levels and X a variate. Thus

```
$eliminate A $fit X$
```

would give the same deviance and the same parameter estimate for X as would be obtained by

```
$fit A+X $
```

In the latter case many estimates for the levels of factor A would be printed, in the former they would not. Furthermore, the working triangle, representing the weighted sums of squares and products matrix, would only have two rows and columns in the first case but many in the second.

In general

```
$eliminate formula1 $fit formula2 $
```

will always give the same deviance as

```
$fit formula1 + formula2 $
```

and the same estimates for *formula2* provided there is no aliasing. If there is aliasing then the parameters that are actually fitted may be different in the two cases. This is because the ELIMINATE directive can essentially force the fit of a high-order term before one of lower order. Consider the case

```
$eliminate A.B $fit 1+A+X $
```

as compared to

```
$fit A.B+1+A+X $
```

In the first case all the parameters corresponding to the A.B term are essentially fitted and the constant term 1, and one level of the main effect A, will be intrinsically aliased. This cannot happen in the second case because terms in a formula are re-ordered by order of interaction. The constant will be fitted and some levels of A.B will be aliased.

Example Consider the following sequence. This is appropriate for a log-linear model analysis of frequencies in a contingency table to assess the affects of factors A and B on the distribution among the categories of the factor C.

```
$eliminate A*B   $ ! This pre-fits 1 + A + B + A.B
$fit A*B*C $
$fit                      - A.B.C $display e $
$fit          - A.C               $display e $
$fit          + A.C - B.C         $display e $
```

The FIT statements above are equivalent to

```
$fit A*B + C + A.C + B.C + A.B.C $
$fit A*B + C + A.C + B.C         $display e $
$fit A*B + C       + B.C         $display e $
$fit A*B + C + A.C               $display e $
```

with the A*B part of the model formula not contributing columns to the model triangle and its parameter estimates suppressed throughout.

If there is no *compound_term* present, then the ELIMINATE directive is unset.

8.2.8 Numerical method: the METHOD directive

To solve the maximum likelihood equations, two numerical methods are provided. The default method is Givens which is accurate and economical in its space requirements. However Givens does not explicitly provide all of the matrix structures required by certain relatively common statistical applications, such as ridge regression, so an alternative numerical method, Gauss–Jordan, is provided; this was the standard numerical method in previous releases of GLIM. Full details of the numerical methods are given in the Modelling Guide [11].

The syntax is:

$method [{ *letter* / * } { *macro_identifier1* / * } [*macro_identifier2*]]

The valid options for *letter* are

J	Gauss–Jordan	{ *macro_identifier1* / * }
G	Givens	

The *macros*, if specified, are called during the fitting process for users to adjust various structures during the iteration (*macro_identifier1*) and at the end of iteration (*macro_identifier2*). These macros allow more specialized techniques, such as the fitting of non-linear models and the use of alternative estimation procedures. See [9] for further details.

The * is used to indicate that an argument is to be left at the previous setting. If the method is set to Givens when a LOAD variate has been set, a combination which is not allowed, then GLIM will unset the LOAD variate and issue the warning

```
-- LOAD variate unset
```

Example To show that the choice of numerical method does make a difference consider the following example of fitting high-degree polynomials. We obtain the exact co-ordinates for the sixth degree polynomial with coefficients all equal to 1:

$$y = 1 + x + x^2 + x^3 + x^4 + x^5 + x^6$$

```
$assign x=1,2,3,4,5,6,7,8$
$calculate x2=x*x : x3=x*x2 : x4=x*x3 : x5=x*x4 : x6=x*x5 $
$calculate y=1+x+x2+x3+x4+x5+x6 $
```

and then fit the sixth degree polynomial using the Givens algorithm

```
$yvariate y $
$acc 9 $
$method G $
$fit 1+x+x2+x3+x4+x5+x6 $dis e $

      deviance =  4.83246144e-21
      residual df =       1

               estimate              s.e.     parameter
      1       1.00000000   1.97034988e-09      1
      2       1.00000000   4.19932089e-09      X
      3       1.00000000   3.23037685e-09      X2
      4       1.00000000   1.18760035e-09      X3
      5       1.00000000   2.25131663e-10      X4
      6       1.00000000   2.12037037e-11      X5
      7       1.00000000   7.84374844e-13      X6
scale parameter 4.83246144e-21
```

This gives a very close fit with every coefficient accurate to 8 decimal places. Repeating the exercise with the Gauss–Jordan method gives

```
$method J $
$fit 1+x+x2+x3+x4+x5+x6 $dis e $
```

```
     deviance =  1.34443346e-15
     residual df =       1

              estimate              s.e.      parameter
     1      1.00000095    1.03927096e-06       1
     2     0.999997854    2.21495293e-06       X
     3      1.00000167    1.70387864e-06       X2
     4     0.999999404    6.26405779e-07       X3
     5      1.00000012    1.18746840e-07       X4
     6      1.00000000    1.11840004e-08       X5
     7      1.00000000    4.13722501e-10       X6
scale parameter 1.34443346e-15
```

This is also accurate enough for most practical purposes, but it illustrates that the Givens algorithm has the edge. Note that the actual numerical results obtained will depend on the particular computer being used.

Of course, explicitly forming and fitting sixth-degree polynomials is very rarely a sensible thing to do, even if it can be done more accurately by one method than another. It is largely to handle this situation that the variate polynomials [8.2.4.6] are provided.

8.2.9 Controlling iteration and other aspects of the fitting process

Since a number of models lead to estimating equations which cannot be solved directly much of the estimation involves iteration to obtain more and more accurate approximations to the solutions. The frequency and convergence criteria of this iterative process have to be controlled and for special applications it may be useful to modify the actual process itself. The following directives are provided for these purposes.

8.2.9.1 Iteration and convergence — the `CYCLE` *and* `RECYCLE` *directives*

The directives `CYCLE` and `RECYCLE` control the iterative process and associated output.

The syntax is:

`$cycle` or `$recycle` [*integer1* [*integer2* [*real1* [*real2*]]]]

Integer1 specifies the maximum number of iterations, and *integer2* specifies how often output of the deviance is to occur. GLIM supplies a default value of 10 for the maximum number of iterations. A zero or null *integer1* implies use of the default value and a zero or null *integer2* gives output at the final iteration only. Iteration stops if convergence is achieved before the maximum number of iterations specified.

The third and fourth items, *real1* and *real2*, set tolerances for the fitting algorithm. *Real1*, the convergence criterion, is used to determine when convergence has occurred; increasing/decreasing it will tend to decrease/increase the number of iterations and hence the accuracy of the estimates obtained and the time taken. *Real2* is the aliasing tolerance

[8.2.6]. All four values are available as system scalars %cyc, %prt, %cc, %tol respectively. The settings remain in force until changed by a subsequent CYCLE or RECYCLE statement.

Examples

```
No directive (default)
      !       implies     $cycle 10 1000 1.0e-4  1.0e-10

$cycle $
      !      implies     $cycle 10 1000 1.0e-4  1.0e-10

$cycle  0  0 $
      !       implies     iteration until convergence or
      !                   up to a maximum of 10 cycles
      !                   with printing at the end only.

$cycle  20  2  1.0e-5 $
      !       implies     iteration for up to 20 cycles
      !                   with printing at every 2nd
      !                   cycle and convergence criterion
      !                   set to 1.0^-5 .
```

The RECYCLE directive differs from CYCLE in that it causes iterations to start from the set of fitted values produced by the last fit. That is, it sets the initial values variate to the system vector %fv from the previous fit. Its syntax is identical with that of CYCLE and it may be used, for example, to continue an iteration begun by a CYCLE statement or to speed up convergence when the current set of fitted values is expected to be a specially good starting point for fitting the next model. The default settings are restored by either directive without arguments and $cycle $ by itself will cancel any previous RECYCLE directive and cause the initial values vector to revert to the observed values of the *y*-variate.

8.2.9.2 *Initial values for iterative fits: the INITIAL directive*

The iterations involve updating the fitted values, held in a system vector %fv, using the values of %fv from the previous iteration. To begin this process starting values are required. By default these are taken to be the observed values of the *y*-variate (or some simple transformation). On occasion it is useful to use other starting values. RECYCLE [8.2.9.1] uses the values of %fv from the previous fit to continue iteration when there is evidence that the fitting process has not yet converged, or the previous fitted values would make good starting values. The INITIAL directive is provided to allow the user complete freedom to specify any starting values felt appropriate. This is particularly useful in the case of non-linear fits.

The syntax is:

`$initial` [*vector1* [(*vector2*)]] [*macro*]

Vector1 specifies a vector of values to be used as the starting values for the fitted values (`%fv`) vector. It may be an indexed variate, in which case *vector2* must specify the index variate. The *macro* if present specifies the `OWN` initial macro which, if specified, is called at the start of a user-defined fit [9.2.1].

8.2.9.3 *Changing the Deviance origin: the* `BASELINE` *directive*

With certain non-standard applications of GLIM, the deviance obtained is not correct for calculating the standard errors of the estimates. Adding the appropriate amount to the deviance and degrees of freedom to remedy this is straightforward with the `BASELINE` directive.

The syntax is:

`$baseline` [*real1* [*real2*]]

Real1 specifies the deviance increment and *real2* the degrees of freedom increment. The increment can be positive or negative.

8.2.9.4 *Loading the working triangle before solution: the* `LOAD` *directive*

Ridge regression depends on modifying the fitting process in such a way as to suggest that there is more information available to estimate the parameters than there actually is. This is achieved by adding a load vector to the diagonal of the working triangle before solving the maximum likelihood equations to obtain the parameter estimates. The `LOAD` directive is provided to make this easy and to provide general access to the matrix at this point in the fitting process.

The syntax is:

`$load` [*identifier*] [*macro_identifier*]]

Identifier specifies a vector of values to be added to the diagonal of the working triangle and the *macro_identifier* specifies a macro which can be used to modify the working triangle on each iteration. Note that a vector of values can only be specified when the Gauss–Jordan numerical method is in use. If Gauss–Jordan is not the method setting then use of a vector identifier will cause a switch to that setting and the following warning will be issued

```
-- METHOD changed to Gauss-Jordan
```

If the model being fitted contains no parameters, that is, it is the null model, there is no working triangle. If there is a `LOAD` variate set GLIM will issue the warning

```
-- Null fit - LOAD variate ignored
```

8.3 Displaying and saving the results of a fit

There is very little automatic output from the GLIM model fitting process. It is assumed that a single fit is generally part of a fuller analysis and that the user is the best judge of what output is required at any particular stage. This may seem uncomfortable to those users preferring a 'black box' producing something close to a final report, but analyses are, in practice, very rarely routine applications of standard procedures with a standard results format.

The relationship between the original problem that generated the data and the appropriate analytical strategy will often mean that the fits of several models should be compared without their parameter estimates being needed. Assumptions made by the analysis need to be clear and tested. Finally the answer to the practical question may require that the results from one or more models are combined and/or modified in some way to give the most informative output for a final presentation.

The appropriate output for all or part of such an overall analysis cannot, in general, be identified in advance so the philosophy of GLIM is to provide efficient and comprehensive access to all the ingredients that may be required and the tools to manipulate them in any way the user chooses.

8.3.1 The `FIT` output

A fit produces output of the form:

```
[ scaled] deviance =  real [ at cycle integer ]
     residual df =  integer
```

where the word '`scaled`' occurs if the scale parameter [8.1.5] has been assigned a prior value. Models with Normal error and identity link can be fitted in one cycle, that is, with a non-iterative fit, and in this case the cycle information is suppressed. All other models require an iterative fit, but if convergence is not achieved after a certain maximum number of cycles the warning

```
-- (no convergence yet)
```

is printed. The results at this point will at best be only approximately correct. Further iterations of the same model may be achieved by

```
$recycle $fit . $
```

If this problem is anticipated the maximum number of cycles (default 10) can be changed by the `CYCLE` or `RECYCLE` directives [8.2.9.1]. Non-convergence, in which the deviance though failing to reach stability, is still falling in each iteration, is distinguished from divergence in which the deviance begins to increase. Divergence produces the warning

```
-- (iterations diverged)
```

and implies a complete failure of the fitting process. Such failures are normally most uncommon, but may occur if an unrealistic offset [8.1.6] is imposed, or if a large number of units have their fitted values held at the limit (Modelling Guide [11.3.4]).

Attempts to fit models where the fitted values are extreme (zero with Poisson errors, and zero or the denominator with binomial errors) may produce a change in the number of non-aliased parameters between iterations (Modelling Guide [11.3.4]). Iteration then stops with the warning

```
-- (change in df)
```

If the fitted values for particular units fall outside the permissible range for a particular `ERROR` distribution during a fit they are set equal to the appropriate limiting value and the following warning is output

```
-- (unit (s) held at limit)
```

There are several other warning messages issued under special circumstances. They are described in full in the Reference Guide [16].

When the only change to the model from the previous fit is that the model formula has been incrementally changed, indicated by a `FIT` directive with a structure formula beginning with an operator, then the change in deviance and degrees of freedom from the previous fit are also printed. If some observations have zero prior weight and are therefore weighted out of the analysis, then the number of effective observations is also printed.

This is particularly convenient for sequences such as

```
$fit A*B        ! this is equivalent to 1 + A + B + A.B
:     - A.B
:     - B  $
```

where a model comparison to test interaction precedes tests of the main effects of both `A` and `B` together. Each model is nested within that preceding it so the change in deviance is a measure of the contribution to the fit of the terms removed.

Note, of course, that sometimes increments will be almost uninterpretable and not useful. For example, in the sequence

```
$fit A*B
:      -A.B
```

```
:     - A
:   + A - B  $
```

The test of interaction is followed by model comparisons for separate and independent tests of the main effects of `A` and `B`. The last model is not nested inside the one before it so that change in deviance has no useful interpretation.

If the `BASELINE` directive has been used to give non-zero settings to either the baseline deviance or the baseline degrees of freedom, then the message '`with baseline adjustment`' is produced following the deviance.

8.3.2 The `DISPLAY` directive

In most cases after fitting a model to some data, the user will probably require more than the single measure of goodness of fit given by the deviance. The `DISPLAY` directive with form:

`$display` *letters*

allows various quantities associated with the fitted model to be displayed. Note that (except for options `I`, `L`, `M` and in certain circumstances `P`) the display will be inhibited if any changes are made to the data or model after the fit has been made. This is to ensure that out-of-date information becomes unavailable. The letters in the directive may be any subset, in any order, of:

```
A C D E I L M O P R S T U V W
```

The two most commonly required displays, those of the parameter estimates `E` and the residuals `R`, are introduced first

8.3.2.1 Option E : estimates of parameters, etc.

The parameter estimates are listed in parallel with their standard errors, and parameter names. Assuming `ALIAS` [8.2.6.1] is set to `ON` (the most useful and the default situation), intrinsically aliased parameters [8.2.6] are omitted from the output completely and must be taken as zero in any calculations involving parameter estimates. If an `ELIMINATE` statement has been used, to pre-fit part of the model and eliminate it from the output, information on the eliminated part of the model will also be displayed.

Extrinsically aliased parameters are also set to zero; they are listed with the value zero in place of an estimate and `aliased` printed in the standard error column. The standard errors of the parameters are calculated using the `SCALE` setting, as described in [8.1.5], whose value is printed below the list.

8.3.2.2 Option R : residuals, fitted values, etc.

This display lists, in parallel, the observed values of the *y*-variate, the fitted values from the model, and the Pearson residuals. These residuals are obtained by scaling the differences (observed values - fitted values) to allow for prior weights, known values of the scale parameter, and the fact that (except for the Normal distribution) the variance of an observation varies with its expected value. Thus,

$$\text{Residual} = (\text{Observed} - \text{Fitted}) * \sqrt{\frac{\text{Prior weight}}{k * \text{Variance function}}}$$

where k = scale parameter if set, or 1 otherwise. (See the Modelling Guide [11.2.1] for the definition of the variance function.) Note that, as a consequence of this definition, if observations are excluded from a fit by being given zero weight, then their residual will appear as zero. Observations with zero weights are identified in the output by parentheses around its observation number. For the listing of subsets of observations only, see the `W` option. For a model with binomial error, the binomial denominator is also listed in a separate column.

8.3.2.3 Option V : (co)variances of estimates

This option gives the (co)variances of non-extrinsically aliased estimates as a lower triangle with numbering of parameters as in option `E` or `P`, using scale parameter values as defined in [8.1.5].

8.3.2.4 Option C : correlations of estimates

This display prints the correlations of the non-extrinsically aliased estimates as a lower triangle; the parameters are numbered as in the `E` option.

8.3.2.5 Option S : standard errors of differences of estimates

This option provides, in lower triangular form, the standard errors of differences of the parameters in the current model. The ordering of the rows is as in the `E` option. Note that the standard errors of factors produced by option `E` are for the comparison of each non-aliased effect of a factor with the aliased effect which has been set to zero.

8.3.2.6 Option W: residuals, fitted values, etc., for included observations

This produces a display identical to that from the `R` option, except that all units for which the system vector `%re` has been assigned zero values by the user are omitted. In particular, if a prior-weight vector, `W` say, has been declared to restrict fitting to a subset of units [8.1.4], then

```
$calculate %re = W $
```

followed by use of this option will cause the listing to be restricted to that subset. Alternatively, a list of residuals and fitted values corresponding to large residuals (> `c`) can be obtained by

```
$number c= .... $calculate %re =(%abs(%rs) > c)$display w$
```

8.3.2.7 Option L : linear model

This option causes the current linear model to be listed as a sum of simple terms. Thus a model specified as `A * B + X` will appear as:

```
Linear model:
terms: 1 + A + B + X + A.B
```

The `ELIMINATED` model term, if set, is also given.

8.3.2.8 Option M: the current model

This option displays the current model specification, including the number of observations available for the fit taken from the (indexed) length of the *y*-variate, the prior weight (if set), the offset (if set), the probability distribution of the *y*-variate, the link function, the scale parameter, the terms eliminated (if there are any), and the terms in the linear model (see option `L`).

Example

```
$slength 20
......
$yvariate Y $fit X $
$display m $
```

gives

```
 Current model:

   number of observations in model is 20

   y-variate  Y
   weight     *
   offset     *


   probability distribution is NORMAL
              link function is IDENTITY
              scale parameter is to be estimated by the mean
deviance

 linear model:
     terms: 1+X
```

A user-defined error and link will be identified by the `OWN` macro names and active `INITIAL`, `LOAD`, and `METHOD` macros to modify the fitting process are listed (see [9]).

Any index vectors associated with the *y*-variate, binomial denominator, offset, weight, and initial values vectors will also be displayed. Vectors currently unset are denoted by an asterisk.

8.3.2.9 *Option P: the parameters*

This lists the parameters in the current model in the order in which they will be fitted. These are available after a `TERMS` directive or a `FIT`.

8.3.2.10 *Option O: fitted orthogonal polynomials*

This lists details of orthogonal polynomials included in the model. For each variate polynomial term present in the model, it displays the coefficients of the powers of the explanatory variate for each orthogonal polynomial from zero up to that explicitly specified in the model.

It also gives the coefficients of the equivalent polynomial in powers of the explanatory variate that would have been obtained if the model had been specified in terms of the powers of the explanatory variate.

8.3.2.11 *Option I: information on the algorithm*

This option gives details of the settings for method, aliasing, the number of iterative cycles, tolerances, and convergence criterion.

8.3.2.12 *Option A: estimates and standard errors with aliasing*

This lists estimates, standard errors, and parameter names, in the same format as for option `E`, except that all parameters are listed, whether intrinsically aliased, extrinsically aliased or unaliased. As for options `E` and `P`, the first column gives the position of each parameter in the model matrix and thus in the vector of parameter estimates, and so on. Intrinsically aliased parameters are labelled with a 0 in this column as they do not contribute to the model matrix.

8.3.2.13 *Option U: estimates and standard errors without aliasing*

This gives the same display as option `E`, except that only unaliased parameters are listed.

8.3.2.14 *Option D: the deviance and degrees of freedom*

This option prints the (scaled) deviance and the degrees of freedom of the current model. If a `BASELINE` directive has been used this will also display the baseline deviance and degrees of freedom values which have been added.

8.3.2.15 *Option T : the working matrix*

Gauss–Jordan algorithm: This option prints in lower triangular form the current working matrix used by the fitting algorithm. The last element (except when `LOAD` is in use) gives the generalized Pearson chi-squared statistic from the previous iteration, and the remainder of the last row gives the parameter estimates. The rest of the matrix is the generalized inverse of the weighted sums of squares matrix. If the diagonal element is non-zero the row contains the asymptotic (co)variances of the estimates without the scale parameter [8.1.5]. If the diagonal element is zero (that is, the corresponding parameter is aliased) the row contains the coefficients of the linear dependency which produced the aliasing.

Givens algorithm: The description of the working matrix in the Givens case is given in the Modelling Guide [11]

8.3.3 Warnings resulting from changes to elements of the model

If any of the model components: error, link, *y*-variate, offset, prior weight, binomial denominator, initial values, load, baseline, terms formula, eliminate formula, or a macro involved in a user-defined model specification, are changed then GLIM issues the warning

```
-- model changed
```

If a vector in the terms formula, eliminate formula, or one set as a *y*-variate, weight, offset, binomial denominator, initial value, or load, or one specified as an index to any of these has had its length changed by an `ASSIGN` statement GLIM issues the following warning

```
-- change to length of model vector
```

If a factor is redeclared with a different reference category [3.1.4], thus changing the parametrization of the model, GLIM issues the following warning

```
-- change to parametrization of the model
```

If a variate involved in the model formula is redeclared as a factor or if a scalar defining the order of a polynomial in the formula is changed in value then the number of parameters in the model is changed and GLIM issues the following warning

```
-- change to number of parameters in the model
```

If a vector in the terms formula, eliminate formula, or one set as a *y*-variate, weight, offset, binomial denominator, initial value, or load, or one specified as an index to any of these has had its values changed by a directive then this affects the model and GLIM issues the following warning

```
-- change to data values affects model
```

If there are no parameters in the model and the `DISPLAY` or `EXTRACT` directives are used with options expecting parameters they are ignored and GLIM issues the following warning

```
-- no parameters in the model
```

If the `O` option to the `DISPLAY` directive for examining orthogonal polynomials in the model is used and there are none present GLIM issues the following warning

```
-- no orthogonal polynomials present in the model
```

If the *y*-variate is changed in a way affecting the number of units GLIM issues the following warning

```
-- number of units changed
```

8.4 Saving the results from a fit

In order to perform further calculations on the results of a fit, or to plot the residuals against the fitted values, and so on. the user must be able to access the various components produced by the fitting algorithm. GLIM allows two modes of access, one directly via the system vectors and system scalars immediately, and the other indirectly via system structures only available through the `EXTRACT` and `PREDICT` directives.

8.4.1 System structures

8.4.1.1 Scalars

The full list of system scalars is given in the Reference Guide [14.2.2]. Those holding results from the most recent fit are:

`%dv` The deviance.

`%df` The degrees of freedom.

`%sc` The value of the scale parameter, if set; otherwise the estimated value of the scale parameter.

`%nin` The number of valid observations included in the fit with positive prior weights.

`%itn` The number of cycles that were required for the current model to converge. During a user-defined fit [9], it contains the current iteration number.

`%x2` The generalized Pearson chi-squared statistic. This is the sum of squared Pearson residuals from the fitted model.

`%cv` The state of the current model fit:

−1	no model yet fitted;
0	model fitted and converged;
1	model fitted, but not yet converged;
2	model fitted, but failed through losing degrees of freedom;
3	model fitted, but failed because of divergence;
4	model fit failed due to invalid linear predictor or fitted value.

`%pl` The number of parameters not intrinsically aliased in the current model. The length of the vectors `%pe`, `%se`, `%al`, and `%sb`.

`%ml` The number of elements in the lower triangle of the variance–covariance matrix for the non-intrinsically aliased parameters.
The length of the vector `%vc`.

8.4.1.2 Vectors

The full list of system vectors is given in the Reference Guide [14.3]. Those holding results from the most recent fit and directly accessible are:

`%fv` The fitted values for each value of the *y*-variate vector whether it was weighted out of the fit or not.

`%lp` The linear predictor.

`%eta` Normally holds the same values as `%lp`, but in certain user-defined models will contain different values (see [9] for details).

`%rs` The Pearson residuals as displayed by the statement `$display r$`.

In addition, there are three high-precision vectors whose values can only be accessed via the `CALCULATE` directive. These are:

`%wvd` The working variate (or adjusted dependent variate).

`%wtd` The iterative weights variate.

`%tri` The working triangle stored as a vector, as displayed by the statement `$display t$`.

8.4.1.3 System pointers

Pointers are GLIM system identifiers that point to another vector or scalar. For example, the pointer `%yv` points to the values of the current *y*-variate. Pointers can be unset — if they are unset and are used in a GLIM statement an error message is produced. Note that when a pointer identifier is used in a directive, the value of the pointer is substituted before the directive is executed. So, for example

```
$yvariate Y$delete %yv$
```

will delete the variate `Y`; and

```
$print *n %yv$
```

will print the identifier `Y`. The available system pointers are:

`%yv` points to the current *y*-variate;

`%pw` points to the current prior-weights vector, if set; if no prior weights vector is declared, points to the real value 1.0;

`%os` points to the current offset vector, if set; if no offset is declared, `%os` points to the real value 0.0;

`%bd` points to the current binomial denominator vector, if set;

`%iv` points to the current initial values vector, if set;

`%lo` points to the current load vector, if set.

In addition, further system pointers are available which point to the index vectors of some of the above quantities. See the Reference Guide [14.4] for further details.

8.4.2 The EXTRACT directive

This directive is used after a fit to cause values resulting from the fitting process to be saved in the system vectors specified (see the Reference Guide [14.3.2.2] for full details). The syntax is

`$extract` *identifier* [(*term*)] [*identifier* [(*term*)]] $

where *identifier* is one of `%pe`, `%se`, `%vc`, `%vl`, `%ft`, `%cd`, `%lv`, `%wt`, `%va`, `%di`, `%dr`, `%wv`, `%dm`, `%al`, or `%sb`.

If the identifier is `%sb`, then and only then, should *term* be supplied. The *term* then should specify a model term in the current model.

The system vectors that can be `EXTRACT`ed are:

`%pe` The vector of parameter estimates produced by the most recent fit.

`%se` The standard errors of the parameter estimates

`%vc` The lower triangle of the variance–covariance matrix of the parameter estimates, as displayed by the statement `$display v$`. It has length `%ml`.

`%vl` The variances of the linear predictors for each observation.

`%ft` Codes indicating the status of each observation, represented by the values of the *y*-variate, in the fit. The codes are:

- 0 observation present in the fit;
- 1 observation weighted out of fit;
- 2 observation weighted out and has an invalid binomial denominator or *y*-variate; fitted values or probabilities can be calculated;
- 3 observation weighted out and has an invalid model vector; fitted values cannot be calculated.

`%cd` Cook's statistics for each observation from the current fit.

`%lv` The leverage values for each observation from the current fit.

`%wt` The iterative weights for each observation generated during the fit.

`%va` The variance function values for each observation generated during the fit.

`%di` The deviance increment values for each observation generated during the fit.

`%dr` The values for each observation of the derivative of the link function generated during the fit.

`%wv` The values for each observation of the working variate generated during the fit.

`%dm` The model matrix for the current model. This is stored as an array of size `%nu` × `%pl`.

`%al` The aliasing structure of the parameter estimates for the current model.

`%sb` The positions in the vector of parameter estimates `%pe` of those parameters (selected β's) representing the part of the model specified by *term*.

Examples

```
$extract %pe %se $
      ! to access the estimates and their
      ! standard errors
$extract %cd %lv $
      ! to access the Cook's statistics
      ! and leverage values for diagnostic plots
$fit sex+year+exposure+year.exposure $
$extract %sb (year.exposure) $
      ! to obtain an indicator variable of the year
      !.exposure interaction terms in the %pe vector
$extract %pe %se$
$pick PEYREX,SEYREX %pe,%se %sb$
      ! pick out the parameter estimates and
      ! standard errors.
```

8.4.3 The PREDICT directive

This directive is used to obtain fitted values, linear predictor values, and the variances of the linear predictors for specified values of the explanatory variates and factors in the current model. The predicted quantities are put respectively into the system vectors `%pfv`, `%plp`, and `%pvl`. The syntax is

`$predict` [(*option list*)] [*item*] [[,] *item*] $

The only option is

`style` = *real*

where if *real* is positive a brief message indicating the assumed values of all the model vectors is output and the values of `%pfv`, `%plp`, and `%pvl` are displayed. If *real* is zero only the message is displayed; if *real* is negative nothing is displayed. The default if *real* is not specified is assumed to be zero and the predicted values are not displayed.

Item takes the form

LHS = *RHS*

where

(1) *LHS* is a variate, factor, or array (optionally indexed) corresponding to a model component in the current model formula

and

(2) *RHS* is an identifier or real number specifying values for *LHS*. The valid combinations of *LHS* and *RHS*, using the same identifier conventions as in [8.7.2] are listed in the following table.

Model component	Value of *LHS*	Valid values of *RHS*	Default if *RHS* not specified
X	X	*real*, *vector*	0.0
A	A	*real*, *vector*	The reference category
Q	Q	*vector*, *array*	A row vector of zeros
X(IND)	X or X(IND)	*real*, *vector*	Values taken as X if present, otherwise 0.0
A(IND)	A or A(IND)	*real*, *vector*	Values taken as A if present, otherwise the reference category
Q(IND)	Q or Q(IND)	*vector*, *array*	Values taken as Q if present, otherwise a row vector of zeros
X<*integer*>	X	*real*, *vector*	0.0
A<*integer*>	A	*real*, *vector*	The reference category
<*integer*>	%units	*real*, *vector*	%nu+1
binomial denominator	X or %bd	*real*, *vector*	1.0
offset	X or %os	*real*, *vector*	0.0

In the simplest case, the model will consist of model components of the form X (variate) or A (factor). Specifying *reals* on the RHS for the sequence of vectors in the model will produce a set of single predicted values, and the system vectors %pfv, %plp, and %pvl produced will be of length 1. If *vectors* are specified instead of *reals* on the RHS, the predicted vectors will be of the same length as these vectors. If a mixture of *reals* and *vectors* is used, then the *reals* are assumed to represent constant vectors.

Model components in the model, but not specified in the PREDICT statement, will take a default value as indicated in the table above. Model components not in the model but specified in the PREDICT statement are ignored.

Examples

(i)
```
$fit x $
$assign xp = 0,1...5 $
$predict (style=1) x = xp  $
```

will derive and display

Predicted linear predictor	Predicted fitted values	Predicted variance of the linear predictor
%plp	%pfv	%pvl
value	*value*	*value*
..	..	..
..	..	..
..	..	..
value	*value*	*value*

The system vectors %pfv, %plp, and %pvl can then be used to plot a fitted line with confidence limits over the chosen range of x-values unrestricted by the values occurring in the data.

(ii)
```
$fit A*X
$ass xp = 0,5...50 $
$predict  X = xp   A = 1   $calculate YF_A1=%pfv $
$predict  X = xp   A = 2   $calculate YF_A2=%pfv $
$predict  X = xp   A = 3   $calculate YF_A3=%pfv $
```

This model represents a set of straight line trends with slopes and intercepts depending on the level of the factor A. The full sequence generates values for %plp, %pfv, and %pvl over the X range of 0 to 50 for each of the A values 1, 2, and 3. The %pfv values are stored in the variates YF_A1, YF_A2, and YF_A3 to avoid subsequent PREDICT statements overwriting the %pfv values before they are used. The three lines can then be plotted to illustrate how their slopes and intercepts differ. If the variances of the linear predictor, %pvl, had also been stored the confidence limits for each fitted line could also have been obtained.

(iii)
```
$fit A + X<2> $
$predict A=4 X=3 $
```

This fits a 'parallel' set of quadratic curves modelling the relationship between the y-variate and X in the categories representing the levels of A. It then predicts %pfv, %plp, and %pvl for the fourth level of A at X = 3.

(iv)
```
$fit (AGE + GENDER)(INDX) $
$assign AGE_SEQ = 20,25...60 $
$predict AGE=AGE_SEQ GENDER=2 $
```

This fits a model to describe the relationship between the y-variate and the values of AGE and `GENDER` as indexed by `INDX` . It then predicts and outputs the values of the `%pfv`, `%plp`, and `%pvl` for `GENDER` level 2 and `AGE` values from 20 to 60 in steps of 5.

(v)
```
$error b n $
$fit X + X(LAG) $
$predict X=3 X(LAG)=5 %bd=100 $
```

This fits a **lagged** model to the relationship between `X` and the *y*-variate and then obtains predicted values for the single point where `X` = 3 and `X(LAG)` = 5. Note that the setting of `%bd` to 100 in this example will result in `%pfv` containing the predicted percentage; other values can be used. For example, `%bd` = 1 will give the predicted probability (the default).

9 User-defined models

Introduction

The model specification described in [8] allows those error and link specifications that provide models which are the most commonly used in practice. The specification of a GLM, however, is more general than this, as explained in the Modelling Guide [11]. GLIM has facilities for specifying any model within this more general framework, by users specifying their own error distribution or their own link or both, though the fitting procedure is slightly less efficient than for the standard models. These models are referred to here as **OWN models.**

At the heart of the standard model fitting procedure is an iteratively re-weighted least squares (IRLS) algorithm as described in the Modelling Guide [11]. Many models that are not strictly within the GLM framework can, nevertheless, be fitted by maximum likelihood using an IRLS algorithm which is a variant of that used in GLIM. The algorithmic differences are that particular system structures which GLIM computes on the assumption of the GLM structure will need to be modified by the user. GLIM also provides a mechanism to achieve these modifications in a straightforward way. Thus GLIM can still be used when the model is from a class of models wider than that of the GLM. These extensions are referred to as **OWN algorithms**. Either or both of an OWN model and an OWN algorithm may be specified for any fit.

9.1 The components of an OWN model

When the user wishes to fit an OWN model, that is, a GLM that is not a standard one, the model specification is exactly as in [8] except for the `ERROR` and/or `LINK` declarations. In this situation user macros will replace the standard computations of particular system structures.

9.1.1 The OWN error specification

In finding the maximum likelihood estimates of the parameters in a GLM the only information required of the error distribution is the specification of the variance function, that is, the variance of an observation as a function of its mean. Also in assessing models we require the computation of the **deviance** which also depends on the error distribution. Thus for a user OWN error specification a macro must be set up that calculates the following system vectors:

`%va` The variance function for each unit, usually as a function of the current fitted values in `%fv` and possibly involving other quantities. `%va` must not include any scale parameter or prior weights, as any `SCALE` or `WEIGHT` settings in force are already used by the fitting algorithm.

`%di` The *deviance increment*, which is the contribution to the deviance for each unit. `%di` must similarly not contain any scale parameter or prior weights, as any `SCALE` or `WEIGHT` settings are taken care of by the fitting algorithm.

The user then declares that this macro should be used in model fitting by the statement:

`$error own` *macro*

The keyword `OWN` may be abbreviated to `O`. The declaration of an own error will leave the link function unchanged but will reset the scale parameter so that it is estimated rather than fixed.

The following checks are carried out:

- The system vectors `%va` and `%di` exist after the macro has been called, and these vectors have the correct length (`%nu`).
- The value of `%va` for each element must be positive.
- The sum of the values of `%di` plus the baseline deviance, if set, must not be negative.

Specification of any keyword option other than `O` to the `ERROR` directive will unset both the user-defined OWN error macro and its associated link function, replacing it with the error specified, and the default link appropriate to that distribution.

Example The Poisson distribution can be declared directly, but here we illustrate how to fit it using a user-defined error.

The Poisson distribution with mean μ has probability function

$$f(y \mid \mu) = \frac{\mu^y e^{-\mu}}{y!} \qquad \text{with } \mathrm{var}(y) = \mu$$

The log-likelihood for n observations $y_1, y_2, \ldots, y_i, \ldots, y_n$ where the ith observation has mean μ_i is

$$\log l = \sum_{i=1}^{n} y_i \log(\mu_i) - \mu_i - \log(y_i!)$$

and the deviance is therefore

$$-2(\log l_{\text{current}} - \log l_{\text{full}}) = y_i \log(y_i / \hat{\mu}_i) - (y_i - \hat{\mu}_i)$$

The OWN error macro for the Poisson distribution would therefore be

```
$macro POISSON !
$calculate  %va=%fv $!
$calculate  %di=2*(%yv*%log(%yv/%fv)-(%yv-%fv)) $!
$endmac
```

and it is declared by the GLIM statement

```
$error own POISSON$
```

9.1.2 The OWN link specification

The information required by the algorithm that is governed by the link function is the functional relationship between the fitted values and the linear predictor and the derivative of this function with respect to the fitted values. Thus for a user-defined link function a macro must be provided that calculates these two quantities. The user then specifies that this macro should be invoked in place of the standard system computations by the statement:

`$link own` *macro* `$`

Specifically this macro should calculate the following two system vectors:

`%fv` as a function of `%lp` (or strictly `%eta`, see [9.2.2]); the calculation must not include any offset vector, as any `OFFSET` settings in force are used by the fitting algorithm.

`%dr` the derivative $d\eta/d\mu$ as a function of the current values of `%fv` or `%eta`.

Specification of any keyword option other than `O` to the `LINK` directive will unset the user-defined link function, but will leave the current error setting unchanged. The following checks are carried out:

- The system vectors `%lp`, `%dr`, and `%fv` exist after the macro has been called, and these vectors have the correct length (`%nu`).
- The value of `%dr` for each element must be non-zero.

For standard models the iterative process is initialized by calculating starting values for `%lp` from the values of the *y*-variate. For own links this cannot be done automatically, so the user **must** initialize `%lp` before a fit can be done. The user can set up an initialization macro that will be invoked automatically using the `INITIAL` directive:

```
$initial macro$
```

Example As an illustration the own link calculations for the log link

$$\eta_i = \log(\mu_i)$$

would be:

$$\mu_i = \exp(\eta_i)$$
$$\frac{\partial \eta_i}{\partial \mu_i} = \frac{1}{\mu_i}$$

and so the macro would be

```
$macro  LOGLINK !
$calculate  %fv=%exp(%eta) $!
$calculate  %dr=1/%fv $!
$endmac
```

A suitable initialization macro would be:

```
$macro LOGINIT !
$calculate %eta=%log(%yv+0.5) $!
$endmac
```

The macros would be declared and a null model fitted by the following statements (note that the error distribution needs to be declared before the link; the Poisson distribution has been used here but any distribution could be specified):

```
$error p$
$link own LOGLINK$
$initial LOGINIT $
$fit $
```

9.2 Components of an OWN algorithm

When a user wishes to fit models outside the class of GLMs or wishes to modify the algorithm in some way, the above facilities of OWN models may not be sufficient. In such situations it will be necessary to modify certain system structures at various stages of the iterative cycle other than those at which OWN link or error macros would be invoked. Four other stages of the iterative cycle can now be specified as points at which user macros will be called. These are shown in Figure 9.1 at the end of this chapter. The descriptive names given to these macros in the following are only meant to indicate the most likely usage of the particular modifications. There is no restriction on what the user modifies at any stage in the iterative cycle subject to the condition that the structure has values at that stage. GLIM continues to compute the values of system structures as for the standard algorithm. There is ***no*** insistence, as there is in the OWN error and link, that users ***must*** calculate the values of certain system structures, rather that users ***may*** modify whatever structure they wish in order to produce the desired effect. The four macros will be referred to as the **own initialization, triangle, predictor,** and **weight** macros. Notice must be taken of the order of stages in the iterative cycle. For example, computing `%eta` in the **own triangle** macro would have no effect as `%eta` would be recalculated at the next stage.

9.2.1 The OWN initialization macro

This macro was introduced above in the discussion of OWN links [9.1.2]. It is called once in each fit after the model formula has been read and all initial checks have been performed, but before the system vectors `%pe`, `%se`, `%al`, and `%sb` from the previous fit have been deleted or overwritten. It is specified via the `INITIAL` directive:

```
$initial macro $
```

A typical use of the macro is to calculate suitable starting values for `%lp` where an OWN link macro is in force.

9.2.2 The OWN triangle macro

This macro is called after the working triangle has been formed, but before it has been inverted to produce the latest values of `%lp` (this is equivalent to re-estimating the parameters).

The main purpose of this macro will therefore be to modify the values of the working triangle, `%tri`. Normally, the user will need to ensure that the algorithmic method is set to Gauss–Jordan; the working triangle at this stage in the iteration will then contain the SSP matrix. However, it is also possible to write a macro to manipulate values of the triangle produced by the Givens algorithm.

The `LOAD` directive is used to specify this macro:

```
$load macro
```

This *macro* will be called at this stage in each iteration and will usually directly modify `%tri`. (Note that `%tri` is a high-precision system vector and it can only be accessed or modified through the `CALCULATE` directive.) If the `LOAD` directive is used with a vector, that is,

```
$load vector
```

then the values held in the *vector* will be added to the ***diagonal*** of the working triangle at this stage. If both *macro* and *vector* are specified then the macro is called first and then the values of the vector are added to the diagonal. If no arguments to the `LOAD` directive are specified, GLIM will unset both the vector and the macro.

Example To perform ridge regression in the correlation form (that is, in which all explanatory variables have been standardized to have unit variance) we require a constant to be added to the diagonal of the SSP matrix. This can be achieved by using the following OWN triangle macro, which assumes that the constant is held in the ordinary scalar %d:

```
$macro  RIDGE !
$assign IDIAG=1,...,%pl $calc IDIAG=%cu(IDIAG) $!
$calculate  %tri(IDIAG)=%tri(IDIAG)+%d $!
$endmac

$method J$
$load RIDGE $
```

Note that in this simple example, specifying a vector with all values equal to %d in the LOAD directive would achieve the same result:

```
$variate %pl CONST
$calculate CONST=%d$
$load CONST$
```

9.2.3 The OWN predictor macro

The OWN predictor macro is called immediately after the linear predictor vector %lp (defined as $\mathbf{X\beta}$) has been calculated and before the link function is used to compute %fv and %dr.

One primary use of this macro is for non-linear models. Consider a non-linear model for which $\boldsymbol{\eta}$ is a general function

$$\boldsymbol{\eta} = \boldsymbol{\eta}(\mathbf{X}, \boldsymbol{\beta})$$

Then the IRLS algorithm is still applicable with $\mathbf{X}$ replaced by $d\boldsymbol{\eta}/d\boldsymbol{\beta} = \mathbf{D}$ (see the Modelling Guide [11.9.3]). However GLIM would then calculate the linear predictor as follows:

$$\boldsymbol{\eta} = \mathbf{D}\boldsymbol{\beta}$$

and this is incorrect.

Thus in fitting non-linear models the user must specify a macro which will calculate $\boldsymbol{\eta}$ in the correct way from $\mathbf{X}$ and the updated values of $\boldsymbol{\beta}$. In order to distinguish between the incorrect linear predictor, $\mathbf{D}\boldsymbol{\beta}$, and the own predictor, $\boldsymbol{\eta}$, a separate system structure, %eta, becomes available. Thus %lp will hold the values of $\mathbf{D}\boldsymbol{\beta}$, and %eta, which will be calculated or modified by the user in the OWN predictor macro, should contain $\boldsymbol{\eta}$. Note that some, if not all, of the matrix of derivatives $\mathbf{D}$ will change at each iteration and will also have to be recalculated in the macro. If no OWN predictor macro is set, `%eta` and `%lp` are synonyms and point to the same system values.

The OWN predictor macro can be used for applications other than non-linear models. In this case `%eta` and `%lp` will still refer to separate quantities, but GLIM will ensure that both hold the values of the linear predictor. However, note that `%eta` and `%lp` have special definitions when an offset is set (see below).

The OWN predictor macro is defined via the second argument to the `METHOD` directive as follows.

```
$method  * macro
```

Note that the first argument, which sets the algorithmic method, should be specified. The asterisk is used to leave the algorithmic setting unchanged.

The meaning of offset is ill-defined for non-linear models so that when an OWN predictor macro is in force `%lp` ***never*** includes the offset. However, in order that the algorithm gives correct results for standard generalized linear models `%eta` is computed including the offset (that is, `%eta=%lp+%os`). Note that `%os` can be safely used in a `CALCULATE` statement even when there is no offset as it then takes the value zero. All links are still available when using own predictors allowing the user to specify the most efficient numerical algorithm. Note that the user may have to give initial values for ***both*** `%eta` and `%lp` for the algorithm to converge.

Example Consider the model of simple similar action in bioassay. This is a standard binomial model with usually a probit or logit link, but with own predictor given by:

$$\eta = \beta_0 + \beta_1 \log(x_1 + \beta_2 x_2)$$

where x_1 and x_2 are the doses of two drugs.

The quantities $\partial\eta / \partial\beta_0$, $\partial\eta / \partial\beta_1$ and $\partial\eta / \partial\beta_2$ need to be calculated.

$$\frac{\partial\eta}{\partial\beta_0} = 1$$

$$\frac{\partial\eta}{\partial\beta_1} = \log (x_1 + \beta_2 x_2)$$

$$\frac{\partial\eta}{\partial\beta_2} = \frac{\beta_1 x_2}{x_1 + \beta_2 x_2}$$

If we refer to $\partial\eta/\partial\beta_1$ as `LPOT` and $\partial\eta/\partial\beta_2$ as `RELPOT`, then the model formula `1+LPOT+RELPOT` will specify the form of **D** and the OWN predictor macro would then be

```
$macro  SIMACTION !
$extract  %pe $!
$calculate  POT=X1+%pe(3)*X2 $!
```

```
$calculate  RELPOT=%pe(2)*X2/POT $!
$calculate  LPOT=%log(POT) $!
$calculate  %eta=%pe(1)+%pe(2)*LPOT $!
$endmac
```

Assuming the y-variate is stored in a vector R and the binomial denominator in a vector N and a probit link, the model would be fitted by the statements:

```
$yvariate R$error b N$link p$
$method  *  SIMACTION $
$fit 1+LPOT+RELPOT $
```

A fairly complex initialization procedure would be required to ensure a high chance of convergence. Note that since $\boldsymbol{\eta}$ is a non-linear function of the β's, `%vl` is only a first-order approximation to var($\boldsymbol{\eta}$) and may need to be used with caution.

9.2.4 The OWN weights macro

This macro is called at the end of each iteration, after new values of the iterative weights and working vector have been formed but before the test for convergence and the formation of the working triangle. Its main use is therefore to modify the iterative weights and working variate. Thus, for example, this macro may be used to modify the fitting algorithm. The algorithm used by GLIM to fit models is based on Fisher scoring in which the ***expected*** information matrix is used in a Newton–Raphson algorithm (see the Modelling Guide [11]). This results in a particularly simple form for the iterative weights and working variate. If the user wishes to use an algorithm other than Fisher scoring, for example using the ***observed*** information matrix, then the iterative weights in IRLS will need to be changed, with a corresponding modification to the working variate. Appendix 1 of Aitkin et al. (1989) contains the mathematical details. The structures to be modified are the double precision structures: `%wtd` (iterative weights) and `%wvd` (working variate); these are used directly by GLIM to form the working triangle at the next iteration.

An OWN weights macro is declared by setting the third argument of the METHOD directive:

```
$method  *  *  macro
```

If there is no third argument for the METHOD directive, then the OWN weights macro is undeclared. Note that the first argument to the METHOD directive sets the algorithm and must always be present, and the second argument sets the OWN predictor macro. Settings are left unchanged by using asterisks as arguments.

Example In order to use the observed information matrix for a binomial distribution with complementary log-log link, the OWN weights macro would be

```
$macro  BINOCLL !
$calculate P=%fv/%bd : T=%log(1-P)/P $!
$calculate W=%wtd+(%yv-%fv)*(T+(1-P)*T**2)  $!
$calculate  %wvd=(%wvd-(%eta-%os))*%wtd/W $!
$calculate  %wtd=W $!
$endmac

$method * * BINOCLL$
```

A secondary use of this macro is to modify permanently the internal stored value of the deviance. This may be achieved by altering the contents of the system scalar `%dv` from within the OWN weights macro. After this macro has been called, GLIM will then check if a change has taken place in the contents of the system scalar `%dv`, and will alter the internal quantities appropriately. Note that this modification will also affect the estimate of the scale parameter if the `SCALE` setting has been set to `deviance`.

9.3 Information on user-defined fits

The `M` option to the `DISPLAY` directive provides information on the current settings of all OWN model macros.

9.4 Deletion of OWN model macros

All OWN model macros may be deleted. For the OWN triangle, OWN predictor, OWN weights, and OWN initialization macros, the effect of deleting the macro is to unset the relevant option. For the OWN error and OWN link macros, the error or link setting will remain defined as an OWN error or link, but with the macro undefined.

9.5 Limitations to user-defined fits

User-defined fits in GLIM are defined as any call of the `FIT` directive which calls a user-defined macro as part of the fitting procedure. In GLIM there are limits to what a user-defined macro can do; these limits are necessary to allow the model fitting procedure to be well defined.

9.5.1 Invalid directives

The following directives are not allowed in an OWN macro.

```
FIT       ELIM      TERMS     YVAR      ERROR     LINK
WEIGHT    OFFSET    INITIAL   LOAD      ALIAS     CYCLE
RECYCLE   METHOD    DISPLAY
```

However, the effect of many of these directives may be achieved in a different way. It is still possible to change the y-variate values, for example, during a fit by changing the values of the *y*-variate through the `CALCULATE` directive rather than the name of the *y*-variate through the `YVARIATE` directive.

The `EXTRACT` and `PREDICT` directives are allowed, but the quantity to be extracted (or the current parameter estimates in the case of `PREDICT`) must be available at the time of extraction. For example, the parameter estimates are not available until the triangle has been formed and inverted for the first time, and are additionally not available on any iteration between the formation of the triangle and its inversion.

9.5.2 Change in number of parameters in the model

During a user-defined fit, a change in the number of parameters in the model will cause an error to be generated. This might be caused by a `FACTOR` or `GROUP` statement changing the type of an identifier from a variate to a factor, or a change in the order of an orthogonal polynomial term. Only vectors and scalars used in the model formula are checked.

9.5.3 Change in length of a model vector

The change of length of any vector used in the model, either in the model formula or in the model definition, will cause an error.

9.5.4 Change of values of a model vector

There is no restriction on the change of the values of a vector used in the model except in the following cases:

(a) change of the value of a factor which is used in the model such that the new factor value is invalid (outside the valid defined range for that factor);

(b) change in the value of a vector which is used as an index vector in the model, such that the index is invalid.

9.5.5 Change of parametrization of a model factor

This can occur when the reference category of a factor changes during a fit. This is allowed, as the reference category of all factors is only used once at the beginning of each fit. Any change in parametrization will therefore be ignored until the fit is completed, and the parameter estimates will therefore remain unchanged.

9.5.6 Alteration of OWN macros

An OWN macro is not allowed to alter the contents of another OWN macro or itself while a fit is in progress.

9.5.7 Use of the PREDICT directive

The `PREDICT` directive can be used for most user-defined fits. However, where an OWN predictor macro has been used to fit a non-linear model, the `PREDICT` directive will usually give incorrect results as `%eta` and `%lp` will not, in general, be equivalent, and `%pfv` is computed from `%plp` rather than the corresponding values of η.

Figure 9.1: The GLIM Iterative Cycle

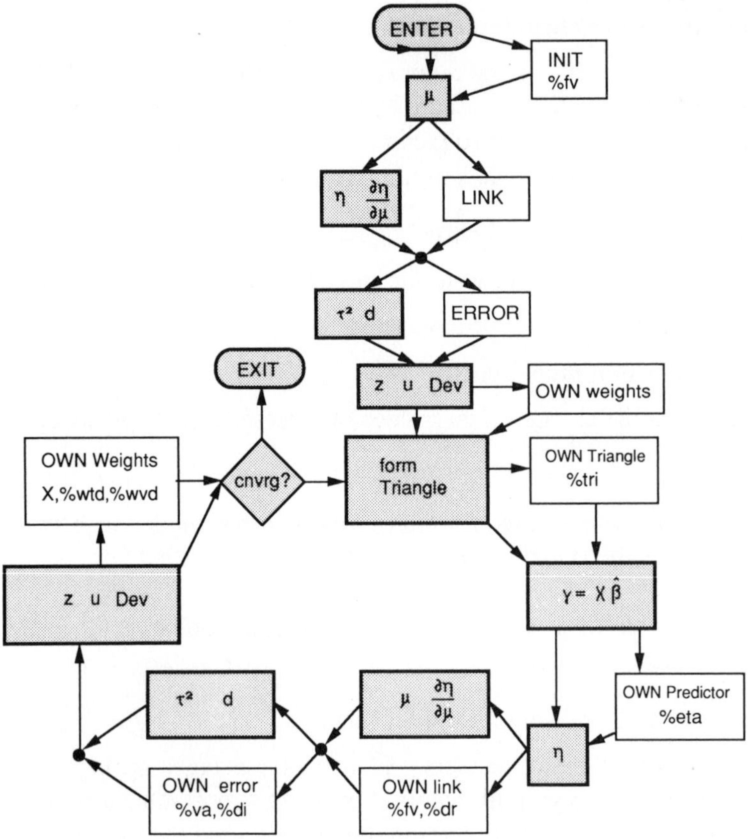

Appendix A The macro library

A.1 Introduction

The GLIM macro library was first made available as a standard facility in Release 3.77. It provided a convenient method of access to many commonly used GLIM macros, some of which had been previously published in the GLIM Newsletter. The macros have been chosen for their usefulness either in providing extra facilities or by extending the range of data analysis and modelling which can be performed with the GLIM system. The library now uses many new features of GLIM4 and is not suitable for use with previous releases of GLIM. The current contents of the macro library are now available on-line during an interactive GLIM session.

The contents of the macro library have changed from GLIM 3.77. Where new facilities in GLIM4 have made the use of a previous GLIM 3.77 macro redundant, that macro has been deleted. For example, the macros `CHIP`, `FPROB`, `TPROB`, and `IGAM` have been deleted from earlier releases and their roles replaced by the functions `%chp`, `%fp`, `%tp`, and `%gp`, respectively.

A.2 Structure of the macro library

The macro library consists of a number of subfiles, the first of which contains an index to the macros available. Each subfile contains one or more macros. Within each subfile, apart from the first, there will be one or more main macros, and possibly many other subsidiary macros which are called by the main macro(s).

The subfiles are currently arranged into four sections:

- data description, exploration, and display;
- robust statistics;
- model checking;
- survival analysis.

The current contents of the macro library can be obtained in several ways. This information can be obtained interactively during a GLIM session by echoing the contents of the first subfile, `INFORM`, as it is read in by GLIM. The GLIM statements needed to do this are:

```
$echo on $
$input %plc 80 INFORM $
```

followed by the statement `$echo off$` once the subfile `INFORM` has been read.

Alternatively, the `MANUAL` directive can be used to display information about the macro library. The statement

```
$manual library $
```

will display summary information on the contents of the macro library. A detailed description of a particular macro can be obtained by including the macro identifier in the `MANUAL` directive

```
$manual library macro_identifier $
```

For example

```
$manual library WEIBULL $
```

will display information about the use of the macro `WEIBULL`.

A.3 Reading the contents of a macro library subfile into a GLIM session

The GLIM macro library has been automatically assigned to a specific FORTRAN channel. In general, this channel number will vary over different machine ranges and installations of GLIM. The macro library channel number for the local installation may be found by issuing the statement

```
$environment c $
```

A system scalar, `%plc`, is also available and contains the channel number to which the macro library has been assigned.

The contents of a subfile can be read in by issuing the statement

```
$input %plc 80 subfile name $
```

For example, the subfile containing the Box–Cox transformation macros is called `BOXCOX`. The macros contained in this subfile can be read into GLIM by using the statement

```
$input %plc 80 BOXCOX $
```

The contents of more than one subfile can be read in by issuing a series of `INPUT` directives. For example

```
$input %plc 80 WEIBULL $
$input %plc 80 BOXCOX $
```

or, more simply,

```
$input %plc 80 NORMAC BOXCOX $
```

Note that GLIM reads the macro library sequentially. Therefore, for greatest efficiency, the subfile names should be specified in the order in which the subfiles are stored.

The contents of a subfile may be displayed as it is being read in by ensuring that the `ECHO` facility is switched on before reading the subfile. For example

```
$echo on $
$input %plc 80 BOXCOX $
$echo off $
```

Each subfile contains a short description of its contents together with condensed documentation on the main macros contained within the subfile. This information is stored at the start of each subfile. The `ECHO` facility therefore provides a way of obtaining on-line documentation on the various macros within the library.

A.4 Passing data to a GLIM library macro

Data can be passed to a macro in many ways. Some ways of passing data are outlined below; each of these methods has been used in at least one macro in the macro library. The on-line documentation indicates the method or methods required for each macro.

A.4.1 Formal arguments

Formal arguments to a macro are specified either by using the `ARGUMENT` directive before the macro is used or, alternatively, by specifying the arguments as part of the `USE` directive. For example, if a macro `M` needs two formal arguments, the first a vector and the second a scalar, then the macro is used with the first argument set to the vector `V` and the second argument set to the scalar `%a` as follows :

either

```
$argument M V %a $
$use M $
```

or

```
$use M V %a $
```

A.4.2 Macro arguments

Macro arguments to a library macro are a convenient way of passing text information to the macro. The text might be a model formula, a variate name, or an expression. Macro arguments are simply macros of the required names which have been set up by the user and which contain the required information stored as text. The user needs to declare all macro arguments before calling the library macro. Examples of suitable declarations are:

```
$macro MODEL AGE*REGION $endmac
$macro ARG1 V $endmac
```

A.4.3 Scalar arguments

Sometimes certain scalars may need to be set before the macro is called. These are referred to as scalar arguments.

A.4.4 Macro prompting

Certain macros may prompt the user for information while the macro is being executed. These macros use the `GET` directive [7.6] to prompt for text strings, real numbers, or arguments according to the requirements of the macro.

A.5 Macro library conventions

A.5.1 Locally declared vectors

Many macros in the macro library need to declare vectors local to that macro. Usually, the required vectors are declared as local structures to the macro, and are thus unavailable to the user on exit from the macro. However, some of these vectors may be useful to the user. In the macro library, a convention has been adopted that all vectors declared within a macro have names which end with the underline symbol _. This convention minimizes the possibility of the names of these locally declared vectors clashing with the names of any vectors set up by the user. The user should therefore avoid choosing names for user-identifiers which end with the underline symbol.

All locally declared vectors are deleted on successful completion of the macro, unless the documentation specifies otherwise. Some macros have the option to keep certain useful vectors in the workspace.

A.5.2 Scalars

Most macros in the GLIM macro library use the macro library scalars (`%z1`, `%z2`, ..., `%z9`) for temporary storage of scalars. Some macros may additionally use some ordinary scalars (`%a`, `%b`, ..., `%z`) and these may contain scalar values potentially useful to the user. If any ordinary scalars are used by a library macro, then this is documented, together with their contents, if useful.

A.5.3 Space recovery

Library macros occupy part of the workspace and it may be necessary to reclaim this space after the macro has been used. To facilitate this, some subfiles contain an additional macro called `DELETE`; the text of this macro is, of course, different for each subfile. After execution of the required library macro the statement

```
$use DELETE $
```

will delete all identifiers from that subfile, with the exception of the macro `DELETE` itself. If a different subfile were then to be read in, `DELETE` could be left, as it would then be overwritten; otherwise it could be deleted.

Note that the statement

```
$print DELETE $
```

is a further way of obtaining the list of macro identifiers used in a subfile.

A.6 Error messages and reporting

Many library macros produce their own specific error and information messages. All macros may fail with a standard GLIM error message in certain circumstances. Common causes of failure are:

(a) exceeding the maximum number of identifiers;

(b) exceeding the size of the workspace;

(c) failure to set or to correctly specify all necessary arguments to a macro.

Some library macros use the `OUTPUT` directive to switch off unnecessary output; for example, iterative macros using the `FIT` directive. If such a macro should fail (or if the user should break-in to it) then, when control returns to the user, output to the primary output channel may still be switched off, and the user will seemingly get no response when further GLIM statements are entered. If this should happen, the statement

```
$output %poc $
```

should be issued to set the output back to its default setting.

A.7 Future development of the macro library

Updates to the macro library will be produced periodically and details of the updates will be published in the GLIM Newsletter. The updates can be obtained either on diskette or on magnetic tape from NAG Ltd. for an additional fee.

Submissions to the macro library are welcome, and potential authors should contact the macro library editor via NAG Ltd. to obtain submission guidelines and to submit macros for refereeing.

Appendix B
Differences between GLIM 3.77 and GLIM4

B.1 Introduction

In general, GLIM4 is compatible with GLIM 3.77, and GLIM 3.77 programs will, except for a few incompatibilities, run in GLIM4. A full list of incompatibilities is given in Appendix C.

B.2 The language

Identifier names may now be up to eight characters long. Note that this might cause a working GLIM 3.77 program to fail; for example:

```
$assign VECTOR=1,2,3,4 $calculate temp=%sqrt(VECT)$
```

would be valid in GLIM 3.77 but not in GLIM4. Strings may now contain the directive symbol `$`, and must be terminated by a single `'`.

B.3 Changes to GLIM 3.77 directives

B.3.1 ACCURACY

It is now possible to specify the maximum number of decimal places to be printed.

B.3.2 ALIAS

The keyword `ON` or `OFF` may now be optionally specified. The aliasing setting may be displayed [8.2.6.1].

B.3.3 ARGUMENT

Keyword arguments are allowed in the body of a macro. Arguments may be unset by using the - operator [7.4.3].

B.3.4 ASSIGN

Assign lists (for example, of the form `2,4...10`) are now allowed on the right-hand side of the assignment. Spaces may be used as a separator between items as well as commas. If a vector appears on both the LHS and RHS of the assignment, it now does not have to be previously defined [5.1].

B.3.5 CALCULATE

All calculation is now carried out in high precision. Indexed vectors on the LHS are initialized to zero before assignment. A wide range of new mathematical and statistical functions are available.

New functions are also provided for string comparison and for determining the attributes of an identifier [5.4].

B.3.6 CYCLE

The aliasing default setting has changed [8.2.9.1].

B.3.7 DATA

A list may be specified as an alternative to the identifiers. In conjunction with the `READ` or `DINPUT` directives, this directive may be used to read a subset of identifiers from a dataset [3.2.6].

B.3.8 DELETE

Deletion of a macro now causes deletion of all its local variables. Deletion of a list deletes the list but not the list elements. Deletion of an identifier in a list deletes the list as well as the identifier [4.6.1].

B.3.9 DINPUT

Filenames in addition or as an alternative to a channel number may now be specified [3.2.7].

B.3.10 DISPLAY

The output has been improved. New options give access to information on the algorithm, the parameters to be estimated, and the definition of any orthogonal polynomial variates present in the model [8.3.2].

B.3.11 DUMP

Filenames in addition or as an alternative to a channel number may now be specified [4.8].

B.3.12 ECHO

The keyword `ON` or `OFF` may now be optionally specified. The directive now affects the transcript output by default [4.2.8].

B.3.13 EDIT

A list can now be specified [5.3].

B.3.14 ENDMAC

The directive can no longer be confused with the `END` directive [7.1].

B.3.15 ENVIRONMENT

The optional channel number has been withdrawn. The format of the output has been enhanced for the `C`, `D`, `I`, `S`, and `U` options. New options provide information on the format, data list, argument settings, maximum and minimum function values, and graphical settings and facilities [4.5].

B.3.16 ERROR

The binomial denominator may be indexed. The inverse Gaussian distribution has been added. User-defined distributions are now specified via this directive [8.1.2].

B.3.17 EXTRACT

A wide range of quantities may now be extracted following or during a fit. These include leverage values, the iterative weights, the model matrix, and subsets of parameter estimates [8.4.2].

B.3.18 FACTOR

A reference level may be specified for a factor [3.1.4].

B.3.19 FIT

The model fitting syntax has been extended, with indexing, orthogonal polynomials, the exponentiation operator, and variate by variate interactions. It is now possible to fit a model with no estimated parameters. The system scalar `%gm` is no longer valid in model formulae (`1` should be used) [8.2].

B.3.20 FORMAT

For fixed format, some limited syntax checking now takes place when the format is specified. I format is automatically translated into F format [3.2.1]. The current format setting is available through the `ENVIRONMENT` directive [4.5].

B.3.21 HELP

This directive no longer controls the automatic output of help messages to the screen (this has been superseded by the `BRIEF` directive). Its sole purpose now is to print the help message if the brief setting is off, and otherwise to direct users to the new `MANUAL` directive [4.4.4].

B.3.22 HISTOGRAM

Lists can be specified for the *y*-vectors and weights [6.4].

B.3.23 INPUT

Filenames in addition or as an alternative to a channel number may now be specified [4.2.2].

B.3.24 LINK

All links are now valid for all error distributions. There is one new link, the inverse quadratic. User-defined links are now specified via this directive [8.1.3] [9.1.2].

B.3.25 LOOK

A list of identifiers may now be specified [6.1.1].

B.3.26 MACRO

Keyword arguments and local identifiers can now be specified. Macros can be edited and debugged using new directives [7].

B.3.27 OFFSET

The offset may be indexed. The system pointer `%os` now points to the value zero if no offset is defined [8.1.6].

B.3.28 OUTPUT

Filenames in addition or as an alternative to a channel number may now be specified [4.3.3].

B.3.29 PAGE

The keyword `ON` or `OFF` may now be optionally specified [4.4.2].

B.3.30 PASS

Now allows keywords to be passed as an alternative to the integer. A list of vectors, arrays, and scalars may now be passed to a FORTRAN subroutine. An option specifies whether to carry out transposition of GLIM arrays to FORTRAN arrays and vice versa [4.10.4].

B.3.31 PAUSE

The directive, if implemented, now allows users to issue operating system commands from within GLIM [4.10.1].

B.3.32 PLOT

The output has been enhanced. Titles and axis labels may be specified. User-defined cut-points for the axis labelling can now be given. Multiple *x*-*y* plots can now be superimposed The maximum limit for the number of *y*-vectors has been removed [6.5].

B.3.33 PRINT

The output from the `PRINT` directive can be redirected to a macro. Vectors may now be indexed. Lists may be printed [6.2].

B.3.34 READ

The directive now takes an optional temporary data list [3.2.3].

B.3.35 RECYCLE

The directive now affects the setting of the `INITIAL` directive [8.2.9.2].

B.3.36 REINPUT

Filenames in addition or as an alternative to a channel number may now be specified [4.2.3].

B.3.37 RESTORE

Filenames in addition or as an alternative to a channel number may now be specified [4.8].

B.3.38 REWIND

Filenames in addition or as an alternative to a channel number may now be specified [4.2.4].

B.3.39 SCALE

A new facility controls whether the mean deviance or the mean chi-squared statistic is used where the scale parameter is to be estimated [8.1.5].

B.3.40 SET

New phrases control the default graphics device, and whether or not the constant term is included in the model formula by default [4.7].

B.3.41 SORT

Multiple source vectors and destination vectors may be specified. Multiple keys are also provided. Lists may be used in the directive. A new option controls the sort direction [6.3].

B.3.42 SSEED

Now resets the system scalars %s1, %s2, and %s3 to the current values of the random number seed after each random number has been generated [5.4.6].

B.3.43 TABULATE

A new statistic WEIGHT returns the total weight, and a new phrase COUNT (as an alternative to WITH) specifies a weight vector to be treated as a frequency count. Multiple statistics are now allowed in a single TABULATE statement. Output has default labelling using the names of the FOR variables and names of the statistics with multiple statistics [6.10].

B.3.44 TPRINT

An alternative separator, /, used in the dimension list specifies which variables are to be used to label the rows and which label the columns. The dimension list may be omitted when outputting arrays. Vectors may be printed without a dimension list [6.9].

B.3.45 TRANSCRIPT

The transcript output is now affected by the OUTPUT, ECHO, BRIEF, WARN, and VERIFY directives. The transcript setting can now be specified relatively as well as absolutely. A scalar may also be specified to control the transcript setting [4.3.1].

B.3.46 UNITS

This directive is now a synonym for the SLENGTH directive, and determines only the standard length of vectors, and not the number of units in a fit. The directive no longer affects the settings of the ERROR, LINK, WEIGHT, SCALE, and OFFSET directives [8.1.1].

B.3.47 USE

See the enhancements for the ARGUMENT directive [7.4.3].

B.3.48 VARIATE

A list may now be given to supply the identifiers [3.1.3].

B.3.49 VERIFY

The keyword ON or OFF may now be optionally specified. The directive now affects the transcript output by default [4.4.4].

B.3.50 WARN

The keyword ON or OFF may now be optionally specified. The directive now affects the transcript output by default [4.4.4].

B.3.51 WEIGHT

The prior weight may be indexed. The system pointer %pw now points to the value 1 if no prior weight is defined [8.1.4].

B.3.52 WHILE

The test scalar is now rounded to the nearest integer before testing for equality to zero [7.3.3].

B.3.53 YVARIATE

The *y*-variate may be indexed. The YVARIATE directive now determines the number of units in the fit [8.1.1].

B.4 New directives

B.4.1 ARRAY

Declares multi-dimensional real valued structures [3.1.5].

B.4.2 BASELINE

Allows constant quantities to be added to the deviance and degrees of freedom before display [8.2.9.3].

B.4.3 BRIEF

A switch to determine whether or not extended help messages are produced following a fault [4.4.4].

B.4.4 CLOSE

Closes a file or channel [4.2.7].

B.4.5 DEBUG

A switch, determining whether or not macros currently being executed are debugged, whereby a macro is executed statement by statement with user intervention allowed between each statement [7.8.2].

B.4.6 EDMAC

Initiates the macro editor [7.10].

B.4.7 ELIMINATE

Specifies a model term which will be pre-fitted, that is, included in the model but not contributing to the SSP matrix [8.2.7].

B.4.8 FAULT

Calls the fault mechanism with a user-defined help message [7.7.4].

B.4.9 GET

Allows user prompting for scalars, strings, and identifiers [7.6].

B.4.10 GFACTOR

Allows the construction and assignment of regular patterned factors [3.2.8].

B.4.11 GRAPH

Produces high-quality line and scatter plots on a graphical device [6.6.5].

B.4.12 GSTYLE

Sets the colour, line type, and symbol style of one or more pens for graphical output [6.6.4].

B.4.13 GTEXT

Writes text to a graphical device [6.7.1].

B.4.14 INITIAL

Sets the initial values vector, which provides starting values for the fitted values. Also sets the OWN initial macro [8.2.9.2].

B.4.15 JOURNAL

A switch, determining whether or not a copy of the primary input is kept in a journal file [4.3.2].

B.4.16 LAYOUT

Determines the layout of plots on a graphical device [6.8].

B.4.17 LIST

Defines or modifies a structure containing a list of identifiers [3.3].

B.4.18 LOAD

Sets the load vector, which, if set, will be added to the diagonal of the SSP matrix when the method is Gauss–Jordan. Also defines the OWN triangle macro [8.2.9.4].

B.4.19 MANUAL

Displays a selected page of the on-line GLIM manual [4.11.1].

B.4.20 METHOD

Sets the model fitting algorithm to be either Givens or Gauss–Jordan. Also defines the OWN predictor and OWN weights macros [8.2.8].

B.4.21 NEWJOB

Replaces the END directive in GLIM 3.77. Starts a new job, with optional closing of all secondary input and output files [4.9].

B.4.22 NEXT

Moves to the next directive when debugging a macro [7.8.2].

B.4.23 NUMBER

Declares user scalars and optionally assigns values to them [3.1.1].

B.4.24 OPEN

Opens a file or channel [4.2.7].

B.4.25 PICK

According to a selection vector, selects a subset of elements from a set of vectors and places them in a set of new vectors or shortened existing vectors [5.5].

B.4.26 PREDICT

From the current fitted model, predicts the fitted values, the linear predictor, and its variance, given values for some or all of the model identifiers [8.4.3].

B.4.27 SLENGTH

Sets the default standard length to which all vectors will be set if not otherwise defined [3.1.2].

B.4.28 TERMS

Sets the current model formula without fitting it, allowing the parameters to the examined and the design matrix to be extracted [8.1.7].

B.4.29 TIDY

Deletes all the identifiers in a list, or deletes the values, type, and attributes of all local variables in a macro [4.6.2].

B.5 GLIM 3.77 directives and features no longer available

B.5.1 END

The END directive has been replaced by the NEWJOB directive [4.9].

B.5.2 OWN

The OWN directive has been replaced by relevant options to the ERROR, LINK, METHOD, LOAD, and INITIAL directives [9].

B.5.3 %gm

The system identifier %gm is no longer available in a model formula. The character 1 should be used instead to represent the constant term [8.2.4.1].

B.6 Changes for interactive use

Prompts for further input now include the prompt

```
STRING?
```

when a string is still to be terminated. Special prompts are provided for the macro editor and the macro debugger. User-defined prompts for numeric values, strings, and identifiers can be specified. GLIM will now pause by default after every high-quality graph has been drawn to the screen, and after a page of output has been produced by the `MANUAL` directive.

B.7 Other enhancements

The list of system scalars has been extended. These relate mainly to model fitting, including the iteration number and the convergence state.

The list of system vectors has been substantially extended. It is now possible to gain access to most quantities used in the model fitting process, usually via the `EXTRACT` directive.

Run-time error detection has been improved, with floating-point overflow detection available on most implementations. FORTRAN input–output errors are now trapped and a text error message is produced. A start-up or initialization file may be used on some systems [4.2.1]. This file, whose name is system dependent, is read before the start of every job.

Appendix C Incompatibilities between GLIM 3.77 and GLIM4

Programs written for GLIM Release 3.77 can be run without change using GLIM4, with the following exceptions (note that references here refer to the Reference Guide).

(1) Strings in GLIM4 may now contain the directive and string symbols [13.2].

(2) The `END` directive has been replaced by the `NEWJOB` directive [15.54].

(3) The system scalars `%s1`, `%s2`, and `%s3`, which are the seeds for the standard random number generator, have changed their meaning.

(4) The `OWN` directive has been withdrawn, and has been replaced by better facilities in the `ERROR`, `LINK`, `INITIAL`, `LOAD`, and `METHOD` directives, [15.25], [15.45], [15.41], [15.47], [15.53].

(5) The test scalar in the `WHILE` directive is now rounded to the nearest integer before the test for equality to zero is carried out [15.95].

(6) The system scalar `%nu` now refers to the ***number of units in the fit*** rather than the standard length of vectors. In GLIM 3.77 these concepts were equivalent; in GLIM4 they have been separated, with a new system scalar `%sl` representing the latter quantity.

(7) User identifiers, including subfile names, may now be up to eight characters, rather than four characters in length. This will affect compatibility in the following manner. In GLIM 3.77 it was possible to define the name of an identifier using a name with more than four characters, then refer at a later stage to that identifier using the first four characters of its name. Programs using this facility will no longer work.

(8) The system scalar `%gm` may no longer be used as a synonym for `1` (the intercept or constant term) in model formulae.

(9) The optional channel number in the `ENVIRONMENT` directive is no longer available [15.24].

(10) The system vectors `%wt` and `%wv` are no longer available by default following a fit, and now need to be extracted or accessed in extra precision through the system vectors `%wvd` and `%wtd`.

(11) The `HELP` directive no longer controls the switch determining whether help messages are displayed following an error. This function is now carried out by the `BRIEF` directive [15.7].

(12) The o and i options to the TRANSCRIPT directive have changed their meaning, and no longer control secondary input and output. The TRANSCRIPT setting is now affected by the setting of the OUTPUT, ECHO, BRIEF, VERIFY, and WARN directives. The contents of the transcript setting scalar %tra have changed [15.59], [15.18], [15.7], [15.92],[15.93].

(13) Any identifiers which have been extracted using the EXTRACT directive are now deleted at the start of a new fit.

(14) The UNITS directive no longer resets the prior weight, offset, error, link, or scale settings. In addition, if the y-variate is changed and the length of one or more of the prior-weight vector, the offset, or the binomial denominator differs from the new number of units, then they are now unset.

(15) The model triangle displayed by the T option to the DISPLAY directive has a different meaning when the default algorithmic method (Givens) is used. The GLIM 3.77 triangle may be obtained by setting the algorithmic method to Gauss–Jordan.

(16) For iterative fits, the value given in the system scalar %x2 (the generalized Pearson chi-squared statistic) is now calculated from the current fitted values, and not the fitted values from the previous iteration. (When no ELIMINATE formula is set, the chi-squared statistic from the previous iteration may be obtained from the last element of the working triangle.)

(17) In the TABULATE directive, if the WITH phrase is present and comes before the FOR phrase, and one or more of the vectors present in the FOR phrase are variates, then the lengths of the output vectors may differ. GLIM4 will ignore values of FOR variates that have zero values in the weight vector in determining the length of the output vectors [15.84].

(18) A directive symbol must terminate the READ directive in fixed-format read, before the in-line data is encountered. In GLIM 3.77 the directive symbol was not mandatory.

(19) The system scalar %pl is now defined to be the parameter length throughout the model fitting process. In GLIM3.77, %pl took the value of 0 in the first iteration of a fit.

Apart from these minor incompatibilities with GLIM 3.77 there are many areas where the output from the two releases is different (such as the PLOT directive) or the syntax has been enhanced. These will not affect the operation of GLIM 3.77 programs.

II
The Modelling Guide

10 Guide to statistical modelling with GLIM

Introduction

This chapter gives a description of the general principles of statistical modelling using the GLIM package. A theoretical treatment is deferred to Chapter 11. Standard models that are subsumed in the GLM family and which can be routinely analysed with GLIM are treated first. Later sections describe many examples of non-standard models that can also be analysed with GLIM. The user may need to refer to the User Guide for more detail. Case studies of analyses for many of these models are given in Chapter 12.

10.1 Statistical modelling and GLIM

The GLIM package is capable of a wide range of descriptive data analysis tasks. However, the package is really designed to give the user access to all the tools necessary for **statistical modelling**.

Before embarking upon a more formal description of what we mean by the term **statistical modelling**, we first summarize the essential characteristics which determine the range of the statistical modelling which can be done in GLIM.

Firstly, GLIM is concerned with modelling **observations** of characteristics of individuals or groups which are assumed to have a random element (typical data might be the numbers in a sample of electors voting for particular political parties, or the numbers of patients whose symptoms are alleviated by different drugs, the yields of some crop under different treatments, the times between certain events, and so on). The set of observations will be a sample representing some population. The variable characteristics are assumed to respond to mechanisms involving the various circumstances of the different individuals. For this reason they are termed **response variables**. The aim of the modelling is to explain how the variation in the observed values of the **response variable** is explained by differences in circumstances between individual cases or groups of cases. These differences may result from nature (gender, age), nurture (social environment, education), exposure (diet, pollution, media) or treatment (drugs, fertilizers, educational techniques). They are represented by **explanatory variables**.

Secondly, GLIM is primarily designed to analyse data where the main variable of interest, the **response variable**, is one-dimensional (univariate). This implies that the package can be used to fit models such as regression, logistic regression, and the analysis of variance or covariance, but it is not suitable for techniques such as multivariate analysis of variance or factor analysis.

GLIM is largely based on the assumption that there is a random sample of independent observations, the variation among which is to be modelled. However, there are certain important problems where the observations may not be statistically independent which can be analysed in GLIM. This is the case when the likelihood (joint density) of the observations can be expressed as a product of two functions; the first being a function of the data alone and the second having the form of a likelihood of independent observations. A very important situation involving non-independent observations which can be analysed this way in GLIM is when the data consists of frequencies in a **multi-way contingency table**. For such data, in the form of **counts**, log-linear models are often most appropriate and GLIM makes the fitting of such models relatively straightforward. Datasets of this type (often called **categorical**) arise in many practical situations; for example in social surveys, opinion polls, and so on, where the counts are the number of people who belong to some category (for example the number of women under 30 in a sample who are married and express preference for a particular political party). The analysis of log-linear models in GLIM is discussed further in [10.6.4.1].

One of the special features of GLIM is that it offers much more than the standard models. The following sections will define many of the powerful statistical models which can be modelled in the Generalized Linear Model (GLM) framework; it will be shown how a GLM is based upon a few basic concepts which, when combined together, give a very general framework encompassing many of the most frequently used techniques of statistical analysis. Moreover, the statistician can use the GLIM framework to control the specification of the GLM elements and thus develop 'macro' techniques which can be applied to a range of new problems. For example, GLIM macros have been developed to allow the relatively easy analysis of censored survival data using Cox's regression or methods assuming Weibull or Extreme Value parametric distributions for survival times. This flexibility and access to the structures and algorithms of GLIM is so far-reaching that there is almost no limit to the range of applications possible. Previous versions of the package have already led to the development of a large number of useful user-defined modelling methods. Section [10.6.10] discusses methods for correlated observations usually requiring a multivariate approach. Two examples are given in [12.4.2] of non-linear modelling problems; the first involving fitting an hyperbola and the second the logistic curve with unknown asymptotes that arises in radio-immuno assay problems. The range of user-defined GLIM analyses is very wide and increasing. However it is not possible to cover all of them in this manual. New data analytic methods based upon GLIM are continually being developed and published in the statistical journals, and promulgated via the GLIM Newsletter. Nonetheless these examples illustrate that the GLIM package provides an excellent environment for the development and implementation of new methods of data analysis within the overall framework of generalized linear modelling. This manual introduces the tools with which it is possible to develop new techniques. Further information on the relevant theory may be obtained from the many comprehensive theoretical texts on GLMs. A number of books on GLMs are listed in the Bibliography; further useful references may be found in learned journals, particularly the Royal Statistical Society's Applied Statistics journal. The GLIM Newsletter is recommended for users who wish to keep up to date with the latest developments in applying GLIM to new and unusual problems.

10.2 Statistical models as a structure for observations

Univariate analyses assume that the practical question can be answered by explaining variation in a single variable of interest. This variable is conveniently considered as an outcome response variable. Such variables which have probability distributions associated with them are termed random variables or variates. The information on the variable is collected as observations which are considered to be realizations of the underlying random variable. Other information will probably have been collected, namely a set of accompanying observations on variables which will be used to explain the variation in the response variable. GLIM analyses assume that these explanatory variables can be combined in some (essentially linear) way to represent the systematic component in the variation of the response variable. A model can then be postulated relating the response variable to the systematic component and specifying an appropriate assumption about the distribution of the random variable underlying the responses.

Thus, underlying the concept of a statistical model for variation in a random variable is the idea that the variable under investigation has a definite structure. It is assumed that this structure can be used to explain the values actually obtained and, often more importantly, that it can also be used to predict future values. The structure provides a description of the population from which the data values are considered to be sampled. We are implicitly assuming that this structure will be mirrored in whatever sample we obtain. Effectively the assumption is that the values of the variable of interest can be expressed in terms of other explanatory variables. The usual analyses assume that the explanatory variables have fixed values. Only with this assumption can a model involving them be considered as a systematic component of variation in the response variable. In addition to the systematic components, we define the random component of the structure; this is a description of the assumed probability distribution of the observed values of the variable of interest; that is, the distribution of the responses.

Thus the basic definition of a GLM has a three part specification:

(i) A **random component**; this is essentially a description of the assumed underlying probability distribution of the (single) variable of interest. The basic formulation assumes that the observed values of this random variable are independently distributed.

(ii) A **systematic component**; this is generally a linear combination of the explanatory variates.

(iii) A **link** between the random and systematic components.

We return to this formulation more formally in Chapter 11.

It may be helpful to note that in classical regression and analysis of variance (ANOVA), statistical models have often been represented in a slightly different way; in particular, as having a **response variable** which consists of two parts:

(i) a **systematic component** (sometimes referred to as a **deterministic component** or as the **signal**).

(ii) a **random error component** (sometimes also referred to as the **error** or **noise**).

Traditional formulations of such models have often been presented in the following additive form

Random variable of interest = Systematic component + Random error.

However, other forms are possible; for example, some models involve the multiplicative error form where

Random variable of interest = Systematic component x **Random error.**

The traditional approach is evidently closely related to the GLM formulation but is less general and can lead to confusion when considering models other than those assuming Normal response distributions. We therefore use the GLM approach throughout; that is, we concentrate on the distribution of the random component, not on the distribution of the random error. However it may be of interest to note that some echo of the traditional approach lingers in the GLIM language, in that GLIM uses the directive `ERROR` to specify the distribution of the random component.

In concluding this section, it may also be noted that a more general theory of statistical modelling does not necessarily assume that (all) the explanatory variables are fixed. Thus it may be assumed that some of the explanatory variables are themselves random variables. An increasingly important class of models of this type are the multi-level models (Goldstein, 1987). GLIM is not really designed to offer a user-friendly output for models with more than one random component, although some of these models can be analysed effectively in GLIM (see [12.1.2.6]).

10.3 Comparing statistical models

A key aspect of statistical modelling is that of comparison of a sequence of models, to ascertain whether one particular model is better supported by the data than another. Given adequate models this strategy allows tests to be constructed for parameters representing some of the explanatory variables while allowing for the potentially confounding effects of others. There are various strategies for model selection and it is quite possible to obtain more than one model which provides an adequate explanation of the data. However, a basic idea behind statistical modelling is that we can partition the total variability which we have observed in the variable of interest into the portion attributable to the systematic component and the portion attributable to the random component. We then typically seek to compare models, using some criteria of **goodness of fit**. A popular criterion for preferring one model to another is that it **explains** a larger proportion of the total variability of the data. That is to say, in comparing two models, we prefer the one where the variation due to the systematic component is larger (or equivalently, where the variation due to the random component is smaller). In pursuing this aim, we may in principle consider a random variable to be representable by a combination of any number of components, but for most practical purposes simple structures suffice and, mathematically and computationally, they are certainly more tractable.

A general aim of statistical modelling can be thought of as an attempt to explain as much of the variability of the data as possible using as simple a structure as possible. Inherent in this is the underlying concept of **parsimony**; that is, we look for simple models which adequately describe the data rather than more complicated models which explain almost all the variability. Statistical modelling involves the estimation of unknown **parameters** in a model; parsimony implies preferring models with less parameters to models with more. (A supporting argument in favour of this principle is that prediction is generally more reliable when using models with fewer parameters rather than models with more.) Thus, when comparing two models, one of which is **nested** within the other (that is, where the linear structure of the simpler model is contained within the linear structure of the more complex model), the more complex model will necessarily explain at least as much of the variability in the response as does the simpler model. But we shall wish to ascertain whether the extra parameters of the more complex model significantly add to this explanation. A measure of goodness of fit facilitates this decision. In GLIM, the measure used, the **deviance**, is a measure of **lack of fit** (that is, it measures the variation due to the random component) and we would thus wish to **minimize** it. We discuss this in more detail in section [10.6] but, firstly, it is necessary to introduce some notation.

10.4 An introduction to the notation of generalized linear models

The random variable under consideration, is called the **response** variable or, sometimes less satisfactorily, the **dependent** variable. In this manual we will only refer to the variable of interest as the response variable; it will be denoted by Y and its sample values by y_i (i = 1, ..., n).

The components will be denoted by other Latin and Greek letters, the latter being reserved for those with unknown values.

A typical structure will consist of linear combinations of the explanatory variables, of the following form

$$\beta_1 x_{i1} + \beta_2 x_{i2} + \ldots + \beta_p x_{ip} .$$

Here x_{i1}, x_{i2}, ..., x_{ip} are the explanatory variables. The β_1, β_2, ..., β_p are the unknown **parameters** to be estimated by GLIM in order to **fit** the model.

10.4.1 The types of response variables in a GLIM analysis

GLIM is concerned with models in which the user seeks to explain the variation in a **response** (or **y**) variable in terms of variation in certain **explanatory** variables. In addition to this response–explanatory distinction, there is the classification of a variable into different **data types**.

The different data types of response variates (the random component) include

Type	Examples
(i) Continuous	Yield from an agricultural experiment
	Breaking strain
	Concentration of a chemical
	Tensile strength of a metal alloy
	Nutrient intake of a human respondent
(ii) Count	Number of accidents in a traffic survey
	Number of people voting for a particular party in an election
	Number of bacterial colonies on an agar plate
	Word frequency in a piece of text
(iii) Binomial count (this covers proportions and rates)	Number responding (y) out of m subjects exposed in an assay
	The number (y) of voters preferring a particular political party out of a sample of m people interviewed.
(iv) Binary count	Whether a person is ill or not.
	Yes/no data, y [= 0 or 1] out of m [= 1].

10.4.2 The data types of explanatory variates

The different **data types** of explanatory variates include

Type	Examples
(i) Continuous	Age of respondents
	Weight
	Hardness of a metal alloy
(ii) Categorical (Nominal) (factor)	Block factor in a randomized block experiment
	Sex
	Marital status
	The variety names in a plant variety trial
(iii) Categorical (Ordinal)	Education (school, university)
	Severity of symptoms for a patient

It is important to note that GLIM uses the name **factor** for a nominal (categorical) explanatory variable which can take only a finite set of possible values; these possible values are called **levels**. A factor can be used to divide the data values into disjunct subsets indexed by its levels. Thus in a social survey, the data may be categorized by the gender of the respondents, that is, by a factor with levels 1 and 2, corresponding to male and female. Similarly, in a randomized-block experiment with 4 blocks, the block factor has levels 1, 2, 3, 4 and these index the plots by giving the block number within which each one falls.

10.4.3 The data matrix

The basic data structure is a **data matrix**. This is a two-dimensional structure involving the variables of interest. **Units** is the neutral name adopted for the plots of an agricultural experiment, the patients of a medical survey, the agar plates in an assay, the number of categories into which people are classified, and in general the number of observations of the response variable.

Conventionally, the units define the rows of the data matrix. The columns are the variables, each of which may be of any of the types described above in [10.4.1] and [10.4.2]. GLIM regards all the columns as **vectors**; however factors must be declared explicitly, and GLIM adopts the convention that their levels must be coded as 1, 2, 3, ...

A data matrix is defined from the number of rows (units), and the names and types (variate or factor) of the variables in each column. Indeed, GLIM does not store the data matrix as a matrix but rather as a set of named columns, each column referring to a vector (a variate or a factor). We illustrate this below.

10.4.4 Tables as data

The original data from an experiment or survey will often have undergone tabulation before an analysis by model fitting begins. A multi-way table is indexed by a set of factors, with typically one cell for each combination of factor levels. Tables can be used as data for a GLIM analysis by turning them into a data-matrix form.

For example, the following table gives the number of students recruited to a Polytechnic Statistics course, indexed by gender (two levels) and by entry qualifications (coded as Standard, Access or Non-standard; that is, three levels) and by whether they are over or under 21 on entry (two levels). (It may be of interest to note that being over 21 on entry is the formal definition of a 'mature' student in the UK.)

	Male		Female	
	Under 21	Over 21	Under 21	Over 21
Standard	20	21	10	4
Access	0	3	0	1
Non-standard	6	8	0	1

This data can be represented as a data matrix as follows

Qualif	Gender	Age	Y
1	1	1	20
1	1	2	21
1	2	1	10
1	2	2	4
2	1	1	0
2	1	2	3
2	2	1	0
2	2	2	1
3	1	1	6
3	1	2	8
3	2	1	0
3	2	2	1

Note that the response or y-variable is the number of students in each category. This is the variable that GLIM treats as the response variable in the analysis of a multi-way table.

There are therefore 3 x 2 x 2 = 12 **units**, indexed by three variables. We would treat gender as nominal (that is, a **factor** in GLIM notation). We would probably also treat qualification as nominal, as there is no inherent ordering in these categories. Age, on the other hand, could be treated as purely nominal but, in some applications, might be treated as ordinal.

More generally, suppose a three-way table of variable `Y` is indexed by factors `A`, `B`, and `C` with 2, 2, and 3 levels respectively; we may represent this in data-matrix form in the following way:

A	B	C	Y
1	1	1	y_1
1	1	2	y_2
1	1	3	y_3
1	2	1	y_4
1	2	2	y_5
1	2	3	y_6
2	1	1	y_7
2	1	2	y_8
2	1	3	y_9
2	2	1	y_{10}
2	2	2	y_{11}
2	2	3	y_{12}

The indexing is shown in **standard order**, with `A` changing slowest and `C` fastest. In fact, any order will do for GLIM, although it is more convenient to use standard order as GLIM provides a very succinct command for generating the levels of factors arranged in standard order (see `GFACTOR`).

10.5 Using GLIM to fit and compare models

10.5.1 Matching models to data

GLIM is primarily a tool for fitting a certain class of models (GLMs) to data. The modelling process may be thought of as one in which the data

$$y_1, y_2, \ldots, y_n \text{ say,}$$

are matched by a set of **theoretical values**

$$\mu_1, \mu_2, \ldots, \mu_n \text{ say.}$$

For a good model the μ's must have the following properties:

(i) they are all derived from a small number of basic quantities called **parameters**, and

(ii) the resulting set of μ's is close to the original data, the y's. Thus the μ's are highly patterned, and, therefore, we hope, easier to understand and interpret than the y's, which will be 'rough' by comparison.

The model fitting process involves two basic decisions:

(i) the choice of the relation between the μ's and the underlying parameters of the model, and

(ii) the choice of a measure of discrepancy which defines how close a given set of μ's is to the data.

The first choice relates to the systematic component of the model, and the second is governed by assumptions we make about the random component. The latter is a statistical description of that part of the variation which the systematic component does not account for.

In GLMs the systematic component is assumed to take the following two-stage form:

For each of our i units, $i = 1,\ldots,n$:

(i) The **linear predictor**, $\eta_i = \Sigma\beta_j x_{ij}$, is a linear combination of explanatory variables x_{ij} with parameters β_j, and

(ii) η_i is related to μ_i by the **link function**

$$\eta_i = g(\mu_i) \ .$$

The basic assumption is that the g function is a **known** function; for example, $\eta_i = \mu_i$ or $\eta_i = log\ (\mu_i)$.

Note that we assume that there are no unknown parameters to estimate in the g function. (This assumption can be relaxed in advanced use of GLIM but we shall restrict ourselves to the case where g is fully known.)

10.5.2 The distribution of the random component

The basic assumption in a GLM is that the random component is such that each y_i, $i = 1, \ldots, n$, follows a probability distribution which is a member of the **exponential family**, with the mean of the distribution given by μ_i . This exponential family (see [11.2]) includes the Normal distribution (suitable for many continuous y-variables), the Poisson distribution (suitable for models for counts and for contingency tables), the binomial distribution (suitable for models for binary and bounded count data), plus the gamma and inverse Gaussian distributions (applicable to certain forms of continuous data). Moreover, this list is not exhaustive as there are other probability distributions within this family and many others outside it which can be modelled in GLIM by the use of the macro facilities (see, for example, Jørgensen (1987)).

10.5.3 Fitting models by minimizing the deviance

The second essential aspect of GLIM's fitting procedure is to minimize a measure of discrepancy, called the **deviance** in GLIM, between the observed data and the corresponding fitted values. Thus GLIM fits a model by choosing as estimates of the β's those values which give μ's that minimize the **deviance**. The actual form of the deviance depends upon which member of the exponential family GLIM has been instructed to use. Thus, with the classical assumption of Normally distributed random components, the deviance is the well-known **residual sum of squares**. For other assumptions about the random component, the deviance takes a different form depending on the distributional assumption. The GLIM user does not need to know the exact formula for the deviance, nor how GLIM finds the fitted μ's which minimize this deviance, as GLIM does all the calculations internally. Nevertheless, the formula for the deviance and the fitting technique are presented later ([11.4.2],[11.3.1]) for users who wish to delve deeper into the theory behind GLIM.

If the response variable (the random component of the model) is assumed to be Normally distributed, then the technique of minimizing the deviance is the same as minimizing the residual sum of squares and thus is equivalent to the method of **least squares**. In general this technique is the same as finding the **maximum likelihood estimates** of the parameters β_j (and hence of the μ_i). Thus, GLIM can be thought of as a program for maximum likelihood estimation in generalized linear models.

10.5.4 The specification of a GLM

To summarize the above, the specification of a GLM requires

(i) the terms to be included in the linear predictor;

(ii) the link function connecting the linear predictor to the theoretical values;

(iii) the distribution of the random component.

GLIM is then able to compute the minimum deviance/maximum likelihood estimates of the unknown parameters in the linear predictor and to display the value of the minimized deviance, together with its associated degrees of freedom (the latter being given by the number of units minus the number of unique parameters estimated).

10.5.5 Estimation of the scale parameter

In a GLM it is assumed that the underlying probability distribution for the random component comes from the **exponential family**. For some members of the family, such as the gamma distribution, the Normal distribution, or the inverse Gaussian distribution, the probability distribution involves two (generally) unknown parameters. One of these parameters, the mean μ, is related to the parameters β in the linear predictor and the other is the **scale** parameter ϕ, [11.2.1]. It will also need to be estimated. One of the advantages of a GLM is that the β's and hence the μ's can nevertheless be estimated without needing to estimate ϕ (see section [11.3.1]). Indeed, we shall show below that the scale parameter ϕ can be estimated after the estimate of the mean μ has been calculated.

Other members of the exponential family, for example, the binomial and Poisson distributions, do not have a scale parameter and hence are completely specified once we know the mean of the distribution. This implies that the variance of the distribution is fixed as soon as the mean has been specified. Thus, if the distribution of y_i involves only one unknown parameter (for example, as in the binomial or Poisson cases) then the scale parameter ϕ can usually be taken to have the value 1 and there is no extra estimation problem.

To clarify the estimation procedure, we may note that in practice we have n observations y_i, $i = 1, \ldots, n$, each y_i being associated with a corresponding $\mu_i = E(y_i)$. We may need to assume that there is an unknown scale parameter ϕ, which is constant for all $i = 1, \ldots, n$. Whether we assume an unknown scale or not, we can use GLIM to estimate the corresponding μ_i, $i = 1, \ldots, n$ without the need to estimate the scale ϕ.

In the case of an unknown scale ϕ, we need to estimate it, particularly if we wish to carry out any significance tests. Provided that a reasonably well-fitting model has been chosen, the usual approach is to estimate the scale ϕ from the mean residual deviance for that model. Thus, if a reasonably well-fitting model has been fitted, with deviance D and associated degrees of freedom d, then we estimate ϕ by the mean residual deviance;

$$\text{that is, } \hat{\phi} = D/d\,.$$

Both D and d are output by GLIM. (In fact, $d = n - r$, where n is the number of response observations in the fit and r is the number of (non-aliased) parameters fitted [10.5.10].

The estimate $\hat{\phi}$ is a moment estimate. This is, in fact, the estimate used to estimate variance components in the analysis of variance where D would be the sum of squares for a source of variation and ϕ the variance component.

More generally we may estimate ϕ by any reasonable estimate (see [11.3.3] for a fuller discussion) but we generally prefer the mean residual deviance for simplicity.

10.5.6 Overdispersion in the response variable

We have noted that, although, in general, members of the exponential family contain a scale parameter, some members have only a single parameter, the mean of the distribution. Thus the binomial, multinomial, and Poisson distributions are completely specified when we know the mean; in particular, the variance is fixed once the mean is specified. However in some situations **overdispersion** can arise in which the variance of the response variable is greater than we would expect for the particular distribution. Consider a binomial sampling situation in which we ask samples of electors from different regions of the country whether they agree that the Prime Minister is performing well. We may have used cluster sampling in which each sample was taken from a particular small geographic area within a region. If there is variation in opinion between areas within a region then taking the cluster sample as representative of opinion in the region as a whole will result in the overall variance of the response being greater than that of a pure binomial. This situation is interrelated with that of **omitted variables** in which an important explanatory variable is omitted from the model, resulting in a poor fit. If there were an explanatory variable that could be used to model the difference between the opinion in the sample area and the opinion in the corresponding region as a whole, then inclusion of this variable in the model would remove the overdispersion. Thus great care should be taken to identify the possible sources of overdispersion, as omitted variables can result in distorted conclusions about the importance of explanatory variables that are strongly related to the omitted variables.

Similarly, overdispersion can occur in modelling data which follows a Poisson distribution. For example, with a clustered sampling mechanism, as above. Or when sampling a Poisson process (Cox, 1967) over an interval whose length is not fixed, but is itself random. A further possibility is that the underlying model assumes the data to be Poisson distributed with some unknown mean which is itself a random variable with some distribution; this gives a continuous mixture for the observed data. Similarly, some models for multinomial data can be handled by a Poisson assumption for the mean, but the analysis requires modification of the assumption of known dispersion.

Models with overdispersion can be handled in a straightforward manner in GLIM by first estimating the unknown means μ_i (a procedure which does not require knowledge of dispersion [11.3.1]) and then estimating the scale of overdispersion from the residual **deviance** (see [12.2.1.4]).

10.5.7 Comparing models: the analysis of deviance

Having fitted a model in GLIM, it is necessary to determine how well the data support the model.

We here assume that the 'error' and link specifications are satisfactory (see [11.6] for further discussion of this point), and concentrate our attention on the linear structure. Thus for the rest of this section the term 'model' refers only to the linear structure of a model.

In general, statistical modelling involves model selection. To carry out such a selection we need to compare any two competing models. As shown in more detail in [11.4.2], we can carry out a likelihood ratio test to compare two nested models by considering the ***difference between their scaled deviances***. If the scale parameter is known the scaled deviance, $S(c,f)$, and the degrees of freedom, d, are output by GLIM after fitting a model. The scaled deviance compares the current model, c, to the **full model**, f, defined as a model that would fit the data exactly. To compare two models, c_1 and c_2, where c_1 is nested within c_2, we calculate the difference, $S(c_1,c_2)$, between their scaled deviances:

$$S(c_1,c_2) = S(c_1,f) - S(c_2,f)\,.$$

The (non-negative) difference in scaled deviances has, asymptotically, a chi-squared distribution, with degrees of freedom given by the difference $d_1 - d_2$ between the degrees of freedom of the two models c_1 and c_2 being compared; that is,

$$S(c_1, c_2) \sim \chi^2_{d_1 - d_2}.$$

A statistically significant value for $S(c_1, c_2)$ indicates that c_2 is a better model than c_1 or equivalently that the term(s) omitted in model c_1 are significant.

The scaled deviance, $S(c,f)$, provides a general test of adequacy of the model since under the assumption that all significant terms are included in c we have:

$$S(c,f) \sim \chi^2_d.$$

If this scaled deviance is significant then it indicates that a significant term has been omitted from the model though it does not indicate which term.

10.5.8 Comparing nested models with unknown scale

For the case of unknown scale, however, a different method is needed. In this case GLIM will output the (unscaled) deviance, $D(c,f)$, which is related to $S(c,f)$ by:

$$S(c,f) = \frac{D(c,f)}{\phi}.$$

Defining the difference in deviances, $D(c_1, c_2)$, by

$$D(c_1, c_2) = D(c_1, f) - D(c_2, f)$$

then to perform the previous test we would require $D(c_1, c_2)/\phi$. Since ϕ is unknown we need an alternative test statistic. A simple procedure is to divide the difference between the deviances by an estimate of the scale, where the estimate of the scale is obtained either from the **maximal model** (the most complex model appropriate to the data) or, alternatively, from the more complex of the two models being compared; that is, from the model with fewer degrees of freedom.

Thus we first obtain an estimate $\hat{\phi}$ of ϕ. Then the test statistic to compare two nested models is $D(c_1, c_2)/\hat{\phi}$, where $\hat{\phi}$ is the mean residual deviance:

$$\hat{\phi} = \frac{D}{d_3}$$

and where D and d_3 are the deviance and degrees of freedom of the model used to estimate ϕ (note that $d_3 = d_2$ if the more complex of the two competing models is used to estimate the scale).

The statistic $D(c_1, c_2)/\phi$ contains the estimate $\hat{\phi}$ so its distribution is no longer asymptotically chi-squared. A common approach is to assume an approximate F distribution; that is

$$\frac{D(c_1,c_2)}{(d_1-d_2)\,\hat{\phi}} \sim F_{d_1-d_2,\,d_3}\ .$$

It may be noted that the estimate $\hat{\phi} = D/d_3$ is in general inconsistent as the number of units tends to infinity and so other estimates such as the modified profile likelihood or generalized Pearson estimates might be used to estimate ϕ (see [11.3.3] for more details).

10.5.9 The GLIM syntax for defining the linear structure

The GLIM syntax for defining the linear structure is very convenient. A full discussion is given in the User Guide [8.2]. We briefly introduce the concepts here.

The linear structure is defined by a model formula; it contains all the information implicit in a formula such as $\Sigma\beta_j x_{ij}$ but without the need for unpleasant summations with subscripts.

(i) For example, to fit a linear structure consisting of a variate X with an intercept included, we merely use the statement

```
$fit  X
```

This instructs GLIM to fit the linear structure $\beta_0 + \beta_1 x_i$.

(ii) Similarly, to fit a linear structure involving two variates X1 and X2, with intercept, we specify

```
$fit  X1+X2
```

This corresponds to the linear structure $\beta_0 + \beta_1 x_{i1} + \beta_2 x_{i2}$.

(iii) The command

```
$fit  X1+X2-1
```

will fit the variates X1 and X2 but exclude the intercept (that is, exclude a variate consisting of a column of 1's). This fit corresponds to the linear structure

$$\beta_1 x_{i1} + \beta_2 x_{i2}.$$

(iv) The syntax is straightforward if nominal variates are to be fitted. Thus having defined **factors** (see the User Guide [3.1.4]) A and B, we can fit an additive linear structure by

```
$fit A+B
```

A representation of the equivalent linear structure could be

$$\beta_0 + \beta_j^{\mathrm{A}} + \beta_k^{\mathrm{B}}$$

corresponding to levels j, $j=1,\ldots,J$ of A and k, $k=1,\ldots,K$ of B.

(v) We can include variates with statements such as

```
$fit A+X
```

which allows for different intercepts for the levels of A but with a common slope. The linear structure formulation for this GLIM syntax is

$$\beta_0 + \beta_j^{\mathrm{A}} + \beta^{\mathrm{X}} x_i$$

where we understand that the β^{A} are different from β^{X}.

(vi) Linear structures involving **interactions** between factors are fitted by commands such as

```
$fit A+B+A.B
```

This last statement can be written more succinctly as `$fit A*B`; that is, GLIM recognizes `A*B` as equivalent to `A+B+A.B`.

An algebraic representation of the linear structure is

$$\beta_0 + \beta_j^{\mathrm{A}} + \beta_k^{\mathrm{B}} + \beta_{jk}^{\mathrm{A.B}}.$$

(vii) A continuous variate with different slopes for the differing levels of A can be included by a statement such as

```
$fit A+A.X
```

The mathematical form could be written as

$$\beta_0 + \beta_j^{\mathrm{A}} + \beta_j^{\mathrm{A.X}} x_i$$

where we understand that the β^{A} are different from the $\beta^{\mathrm{A.X}}$.

10.5.10 Estimating the parameters: aliased parameters

GLIM finds the maximum likelihood estimates $\hat{\beta}_j$ of the parameters β_j, $j = 1, \ldots, p$ by minimizing the deviance, that is, by maximizing the likelihood. This enables GLIM to calculate an estimate of the linear predictor η_i and hence to estimate the unknown means μ_i of the random responses y_i, for $i = 1, \ldots, n$. The method is discussed further in section [11.3.1].

If the explanatory data consists of continuous variates alone, it is likely that the **model** (or **design**) **matrix** $\mathbf{X} = [x_{ij}]$ will be of full rank (that is, rank($\mathbf{X}$) = p) and the β_j will be well defined. But if the explanatory variates involve factors, there is an essential indeterminacy in the definition of the β_j. We cannot uniquely estimate a parameter for each level of a factor as there is a linear dependence between the columns of the design matrix; that is, the design matrix is of rank $r < p$. When there is a linear dependence between the columns of $\mathbf{X}$ we say that the β_j corresponding to these columns are **aliased**. The parametrization is then arbitrary as there are many sets of β_j which satisfy the maximum likelihood equations. All these sets of β_j give the same deviance and the same estimates of μ_i but some extra constraints are needed to ensure that the set of β_j is uniquely specified.

One possible set of constraints is to demand that factor effects sum to zero, but this is not computationally simple and, with 'unbalanced' designs or designs with covariates, may produce parameter values that are difficult to interpret.

A computationally simpler and more general method is to evaluate the parameter estimates sequentially and on finding one which is aliased with preceding estimates to drop it from the model, which is equivalent to assigning the value zero to it. Thus we implicitly impose $p - r$ constraints and can estimate the remaining r of the estimates $\hat{\beta}_j$ under these constraints. Such a procedure is to some extent arbitrary but knowledge of the experimental design or of the model matrix should enable us to re-parametrize as necessary. By default, GLIM adopts this procedure, with some additional features, which may be summarized as follows.

Unless otherwise instructed, GLIM fits a parameter corresponding to the unit vector in the model matrix; this is known as `1` in GLIM terminology (formerly referred to as the 'grand mean', `%GM`, in earlier versions of GLIM). Provided that this constant term is in the model, GLIM then takes the first level of each factor as the reference point from which other parameter estimates are calculated; in other words, if it is impossible to get parameter estimates for all levels of a factor, then by default GLIM will intrinsically alias the first level of the factor; that is, it sets the parameter estimate corresponding to the first level of the factors to zero (rather than, for example, the sum being zero). If there is no constant term in the model, all the parameters of the first factor in the structure formula are estimated and subsequent factors follow the previous rule that the first level is taken as the reference value. However, as these choices are somewhat arbitrary, GLIM allows you (through the `FACTOR` directive) to choose a different level to be the reference category of each factor (see the User Guide [3.1.4]). Note, however, that the constraint of the sum being zero is ***not*** available in GLIM. If a particular form of parameter estimates is required, these can be obtained in GLIM by specifying the model matrix.

The term **aliasing** is applied more generally than in the above sense. We say that parameters $\beta_1, \ldots, \beta_l$ are **aliased** if their corresponding columns $\mathbf{x}_1, \ldots, \mathbf{x}_l$ are linearly dependent. This can occur in two ways. **Intrinsic aliasing** occurs as discussed above. Thus, if we try to estimate a parameter for each level of a factor, the linear structure has an inherent redundancy, whatever data we collected. **Extrinsic aliasing**, however, occurs when some quirk of the data causes a linear dependence amongst the columns of the design matrix (this can, for example, occur when one explanatory variate happens to be a linear combination of the others). This does not imply that the corresponding population parameters are aliased; however, the observed data does not provide us with sufficient information to estimate all the parameters. GLIM automatically checks to see if there is intrinsic or extrinsic aliasing; intrinsic aliasing is detected by the program before any fitting takes place, and no attempt is made to estimate intrinsically aliased parameters. Similarly GLIM reports parameters which are extrinsically aliased but, of course, it cannot estimate them.

The detection of extrinsic aliasing by GLIM is based upon a numerical check of the size of the multiple correlation between the columns of the design matrix under consideration and the columns already fitted. A correlation 'near' 1.0 triggers the program to conclude that there is extrinsic aliasing. GLIM has a default value of 10^{-10} used to decide how near to 1.0 this correlation should be to conclude that aliasing exists. This default value can be changed by use of the `CYCLE` directive. For a fuller discussion of aliasing, see [11.3.2].

10.5.11 Relaxing the exponential family assumption: quasi-likelihood

It may be noted that it can be helpful in some applications of GLMs to be able to simplify assumption (iii) of Section [10.5.4]; that is, to be able not to make any strong distributional assumption about the random component. Thus, noting that knowledge of the variance function determines which member of the exponential family is being used, we are led to the idea of assuming only the form of this variance function, rather than the full distributional form. The corresponding distribution will then, in general, not be a member of the exponential family but valuable estimates can still result. This technique is known as **quasi-likelihood** (Wedderburn, 1974b); it can be convenient when the form of the mean–variance relationship is known (for example, when $\text{var}(y) = \mu + \mu^2$) but where it is difficult or not appropriate to make a firm assumption about the distribution of the y's (see, for example, Nelder, 1985). For a fuller discussion, see [11.8].

10.6 Examples of GLMs

It follows from Section 10.5 that we can describe specific instances of GLMs merely by defining the types of the response and explanatory variables (the data side) and the link function and the distribution (the model side).

We now illustrate this by summarizing some of the models which can be handled effectively by GLIM. It is to be stressed that this list is not exhaustive but rather is intended to give some indication of the wide range of possible analyses for which GLIM can be used.

10.6.1 Continuous response variates

10.6.1.1 Classical regression

The classical regression model can be thought of as having:

(a) independent observations;
(b) a continuous response variable (the random component);
(c) continuous explanatory variables (to be linearly combined to give the linear structure);
(d) the identity link function, that is, $\eta = \mu$;
(e) a Normally distributed response variable.

In traditional algebraic form this can be written:

$$y_i = \Sigma_j \beta_j x_{ij} + \epsilon_i,$$

where $E(\epsilon_i) = 0$ and the ϵ_i are independent and Normally distributed errors. From the formulation above, it can be seen that this model is simply a special case of a GLM (see [12.1.1]).

10.6.1.2 Analysis of variance and covariance, and the general linear model.

Similarly, traditional analysis of variance and covariance models (ANOVA and ANCOVA) fit equally well into the GLM framework. The difference between these and regression models is that the explanatory variates include factors.

In the classical general linear model (note that this is not the same as the **generalized** linear model), the dependent variable is considered to be the sum of several systematic components and one random component, the latter having a Normal distribution. The classical linear model is traditionally written (Rao, 1973) in the form:

$$y_i = \Sigma x_{ij}\beta_j + \epsilon_i \text{, where } \epsilon_i \text{ is distributed as } N(0,\sigma^2).$$

The x_{ij} may be the values of explanatory variables or they may be 'dummy' variables indicating the presence or absence of an effect (a set of such variables defining a **factor**) or they may be a mixture of both. The ϵ_i represent independent variables which have a random, distorting effect on the observations, the magnitude of which effect can be measured by their variance. Thus the structure of y_i is expressed as the sum of a linear combination of systematic components and one random component.

This can evidently be expressed in the GLM formulation, except that the explanatory variates are not restricted to being only continuous or only categorical (see [12.1.2]).

10.6.2 Transformation to Normality

Statistical analysis of response variate data is often based upon the use of an assumption that they are Normally distributed (with constant variance). If the data do not support this assumption, it is possible to transform the response variate to achieve a better approximation to constant variance. Various techniques can be suggested to achieve the most appropriate transformation; [12.1.1.4] gives an illustration of a simple method of finding a suitable transformation. However, within a GLM formulation, it is generally more appropriate to use a member of the exponential family to model the response (see [12.4.1.2]), as not only will the mean–linear predictor relation be maintained but it is also possible to have exact distributional assumptions.

Thus the commonly used Box–Cox data transformation applied to the response variate y is given by

$$y^* = (y^t - 1)/t, \quad t \neq 0$$

$$y^* = \log y \quad , \quad t = 0.$$

This transformation gives a variance stabilizing effect corresponding to an assumption that

$$\text{var}(y) = c\mu^{2(1-t)}, \text{ for some suitable constant } c.$$

For example:

$t = -0.5$: applying the inverse square root data transformation and treating the transformed data as Normally distributed corresponds to an assumption that the original data comes from the inverse Gaussian member of the exponential family (that is, that $\text{var}(y) = c\mu^3$, for a suitable constant c).

$t = 0.0$: applying the log data transformation and treating the transformed data as Normally distributed corresponds to an assumption that the original data is from the gamma member of the exponential family (that is, that $\text{var}(y) = c\mu^2$).

$t = 0.5$: applying the square root data transformation and treating the transformed data as Normally distributed corresponds to an assumption that the original data comes from the Poisson member of the exponential family ($\text{var}(y) = c\mu$).

In Section 11.6 we discuss the assessment of the suitability of choice of both 'error' and link. In general, in a GLM analysis, it is desirable to use an appropriate distributional assumption for the random component, rather than transforming the response variate in an attempt to obtain approximately constant variance (and approximate Normality). However, when there is no prior knowledge about the form of the distribution, the transformation approach can provide a useful guide as to what is an appropriate assumption.

10.6.3 Methods for binary and binomial response variates

Suppose the data consists of a dichotomous response (success/failure, yes/no, or died/survived, and so on) or of an observed bounded count of the form 'y successes from m trials'. Sometimes, such data is expressed in the form of **proportions**; that is, y/m. Typically the data is to be explained by some explanatory variables such as age of respondent, sex of respondent, dose, and so on. Clearly, the predicted probabilities (of success) must lie between zero and one and, moreover, the analysis should make use of the assumption that the responses are from an underlying **binomial** or **binary** distribution. A special form of regression model is required. For example, an experiment in which it is noted that there are increasing proportions (of, say, patients with alleviation of symptoms) with increasing dose leads to an S-shaped curve relating the proportions to the dose. It would be possible to transform the response variable and fit the resulting variable using least squares (that is, with an assumption of approximate Normality of the transformed variate), but it is preferable to model the binomial parameters as a function of parameters which are linear on some scale. We here briefly review this approach.

The original development of the modelling of proportions was in the context of bioassay, with the modern method of **probit analysis** dating back to Bliss (1935). Probit analysis is typically applied to toxicology experiments (see [10.6.3.3], below); however, although it remains popular in this field of application, the related **logistic regression** model is perhaps of wider application. We now describe a general model which includes both the probit and logistic formulations (see [12.2]).

10.6.3.1 Tolerance distribution for binomial response data.

We assume that the (discrete) dependent variable y_i is the number of successes out of m_i independent trials, where π_i, the probability of success in a trial, varies with i, $i = 1, \ldots, n$. Hence the response variate y_i may be assumed to follow a binomial distribution; that is,

$$y_i \sim B(m_i, \pi_i).$$

Note, moreover, that the binary case is merely a special case of the binomial, corresponding to $m_i = 1$, so is included in our discussion.

The probability of 'success' π must lie in [0,1] so it is convenient to assume we can model it by

$$\pi = \int_{-\infty}^{t} f(s)\, \mathrm{d}s, \text{ where } f(t) \text{ is a probability density function}$$

(that is, f integrates to unity over $(-\infty,\infty)$ and is non-negative).

The probability density function f is often referred to as the **tolerance distribution.** The above formulation states that π represents the probability of some random variable (the tolerance of some individual) being less than some value t (typically in a toxicology experiment this would be the dose or log(dose)).

The form of analysis depends upon the form chosen for the density f. The probit model assumes f to be a Normal density, with mean and variance ξ and ν^2, respectively. The logistic formulation is now more widely used. Logistic regression assumes a logistic tolerance distribution. In this case the density f is given by the logistic density

$$f(t) = \frac{\exp(t)}{[1+\exp(t)]^2}, \qquad -\infty < t < \infty.$$

10.6.3.2 Logistic regression

Logistic regression is widely applied where the data consists of a dichotomous response (success/failure, yes/no or died/survived, and so on) or of the form 'r successes from m trials'. Typically the response is to be explained by some explanatory variates such as age of respondent, sex of respondent, and so on. Logistic regression is a popular approach for such data.

For $i = 1, \ldots, n$, we assume $y_i \sim B(m_i, \pi_i)$, where $\eta_i = \Sigma \beta_j x_{ij}$ is a systematic component which linearly combines the explanatory variates x_{ij} (which may be continuous, nominal, or ordinal).

We assume that

$$\pi_i = \int_{-\infty}^{\eta_i} \frac{\exp(s)}{(1+\exp(s))^2}\, ds = \frac{\exp(\eta_i)}{1+\exp(\eta_i)}$$

and hence

$$\mu_i = m_i \pi_i = m_i \left(\frac{\exp(\eta_i)}{1+\exp(\eta_i)} \right).$$

It follows that

$$\eta_i = \log \left(\frac{\mu_i}{m_i - \mu_i} \right),$$

which we refer to as the **logit link**.

Thus in GLIM terminology a logistic regression model can thus be thought of as having:

(a) independent responses;

(b) response variable: a non-negative count bounded above by a number m (where m can be 1);

(c) a logit link function, that is, $\eta = \log [\mu/(n-\mu)]$;
(d) a binomially distributed response variable.

For an application of this model see [12.2.1].

10.6.3.3 Probit analysis

One of the first uses of regression-type models applied to binary or binomial data was in **bioassay** (see, for example, Finney (1977)). A typical bioassay problem is where a number of animals or humans are given varying doses of some drug and the proportion of 'successes' is recorded for each dose level. Here 'success' might mean the alleviation of symptoms. In a toxicity trial, it could be a somewhat unfortunate terminology for the death of an experimental animal.

In the probit analysis approach, it is assumed that an explanatory variable, the tolerance, has a Normal distribution over the experimental units such that if a unit i is given a 'dose' x_i, then the probability of its tolerance being less than x_i (that is, the probability of a successful trial) is

$$\pi_i = \int_{-\infty}^{x_i} \frac{1}{\sqrt{2\pi\nu^2}} \exp\left[-\frac{1}{2}\frac{(t-\xi)^2}{\nu^2}\right] \mathrm{d}t = \Phi[(x_i - \xi)/\nu]$$

where $\Phi(.)$ is the standard Normal probability integral and ξ is the dose giving a 50 percent response.

Thus $\mu_i = \mathrm{E}(y_i) = m_i\, \Phi(\eta_i)$, which can be written in the form

$$\eta_i = \Phi^{-1}(\mu_i/m_i).$$

This form of link between the linear predictor η_i and the binomial responses with means μ_i is called the **probit link**.

Note that

$$\eta_i = (x_i - \xi)/\nu = \beta_0 + \beta_1 x_i$$

so estimation of β_0, β_1 provide estimates of ν and ξ.

In GLIM terminology, a probit analysis has

(a) independent responses: y;
(b) response variable: a non-negative count y bounded above by a number m (where m can be 1);
(c) explanatory variable: continuous (for example, dose);

(d) link function: probit (inverse of the Normal probability integral);
(e) distribution: binomial.

10.6.4 Count data as the response variable

The last section discussed response data in the form of non-negative counts which are bounded above; however, some count data is not evidently of this bounded form. In such cases, it is appropriate to consider treating the responses as following the Poisson distribution. This distribution is again a member of the exponential family and thus amenable to analysis as a GLM.

10.6.4.1 Contingency tables: log-linear models for categorical data.

A (multi-way) contingency table is a table of data consisting of **counts**; these counts (frequencies) correspond to the number of **individuals** classified into one of several distinct and exhaustive categories, usually referred to as the **cells** of the table. A typical example would be respondents in a social survey classified by the three factors *social class, age,* and *gender* to form a three-way table. In general, there could be a larger number of classifying variables. Our initial discussion concentrates on the case where these are nominal (factors); however, some of these classifying variables might be ordinal and we return to this in Section [10.6.4.2]. For a thorough treatment of these models see Agresti (1990).

The individuals are assumed to be sampled from some population. Various sampling mechanisms can be applied to obtain the data; the simplest is when a fixed number of individuals have been chosen at random. This implies that the count (frequency) in each cell follows a multinomial distribution, with unknown probabilities which we need to estimate. There is a slight complication to consider, namely that as the sum of the counts is fixed in advance, the responses are not strictly independent. However, it is not too difficult to show that, provided that an appropriate link function is used, the likelihood of the (statistically dependent) multinomial response variates is proportional to the likelihood of a GLM of independent Poisson variables with means μ_i. Thus a GLM analysis of multinomial data is available by treating the responses as independent Poisson variables; this is the so-called **Poisson trick**.

An analysis of a multi-way contingency table is based upon the assumption that the μ's, the systematic components, may themselves have a structure. Multinomial sampling assumes that $\sum y_i = m$ is fixed (where we sum over all cells of the table; that is, over all units).

As an illustration, consider a two-way table in which it is assumed that rows (R) and columns (C) are independent factors.

Let y_i be the count in cell (rc), corresponding to $R = r$ and $C = c$. Then, assuming independence,

$$\mu_i = \mathrm{E}(y_i) = m\ \pi_r^{\ R}\ \pi_c^{\ C},$$

where $\pi_r^{\ R}$ is the probability of a random individual being classified into row r and $\pi_c^{\ C}$ is the probability of a random individual being classified into column c.

Hence, if we define $\eta_i = \log \mu_i$ then

$$\eta_i = \log m + \log \pi_r^R + \log \pi_c^C.$$

The right-hand side of this equation can be rewritten in the more familiar form

$$\eta_i = \alpha + \alpha_r^R + \alpha_c^C,$$

giving a formula as would be used in a two-way analysis of variance without interaction.

It can thus be seen that we can write

$$\eta_i = \Sigma\, \beta_j x_{ij},$$

for suitably chosen nominal explanatory variates x_{ij}, where $\eta_i = \log \mu_i$.

Thus the independence model for a two-way contingency table can be written in the form of a linear structure, with the **log link** between the linear structure and the means of the counts. Moreover, the use of the log link ensures that the estimated μ's will sum to m, so validating the use of the Poisson trick. Hence, a model of independence in a two-way contingency table can be fitted as a GLM by assuming that the response variates are distributed as independent Poisson variates, with the log link used to link the mean of the y's to the linear structure.

This approach can be generalized; log-linear models for general multi-way contingency tables can be written in the GLM form, with the log link and with an appropriate linear structure. Moreover, the model definition syntax of GLIM is particularly well suited to the definition of such models (see [12.2.1.3]). A GLIM analysis assumes that hierarchical models are to be fitted; that is, the existence of a high-order interaction necessarily implies the inclusion of all lower-order terms marginal to it. This is the usual approach in log-linear modelling. Other sampling schemes, such as quota sampling, which result in product multinomial likelihoods, can be handled similarly by GLIM, by fixing appropriate margins. Thus if the quota sampling predetermines the values for a margin as defined by a term in the model then this term must be included in any fitted model formula. For example, if the sampling scheme specifies, in advance, the total numbers of men and women that will be sampled then a term `SEX` must be included in the model formulae. Special cases of the hierarchical log-linear models, such as the graphical and decomposable models, can also be fitted easily in GLIM (see [10.6.4.5]).

In summary, a log-linear model applied to a contingency table can be analysed as a GLM with:

(a) independent responses;
(b) response variable: a count;
(c) explanatory variables: nominal (factors) or ordinal variables;
(d) link function: log, that is, $\eta = \log \mu$;
(e) distribution: Poisson.

10.6.4.2 Log-linear models for ordinal data

A traditional analysis of a multi-way contingency table assumes that the explanatory variates are nominal; that is, factors in the GLIM language. However, there are many occasions, especially in the social sciences, where an ordering can be placed on the levels of the explanatory variates (such as disagree, indifferent, agree), so an ordinal analysis is required (Agresti, 1984). This then necessitates the use of both factors and variates as explanatory variates.

The GLIM formulation of log-linear modelling for ordinal data is therefore exactly as in [10.6.4.1], except that the explanatory variates are either continuous or nominal (that is, variates or factors in GLIM terminology). Provided that the scale of the ordered variates is assumed known (this scale is often assumed to be the integers 1, 2, 3, ...), your GLIM analysis proceeds exactly as in Section [10.6.4.1]).

The case of unknown scale in an ordinal analysis is more difficult. It is possible to estimate the scale if there is sufficient data (Becker, 1990). However this requires advanced use of GLIM as it needs use of GLIM's PASS facility (see the Reference Guide [17]).

10.6.4.3 Log-linear models where one factor is a binary response.

In some analyses of log-linear models, one factor is of prime interest; that is, may be considered to be a response variate. If this factor has only two levels, the data can be reformulated as a binomial sampling experiment. A logistic regression analysis [10.6.3.2] is then appropriate. Alternatively, the data can still be analysed as a log-linear model.

If a log-linear approach is used, it is necessary to consider if the sampling mechanism was essentially binomial. If so, the log-linear analysis should preserve this property; in other words, the sum of fitted yes's and no's in each cell of the table should equal the sum of observed yes's and no's in that cell. This is achieved in GLIM by **fixing the margins** for all the explanatory factors; that is, ensuring that any model fitted includes terms corresponding to those marginal totals that are predetermined by the sample design.

As an example, suppose we have some data on voting preferences for the two principal political parties in the UK, classified by gender and social class. Then, if the data were collected as a **quota sample** in which a fixed number of men/women of differing social classes recorded their preference for Conservative or Labour, then the data can be considered as binomial. However, an appropriate analysis via log-linear modelling can be achieved by always including the term `GENDER*CLASS` in any model fitted.

Typical model formulae might then be

`VOTE + GENDER*CLASS` (corresponding to fitting the constant term only in the equivalent logistic analysis)

`VOTE*CLASS + GENDER*CLASS` (corresponding to fitting the main effect of `CLASS` in the equivalent logistic analysis)

VOTE*(CLASS+GENDER) + GENDER*CLASS (corresponding to fitting the additive model CLASS + GENDER in the equivalent logistic analysis)

It is also possible to analyse a contingency table as a log-linear model, treating one factor as a response but without assuming a binomial sampling mechanism; that is, without insisting on fixing the margins in the subtable of explanatory factors. However, it is necessary to be careful when considering the results of such an approach, as the resulting models may not all be interpretable. For example, a model implying the **conditional independence** of gender and social class given voting preference might be deemed to give an acceptably small scaled deviance. But attitudes to political parties can easily change and so we cannot accept conditional independence of two demographic variables GENDER and CLASS given VOTE unless the gender and vote are also **marginally independent**. See Edwards and Kreiner (1983) for a fuller discussion.

10.6.4.4 Multinomial response models for count data

A multinomial logit model is a model for polytomous data; that is, data where some response variable falls into one of *r* different categories. This is exactly the same as the model of [10.6.4.1], except in the multinomial case we have more than two categories for the response variate. Again this type of data can be analysed easily in GLIM by fixing the appropriate subtable of the explanatory variates if we assume that the **multinomial logit link** formulation is appropriate (Aitkin et al., 1989). However, more general models for multinomial responses cannot be handled so easily in GLIM, or at all. One such model which can be analysed via GLIM is the proportional odds model (Hutchinson, 1985); more complex models suggested in McCullagh and Nelder (1989) do not appear amenable to easy analysis by GLIM.

10.6.4.5 Multivariate graphical models for count data.

Many log-linear analyses can be carried out in GLIM; see for example, Lindsey (1989). An important class of the log-linear models are the **graphical log-linear models** (Whittaker, 1990). These are, in one sense, merely special models amongst the more general class of hierarchical log-linear models; the importance of graphical models is that they can be interpreted in terms of conditional independence statements. Graphical log-linear models for count data can be fitted easily in GLIM, as can the more specialized class of decomposable (direct) models. GLIM is an effective way of fitting graphical and decomposable models, and the user is greatly assisted by GLIM's model definition syntax. However, it should be noted that GLIM uses the general-purpose iteratively re-weighted least squares algorithm to fit all its models. For tables in many dimensions, the graphical models (and more specifically the decomposable models) should preferably be analysed using an iterative proportional scaling algorithm. Thus, GLIM can be expected to be quite slow for tables with a large number of dimensions. Moreover, care should be taken about asymptotic assumptions about the deviance for large sparse tables (tables with many observed zero counts). Model zeros can cause GLIM to give erroneous degrees of freedom; see [11.3.4.1].

10.6.4.6 Square tables for count data

In some applications, for example, in considering social mobility, an analysis of a square table (that is, a two-way table with rows and columns classified by the same type of factor) is required which takes into account the distinctive structure of the table. Models for symmetry, quasi-symmetry, quasi-independence, and marginal homogeneity (Bishop et al., 1975) can be fitted in the GLIM framework by suitable choice of the explanatory variates x_{ij} (see Firth and Treat, 1988; de Falguerolles, 1989).

10.6.5 Models for continuous data assuming gamma distribution

Some data can be best modelled by an assumption that the variance of the response is proportional to the square of the mean of the response; this is sometimes referred to as the coefficient of variation being constant. A traditional approach to this type of problem was to take logs of the response variable, to attempt to preserve variance homogeneity. With the availability of GLIM, this transformation approach can be avoided. This may be particularly helpful if it is desired to keep the original scale of the observations.

The member of the exponential family with constant coefficient of variation is the gamma distribution. An example where this assumption might be a reasonable first attempt is the example in [12.4.1.2]. A special case of the gamma distribution is the (negative) exponential distribution (not to be confused with the much more general exponential family). As GLIM can analyse data which follows the gamma distribution, it can similarly also model data for exponentially distributed data.

The chi-squared distribution is a special case of the gamma distribution; thus data following the chi-squared distribution can be analysed in GLIM, by use of the gamma assumption. For example, when carrying out the well-known technique of analysis of variance, various (corrected) sums of squares of observations are calculated. If the underlying observations are assumed to come from a Normal distribution, these (corrected) sums of squares will be proportional to a chi-squared distribution, with appropriate degrees of freedom. Thus traditional tests of assumptions of variance homogeneity can be carried out in GLIM [12.1.2.4].

10.6.6 Models for survival data, with or without censoring

A common form of medical study is one in which a group of patients is followed up after medical treatment. The variable of interest is the survival time of such patients and the object of the study is to determine which characteristics of the patient or treatment affect the length of the survival time. A frequent problem with such data is that at the end of the study period some of the patients may still be alive and thus there is not a precise measurement of their survival time but only that it exceeded the length of the study period. Such observations are said to be **censored.** In this section we discuss some popular models for censored survival data.

10.6.6.1 Survival times with the exponential distribution.

Suppose we have censored survival data with the assumption that survival time has the exponential distribution.

The hazard function $h(t)$ is defined by

$$h(t) = \frac{f(t)}{[1 - F(t)]} \quad ,$$

where $f(t)$ and $F(t)$ are the probability density and distribution functions of the survival times.

If we assume that the survival time t_i of an individual can be modelled by

$$h(t_i) = \exp\,(\Sigma \beta_j x_{ij})$$

then we are effectively assuming that the survival times are exponentially distributed with density

$$f(t_i) = \exp\,[\,(\Sigma \beta_j x_{ij}) - t_i \exp\,(\Sigma \beta_j x_{ij})]\,, \quad t_i > 0.$$

Then, on the assumption of random censoring, it can be shown that the likelihood L of the observed survival times is proportional to the likelihood for n independent Poisson variates with means μ_i, where

$$\mu_i = \exp\,(\,\log t_i + \Sigma \beta_j x_{ij}\,)$$

and the response variate is taken to be a vector with a value 0 for censored observations (survivors) and 1 for uncensored cases.

The linear structure is effectively a **log-linear** model for the μ_i; maximization of the likelihood L may be obtained by stipulating that log t_i is a known function in the linear structure and hence has no associated parameter. This is achieved by use of the `OFFSET` directive of GLIM.

In summary, censored survival data can be modelled by:

(a) independent responses;
(b) response variate: 0 (censored), 1 (uncensored);
(c) log link;
(d) offset: log t, for survival times t;
(e) Poisson distribution for the random component.

10.6.6.2 Extreme value and Weibull distributions

The Extreme Value and Weibull distributions are used to model survival times, breaking strengths, and so on. These are not standard GLM models; these probability distributions do not fit immediately into the exponential family which GLIM uses. However, these distributions can be easily fitted in GLIM by various methods using GLIM macros (Aitkin and Francis, 1980; Roger and Peacock, 1982; Roger 1985). The analysis can take into account both right and left censoring; that is, either the case where recording of data stops before the deaths of all of a group of individuals or, less usually, where the observed time is greater than the actual survival time.

10.6.6.3 Logistic and log-logistic distributions

The logistic distribution is used to model survival data, especially where left and right censoring are present. The logistic distribution is similar to the Normal distribution but with 'heavier' tails; the log-logistic is sometimes preferred in the same way that the log-Normal distribution is sometimes used instead of the Normal. Both distributions can be fitted easily in GLIM by use of GLIM macros (Roger and Peacock, 1983; Roger, 1985).

10.6.6.4 The Cox proportional hazards model

The proportional hazards model developed by Cox (1972) is widely used to model survival data in medical applications; for example, in trials of treatments for cancer patients. This model makes few assumptions and can be considered to be **semi-parametric**. GLIM is able to fit this model relatively easily, using macros, the `ELIMINATE` directive, and model indexing (see the User Guide [8.2]). An example of this form of analysis is given in [12.3.2.2].

10.6.7 Distribution mixtures

It is possible to imagine problems where the underlying probability distribution consists of a combination of a (finite) number of different distributions; an example might be data on the heights of people, where the data for men and women have been combined. Clearly, this can be thought of as the mixture of two different components, one for the men, the other for the women. When confronted with such data, various special techniques have been advocated (see, for example, Titterington et al., (1985)). A general technique for estimation of the components of a mixture and the mixing proportions is to use the **EM algorithm** approach. However, this estimation problem can alternatively be handled directly in GLIM by means of its macro facility, by the use of **composite links** [11.7]. The GLIM approach, via composite links, allows the fitting of mixtures of different distributions, or truncated distributions. Furthermore, explanatory factors such as social class can be incorporated in the analysis. It is simple to constrain certain parameters in the model and hence to test whether they are the same for each level of a factor of interest (see Scallan and Evans, (1989)).

10.6.8 Generalized additive models

In some applications, it may not be appropriate to assume the data can be modelled linearly by the use of explanatory variates such as `WEIGHT` or `AGE`. Rather the predictor might require terms such as `WEIGHT*WEIGHT`, or log(`AGE`) or some much more complicated functional form. The appropriate functional form may be difficult to guess in the absence of a theoretical prediction. In such cases, the data themselves may be used to determine the form of dependence between the response variate and the explanatory variate, without imposing a rigid parametric assumption about the form of this dependence, simply requiring it to be smooth. Different forms of smoothing are possible. Moreover, it is not necessary to abandon entirely the parametric framework as a semi-parametric model may be used, in which the linear predictor is a sum of both parametric and non-parametric terms.

These models have been developed by Green and Yandell (1985), who call them semi-parametric generalized linear models, and by Hastie and Tibshirani, in a series of papers culminating in a book (1990), who term them generalized additive models (GAM). The approaches are similar, except that the former base their model fitting on a **penalized likelihood** principle, while the latter usually define their procedures operationally by means of selected scatter plot smoothers. In these models, some explanatory terms βx in the linear predictor are replaced by arbitrary functions $f(x)$, estimated by smoothing relative to x. This model is evidently closely related to the GLM concept and, indeed, GAM models can be analysed in GLIM, although it is usually necessary to use the (somewhat sophisticated) PASS facility.

10.6.9 Taguchi type methods: simultaneous modelling of mean and variance

Section [10.6.5] discussed the modelling of chi-squared distributed data in GLIM. It is more generally possible with heterogeneous Normally distributed data to think of modelling both the mean and variance simultaneously. Thus, having estimated the means of the data, the variances (assumed to be varying) can be modelled by a chi-squared assumption, using the same or different explanatory variates; the new variance estimates can then be used in finding new estimates of the means, and so on, to convergence. This technique is similar to the Taguchi methods and is discussed by Aitkin (1987).

More generally still, it is possible in GLIM to model both mean and dispersion for non-Normal data. This is a fairly advanced use of GLIM, requiring some knowledge of extended GLM methodology (see Gilchrist, 1987; McCullagh and Nelder, 1989).

10.6.10 Multivariate observations

The basic definition of a GLM uses the assumption that the data are independent. We have noted that this is not entirely necessary as we have observed that multinomial data can be modelled in GLIM. Indeed this is one of the major uses of the package. Users of GLIM have also applied the package to analyse other models where the data are assumed to be correlated. The analysis of time series is an area where dependent data arise naturally. GLIM can be used effectively to model some time series data; the autoregressive model with Normally distributed data is the most amenable to a GLIM approach. (Note, of course, that using GLIM allows the autoregressive model to have ***covariates***). Other multivariate models with Normal data can be fitted in GLIM, although it would only be the package of choice for a limited number of such models. Examples where GLIM can be used to good effect for correlated responses y_i are given by Scallan (1985) and by Forcina (1986); see also Gilchrist (1987).

Models with non-Normal assumptions can arise naturally as process driven models (Cox, 1981). For example, in a seed trial, the number germinating, of those so far not germinated, is recorded; this is a case where the conditional distribution at a given time t is specified as a function of past observations. Models with non-Normal dependent structure can also arise as parameter driven models, whereby the dependence is introduced through some latent process. For example, in Zeger (1988), a time series of counts is analysed as a GLM by assuming the covariance matrix of the observations can be approximated by a form amenable to a technique of quasi-likelihood [11.8] (McCullagh and Nelder, 1989).

Although GLIM is not primarily designed for multivariate data, as the above examples illustrate, it can be and is increasingly being used to model specific examples of such data.

10.6.11 Ad-hoc models

In many experimental situations (especially in the physical sciences) the statistician may know that a specific functional relationship exists between the variables. An example might be

$$E(y_i) = \mu_i = f(\alpha + \beta x_i)$$

where $f(.)$ is a known function (such as, $f(\theta) = \theta^k$). One method of analysis would be to write $U = f^{-1}(Y)$ so that we have, approximately,

$$E(u_i) \cong \alpha + \beta x_i$$

but if $y_i \sim N(\mu_i, \sigma^2)$, then the distribution of u_i is either unknown or difficult to specify, and to approximate it by the Normal distribution (thereby losing any constant-variance property that y_i possessed) is to depart even further from the known model.

A simpler, exact analysis is to treat y_i as $N(\mu_i, \sigma^2)$ and, with $\eta_i = \alpha + \beta x_i$, to define the link function as $g(\mu) = f^{-1}(\mu)$ or $\mu = f(\eta)$, whereupon the results and algorithms for all GLMs are now applicable to this case. The GLIM specification and fitting of such models is described in the User Guide [9].

10.7 How to set up a generalized linear model in GLIM

In this section we give an informal summary of the main steps in setting up a generalized linear model in GLIM. This is intended not as a comprehensive guide but rather as an introduction to the way that GLIM can be used to set up the models described in the earlier sections of the Chapter.

The first step is to determine if the data are amenable to a GLIM analysis; perhaps the data fits into one of the examples so far discussed or those given in detail in [12]. If not, the requirements of a GLM will need to be considered:

(i) What is the response variable?

(ii) Are the responses independent? If not, are they in a form which GLIM can nevertheless handle (for example, multinomial data)?

(iii) Can it be assumed that the random component is from some member of the exponential family (for example, Normal, Poisson, gamma, negative exponential, inverse Gaussian, binomial). Or can GLIM nevertheless be used to model the distribution by some `OWN` macros?
Note that GLIM will, by default, assume use of the Normal distribution unless otherwise specified.

The mean–variance relation which is specified by the choice in the `ERROR` directive is as follows:

Distribution	Mean–variance
Normal (Gaussian)	$\text{var}(y) = c$, for a constant c
Poisson	$\text{var}(y) = c\mu$,
Gamma	$\text{var}(y) = c\mu^2$,
Inverse Gaussian	$\text{var}(y) = c\mu^3$,
Binomial	$\text{var}(y) = c\mu(m-\mu)/m$

Note that c is usually taken to be equal to 1.0 in the Poisson and binomial case; see [10.5.6].

(iv) What is the functional relationship between the mean of the data and the linear structure; that is, what is the link function? Again, if GLIM is not informed otherwise, it will assume the use of the canonical link which exists for each form of distribution. The canonical links are as follows:

Distribution	Canonical/default link
Normal (Gaussian)	Identity
Poisson	Log
Gamma	Reciprocal
Binomial	Logit
Inverse Gaussian	Inverse quadratic

(v) What are the explanatory variates? Are they categorical? If they are categorical, should they preferably be treated as ordinal?

(vi) How might the explanatory variables be combined (linearly) to give possible linear structures to explain the responses?

If users have got this far, then it would appear that they can set up their analyses as GLM's. The GLIM package has a succinct language to enable this to be done (described in the User Guide [8]). Essentially, the answers to (i) to (vi) determine a set of GLIM **directives** which inform the package of the model which is required. GLIM can then be told to **fit** the model; that is, to estimate the unknown parameters (the β's) in the linear structure. GLIM will output its measure of the goodness of fit of the chosen model, together with its degrees of freedom. The goodness of fit measure is called the **deviance** by GLIM. The deviance which GLIM outputs will be called a **scaled deviance** or a **deviance**. Scaled deviances occur with the binomial and Poisson distributions. For other cases, GLIM outputs the **(unscaled) deviance** but then there is still the need to estimate the unknown scale for the data in order to **test** the acceptability of the linear structure. Typically, the unknown scale is estimated from the **mean residual deviance** for some appropriate maximal model (see section [10.5.5]). Significance tests can then be carried out, eliminating terms from the linear structure, to obtain a parsimonious representation of the data in terms of the explanatory variates. The adequacy of the assumptions (iii) and (iv) can be tested and, if necessary, modified. Changing the link or 'error' necessitates the rechecking of the linear structure and may subsequently lead to a different choice of best fitting linear structure.

The above is a brief introduction to how GLIM can be used (usually interactively) to set up and fit a statistical model. Model selection is often a key element in model building, as is the testing of the acceptability of the model once it has been estimated. We return to these in Sections [11.4] and [11.6].

11 The theory of generalized linear models

Introduction

In Chapter 10 it was observed that the standard (univariate) statistical models can be thought of as being special cases of a more general structure, called a generalized linear model (GLM).

This chapter introduces the underlying theory behind a GLIM analysis by defining more rigorously what is meant by a generalized linear model (GLM) and by outlining the algorithmic basis of the model fitting procedures of GLIM.

11.1 An outline of the generalized linear model as a comprehensive class of structures

The examples of Chapter 10 show a common pattern.

(a) The response variates y_i are independent and follow a distribution which is a member of the **exponential family** with mean μ_i. This family includes such well known distributions as the Normal, the binomial, the Poisson, the χ^2 (chi-squared), the gamma, the negative exponential and the inverse Gaussian.

(b) The explanatory variables enter as a linear combination of their effects

$$\sum_j \beta_j x_{ij} = \eta_i$$

where η_i is called the **linear predictor.**
We may recall that the x_{ij} are here assumed to be known values of the **explanatory variates**. They will be used to explain the variation in the response y_i; this is achieved by estimating the unknown **parameters** β_j.

(c) The mean of the y_i, which is denoted by μ_i, is functionally related to the linear predictor η_i through the (known) function $g(.)$ which we call the **link function**; that is, we assume

$$\eta_i = g(\mu_i).$$

In the contingency table example [10.6.4.1], $g(\mu) = \log\mu$, while in the linear model example [10.6.1.2], $g(.)$ is reduced to the identity function so that $\mu = \eta$. For a Probit analysis [10.6.3.3], $g(\mu) = \Phi^{-1}(\mu/m)$, whilst for a logistic model [10.6.3.2], $g(\mu)=\log[\mu/(m - \mu)]$.

We note that the superficial dissimilarities of the widely used models discussed in the first chapter may disguise their common underlying structure. An important aspect of a GLM analysis is that the ideas and techniques acquired in connection with one type of model can be carried over to other models within this class of structures. Thus, the class not only represents a compact and useful characterisation of standard models but also offers the opportunity of choosing from this wider class the particular structure that reflects most closely the process being modelled.

Any structure which satisfies the definitions given in the next section is termed a generalized linear model (GLM). Such models were initially proposed by Nelder and Wedderburn (1972); there have been many subsequent books and papers on both GLM theory and applications. An annotated list of some relevant publications is given in the Bibliography. Note that publications prior to 1992 will use GLIM3 syntax; this is generally not a problem for GLIM4 users. Any minor lack of compatibility can be resolved by referring to Appendix C of the User Guide.

11.2 The mathematics of a generalized linear model

This section defines the basic mathematical structure of a GLM.

The approach here is directly oriented to GLIM. It is based upon the concepts discussed in [11.2.1], namely the **exponential family** of independent probability distributions for the responses y_i, $i = 1, 2, ..., n$, the **linear predictor** $\sum_j \beta_j x_{ij} = \eta_i$, and the **link function** $g(.)$ which links η and μ. There is an even more general framework which is discussed in [11.9.3].

11.2.1 The error structure

The probability density function of y_i is assumed to follow the form of the **exponential family** of distributions

$$p(y_i) = \exp\left\{ \frac{y_i\theta_i - b(\theta_i)}{a_i(\phi)} + c(y_i, \phi) \right\}$$

for suitable choice of a_i, b, and c. (Note that ϕ, termed the **scale parameter**, is assumed constant for all i. Moreover, the function $a_i(\phi)$ will be assumed to be of the form ϕ/w_i where the w_i, the **prior weights**, are known. In this manual we deal only with this form for $a_i(\phi)$).

This formula for $p(y_i)$ is somewhat unusual at first sight. It is, in fact, sufficiently general to encompass a wide range of standard distributions. However, it is also sufficiently restrictive to allow the development of a powerful and stable algorithm which can be used in the numerical fitting of models to data assumed to follow this form of density.

We shall find it useful to note that, if we assume the above form for the density of y_i, then the mean and variance of y_i can be expressed in terms of θ and ϕ as follows:

$$\mu_i = E(y_i) = b'(\theta_i)$$

and $\quad \text{var}(y_i) = a_i(\phi)\, b''(\theta_i),$

where primes denote differentiation with respect to θ.

For example, the Normal distribution is obtained by setting $a_i = \phi$, which we would more usually represent by σ^2 (that is, $a_i = \phi = \sigma^2$),

$$b(\theta_i) = \tfrac{1}{2}\,\theta_i^2$$

and

$$c(y_i\,,\phi) = -\tfrac{1}{2}\,[\log(2\pi\phi) + y_i^2/\phi].$$

Section [11.9.1] presents a fuller list of the forms of the functions a_i , b, and c for the well-known distributions.

The function c tends to be somewhat complicated, as above, but we shall see that, fortunately, its form is usually of little concern to us.

It can be convenient to write $b''(\theta_i) = \tau_i^2$, which we call the **variance function**. It is a function of μ_i only. With this assumption, we can therefore write

$$\text{var}(y_i) = \sigma_i^2 = \phi\tau_i^2/w_i\,.$$

An important consequence of the assumption of the exponential family is that the membership of the family is entirely specified by the mean–variance relation; that is, by the variance function. For example, a constant variance implies a Normal distribution, variance equal to the mean implies a Poisson distribution, variance proportional the square of the mean signifies a gamma distribution, and variance proportional to the mean cubed applies to the inverse Gaussian (see [11.9.1]).

11.2.2 The linear predictor

The role played by other variables in the structure of each observation is expressed as a linear sum of their effects for the observation, called the linear predictor, η_i where

$$\eta_i = \sum_{j=1}^{p} \beta_j x_{ij}$$

or

$$\boldsymbol{\eta} = \mathbf{X}\,\boldsymbol{\beta}$$

in matrix notation, where the x_{ij} are known and the β_j are (usually) unknown parameters. The matrix $\mathbf{X}$, of order n by p, is called the **model or design matrix.** The right-hand side of the equation is called the **linear structure.**

If x_{ij} is the value of a quantitative covariate then β_j scales x_{ij} to give its effect on η_i and if an x_{ij} is a 'dummy' variate representing the presence or absence of a level of a factor then β_j is the effect of that factor level;

Such structures are widely used in linear models, such as for the analysis of data from designed experiments, and will be familiar to many readers. A very useful, shorter notation, very suitable for use with computer programs, was first devised by Wilkinson and Rogers (1973). Such Wilkinson and Rogers model specifications are often known as model or structure formulae, though strictly they refer only to the linear structure of the model. GLIM makes heavy use of the Wilkinson and Rogers notation [10.5.9] for specification of the linear structure and, as in common usage, we will refer to the Wilkinson and Rogers linear structure formula as the model formula, when no ambiguity can result.

11.2.3 The link function

The relationship between the mean of the *i*th observation and its linear predictor is given by the **link function** g_i:

$$\eta_i = g_i(\mu_i)$$

where the g_i are assumed monotonic and differentiable.

We define h_i , where $\mu_i = h_i(\eta_i)$, as the **inverse of the link function.**

Although each observation could in principle have a different link function, this is rare in practice. Hence, in this manual the subscript is dropped from the function. Thus we assume $\eta_i = g(\mu_i)$, $i = 1, ..., n$. (You can re-introduce the idea of a different link for each observation by defining your own GLIM program through use of GLIM's facility for user-defined models. For example, Bennett and Whitehead (1981) use this technique to fit the univariate logistic distribution in GLIM).

It may be noted that our mathematical formulation does not necessarily assume that g and h are known functions. However, ***the GLIM program does assume that g and h are known functions***; hence GLIM does not immediately allow for estimation of extra parameters in these link functions. However, there are occasions when it is desirable to estimate the link function, assuming it to belong to some parametric family. Several approaches can be used to do this. A basic technique is to fix the unknown parameters and to search for the best values; an auxiliary variable technique is described in Pregibon (1980); a stable, albeit somewhat complicated, two-stage algorithm is given by Scallan et al., (1984).

11.2.4 Summary of the properties of a GLM

A particular GLM can be identified by specifying the error distribution of the random component, the terms of the linear predictor and the function linking the mean of the random component to the linear predictors. It may be helpful at this stage to note some types of models which do not fit into the above framework:

(i) Models with more than one random component (although they can be handled by GLIM to a certain extent if they have Normal errors, see [12.1.2.6]).

(ii) Probability distributions such as the Cauchy distribution which do not belong to the exponential family. Note, however, that some distributions which are not members of the exponential family, such as the Extreme Value and Weibull distributions, the logistic distribution and the von-Mises distribution can be used as the random component in a GLIM analysis. However, they are outside our basic framework and so they will, in general, require additional user control by the use of GLIM's facility for fitting user defined models.

(iii) Non-linear structures such as $\eta_i = \sum_j \beta_j \exp(k_j x_{ij})$, unless the k_j are assumed known. Again, however, it may be possible to use GLIM to estimate the parameters of such a model, by extra user-controlled iterative procedures.

Note that even keeping strictly within the GLM framework, we still have a wide range of choice of models. Thus the strict GLM framework presents a unified view of many analyses which were previously considered as distinct; from this manual you will learn primarily how GLIM can be used to analyse data from such models. However, you will also be directed as to how GLIM can be used for a wider range of problems, some not strictly within the basic GLM paradigm but rather within the more general theory of GLMs as, for example, discussed by McCullagh and Nelder (1989).

11.3 Inference questions in GLMs

Given a set of sample data we can now describe, in terms of a GLM, a theoretical population to which the data are believed to belong. Two questions arise:

(i) Do the data support the belief that the proposed GLM is a reasonable description of the population?

(ii) Since the parameters of the GLM are unknown, what values do the data suggest for them?

The questions are connected but for convenience they will be dealt with separately. The second question will be considered now and the first will be considered in the following section [11.4].

11.3.1 Estimation of the linear parameters

Our approach is based upon the model formulated in Section [11.2]; in other words, we assume that we have independent observations from an exponential family of distributions with a linear predictor related to the mean of the data through a known link function. The method of maximum likelihood is used to estimate the linear parameters β_j and hence the linear predictors η_i and the fitted values μ_i .

We assume for the moment that the model matrix $\mathbf{X}$ is of full rank. We consider the set of β_j, $j = 1, \ldots, p$ as a vector $\boldsymbol{\beta}$. (We shall write vectors and matrices in **bold** type.)

11.3.1.1 Maximising the likelihood function

Considering the likelihood L as a function of $\boldsymbol{\beta}$ we may, in general, find the maximum of L by finding the maximum of $\log L$; this may be achieved by solving

$$\frac{\partial \log L}{\partial \beta_j} = 0, \text{ for } j = 1,\ldots,p.$$

We may note that $\log L = l = \sum_{i=1}^{n} l_i$, where $l_i = \log p(y_i)$, with $p(y_i)$ as in [11.3.1].

$$\text{Now} \quad \frac{\partial l_i}{\partial \beta_j} = \frac{\partial l_i}{\partial \theta_i}\frac{\partial \theta_i}{\partial \mu_i}\frac{\partial \mu_i}{\partial \eta_i}\frac{\partial \eta_i}{\partial \beta_j}$$

$$= \frac{y_i - b'(\theta_i)}{a_i(\phi)} \frac{1}{b''(\theta_i)} \frac{1}{g'(\mu_i)} x_{ij},$$

where primes denote differentiation.

If we set $d_i = g'(\mu_i) = \partial\eta_i/\partial\mu_i$ (which is, in general, a function of $\beta_1,\ldots,\beta_p$) and, furthermore, if we write

$$u_i = \frac{1}{\sigma_i^2 d_i^2} = \frac{1}{a_i(\phi)\, b''(\theta_i)\, d_i^2}$$

$$\text{then} \quad \frac{\partial l}{\partial \beta_j} = \sum_i \frac{\partial l_i}{\partial \beta_j}$$

$$= \sum_i (y_i - \mu_i)\, u_i\, d_i\, x_{ij} \quad . \tag{11.1}$$

To find the maximum likelihood estimates we need to solve

$$\frac{\partial l}{\partial \beta_j} = 0, \text{ for } j = 1, \ldots, p.$$

In general, this is a non-linear problem, as μ, u, and d are, in general, functions of $\beta_1,\ldots,\beta_p$.

11.3.1.2 Newton–Raphson and Fisher-scoring

The **Newton-Raphson** approach to solving these equations would be to set up an iterative scheme for the vector $\boldsymbol{\beta} = (\beta_1,\ldots, \beta_p)'$, as follows:

$$\boldsymbol{\beta}^{(n)} = \boldsymbol{\beta}^{(n-1)} + \boldsymbol{\delta\beta}^{(n-1)}$$

where

$$\boldsymbol{\delta\beta}^{(n-1)} = -\left[\frac{\partial^2 l}{\partial \boldsymbol{\beta}\, \partial \boldsymbol{\beta}'}\right]^{-1} \frac{\partial l}{\partial \boldsymbol{\beta}}$$

The second derivatives are usually complicated to calculate and may be replaced by their expectations (the **Fisher-scoring** technique):

Firstly we note that

$$\mathrm{E}\left(\frac{\partial^2 l}{\partial \beta_j \partial \beta_k}\right) = -\mathrm{E}\left(\frac{\partial l}{\partial \beta_j}\frac{\partial l,}{\partial \beta_k}\right).$$

Using the result that $\mathrm{E}(y_i - \mu_i)^2 = \sigma_i^2$, we can deduce from eqn (11.1) that

$$\mathrm{E}\left(\frac{\partial l_i}{\partial \beta_j}\frac{\partial l_i}{\partial \beta_k}\right) = \sigma_i^2 u_i^2 d_i^2 x_{ij} x_{ik}$$

Hence $$\mathrm{E}\left(\frac{\partial l_i}{\partial \beta_j}\frac{\partial l_i}{\partial \beta_k}\right) = u_i x_{ij} x_{ik}$$

so

$$-\mathrm{E}\left(\frac{\partial^2 l}{\partial \beta_j \partial \beta_k}\right) = \sum_i u_i x_{ij} x_{ik}$$

Thus $\boldsymbol{\delta\beta}^{(n-1)}$ is the solution of a matrix equation, of which the typical (row j, column k) elements are:

$$\left[\sum_i u_i x_{ij} x_{ik}\right]\left[\delta\hat{\beta}_j^{(n-1)}\right] = \left[\sum_i (y_i - \mu_i) u_i d_i x_{ik}\right] \tag{11.2}$$

Although one could solve these equations, it is more convenient to note firstly that

$$\hat{\eta}_i = \sum_j \hat{\beta}_j x_{ij};$$

hence typical elements of this expression expressed as a matrix equation may be written

$$\left[\sum_i u_i x_{ij} x_{ik}\right]\left[\hat{\beta}_j^{(n-1)}\right] = \left[\sum_i u_i x_{ik} \hat{\eta}_i\right], \tag{11.3}$$

Where $\hat{\eta}_i$ is evalued at the (n–1)th iteration.
Summing eqns (11.2) and (11.3) we obtain a matrix equation with typical entries (row j, column k) given by

$$\left[\sum_i u_i x_{ij} x_{ik}\right]\left[\hat{\beta}_j^{(n)}\right] = \left[\sum_i u_i\{\hat{\eta}_i + d_i(y_i - \mu_i)\}x_{ik}\right] \tag{11.4}$$

We now define the **working variate** z_i by

$$z_i = \hat{\eta}_i + d_i(y_i - \mu_i) .$$

Then our eqn (11.4) can be written as

$$(\mathbf{X}'\mathbf{U}\mathbf{X})\hat{\boldsymbol{\beta}}^{(n)} = \mathbf{X}'\mathbf{U}\mathbf{z}$$

where $\mathbf{X}'$ is the transpose of the model matrix $\mathbf{X}$ and $\mathbf{U}$ is a diagonal matrix of the u_i. This can be written more familiarly as

$$(\mathbf{X}'\mathbf{V}^{-1}\mathbf{X})\hat{\boldsymbol{\beta}}^{(n)} = \mathbf{X}'\mathbf{V}^{-1}\mathbf{z}$$

which is a weighted least squares equation with the working variate $\mathbf{z}$ as response variate, where $\mathbf{V}$ and $\mathbf{z}$ are evaluated at the previous iteration.

11.3.1.3 Iteratively re-weighted least squares (IRLS)

We can summarize the last sections by noting that, to maximize for $\boldsymbol{\beta}$, we obtain an equation of the form

$$\mathbf{A}\mathbf{b} = \mathbf{c} \tag{11.5}$$

where $\mathbf{b} = (b_i,\ldots,b_p)' = \hat{\boldsymbol{\beta}}$ is the maximum likelihood estimate of $\boldsymbol{\beta}$; both $\mathbf{A}$ and $\mathbf{c}$ are, in general, functions of the $\hat{\boldsymbol{\mu}}$, the estimate of $\boldsymbol{\mu}$, which is unknown since it is a function of $\mathbf{b} = \hat{\boldsymbol{\beta}}$.

A direct solution is not generally possible but we may solve iteratively. We take the observations (or some minor modification of the observations) as initial estimates of $\boldsymbol{\mu}$, solve for $\mathbf{b}$, use these values of $\mathbf{b}$ to calculate a second estimate of $\boldsymbol{\mu}$ and again solve for $\mathbf{b}$. We continue this until $\mathbf{b} = \hat{\boldsymbol{\beta}}$ converges to the maximum likelihood estimate (MLE). (See [11.5.6] for a discussion of non-convergence, and so on.)

We may note again that $\mathbf{A} = \mathbf{X}'\mathbf{V}^{-1}\mathbf{X}$ and $\mathbf{c} = \mathbf{X}'\mathbf{V}^{-1}\mathbf{z}$

where

$$\mathbf{V} = \text{diag}\left\{\tau_i^2\left(\frac{\mathrm{d}\eta_i}{\mathrm{d}\mu_i}\right)^2\frac{\phi}{w_i}\right\} = \text{diag}\left\{\frac{1}{u_i}\right\}, \text{ say}$$

(the u_i, excluding ϕ if unknown, are known as the **iterative weights**) and $z_i = \eta_i + (y_i - \mu_i)(\mathrm{d}\eta_i/\mathrm{d}\mu_i)$.

We have noted that we have obtained the same equations as for a weighted least-squares regression for a random variable $\mathbf{z}$ with $\mathbf{E}(\mathbf{z}) = \boldsymbol{\eta}$, var($\mathbf{z}$) = $\mathbf{V}$. At each iteration, $\mathbf{z}$,

$\boldsymbol{\eta}$ and $\mathbf{V}$, and the weights u_i will be updated. Thus, our algorithm for obtaining the maximum likelihood estimates is often referred to as **iteratively re-weighted least squares;** ([11.9.3] gives a comprehensive account of this more general technique).

11.3.1.4 The variance-covariance matrix of the parameter estimates

The variances and covariances of the parameter estimates may be estimated by noting that eqn (11.5) has the solution

$$\mathbf{b} = \mathbf{A}^{-1}\,\mathbf{c} \tag{11.6}$$

where $\mathbf{b}$ is the MLE of $\boldsymbol{\beta}$. From general estimation theory we know that MLEs are asymptotically unbiased. By the same theory, the dispersion matrix (the matrix of variances and covariances) of the parameter estimates is asymptotically given by

$$\text{var}(\mathbf{b}) \cong (\mathbf{X}'\mathbf{V}^{-1}\mathbf{X})^{-1}$$

and estimates of $\boldsymbol{\eta}$ and $\boldsymbol{\mu}$ are given by $\mathbf{Xb}$ and $h(\hat{\boldsymbol{\eta}})$, respectively.

11.3.1.5 The mean–variance relation and quasi-likelihood

It is of interest that, in the above algorithm, the distributional assumption [11.2.1] is introduced via the mean–variance relation; that is, it depends upon $\text{var}(y_i) = a_i(\phi)b''(\theta_i)$ but not upon higher moments of the distribution. This leads us to consider the **quasi-likelihood** method [11.8] which assumes a (generally non-standard) mean–variance relation as a basis for estimation.

11.3.2 Aliasing

If we now relax the restriction that $\mathbf{X}$ be of full rank p and allow it to have rank $r < p$, eqn (11.5) may no longer have a unique solution. Instead we write

$$\mathbf{b}^0 = \mathbf{A}^-\,\mathbf{c} \tag{11.7}$$

where $\mathbf{A}^-$ is any generalized (or g-) inverse matrix of $\mathbf{A}$, that is, $\mathbf{A}^-$ satisfies $\mathbf{AA}^-\mathbf{A} = \mathbf{A}$. (See, for example, Searle (1971), or Pringle and Rayner (1971) for a more detailed treatment.) Such a g-inverse is not necessarily unique, but for any $\mathbf{A}^-$ it can be shown that $\mathbf{b}^0$ is a solution of (11.5).

For statistical purposes the particular solution chosen is irrelevant since, for example, $\hat{\eta}$, and thus $\hat{\boldsymbol{\mu}}$, can be shown to be the same whichever value of $\mathbf{b}^0$ we choose. We emphasize that our choice of a particular $\mathbf{b}^0$ does not change the model we are fitting but merely determines our manner of expressing the linear structure. However, from an algebraic point of view we note that eqn (11.5) imposes only r independent constraints on the p estimates $\mathbf{b}^0$, so that if $r < p$ there will be many sets of values of $\mathbf{b}^0$ satisfying the equation. We can add, in simple cases, some extra constraints (for example that factor effects sum to zero) so that the solution becomes unique; in more complex cases, however, the form of such constraints is not obvious (for example in a complex unbalanced design).

Now $\mathbf{X}$ is of less than full rank when one or more of the columns of $\mathbf{X}$ can be formed from linear combinations of other columns of the matrix. When the set of columns $\mathbf{x}_i,\ldots,\mathbf{x}_j$ form a linearly dependent set then one or more of the parameters $\beta_i,\ldots,\beta_j$ corresponding to these columns are said to be aliased, with the following implications.

If β_1, β_2 and β_3 are parameters, $\mathbf{x}_1$, $\mathbf{x}_2$ and $\mathbf{x}_3$ being the corresponding columns of the model matrix, and if $\mathbf{x}_3 = \mathbf{x}_1 + \mathbf{x}_2$ then knowledge of the values of $\mathbf{x}_1$ and $\mathbf{x}_2$ (say) implies knowledge of $\mathbf{x}_3$ so that $\mathbf{x}_3$ and its parameter β_3 are then redundant in any explanation of the structure of the observations. The composition of the model matrix shows that the data give no further information on β_3 once the information on $\beta_1$and β_2 has been removed. Of the three parameters only two are needed so that the solution is adequately expressed as

$$\hat{\boldsymbol{\eta}} = b_1\mathbf{x}_1 + b_2\mathbf{x}_2$$

for suitable estimates b_1 and b_2. We may, of course, use three parameters

$$\hat{\boldsymbol{\eta}} = a_1\mathbf{x}_1 + a_2\mathbf{x}_2 + a_3\mathbf{x}_3$$

by defining

$$a_1 = b_1,\ a_2 = b_2,\ a_3 = 0$$

or with a different parametrization by

$$\begin{aligned} a_1 &= (2b_1 - b_2)/3, \\ a_2 &= (2b_2 - b_1)/3, \\ a_3 &= (\,b_1 + b_2)/3. \end{aligned}$$

GLIM uses the first kind of parametrization while the second is often known as 'the usual constraints', giving

$$a_1 + a_2 + a_3 = 0.$$

Though the estimate of η is the same in both cases the manner of expressing it is different. Thus aliasing among a set of parameters implies that less than the whole set is needed to specify the structure but that if the whole set is used we have extra freedom in our choice of parameter values.

Aliasing of parameters can arise in two ways. If the specification of the linear structure of the population is redundant whatever the model matrix then the aliasing is said to be **intrinsic**. Such redundancy occurs, for example, when both an overall mean and effects for all levels of a factor are included in the model. It is easily seen that the sum of the model matrix columns for the factor effects is identical to that of the column for the mean. Intrinsic aliasing is a result of the parametrization of the population and will occur regardless of the sample we obtain. In contrast, **extrinsic aliasing** is said to occur when by some quirk of the data there is linear dependence among columns; for example, when one covariate is a linear combination of other covariates or when there are no observations for some level of a factor (resulting in a zero column for the corresponding dummy variable). This does not necessarily imply that the corresponding population parameters are intrinsically aliased, but only that the sample data do not contain enough information to estimate all the parameters.

Given that aliasing has occurred among the parameter estimates we then face a choice of possible parametrizations. To narrow the choice we must add extra constraints on the parameter estimates; $(p - r)$ independent constraints would give a unique solution. One possible set of constraints is to demand that factor effects sum to zero, but this is not computationally simple and, with 'unbalanced' designs or designs with covariates, may not be very helpful.

As discussed in [10.5.10], a computationally simpler and more general method is to evaluate the parameter estimates sequentially and on finding one which is aliased with preceding estimates to drop it from the model, which is equivalent to assigning the value zero to it. Thus we implicitly impose $p - r$ constraints and can estimate the remaining r of the estimates b_j under these constraints. Such a procedure is to some extent arbitrary but knowledge of the experimental design or of the model matrix should enable us to re-parametrize as necessary. GLIM adopts this procedure with some additional features. We have introduced these in [10.5.10] but we briefly summarize them here for completeness.

Unless otherwise instructed, GLIM fits a parameter corresponding to the unit vector in the design matrix; this is known as `1` in GLIM terminology (referred to as the '**grand mean**' `%GM` in earlier versions of GLIM). Provided that this constant term is in the model, GLIM then takes the first level of each factor as the reference point from which other parameter estimates are calculated; in other words, if it is impossible to get parameter estimates for all levels of a factor, then by default GLIM will intrinsically alias the first level of the factor; that is, it sets the parameter estimate corresponding to the first level of the factors to zero (rather than, for example, the sum being zero). However, as this choice is somewhat arbitrary, GLIM allows the user (through the `FACTOR` directive) to choose a different level to be the reference category of each factor (see the User Guide [3.1.4]). The convention of the sum being zero is ***not*** available in GLIM. If you wish to have that particular form of parameter estimates, you can obtain these in GLIM by specifying your own design matrix. Alternatively, you might wish algebraically to carry out a transformation from one set of parameters to the other, although this can be tedious for large designs as it effectively uses a Mobius transformation (A possible method is indicated by Leimer and Rudas (1989)). In fact, in some applications, it is the fitted values which are more important than the parameter estimates. In such cases, the effect on the fitted values of a term in the linear structure can be well illustrated by tabulating the fitted values with and without the term in question.

We conclude this section by noting the effect of aliasing on the dispersion matrix of the parameter estimates.

Referring again to eqn (11.6) of [11.3.1.4], it can be shown that if $\mathbf{A}^-$ is the g-inverse which gives such a solution and is also reflexive, then

$$\text{var}(\mathbf{b}) \cong \mathbf{A}^-,$$

although it should be noted that the rows and columns corresponding to parameters set to zero will be zero, indicating that such parameters have not been estimated and so have zero variances. Since the matrix $\mathbf{A}^-$ contains the scale parameter ϕ as a multiplier it is necessary, when ϕ is unknown, to estimate it, as indicated below.

11.3.3 Estimation of the scale parameter

If the distribution of y_i involves only one unknown parameter (such as the binomial or Poisson cases) then the scale parameter ϕ can usually be taken to have the value 1.0 and there is no estimation problem. Otherwise, we will need to estimate ϕ from the data: several alternative estimators have been proposed and a complete picture of their relative merits is not yet available. One possibility is to estimate ϕ by maximum likelihood.

This, for example, is appropriate and convenient for the gamma case, as GLIM provides the gamma, digamma and trigamma functions which turn out to be needed. (Note, however, that the MLE estimate for the gamma scale should preferably be transformed to give smaller bias and mean square error; see Anderson and Ray (1975) and Gilchrist (1981) for a fuller discussion). In general, though straightforward in theory, the use of maximum likelihood to estimate ϕ can be computationally cumbersome as no explicit estimators exist.

To derive a simpler estimate we must anticipate results presented in the next section where a quantity S is shown to have (possibly approximately) a χ^2_{n-r} distribution under a model with r independent linear parameters.

It is further shown that $S = D/\phi$ where D, known as the **deviance**, is a statistic which can be computed from the data. If we calculate D under a reasonable model then, since $\text{E}(S) = n - r$, we may estimate ϕ by

$$\hat{\phi} = D/(n-r)$$

on the grounds that under a reasonable model S will be relatively close to its expected value. This is, in fact, the method used to estimate variance components in the analysis of variance where D would be the sum of squares for a source of variation and ϕ the variance component.

More generally we may use this idea with D replaced by any 'reasonable' measure of discrepancy for the model. Another important measure of discrepancy is the generalized Pearson statistic, which takes the form

$$X^2 = \Sigma \frac{(y-\hat{\mu})^2}{V(\hat{\mu})} ,$$

where $V(\hat{\mu})$ is the estimated variance function for the distribution being used. For the Normal distribution, X^2 is again the residual sum of squares, while for the Poisson or binomial distributions it is the well-known Pearson X^2 statistic.

In tests of hypotheses, the use of (scaled) deviances is probably preferred, as the deviance is additive for nested sets of models, whereas the Pearson X^2 is not. The generalized Pearson statistic and the deviance, each after division by ϕ, have exact chi-squared distributions for models assuming Normal errors (when the model is true), whilst asymptotic chi-squared results are available for other distributions. However, the corresponding estimates $\hat{\phi}$ are, in general, inconsistent in the limit as $n \to \infty$ which casts some doubts upon their validity for general use. Some authors suggest that the modified profile likelihood estimate (Barndorff-Nielsen, 1988) or the generalized Pearson statistic might be preferred (see Jørgensen, 1987).

11.3.4 Implications of non-convergence

The algorithm underlying the fitting of iterative GLMs is generally very robust, converging quite rapidly. However, it can sometimes fail to converge, and can do so in two quite distinct ways which we now discuss.

11.3.4.1 Infinite parameter values

Consider a two-way table of counts in which an entire row consists of zeros. If we fit a model with row and column terms using a Poisson error and the link function $\mu = \exp(\eta)$, then the fitted values for the row of zeros should also be zero, implying that the linear predictors must be minus infinity, which in turn implies that the parameter estimate corresponding to that row must be minus infinity. In practice the algorithm will produce larger and larger negative values for the estimate until the maximum number of cycles set for the iteration is reached. The deviance and the parameter estimates will be approximately correct, and the parameters whose estimates should tend to minus infinity can be found by inspection from their large negative values. Models with binomial errors can also produce infinite parameter values corresponding to $\mu = 0$ or m.

When infinite parameter estimates can be detected in advance (for example, a complete row of zeros in a table) it is advisable to omit the appropriate data values from the fit (which may be done by assigning them prior weights of zero). Their inclusion may distort the deviance because the expected contribution from those data values is zero.

In practice, a sparse contingency table will have many cells which are observed to be zero. In such cases, it is necessary to find out whether there are zeros in any of the marginal configurations defined by the (log-linear) model in question. GLIM's numerical algorithm will produce the correct fitted values (the estimates of the μ's) but it will not automatically take into account the fact that some of the estimated cell frequencies are identically zero. Thus, in such cases, although GLIM produces the correct MLEs of the

μ's, some of the parameter estimates and their standard errors will be incorrect. Moreover, the degrees of freedom will be wrong. The occurrence of this problem is easily identified, as the standard errors of some of the parameter estimates will be very large compared to the estimates themselves. Thus, in carrying out such an analysis, it is recommended that the user identify the occurrence of zeros in any of the margins defined by the log-linear model and weight out these observations; that is, identify fitted values near zero and constrain these cells which should be zero to have fitted values equal to zero. The degrees of freedom, parameter estimates and their standard errors will then be correctly output by GLIM. It may be appropriate to note further that it is evident that **structural zeros** (for example, pregnant men) should be weighted out of all analyses of the data. For a full discussion of the problem of zeros in sparse contingency tables, see Aston and Wilson (1984).

11.3.4.2 Divergence

If the deviance, instead of falling between successive cycles, more than doubles, then iteration stops with a comment that divergence has occurred. No instances of divergence are known to have occurred when the link function is that giving sufficient statistics for the parameters, and divergence is generally most uncommon. Some cases have arisen, for example, with gamma errors and the identity link; in one the fit oscillated indefinitely between two positions, and in another, a perfect fit produced two consecutive deviances, both extremely small, the second being accidentally more than twice the first. This latter does not, of course, constitute a genuine divergence.

If a divergence should occur it is worth repeating the fit with `RECYCLE`, starting from the fitted values produced by a similar, but different, model. Convergence may then follow.

11.3.5 'Saw-toothing'

It may happen in the course of an iterative fit that the fitted weights attached to the observations become such that the amount of information about one of the parameters becomes very small. Such a state is not distinguishable numerically from the linear dependencies that underlie extrinsic aliasing [11.3.2]. The result is that a parameter which began by not being aliased appears suddenly to become aliased, and so is dropped from the model. If iteration is allowed to continue with the parameter set to zero, the deviance, which has so far been falling cycle by cycle, suddenly rises sharply, and the parameter is now found to be unaliased again. Further iteration produces further falls in the deviance until apparent aliasing occurs again. Plotting the deviance against the cycle number produces a saw-tooth profile, hence the name for this phenomenon.

Saw-toothing will usually occur when parameter estimates are tending to + or –infinity, and results from the limited accuracy of the computer arithmetic. It is detected by a change in the d.f. of the deviance during iteration and results in a warning being printed. The fit for the cycle immediately preceding the jump in deviance will be approximately correct.

11.4 Goodness of fit

Our concern in this section is with the first of the questions raised in [11.3], namely, how well do the data support the model?

We assume that the error and link specifications are satisfactory (see [11.6] for further discussion of this point), and concentrate our attention on the linear structure. Thus for the rest of this section the term 'model' refers only to the linear structure of a model.

11.4.1 The linear structure as a compromise

The linear structure is the sum of the effects of explanatory variables and its make-up expresses the impact of these variables on the response variable. The data give information on which effects have an important effect and which can be neglected. A smaller number of parameters means easier interpretation and, generally, better prediction so our aim is to obtain the best 'trade-off' between the number of variables and their parameters that we must include in the linear structure (keeping the number as small as possible), and the ability of the model to represent the data (keeping the fit as good as possible). It is useful to distinguish five special instances of the linear structure, as follows.

(i) If we include n linearly independent explanatory variates in the linear structure, the MLE of the μ_i are equal to the observations themselves. This is known as the **full model**. GLIM in fact treats all models alike so will estimate parameters for a full model by iteratively re-weighted least squares; in this case, provided the algorithm has enough iterations to converge, the data will be reproduced exactly, but without any simplification of interpretation.

(ii) Conversely, if we propose one common value for the μ_i, then we have a very simple model, known as the **null model**. In most cases the null model will not adequately represent the structure of the data.

(iii) There may, however, be two other, less extreme, limiting models. Certain parameters may have to be in the model (for example, when there are known to be 'block effects' or fixed margins in a contingency table); the model containing only these parameters is termed the **minimal model** since it is the simplest model we wish to consider.

(iv) Conversely, the largest, most complex, model we wish to consider is termed the **maximal model**.

(v) Between these two extremes is the model under investigation, which is called the **current model**.

11.4.2 A measure of goodness of fit: the scaled deviance

Our problem is to determine the usefulness of an extra parameter in the current model, or conversely, the lack of fit induced by omitting that parameter.

GLIM's estimation procedures can be considered as being based upon maximum likelihood. In this spirit, a measure of the reasonableness of a model is the likelihood of the model given the data. By comparing the likelihood of the current model L_c to the likelihood of the full model L_f with the given data, we obtain a measure of the acceptability of the current model relative to that of the full model.

A widely used method of comparing the likelihoods is the (maximum) likelihood ratio test; this is based upon the ratio L_c/L_f. More conveniently, we use the quantity $S(c,f)$ where

$$S(c,f) = -2\log(L_c/L_f).$$

We term S the **scaled deviance**, its arguments denoting the models being compared. It can be seen that large values of S indicate low values of L_c relative to L_f, that is, increasing lack of fit; hence the term deviance (that is, a measure. of the deviation of the data from the fitted model)

If we substitute for L_c and L_f, the p.d.f. given in [11.3.1] with ML estimates of the parameters, we obtain

$$S(c,f) = 2\sum_{i=1}^{n} \frac{y_i(\tilde{\theta}_i - \hat{\theta}_i) + \{b(\tilde{\theta}_i) - b(\hat{\theta}_i)\}}{a_i(\phi)}$$

where $\hat{\theta}_i$ and $\tilde{\theta}_i$ are the MLEs of θ_i under the current and full models respectively. For a particular function $b(.)$ we may sometimes find it more convenient to re-parametrize $S(c,f)$ in terms of the corresponding $\hat{\mu}$ and $\tilde{\mu}$

We may note that, in [11.2.1], we restricted $a_i(\phi)$ to be of the form ϕ/w_i. This enables $S(c,f)$ to be rewritten in the form

$$S(c,f) = D(c,f)/\phi.$$

With this definition, $D(c,f)$ is termed the **deviance** of the current model relative to the full model and ϕ is the **scale parameter**. Note that D is a genuine statistic as it contains no unknowns and so can be computed given the data and the MLEs obtained from the data. The contribution of unit i to the deviance is termed the **deviance increment**, d_i, so that

$$D(c,f) = \sum_i d_i \quad .$$

The deviance (effectively the likelihood ratio test statistic) is in fact a commonly used statistic. For the Normal distribution, $D(c,f)$ is the residual sum of squares under the current model since $b(\theta) = \frac{1}{2}\theta^2$ and under a full model we always have $\hat{\mu}_i = y_i$, while for the Poisson distribution, $D(c,f)$ is the log likelihood-ratio statistic which is often used as a test statistic in the analysis of contingency tables and is closely related to the Pearson 'goodness-of-fit' statistic.

11.4.3 The distribution of the scaled deviance *S*

The following general result is derived by, for example, Kendall and Stuart (1967, p.224): assuming certain regularity conditions, if L_1 is the likelihood of $\mathbf{y}$ for model 1, and L_2 similarly for model 2, and if model 2 is nested in model 1 (that is, the parameter space under model 2 is a subspace of that under model 1) then, if model 2 is correct $S(2,1) = -2 \log(L_2/L_1)$ is distributed as χ^2 with $t_1 - t_2$ d.f., where the t_i are the number of independent parameters estimated under model *i*. The distribution is exact for the Normal distribution and identity link, and for some models with the inverse Gaussian distribution but only approximate for other 'error'/link combinations. (See [11.5.3] for further details.)

A practical difficulty in using such a test is that for two-parameter families *S* is a function of ϕ which is usually unknown. A solution is to fit a maximal model and use the estimate of ϕ suggested in [11.3.3] on the grounds that, having removed systematic variation, the residual variation will be well approximated by χ^2 and will then provide an adequate estimate of ϕ. This estimate may be used to calculate *S* which may then be referred to the appropriate χ^2 tables.

Alternatively we define the mean scaled deviance of model 2 from model 1 as $S(2,1)/(t_1 - t_2)$ and note that ratios of such statistics do not involve the scale parameter. If we are willing to accept such a ratio as approximately distributed as F with the appropriate degrees of freedom then comparison of a calculated ratio with tabulated values provides a test of significance. (For the model with Normal distribution and identity link this distribution is exact.)

11.4.4 Analysis of deviance

The above results enable us to assess the usefulness of the linear parameters of the GLM in explaining the data. If model 1 contains two sets of parameters $\beta_1,\ldots,\beta_t$ and $\beta_{t+1},\ldots,\beta_p$ all linearly independent, and model 2 contains only the second set $\beta_{t+1},\ldots,\beta_p$ the models being otherwise identical, then $S(2,1)$ is distributed asymptotically as a chi-squared distribution with *t* d.f. under model 2. A test of the hypothesis $\beta_1 = \beta_2 = \ldots = \beta_t = 0$ may thus be performed by comparing $S(2,1)$ and the relevant percentage point of the chi-squared distribution with *t* d.f. These percentage points can be obtained using the `%chd` function in `CALCULATE`.

If the parameters $\beta_1,\ldots,\beta_t$ are not linearly independent (that is, are aliased) or not linearly independent of $\beta_{t+1},\ldots,\beta_p$ (that is, are aliased with the other parameters) then the degrees of freedom associated with the χ^2 will be less than *t*; the degrees of freedom measure the difference in dimension of the nested parameter spaces available to the models and, of course, removal of an aliased parameter does not alter the parameter space, though the manner of representation may change.

Consider now the situation given above when model 1 is the full model and model 2 has r independent parameters. Then $S(2,1)$ represents the lack of fit induced by estimating from an r-dimensional parameter space instead of an n-dimensional space. Similarly if model 3 is nested in model 2 and contains t ($< r$) independent parameters we can form $S(3,1)$, and

$$\begin{aligned} S(3,1) - S(2,1) &= -2\log(L_3/L_1) + 2\log(L_2/L_1) \\ &= -2\log(L_3/L_2) \\ &= S(3,2) \end{aligned}$$

which is the scaled deviance for the extra parameters between models 2 and 3, and since these models are nested it is distributed as χ^2_{r-t} when model 3 is correct. Hence if, for a sequence of k nested models each with r_i independent parameters, we form $S(i,1)$, $i = 1,\ldots,$ k then the scaled deviance for changing from (say) model i to a simpler model (say j) is given by $S(j,1)–S(i,1)$. Thus we may build up a table of differences of deviances for a sequence of nested models analogous to the tables of sums of squares used in the analysis of variance. Note that, in general, orthogonality cannot be assumed and so the deviance attributable to a set of parameters will depend upon which other parameters are in the model. We discuss this more fully in the next section.

11.4.5 Marginality and orthogonality

A set of parameters in the linear structure (corresponding, for example, to the main effects of a factor or to the interaction effects of two or more factors) is called a **term**.

If the space spanned by the columns of the model matrix corresponding to a term T is a subspace of the space for a term R then T is said to be **marginal** to R. Thus, by the definition of Section [11.3.2], such a term T is **aliased** with the term R, and including T in a model if R is already included will not alter the deviance, though it may, depending upon the constraints imposed, alter the parametrization. The grand mean is marginal to all factor terms (as can be seen from inspection of a model matrix), main effect terms are marginal to interaction terms and low-order interaction terms to high-order ones that contain them.

If the inclusion of a term A in the model produces the same reduction in the scaled deviance regardless of whether a term B is already in the model or not (in which case the reverse is also true) then A and B are said to be **orthogonal**. The implication is that the extra space spanned by including A in the model is orthogonal to the extra space spanned by B over other terms in the model so that estimates of the non-aliased parameters in A and B are uncorrelated.

When terms A and B are not orthogonal the order in which they are included in the model may be important. Since the relevant spaces are not orthogonal the magnitude of the change in the scaled deviance due to the addition of (say) A will depend on whether or not the term B is already included.

Thus, when speaking of the scaled deviance due to a term A (that is, the reduction in scaled deviance after including A) we must also mention the terms (say T) already in the model, for example, by writing $S(A+T, T)$. It only makes sense to speak of $S(A)$ when A is orthogonal to the other terms in the model or, possibly, when A is orthogonal to a subset of terms, the remainder being considered the minimal model.

11.4.6 t tests

When the deviance associated with a factor A is large enough to merit its inclusion in the model it may still be possible to economize on parameters by reducing the number of levels in A. Thus we may hypothesize that the effects associated with levels i and j of A are the same. Such a hypothesis may be examined in two ways.

By subtracting the scaled deviance for the model with a reduced number of levels for A from the scaled deviance for the model with all levels of A we obtain a statistic whose ratio to the scaled mean deviance under the full model can be referred to the appropriate F tables, as indicated in [11.4.3]. We can constrain effect i to be the same as effect j by, for example, changing column i so that it indicates the presence of effect i in the reduced model, wherever either i or j were previously present. (If effect j is still included it will now be aliased with effect i.)

Alternatively we note that **b**, the maximum likelihood estimator of $\boldsymbol{\beta}$, is asymptotically distributed as a multivariate Normal distribution with mean $\boldsymbol{\beta}$ and dispersion matrix $\mathbf{A}^-$ as given in Section [11.3.2].

Thus under the hypothesis $\mathbf{c}'\boldsymbol{\beta} = 0$ (where prime denotes transposition) the statistic

$$(\mathbf{c}'\mathbf{b})/(\mathbf{c}'\mathbf{A}^-\mathbf{c})^{\frac{1}{2}}$$

will be distributed as t with $(n - r)$ d.f., where r is the rank of $\mathbf{A}$. Note that if $\mathbf{c}$ contains 1 in position i, -1 in position j and zeros elsewhere, and if either i or j (but, of course, not both) refers to an aliased parameter then we are effectively using the statistic b/s.e.(b), where b is the value of the non-aliased parameter, since the aliased parameter has been set to zero.

The F and t method are identical when a single hypothesis is being tested. The second method, however, is unreliable when more than one hypothesis is tested simultaneously. In such a case, there is a possibility that, although the t values for the hypotheses may be individually non-significant, the compound hypothesis that they are true simultaneously (which cannot be tested with a t test) may be highly untenable. This danger increases with increasing correlation between the parameter estimates of the individual hypotheses.

11.4.7 Known parameter values

It is sometimes required to fit a model where some of the β_j in the linear predictor $\eta_i = \sum_j \beta_j x_{ij}$ are fixed in advance. Thus in a simple dilution assay the proportion of fertile tubes π is related to the dilution u by

$$\pi_i = 1 - \exp(-\lambda\, u_i)$$

so that the complementary log-log transformation gives

$$\eta_i = \log(-\log(1 - \pi_i)) = \log \lambda + \log u_i \;.$$

If we write $x_i = \log u_i$ as the covariate and $a = \log \lambda$ as the intercept we have a GLM with

$$\eta_i = a + x_i$$

that is, the slope is fixed at 1. More generally, if a subset of the β_j are fixed, the sum of their contributions to η_i is called an **offset** so that

$$\eta_i = \text{offset} + \sum_j \beta_j x_{ij}$$

where the summation is over the terms for which the β_j are not fixed. In fitting such a model the offset is first subtracted from the linear predictor and the result can then be regressed on the remaining covariates.

11.5 Some notes on approximations

11.5.1 Existence of maximum likelihood estimates

For the classical linear model (in GLIM notation that with Normal distribution and identity link) the likelihood surface is quadratic with respect to the β's and has a unique maximum, except when some covariates in the model are linearly dependent. Such dependence is dealt with by assigning the values of certain (aliased) parameters to zero, and with this convention the maximum likelihood estimates (MLE) are unique. For other GLMs the likelihood surface is not quadratic (though it may be close to being quadratic in the neighbourhood of its maximum); it may have more than one local maximum, or attain its maximum for infinite values of some of the parameters [11.3.4]. Wedderburn (1976) gives a table of the properties of the MLE for the four standard members of the exponential family of distributions handled by GLIM combined with a variety of link functions. In particular the estimates are unique for link functions giving sufficient statistics (the default settings in GLIM) and also for the probit and complementary log-log links with binomial errors. They will also be finite for these cases unless some of the fitted values equal extreme values of the data (zero for Poisson errors, and zero or m for binomial errors). See [11.3.4] for further discussion.

Gamma errors with an identity link give an example where uniqueness cannot be guaranteed. Explicit proof of the existence of more than one maximum in particular cases may be difficult to find. Some empirical checks can be made by starting the fitting process from several different points and checking whether they end by producing the same estimates.

It is possible that the parameter estimates from a fit may take 'impossible' values. For example, when the data are sums of squares whose expectations are linear combinations of variance components, the β's must be non-negative; however their estimates may contain negative values. In such cases it may be useful to re-evaluate the fit of the model with those β's for which b was negative set to zero. GLIM allows this to be done quite simply.

This section has given a practical approach to the possible problems of uniqueness. A more detailed, theoretical discussion is given in Verbeek (1989).

11.5.2 The chi-squared approximation to the scaled deviance

The assessment of the goodness of fit of a GLM requires that the scaled deviance, with its associated d.f., be matched against a theoretical distribution which represents its sampling distribution if the model is true. For the linear model with the Normal distribution this distribution is chi-squared with the appropriate d.f. Note that the scale factor, σ^2, is assumed known; if it is not known then the deviance may be divided by an independent estimate of σ^2, if one is available, in which case the chi-squared distribution is replaced by an F distribution. For all other GLMs the scaled deviance is known to be distributed as chi-squared only asymptotically, and rather little is known about how good the asymptotic approximation is for small sets of data. It seems that the approximation may be better for the difference of two deviances, which expresses the effect of adding a term to a model, than for an absolute deviance expressing the goodness of fit of a single model. Care is needed particularly with binary data, where it is known that the absolute deviance is completely uninformative about the goodness of fit. McCullagh (1985) gives GLIM macros which can help to evaluate the distribution of the Pearson chi-squared statistic for data with Poisson or binomial errors.

With data on counts or proportions where very large samples are involved, the deviance is often larger than expectation, perhaps much larger, even for models which fit well, as judged by the closeness of the fitted and actual values. This happens because with very large samples very small deviations from the model can be detected, deviations so small as to be of no practical importance, although they produce statistically significant results.

A further approximation is involved with discrete data, because the exact distribution of the deviance is discontinuous. Again little firm advice can be given, except that distortions will be worse in the extreme tails of the distribution. As with t statistics, 'exact' p-values should not be attached to scaled deviances from iterative fits, or to ratios of deviances, and the corresponding χ^2 and F distributions should be regarded only as general guides in assessing goodness of fit.

11.5.3 Normality assumptions in t tests on parameters

The use of the t distribution when parameter estimates are compared with their standard errors, as outlined in [11.4.6], is exact for the classical linear model, but is otherwise justified only by asymptotic theory. No general results are known about the adequacy of this approximation for all the other models covered by GLIM, so the standard errors provided must be regarded as only a general guide to the accuracy of the estimates, no attempt being made to provide 'exact' p-values for significance tests. Thus a parameter estimate less than its standard error will usually be insignificant, and one more than 3 times its standard error usually significant; for a better test the parameter concerned should be omitted from the model and the change in deviance assessed (see [11.4.4]). Note also the warning in [11.4.6] about the effect of correlations of estimates on the behaviour of individual t values. When correlations of estimates are small, individual t values can be a useful guide to the importance of different parameters in a model. When correlations are substantial they can be misleading and are no substitute for information from the deviances obtained from fitting various submodels.

11.5.4 Assumption of unit scale parameter

Unlike the Normal and gamma distributions, the Poisson and binomial have no adjustable scale parameter, so that the scaled deviance is calculated with an a priori scale parameter of 1.0. It is not uncommon to find in practice, for example in bioassay, that the residual variation is larger than that expected from binomially distributed data. It is often proposed that the variance function should be amended in such cases to be proportional to the binomial variance, rather than equal to it. The scale parameter is then estimated from the mean deviance (that is, deviance/d.f.) as with the Normal and gamma distributions, and the parameter estimates remain unchanged. Such a model can be justified to some extent by the idea of **quasi-likelihood** [11.8], but its introduction brings in a further element of uncertainty in the inferences made from the model.

With data on counts the variance function may similarly be postulated to be proportional to, rather than equal to, the mean. As an illustration of how this might occur, consider the case where the distribution of the responses was actually negative binomial rather than Poisson so that

$$\mathrm{var}(y) = \mu + \mu^2/k.$$

Then, over a reasonable range of μ, the right-hand side could be approximated by

$$\mathrm{var}(y) = \phi\mu \text{ with } \phi > 1.$$

Thus the use of a GLIM Poisson 'error' combined with an empirical scale parameter could approximate a model with responses which follow a negative-binomial errors distribution. Fitting is carried out exactly as for the Poisson case, except that the mean deviance is used to estimate ϕ and the asymptotic co-variances are adjusted using the estimated scale. (Note that GLIM allows user-defined models, so that a model with a negative binomial variance function with known k could, in fact, be fitted exactly).

11.6 Model checking

If a model is to represent the data adequately it must (i) model the mean–variance relationship adequately and (ii) produce additive effects on the appropriate scale as defined by the link function. A parsimonious model is also required, in which the number of parameters needed for an adequate fit is as small as possible; this means in particular that when the data are cross-classified by a number of factors a parsimonious model will minimize the number of interaction terms needed.

The choice of a particular combination of error and link may result from prior considerations, examination of the data themselves, or more usually a mixture of the two.

11.6.1 Prior considerations

Normal, gamma, and inverse Gaussian distributions are associated with continuous, or effectively continuous, measurements. Many measured quantities are essentially positive, whereas the Normal distribution extends to minus infinity. The assumption of Normality may nevertheless be justifiable provided that zero is well removed from the range of values taken by the observations. Alternatively we can transform the data and consider Normality as applying to some function of the data rather than to the data themselves.

For example, if y is positive, then $0 < y < \infty$. Hence $-\infty < \log(y) < \infty$. Thus for y which is necessarily positive, the assumption of Normality might better be applied to $\log(y)$. As discussed in [10.6.2], there is a close relation between such data transformations and use of different variance functions. Thus an alternative approach to use of a data transformation is to assume a suitable form for the distribution of y; for example, to assume a gamma (or inverse Gaussian) distribution for y, for which the range is positive. The use of such an appropriate error distribution is perhaps to be preferred, particularly as it has the advantage of preserving the original mean–linear predictor relation.

The Poisson distribution applies to data in the form of counts, or averages of counts. However the Poisson variance function may be usefully applied to continuous data if the variance appears to be proportional to the mean.

The binomial distribution applies to data in the form of proportions. The probit and logit link functions usually give very similar fits unless some expected values are very close to the extreme proportions of 0 or 1. Both these links are symmetrical about proportion $\frac{1}{2}$. The complementary log-log link is not symmetrical. It is required for dilution assays [11.4.7] and may be more generally useful for analysing data on disease incidence (see [12.4.1.1]).

11.6.2 Examination of the data

Checking the mean–variance relationship is itself an iterative process because the means must themselves be estimated by fitting a model. A graphical check can be made by plotting the standardized residuals against the fitted values. The scatter should show no obvious trend. The Wedderburn (1974a) example uses a non-standard mean–variance relationship. Pregibon (1981) discusses model-checking techniques for data in the form of proportions; many of his methods can be adapted quite simply to other data types and models.

The possibility of gross errors in the data should also be borne in mind. They may appear (though not always) with large residuals, or the models being fitted may appear to require many high-order interactions. Graphical inspection is always a powerful tool in the detection of gross errors, and their possible existence should always be considered in any model fitting.

It may be required to modify the link function by introducing other, non-linear, parameters into the model. Example [12.4.1.2] illustrates the introduction of an origin into an inverse link. See Copenhaver and Mielke (1977) for a generalisation of bio-assay links which include the logit and probit as special cases. Pregibon (1980) discusses goodness-of-link tests, where the link function contains unknown parameters that are required to be estimated. See also Scallan *et al.* (1984) and McCullagh and Nelder (1989, Chapter 11) for the inclusion of non-linear parameters in GLMs (Chapter 10).

In the following sections, we outline some of the basic tools of model diagnostics. In general, a range of informal and formal techniques should be tried, attempting to find extreme points and/or systematic variation from the fitted model. Model checking is an interactive process. We must first find a linear predictor, for an assumed link function and variance function. We can then try to discover unusual points and possibly eliminate them. A test of the adequacy of the assumed link might then be carried out; unfortunately, if the link is then changed, the 'best' linear predictor may change. A test of the adequacy of the variance function is desirable but, again, changing the variance function may well change the best form of link and, indeed, the best linear predictor.

Model checking is as much an art as a science. Different points may be 'unusual' when we assume different link/variance function combinations. A link test may indicate a poorly fitting link because of a missing interaction or a wrongly scaled covariate, or because of some extreme points. Procedures to isolate extreme points work best if these points are isolated; a small group of such points may be difficult to identify. Thus, in general, there are no hard and fast rules to follow. The procedures below can be helpful but are not exhaustive. McCullagh and Nelder discuss general techniques of model checking (Chapter 12); also see Williams (1987) for a review of GLM methods.

11.6.2.1 Residual plots

Residual plots provide a means of detecting departures from the fitted model. There are a variety of plots which can be considered and, indeed, a variety of ways in which the residuals can be defined. Before discussing these questions, we first summarize some of the most useful plots (see Atkinson, 1985).

(1) A plot of the residuals against (transformed) fitted values. If the variance seems to increase (or decrease) with the (transformed) fitted values, then a different distribution for the response may be appropriate. (Here the required transformation of the fitted values is equivalent to the variance stabilizing transformation for the assumed distribution of the responses).

For the Normal distribution, plot the residuals against $\hat{\mu}$.

For the Poisson distribution, plot the residuals against $2\sqrt{\hat{\mu}}$.

For the gamma distribution, plot the residuals against $2\log\hat{\mu}$.

For the inverse Gaussian distribution, plot the residuals against $-2\,\hat{\mu}^{-\frac{1}{2}}$.

For the binomial distribution, plot the residuals against $2\,\sin^{-1}\sqrt{\hat{\mu}/m}$. The `%angle` function can be used [14.6.1].

(The constant multipliers are not essential but are included by analogy with the Normal case; see, for example, Nelder and McCullagh, 1989, p.399).

The plots should display a mean zero and constant range. No systematic deviations should be observable. This form of plot is unlikely to be helpful for binary data.

(2) A plot of the residuals against each explanatory variate. No clear pattern should be seen. A curvilinear relationship suggests that a higher-order interaction has been omitted.

(3) A Normal (or half-Normal) plot of the residuals [11.6.4].

(4) A plot of residuals against time (this can be a useful idea even if time is not explicitly included in the explanatory variates).

(5) Partial residual plots; here the partial residuals are plotted against each explanatory variate.

11.6.3 Definitions of residuals

Various approaches have been suggested for defining residuals. Unfortunately, there is not a uniform terminology for such residuals. We essentially follow the terminology of McCullagh and Nelder (1989) to define some of the more popular residuals. See Atkinson (1985) for a fuller account. The so-called **Pearson residuals** have traditionally been used in model diagnostic procedures. GLIM therefore provides these Pearson residuals by default; however, other approaches to defining residuals are becoming increasingly widely used. We introduce some of these concepts and give a brief introduction to their use in model checking.

11.6.3.1 Modified Pearson residuals (the residuals of GLIM)

The simplest approach (as output by default with GLIM) is the (Modified) Pearson residual r^P. For each unit i, we have:

$$r^P_i = \frac{y_i - \hat{\mu}_i}{\sqrt{\dfrac{k}{w_i}\,V(\hat{\mu}_i)}} ,$$

where k = scale parameter if set or 1 otherwise. The 'modification' here introduced is the factor k/w_i, which ensures that the denominator is a reasonable estimate of var (y_i).

For example, for the Normal distribution,

$$r^P_i = y_i - \hat{\mu}_i \ .$$

Similarly, for the Poisson distribution,

$$r^P_i = \frac{(y_i - \hat{\mu}_i)}{\sqrt{\hat{\mu}_i}} \quad .$$

Here the $\hat{\mu}_i$ are the GLIM fitted values (stored as `%fv`). The r^P_i are available in GLIM as `%rs`.

11.6.3.2 Standardized residuals

The (modified) Pearson residual suffers from the obvious disadvantage that it does not take into account that the $\hat{\mu}_i$ are merely estimates of μ_i and hence are correlated with the responses y_i. The estimated variance should ideally take into account this correlation. It is therefore desirable to adjust the (modified) Pearson residual by dividing it by a factor $\sqrt{(1-h_i)}$ which compensates for the correlation between y_i and $\hat{\mu}_i$.

Thus we define the standardized Pearson residual

$$r^{PS}_i = \frac{r^P_i}{\sqrt{1-h_i}}$$

$$= \frac{y_i - \hat{\mu}_i}{\sqrt{kV(\hat{\mu}_i)(1-h_i)/w_i}} \quad .$$

The calculation of this estimate requires knowledge of the h_i. These quantities are, in fact, the diagonal entries in the **hat-matrix**, $\mathbf{H} = \mathbf{V}^{-\frac{1}{2}}\mathbf{X}\ (\mathbf{X'V}^{-1}\mathbf{X})^{-1}\ \mathbf{X'V}^{-\frac{1}{2}}$. The h_i may be **extracted** in GLIM as `%lv`, which enables the user to 'standardize' the GLIM residuals.

11.6.3.3 Studentized residuals

In residual plots, we are generally interested in the pattern rather than the size of the plots so it is usually not necessary to scale the residuals by an estimate of the unknown scale parameter ϕ. However, following McCullagh and Nelder (1989), scaled residuals can be defined as **Studentized residuals**. For the standardized (modified) Pearson residuals, the Studentized (scaled) version is:

$$r^{PS'}_i = \frac{r^{PS}_i}{\sqrt{\hat{\phi}}}$$

$$= \frac{r^P_i}{\sqrt{\hat{\phi}(1-h_i)}}$$

$$= \frac{y_i - \hat{\mu}_i}{\sqrt{k\hat{\phi}V(\hat{\mu}_i)(1-h_i)/w_i}} .$$

11.6.3.4 Deviance residuals

The deviance plays a central role in inferential aspects of generalized linear modelling. A deviance residual r^D can be defined as

$$r^D_i = \text{sgn}(y_i - \hat{\mu}_i)\sqrt{d_i}$$

where sgn() indicates that the deviance residual is taken as positive if $y_i - \hat{\mu}_i > 0$, and negative if $y_i - \hat{\mu}_i < 0$. Here the d_i are the deviance increments as given in general form in [11.4.2] and in specific form in [11.9.1]. The values of the d_i may be extracted in GLIM as `%di`.

11.6.3.5 Standardized, Studentized deviance residuals

As with Pearson residuals, it is better to standardize the deviance residuals namely by defining

$$r^{DS}_i = \frac{r^D_i}{\sqrt{1-h_i}}$$

$$= \frac{\text{sgn}(y_i - \hat{\mu}_i)\sqrt{d_i}}{\sqrt{1-h_i}}$$

It may again be convenient to divide these standardized residuals by an estimate of ϕ, to give the Studentized version

$$r^{DS'}_i = \frac{r^D_i}{\sqrt{\hat{\phi}(1-h_i)}}$$

11.6.3.6 Other residuals

Other residual plots can prove useful. For example, Williams (1987) suggests a form of residual which combines the attractive properties of Pearson and Deviance residuals. So-called added variable or partial residual plots can also be helpful in assessing the evidence for the addition or removal of a variable (See Atkinson, 1985).

11.6.4 Normal and half-Normal plots

Normal (or quantile–quantile) plots provide a useful way of displaying residuals, particularly as a check of the Normality assumption of the model. The residuals are sorted into ascending order and are plotted against the expected order statistics of a Normal sample; a linear plot should result. The expected order statistics can be calculated in various ways. McCullagh and Nelder (1989, p.407) suggest that, for a Normal plot, these are calculated as

$$\Phi^{-1}\left(\frac{i-\frac{3}{8}}{n+\frac{1}{4}}\right)$$

For the half-Normal plot, the sorted ***absolute*** values of the residuals are plotted against

$$\Phi^{-1}\left(\frac{n+i+\frac{1}{2}}{2n+\frac{9}{8}}\right)$$

The half-Normal plot can be particularly helpful in detecting non-linearity in the Normal plot. For non-Normal data, transformation of the residuals to approximate Normality may be necessary. The 'Anscombe' residuals are derived for this purpose (see McCullagh and Nelder (1989)), but deviance residuals are often remarkably similar in value. It should, however, be noted that count data (be it Poisson or binomial) will show distortions if there are many zeros (or ones in the case of Binary data). Such data will tend to produce a large number of residuals near zero, which may distort the plot.

11.6.5 Influence and leverage: Cook's distance

Points with large residuals are certainly worthy of further investigation. However, there can be points with small residuals which nevertheless have a marked effect upon the parameters of our linear predictor. In order to consider this phenomenon, we introduce the concepts of **leverage** and **influence**. We do so by reference to the simple (Normal distribution) regression model with one explanatory variate x, say.

Fig. 11.1

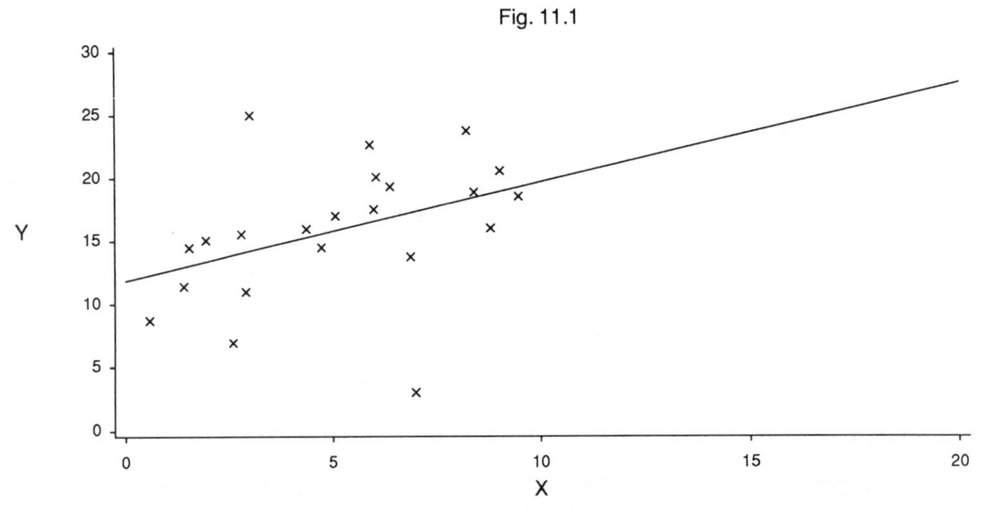

It is quite possible that all the values of the explanatory variate x are spread reasonably evenly over all the range of x (see Fig. 11.1). However, it might be the case that one value of the explanatory variate x is a long way from the rest of the values. This isolated value has a clear chance of being important in the fit; such a point is said to have a high **leverage**. Indeed, we could define the leverage of a point (unit) as a measure of how far the x-value of the point is away from the average of the rest of the x-values. Thus, more generally with p explanatory variates, we define the leverage of a p-dimensional point in the design space as a measure of how far the point is away from the centroid of the other points in the design space.

Having a high leverage does not necessarily imply that the point in question will have any marked effect upon the regression line. It can do (Fig. 11.3), or it may not (Fig. 11.2). This depends upon the magnitude of the observed response for the point in question, not just on the explanatory variates. If the removal of the unit would greatly change the regression line, then it is said to be **influential**. The extreme point in Fig. 11.2 is not influential, whereas the extreme point in Fig. 11.3 is influential.

Fig. 11.2

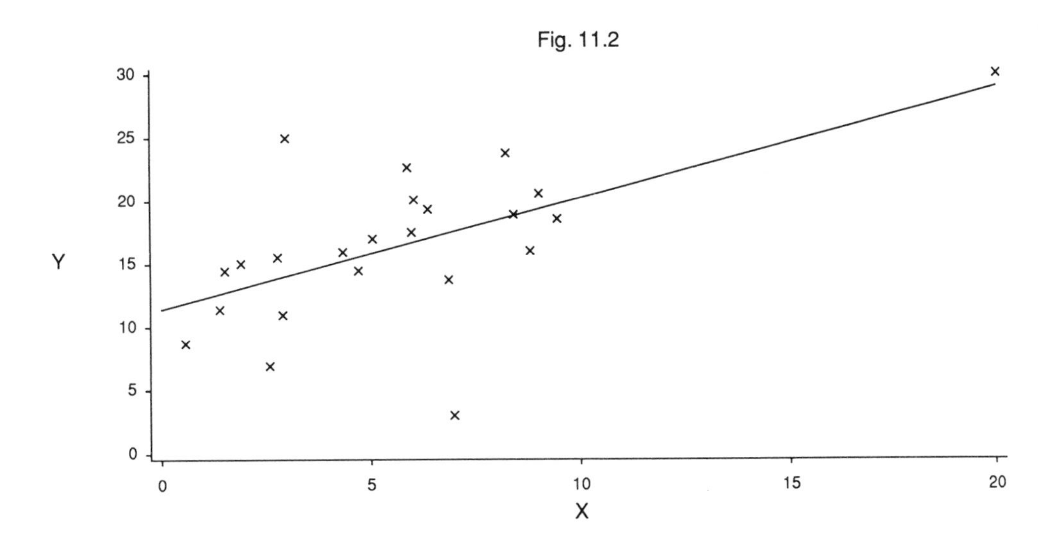

Fig. 11.3

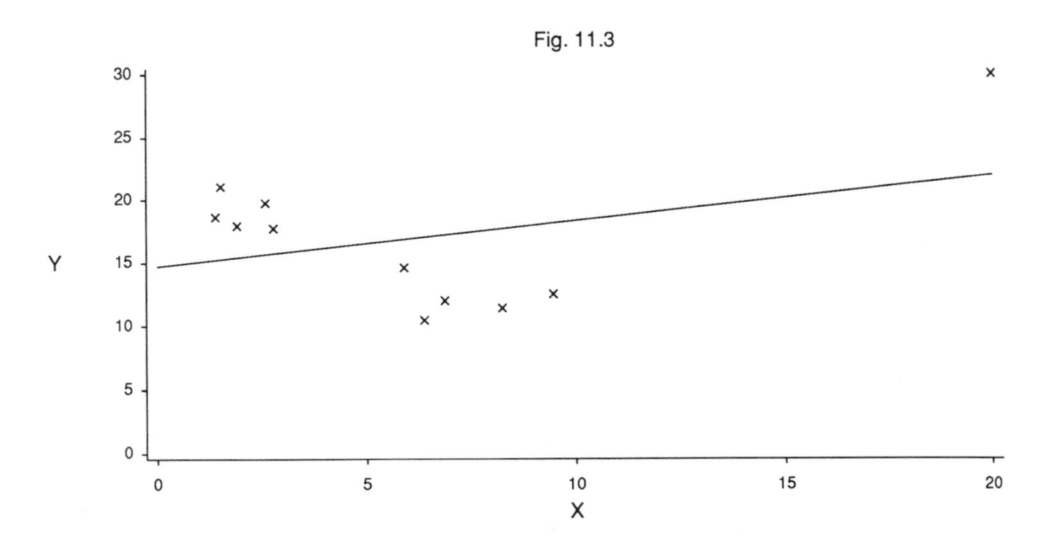

A measure of leverage and influence is required. In the ordinary (Normal distribution) regression case, leverage can be defined as the distance of a p-dimensional point in the design space from the p-dimensional centroid of the other points. Thus, in this case, we can show that we can define the leverage h_i of a given point in the design space as the ith diagonal entry of the **hat-matrix**, $\mathbf{X(X'X)^{-1}X'}$. Generalizing this to generalized linear models, the corresponding hat-matrix is

$$\mathbf{H} = \mathbf{V}^{-\frac{1}{2}}\mathbf{X}(\mathbf{X}'\mathbf{V}^{-1}\mathbf{X})^{-1}\mathbf{X}'\mathbf{V}^{-\frac{1}{2}},$$

and we define the leverage of a point with suffix i in the design space as the ith diagonal entry of this matrix.

Before moving on to define a measure of influence, we may note that the sum of the diagonal elements of $\mathbf{H}$, namely trace($\mathbf{H}$), is given by trace($\mathbf{H}$) = p. Thus an average value of h_i is p/n, which enables us to have some idea as to whether a point has a large leverage or not. Typically, a leverage greater than $2p/n$ might be taken to warrant further investigation.

Leverage alone cannot tell us if our linear predictor is being affected strongly by a given unit. A measure which does this is **Cook's distance.**

Cook (1977) introduced the following statistic c_i as a measure of influence:

$$c_i = \frac{h_i}{p\,(1-h_i)} \, (r^{PS'}_i)^2.$$

The values of c_i can be extracted in GLIM as `%cd`. In passing, we may note that we could use an alternative definition based upon the deviance residuals, namely

$$c^D_i = \frac{h_i}{p\,(1-h_i)} \, (r^{DS'}_i)^2 .$$

A search for influential points is carried out by looking for large values of Cook's distance. Unfortunately, no clear rules can be given for what constitutes a large value of c_i. Nevertheless, provided that the sample size is not too large, a simple index plot of the statistic against case number can be useful in determining which are the largest values. Alternatively, the c_i can be sorted, whilst preserving the case number, so that the largest can be found. Cases with large values of c_i can then be weighted out of the analysis and a test carried out of the change in scaled deviance. Significant changes in scaled deviance lead us to consider the status of the case in question.

11.6.6 Testing link functions

The fitting of a generalized linear model is an iterative process. It is first necessary to decide upon an initial variance function and an initial link function. Having found an appropriate linear predictor, it is then desirable to use it to check the validity of the original assumption regarding the form of link. Since application of an inappropriate link function would result in a non-linear η, a simple, though not very powerful, technique is to assess the change in fit by adding $\hat{\eta}^2$ as an extra covariate. Pregibon (1981) considered the link function as being embedded in some parametric family of links and showed how a test of the link function could be carried out by fitting auxiliary variables.

Pregibon's technique essentially finds a first step approximation to the maximum likelihood estimates of the extra parameters in the link function. However, for reasonably sized problems, there is no reason why we should not search for the full MLEs of the extra parameters. An approach well suited to GLIM is that of **profile likelihood.** The link function $\eta = g(\mu)$ is considered to belong to a parametric family; for example, it might be assumed that $\eta = g(\mu, a, b)$, for some unknown parameters a, b. The model with given variance function and linear predictor can be fitted for different values of the parameters a and b and the overall maximum likelihood (minimum scaled deviance) can be found by inspection. Interval estimates can then be found for the parameters of interest and a test carried out of the validity of the original assumption that $\eta = g(\mu)$.

As a simple illustration of this technique for testing the adequacy of either the reciprocal link, the identity link or the log-link, we might assume that the link function is a member of the one parameter Box–Cox family, for which

$$\eta = (\mu^a - 1)/a, \quad a \neq 0,$$
$$\eta = \log \mu, \qquad a = 0.$$

The minimum scaled deviance (maximum likelihood) estimate of a can then be found by plotting the deviance (or likelihood) as a function of a, as illustrated in Fig. 11.4.

Fig. 11.4

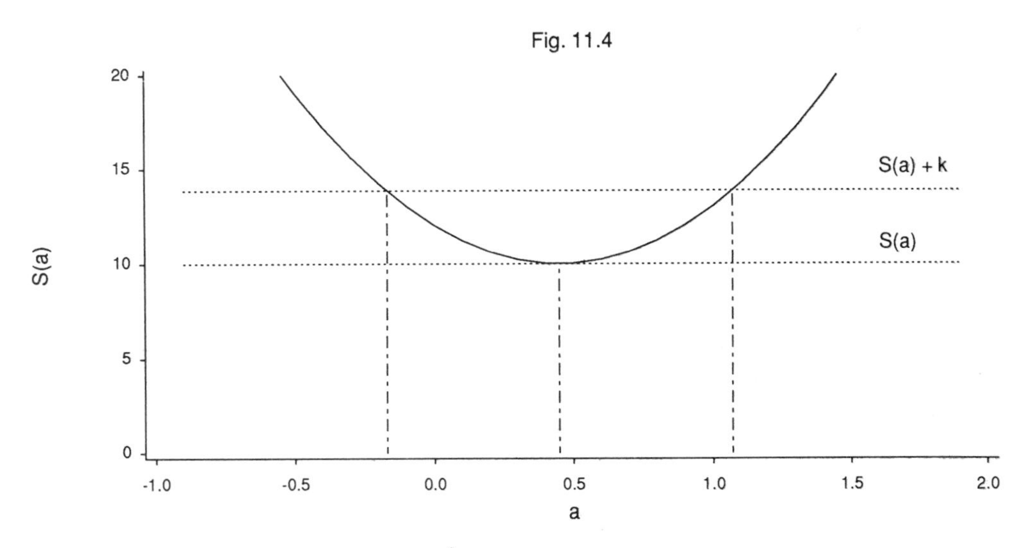

The maximum likelihood estimate $\hat{a}$ can be obtained by visual inspection of this graph. Using a chi-squared approximation to the distribution of the scaled deviance S, a 95 per cent interval estimate for the parameter a is then given by the set of values of a such that

$$S(a) - S(\hat{a}) < 3.84.$$

For cases where the scaled deviance equals the deviance (such as the Poisson distribution), this interval estimate is

$$\{ a \mid D(a) < D(\hat{a}) + 3.84 \}.$$

For examples where the scale ϕ needs to be estimated, we might use

$$\hat{\phi} = D(\hat{a})/d$$

where d is the degrees of freedom of the model being fitted. This then gives the approximate 95 per cent interval estimate of a as

$$\{ a \mid D(a) < D(\hat{a})(1 + 3.84/d) \} .$$

11.6.7 Checking the variance function

A plot of the residuals against fitted values can provide a visual check of the validity of the assumed mean–variance relation. The fitted values might be first transformed as discussed in [11.6.2]. The absolute value of the residuals might be helpful in looking for a trend in the residuals. A positive slope would indicate that the assumed variance function does not vary enough with increasing μ, so a variance function with a higher power of μ might be tried. Similarly, with a negative trend, a lower power should be tried.

As discussed in [10.6.2], a variance stabilizing data transformation might be applied to the data, perhaps estimating the best such transformation from some appropriate family (such as the Box–Cox family of data transformations). Such a transformation simultaneously changes both the variance and the link and it is indeed fortuitous if both can be successfully transformed. Thus, although such data transformations can be informative in helping to identify an appropriate form for the variance function, it is generally preferable to treat the variance function and the link separately. Formal model checking of the variance function can be achieved by embedding the variance function in some family. For example, we might assume $V(\mu) = \mu^{\alpha}$ for some unknown α, and observe how the fit varies for different values of α. This technique requires the use of **extended quasi-likelihood**, as discussed in [11.8.1]. See also McCullagh and Nelder (1989, p.400).

11.7 Composite link functions

The discussion of [11.2.3] defined a key element of a GLM as being the link function $\eta_i = g_i(\mu_i)$. This concept can be extended by allowing μ_i to relate to more than one of the η_i. This so-called **composite link** technique, first suggested by Thompson and Baker (1981), allows a wide range of models to come into the GLM framework.

In the simplest form of composite link, we might assume that the vector $\boldsymbol{\mu}$ is related to the vector $\boldsymbol{\eta}$ by the matrix equation

$$\boldsymbol{\mu} = \mathbf{C}\boldsymbol{\gamma}$$

where $\boldsymbol{\gamma} = g(\boldsymbol{\eta})$, $\boldsymbol{\eta} = \mathbf{X}\boldsymbol{\beta}$ and $\mathbf{C}$ is a known n x m matrix.

Thompson and Baker showed that the weighted least squares technique of [11.3.1.3] could be used to find the maximum likelihood estimates of the μ_i for this formulation, by defining an iterative model matrix **CHX** and working variate **CHz**, where $\mathbf{H} = [\partial\boldsymbol{\gamma}/\partial\boldsymbol{\eta}]$. The required weight matrix in the iterative procedure is given by $\mathbf{U} = \text{diag}(1/\text{var}(y_i))$.

The composite link technique thus requires the re-calculation of the model matrix at each iteration, which has restricted its use in GLIM to models in which **C** has a comparatively simple form.

Thompson and Baker introduce composite links by means of a genetics example; the data consist of phenotype counts which are the sums of frequencies of subsets of the cells of a table of genotype frequencies, where this table of genotype frequencies cannot be observed directly. Burn (1982) discusses how a wider range of estimation problems in genetics can be similarly handled. Thompson and Baker discuss how grouped Normal data (that is, histogram type data) can be legitimately analysed via composite links; the composite link is needed as the data (the frequencies in each cell) can be expected to depend upon the two end-points of the respective cells. Scallan and Evans (1989) show how the composite link approach can be used to handle the estimation of the parameters of truncated distributions.

A natural although more complicated formulation allows the matrix **C** to contain extra parameters. Thus the estimation procedure is a composite link involving extra parameters not contained in the usual linear predictor. Such a model might be referred to as a ***parametric composite link***. One application of such parametric composite links is for the analysis of grouped data where the data are assumed to come from a finite distribution mixture (Scallan and Evans, 1992). In this case, the expected frequency (count) in each cell depends upon the end-points of the cell, as with Thompson and Baker's original formulation. However, the cumulative frequency distribution which determines the probability of an observation being classified into a given cell is now a finite distribution mixture, so involves mixing parameters as well as the standard parameters of each component of the mixture. For example, if the underlying components each have two parameters m_i and v_i (for example, location and scale), then the cumulative frequency distribution $F(x)$ might be defined as

$$F(x) = \sum_{i=1}^{q} a_i F_i(x, m_i, v_i),$$

where the a_i, $i = 1,...,q$ are mixing parameters such that $\sum_i a_i = 1$.

The expected frequencies can be expressed in the composite link form, with the **C** matrix being a function of the unknown a_i.

11.7.1 Non-independent Normal responses

A further application of composite links enables the modelling of response data which is Normally distributed but correlated. Thus the use of composite links allows us to move away from the assumption that the likelihood of the responses can be written as the product of two functions, one of which is a function of the data alone and the other having the form of a likelihood of independent observations; more specifically, by transforming the (non-independent) Normal responses to obtain independent transformed data. However, the data transformation induces an equivalent transformation of the mean of the data. This then implies a composite link between the mean of the new responses and the linear predictor.

This approach, in theory, applies quite generally to correlated Normal responses but it is really only tractable for simple forms of covariance matrix. For example, it has been applied by Scallan (1985) to the fitting of first-order autoregressive processes and by Forcina (1986) to the case where the eigenvectors of the correlation matrix are independent of the unknown parameters; see also Gilchrist (1987). It is also tempting to try to apply this technique to correlated responses which are not Normally distributed. There are, however, difficulties with this, as we shall see in Section [11.8.2]. The technique effectively involves an extra iterative procedure to estimate the unknown parameter(s) in the covariance matrix (for example, the parameter ρ in the first order autoregressive process). This implies that we need to amend the covariance matrix of the estimates of the β's as output by GLIM. A suitable method is given in Richards (1961).

11.8 Quasi-likelihood

In [11.3.1.5], it was noted that GLIM's fitting algorithm makes use of the mean and the variance of the responses y_i but does not use any of the higher moments of the distribution of the y_i. This leads to the idea of an estimation procedure which depends only upon the mean and variance of the responses.

Wedderburn (1974b), seems originally to have suggested this so-called **quasi-likelihood** approach; a discussion of quasi-likelihood and GLIM is given in Nelder (1985).

To derive the quasi-likelihood approach, it is noted again that the original GLM density is of the form

$$l_i = \log p(y_i) = \frac{y_i\theta_i - b(\theta_i)}{a_i(\phi)} + c(y_i, \phi) \tag{11.8}$$

for suitable choice of a_i, b, and c.

It thus follows that for this density

$$\frac{\partial l_i}{\partial \mu_i} = \frac{\partial l_i}{\partial \theta_i} \frac{\partial \theta_i}{\partial \mu_i}$$

$$= \frac{y_i - b'(\theta_i)}{a_i(\phi)} \frac{1}{b''(\theta_i)}$$

$$= \frac{y_i - \mu_i}{\text{var}(y_i)}$$

where primes denote differentiation.

In the quasi-likelihood approach we may assume more generally that we have a **quasi-likelihood** Q defined by

$$Q(y, \mu) = \int_y^\mu \frac{y - \mu}{\phi V(\mu)} \, d\mu, \quad \text{where } \text{var}(y) = \phi \, V(\mu),$$

for some scale/dispersion parameter ϕ. (We have omitted any prior weight, for convenience. If they exist, they can be re-introduced wherever V occurs.)

Of course, if a distribution exists within the exponential family with variance function $V(\mu)$, then the quasi-likelihood Q is effectively equivalent to a standard (log) likelihood function (differing only by a function of the y_i alone). However, more generally, we see that we need only specify the first two moments μ and $\text{var}(y) = \phi V(\mu)$ and maximize Q to obtain the so-called **maximum quasi-likelihood (MQL)** estimates.

Such MQL estimates have properties which are similar to those of the maximum likelihood estimates (Firth, 1987; Godambe and Heyde, 1987). However, assuming a general form of variance will not correspond to a true density of the form of eqn (11.8) above.

The function $Q(y, \mu)$ is referred to as the (log) quasi-likelihood for μ based upon the data y. We (initially) assume the components are independent, so the QL for the whole data is

$$Q(\mathbf{y}, \boldsymbol{\mu}) = \sum_{i=1}^{n} Q(y_i, \mu_i).$$

The corresponding quasi-deviance for each observation is usually written as $D(y_i, \hat{\mu}_i)$ where

$$D(y_i, \hat{\mu}_i) = -2 \, \phi \, Q(y_i, \hat{\mu}_i)$$

$$= 2 \int_{\hat{\mu}_i}^{y_i} \frac{y - \mu}{V(\mu)} \, d\mu.$$

The total quasi-deviance $D(\mathbf{y},\hat{\boldsymbol{\mu}}) = \sum_i D(y_i,\hat{\mu}_i)$ is a genuine statistic, as it contains no unknown parameters; that is, it is a function of the y_i and the $\hat{\mu}_i$, but not of the unknown scale ϕ.

MQL estimates can be obtained from GLIM by supplying GLIM with the appropriate variance function via the `ERROR` directive. A user-defined quasi-deviance must also be specified. One slight difference from ML estimation is that, in MQL, the scale ϕ should certainly be estimated from the generalized Pearson statistic X^2, rather than from the mean residual (quasi-) deviance. Thus ϕ is estimated by

$$\hat{\phi} = \frac{\sum_i \{(y_i - \hat{\mu}_i)/V_i(\hat{\mu}_i)\}}{n-p}$$

$$= \frac{X^2}{n-p} \, .$$

11.8.1 Estimation of the variance function: extended quasi-likelihood

The theoretical discussion has considered models where the variance function is assumed to be of a known form. In general, this will not be true, although an educated guess at a suitable form may be made and used to form initial estimates of the means μ_i, $i = 1, \ldots,$ n. The residuals from such a fit can be used to assess the acceptability of the chosen mean–variance relation [11.6]. Alternatively the mean–variance relation could be estimated by a method known as **extended quasi-likelihood** (Nelder and Pregibon, 1987). Essentially, this method is quasi-likelihood with a modification to allow for the comparison of the quasi- deviances which are here calculated from distributions with different variance functions (it is, for example, clear that deviances from, for example, Normally distributed data cannot be directly compared with deviances from gamma distributed data, and so on). In other words, the extended quasi-likelihood must have the properties of a log-likelihood when considered as a function of ϕ.

The quasi-likelihood Q is therefore **extended** by defining Q^+ as

$$Q^+ = Q - \tfrac{1}{2} \log \{2\pi\phi V(y)\}.$$

(Prior weights have been omitted for convenience; they can be re-introduced wherever V occurs).

It may be observed that the definition of Q^+ has essentially used a Normal-like approximation to the overall density, but with $V(\mu)$ replaced by $V(y)$. For example, for gamma distributed data, $V(y) = y^2$. This **extended quasi-likelihood** Q^+ is, in fact, the saddle point approximation to the members of the exponential family, provided the distribution exists. It ensures that Q^+ behaves as a quasi- likelihood when treated as a function of ϕ.

In order to compare two variance functions, we need to define an **extended quasi-deviance** D^+. Multiplying Q^+ by –2 and noting that

$$D = 2\int_{\mu}^{y} \frac{y-\mu}{V(\mu)}\, d\mu$$

gives

$$D^{+}(\mathbf{y},\hat{\mu}) = \frac{D(\mathbf{y},\hat{\mu})}{\phi} + \sum_{i=1}^{n} \log(2\pi\phi V(y_i)).$$

$$= \frac{\Sigma\, D(y_i,\hat{\mu}_i)}{\phi} + \Sigma \log(2\pi\phi V(y_i)).$$

We may note that Nelder (1985) suggests replacing y by $y + \frac{1}{6}$ in $V(y)$ for discrete distributions to avoid problems when y=0.

A possible approach to testing variance functions for continuous observations (responses) is to use the family $V(\mu) = \mu^{\alpha}$ (see, for example, Jørgensen (1987) and the discussion of that paper).

Simultaneous modelling of the mean and variance is a natural extension of extended quasi-likelihood; this is similar to the increasingly popular **Taguchi methods**. A brief summary of a GLIM oriented approach is as follows.

For Normally distributed data, the residual sum of squares is known to be chi-squared distributed. Thus it is simple to estimate simultaneously the mean μ_i and the variance σ_i^2 of Normally distributed responses by alternating between estimation of the μ_i and σ_i^2, using Normal errors for the data y_i and gamma (that is, scaled chi-squared) errors for the residual sum of squares. A separate linear predictor and link can be used for μ_i and σ_i^2. It is well known and convenient to note that the sample mean and sum of squares are independent. Implementation of the approach in GLIM for Normal responses is discussed by Aitkin (1987).

An obvious generalization of this approach for general (non-Normal) data is to model the mean and variance simultaneously by a similar alternating procedure with the means μ_i being modelled given the dispersion ϕ_i and dispersion ϕ_i being modelled with a gamma quasi-likelihood using the deviance components from the mean fit. Some fine tuning of the gamma assumption is desirable to allow for the non-Normality of the underlying data; see, for example, McCullagh and Nelder (1989, p.361).

11.8.2 Quasi-likelihood for dependent observations

Our theoretical discussion has so far required the assumption of the independence of the observations y_i, $i = 1,...,n$, or, at least, that the likelihood can be written in a form which can be treated as equivalent to a likelihood of independent data.

Quasi-likelihood seems to offer one way of allowing dependence in a GLM framework. Thus, we might assume that we have a vector $\mathbf{y}$ of observations, such that the variance–covariance matrix of $\mathbf{y}$ is given by $\phi\mathbf{V}(\boldsymbol{\mu})$, where $\mathbf{V}$ is a real, symmetric positive definite n x n matrix of known function $v_{ij}(\boldsymbol{\mu})$.

The quasi-likelihood might be defined as a function Q satisfying

$$\frac{\partial Q}{\partial \boldsymbol{\mu}} = \mathbf{V}(\boldsymbol{\mu})^{-}(\mathbf{y} - \boldsymbol{\mu})/\phi,$$

where $\mathbf{V}(\boldsymbol{\mu})^{-}$ is some generalized inverse of $\mathbf{V}(\boldsymbol{\mu})$. However, it should be noted that, for correlated y_i, this differential equation will not necessarily lead to a well-defined quasi-likelihood. Put more mathematically, we note that the quasi-estimating equations for $\boldsymbol{\beta}$ are obtained by differentiating $Q(\mathbf{y},\boldsymbol{\mu})$ to obtain $\mathbf{U}(\boldsymbol{\beta}) = \mathbf{O}$, where

$$\mathbf{U}(\boldsymbol{\beta}) = \mathbf{D}'\mathbf{V}^{-1}(\mathbf{y} - \boldsymbol{\mu})/\phi,$$

where $\mathbf{D}$ is an n x p matrix with elements given by $D_{ij} = \partial\mu_i/\partial\beta_j$.

But if this score vector $\mathbf{U}(\boldsymbol{\beta})$ is to be the gradient vector of a quasi-likelihood or log-likelihood, then it is necessary and sufficient that the derivative matrix of $\mathbf{U}(\boldsymbol{\beta})$ be symmetric with respect to $\boldsymbol{\beta}$ (Borre and Lauritzen, 1989). This condition debars many choices of $\mathbf{V}(\boldsymbol{\mu})$ if a true quasi-likelihood is required. For example, the simple case of two correlated Poisson variates y_1, y_2 with $E(y_i) = \mu_i$, $\text{var}(y_i) = \mu_i$ and $\text{cov}(y_i,y_j) = \rho\sqrt{(\mu_i\mu_j)}$ does not lead to a symmetric form unless $\rho = 0$; that is, unless they are independent (which rather defeats the purpose).

It may be concluded that some caution is required when using quasi-likelihood with dependent y_i. For example, it is tempting to assume the form $\mathbf{V}(\boldsymbol{\mu}) = \mathbf{V}^{\frac{1}{2}}\, \Sigma\, \mathbf{V}^{\frac{1}{2}}$ but this will not generally give a well-defined quasi-likelihood. Nevertheless, it is quite possible to obtain estimates with this form of $\mathbf{V}$; however, such estimates then require further justification in use. Indeed, Liang and Zeger (1986) and Zeger and Liang (1986) use this sort of approach for binomial data, giving an alternative rationale for the validity of their estimates.

McCullagh and Nelder (1989, p.335) give a form for $\mathbf{V}^{-1}$ which guarantees that $\mathbf{U}(\boldsymbol{\beta})$ has a symmetric derivative matrix when differentiated with respect to $\boldsymbol{\beta}$. Their condition can sometimes be verified for certain covariance functions but it would seem to be of limited general applicability.

11.9 Technical considerations

11.9.1 The exponential family

This section gives the form of the constants a_i, b, and c in the probability density function

$$p(y_i) = \exp\left\{\frac{y_i\theta_i - b(\theta_i)}{a_i(\phi)} + c(y_i,\phi),\right\}$$

as introduced in [11.2.1].

The **scale parameter**, ϕ, is assumed constant for all i. Moreover, the function $a_i(\phi)$ will be assumed to be of the form ϕ/w_i where the w_i, the **prior weights**, are known.

We have noted that, if we assume the above form for the density of y_i, then the mean and variance of y_i can be expressed in terms of θ and ϕ as follows:

$$\mu_i = \mathrm{E}(y_i) = b'(\theta_i)$$

and

$$\mathrm{var}(y_i) = a_i(\phi)\, b''(\theta_i) = \frac{\phi}{w_i} V(\mu_i) = \frac{\phi}{w_i} \tau_i^2$$

where $b''(\theta_i) = V(\mu_i) = \tau_i^2$ is the **variance function**. It is a function of μ_i only.

This section also shows the **canonical link** (the default link in GLIM) for each distribution. The canonical link is defined by $\theta_i = \eta_i$, where θ_i is the canonical parameter. Also included is the form of the (scaled) deviance for each distribution.

The following sections discuss the Normal distribution, the Poisson distribution, the gamma distribution, the inverse Gaussian distribution, and the binomial distribution.

11.9.1.1 The Normal/Gaussian distribution

(i) The y_i, $i = 1, \ldots, n$ are assumed to be independent and Normally distributed with means μ_i, $i = 1, \ldots, n$ and variance σ^2/w_i, $i = 1, \ldots, n$.

(ii) This may be written

$$y_i \overset{\text{ind}}{\sim} \mathrm{N}(\mu_i, \sigma^2/w_i),\ i=1,\ldots,n.$$

(iii) The (log) density may be written

$$\log f(y_i) = \frac{w_i}{\phi}\left(y_i\mu_i - \tfrac{1}{2}\mu_i^2\right) - \frac{1}{2}\left(\log\frac{2\pi\phi}{w_i} + \frac{w_i y_i^2}{\phi}\right).$$

(iv) The range of y_i is $(-\infty,\infty)$.

(v) The scale (dispersion) parameter $\phi = \sigma^2$.

(vi) The canonical/natural parameter is $\theta_i = \mu_i$.

(vii) The cumulant function $b(\theta_i) = \frac{1}{2}\theta_i^2$.

(viii) $$c(y_i, \phi) = -\frac{1}{2}\left(\log\frac{2\pi\phi}{w_i} + \frac{w_i y_i^2}{\phi}\right)$$

(ix) $\mu_i(\theta_i) = \mathrm{E}(y_i;\theta_i) = \theta_i$.

(x) The canonical/default link is $\theta_i = \eta_i$; that is, $\mu_i = \eta_i$.

(xi) The variance function $V(\mu_i) = 1.0$.

(xii) The deviance $D = \sum_i w_i (y_i - \hat{\mu}_i)^2$.

11.9.1.2 The Poisson distribution

(i) The y_i, $i = 1,\ldots,n$ are assumed to be independent and Poisson distributed with means μ_i, $i = 1,\ldots,n$.

(ii) This may be written

$$y_i \overset{\text{ind}}{\sim} P(\mu_i),\ i = 1,\ldots,n,$$

or, alternatively, as

$$P[Y_i = y_i] = \frac{e^{-\mu_i} \mu_i^{y_i}}{y_i!} \quad ,\ i=1,\ldots,n, \text{ for independent } Y_i.$$

(iii) The (log) density can be written

$$\begin{aligned} \log f(y_i) &= \log P[Y_i = y_i] \\ &= y_i \log \mu_i - \mu_i - \log y_i! \end{aligned}$$

(iv) The range of y_i is : (0, 1, 2, …); that is, the non-negative integers.

(v) The scale (dispersion) parameter is 1.0.

(vi) The canonical/natural parameter is $\theta_i = \log \mu_i$.

(vii) The cumulant function $b(\theta_i) = \mu_i = \exp(\theta_i)$.

(viii) $c(y_i, \phi) = -\log y_i!$

(ix) $\mu_i(\theta_i) = \mathrm{E}(y_i; \theta_i) = \exp(\theta_i)$.

(x) The canonical/default link is $\theta_i = \eta_i$; that is, $\log \mu_i = \eta_i$.

(xi) The variance function $V(\mu_i) = \mu_i$.

(xii) The scaled deviance $S = \sum_i w_i \left\{ y_i \log \frac{y_i}{\hat{\mu}_i} - (y_i - \hat{\mu}_i) \right\}$.

11.9.1.3 The gamma distribution

(i) The y_i, $i=1,...,n$ are assumed to be independent and gamma distributed with means μ_i, $i=1,...,n$ and variances $w_i\mu_i^2/\phi$.

(ii) This may be written

$$y_i \overset{\text{ind}}{\sim} G(\mu_i, \frac{\phi}{w_i}),\ i = 1, ..., n.$$

(iii) The (log) density can be written

$$\log f(y_i) = \frac{w_i}{\phi}\left\{y_i\left(-\frac{1}{\mu_i}\right) - \log\mu_i\right\} + \left\{\frac{w_i}{\phi}\log\left(\frac{w_i y_i}{\phi}\right) - \log y_i - \log\Gamma(w_i/\phi)\right\}$$

(iv) The range of y_i is $(0,+\infty)$; that is, the positive real line.

(v) The scale (dispersion) parameter is ϕ.

(vi) The canonical/natural parameter is $\theta_i = -1/\mu_i$.

(vii) The cumulant function $b(\theta_i) = \log\mu_i = -\log(-\theta_i)$.

(viii) $c(y_i, \phi) = \frac{w_i}{\phi}\log\left(\frac{w_i y_i}{\phi}\right) - \log y_i - \log\Gamma\left(\frac{w_i}{\phi}\right)$

(ix) $\mu_i(\theta_i) = \mathrm{E}(y_i;\theta_i) = -1/\theta_i$.

(x) The canonical/default link is $\theta_i = \eta_i$; that is, $-1/\mu_i = \eta_i$, which is equivalent to a reciprocal link.

(xi) The variance function $V(\mu_i) = \mu_i^2$.

(xii) The deviance $D = \sum_i w_i\left\{-\log\left(\frac{y_i}{\hat{\mu}_i}\right) + \frac{y_i - \hat{\mu}_i}{\hat{\mu}_i}\right\}$

11.9.1.4 The inverse Gaussian distribution

(i) The y_i, $i = 1,...,n$ are assumed to be independent and distributed as the inverse Gaussian distribution with means μ_i, $i = 1,...,n$ and variances $\sigma^2\mu_i^3/w_i$.

(ii) This may be written

$$y_i \overset{\text{ind}}{\sim} IG\left(\mu_i, \frac{\sigma^2}{w_i}\right),\ i = 1,...,n.$$

(iii) The (log) density may be written

$$\log f(y_i) = \frac{w_i}{\phi}\left\{ y_i\left(-\frac{1}{2\mu_i^2}\right) + \frac{1}{\mu_i}\right\} - \frac{1}{2}\left\{\log\left(\frac{2\pi\phi y_i^3}{w_i}\right) + \frac{w_i}{\phi y_i}\right\}.$$

(iv) The range of y_i is $(0, \infty)$.

(v) The scale (dispersion) parameter $\phi = \sigma^2$.

(vi) The canonical/natural parameter is $\theta_i = -\dfrac{1}{2\mu_i^2}$.

(vii) The cumulant function $b(\theta_i) = -1/\mu_i = -\sqrt{-2\theta_i}$

(viii) $c(y_i, \phi) = -\dfrac{1}{2}\left\{\log(2\pi\phi y_i^3 w_i) + \dfrac{w_i}{\phi y_i}\right\}$.

(ix) $\mu_i(\theta_i) = E(y_i; \theta_i) = (-2\theta_i)^{-\frac{1}{2}}$.

(x) The canonical/default link is $\theta_i = \eta_i$; that is, $-1/(2\mu_i^{\,2}) = \eta_i$, which is equivalent to the inverse quadratic link.

(xi) The variance function $V(\mu_i) = \mu_i^3$.

(xii) The deviance $D = \sum_i w_i \left\{\dfrac{(y_i - \hat{\mu}_i)^2}{\hat{\mu}_i^2 y_i}\right\}$.

11.9.1.5 The binomial distribution

(i) The y_i, $i=1,\ldots,n$ are assumed to be independent and binomially distributed, each y_i corresponding to m_i trials, with probability p_i of success for each trial, $i=1,\ldots,n$.

(ii) This may be written

$$y_i \overset{\text{ind}}{\sim} B(p_i, m_i) \quad , i = 1, \ldots, n,$$

or, alternatively, as

$$P[Y_i = y_i] = {}^{m_i}Cy_i \; p_i^{\,m_i}(1 - p_i)^{m_i - r_i} \quad , i = 1, \ldots, n,$$

with the y_i being independent.

(iii) The (log) density can be written

$$\log f(y_i) = y_i \log\left(\frac{p_i}{1-p_i}\right) + m_i \log(1-p_i) + \log {}^{m_i}Cy_i$$

(iv) The range of y_i is $(0, 1, 2, ..., m_i)$.

(v) The scale (dispersion) parameter is 1.0.

(vi) The canonical/natural parameter is $\theta_i = \log\left(\frac{p_i}{1-p_i}\right)$.

(vii) The cumulant function $b(\theta_i) = \mu_i = m_i \log(1+\exp(\theta_i))$.

(viii) $c(y_i, \phi) = -\log {}^{m_i}Cy_i$.

(ix) $\mu_i(\theta_i) = E(y_i; \theta_i) = m_i \dfrac{\exp(\theta_i)}{1+\exp(\theta_i)}$.

(x) The canonical/default link is $\theta_i = \eta_i$; that is, $\log\left(\frac{\mu_i}{m_i-\mu_i}\right) = \eta_i$.

(xi) The variance function $V(\mu_i) = m_i p_i(1-p_i) = \dfrac{\mu_i(m_i-\mu_i)}{m_i}$

(xii) The scaled deviance $S = \sum_i w_i \left\{ y_i \log\left(\frac{y_i}{\hat{\mu}_i}\right) + (m_i-y_i)\log\left(\frac{m_i-y_i}{m_i-\hat{\mu}_i}\right)\right\}$.

11.9.2 The numerical methods

The vast majority of GLIM applications require no knowledge or manipulation of the numerical method that is used and details should, as far as possible, remain hidden from the user. Nevertheless there are some cases where more understanding is necessary to make the best use of GLIM.

The basic ideas can be quite easily explained. If the model matrix is $\mathbf{X}$, the response variate $\mathbf{y}$, and the parameters $\boldsymbol{\beta}$, consider a single iteration and, for simplicity of exposition, ignore the iterative weight (for the weighted version, with diagonal weight matrix $\mathbf{U}$, replace $\mathbf{X}$ by $\mathbf{U}^{\frac{1}{2}}\,\mathbf{X}$ throughout in the following). Familiar theory leads to the normal equations

$$\mathbf{X'X}\boldsymbol{\beta} = \mathbf{X'y}$$

which can then be solved for the parameter estimates. The problem with straightforward applications of this theory is that $\mathbf{X'X}$, and subsequently $(\mathbf{X'X})^{-1}$, are explicitly formed and hence if $\mathbf{X}$ is at all ill-conditioned $(\mathbf{X'X})^{-1}$ can be very poorly determined.

A better method than explicitly forming the sums of squares and products is to perform some cancellation analytically in the normal equations by first expressing the model matrix in the form $\mathbf{X} = \mathbf{QR}$ where $\mathbf{Q'Q} = \mathbf{C}$, a diagonal matrix, and $\mathbf{R}$ is upper triangular with unit diagonal. The normal equations then reduce to

$$\mathbf{X'X\beta} = \mathbf{R'Q'QR\beta} = \mathbf{R'CR\beta} = \mathbf{R'Q'y}$$

so that, for the non-singular portions of $\mathbf{R}$ and $\mathbf{C}$ corresponding to the unaliased parameters,

$$\mathbf{R\beta} = \mathbf{C}^{-1}\mathbf{Q'y}$$

There are many ways of doing this. The well-known Householder and Gram–Schmidt methods involve explicitly storing the whole model matrix. Using Givens rotations however it is possible to restrict storage to exactly the same sized working triangle as is required for methods that explicitly form and invert $\mathbf{X'X}$, while getting the benefit of improved numerical stability, fully comparable to that obtainable from Householder or Gram-Schmidt. A detailed description of the implementation of the Givens method is given in Gentleman (1974).

Previous versions of GLIM explicitly formed the normal equations into a triangular structure and then used an adaptation of the Gauss–Jordan method to solve the normal equations and overwrite $\mathbf{X'X}$ by its inverse. Diagrammatically the state of the working triangle after doing this is as follows

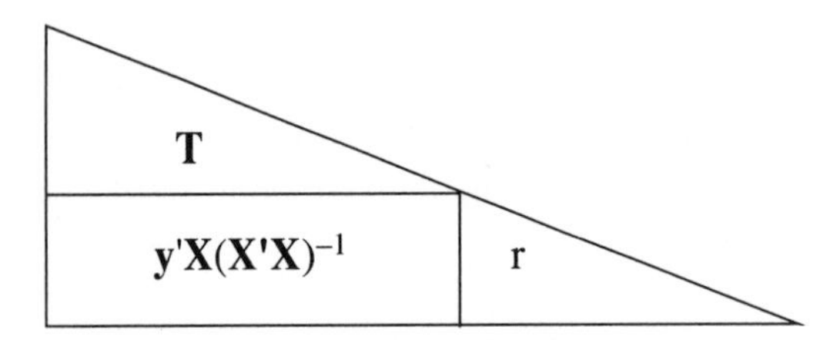

where $\mathbf{T}$ is the lower triangular part of $-(\mathbf{X'X})^{-1}$ and $\mathrm{r} = \mathbf{y'y} - \mathbf{y'X(X'X)^{-1}X'y}$.

It will be seen that, with the Gauss–Jordan method, every element of the working triangle has an immediate interpretation in terms of the linear model. The parameter estimates are explicitly given by $\boldsymbol{\beta}' = \mathbf{y'X(X'X)}^{-1}$, the $\mathbf{T}$ section is, apart from a constant multiplier, the covariance matrix of the parameter estimates, while the single element, r, at bottom right is, for the least squares case, the residual sum of squares.

Using the Givens method the working triangle is as follows, where the matrices $\mathbf{Q}$, $\mathbf{R}$, and $\mathbf{C}$ are those defined above.

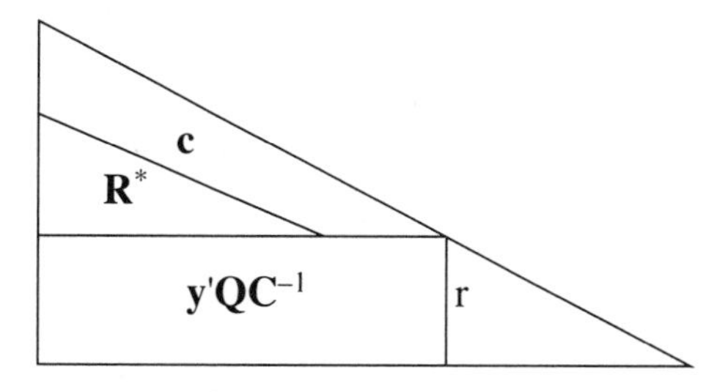

where $r = \mathbf{y'QC^{-1}Q'y}$
$\mathbf{R}^*$ is $\mathbf{R'}$ with the unit diagonal removed, and
$\mathbf{c}$ is the diagonal of $\mathbf{C}$.

It turns out that, using this representation, the linear predictor $\mathbf{X\beta}$ can be calculated without explicitly forming $\boldsymbol{\beta}$. This has advantages in speed and accuracy but means that additional work has to be done behind the scenes if the user wants to extract the parameter estimates or covariances.

Apart from questions of accuracy both methods have their advantages for users with special requirements. For example, the Gauss-Jordan method is simpler and more flexible for such techniques as stepwise or ridge regression.

11.9.3 IRLS for general regression models

As explained in [11.3], GLIM, and in particular the IRLS algorithm that forms the heart of the `FIT` directive, can be used to fit much more general regression models than just those that fall within the framework of generalized linear models. In this section, a brief derivation is given of the IRLS algorithm for maximum likelihood estimation in this more general class. The theory here provides an explanation for the success that creative users have had in fitting non-GLM models in GLIM, which have provided so many contributions to the GLIM Newsletter over the years. Almost all of these non-standard methods fall into the framework of this section. This section is intended to help future users make use of GLIM's facilities for fitting user-defined models to derive their own analyses.

For more detail about these ideas, consult the papers by Jørgensen (1983) or Green (1984), who extend this treatment in different directions.

Suppose that we have a probability model for the distribution of a response variable $\mathbf{y}$, which depends on a vector of explanatory variables $\mathbf{x}$. The models considered here are those that can be decomposed into a systematic and a random component as follows.

The random component: The log-likelihood function l for $\mathbf{y}$ depends on the explanatory variables $\mathbf{x}$ and the unknown parameters, $\boldsymbol{\beta}$ say, only through the values of a finite-dimensional vector of predictors $\boldsymbol{\eta}$. l is thus a specified function of $\mathbf{y}$ and $\boldsymbol{\eta}$.

The systematic component: The predictor vector $\boldsymbol{\eta}$ is a prescribed, deterministic, function of the explanatory variables $\mathbf{x}$ and the unknown parameters $\boldsymbol{\beta}$.

When modelling data in this way, the statistician has considerable freedom in how the predictors are chosen. For example if the model is a standard GLM, the predictors might be the expectations $\boldsymbol{\mu}$, the linear predictors $\mathbf{X}\boldsymbol{\beta}$, or indeed any other parametrization of the error distribution that is convenient.

Once the data $\mathbf{y}$, $\mathbf{x}$ are observed, the process of model-fitting by maximum likelihood is simply that of maximising the composite log-likelihood function $l(\boldsymbol{\eta}(\boldsymbol{\beta}))$ (we are suppressing $\mathbf{y}$ and $\mathbf{x}$ from the notation for clarity), and it is this numerical calculation that the IRLS algorithm is well-adapted to.

It is required to solve the maximum-likelihood equations

$$\frac{\partial l}{\partial \boldsymbol{\beta}} = \mathbf{0} \ .$$

But in this framework,

$$\frac{\partial l}{\partial \boldsymbol{\beta}} = \mathbf{D}' \frac{\partial l}{\partial \boldsymbol{\eta}} \ ,$$

where $\mathbf{D}$ is the matrix of derivatives with

$$D_{ij} = \frac{\partial \eta_i}{\partial \beta_j} \ .$$

Except in the most simple linear/Normal models, these equations cannot be solved explicitly, so that iteration is needed. The most familiar approach is the Newton–Raphson algorithm:

$$\boldsymbol{\beta}_{\text{new}} = \boldsymbol{\beta} + \mathbf{H}^{-1} \frac{\partial l}{\partial \boldsymbol{\beta}}$$

where $\mathbf{H}$ is the matrix of negative second derivatives

$$\mathbf{H} = \left[- \frac{\partial^2 l}{\partial \beta_i \partial \beta_j} \right] = \mathbf{D}' \left[- \frac{\partial^2 l}{\partial \eta_i \partial \eta_j} \right] \mathbf{D} + \frac{\partial l}{\partial \boldsymbol{\eta}} \left[- \frac{\partial^2 \boldsymbol{\eta}}{\partial \beta_i \partial \beta_j} \right] .$$

In statistical model fitting it has been customary since the time of Fisher to replace these second derivatives by their expected values (at the current parameter values). But then, use of the standard identities for expectations of derivatives of log-likelihoods (see Kendall and Stuart (1967)) allows us to replace $\mathbf{H}$ by $\mathbf{D}'\mathbf{W}\mathbf{D}$, where

$$W_{ij} = \mathbf{E}\left(- \frac{\partial^2 l}{\partial \eta_i \partial \eta_j} \right) .$$

$$\boldsymbol{\beta}_{\text{new}} = \boldsymbol{\beta} + (\mathbf{D'WD})^{-1}\,\mathbf{D'}\,\frac{\partial l}{\partial \boldsymbol{\eta}}$$

or equivalently

$$\boldsymbol{\beta}_{\text{new}} = (\mathbf{D'WD})^{-1}\,\mathbf{D'W}\,(\mathbf{D}\boldsymbol{\beta} + \mathbf{W}^{-1}\,\frac{\partial l}{\partial \boldsymbol{\eta}}\,)$$

$$= (\mathbf{D'WD})^{-1}\,\mathbf{D'Wz}\ \ ,\ \text{say.}$$

So, as claimed earlier, even in this very general framework, the maximum likelihood estimation can be reduced to use of the IRLS algorithm.

To make use of these ideas in constructing a new model fitting procedure in GLIM, this updating equation is compared with that derived earlier for the standard GLM models [11.3.1.2]. In that context:

the derivative matrix **D**	was the model matrix **X**
the iterative weight matrix **W**	was **U**
the working variate **z**,	
now $\mathbf{D}\boldsymbol{\beta} + \mathbf{W}^{-1}\,\partial l/\partial\boldsymbol{\eta}$	was $\eta + (\mathbf{y} - \mu)\,\partial\eta/\partial\mu$.

Any regression model falling into this framework can thus be fitted in GLIM if the system structures **X**, **U**, η, μ, and $\mathrm{d}\eta/\mathrm{d}\mu$ are appropriately set to give the above correspondence. The only major restriction is that in the current version of GLIM, only diagonal weight matrices are allowed. Except in very special cases, therefore, or by making subtle use of the PASS facility, it is necessary that $\mathbf{W} = \mathbf{E}(-\partial^2 l/\partial\eta\partial\eta')$ be diagonal. Effectively this means that the log-likelihood is a sum of terms, each involving only one component of η.

12 Examples

Introduction

This chapter of the Modelling Guide is intended to give a set of more or less complete analyses covering a range of practical problems for which GLIM is useful. The examples are also intended to convey how the various facilities of GLIM, especially the high–resolution graphics, may be used with practical problems, and to provide a useful reference for users with similar problems.

The examples given here are designed to cover the most common problems that arise for which univariate modelling is appropriate. They are grouped, mainly, according to the type of response variable of which the main types of interest are continuous (measured), dichotomous (two categories), and counted (frequencies). However there is some overlap so it was not possible to keep rigidly to this scheme.

The continuous response variable section examples are almost entirely restricted to the case where the response variable has a Normal or Gaussian distribution. However the general approach is much the same for other continuous distributions and there is a specific example using the gamma distribution so this should not be a limitation. The dichotomous response variable section is mainly restricted to classical and conditional logistic regression problems, but again other approaches such as probit analysis follow much the same lines with minor changes in the model specification. The counted response variable section includes an example of a log-linear modelling analysis of a contingency table, but with one dimension of the table representing a categorical outcome. This is a common situation and one in which it is much easier to understand the complex manoeuvres required for an analysis using frequencies as a proxy for underlying categorical response variables. This means that there is no example specifically covering the wide class of applications involving the analysis of contingency tables where there is no single response dimension. However these tend not to follow standard patterns so it is difficult to choose a generally informative example. Users with interests in this area are recommended to consult the specialist literature (Lindsey, 1989; Aitkin et al, 1989; Agresti, 1984; Whittaker, 1990). The non-linear applications are simply to illustrate what is possible in GLIM using the very flexible facilities for users to specify their own models that are now available. Quite complex problems can be handled, but it becomes progressively more difficult and specialist packages are likely to be much more efficient for general use.

There is still a wide range of applications for which GLIM is likely to prove useful which is not covered by the examples here. GLIM also provides a very flexible working environment for implementing new techniques with very powerful macro facilities. There is a comprehensive set of functions available including a range of probability distributions and their inverses. This with the graphical facilities makes it possible to display standard distributions very effectively for teaching and other purposes. It also means that tables may be produced very simply using GLIM, for use with students thus avoiding the copyright problems of copying published tables. The random number generation facilities together with the distribution functions make a wide range of simulation studies practicable.

Some of the methods and applications for which examples have not been provided and where GLIM can or may prove useful are: penalised regression, ordered categorical outcomes, association and graphical models, repeated measures with serial correlation, survival data analysis with time dependent covariates, discrimination, estimating the parameters of distributions from a single frequency distribution, generalized additive models, composite links, latent class analysis, Taguchi methods, social mobility and gravity models, `bootstrap' methods, Gibb's sampling techniques, and quasi–likelihood models.

Papers and reports on these and other topics may be found in the GLIM Newsletters and in the proceedings of the four GLIM conferences.

12.1 Continuous *y* variables

12.1.1 Continuous *x* variables: classical regression

12.1.1.1 A straight line regression model

Suppose the 12 data points are:

x	y
8	59
6	58
11	56
22	53
14	50
17	45
18	43
24	42
19	39
23	38
26	30
40	27

The simplest way to input them to GLIM is by using the `READ` directive after first declaring two 12–element long vector identifiers

```
$var 12 x y $read x y
     8    59
     6    58
    11    56
    22    53
    14    50
    17    45
    18    43
    24    42
    19    39
    23    38
    26    30
    40    27
```

A scatter plot indicates the form of the relationship

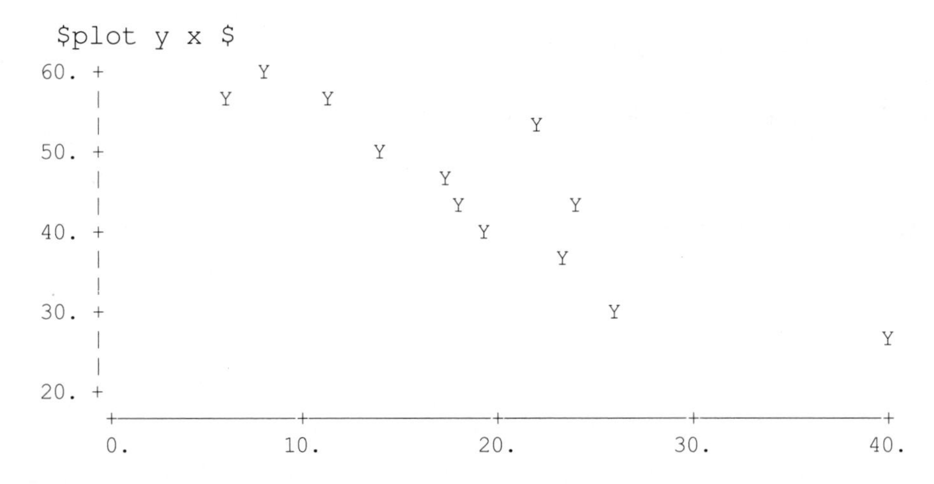

There appears to be a steady, downward, linear trend. A straight line model of the form $y = a + bx$ can be used to describe and quantify this trend. The two parameters, a representing the intercept and b representing the slope of the line, are estimated by fitting the model, assuming the variation in y follows a Normal distribution, as follows

```
$yvar y $
$fit 1 + x $dis e $
   deviance =  273.84
residual df =    10
           estimate        s.e.     parameter
     1       64.25         3.603      1
     2      -1.013         0.1722     X
scale parameter 27.38
```

The deviance is the sum of the squared deviations of the points from the fitted line and the scale parameter is the residual variance of y about the model obtained by dividing the deviance by the degrees of freedom. The estimate of a is the first parameter estimate labelled `1` and the estimate of the slope b is the second parameter estimate labelled `X`.

The values predicted from this model for x–values over the whole range 0,50 in steps of 5 are obtained and plotted with the observed values by

```
$ass xf=0,5...50 $
$predict (s=-1) x=xf $
$cal yf=%pfv $
```

The data points and how closely the model fits can be seen with a plot

```
$plot y;yf x;xf 'x.' $
```

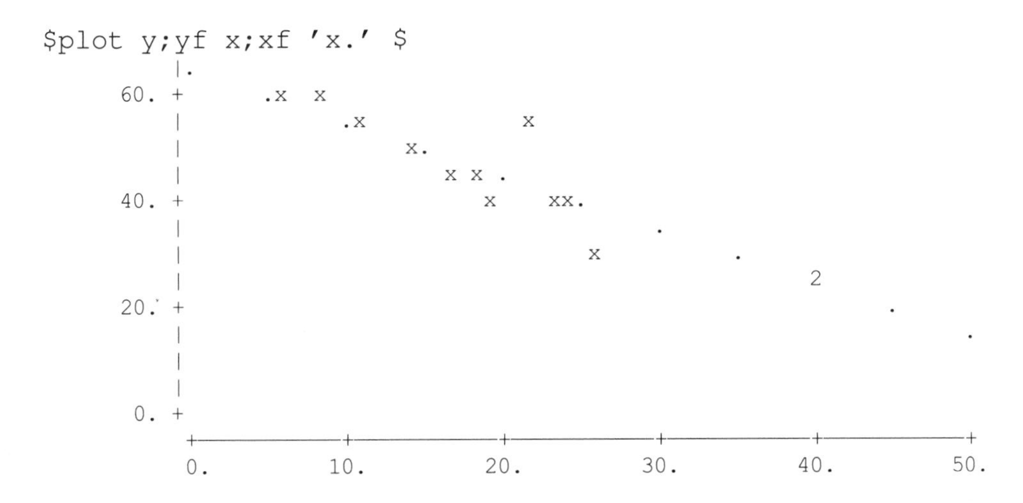

A high–resolution graphics plot can be obtained by

```
$gra ( t='Fig. 12.1  Data points with Fitted Straight Line'
       v='y - axis' h='x - axis' x=0,50,10 y=0,80,10 )
        y;yf x;xf  3,12 $
```

For presentation purposes a *key* and text labelling can be added, with the `GTEXT` directive. How to do this is shown in [12.4.3.1].

It is usual to test whether the trend seen in the sample reflects a real non-horizontal trend in the population it represents. This is done by performing a t test comparing the estimated regression coefficient with zero.

The t test with the two-tail probability of values this far from zero by chance is obtained by `EXTRACTING` the parameter estimate and standard error vectors and using

Fig. 12.1 Data points with Fitted Straight Line

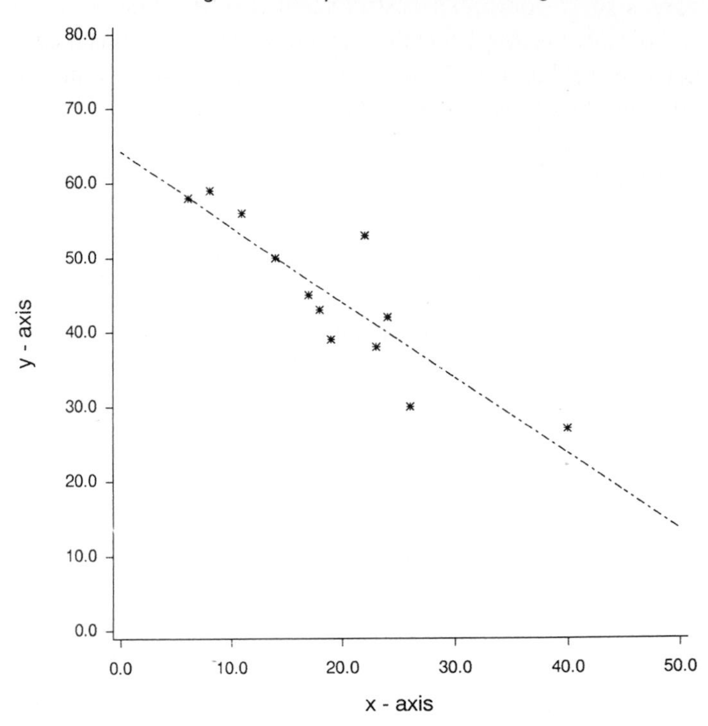

NUMBER variables and the t distribution cumulative probability function for the calculations as follows

```
$extract %pe %se $
$number b t p seb $
$cal b=%pe(2) : seb=%se(2)
  :  t=b/seb : p=2*(1-%tp(%abs(t),%df)) $
$print ;' t value = 't'  p(values this far from zero) = 'p$
```

```
 t value =    -5.884  p(values this far from zero) =    0.0002
```

A 95 per cent confidence interval on the estimated slope is obtained using the t distribution deviate function as follows

```
$number low high tval $
$cal tval=%td(0.975,%df)
  :  low= b - tval*seb: high=b + tval*seb$
```

These can be printed by

```
$pr low high $
  -1.397 -0.6294
```

or with appropriate explanatory text useful for later reference or passing to the customer

```
$pr ; 'The estimated slope is: 'b
' units change in y for a unit increase in x'
;'  with 95% Confidence limits of 'low'  to  'high $

The estimated slope is:   -1.013 units change in y for a unit increase in x
  with 95% Confidence limits of   -1.397  to   -0.6294
```

This is the complete analysis for a simple regression although it might be useful to obtain confidence limits for the fitted line. This is covered in a later example.

12.1.1.2 Simple linear regression to test association

A study on the health effects of pollution by Forni and Sciame (1980) produced data on chromosome abnormality rates and exposure to lead. The main question was whether exposure to lead and hence absorption of lead causes an increase in chromosome abnormalities.

The subjects in the study were female workers in an Italian battery factory and the complete dataset consisted of observations on each of the eight variables

age	age in years;
exp	period of exposure to lead;
pbb	blood lead levels;
alad	delta-aminolevulinic acid dehydratase;
ep	erythrocyte protoporphyrin;
hct	haematocrit;
gap	percentage of cells with gaps in the chromosomes;
abn	percentage of cells with chromosome abnormalities;

The data are read into GLIM by declaring the number of units observed, listing the names of the variables in the dataset and directing GLIM to read as follows

```
$units 30  $
$data  age exp pbb alad ep hct gap abn  $
$read
40    0.0   22   44.0    45   43   2    2
33    0.0   23   30.0    28   42   2    5
36    0.0   23   49.0    32   46   2    4
```

```
35   0.0  24  40.0   26  42  2   5
48   0.0  25  30.5   27  44  2   4
29   0.0  26  45.0   25  41  1   5
39   0.0  27  35.2   21  44  1   6
34   0.0  30  30.0   22  40  1   3
28   0.0  31  32.5   22  40  0   0
42   0.0  32  31.4   40  45  2   6
44   0.0  34  39.8   26  46  2   4
39   0.0  37  25.8   43  46  3   4
42  23.0  24  23.0   57  40  3   5
38  15.0  25  48.0   43  41  2   8
46   6.0  31  33.2   50  45  1   7
30   6.0  36  28.0   73  39  1   4
27   3.0  36  31.8   59  36  1   5
38  10.0  36  17.8   38  46  4   9
38   6.0  38  27.8   35  40  4   6
40  13.0  39  22.0   76  41  1   7
35   8.0  42  22.8   55  42  0   1
38   5.0  45  15.4   -1  -1  2   7
50  13.0  46  13.6   -1  -1  4  22
42   5.0  49  14.3   70  41  1  11
40   3.0  49  23.2   87  40  3   4
32   4.0  49  14.7   39  40  1   5
37  13.0  49  23.0   25  44  3   8
44  12.0  50  18.3   -1  -1  1   6
44  23.0  54  14.5  101  43  2   6
29   4.5  59  19.3   -1  -1  1  10
```

The `-1`'s have been inserted in the data where values could not be obtained or were unknown, to indicate that a value is missing. The individual concerned should not be used in analyses using that variable.

One measure of the exposure to lead is the level found in the blood. We need to model how the mean number of abnormalities per 100 cells `(abn)` changes as the blood levels of lead `(pbb)` change. The first step is to investigate the relationship graphically using a scatter diagram obtained by plotting `abn` against `pbb`.

```
$plot (x=0,60 y=0,30) abn pbb 'x' $
```

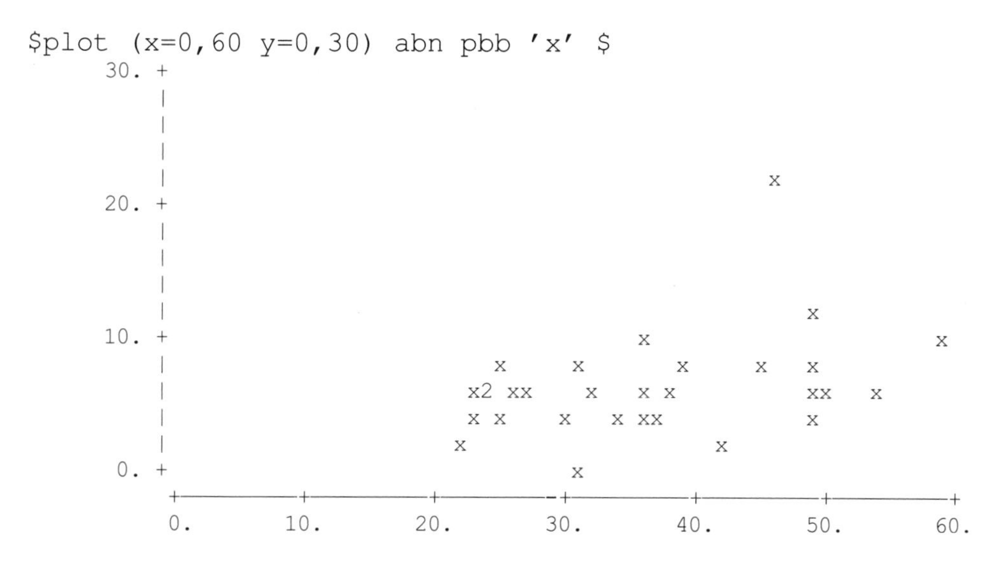

Alternatively a high–resolution plot can be produced with

```
$graph (t='Fig. 12.2 Chromosome abnormalities by Lead levels'
       v='Chromosome abns/100 cells' h='Blood lead levels'
       xscale=0,60,10 yscale=0,30,10 )
       abn pbb  1 $
```

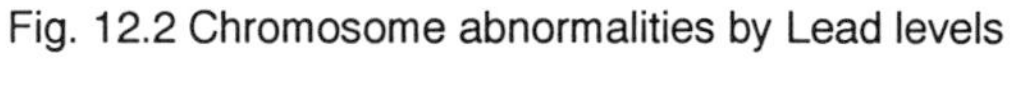

Fig. 12.2 Chromosome abnormalities by Lead levels

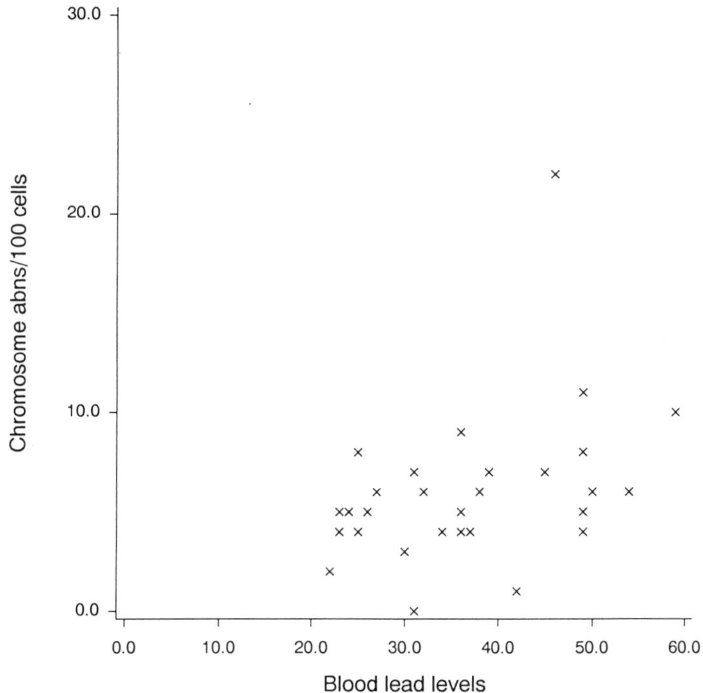

There appears to be some tendency for the number of abnormalities to increase with lead levels. We can examine this more closely by grouping the data on the `pbb` scale and obtaining the group `abn` means using the `GROUP` directive.

```
$ass cut_pts=20,30,40,50,60 $
$group gpb=pbb intervals cut_pts  $

$tab the abn mean;dev;weight for gpb into mean;sd;n by grp $
$tprint mean;sd;n grp $
  GRP      1        2        3        4
 MEAN    4.889    5.000    8.286    7.333
   SD    1.616    2.408    6.824    2.309
    N    9.000   11.000    7.000    3.000
```

This shows that, on average, the mean `abn` levels do increase as the `pbb` levels increase. This is seen more clearly in a plot against the midpoints of each group calculated from the group code variable `grp`.

```
$cal gx=grp*10+15 $
$plot mean gx $
```

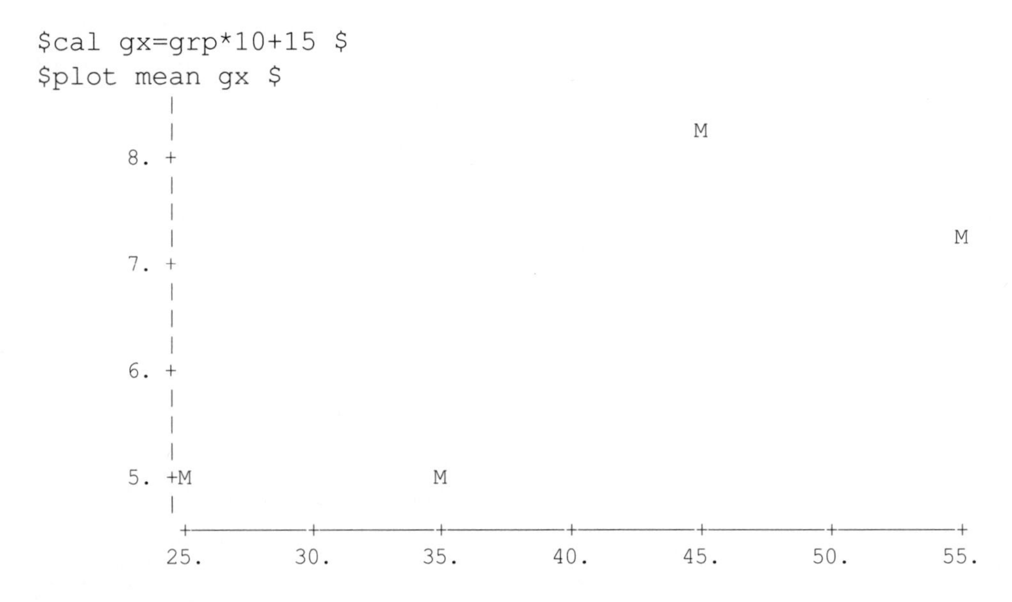

Returning to the raw data to fit the straight line regression model with `abn` as the *y* variable we need

```
$yvar abn $
$fit 1+pbb $dis e $
   deviance =  368.80
residual df =    28
          estimate        s.e.     parameter
     1      0.8979       2.390      1
```

```
     2       0.1394      0.06315      PBB
scale parameter 13.17
```

By default the ERROR has been assumed to be Normal and hence the deviance is simply the residual sum of squares about the regression. The scale parameter, for the Normal ERROR case, is the deviance divided by the degrees of freedom, that is, the usual estimate of the residual variance.

The estimate of the intercept a is identified by the symbol 1 which is the coefficient of the parameter in the model. Unless the SET directive has been used to set the NOCONSTANT option this will be supplied automatically. The slope b is identified by the name of the x variable and it indicates that, on average, the number of abnormalities/100 cells increases by 0.1394 for each unit increase in pbb.

The crucial question is whether this non-zero regression coefficient represents a genuine relationship between blood lead levels and numbers of chromosome abnormalities in the population of such women exposed to lead in this way. This is tested with a t test

t value = $b/\text{se}(b)$ = 2.21 on 28 d.f.

This is greater than the 97.5 centile of the t distribution with 28 d.f., that is,

```
$cal  %td(0.975,28) $
           2.048
```

so the regression coefficient is significantly different from zero at the 5 per cent level. The test assumes that the distribution of abn values about the model follows a Normal distribution. This must be investigated. One approach is to inspect a histogram of the residuals, that is, the differences between the observed y–values and those predicted for that observation from the fitted model. These residuals are held in the variable %rs.

```
$histogram %rs 'd' $
[-10.00, -5.00)  2 dd
[ -5.00,  0.00) 14 dddddddddddddd
[  0.00,  5.00) 13 ddddddddddddd
[  5.00, 10.00)  0
[ 10.00, 15.00]  1 d
```

This is not obviously Normal or non-Normal. It is easier to judge from a Normal plot. These are obtained by plotting the ordered residuals against the equivalent quantiles of the Normal distribution. The GLIM sequence for this is

```
$sort res %rs $
```

to sort them, into a new variable, in ascending size order. The sample size is the number of units in the sample n held in the system scalar `%nu`. This is used to get approximations to the $1/n$th, $2/n$th, .., n/nth quantiles of the standardized Normal distribution avoiding the infinite result at n/n with a small correction

```
$cal nqs=%nd((%gl(%nu,1)-0.5)/%nu) $
$plot res nqs $
```

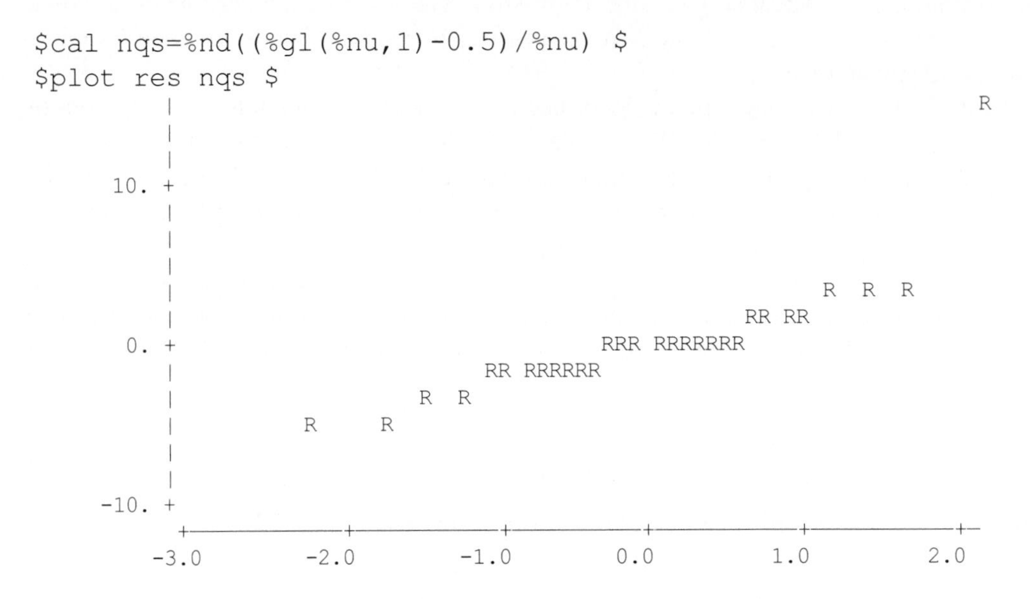

If the sample was much as would be expected from a Normal distribution this plot would be reasonably close to a straight line. In fact it is apart from the largest residual which is much larger than would be expected from the Normal distribution. Plotting the unsorted residuals against the observed values we can see which observation this residual comes from

```
$plot %rs abn $
```

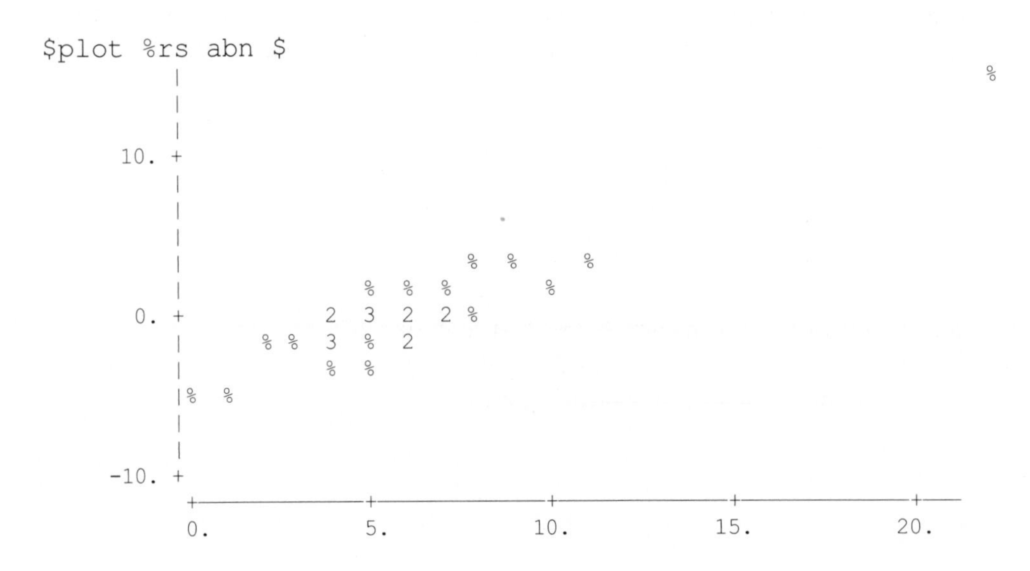

We see that it arises from the extreme observation with an abn value twice as large as any of the others. It is worth investigating the effect of omitting that observation from the analysis. First save the fitted values from this model for a later comparison.

```
$cal fv_all=%fv $
```

Then give that observation zero weight in the analysis to exclude it and fit the model again

```
$cal wt=(abn < 20) $weight wt $

$fit pbb $dis e $
   deviance =  138.86
residual df =    27    from 29 observations
          estimate        s.e.     parameter
     1       2.036       1.503      1
     2     0.09373     0.04005      PBB
scale parameter 5.143
```

This shows how much effect that outlying observation had on the results. The residual variance (the scale parameter) is now very much smaller. The regression coefficient is also smaller, but so is its standard error due to the decrease in the residual variance. The net result is that the coefficient is actually more significant than it was

t value = 0.09373/0.04005 = 2.34

This is greater than the 97.5 centile of the t distribution with 27 d.f. which is

```
$cal %td(0.975,27) $
            2.052
```

The outlying observation had inflated the residual variance more than it had affected the estimate of *b*.

The effect on the Normal plot of the residuals from excluding the outlier can be seen from a weighted analysis plot obtained as follows. First get a vector of residuals from those observations still included and sort them into ascending size order. Then generate a vector, of the same length, of the appropriate quantiles from the Normal distribution.

```
$pick r_in  %rs wt $
$sort r_in $

$number nin  $
$cal nin=%len(r_in) $
```

```
$var nin n_in $
$cal n_in=%nd((%gl(nin,1)-.5)/nin) $
```

Note that this sort of potentially complex calculation sequence, which may be required many times, would usually be written as a general macro. In later examples a variety of such macros are read from library files and used when required. An official GLIM macro library containing a full range of frequently needed and useful macros is supplied with the program.

The Normal plot is then obtained by plotting the residuals against the quantiles.

```
$plot r_in n_in $
```

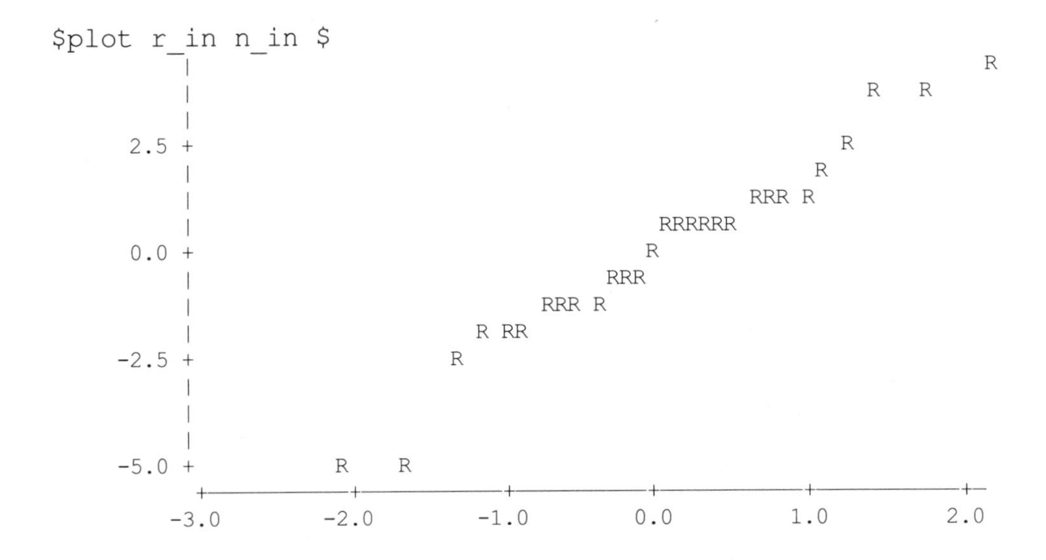

Not exactly a straight line, but better than before and enough to justify the use of the t test.

Finally the ‘aberrant’ point, magnified by a weight of 2, and the models obtained with and without it can best be seen with a graph using

```
$cal wt_out=2*(1-wt) $

$graph (t='Fig. 12.3 Chromosome abnormalities by Lead levels'
       v='Chromosome abns/100 cells' h='Blood lead levels'
       xscale=0,60,10 yscale=0,30,10 pause=no)
       abn/wt;fv_all;abn/wt_out;%fv   pbb  1,10,3,12 $
$gtext ( m=p l=r ) 30/1.2 2 'Two models with and without the
Extreme value' $
```

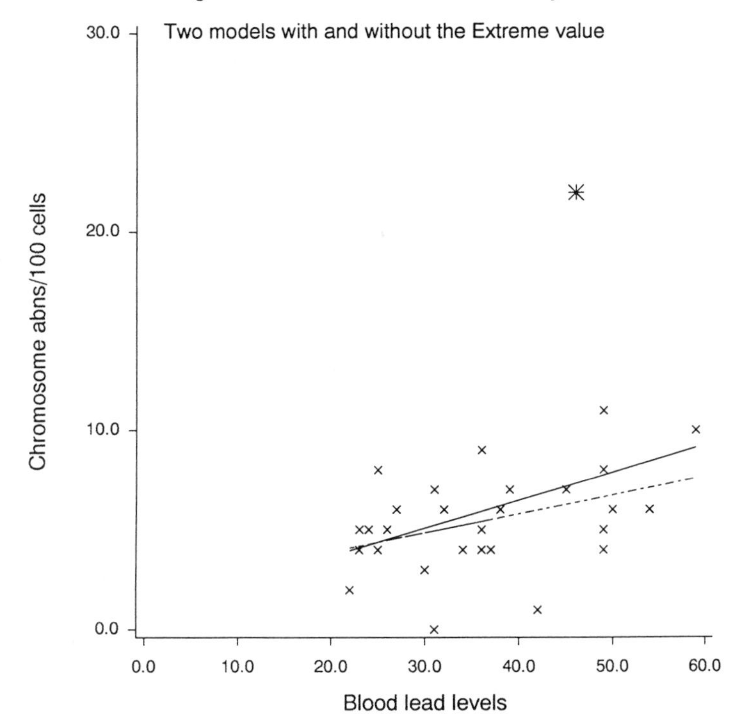

To assess the magnitude of the apparent effect of lead levels on chromosome abnormalities thoroughly we need confidence limits on the regression coefficient. We can obtain these by using the actual values in a calculate statement or more generally by extracting the parameter estimates and their standard errors by

```
$extract %pe %se $
```

The regression coefficient is the second element of `%pe` and its standard error is the second element of `%se`. The variance of the regression coefficient is also available as the second diagonal element, that is, element 3 of the covariance matrix held as a lower triangle in `%vc`. The 95 per cent confidence limits on the `true' regression coefficient representing the mean increase in `abn` per unit increase in `pbb` using the 97.5 centile of the t distribution with 27 d.f. is

```
$number low high tval $
$cal tval=%td(0.975,%df)
    : low=%pe(2) - tval*%se(2) : high=%pe(2) + tval*%se(2) $
$print ;;' 95% CLs for the abn v pbb regression
coefficient';;' Lower = 'low ' Upper = 'high $
```

```
Lower =    0.0116 Upper =    0.1759
```

This is reasonably convincing evidence that the observed association is not due to chance. However, as for all epidemiological evidence from observational studies, it needs the support of similar findings from other studies before the association could be considered strong evidence of a causal relationship. In practice it would also be necessary to consider the possibly biasing (confounding) effects of other variables such as age and exposure. To keep this example simple the appropriate techniques for that are introduced later.

The checks on the model assumptions here are relatively superficial. If there was thought to be a 'correct' model to be deduced and used for prediction, or precise inferences on the algebraic form of the model needed to distinguish between possible mechanisms underlying the relationship, more rigorous methods would be necessary. Appropriate techniques will be discussed in subsequent examples.

In this type of situation establishing whether or not there was any sort of trend and its magnitude will generally be sufficient to answer the epidemiological question. Further questions, on whether the relationship is causal among other things, will generally require data from different types of study. Nonetheless even in pragmatic modelling of this sort one would have to test the assumption of linearity by investigating whether a curved relationship would fit the data better. Appropriate modelling approaches for doing this are discussed in later examples.

This analysis has concentrated on the relationship between chromosome abnormality rates and blood lead levels. However the dataset can obviously be used to investigate other questions on the effects of lead and how they may vary with age and duration and exposure.

12.1.1.3 Simple linear regression to obtain a model for prediction

The hardness of a metal alloy is easier to measure than tensile strength although the latter is the most important characteristic. It would be very useful if hardness could be used for reasonably accurate predictions of tensile strength.

Data were obtained on the hardness and tensile strengths of 20 samples of an alloy produced under a variety of conditions. A model of the relationship between the two variables and measures of how accurately it can be used to predict tensile strength from hardness measurements is needed. The data are taken from Brookes and Dick (1951).

The data are input by

```
$un 20 $data hard tensile $read
 52 12.3   54 12.8
 56 12.5   57 13.6
 60 14.5   61 13.5
 62 15.6   64 16.1
 66 14.7   68 16.1
 69 15.0   70 16.0
```

```
71 16.7   71 17.4
73 15.9   76 16.8
76 17.6   77 19.0
80 18.6   83 18.9
```

A plot of `tensile` against `hard` shows that there is a strong relationship which appears to be linear.

```
$plot tensile hard $
```

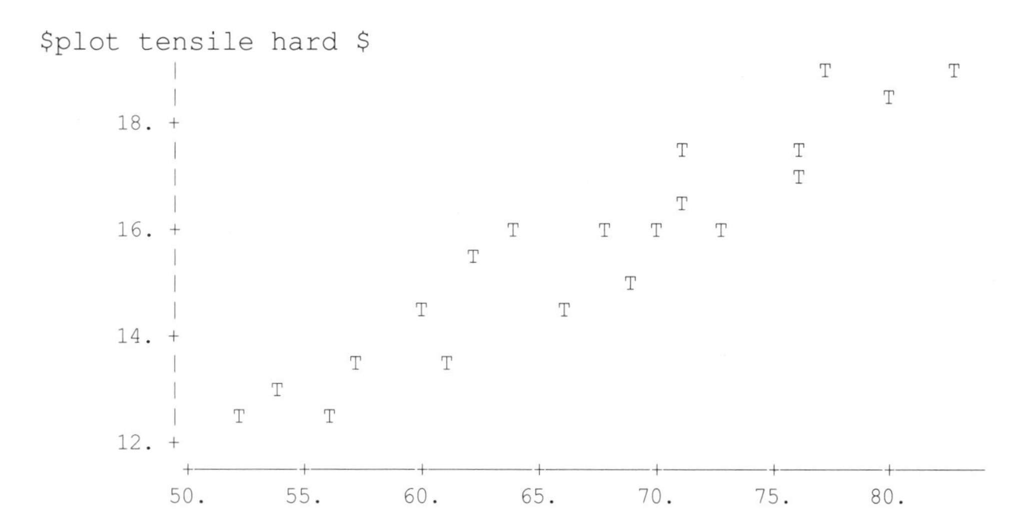

However rather than assume linearity it is appropriate to test for curvature immediately. We do this most simply by fitting a quadratic term, that is, we put a variable with values equal to those of `hard` squared in the model.

```
$cal hd2=hard*hard $
```

Then declare `tensile` as the y variable and fit the model

```
$yvar tensile $
$fit 1+hard+hd2 $dis e $
   deviance =    9.5817
residual df =   17
          estimate          s.e.     parameter
     1     -0.5971         9.898      1
     2      0.2666        0.2983      HARD
     3  -0.0003611      0.002221      HD2
scale parameter 0.5636
```

The coefficient of the `hd2` term is considerably less than its s.e. so it appears that the assumption of linearity is adequate. We must also test the assumptions of constant variance and Normality. The first can be done visually by plotting the residuals against the fitted values.

```
$plot %rs %fv 'r' $
```

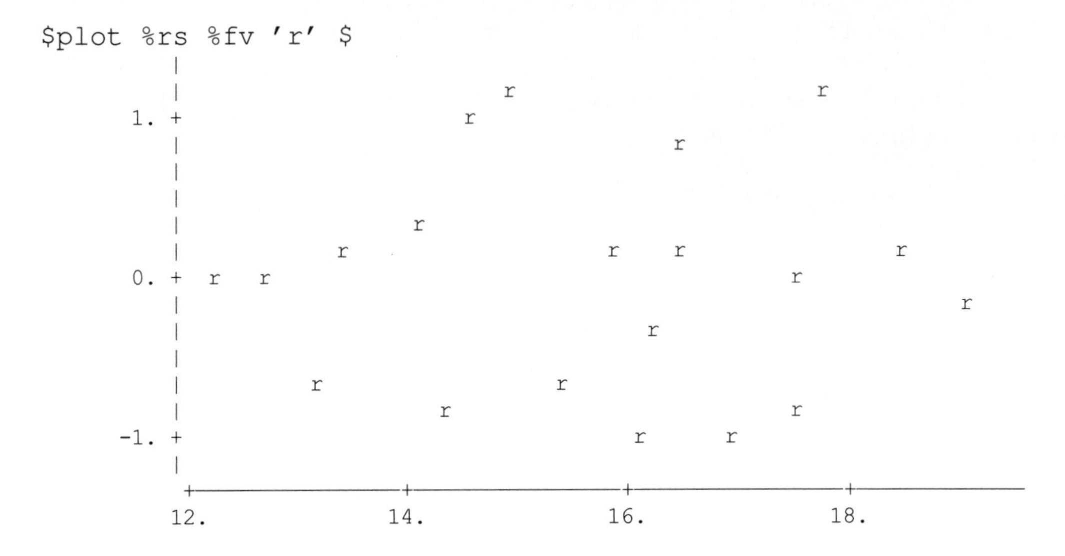

There is no evidence of the vertical spread of the residuals increasing with increasing y values. We test for Normality using a quantile–quantile plot as in [12.1.1.2] using a previously stored macro.

```
$use crep $
```

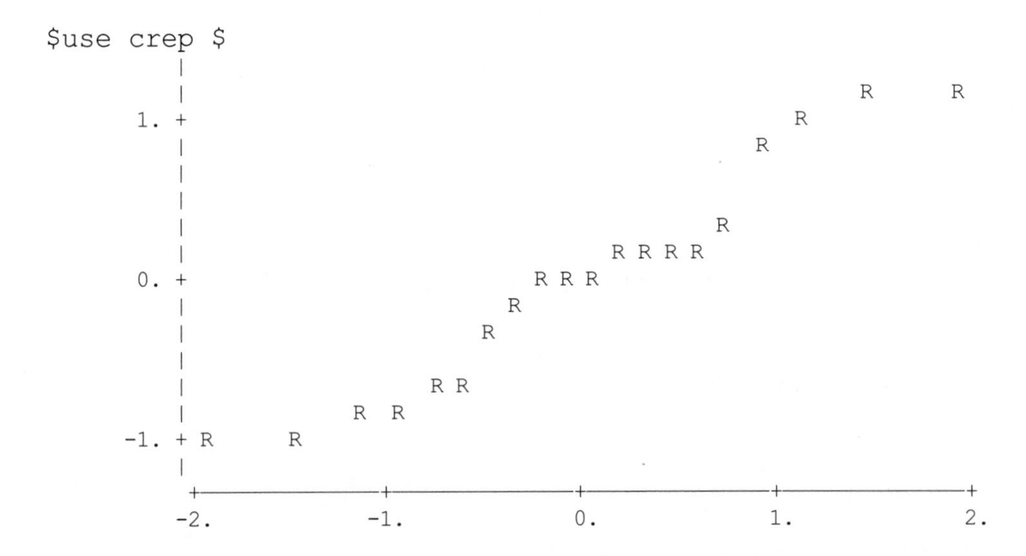

There is a slight indication of oscillation about the required straight line. This could be due to the way the data values are collected or presented. For example, rounding errors might be involved. In practice this would be investigated with the originator of the data. Otherwise it is not untypical of the sort of plots arising from adequately Normal data. The assumptions all look reasonable so we can progress to fitting the simpler straight line model.

```
$fit 1+hard $dis e $
   deviance =   9.5966
residual df =  18
          estimate         s.e.     parameter
     1      0.9978        1.278      1
     2      0.2182      0.01884      HARD
scale parameter 0.5331
```

It is wise to check that this simplification hasn't made the assumptions less reasonable

```
$plot %rs %fv 'r' $
```

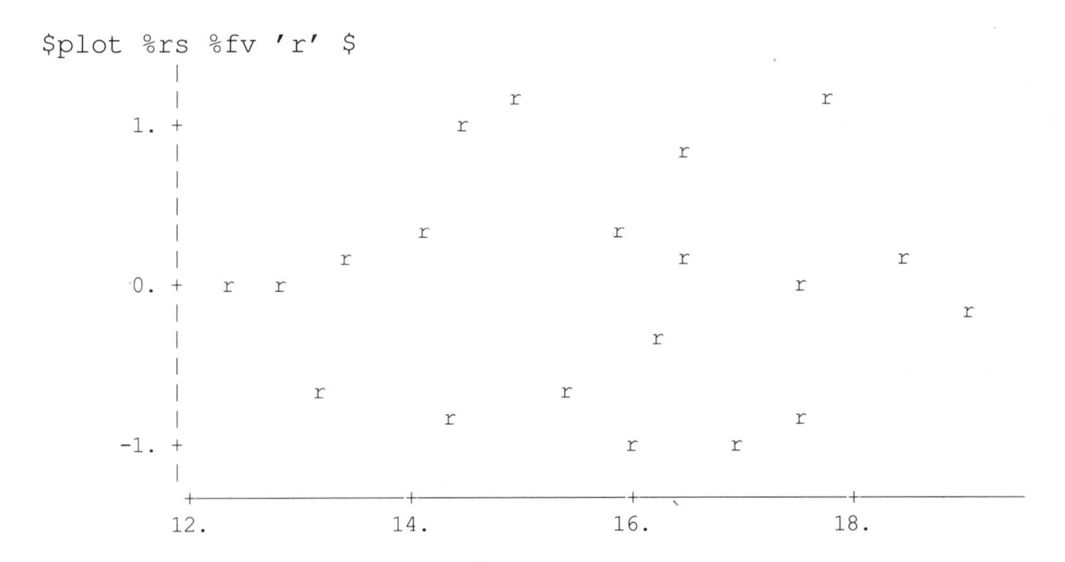

There is still no evidence of the residual variances increasing with the *y* variable

```
$use crep $
```

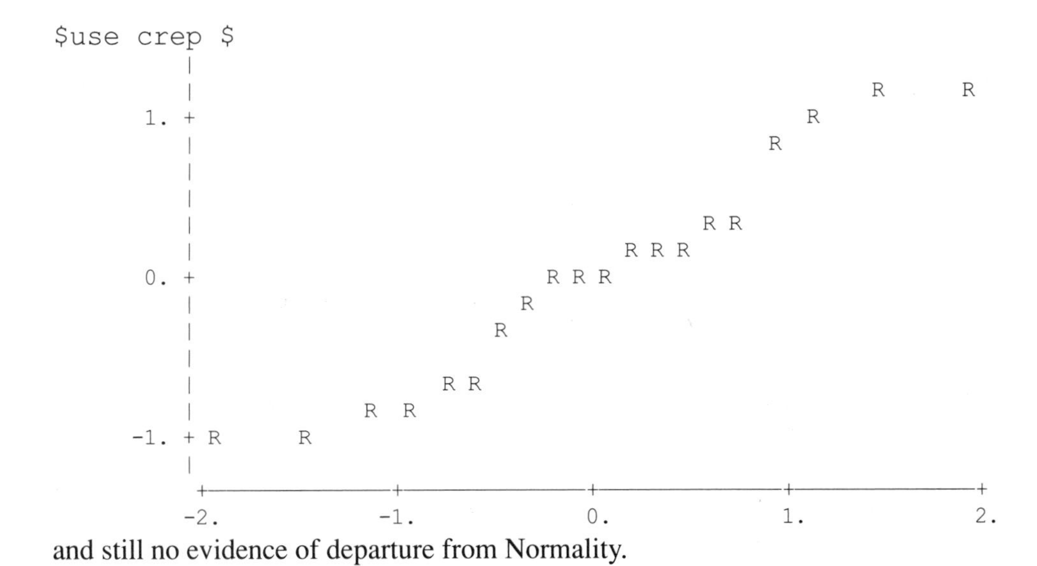

and still no evidence of departure from Normality.

The coefficient of `hard` is well over twice its s.e. and hence highly significant, but this is not the main interest. We need to assess how close this fitted line is to the `true' relationship assumed linear. We need confidence limits, or more precisely a confidence region, for the population line. That will tell us how accurately we could predict, using this fitted linear relationship, the average tensile strength to be expected for alloys of a given hardness. Subsequently we may also need to estimate how accurately we can predict the tensile strength of individual samples.

For the 95 per cent confidence region for a fitted regression line, that is, the predicted mean values for the conditional distributions, we need access to the variances of the fitted values which, in the case of a Normally distributed response variable, are the same as the linear predictors. These are held in system vector `%vl` which contains the lower triangle of the variance-covariance matrix of the linear predictors.

```
$extract %vl $
```

Then calculate the 95 per cent confidence limits for the fitted values using the 97.5 centile of the t distribution with 18 d.f.

```
$cal tval=%td(.975,18) $
$cal low=%fv - tval*%sqrt(%vl)  : high=%fv + tval*%sqrt(%vl) $
$plot ( x=50,90 ) low;high;%fv hard '^v.' $
  20.0 +                                                     v
       |                                                 v   .
       |                                         2 v     .   ^
  17.5 +                                     v   4 2     ^
       |                              v2 4   2
       |                          v  3 2^ 2
  15.0 +                   vv  3  2
       |         v v    2 22
       |     v  v  . .    ^
  12.5 +     .  2  ^ ^
       |     ^
       |
  10.0 +
       +-----------+-----------+-----------+-----------+
      50.         60.         70.         80.         90.
```

The high–resolution version of this, using macros to store the title strings, is obtained by

```
$mac title Fig. 12.4 Data, Fitted Line and 95% CLs $end $
$mac xlabel Hardness $e $
$mac ylabel Tensile strength $e $

$gra ( t=title h=xlabel v=ylabel x=50,90,10 y=0,30,10)
       tensile;low;%fv;high hard  3,12,10,12 $
```

Fig. 12.4 Data, Fitted Line and 95% CLs

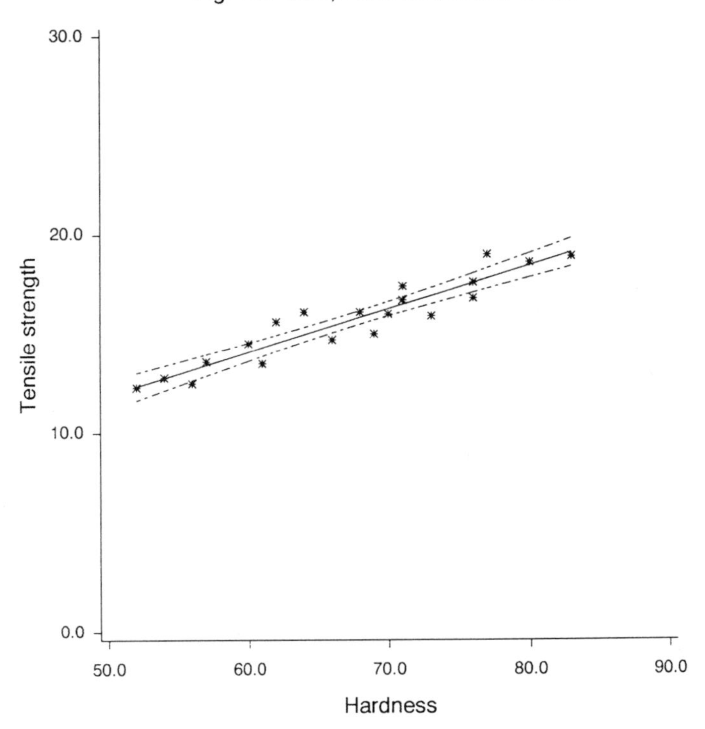

To see the upper limits on the percentage errors likely in using this relationship to predict mean tensile strengths from hardness measurements:

```
$cal perr=100*tval*%sqrt(%vl)/%fv $
$look hard low %fv high perr $
      HARD     LOW      %FV     HIGH     PERR
  1   52.00    11.65    12.34   13.04    5.639
  2   54.00    12.15    12.78   13.41    4.917
  3   56.00    12.65    13.21   13.78    4.265
  4   57.00    12.90    13.43   13.97    3.966
  5   60.00    13.64    14.09   14.54    3.184
  6   61.00    13.88    14.31   14.73    2.964
  7   62.00    14.12    14.52   14.93    2.768
  8   64.00    14.59    14.96   15.33    2.453
  9   66.00    15.05    15.40   15.74    2.253
 10   68.00    15.49    15.83   16.18    2.174
 11   69.00    15.70    16.05   16.40    2.178
 12   70.00    15.91    16.27   16.63    2.208
 13   71.00    16.11    16.49   16.86    2.262
 14   71.00    16.11    16.49   16.86    2.262
```

```
 15    73.00    16.51    16.92    17.33    2.426
 16    76.00    17.09    17.58    18.06    2.765
 17    76.00    17.09    17.58    18.06    2.765
 18    77.00    17.28    17.80    18.31    2.893
 19    80.00    17.84    18.45    19.06    3.298
 20    83.00    18.40    19.11    19.81    3.715
```

To assess the accuracy of predictions for individual observations we need to use the variances of the linear predictors increased by adding the residual variance, which is the scale parameter held in the scalar variable `%sc`. Thus

```
$cal low=%fv - tval*(%sqrt(%vl+%sc))
: high=%fv + tval*(%sqrt(%vl+%sc)) $
$plot ( x=50,90 ) low high %fv hard '^v.' $
```

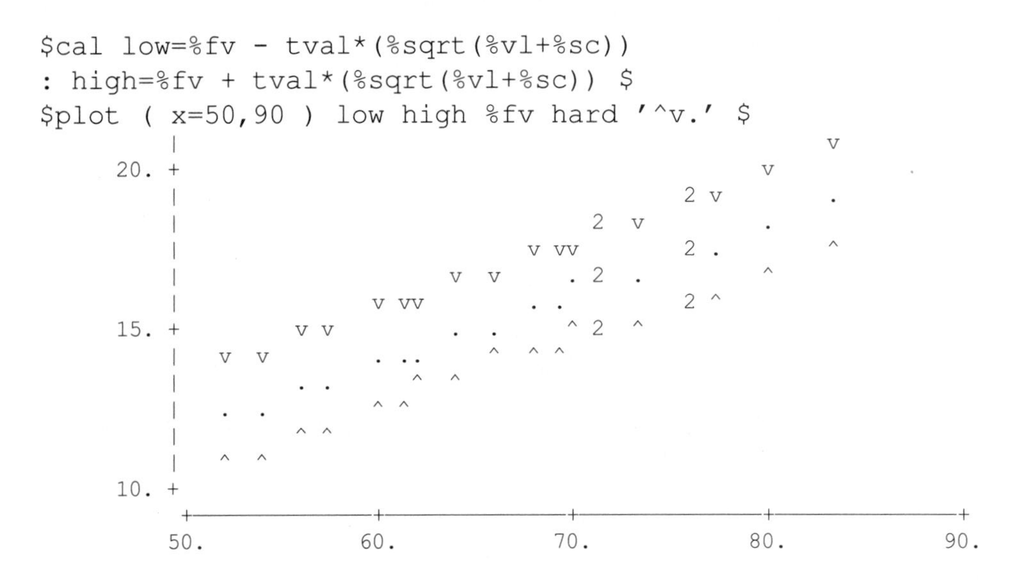

The `GRAPH` equivalent is

```
$mac title2 Fig. 12.5  Data, Fitted Line and 95th Centiles
$end $
$gra ( t=title2 h=xlabel v=ylabel x=50,90,10 y=0,30,10)
       tensile;low;%fv;high hard  3,12,10,12 $
```

To see the range of percentage errors in these predictions:

```
$cal perr=100*(high-%fv)/%fv $
$look hard low %fv high perr $
      HARD     LOW     %FV     HIGH     PERR
  1   52.00   10.66   12.34   14.03   13.648
  2   54.00   11.12   12.78   14.44   12.973
  3   56.00   11.58   13.21   14.85   12.367
  4   57.00   11.81   13.43   15.06   12.089
  5   60.00   12.49   14.09   15.69   11.345
```

6	61.00	12.71	14.31	15.90	11.125
7	62.00	12.94	14.52	16.11	10.919
8	64.00	13.38	14.96	16.54	10.544
9	66.00	13.82	15.40	16.97	10.215
10	68.00	14.26	15.83	17.40	9.930
11	69.00	14.48	16.05	17.62	9.802
12	70.00	14.69	16.27	17.84	9.684
13	71.00	14.91	16.49	18.07	9.575
14	71.00	14.91	16.49	18.07	9.575
15	73.00	15.34	16.92	18.51	9.383
16	76.00	15.97	17.58	19.19	9.155
17	76.00	15.97	17.58	19.19	9.155
18	77.00	16.18	17.80	19.41	9.092
19	80.00	16.80	18.45	20.10	8.945
20	83.00	17.41	19.11	20.80	8.847

Fig. 12.5 Data, Fitted Line and 95th Centiles

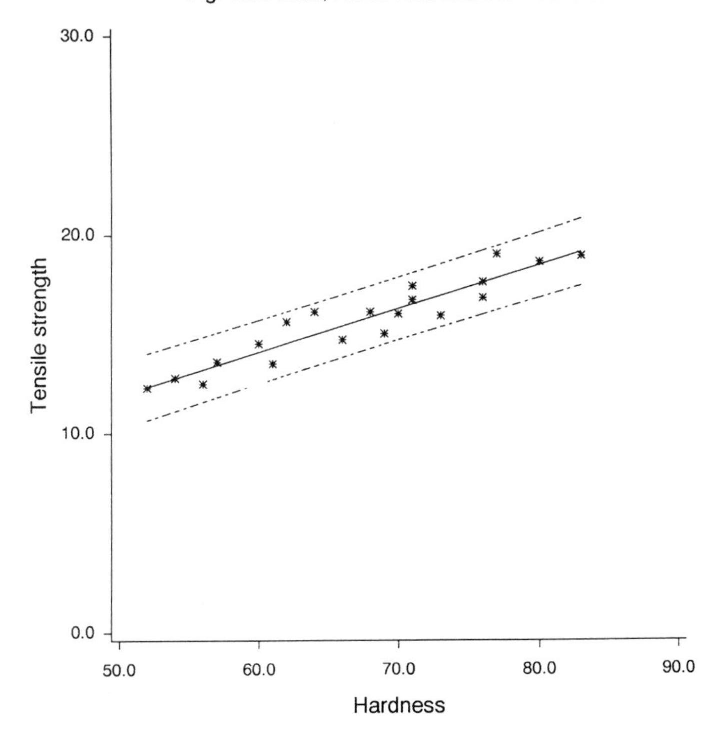

It appears that the relationship is reasonably accurate for predicting mean values, but that there is rather too much residual variation in individual values for it to be accurate for single observations. Whether it is accurate enough or more extensive data are needed to increase the precision of the estimated model parameters and hence the predictions is a matter for the originators of the data to decide.

12.1.1.4 Regression with non-constant variance

A study of foetal growth in Brazil included a set of ultra-sound measurements of the head to coccyx length in millimetres (`htc`) of the foetuses in 83 women between 7 and 16 weeks pregnant. Despite the cross–sectional nature of the data the clinicians wished to use them to estimate foetal growth curves. They wished to estimate rate of growth and obtain reference curves that they could use to:

(i) estimate the range of sizes to be expected at a particular week of the pregnancy so they could judge whether a foetus was particularly large or small for its age

and

(ii) estimate from the size of a foetus how long the mother had been pregnant.

There could be a growth mechanism that might be used to deduce a `correct' model for these data. However there are so many potential sources of error and possibly bias that it is much better to take a pragmatic approach and attempt to find an algebraically simple model that fits the data adequately. (Source: Swan and Gomes, 1989)

```
$units 83 $
$read  week htc
  7  1.4  7  1.1  8  1.0  8   2.1  8  1.4  8  1.6
  8  2.4  8  2.0  8  1.7  8   2.2  8  1.1  8  2.8
  8  1.5  8  2.0  8  1.5  8   1.2  9  2.2  9  2.1
  9  2.2  9  2.0  9  1.8  9   2.6  9  2.5  9  2.7
  9  1.1  9  2.6  9  2.4  9   2.8  9  2.0  9  2.8
  9  2.7  9  2.5  9  2.2  9   2.0 10  2.6 10  3.0
 10  1.7 10  2.1 10  3.8 10   2.5 10  1.5 10  3.9
 10  6.4 10  3.5 10  3.6 10   1.8 10  1.9 10  1.0
 10  3.1 11  3.9 11  2.6 11   5.1 11  3.9 11  2.3
 11  4.1 11  5.3 11  4.7 12   6.2 12  4.7 12  5.7
 12  5.5 12  5.2 12  4.2 12   4.6 12  3.7 12  7.3
 12  5.8 12  6.0 12  4.9 12   3.5 13  3.4 13  5.2
 13  3.5 13  6.8 13  6.0 13   6.3 13  2.7 13  2.2
 13  4.2 13  4.6 14  5.0 15   8.0 16  8.9
```

The first step is to plot `htc` against `week` of pregnancy

```
$plot htc week 'x' $
```

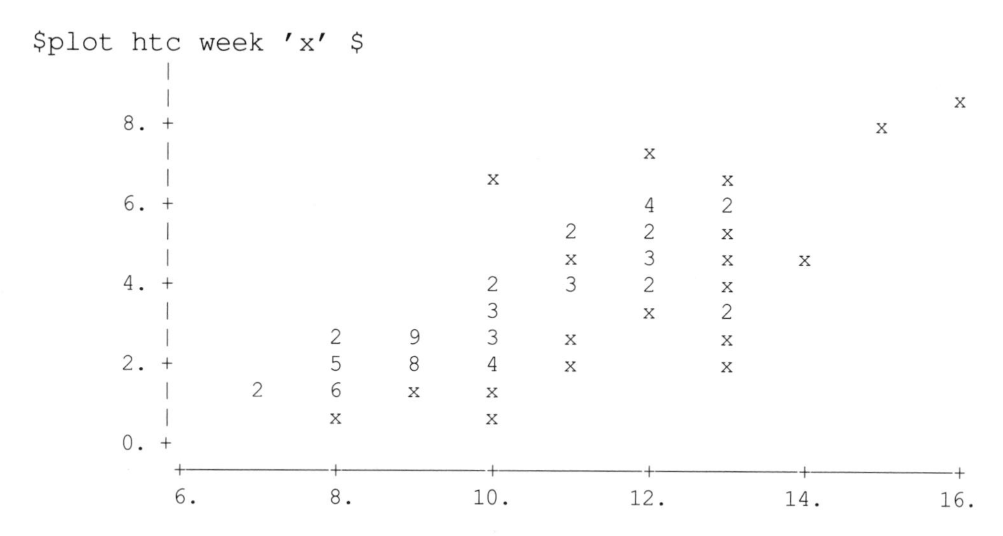

The form of the trend in the data is not immediately obvious although there seems to be some evidence of a slight upward curve. This can be allowed for by including a quadratic term in the model. However the assumption of Normality and constant variance must be thoroughly investigated.

The best strategy is to fit a quadratic regression model and use the residuals to check the assumptions. First identify the y variable and calculate a variable for `week` squared by

```
$yvar htc $
$cal week2=week*week $

$fit week+week2 $dis e $
   deviance =  92.484
residual df =  80
          estimate        s.e.      parameter
     1      -1.430       3.119       1
     2       0.1967      0.5882      WEEK
     3       0.02475     0.02707     WEEK2
scale parameter 1.156
```

The coefficient of `week2` is actually smaller than its s.e. so there is not much evidence of curvature in the relationship. Testing the Normality of the `htc` distributions about the curve is achieved [see 12.1.1.2] by

```
$use crep $
```

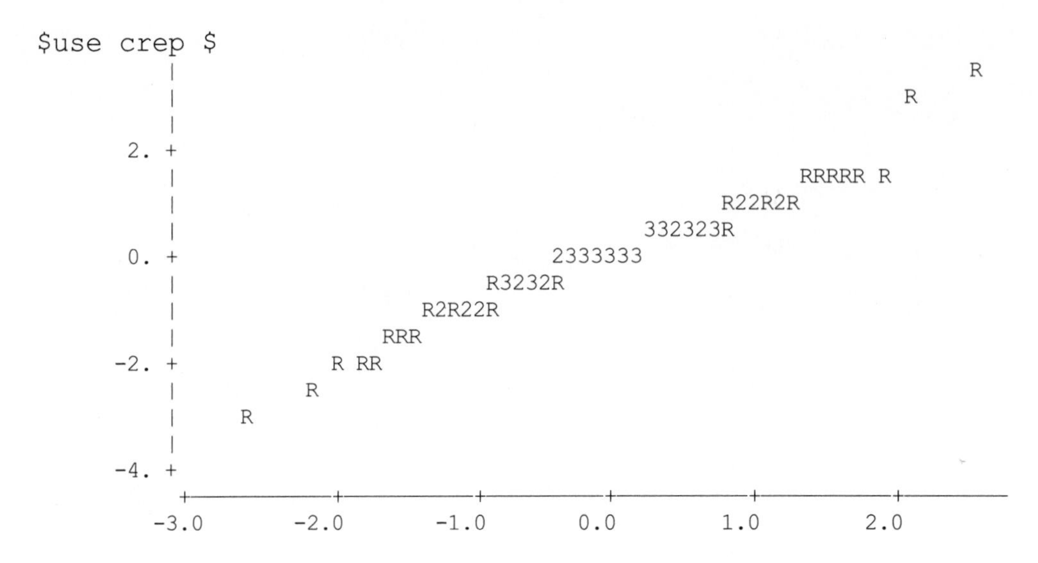

There is some curvature in this plot suggesting that the error distribution is not actually Normal although it is close enough for most practical purposes. The constant variance assumption is more important. We can investigate that by plotting the residuals against `week`.

```
$plot %rs week 'r' $
```

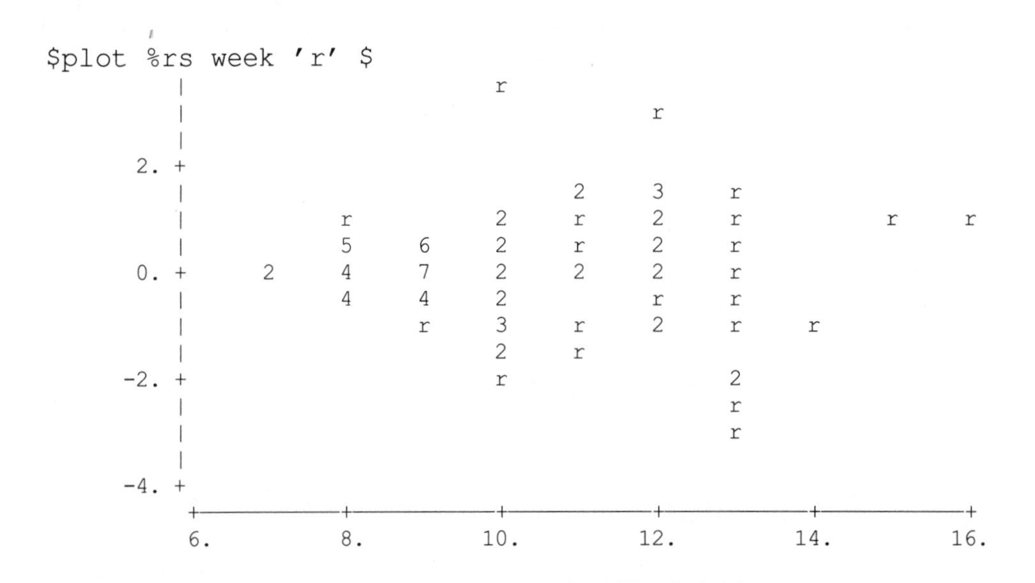

It seems fairly clear from this plot that the spread of the residuals is increasing with `week`. This means that the variance of `htc` is increasing as the mean `htc` increases. A tabulation of the means and standard deviations by week confirms this.

```
$tab the htc mean;dev;weight for week into mean;sd;n by week_ $
$tprint (s=1) n;mean;sd week_ $
```

WEEK_	7.000	8.000	9.000	10.000	11.000	12.000	13.000
N	2.000	14.000	18.000	15.000	8.000	13.000	10.000
MEAN	1.250	1.750	2.289	2.827	3.987	5.177	4.490
SD	0.2121	0.5215	0.4283	1.3360	1.0869	1.0639	1.5673

WEEK_	14.000	15.000	16.000
N	1.000	1.000	1.000
MEAN	5.000	8.000	8.900
SD	0.0000	0.0000	0.0000

The standard deviations of zero from single observations provide no useful information so they should be weighted out:

```
$cal w_sd=(n>1) $
$plot sd/w_sd mean $
```

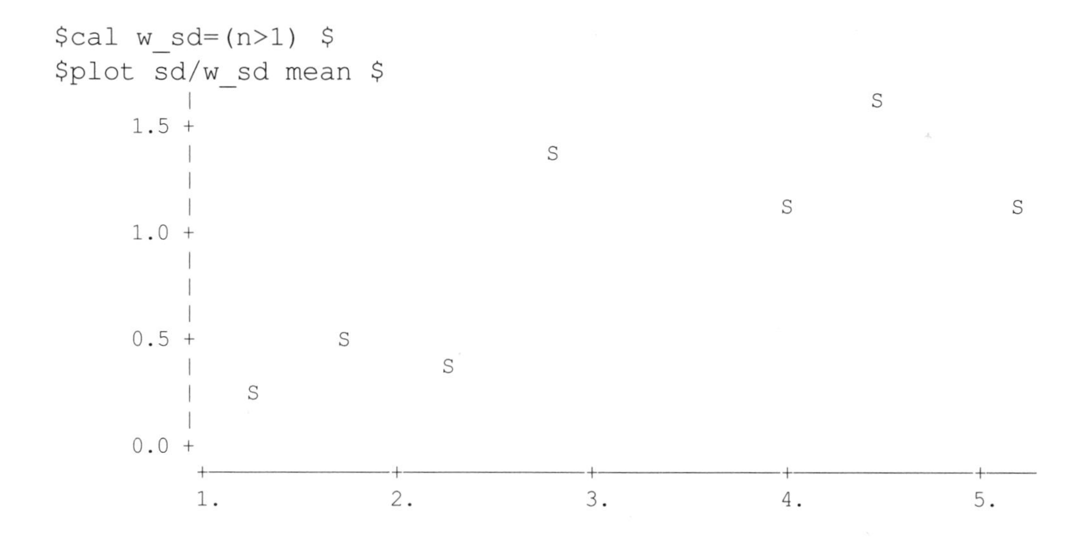

The standard deviations, on average, are increasing with the mean. This suggests that another error distribution with an appropriate mean–variance relationship might be more appropriate. For example, the gamma distribution has the mean proportional to the standard deviation. However it does not work very well with this dataset. A more traditional approach is much more effective. This involves assuming that the response y follows a log-Normal distribution and using a log transformation of the y variable to achieve Normality. If the y mean is proportional to the standard deviation this will have the effect of stabilizing the variance (Box and Cox 1964).

The general technique for identifying variance stabilizing transformations is to plot the logarithms of the standard deviations against the logarithms of the means. A straight line trend of slope b implies the need for the original variables to be transformed by raising them to the power of $1 - b$ (or to logarithms if b=1). The plot, which excludes values based on n's of 1, is

```
$cal lm=%log(mean) : lsd=%log(sd) $
$plot lsd/w_sd lm 'x' $
```

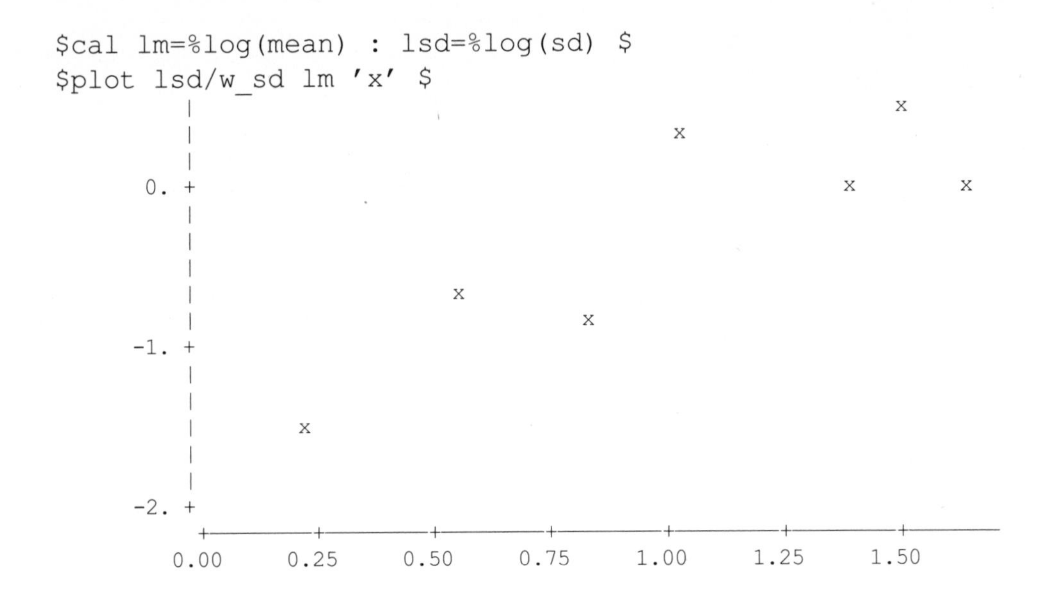

The slope of this line can be obtained by performing a weighted regression of the log(standard deviations) on the log(means)

```
$cal w10=n-1 $
$weight w10 $
$yvar lsd $fit lm $dis e $
   deviance =  8.4734
residual df =  5        from 7 observations
           estimate        s.e.     parameter
     1       -1.300      0.4441      1
     2        1.011      0.3852      LM
scale parameter 1.695
$weight $
```

The slope of nearly 1 (in fact 1.011) suggests that taking the logarithms of the `htc` values will stabilize the variance sufficiently.

Taking logs of `htc` we get

```
$cal lhtc=%log(htc) $
$plot lhtc week $
```

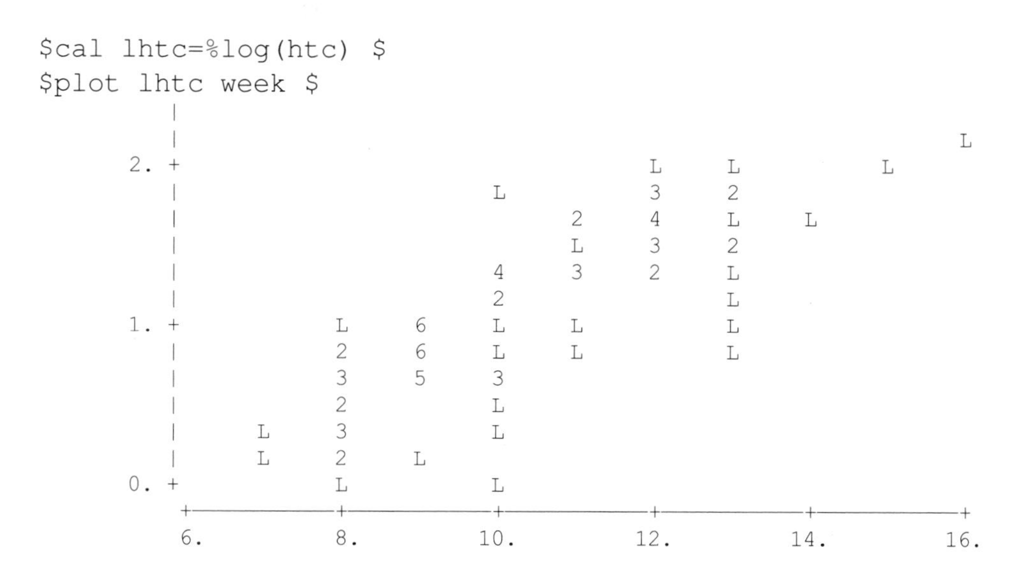

It appears that the transformation has modified the form of the trend which now shows hardly any curvature. Patterns of change in the spread about a trend are not easy to visualize, but the variation does not appear to be changing in any systematic way with `week`. We can examine this more closely by obtaining the mean and standard deviation of the transformed variable grouped in weeks

```
$tab the lhtc mean;dev;weight for week into mean;sd;n by
week_$ $tprint (s=1) n;mean;sd week_ $
```

WEEK_	7.000	8.000	9.000	10.000	11.000	12.000	13.000
N	2.000	14.000	18.000	15.000	8.000	13.000	10.000
MEAN	0.2159	0.5177	0.8075	0.9397	1.3457	1.6242	1.4417
SD	0.1705	0.3031	0.2223	0.4659	0.3035	0.2100	0.3754

WEEK_	14.000	15.000	16.000
N	1.000	1.000	1.000
MEAN	1.6094	2.0794	2.1861
SD	0.0000	0.0000	0.0000

```
$plot sd/w_sd mean $
```

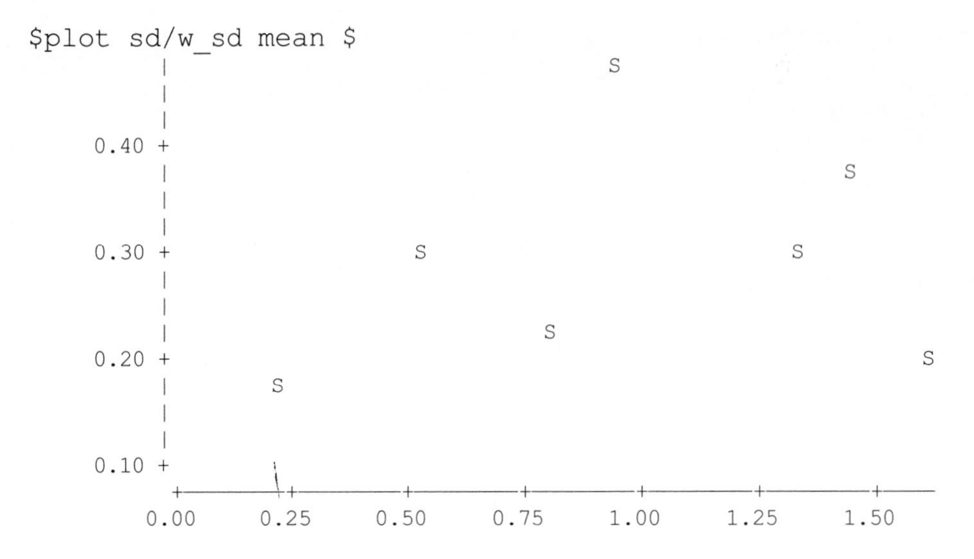

Although it has not disappeared there is less evidence of a systematic tendency for the variance to increase with the mean. For the purposes of this example we will assume the transformation to be appropriate.

To check for curvature on the transformed scale now fit the curved model with `week` and `week` squared to the new y variable:

```
$yvar lhtc $
$fit week+week2 $dis e $
   deviance =     8.5635
residual df =   80
          estimate          s.e.       parameter
     1      -2.540        0.9492        1
     2       0.4794       0.1790        WEEK
     3     -0.01218     0.008237        WEEK2
scale parameter 0.1070
```

The coefficient of the squared term has become negative indicating that any curvature is now downward. However it is still smaller than twice its se and can reasonably be ignored. In this context we might, in fact, reason that whether or not it is significant at some level is not the important question. The need is for a prediction model and the important question is whether predictions will be more reliable and accurate with or without the squared `week` term. However the evidence for curvature is not strong and we will assume that a straight line relationship between log `(htc)` and `week` is appropriate.

```
$fit week $dis e $
   deviance =     8.7974
residual df =   81
          estimate          s.e.       parameter
     1      -1.167        0.1966        1
```

```
     2       0.2163      0.01871        WEEK
scale parameter 0.1086
```

The assumption of Normality is again tested by using the macro

```
$use crep $
```

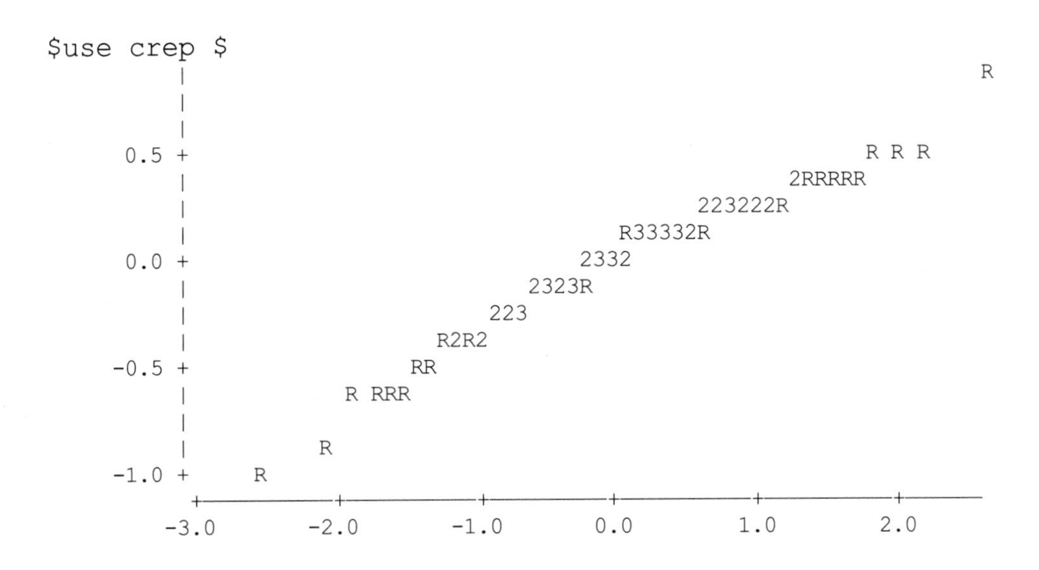

This is not a very straight line, but adequate in most circumstances for us to feel justified in assuming Normality. To check the constant variance assumption a plot of the residuals against the fitted values is necessary.

```
$plot %rs %fv 'x' $
```

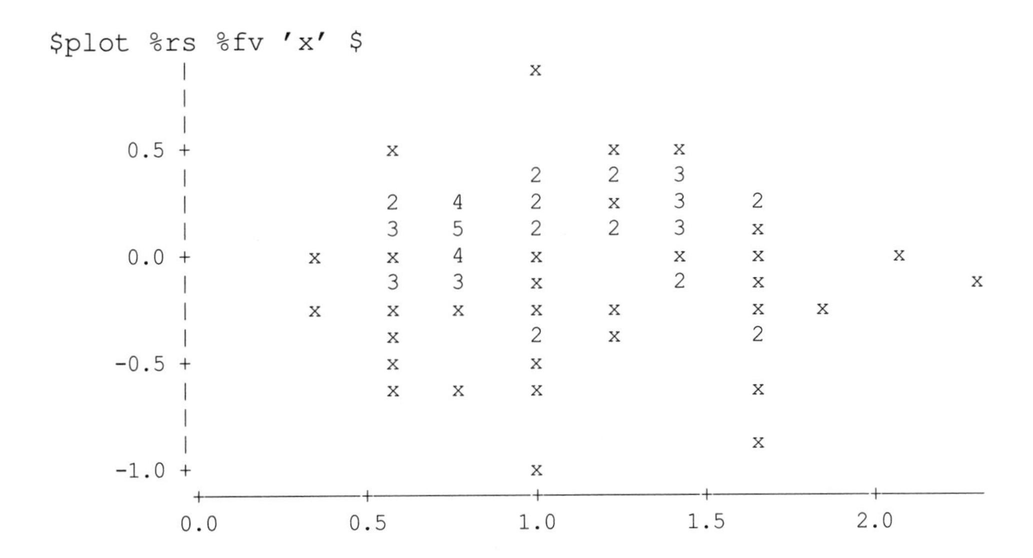

The systematic increase in spread with increasing `htc` values has disappeared confirming that an assumption of constant variance is reasonable.

Now the aim is prediction so we need the fitted line and a confidence region for it. In practice we will also want this transformed back to the original `htc` scale. To plot the fitted values with one point for each week requires a new x variable with the appropriate `week` values. The values for the fitted line are then obtained using `PREDICT` which places the 'predicted fitted values' in the system vector `%pfv`. The variances of these, needed to calculate the confidence limits, are also produced and made available in the system vector `%pvl`.

```
$ass wk=5,6...20 $

$predict (s=-1) week=wk $
$cal fl=%pfv $
$plot fl wk '.' $
       |                                                          .
    3. +                                                      .
       |                                                  .
       |                                              .
       |                                          .
    2. +                                      .
       |                              .   .
       |                          .
       |                      .
    1. +                  .
       |              .
       |          .
       |      .   .
    0. +.
       |
       +---------+---------+---------+---------+---------+---------+
      5.0       7.5      10.0      12.5      15.0      17.5      20.0
```

```
$mac title Fig. 12.6  Data on Log scale with Fitted Line
$end$ $mac xlabel Week of gestation $e $
$mac ylabel Log(Head to Coccyx length) $e $
$gra ( t=title h=xlabel v=ylabel x=5,20,5 y=0,3,1)
        lhtc;fl week;wk  1,10 $
```

To predict the mean `htc` for a given week requires either using this plot on the `log(htc)` scale and taking the exponentials (anti-logs) of the predictions or using the fitted line plotted on the original scale which is also useful for presentation purposes.

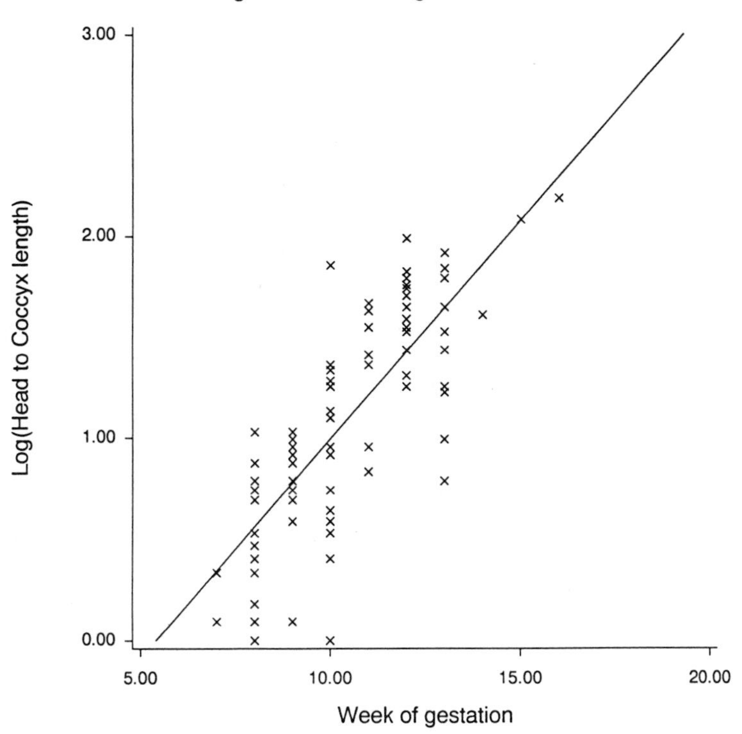

Fig. 12.6 Data on Log scale with Fitted Line

```
$cal fy= %exp(fl)   $plot fy wk '.' $
          |
          |
          |                                                          .
      20. +
          |                                                      .
          |
          |                                                  .
          |
          |
          |                                              .
      10. +                                          .
          |
          |                                      .
          |
          |                                  .
          |                          .   .
          |                  .   .
          |.   .   .   .   .
       0. +
          +------------+-----------+-----------+-----------+-----------+-----------+
         5.0          7.5        10.0        12.5        15.0        17.5        20.0
```

```
$mac title2 Fig 12.7 Data with Fitted Line - original scale $end $
$mac xlabel Week of gestation $e $
$mac ylabel2 Head to Coccyx length $e $
$gra ( t=title2 h=xlabel v=ylabel2 x=5,20,5 y=0,10,2)
       htc;fy week;wk  1,10 $
```

Fig. 12.7 Data with Fitted Line - original scale

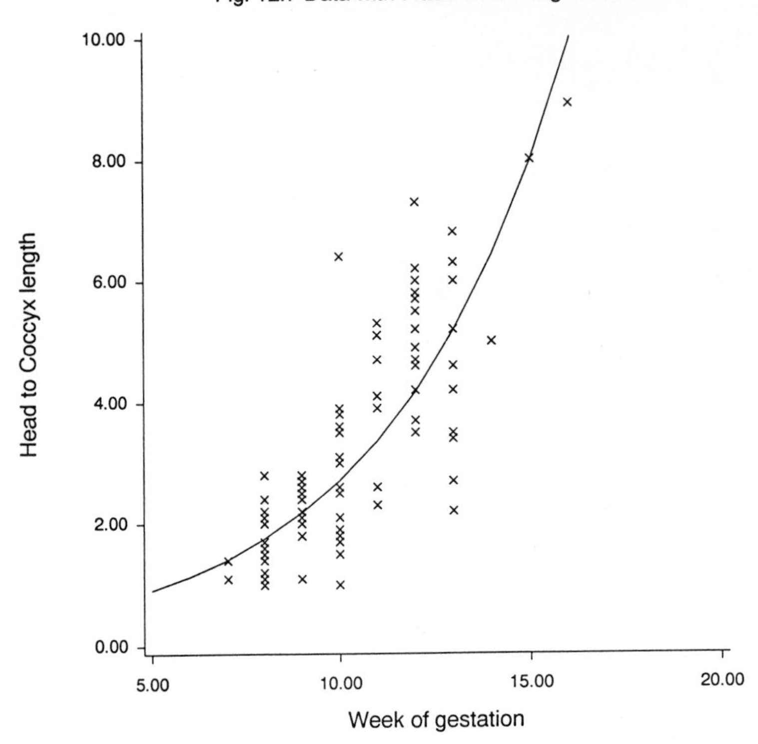

Then calculate the 95 per cent confidence limits for the fitted values using the 97.5 centile of the t distribution with 81 d.f.

```
$number tval $
$cal tval=%td(.975,%df) $
```

```
$cal  low=%pfv - tval*%sqrt(%pvl) : high=%pfv +
tval*%sqrt(%pvl)$
$plot ( x=6,18 ) low high %pfv wk '^v.' $
```

```
         |
         |
      3. +                                                v
         |                                        v   2   .
         |                                    v   .   ^   ^
      2. +                                2   2   ^
         |                        v   3   ^
         |                    2   2
      1. +            v   3   ^
         |        2   2
         |v   3   ^
      0. +2
         |
         |
     -1. +
          +-------+-------+-------+-------+-------+-------+
          6.      8.      10.     12.     14.     16.     18.
```

To see these on the original scale:

```
$cal low=%exp(low) : htcf=%exp(%pfv) : high=%exp(high) $
$plot ( x=6,18 ) low high htcf wk '^v.' $
      I
      I
  30. +
      I
      I
      I
  20. +                                                  v
      I
      I                                             v    .
      I                                        v    .    ^
  10. +                                   v    .    ^
      I                              2    2    ^
      I                    3    3    ^
      Iv   2    3    3    3    3
   0. +2   ^
      +---------+---------+---------+---------+---------+---------+
      6.        8.       10.       12.       14.       16.       18.
```

and using GRAPH

```
$mac title2 Fig. 12.8 Data with Fitted Line - original scale
$end $
$mac xlabel Week of gestation $e $
$mac ylabel2 Head to Coccyx length $e $
$gra ( t=title2 h=xlabel v=ylabel2 x=5,20,5 y=0,10,2)
       htc;low;htcf;high week;wk;wk;wk  1,12,10,12 $
```

The upper limits on the percentage errors likely in predictions of this form are obtained by

```
$cal pelo=100*(1-low/htcf) : pehi=100*(high/htcf-1) $
$look  wk low htcf high pelo pehi $
```

	WK	LOW	HTCF	HIGH	PELO	PEHI
1	5.000	0.7434	0.9179	1.133	19.016	23.481
2	6.000	0.9553	1.1395	1.359	16.170	19.289
3	7.000	1.2259	1.4147	1.632	13.342	15.396
4	8.000	1.5692	1.7562	1.965	10.648	11.916
5	9.000	1.9980	2.1802	2.379	8.356	9.118
6	10.000	2.5161	2.7066	2.912	7.039	7.572
7	11.000	3.1135	3.3601	3.626	7.340	7.921
8	12.000	3.7925	4.1713	4.588	9.083	9.991
9	13.000	4.5797	5.1785	5.855	11.562	13.073
10	14.000	5.5080	6.4287	7.503	14.323	16.717
11	15.000	6.6109	7.9809	9.635	17.166	20.723
12	16.000	7.9258	9.9078	12.385	20.004	25.006

13	17.000	9.4961	12.2999	15.931	22.795	29.526
14	18.000	11.3726	15.2695	20.502	25.521	34.266
15	19.000	13.6161	18.9562	26.391	28.171	39.219
16	20.000	16.2988	23.5329	33.978	30.740	44.384

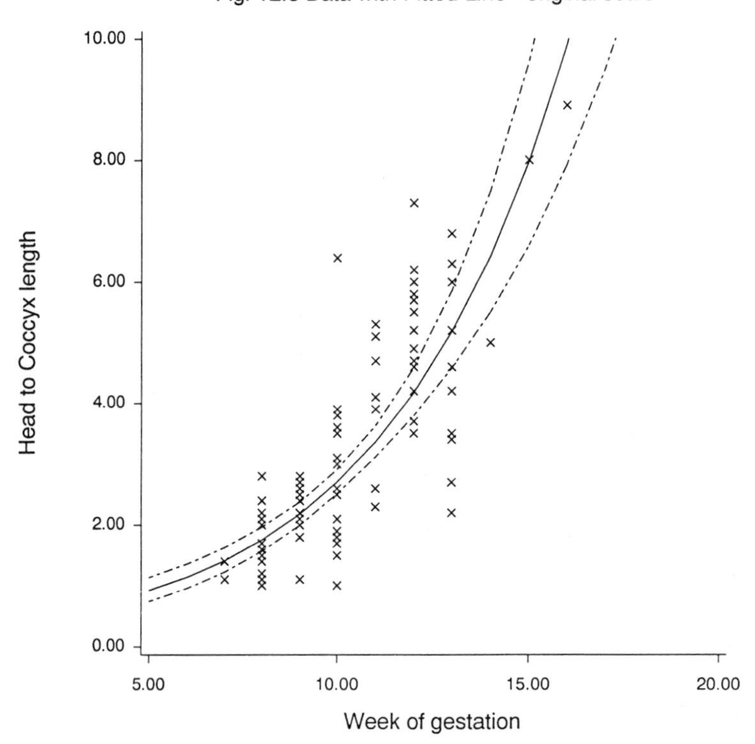

They are not very good percentage errors although as an approximate guide for a clinician monitoring the progress of a pregnancy they may be quite helpful. A clinician may also be interested in the centiles within which an individual's measurement should fall. These are obtained in exactly the same way as above with the variances of the linear predictors increased by the residual variance estimate.

The second aim, to make predictions of the length of gestation from the length of the foetus, that is, from the `htc` value, is a little more difficult. It can be done graphically using the above plot on the log scale and returning to the original scale with an exponential transformation. This involves identifying the x values for where a horizontal line representing a fixed `htc` cuts the fitted line and the confidence limit curves and performing the appropriate transformations. More generally these inverse predictions can be obtained by an equivalent algebraic exercise which involves solving a cumbersome quadratic equation. This can be used to produce the x values consistent with a set of y values from which an appropriate plot or a table can be constructed. As before an exponential transformation must be used to get back to the htc scale. This is rather more conveniently dealt with in a macro, XPRD, which has previously been retrieved from a file. First set a weight variable to indicate the units included in the fit which in this case is all of them.

```
$cal w_in=1  $
```

Then to obtain results for a set of 'observed' htc values e.g. 1.5, 2, 4, 6, 8, 10:

```
$ass ohtc=1.5,2,4,6,8,10 $cal yval=%log(ohtc) $

$use xprd lhtc week w_in yval  $
95% CLs using t =     1.989
  for the range of x values consistent with a set of "observed" y values
       YVAL        XL_        X_       XH_
 1    0.4055     4.035     7.271    10.32
 2    0.6931     5.438     8.601    11.66
 3    1.3863     8.744    11.806    14.96
 4    1.7918    10.631    13.681    16.94
 5    2.0794    11.949    15.011    18.36
 6    2.3026    12.960    16.043    19.48
 Plot of predicted "x" values with CLs against chosen "y" values
      20. +                                                          v
          |                                                    v
          |                                             v            .
      15. +                                     v                  .
          |                                                  .       ^
          |                 v                   .              ^
      10. +            v                                 ^
          |                 .                   ^
          |            .
       5. +                 ^
          |            ^
          |
       0. +
          +------------+------------+------------+------------+------
         0.0          0.5          1.0          1.5          2.0
```

where X_ is the best point estimate and XL_, XH_ are the 95 per cent confidence limits (using t = 1.99) for the week of pregnancy consistent with these specified y values using the fitted straight line model. Plotted against the selected htc values we obtain the curves required by the clinicians.

```
$plot xl_ x_ xh_ ohtc '^.v'$
```

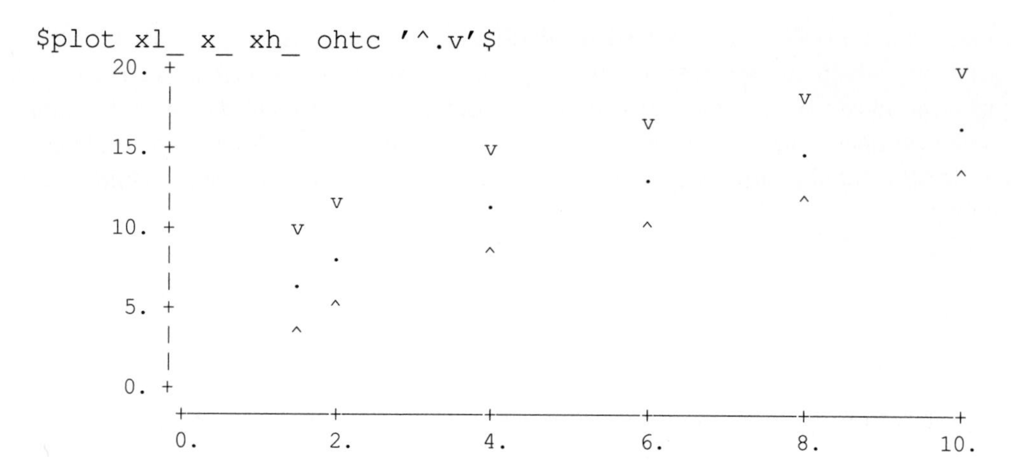

The equivalent graph is obtained by

```
$graph ( t='Fig. 12.9 Gestation prediction from HTC'
         v='Gestation - weeks'
         h='Head to Coccyx length' x=0,10,2 )
         xl_;x_;xh_ ohtc 11,10,11 $
```

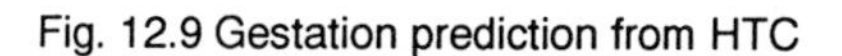

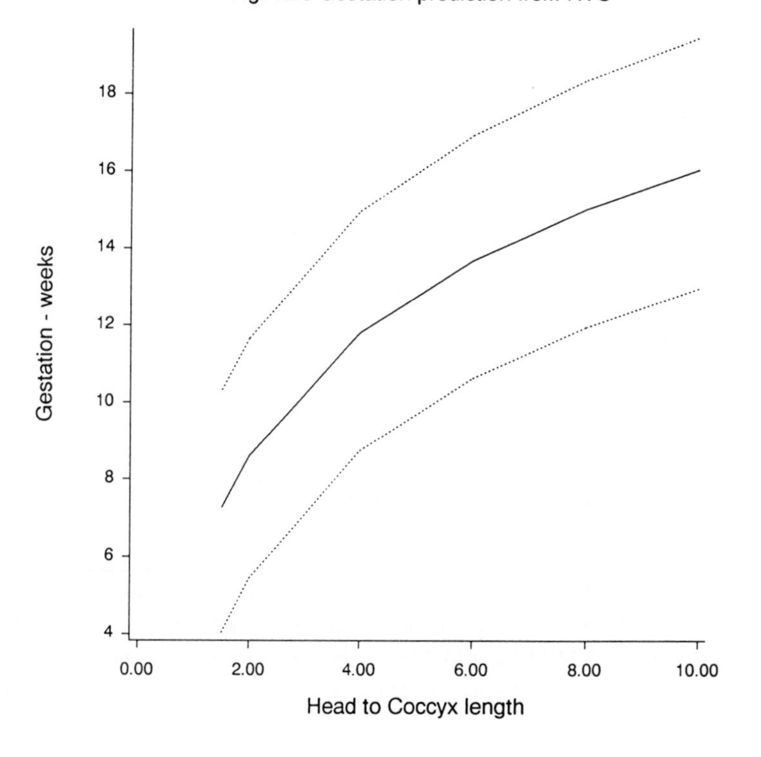

However it is clearly not possible to deduce the week of gestation at all accurately from htc values. The variation among foetuses of the same age is far too great.

12.1.1.5 Multiple regression

These data, taken from Brookes and Dick (1951), arose in an investigation of the relative effects of temperature and calcium concentrations in the early salt phase on the final calcium concentrations in the metal phase in a certain chemical process. The three variables measured were

pcalc - the percentage of calcium in the metal phase of the calcium chloride/sodium chloride/calcium/sodium melt

temp - the melt temperature

and

pcachl- the percentage of calcium in the salt phase of the melt.

```
$units 28 $data pcalc temp pcachl $read
3.02  547   63.7 3.22  550   62.8 2.98  550   65.1 3.90  556    65.6
3.38  572   64.3 2.74  574   62.1 3.13  574   63.0 3.12  575    61.7
2.91  575   62.3 2.72  575   62.6 2.99  575   62.9 2.42  575    63.2
2.90  576   62.6 2.36  579   62.4 2.90  580   62.0 2.34  580    62.2
2.92  580   62.9 2.67  591   58.6 3.28  602   61.5 3.01  602    61.9
3.01  602   62.2 3.59  605   63.3 2.21  608   58.0 2.00  608    59.4
1.92  608   59.8 3.77  609   63.4 4.18  610   64.2 2.09  695    58.4
```

The form of the relationship between these variables will determine a response surface in three dimensions representing the variation observed in the y variable with changes in one or both of the two x variables.

The obvious approach is to regress the y variable pcalc on both variables using product and quadratic terms to check the possibilities of a curved surface or relationship. It appears that the problem is a mixture of a pragmatic attempt to see which of the x variables are most strongly related to the y variable and a precise modelling approach for prediction if the relationship is close enough.

First it is wise to investigate the data graphically before fitting any models because the nature of the data easily seen in a plot may make particular modelling approaches quite inappropriate. It is usual to start by plotting the y variable against the two x variables separately:

```
$plot pcalc temp $
```

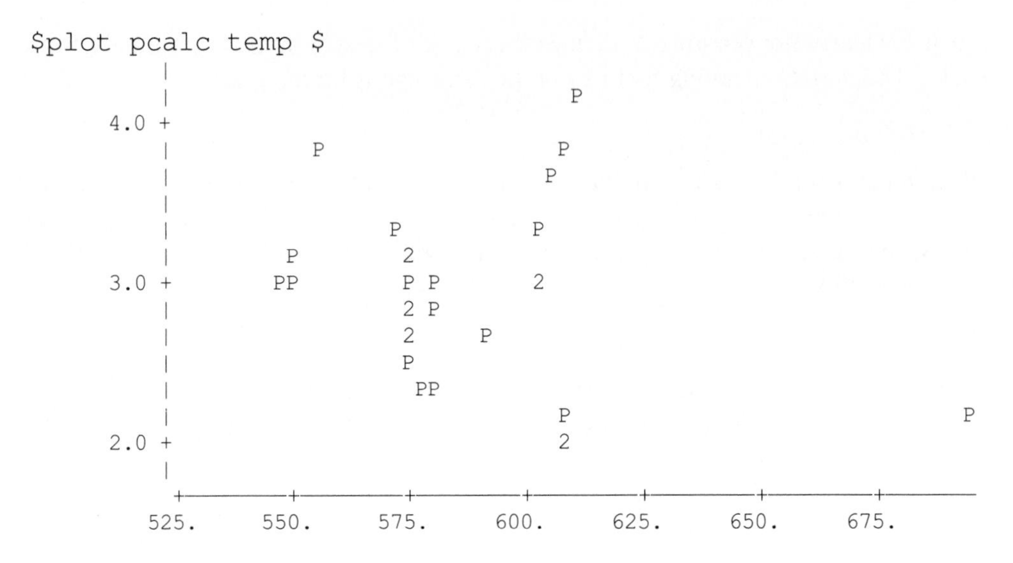

This is not a very illuminating plot; it gives the impression of some sort of relationship, but this is a projection on the `pcalc-temp` plane of points with potentially very different values of the other x variable. If the other x is related to y it could distort this plot in a variety of ways dependent on how it varies with temperature.

```
$plot pcalc pcachl $
```

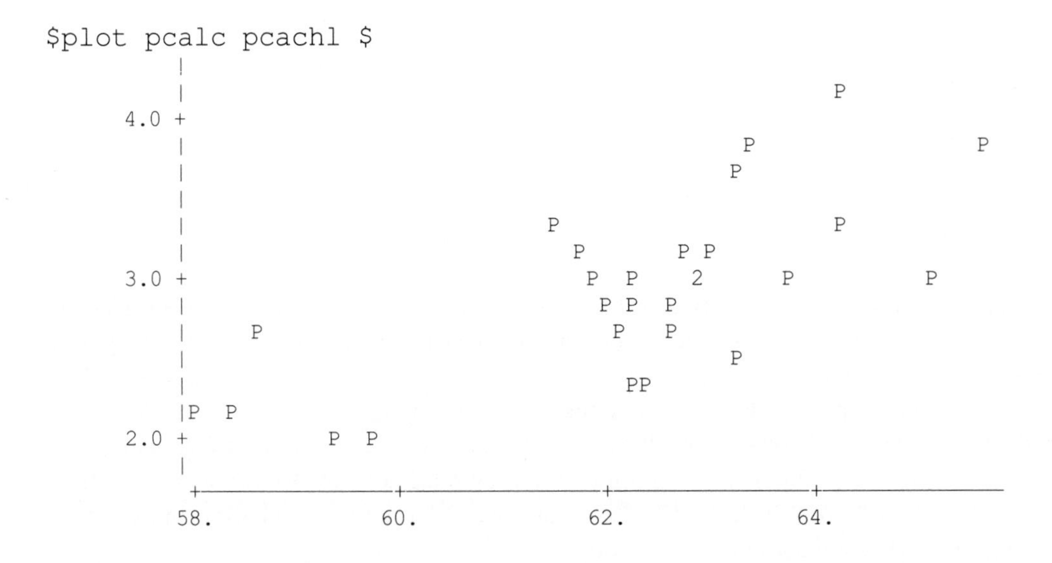

As before there is some pattern, but it is rather difficult to interpret. Now it is necessary to investigate the relationship between the two x variables. This is the essential design of the study and the more closely they are related the lower the sensitivity of the design for estimating and distinguishing their separate effects.

```
$plot pcachl temp $
```

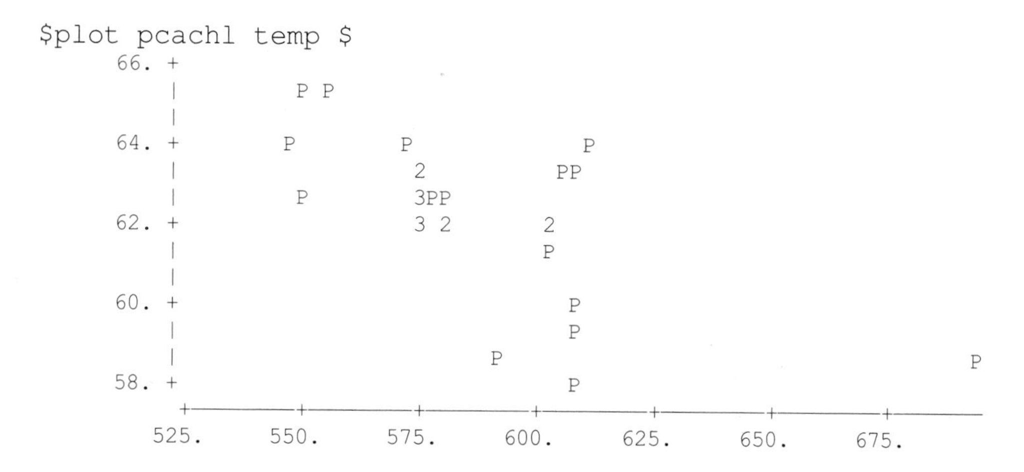

There is clearly a relationship between the two x variables. The higher pcachl values are associated with the lower temperatures. However there is some spread in both directions so the fitted surface will be moderately stable and the s.e.s of the parameter estimates not too excessive.

Obviously the simplest model to interpret and use for prediction would be linear in the x variables as well as in the parameters. However this can only be accepted as an appropriate model if it can be assumed that there is no curvature in the response surface. This assumption ought to be tested first before we even see the simpler model. We will fit a full quadratic model with both x^2 terms and the cross product. First we must obtain variables to represent the squared terms; the product term can be specified as such in the model.

```
$cal pcachl2=pcachl*pcachl : tem2=temp*temp  $
```

Then fit the model with pcalc as the y variable using

```
$yvar pcalc $
$fit temp+pcachl+tem2+temp.pcachl+pcachl2 $dis e $
   deviance =    1.8486
residual df =   22
          estimate          s.e.     parameter
     1       801.2         199.0      1
     2      -1.092        0.3031      TEMP
     3      -15.65         3.746      PCACHL
     4   0.0003176     0.0001071      TEM2
     5     0.07311       0.01892      PCACHL2
     6     0.01162      0.002894      TEMP.PCACHL
scale parameter 0.08403
```

All the product terms coefficients are more than twice their s.e.s. Clearly a planar or flat surface relationship between the y variable and the two x variables is too simplistic. We will carry out a formal test of this after inspecting the residuals to test the assumptions of Normality and constant variance.

```
$hist %rs 'C' $
[-0.600,-0.350)  3 CCC
[-0.350,-0.100)  5 CCCCC
[-0.100, 0.150) 14 CCCCCCCCCCCCCC
[ 0.150, 0.400)  5 CCCCC
[ 0.400, 0.650]  1 C
```

```
$plot %rs %fv 'r' $
         |
         |                   r
     0.5 +
         |                       r                     r
         |                       r   r
         |                       rr  2   r
     0.0 +    rr             r    rr   r r       r      r     r
         |     r                   r
         |      r                              r                                  r
         |
    -0.5 +                       rr
         |                              r
         |
         |
    -1.0 +
          +--------------+--------------+--------------+--------------+----------
         2.0            2.5            3.0            3.5            4.0
```

There is no evidence of the vertical spread of the residuals increasing with increasing y values. Normality is tested using the macro `crep` to obtain a Normal plot as in [12.1.1.3].

```
$use crep $
```

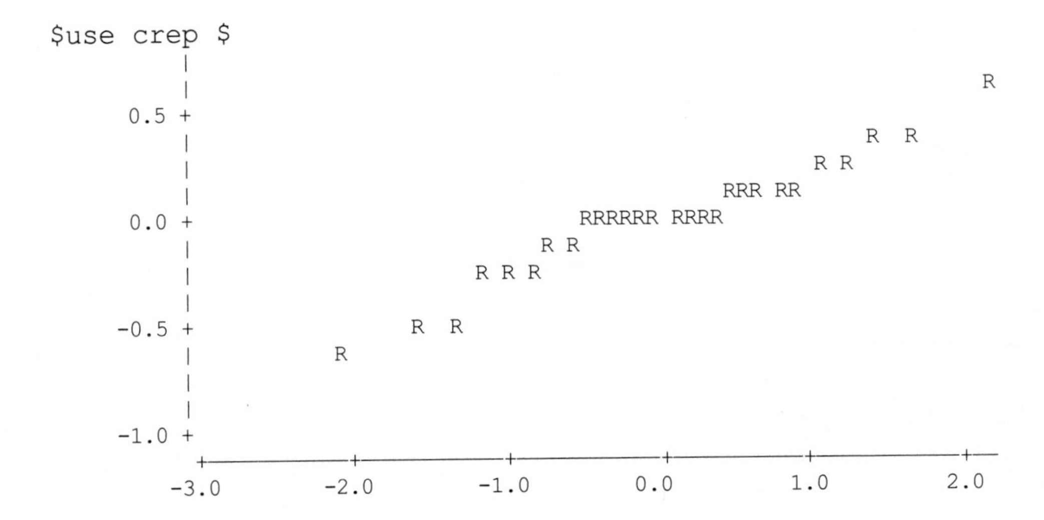

This is a perfectly respectable Normal plot. All the assumptions seem justified except that the model is anything but simple. The relationships seem to have produced a relatively complex response surface.

To obtain an overall test of the apparent curvature we fit the model without the squared and product terms and use the increase in the residual sum of squares (the deviance) to construct an F test. First it is necessary to store the deviance and degrees of freedom from the earlier fit.

```
$number RSS1 df1 s2 df21 F $
$cal RSS1=%dv : df1=%df : s2=%sc $

$fit temp+pcachl  $dis e $
   deviance =   3.6791
residual df =  25
          estimate        s.e.     parameter
     1      -16.63       4.607      1
     2    0.005478    0.003325      TEMP
     3      0.2625     0.05022      PCACHL
scale parameter 0.1472
```

The appropriate F test is then obtained by

```
$cal df21=%df-df1 : RSS2=%dv : F=((RSS2-RSS1)/df21)/s2 $
```

giving

```
$pr '  F ='F'  with 'df21'  and 'df1'  degrees of freedom' ;; $
    F =   7.262  with    3.000  and     22.00  degrees of freedom
```

The probability of a value as large or larger than this is obtained by

```
$cal 1-%fp(F,df21,df1) $
       0.001463
```

So the *p*–value is considerably less than 0.05 and the F–value is highly significant

The more complex model is clearly necessary. To investigate its fit in more detail we need to fit it again and look at some measures of how much influence each of the points had on the fitting process. It will also be useful to explore the shape of the fitted surface with a series of cross–sectional curves.

```
$fit temp+pcachl+tem2+temp.pcachl+pcachl2 $dis e $
   deviance =   1.8486
residual df =  22
```

```
        estimate          s.e.      parameter
1          801.2         199.0       1
2         -1.092        0.3031       TEMP
3         -15.65         3.746       PCACHL
4      0.0003176     0.0001071       TEM2
5        0.07311       0.01892       PCACHL2
6        0.01162      0.002894       TEMP.PCACHL
scale parameter 0.08403
```

There were no very obvious extreme data points in the plots above; however it is possible that there are points, not obvious from a visual inspection of the values, with an undue influence on the form of the best fitting model. The amount of influence a single point has on the model fitting can be measured in a number of ways. Two of the most useful are the leverages which show for each observation its influence on the overall fit of the model and Cook's distances which measure the overall influence each observation has on the parameter estimates. Both of these are available in GLIM after a fit and can be extracted via the system vectors `%lv` and `%cd`. Plotting the measures of influence against temperature will show both the range of values they take and where they are on the temperature scale.

```
$extract %lv %cd $

$plot %lv temp 'i' $
```

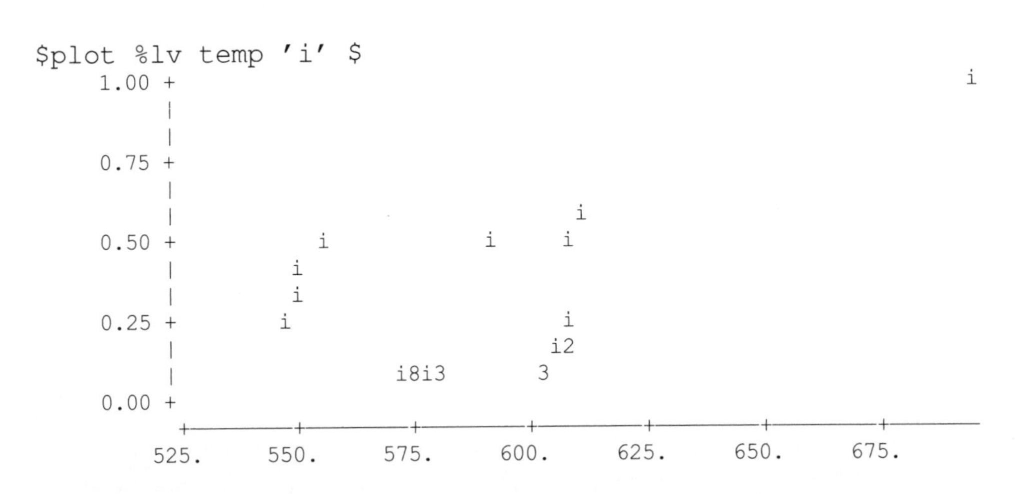

Leverages should be compared with an average of about twice the number of parameters divided by the number of observations, that is, 2*6/28 = 0.43. On this basis it can be seen that a small number of observations have a large share of the leverage. They appear to be those at the extremes of the temperature range and that is as one would expect. However there is one point with a markedly higher temperature than the others which has very much the highest leverage. Cook's distances are also available and can be investigated in the same way:

```
$plot %cd temp 'c' $
        |
        |                                               c
    1.0 +
        |
        |
        |
        |
        |        c
    0.5 +
        |
        |
        |                        c
        |                     c
        |       2
    0.0 +      c      c8c3  c 2c4
         +------+------+------+------+------+------+------
       525.   550.   575.   600.   625.   650.   675.
```

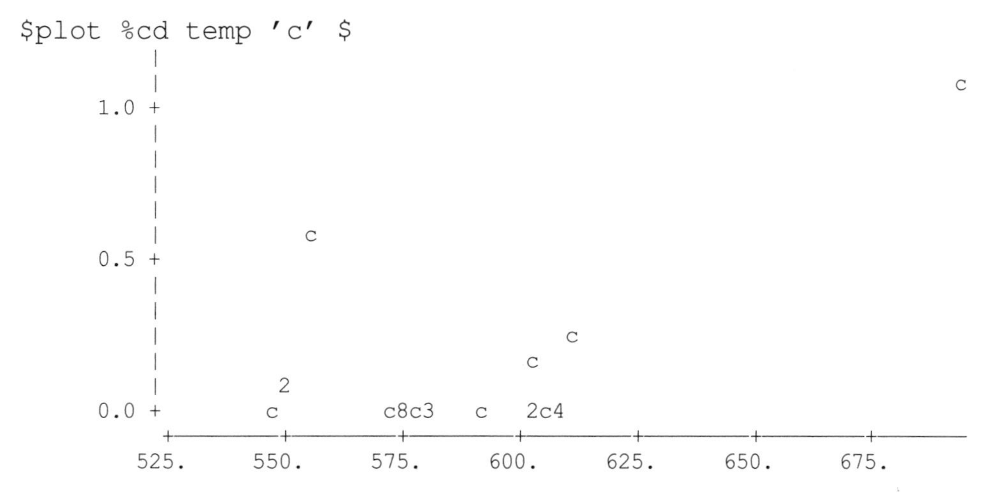

A similar pattern is seen with the observation at the highest temperature standing out most. It is worth checking what effect on the fit it has if this observation is omitted, but before we leave this model let us investigate the shape of the response surface it represents.

The simplest way to look at the shape of the surface is to obtain and plot curves representing cross–sections through the surface in the temperature direction with `pachl` held constant at a selection of values. This can be done using `PREDICT` with values supplied for temperature and `pcachl` and their squares. The product term will be dealt with automatically. The `GFACTOR` directive can be used to generate a vector containing the 20 temperature values 510, 520, ..., 700 repeated for each of the 8 `pcachl` values 58, 59, ..., 65 and a vector containing each of these values repeated for each of the 20 points in each of the 8 cross–sectional curves.

```
$gfactor 160 t 20 p_fac 8 $
$cal t=500+10*t : t2=t*t $
$cal pchl=p_fac+57 : pchl2=pchl*pchl $

$predict (s=-1) temp=t tem2=t2 pcachl=pchl pcachl2=pchl2  $
$plot (x=500,720 y=0,12) %pfv  t  'abcdefgh' p_fac $
        |
        |
        |                                         h
    10. +                                      h    g
        |                                     h   g
        |  a                                h  g    f
        |  b  a                          h    g   f
        |     ba                       h g  g  f    e
        |  c  cb a                 hh  g    f ee  e
     5. +  d  dc b  a         h  h gg  f f    e   d d
        |  2  ed 2  2 2 ah  h g  g ff    e  e dd    c
        |  2  34 4  5 6 54  2 f  2 ee  e d  d  c  c
        |               23  5 5  3 22  2 c  c cb  b b
        |                        b 22  2 2  2 2a  a a
        |
     0. +
         +---------+----------+----------+----------+-----
       500.       550.       600.       650.       700.
```

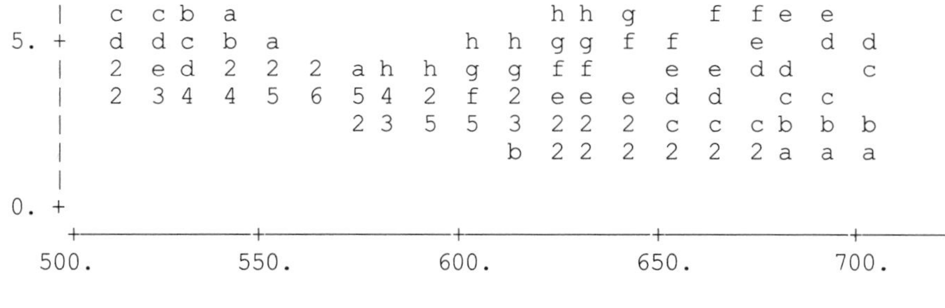

It appears that the surface is concave with metal calcium output initially decreasing as temperature increases until about the middle of the temperature range after which it increases again. At the same time there is a slight, but steady increase with initial calcium chloride concentrations. More detailed appreciation of the form of the surface needs more sections and a high–resolution graph. This can be obtained with the following directives first predicting the points for all the cross–sectional curves in one operation and defining a factor and sequence of pens to ensure that they are each plotted as separate curves.

```
$var 8 pens $cal pens=15 : wts=4/p_fac $

$gra ( t='Fig. 12.10 X-sections through Fitted Surface'
       h='Temp in degrees Centigrade' v='%Calcium in the metal phase'
       x=500,720,40   y=0,10,1 pause=no )
       %pfv   t   pens p_fac  $
$ass xtxt=700,700,700,700,700,700,700
  :  ytxt=1.5,2,3,4.2,5.6,7.1,8.5 $
$gtext (m=p l=r ) ytxt/1.3 xtxt
       'CaChl',' 58',' 59',' 60',' 61',' 62',' 63'$
```

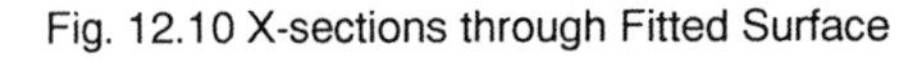

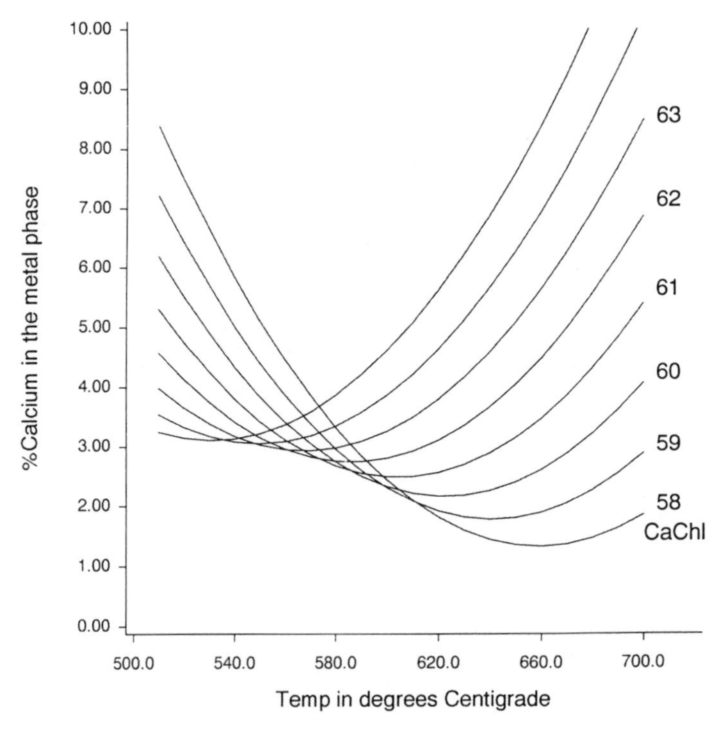

To investigate the magnitude of the effect of the exceptionally influential observation on the fit of the model and the parameter estimates, fit the model again with that observation weighted out. (The observation concerned was the only one with a leverage above 0.75.)

```
$cal wt=(%lv < 0.75) $weight wt $
$fit . $dis e $
   deviance =    1.8386 (change =   -0.01003)
residual df =   21       (change =   -1        ) from 27 observations
          estimate         s.e.      parameter
     1       798.5        203.3        1
     2      -1.138       0.3378        TEMP
     3      -15.14        4.116        PCACHL
     4   0.0003690    0.0001871        TEM2
     5     0.06997      0.02142        PCACHL2
     6     0.01140     0.003023        TEMP.PCACHL
scale parameter 0.08755
```

The deviance has changed by very little so although that observation had the greatest leverage it was not having an undue effect on the fit. The parameter estimates have changed a little and their standard errors have increased, but the general conclusions about the components of the best model are unaffected. Only if a 'best' model was required from this investigation would there be a problem. It could be argued that the most influential observation was incorrect in some way and biasing the estimation process. On the other hand it may be an important and genuine observation, the omission of which would equally bias the estimation. In practice it is almost always most sensible to assume the latter.

12.1.1.6 Linear modelling for curvilinear relationships

As illustrated in earlier examples non-linear functions of the covariates are perfectly acceptable components of linear models. It is the unknown parameters requiring estimation that must be expressible in a linear form. Functions such as $E(y)= a + b \sin(x) + c \exp(-x^2)$ can be fitted quite straightforwardly. However in this example we will restrict our attention to polynomials partly to illustrate the use of orthogonal polynomials.

This is an analysis of the variation in temperature measured hourly over a 24–hour period in Botswana (then known as Bechuanaland) taken in December 1940. The aim is to obtain the simplest polynomial that gives an accurate fit to the observed curvilinear pattern potentially for prediction purposes. The data are taken from Biometrika Tables (Pearson and Hartley, 1962).

We will consider only the model selection part of the analysis and illustrate the use of orthogonal polynomial model specification. The 24 temperatures are input by

```
$un 24 $data temp $read
65.25 69.37 74.44 79 82.72 85.97 88.11 89.67 90.36 90.35 88.92
87.81 84.95 80.41 76.15 74.07 73.13 71.90 70.57 69.08 68.12
67.09 66.71 65.88
```

The time values, held in variable `t`, are generated by

```
$ass t=1,2,...24 $
```

The first thing to do is plot the data

```
$plot temp t  $
```

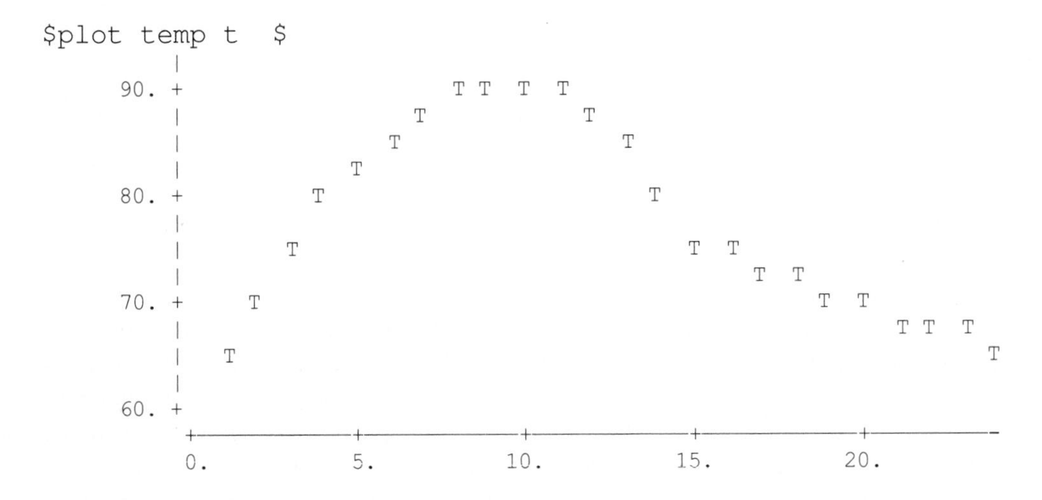

It appears that quite a high degree polynomial will be needed to obtain a well–fitting curve. We will start with a sixth degree polynomial. If we do not use the orthogonal polynomial facility we first need to calculate variables for the powers of *x*. This can be done rather ingeniously using the within expression assignment facility of the `CALCULATE` directive

```
$cal t6=t*t5=t*t4=t*t3=t*t2=t*t  $
```

```
$acc 4 6 $
```

to force fixed format with 6 decimal places

```
$yvar temp $
$fit 1+t+t2+t3+t4+t5+t6 $dis e $
   deviance =  12.575
residual df =  17
          estimate        s.e.     parameter
```

```
      1        62.34         2.209      1
      2        2.233         2.182      T
      3       0.9189        0.7036      T2
      4      -0.1347        0.1015      T3
      5     0.005115      0.007239      T4
      6    -0.000015      0.000250      T5
      7    -0.000002      0.000003      T6
scale parameter 0.7397
```

A graph of the observed and fitted values may be obtained using GRAPH. Firstly, to make it easy to use scaling and labels repeatedly, store them as strings in macros. Pen styles 1 and 10 are set by default to crosses and a solid line.

```
$macro xlabel time in hours since midnight  $end $
$macro ylabel temperature $end $

$gra ( t='Fig. 12.11 24hr Temps Dec 1940 Armoedsvlakte'
       h=xlabel v=ylabel x=0,24,1 ) temp;%fv t  1,10 $
```

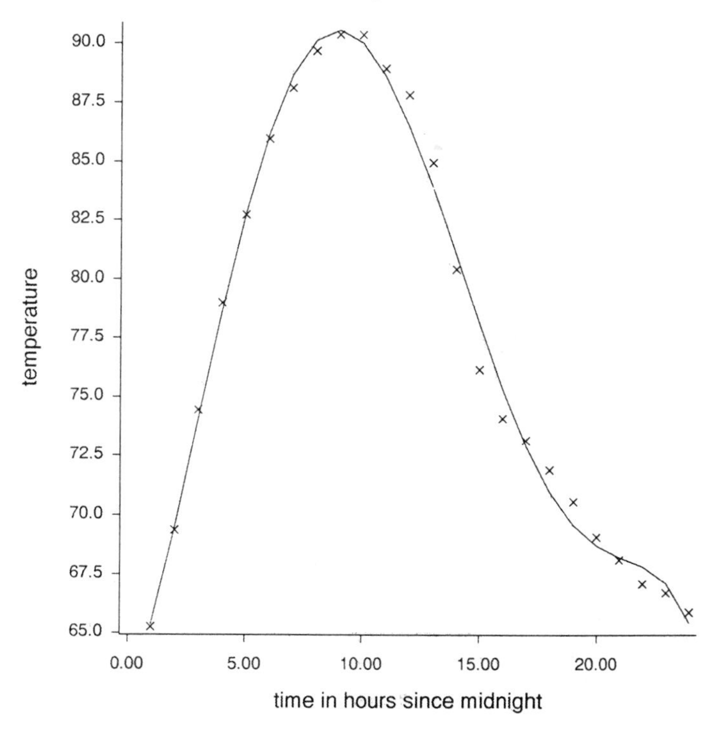

The residuals are inspected, to assess constancy of variance, Normality, and the possibility of outliers, by

```
$plot %rs %fv 'r' $
```

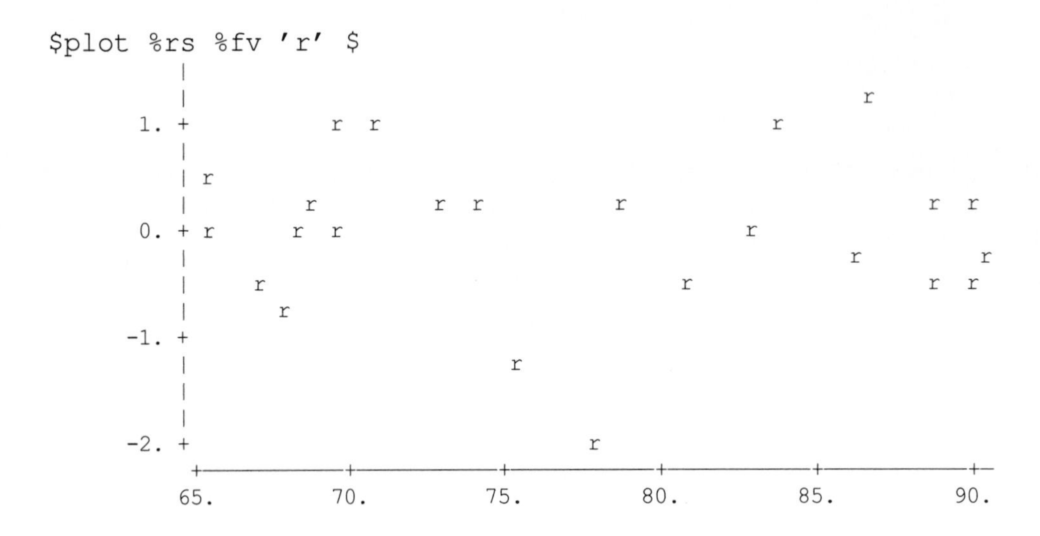

The variation seems reasonably constant.

```
$use crep $
```

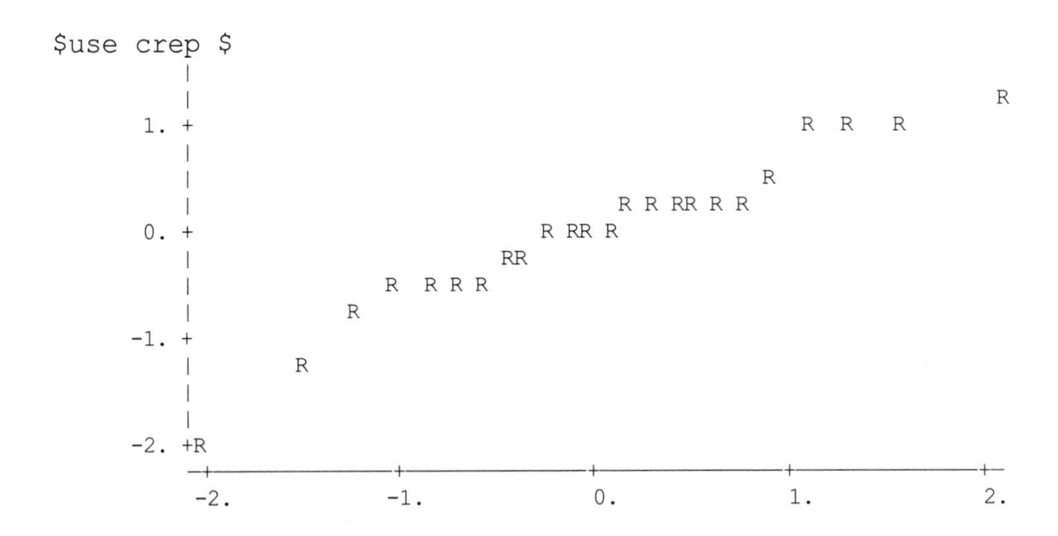

There is some curvature, but no evidence of serious non–Normality.

Turning to the parameters of the fitted model it appears that the sixth power of `t` is not needed since its coefficient is rather larger than its s.e., but we cannot tell anything about the likely fit of lower–order polynomials. We must fit them to find out. First let us calculate the t-test for the coefficient of t^6 and store the residual sum of squares and degrees of freedom for the equivalent F test.

```
$extract %pe %se $
```

To access the stored parameters and their standard errors:

```
$number tval6  Fval RSS6=%dv df6=%df s2=%sc t975_6 F95 $
$cal tval6=%pe(7)/%se(7)  : t975_6=%td(.975,df6) $
```

The t-value testing the coefficient of t^6, its degrees of freedom and the 97.5 centile of the t-distribution are

```
$pr 't-value 'tval6'  with 'df6
    ' df to be compared with +/-'t975_6 $
t-value  -0.4695  with    17.00 df to be compared with +/-   2.110
```

This is far from significant so we re-fit the model without the t^6 term

```
$fit -t6 $dis e $
   deviance =  12.738 (change =  +0.1630)
residual df =  18     (change =  +1      )
         estimate         s.e.      parameter
     1      63.06        1.554        1
     2      1.374        1.164        T
     3      1.222       0.2741        T2
     4    -0.1805      0.02716        T3
     5   0.008466     0.001188        T4
     6  -0.000132     0.000019        T5
scale parameter 0.7077
```

The change in deviance gives an F test for the t^6 term by

```
$number RSS5=%dv df5=%df tval5 t975_5 $
$cal Fval=(RSS5-RSS6)/s2 : F95=%fd(.95,1,df6) $
```

```
$pr ' F test of t^6 term ='Fval' with 1 and'df6' df to be
compared with'F95 $
 F test of t^6 term =  0.2204 with 1 and   17.00 df to be compared with
4.451
```

Note this is equal to the t-value squared:

```
$cal tval6**2 $
        0.2204
```

The t-test for the coefficient of t^5 (parameter 6) is

```
$extract %pe %se $cal tval5=%pe(6)/%se(6)  : t975_5=%td (.975,%df)
```

The t-value testing the coefficient of t^5, its degrees of freedom and the 97.5 centile of the t-distribution are

```
$pr 't-value 'tval5'  with 'df5' df to be compared with
+/-'t975_5 $
t-value   -6.956  with    18.00 df to be compared with +/-   2.101
```

The coefficient of t^5 is more than 6 times its s.e. so clearly this fifth order polynomial, at least, is needed to fit these data adequately.

Now this process required that we calculated variables for all the powers of the explanatory variable t involved in the model specification and fitted the models in sequence until the coefficient of the highest power term was significant. The orthogonal polynomial facility in the model specification algebra allows us to do this rather more efficiently and, although it is not immediately obvious here, more accurately.

```
$fit t<6> $dis e $
   deviance =  12.575
residual df =  17
          estimate        s.e.     parameter
     1       77.50      0.1756      1
     2      -19.34      0.8601      T<1>
     3      -31.57      0.8601      T<2>
     4       20.71      0.8601      T<3>
     5       1.817      0.8601      T<4>
     6      -5.851      0.8601      T<5>
     7     -0.4038      0.8601      T<6>
scale parameter 0.7397
```

Because this is a fit of six orthogonal polynomials of order 1, 2, ..., 6 their estimated coefficients are independent of the other polynomials in the model. This means that the coefficient of t<5> here is the same as would be obtained from the model with t<6> omitted. Because of that the simplest polynomial necessary to fit these data adequately can be deduced from this single fitted model. The t-tests for t^6 and t^5 (and t^4, t^3, and so on) can be obtained from the one model. As a consequence of this they all use the same estimate of residual variance, but it can be argued that this is in fact the most appropriate since the estimate from the maximal model is based on the least assumptions about zero coefficients. The use of the estimate from the current model in the previous approach makes what can be considered unnecessary assumptions. The t-values are

```
$extract %pe %se $
$cal tval6=%pe(7)/%se(7) : tval5=%pe(6)/%se(6) :
t975_6=%td(.975,%df) $
```

The t-values testing the coefficients of t^6 and t^5 have the same degrees of freedom and need to be compared to the same 97.5 centile of the t-distribution. They are

```
$pr 't-values 'tval6' and'tval5' with 'df6' df to be compared
with +/-'t975_6 $
t-values   -0.4695  and   -6.804  with      17.00  df  to  be  compared  with  +/-
2.110
```

They are almost exactly the same as before. The difference is a result of the greater accuracy achieved by the orthogonal polynomial fit and the reduction in the degrees of freedom for the test of t<5> because t<6> is in the model. The fitted curve has not changed and we can see from the plot above that it fits very well in the first part of the 24-hour period, but not so well later. To investigate the curve more closely in that region it would be useful to obtain fitted values for finer divisions of the t axis. Unfortunately using orthogonal polynomials means that we do not have the actual algebraic form of the fitted model immmediately available.

However predicted values from the fitted curve can easily be obtained for any covariate values using the PREDICT directive.

```
$ass times=18.25,18.5,...24  $
$predict t=times $
prediction for the current model with T=TIMES
```

This produces fitted values and their variances for the chosen time values in times and makes them available in the system vectors %pfv and %pvl. These can be used to obtain 95 per cent confidence limits for the fitted line as below and can be plotted to show how well the curve fits.

```
$cal low=%pfv-%td(0.975,%df)*%sqrt(%pvl) :
high=%pfv+%td(0.975,%df)*%sqrt(%pvl)
$plot (x=18,24 y=65,73) temp;%pfv;low;high
t;times;times;times 'oe^v' $
```

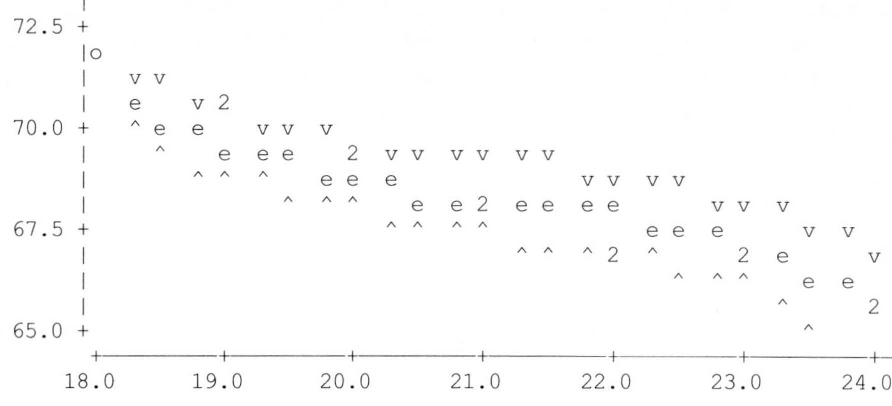

As this is not very clear, higher–resolution graphics are needed to show that the observed 'curve' wanders about rather and is not confined to the region within the confidence limits.

```
$gra ( t='Fig. 12.12 Data with fitted line and 95% CLs'
       h=xlabel v=ylabel x=17.5,24.5,1 y=60,80,10)
       temp;%pfv;low;high t;times;times;times  1,10,12,12 $
```

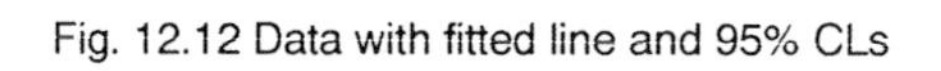
Fig. 12.12 Data with fitted line and 95% CLs

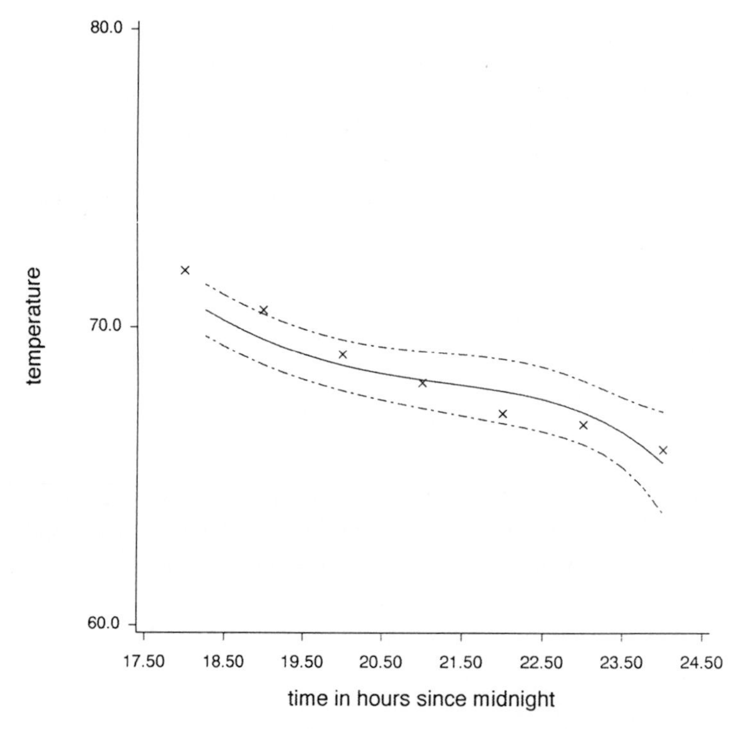

If very precise predictions were needed this would not be adequate. In practice, data from many periods would need to be averaged and the exact form of the relationship might very well be of diminished importance in the face of the day to day variation. In an industrial process on the other hand it might be possible to control the background variation and the precision of the fitted relationship could then be very important.

12.1.2 Qualitative or categorical x variables: group comparisons

12.1.2.1 Comparing two groups: the classical t test

The following data arose from a study to determine the relative merits of high and low protein diets using laboratory rats. Two groups of rats were fed the diets over a period and their weight gains between 28 and 84 days of age recorded (Snedecor, 1956).

```
$un 19 $data group wtgain $read
2 134 2 146 2 104 2 119 2 124 2 161
2 107 2   83 2 113 2 129 2   97 2 123
1  70 1 118 1 101 1  85 1 107 1 132 1 94
```

The `group` variable is a code identifying different categories of individual to be compared. In general it is qualitative in the sense that distances along the scale representing it do not have any particular meaning. Such variables are known as factors in GLIM and generally need to be declared as such together with the number of categories (or levels) they represent.

```
$factor group 2 $
```

We can see how these groups compare by a graph of `wtgain` against `group` using the range 0,3 on the x–xis to keep the `group` codes 1 and 2 centred:

```
$plot (x=0,3) wtgain group $
```

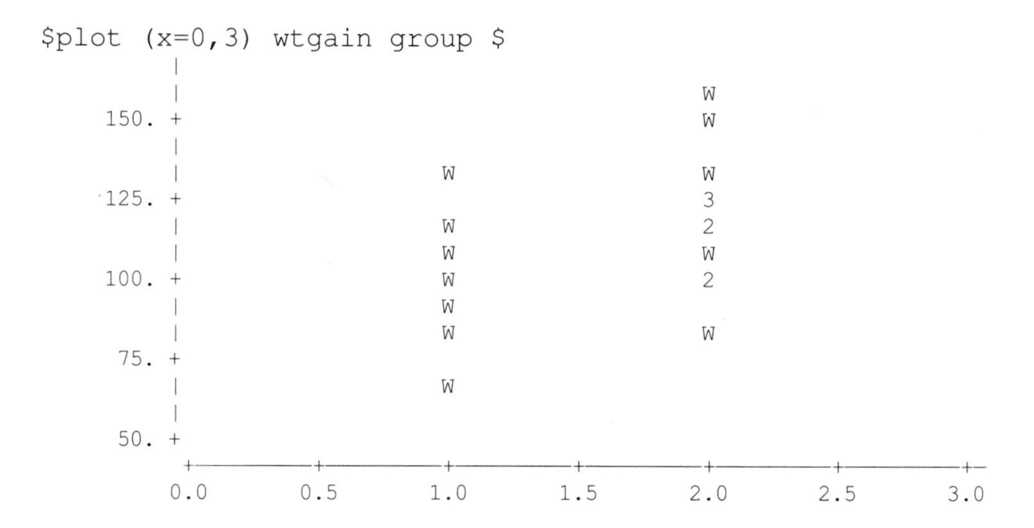

and in higher resolution

```
$gra (title='Fig. 12.13 Weight gains by Diet Group'
     vlab='Wt gain' hla='Diet Group'
     xscale=0,3,1 yscal=20,180,40) wtgain group 1 $
```

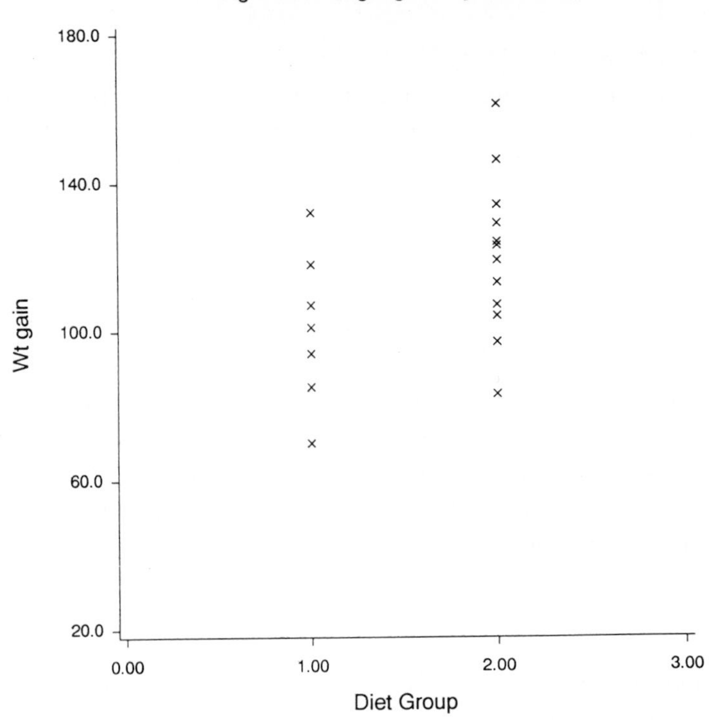

The variances of the two groups appear similar, but the first group seems to have a lower mean. The means and variances are obtained by

```
$tab (style=1) the wtgain mean;var;dev;weight for group by
diet$
```

DIET	1	2
MEAN	101.0	120.0
VARIANCE	425.3	457.5
S.D.	20.62	21.39
TOTAL_WT	7.00	12.00

There is a substantial difference in the means, but the variances are large. The variances are very similar so the assumption of equal variances in the populations represented by these samples is almost certainly justified.

The analysis is achieved quite simply by fitting a linear model describing observations arising from groups with different means. The test can then be performed using the appropriate parameter estimates and their s.e.s. A possible model is

$$y = \text{Group mean} + \text{Random error.}$$

However the question of interest is whether the difference between means is more than could easily occur by chance. For this reason it is useful to parametrize the model in terms of the difference between the means. This is done by default in GLIM; the convention is to identify a reference group and then define the other group effects as differences from that reference mean. That means that the model is

$$y = \text{Reference mean} + \text{Group effect} + \text{Random error.}$$

The package interprets model formulae, constructed using factors, to be of this form with parameter `1` as the reference group mean, and group effects defined to be the differences between the other group means and the reference mean. By default the group represented by factor level 1 is taken as the reference group, although users can change this if they wish. The model in this case is specified in the `FIT` directive by simply providing the factor variable name.

```
$yvar wtgain $
$fit group $dis e $
   deviance =  7584.0
residual df =     17
          estimate        s.e.     parameter
     1       101.0       7.983      1
     2       19.00       10.05      GROUP(2)
scale parameter 446.1
```

The fit of the model to the data can be illustrated with

```
$ass g_pred=1,2 $predict (s=-1) group=g_pred $
$graph (title='Fig. 12.14 Weight Gains by Diet Group'
        vlab='Wt gain' hla='Diet Group' xscale=0,3,1
        yscal=20,180,40)
        wtgain;%pfv group;g_pred 1;10 $
```

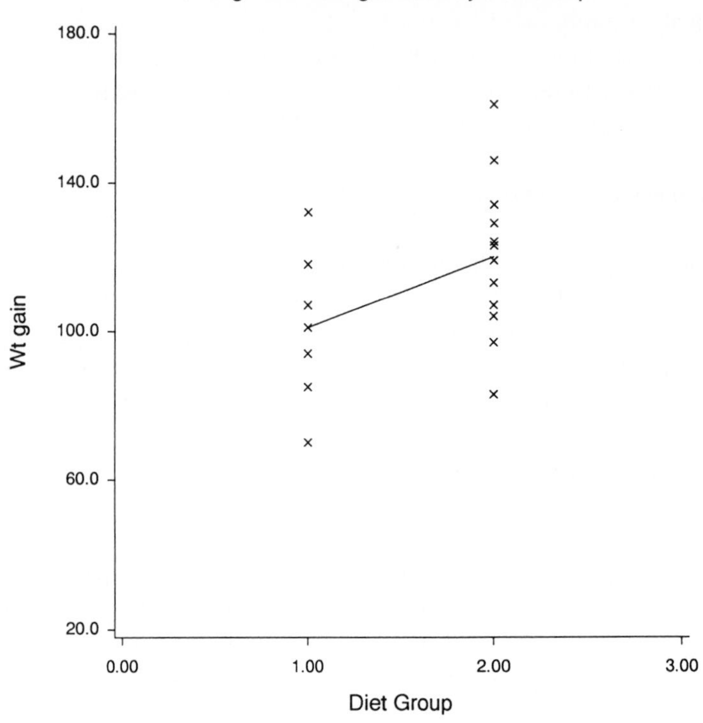

Because group is a factor parameter 1 is the mean of the values in the first category. The second parameter group(2) is the difference between the mean of the values in the second category and the reference mean (parameter 1). The deviance is the residual sum of squares about the model with each group having its own mean, that is, the pooled sum of squares, and the scale parameter (%sc) is the usual pooled variance. The s.e. for group(2) is obtained from the pooled variance, and group(2) divided by its s.e. is identical to the standard two–group t test. Thus

```
$number t df $extract %pe %se $
$cal t=%pe(2)/%se(2)  : df=%df $
$pr ;; ' The t test comparing the two group means is';
'       t = 't' with 'df ;;$
 The t test comparing the two group means is
      t =     1.891 with    17.00
```

This should be compared with the 97.5 centile of the t-distribution with 17 d.f. which is

```
$cal %td(0.975,17) $
           2.110
```

We see it falls short of the two tailed 5 per cent significance level. The exact *p* value representing the probability of values this far from zero in the positive or negative direction occurring by chance from this t-distribution is obtained by

```
$cal 2*(1-%tp(t,17)) $
      0.07573
```

So it is quite a surprising value and would be significant if we were using a 10 per cent or even an 8 per cent level.

Of course this is a rather laborious way to perform such a t test, but assumptions can be tested very easily and the calculations repeated easily after any necessary modifications of the data. However the real argument for this approach is that it points the way to how such problems can be generalized to one–way analyses of variance as in the next section and the completely general cross classifications covered later.

To test the assumption of Normality we use a standard macro.

```
$use crep $
```

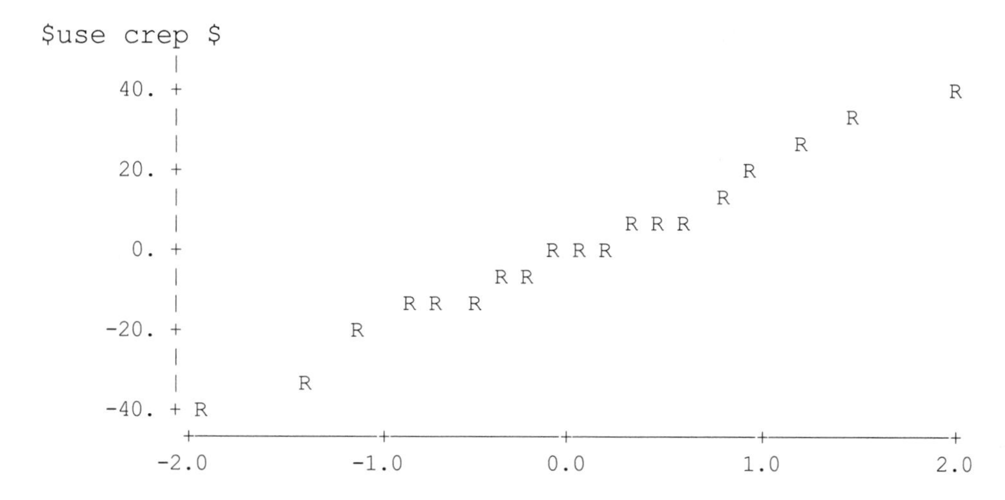

This is a respectable Normal plot which casts no doubt on the assumption of Normality. The conclusion is that there is some suggestion that rats in general would gain weight faster on the high–protein diet than those on the low–protein diet, but it did not reach the conventional 95 per cent level of significance. A repeat of the experiment with larger numbers in the two groups is required.

12.1.2.2 Comparing groups: one way analysis of variance

The following data, taken from Brookes and Dick (1951), are from an investigation of the tensile strength of four rubber compounds. It was designed to estimate their relative strength to identify which was the best and to estimate the measurement error.

Tensile strength in pounds/square inch

Compound A	B	C	D
3210	3225	3220	3545
3000	3320	3410	3600
3315	3165	3320	3580
Defective	3145	3370	3485

It is simplest to put the data into GLIM as if all the values were available using –1 to indicate the value lost through a defective sample

```
$data 16 tens $read
   3210    3225    3220    3545
   3000    3320    3410    3600
   3315    3165    3320    3580
    -1     3145    3370    3485
```

Then define the standard length for vectors to be 16 and obtain values for the factor variable identifying the compound group with the 'generate level' function `%gl( )`.

```
$slen 16 $
$cal comp=%gl(4,1) $
```

This generates the integers 1 to 4 one at a time to give 1 2 3 4 1 2 3 4, and so on.

To ensure that the –1 value is excluded from any analyses a weight variable must be calculated and declared:

```
$cal wt=(tens/=-1) $
$weight wt $
```

First is a plot using the weight vector `wt` to restrict the plot by excluding the –1 value

```
$plot (x=0,5) tens/wt comp $
```

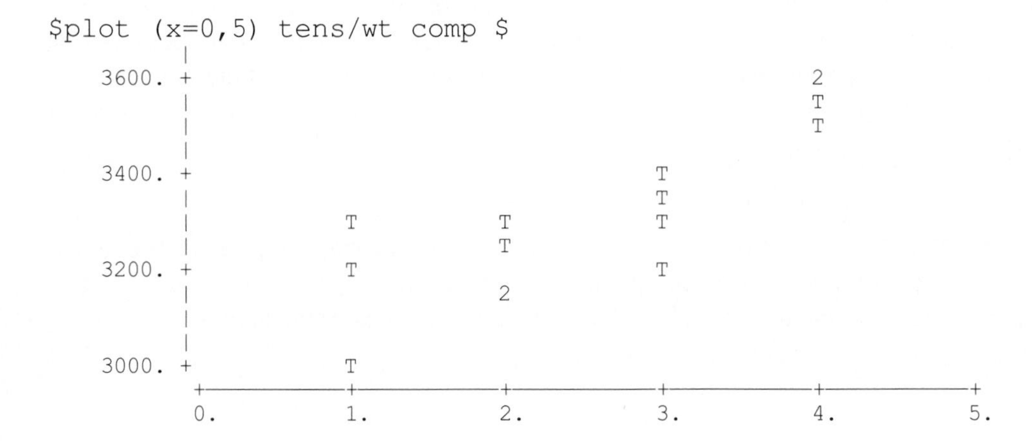

There are quite marked differences between the distributions of the four compounds. Let us identify COMP as a factor with four levels and obtain the table of means using the with keyword and the weight variable to exclude the missing value code of –1:

```
$factor comp 4 $
$tab (s=1) the tens mean;var;dev;weight with wt for comp by
cmpd$
```

CMPD	1	2	3	4
MEAN	3175.	3214.	3330.	3553.
VARIANCE	25725.	6173.	6733.	2542.
S.D.	160.39	78.57	82.06	50.41
TOTAL_WT	3.000	4.000	4.000	4.000

The differences between the means seems large, but the variances seem rather different – that of the first compound is more than four times those of the others. However the numbers are rather small. In fact assuming a test of the first group variance against a pooled variance from the other three groups the comparison would be an F test with 2 and 9 degrees of freedom. That should be a two–tailed F test since it would be a ratio of independent variances so the 95 per cent significance point would be %fd(0.975,2,9), that is, 5.715. Bearing in mind that we chose to test the largest against the mean of the rest it is likely that these differences are within the bounds of chance. Nonetheless they are worrying and a larger study is needed for firm conclusions. However, for this example let us ignore these complications and proceed to fit the model COMP:

```
$yvar tens $
$fit comp $dis e $
   deviance =  97794.
residual df =      11  from 15 observations
          estimate         s.e.     parameter
     1       3175.        54.44      1
     2       38.75        72.01      COMP(2)
     3       155.0        72.01      COMP(3)
     4       377.5        72.01      COMP(4)
scale parameter 8890.
```

and test Normality remembering to allow for the value excluded with the zero weight using a stored macro WREP for weighted residual plots

```
$use wrep $
```

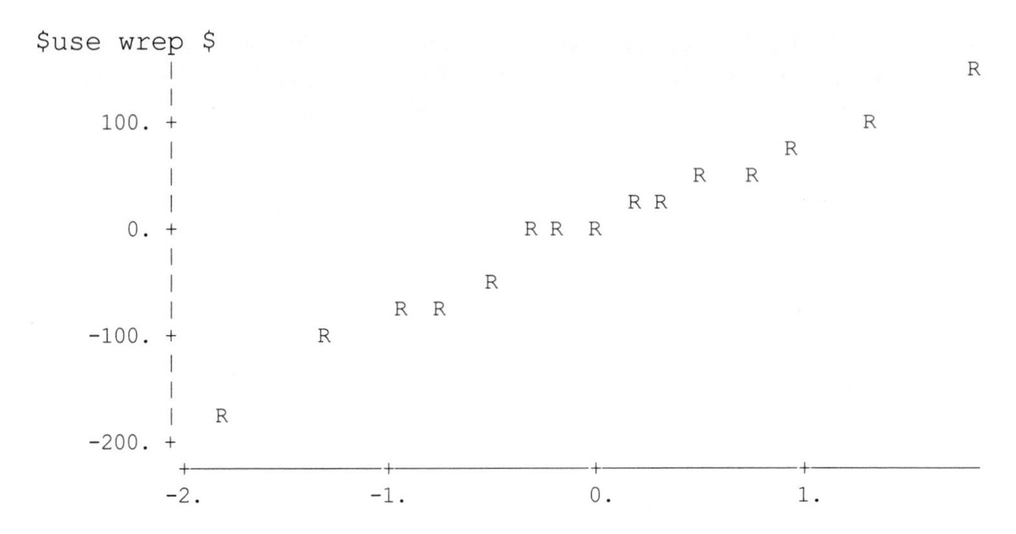

The data certainly seem adequately Normal so we can proceed to interpret the fitted model. First we should clarify the parametrization that GLIM uses here. Since no reference level was specified when `COMP` was declared to be a factor GLIM takes the first level as a reference point for the parametrization. The mean or expected value, for observations in the category represented by that level, is then referred to as parameter `1`. The parameters `COMP(2)`, `COMP(3)` and `COMP(4)` represent the differences between the means for those categories and that for the `COMP=1` category, that is, parameter `1`. That means that tests of individual parameter estimates are simply t tests comparing the first group against each of the others. A comparison of the second and third `COMP` groups cannot be obtained immediately with this parametrization. However it is not appropriate to test the groups in pairs in this way. The proper test is a single test for heterogeneity between all four group means. Remembering that the parameters represent differences from `COMP` group 1 we can see that there appears to be a systematic increase in mean tensile strength from compound 1 to compound 4. If they were ordered in some way it might be appropriate to test for a trend in the four group means, but we have no evidence that they are. We must simply test whether the differences between the four groups as a whole are more than should readily occur by chance. We do this by fitting the model without `COMP`, that is, one overall mean for all four groups and assessing how much worse it fits.

First store the residual sum of squares and degrees of freedom from the earlier fit:

```
$number RSS1 df1 s2 df21 F $
$cal RSS1=%dv : df1=%df : s2=%sc $


$fit 1 $dis e $
   deviance =  421843.
residual df =       14  from 15 observations
         estimate        s.e.     parameter
```

```
     1        3327.        44.82       1
scale parameter 30132.
```

The appropriate F test is then

```
$cal df21=%df-df1 : RSS2=%dv : F=((RSS2-RSS1)/df21)/s2 $
$pr '    F ='F'  with 'df21'  and  'df1'  degrees of freedom';;$
   F =   12.15  with    3.000  and     11.00  degrees of freedom
```

The probability of an F–value as high or higher than this is

```
$cal 1-%fp(F,df21,df1) $
     0.0008156
```

Since this is very much less than 0.01 the F–value is highly significant at the 5 per cent and indeed at the 1 per cent level.

Clearly there are real differences between the compounds and the fourth appears to be the strongest. Confidence intervals for the four means can be obtained from a model if we force the parameters to estimate the means. This is done by removing the constant term from the model, that is, the parameter `1` as follows:

```
$fit comp-1 $dis e $
   deviance =   97794.
residual df =      11  from 15 observations
           estimate         s.e.     parameter
     1        3175.        54.44       COMP(1)
     2        3214.        47.14       COMP(2)
     3        3330.        47.14       COMP(3)
     4        3553.        47.14       COMP(4)
scale parameter 8890.
$extract %pe %se $
$cal %t=%td(0.975,11) : low=%pe - %t*%se : high=%pe + %t*%se $
$look low %pe high $
      LOW      %PE      HIGH
 1   3055.    3175.    3295.
 2   3110.    3214.    3318.
 3   3226.    3330.    3434.
 4   3449.    3553.    3656.
```

Finally the estimate of the measurement error is the pooled variance from the four groups which is 8890psi. In this particular case the customer should be warned that there is some evidence that the variation between samples may differ from compound to compound and the results should be treated with caution.

12.1.2.3 Comparing groups with a covariate: analysis of covariance

The data below were obtained as part of a health survey in the US investigating cholesterol levels in women from two states – Iowa (1) and Nebraska (2). The question here is whether there are real differences in cholesterol levels among women from the two states allowing for age differences. (Source: Snedecor;1956)

```
$un 30 $data state age chol $read
 1 46 181 1 52 228 1 39 182 1 65 249 1 54 259
 1 33 201 1 49 121 1 76 339 1 71 224 1 41 112 1 58 189
 2 18 137 2 44 173 2 33 177 2 78 241 2 51 225 2 43 223
 2 44 190 2 58 257 2 63 337 2 19 189 2 42 214 2 30 140
 2 47 196 2 58 262 2 70 261 2 67 356 2 31 159 2 21 191
 2 56 197
```

The problem is to assess the relationship of cholesterol with age in the two groups and if possible use it to deduce what the difference between the two groups would have been if they had all been the same age. First let us plot cholesterol against age with different symbols for each state.

```
$factor state 2 $
$plot chol age 'in' state $
```

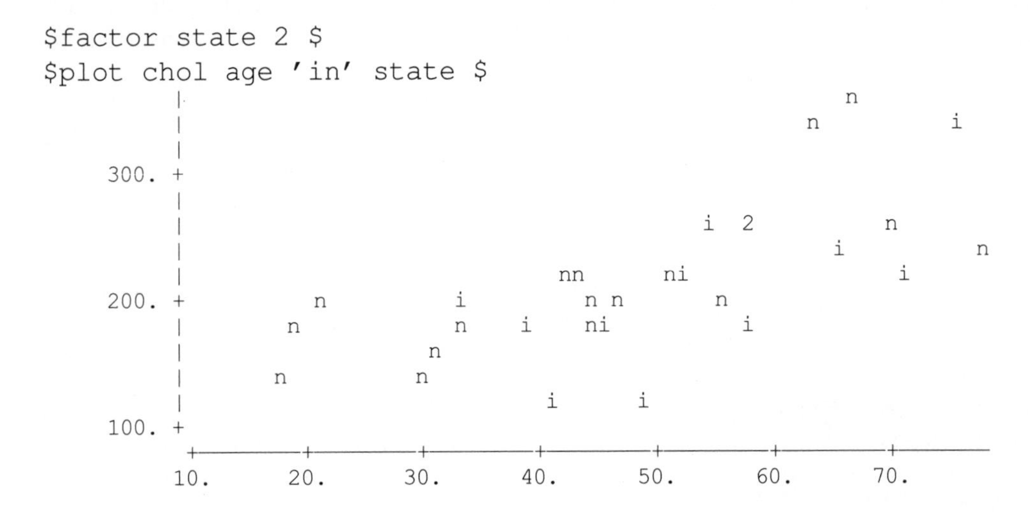

There is a reasonably clear positive association with age, but no clear differences between states. The questions to answer are :

(i) Are the trends with age different in the two states?

If they are, we cannot make any simple statements about cholesterol differences in the two states. We would have to qualify everything according to which age group was involved

If they are not:

(ii) What is the best estimate of the common population age trend and is it non-zero?

and

(iii) What is the best estimate of the difference between women of the same age in the two states and is the difference in the population of such women non-zero?

It looks from the plot as if there may be a curved trend. To investigate this it is necessary first to fit a model with different quadratic trends with age in the two states using the nesting operator / .

```
$cal age_sqd=age*age $
$yvar chol $
$fit state/(age+age_sqd) $dis e $
   deviance =  44293.
residual df =      24
          estimate          s.e.      parameter
     1        359.3         230.0       1
     2       -237.4         239.0       STATE(2)
     3       -9.260         8.685       STATE(1).AGE
     4        1.476         3.025       STATE(2).AGE
     5       0.1139       0.07862       STATE(1).AGE_SQD
     6      0.01139       0.03236       STATE(2).AGE_SQD
scale parameter 1846.
```

It is now necessary to investigate the residuals to check the constancy of the variance and identify potential outliers, the leverages to check for unduly influential points, and the Normal assumptions with a quantile–quantile plot.

```
$plot %rs %fv 'r'  $
```

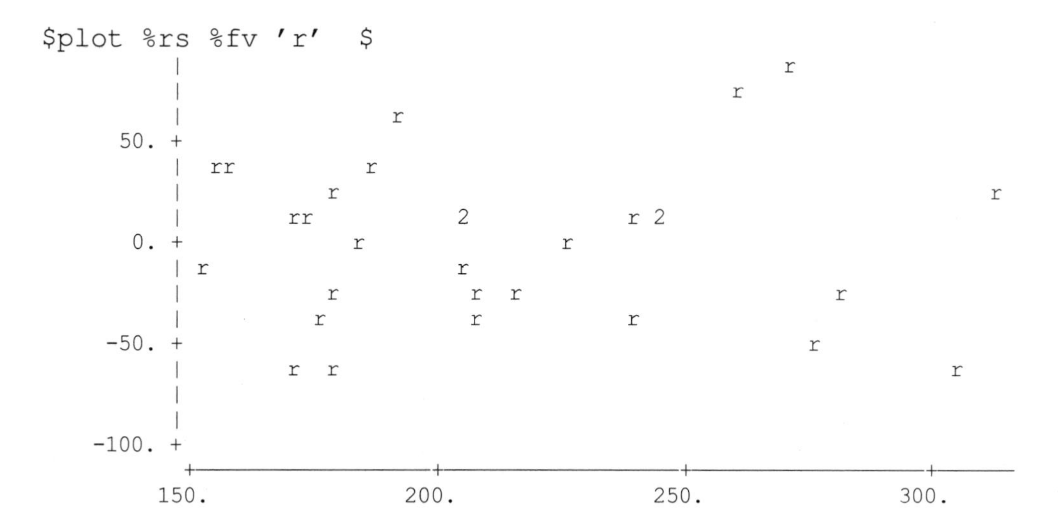

There is no obvious sign of a systematic relationship between the fitted values and the magnitudes or variance of the residuals. Next we obtain an index plot of the leverage values.

```
$extract %lv $
$cal index=%ind(%lv) $

$plot %lv index 'in' state  $
       |              i   i
  0.6 +
       |                                n
       |
       |
  0.4 +
       |                          n
       |     .              i                        n
       |
  0.2 +     i i i i            i                                    n     n
       | i             i     i                                         n
       |                            n n   n n n n n   n n n n      n   n
       |
  0.0 +
       +---------+---------+---------+---------+---------+---------+
       0.        5.        10.       15.       20.       25.       30.
```

It appears that there are three very influential points. We can see which these are by creating a factor to identify those above 0.4.

```
$ass cut=0.4 $group hi=%lv intervals * cut * $
$cal stat_lev=(hi-1)*2+state $factor stat_lev 4 $
$plot chol age 'inIN' stat_lev $
```

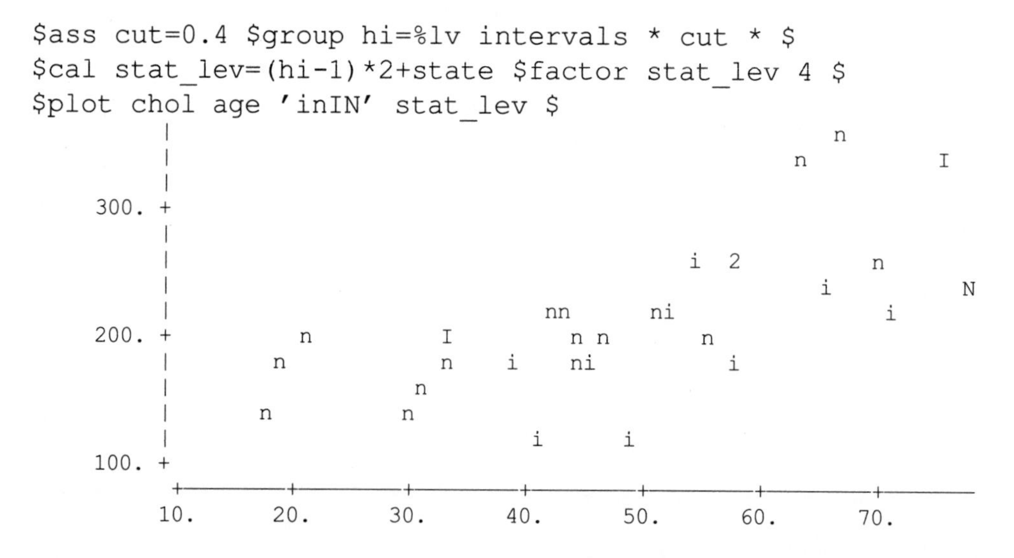

The points with the high leverage values are plotted with upper case letters. Not very surprisingly, the points are the youngest and oldest Iowans and the oldest Nebraskan. It is possible that these points are the main reason for the apparent curvature. This will need to be investigated if the curvature is significant. First the quantile plot to test the Normality of the residuals.

```
$use crep $
```

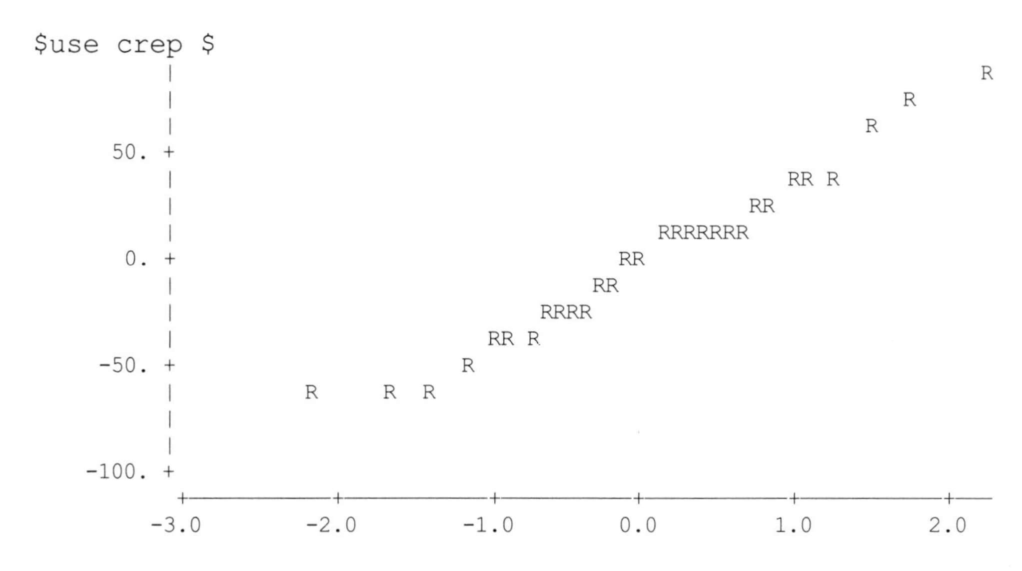

There is a slight indication of a sigmoid or S–shape to this plot, but insufficient evidence of serious non-Normality, certainly not enough to bias the results. The other assumptions seem adequately justified so it is reasonable to proceed; first by testing the apparent curvature. This is done by comparing the fit of the current model with that of a model reduced by omitting the quadratic terms, that is, a model using straight lines to describe the two trends. To compare the models the residual sum of squares and degrees of freedom for this model need to be stored.

```
$number RSS0 df0 $
$cal RSS0=%dv : df0=%df : $
```

Then the non-parallel straight lines model is fitted.

```
$fit state/age $dis e $
   deviance =  48395.
residual df =      26
          estimate        s.e.      parameter
     1       35.81       55.12       1
     2       65.49       61.98       STATE(2)
     3       3.238       1.009       STATE(1).AGE
     4       2.520      0.5783       STATE(2).AGE
scale parameter 1861.
```

The test of curvature is then an F test, giving a p value of

```
$cal 1-%fp((%dv-RSS0)/(%df-df0)/(RSS0/df0),%df-df0,df0) $
          0.3455
```

This is far from significant so the apparent curvature was no more than could easily occur by chance. The straight line model is an adequate representation of these data. In this model the parameters should be interpreted as follows: parameter `1` is the intercept for state 1 and `state(2)` is the difference between that and the intercept for state 2. The two parameters `state(1).age` and `state(2).age` are the two separate slopes for the two states. Since the difference is of the same order as their s.e.s it is unlikely to be significant. We can test this most simply by re-parametrizing the model without the nesting operator / so the difference between the slopes is itself a parameter. This is done with the crossing operator `*`. The parameter representing the difference can then be used to test the difference between the slopes.

```
$fit state*age $dis e $
   deviance =  48395.
residual df =      26
           estimate          s.e.     parameter
     1        35.81         55.12      1
     2        65.49         61.98      STATE(2)
     3        3.238         1.009      AGE
     4      -0.7177         1.163      STATE(2).AGE
scale parameter 1861.
```

The deviance is the same because it is the same model with a different parametrization. However there is now an `age` coefficient, which is the slope of the age trend for state 1 (Iowa) and a parameter for the difference between the two slopes: `state(2).age`. Since this is only half its s.e. there is no evidence for a real difference between the slopes in the populations represented by these samples. This means that a parallel line model is probably adequate to describe the patterns in these data. Before we fit such a model we should store the residual sum of squares and degrees of freedom for the current model for later testing the difference in the fit of the two models.

```
$number RSS1 df1 s2  $
$cal RSS1=%dv : df1=%df : s2=%sc $
```

Now fit the parallel line model and store the parameter estimates and their s.e.s for later calculations.

```
$fit state+age $dis e $
   deviance =  49104.
residual df =      27
           estimate          s.e.     parameter
     1        64.49         29.30      1
     2        28.65         16.54      STATE(2)
     3        2.698        0.4960      AGE
scale parameter 1819.
$extract %pe %se $ass est=%pe : sest=%se $
```

The parameter `1` is still the intercept for the first state and `state(2)` is the difference between the two intercepts. They have changed considerably because the separate lines actually crossed. Averaging the two slopes to obtain the best pair of parallel lines has raised the intercept for state 1 markedly although it has not changed that of state 2 much (remember that is obtained as the sum `1 + state(2)` from both models). The F test of the difference in slope is obtained by

```
$number RSS2 df2 df21 SS21 MS21 vr21  $
$cal df2=%df : df21=df2-df1 : RSS2=%dv : SS21=RSS2-RSS1 :
    MS21=SS21/df21 : vr21=MS21/s2 $
```

The F test for the difference in slopes is

```
$pr '     F ='vr21'  with 'df21'  and  'df1'  degrees of freedom'
;; $
   F =  0.3809  with    1.000  and     26.00  degrees of freedom
```

The *p*–value for this F is

```
$cal 1-%fp(vr21,df21,df1) $
       0.5425
```

Since this is much larger than 0.05 it is far from significant. Clearly the parallel line model is adequate and there is no evidence that such differences as there are between the mean cholesterol levels in the women of the two states change with age. However there does appear to be a very clear age trend with the age coefficient more than 5 times its s.e.

The next question is whether, allowing for age differences, this data provides any evidence for there being a systematic cholesterol difference, on average, between the women of the two states. Assuming the difference to be constant at all ages the best estimate from this data is the `state(2)` parameter from the last model. Since this in turn is less than twice its s.e. the evidence is that once age is taken into account there is no evidence of a difference in the mean level of cholesterol between the two states either.

The questions, in this simple case with just two groups, can be answered by using simple t tests of the appropriate parameters. However a more general approach is to use F tests from appropriate model comparisons as above and it is wise to do this in a systematic way constructing the equivalent of the classical analysis of covariance table. We obtain the model comparison for testing the age trend against the null hypothesis, that is, by dropping `age` from the model and effectively fitting a pair of horizontal lines.

```
$fit -age $
   deviance =  102924. (change =    +53820.)
residual df =       28  (change =        +1 )
```

The parameters are not actually of interest, because we already know it to be an inappropriate model, so we do not need to display them. We just need to store the residual sum of squares and degrees of freedom as before.

```
$number RSS3 df3 SS32 MS32 df32 VR32 $
$cal RSS3=%dv : df3=%df : SS32=RSS3-RSS2 : df32=df3-df2
: MS32=SS32/df32 : vr32=MS32/s2 $
```

Finally fitting the model with coincident lines, that is, allowing an age trend but constraining it to be the same in both states and forcing the mean level at any age to be the same as well:

```
$fit +age - state $
   deviance =  54560. (change =    -48364.)
residual df =     28  (change =         0 )
```

```
$number RSS4 df4 SS42 MS42 df42 VR42 $
$cal RSS4=%dv : df4=%df : SS42=RSS4-RSS2 : df42=df4-df2
: MS42=SS42/df42 : vr42=MS42/s2 $
```

The stored deviances and degrees of freedom can now be used in a macro which prints the appropriate analysis of covariance table.

```
$use anco3123 $
           Model      Residual SS    df   Mean square    Variance Ratio(F)
 ____________________________________________________________________________
   1)  STATE*AGE        48394.86     26     1861.34
   2)  STATE+AGE        49103.91     27
 ________________________________________
(2-1)Due to STATE.AGE     709.05      1      709.05                  0.38
                                                          ( p =   0.5425 )
 ____________________________________________________________________________
   3)  STATE
       (that is -AGE)    102923.97    28
 ________________________________________
(3-2)Due to AGE         53820.05      1    53820.05                 28.91
                                                          ( p =   0.0000 )
 ____________________________________________________________________________
   4)  AGE
       (ie -STATE)      54560.36     28
 ________________________________________
(4-2)Due to STATE        5456.45      1     5456.45                  2.93
                                                          ( p =   0.0983 )
 ____________________________________________________________________________
 ____________________________________________________________________________
```

The *p*–values are the probabilities of F values on 1 and 26 degrees of freedom being as large or larger than those observed. Alternatively to test for significance at the 5 per cent level the variance ratios can be compared with the 95th centile of the F distribution with 1 and 26 degrees of freedom which is

```
$cal %fd(0.95,1,26) $
      4.225
```

These results show that the test of non-parallelism is far from significant. The age trend is very significantly non-zero and there is not really much evidence of a mean difference in cholesterol levels between the states.

This analysis confirms the initial findings that the age trends in the two states are no more different than could easily occur by chance, that the age trend is quite pronounced and could lead to differences being observed between groups of different ages and that although there is a slight difference in mean cholesterol levels after the effects of age have been taken into account this in turn is not more than could relatively easily be explained by chance. Finally, for a complete analysis, it is necessary to obtain confidence limits for the mean cholesterol difference and possibly the age effect. Restricting attention to the difference between the states the estimate with 95 per cent confidence limits is:

```
$cal est(2) : est(2)-%td(0.975,27)*sest(2) :
est(2)+%td(0.975,27)*sest(2) $
       28.65
      -5.288
       62.59
```

Although the difference was not found significant above and the confidence interval includes 0 these limits show that the data are consistent with a true difference between the states of as much as the upper confidence limit of 62.59.

12.1.2.4 Two way classifications

This example uses data from Snedecor (1956). The observations are of average daily weight gain in pigs in a trial of the benefits of vitamin B12 and antibiotic additions to their diet. The results were:

Average Daily Increase in Weight (pounds)

Antibiotics (milligrams/day)	B12 (micrograms/day) 0	5
0	1.30	1.26
	1.19	1.21
	1.08	1.19
40	1.05	1.52
	1.00	1.56
	1.05	1.55

The questions are whether either supplement affects the rate at which the animals increase in weight and by how much.

The data are input and factor variables calculated by

```
$units 12 $data wt_gain $read
   1.30      1.26
   1.19      1.21
   1.08      1.19
   1.05      1.52
   1.00      1.56
   1.05      1.55
$cal B12=%gl(2,1) : A_bio=%gl(2,6) $
$factor B12 2 A_bio 2 $
```

The first step is to look at the means within each treatment combination. These are obtained, with standard deviations and cell frequencies, using

```
$tab the wt_gain mean;dev;weight for A_bio;B12
 into y_mean;y_dev;y_n by Ab_;B12_ $
```

This stores the cell statistics in the three variables `y_mean,` `y_dev` and `y_n` and sets up variables `B12_` and `Ab_`, defining the dimensions of the table, with lengths equal to the number of cells.

The table is printed, with the optional border `(s=1)`, by

```
$tprint (s=1) y_mean;y_dev;y_n  Ab_;B12_ $
            +-------------------------+
      B12_  |     1          2        |
 AB_        |                         |
+-----------+-------------------------+
|  1 Y_MEAN |    1.190      1.220     |
|    Y_DEV  |  0.11000    0.03606     |
|      Y_N  |    3.000      3.000     |
---------------------------------------
|  2 Y_MEAN |    1.033      1.543     |
|    Y_DEV  |  0.02887    0.02082     |
|      Y_N  |    3.000      3.000     |
+-----------+-------------------------+
```

The treatment effects appear as differences between the various means, but notice that one of the standard deviations is rather larger than the rest. The analysis needs an assumption that the true variances in the populations represented by the four groups are the same. Before the analysis is concluded this will have to be investigated in more detail.

The differences between the means, representing the treatment effects, are best illustrated by plotting the observed data points together with the group means against one of the treatment category variables, such as `B12`. This means that the points for the two `B12` categories will be distinguishable because of their horizontal positioning. To identify the points from the two antibiotic categories, which for a given `B12` level will only be distinguishable to the extent that their vertical positioning, or *y* values, differ, the `A_bio` factor can be used to specify different plotting symbols. It is simplest to use `n` for no (antibiotics) and `y` for yes. The means can be identified with the symbol +.

```
$plot (x=0,3 y=0.7,1.7) wt_gain;y_mean B12;B12_ 'ny+'  A_bio; $
```

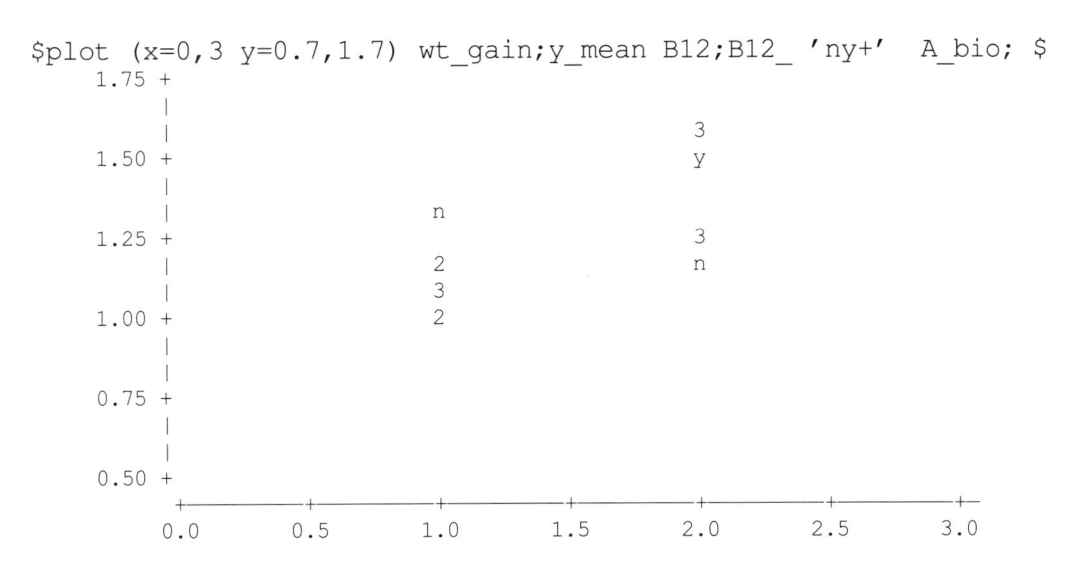

Unfortunately the Vertical scale is too compressed for the groups and means to be distinguished clearly. However it can be deduced, to some extent, from the table of means the effects of the two supplements are dependent on each other. Without antibiotics there is a slight increase in the mean weight gain associated with a `B12` supplement; with antibiotics there is a marked increase. The effects of the two factors interact. This can be seen much more clearly with a graph.

```
$graph (t='Fig. 12.15 Mean Wt Gains by A-biotic and B12 Grp'
        vla='Daily Wt gain'
        hla='B12 dose  - (A-biotic 0:o 40:x)'
        xsc=0,3,1 ysc=0.5,2,0.5 )
        wt_gain;y_mean B12;B12_ 5;1;10;10 A_bio;Ab_ $
```

All the effects of interest can be estimated and tested by combinations of the means and appropriate functions of the variance pooled over the four cells, that is, if the table of means is represented by

Fig. 12.15 Mean Wt Gains by A-biotic and B12 Grp

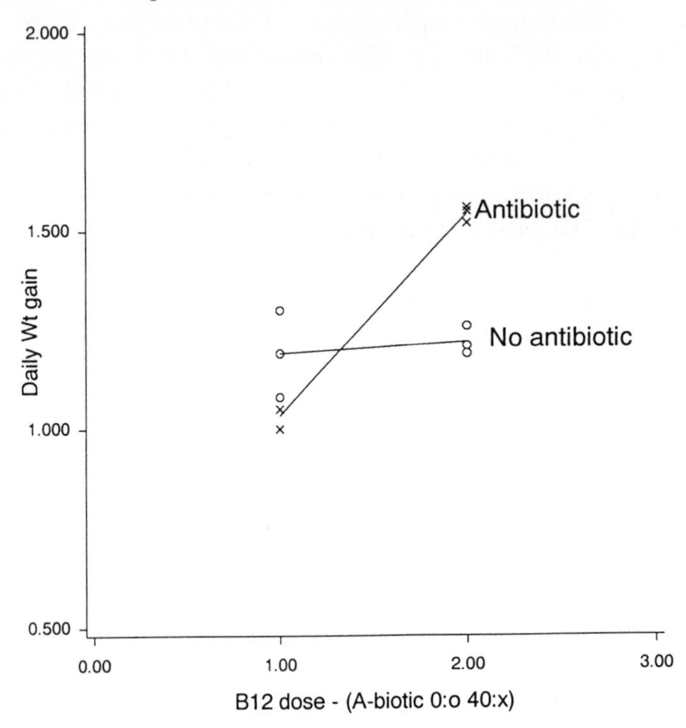

Average Daily Increase in Weight (pounds)

Antibiotics (milligrams/day)	B12 (microgms/day) 0	5
0	m_{11}	m_{12}
40	m_{21}	m_{22}

The interaction effect is the difference between the two B12 effects

$$(m_{12} - m_{11}) - (m_{22} - m_{21}).$$

Divided by its standard error this would give a t test of whether the `true', that is, the population interaction were zero. If it were close to zero the lines joining the means in the above plots would be almost parallel. Consequently tests of interactions can usefully be thought of as tests of parallelism and are closely analogous to such tests comparing regression lines. The overall main effects of B12 and A_bio, which can only be usefully interpreted if the model is 'parallel' and the interaction can be assumed zero, are

$$((m_{12} - m_{11}) + (m_{22} - m_{21}))/2 \text{ for B12}$$

and

$$((m_{21} - m_{11}) + (m_{22} - m_{12}))/2 \text{ for A\_bio.}$$

This approach can be used and the calculations are reasonably straightforward but it quickly becomes very complicated if the numbers in the groups are not equal, when relatively complex weighting is required, and if there are more than two factors or factors with more than two levels. In practice it is usual for one or more of these circumstances to apply so a general model comparison approach leading to an analysis of variance is almost essential. The modelling approach also makes it much easier to test the various assumptions that are necessary whatever the approach taken. The analysis is as follows.

```
$yvar wt_gain $
```

```
$fit B12*A_bio $dis e $
   deviance =  0.029333
residual df =  8
          estimate         s.e.      parameter
     1       1.190      0.03496       1
     2     0.03000      0.04944       B12(2)
     3     -0.1567      0.04944       A_BIO(2)
     4      0.4800      0.06992       B12(2).A_BIO(2)
scale parameter 0.003667
```

The interaction parameter `B12(2).A_bio(2)` is very large compared to its standard error. This parameter estimate divided by the s.e. is the t test described above. However such tests require assumptions of Normality and constant variance over the treatment groups. These need to be tested.

```
$use crep $
```

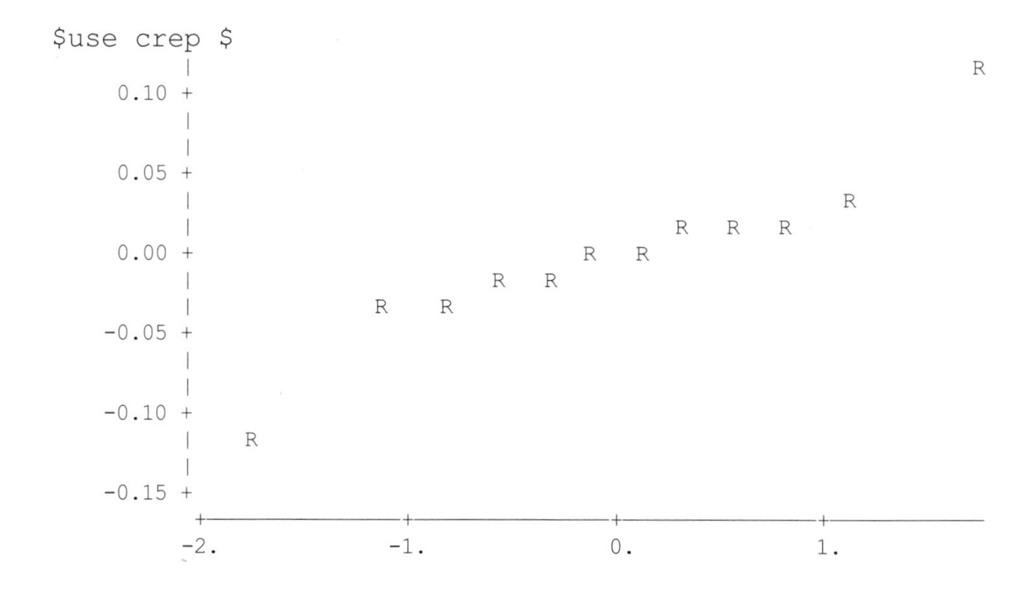

This is not a very tidy plot, but reasonably linear except for the two extreme residuals. It is probably close enough to what should be expected for samples of this size from a Normal population. Now consider the residual variation. A plot of residuals against the index or order number of their observed values should show if there are any extreme values or evidence of changing levels of variation.

```
$cal indx=%ind(wt_gain) $
$plot  %rs indx 'i' $
```

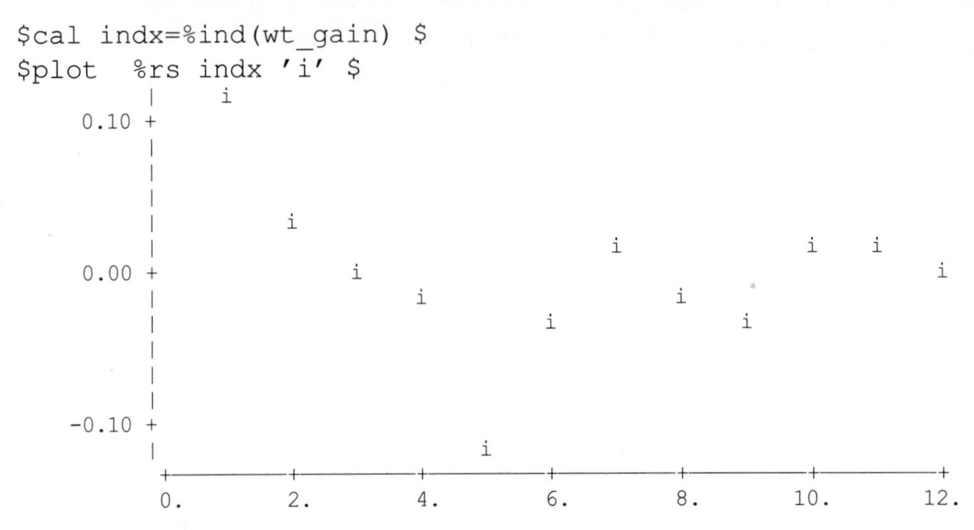

The spread of residuals above and below the zero mean is more or less constant across the plot, but the four separate treatment categories cannot be identified easily. This should be done to investigate the difference in standard deviations observed earlier. First calculate a variable to identify the four treatment combination categories.

```
$cal categ=(B12-1)*2+A_bio $
```

Then plot the residuals against this variable.

```
$plot %rs categ 'r' $
```

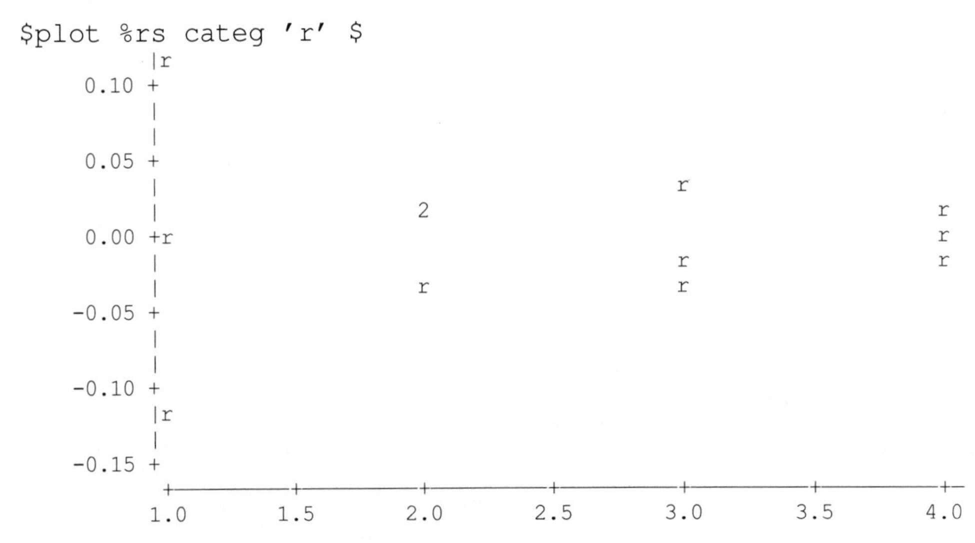

Clearly the spread is considerably greater in category 1 than in the others. Whether the differences observed represent genuine population differences in variance can be tested using Bartlett's test of homogeneity of variance. However it should be noted that Bartlett's test is generally considered to be rather more sensitive to skewness in the underlying distributions than the analysis of variance F tests for which the assumption being tested is required (Snedecor and Cochran, 1967). Nonetheless it is worth illustrating the use of the test macro. The test is based on modelling the variances weighted according to the degrees of freedom as a constant, assuming they are observations from a gamma distribution. First the variances and degrees of freedom need to be obtained for each cell.

```
$cal var=(y_dev)**2 : df=y_n-1 $
```

The technique is conveniently stored as a macro

```
$mac  !  bartlett
!
! Macro for Bartlett's test of homogeneity among several
! variances using the Gamma distribution
!
(arg=var,df  loc=chi,c,p) bartlett $
$yvar var $weight df $error G $scale 2 $
$echo off $trans -o $fit 1 $echo on $trans +o $
$number chi c p $
$cal c=(1+(%cu(1/df)-1/%cu(df))/(3*%df)) $
 : chi=%dv/c : p=1-%chp(chi,%df) $
!
$print
'Bartlett''s approximate Chi-squared test'
' for homogeneity of variance is';'    '
*r chi,6,2 ' on '*i %df,3
' degrees of freedom. The p value is ' *r p,6,4,1,1 $
$endmac $
```

```
$use bartlett var df $
Bartlett's approximate Chi-squared test for homogeneity of variance is
    5.73 on   3 degrees of freedom. The p value is 0.1257
```

It appears that the difference between the variances is quite a way from significance with p = 0.13. Thus although the extra spread in the first group looks quite striking such effects could relatively easily occur by chance in samples of this size. It is reasonable to proceed with the analysis of variance assuming equal population variances although it might be useful to encourage the originator of the data to consider whether or not there are potentially interesting biological reasons for a genuine effect such as this. That would, of course, come under the heading of hypothesis generation and require new data to investigate it.

The next step is to fit an appropriate sequence of models and to compile an analysis of variance table. This can be done automatically within GLIM by storing the residual sums of squares and degrees of freedom for each model to obtain the appropriate components of the F test calculations and completing the calculations at the end of the sequence in a macro to produce the analysis of variance table.

For this purpose it is necessary to create scalar variables to store the residual sum of squares, SS, and the d.f. from the interaction model together with the estimate of the residual variance from this model (identical to the classical pooled variance) which is available as the 'scale parameter'.

```
$number SS1 df1 s1 $
```

Because the Bartlett's test macro fitted a model it is necessary to re-set the Normal ERROR, cancel the weight setting used in the Bartlett test macro and then redeclare wt_gain as the *y* variable. Finally we need to fit the interaction model again to obtain the appropriate deviance and degrees of freedom.

```
$error Normal $weight $
$yvar wt_gain $
$fit B12*A_bio $dis e $
   deviance =  0.029333
residual df =  8
          estimate        s.e.      parameter
     1       1.190     0.03496       1
     2     0.03000     0.04944       B12(2)
     3     -0.1567     0.04944       A_BIO(2)
     4      0.4800     0.06992       B12(2).A_BIO(2)
scale parameter 0.003667
$cal SS1=%dv : df1=%df : s1=%sc $
```

Then fit the main effects model as before

```
$fit B12+A_bio $dis e $
   deviance =  0.20213
residual df =  9
          estimate        s.e.      parameter
     1       1.070     0.07493       1
     2      0.2700     0.08652       B12(2)
     3     0.08333     0.08652       A_BIO(2)
scale parameter 0.02246
```

The sum of squares and d.f. are stored with

```
$number SS2 df2 $
$cal SS2=%dv : df2=%df $
```

Then to test the effect of `A_bio`, we fit a model assuming it to be zero to assess how much worse the model fits

```
$fit  B12  $
   deviance =   0.22297
residual df =  10
```

There is no need for the parameter estimates; as we only need the residual sum of squares and degrees of freedom

```
$number SS3 df3 $
$cal SS3=%dv : df3=%df $
```

Finally we fit a model assuming the B12 effects are zero

```
$fit A_bio $
   deviance =   0.42083
residual df =  10

$number SS4 df4 $
$cal SS4=%dv : df4=%df $
```

The analysis of variance is then obtained (using a macro specific to this example) as

```
$use anco3124 $
           Model      Residual SS    df   Mean square    Variance Ratio(F)
 ----------------------------------------------------------------------------
   1) B12*A_bio             0.03     8       0.0037
   2) B12+A_bio             0.20     9
 ----------------------------------------------
(2-1)Due to B12.A_bio       0.17     1         0.17                  47.13
                                                            ( p =   0.0001 )
 ----------------------------------------------------------------------------
   3)    B12
       (that is -A_bio)     0.22    10
 ----------------------------------------------
(3-2)Due to A_bio           0.02     1         0.02                   5.68
                                                            ( p =   0.0410 )
 ----------------------------------------------------------------------------
   4)   A_bio
       (ie - B12)           0.42    10
 ----------------------------------------------
(4-2)Due to  B12            0.22     1         0.22                  59.65
                                                            ( p =   0.0000 )
 ----------------------------------------------------------------------------
 ----------------------------------------------------------------------------
```

For significance the F tests on 1 and 8 d.f. only need to reach

```
$cal %fd(0.95,1,8) $
        5.318
```

Clearly the interaction term is highly significant. This means that the effect of antibiotics depends on whether the animals were also getting B12 or not. The effects of antibiotics have to be described for each B12 group separately and vice-versa for the effects of B12. A model specification using the nesting operator / will give the antibiotic effect estimators within each of the B12 groups.

```
$fit B12/A_bio  $dis e $
   deviance =  0.029333
residual df =  8
          estimate        s.e.      parameter
     1       1.190     0.03496       1
     2     0.03000     0.04944       B12(2)
     3     -0.1567     0.04944       B12(1).A_BIO(2)
     4      0.3233     0.04944       B12(2).A_BIO(2)
scale parameter 0.003667
```

It appears that without B12, antibiotics result in a decrease in the average daily weight gain of around 0.16 pounds. With B12 the effect is to increase the average weight gain by around 0.32 pounds. 95 per cent confidence limits for these estimates are obtained by EXTRACTing the parameter estimates and their standard errors and calculating the appropriate percentage points of the t distribution as follows.

```
$extract %pe %se $
$cal %t=%td(.975,8) : lo=%pe-%t*%se : hi=%pe+%t*%se $
```

Note that because this was a balanced orthogonal design the standard errors of the two estimates were identical. However in general this will not be the case. The estimates and confidence limits can be printed by

The effect of antibiotics without B12 with 95 per cent confidence limits was

```
$print ; *r %pe(3),8,2 '   '*r lo(3),8,2' to'*r hi(3),8,2 $
   -0.16     -0.27 to   -0.04
```

The effect of antibiotics with B12 with 95 per cent confidence limits was

```
$print ; *r %pe(4),8,2 '   '*r lo(4),8,2' to'*r hi(4),8,2 $
    0.32      0.21 to    0.44
```

Similarly the two separate effects of B12 may be estimated with their confidence limits by re-fitting the model parametrized so the effects of B12 are nested within the A_bio factor

```
$fit A_bio/B12  $dis e $
   deviance =  0.029333
residual df =  8
          estimate        s.e.     parameter
     1       1.190     0.03496      1
     2     -0.1567     0.04944      A_BIO(2)
     3     0.03000     0.04944      A_BIO(1).B12(2)
     4      0.5100     0.04944      A_BIO(2).B12(2)
scale parameter 0.003667
```

It appears that without antibiotics, B12 results in a very small increase in the average daily weight gain of around 0.03 pounds. With antibiotics the effect is to increase the average weight gain by 0.51 pounds. 95 per cent confidence limits for these two estimates are obtained as above by

```
$extract %pe %se $
$cal lo=%pe-%t*%se : hi=%pe+%t*%se $
```

The effect of B12 without antibiotics with 95 per cent confidence limits was

```
$print ; *r %pe(3),8,2 '  '*r lo(3),8,2' to'*r hi(3),8,2  $
    0.03     -0.08 to    0.14
```

The effect of B12 with antibiotics with 95 per cent confidence limits was

```
$print ; *r %pe(4),8,2 '  '*r lo(4),8,2' to'*r hi(4),8,2 $
    0.51      0.40 to    0.62
```

It seems reasonably clear from this data that neither treatment is of much benefit without the other, but together they have a highly significant effect.

12.1.2.5 General cross classifications with covariates

The following data were collected over a period of five years from 156 heart transplant patients to monitor various aspects of patient management as well as survival. (Source: personal communication)

One problem in heart transplant patients under immuno-suppression is that they may suffer kidney damage for various reasons. This investigation was undertaken with the intention of assessing how much management of transplant patients had advanced over time in its ability to restrict renal damage, measured by increased levels of creatinine, allowing for the pathology category of the patient, age, and the cyclosporinD dosage which may all affect the risk of renal damage and hence the levels of creatinine. This analysis is focused on the levels of cyclosporinD and creatinine in the serum at two weeks after transplant.

Sixteen of the patients are women, but they make up only 10 per cent of the sample so they have been omitted from the analysis here to avoid the complication of having to test and allow for possible sex differences in the various analyses.

The variables in the dataset are

`age`	age at transplant in years;
`sex`	1. males; 2. females;
`path`	pathology category coded 1 to 3 with 1 the best;
`wei`	weight (not used in this analysis);
`lvf`	a measure of liver function (not used in this analysis);
`yrtx`	year transplanted (data from 1982 to 1986);
`cr2w`	serum creatinine levels two weeks after transplant;
`cd2w`	serum cyclosporinD levels two weeks after transplant.

```
$un 156 $data age sex path wei lvf yrtx cr2w cd2w $
$read age sex path             yrtx cr2w cd2w $
    38    1   3     55    60    82    273    16.6
    40    1   2     52    73    82    237     7.3
    50    1   1     60    74    82    290     6.3
    53    1   1     97    66    83    321    10.5
     .    .   .   .   .   .   .   .   .   .   .   .
    24    1   2     75    75    86    132     2.7
    30    1   2     50    78    86    280     8.2
```

To check the ranges of values in the data it is useful to look at the distributions of the variables.

```
$hist cr2w $
[  0.0, 50.0)   4 CCCC
[ 50.0,100.0)  33 CCCCCCCCCCCCCCCCCCCCCCCCCCCCCCCCC
[100.0,150.0)  53 CCCCCCCCCCCCCCCCCCCCCCCCCCCCCCCCCCCCCCCCCCCCCCCCCCCCC
[150.0,200.0)  29 CCCCCCCCCCCCCCCCCCCCCCCCCCCCC
[200.0,250.0)  13 CCCCCCCCCCCCC
[250.0,300.0)  12 CCCCCCCCCCCC
[300.0,350.0)   8 CCCCCCCC
[350.0,400.0)   1 C
[400.0,450.0)   2 CC
[450.0,500.0)   0
[500.0,550.0)   0
[550.0,600.0]   1 C
```

The distribution of the outcome measure looks rather skew. It is likely that a transformation will be needed.

```
$hist cd2w $
[ 0.00, 2.00)  1 C
[ 2.00, 4.00)  8 CCCCCCCC
[ 4.00, 6.00) 20 CCCCCCCCCCCCCCCCCCCC
[ 6.00, 8.00) 19 CCCCCCCCCCCCCCCCCCC
[ 8.00,10.00) 23 CCCCCCCCCCCCCCCCCCCCCCC
[10.00,12.00) 30 CCCCCCCCCCCCCCCCCCCCCCCCCCCCCC
[12.00,14.00) 21 CCCCCCCCCCCCCCCCCCCCC
[14.00,16.00) 16 CCCCCCCCCCCCCCCC
[16.00,18.00) 14 CCCCCCCCCCCCCC
[18.00,20.00)  3 CCC
[20.00,22.00)  1 C
[22.00,24.00]  0
```

This covariate seems relatively well behaved with no very extreme values.

```
$hist age $
[ 5.00,10.00)  3 AAA
[10.00,15.00)  7 AAAAAAA
[15.00,20.00)  5 AAAAA
[20.00,25.00)  5 AAAAA
[25.00,30.00)  6 AAAAAA
[30.00,35.00)  9 AAAAAAAAA
[35.00,40.00) 12 AAAAAAAAAAAA
[40.00,45.00) 22 AAAAAAAAAAAAAAAAAAAAAA
[45.00,50.00) 36 AAAAAAAAAAAAAAAAAAAAAAAAAAAAAAAAAAAA
[50.00,55.00) 32 AAAAAAAAAAAAAAAAAAAAAAAAAAAAAAAA
[55.00,60.00) 16 AAAAAAAAAAAAAAAA
[60.00,65.00]  3 AAA
```

There is a wide age range stretching from 5 to 65 although the bulk of patients are over 30 years old.

```
$hist yrtx $
[82.0,82.4)  3 YY
[82.4,82.8)  0
[82.8,83.2) 26 YYYYYYYYYYYYY
[83.2,83.6)  0
[83.6,84.0) 62 YYYYYYYYYYYYYYYYYYYYYYYYYYYYYYY
[84.0,84.4)  0
[84.4,84.8)  0
[84.8,85.2) 62 YYYYYYYYYYYYYYYYYYYYYYYYYYYYYYY
[85.2,85.6)  0
[85.6,86.0)  3 YY
[86.0,86.4)  0
[86.4,86.8]  0
```

The majority of patients in this sample were transplanted between 1984 and 1986. A sensible grouping for the analysis will be pre–, during, and after 1984.

```
$hist path $
[1.00,1.20) 77 PPPPPPPPPPPPPPPPPPPPPPPPPPPPPPPPPPPPPPP
[1.20,1.40)  0
[1.40,1.60)  0
[1.60,1.80)  0
[1.80,2.00) 63 PPPPPPPPPPPPPPPPPPPPPPPPPPPPPPPP
[2.00,2.20)  0
[2.20,2.40)  0
[2.40,2.60)  0
[2.60,2.80)  0
[2.80,3.00) 16 PPPPPPPP
[3.00,3.20)  0
[3.20,3.40]  0
```

Notice that the third and most severe pathology group is rather small.

The relationship of age and cyclosporinD levels with creatinine can be inspected by

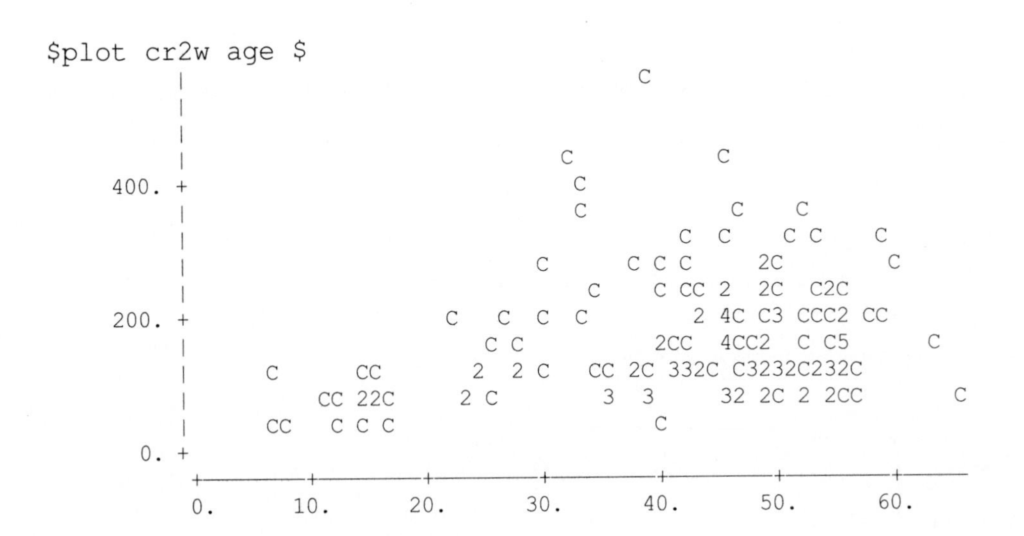

```
$plot cr2w cd2w $
```

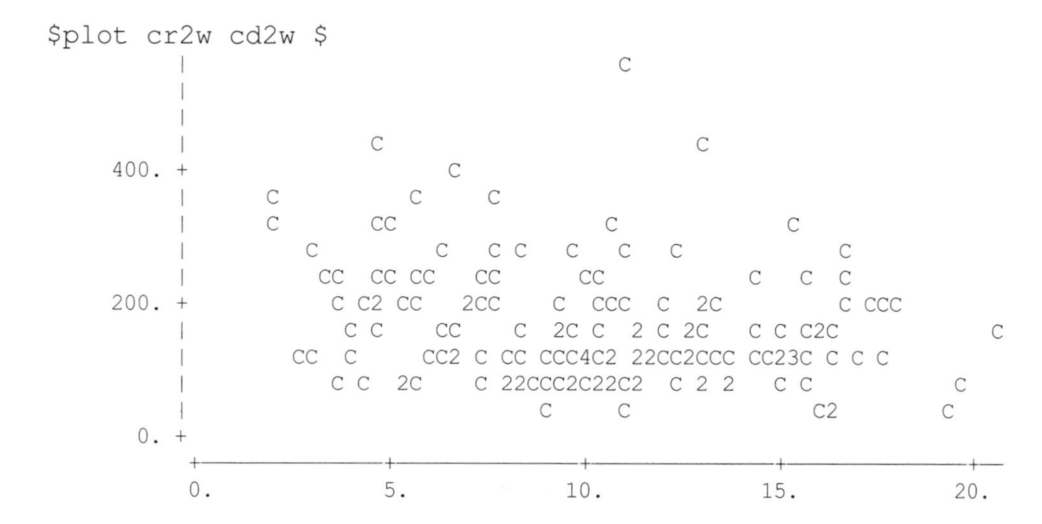

Neither indicates a very clear cut relationship. However it is possible that other factors such as year of transplant have an effect that is obscuring some relationship.

First we need to identify the factor variables and create one for the year of transplant which we will group as:

1. pre–1984; 2. 1984; and 3. 1985 and later.

```
$fac path 3 yrgp 3 $
```

Then assign cut-points for the year of transplant variable and group by

```
$ass iyt=84,85 $group yrgp=yrtx int * iyt * $
```

Calculate a weight variable to exclude the women from this analysis, and graph `cr2w` against `cd2w` using different symbols for the three period groups

```
$cal wt=(sex==1) $weight wt $
```

```
$graph (t='Fig. 12.16 Creatinine by CyclosporinD and Year Tx'
       vla='Creatinine' hla='CyclosporinD'
       ys=0,600,50 xs=0,25,5)
        cr2w  cd2w  1;6;2  yrgp $
```

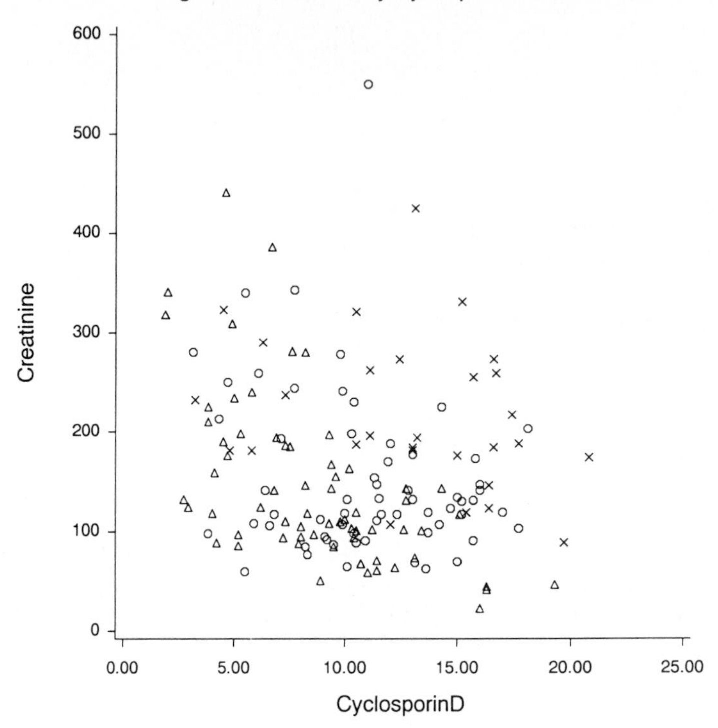

To see the pattern of change over the period 1982–1986 in the creatinine levels two weeks after transplant

```
$tab the cr2w mean;dev;weight with wt for path;yrgp into
m_;sd_;n_ by pth;ygp $
$tpr (s=1) m_;sd_;n_ pth;ygp $
                +-------------------------------+
       YGP |        1         2           3 |
   PTH     |                                |
+------------+-------------------------------+
|    1 M_  |    216.6     149.5      155.9 |
|      SD_ |    86.21     49.45      70.12 |
|       N_ |   17.000    33.000     26.000 |
+--------------------------------------------+
|    2 M_  |    247.0     164.5      126.3 |
|      SD_ |    45.18    120.23      57.26 |
|       N_ |    7.000    21.000     25.000 |
+--------------------------------------------+
|    3 M_  |    175.0     152.6      200.0 |
|      SD_ |    70.88    110.61     199.40 |
|       N_ |    4.000     5.000      2.000 |
+------------+-------------------------------+
```

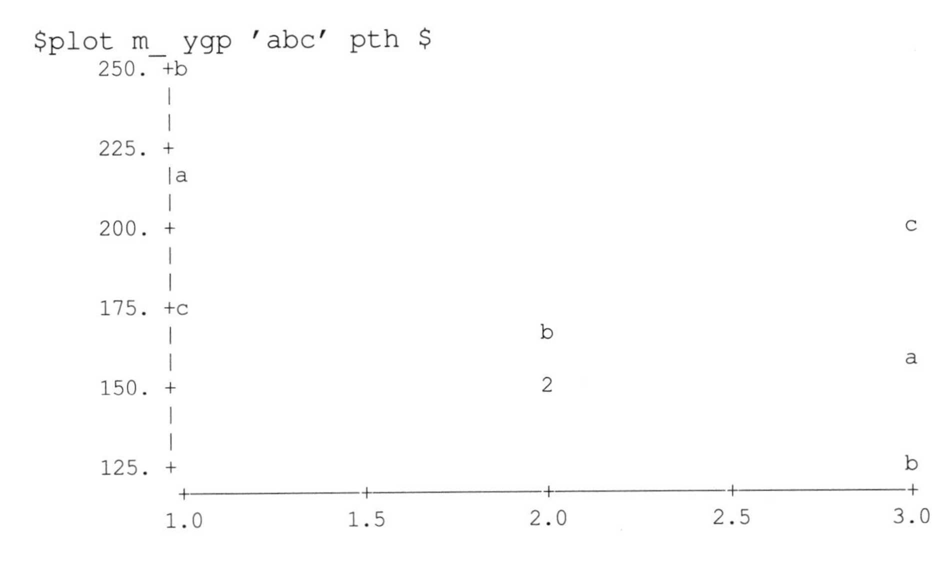

Apart from the third pathology group (the c's) with very small numbers there has been a relatively steady decrease over time in this measure of outcome.

There are two factors, pathology group and year of transplant group, and there are also two covariates, age and cyclosporinD levels. There are many possibilities for interactions of various sorts. The relationships with the two covariates might be non-linear and also interact with each other, requiring product terms in the model. In turn they might interact with the two factors so relationships from year group to year group or from one pathology group to another would not be parallel with each other. In fact, even ignoring the non-linearity problem, the possibilities of this sort are extremely numerous, including six interaction terms involving all the variables taken in pairs, four for the variables taken in threes, and one for the interaction involving all four variables. In this particular case that would require a model with 26 parameters.

A good strategy for coping in such a complex situation is to start with a model that includes all the second–order interactions. If those interactions appear negligible then a standard analysis involving the usual sequence of progressively simplified models can start from there. If there is evidence of more complex relationships then the data can be split into subgroups, for example, by taking the pathology groups separately, and the process repeated separately for each group. This permits a well–structured and searching analysis in complex situations without too many assumptions.

In this case the model would be

```
path +      yrgp +      age +      cd2w
     + path.yrgp + path.age + path.cd2w
                 + yrgp.age + yrgp.cd2w
                            + age.cd2w
    ( + quadratic terms in age and cd2w    )
```

The first part can be written rather more simply as

```
(path+yrgp+cd2w+age)**2
```

because the squared terms such as `path.path` in the model algebra reduce to `path`. For reasons of algorithmic efficiency this also applies to squares of covariates which means that the squared covariate terms to be included in the model have to be separately calculated variables. So to include `age` and `cd2w` squared it is necessary to proceed as follows

```
$cal ag2=age*age : cd2=cd2w*cd2w  $
```

```
$yvar cr2w $
$fit (path+yrgp+cd2w+age)**2+ag2+cd2 $dis e $
   deviance =  637702.
residual df =      118  from 140 observations
           estimate         s.e.       parameter
     1       -303.3         302.8        1
     2        94.29         125.9        PATH(2)
     3        90.51         200.9        PATH(3)
     4        120.2         167.0        YRGP(2)
     5        181.4         183.2        YRGP(3)
     6        14.27         16.64        CD2W
     7        16.67         7.311        AGE
     8      -0.1083       0.05077        AG2
     9     -0.02810        0.4045        CD2
    10       -22.16         43.32        PATH(2).YRGP(2)
    11       -43.40         47.13        PATH(2).YRGP(3)
    12       -11.01         86.37        PATH(3).YRGP(2)
    13        11.54         116.7        PATH(3).YRGP(3)
    14  -0.0005544          4.645        PATH(2).CD2W
    15       -4.954         8.178        PATH(3).CD2W
    16       -3.806         4.600        YRGP(2).CD2W
    17       -9.585         5.713        YRGP(3).CD2W
    18       -1.029         2.021        PATH(2).AGE
    19      -0.2522         2.747        PATH(3).AGE
    20       -2.918         2.887        YRGP(2).AGE
    21       -3.697         3.092        YRGP(3).AGE
    22      -0.3440        0.2100        CD2W.AGE
 scale parameter 5404.
```

Before proceeding any further in assessing the contributions of the various factors and covariates in this complex model it is necessary to check a number of assumptions using the residuals.

```
$plot %rs %fv 'r' $
```

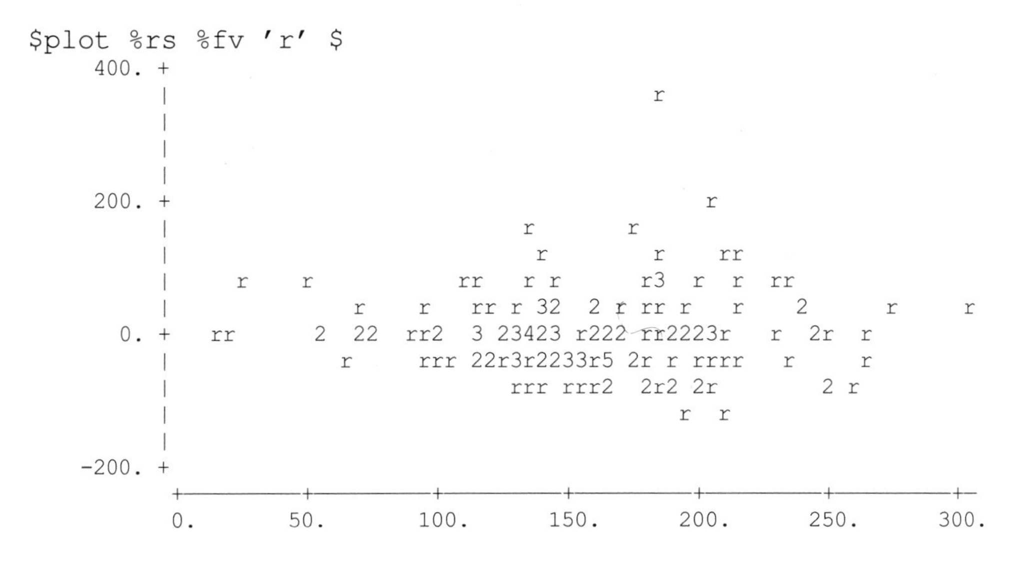

There is evidence of a rather complex and systematic relationship between the residuals and the fitted values. There is some suggestion of a variance increasing with the mean, that is, the fitted values which would suggest a logarithmic transformation, but the plot looks more complicated than just that. There is some suggestion that the larger fitted values are also systematically overestimating the true values and producing a disproportionate number of small or negative residuals. Let us use a macro to obtain a Normal plot allowing for the weighing to exclude female patients.

```
$use wrep $
```

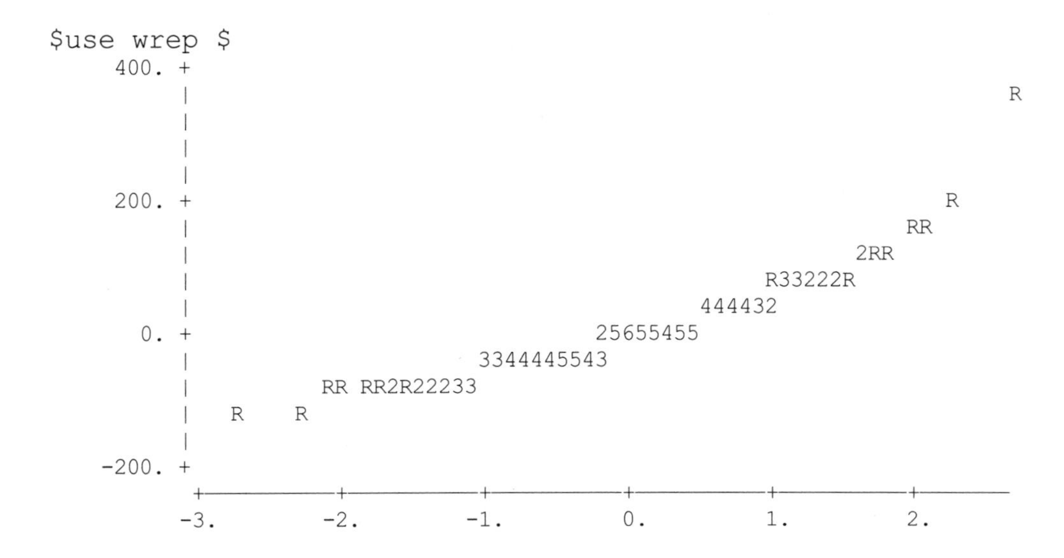

From this we can see that the error distribution is clearly not Normal. The upward curve shows that the residual distribution is markedly skew in the positive direction, that is, a long tail to the right, because it is necessary to go much further along the residual axis than along the Normal axis to reach equivalent quantiles. This pattern also suggests that a logarithmic transformation may be helpful.

```
$cal lcr2=%log(cr2w) $yvar lcr2 $
```

Then fit the model again

```
$fit (path+yrgp+cd2w+age)**2+ag2+cd2   $dis e $
   deviance =    19.707
residual df =   118       from 140 observations
          estimate          s.e.      parameter
     1       2.791         1.684       1
     2      0.4880        0.6998       PATH(2)
     3      0.2316         1.117       PATH(3)
     4      0.3907        0.9284       YRGP(2)
     5      0.8327         1.018       YRGP(3)
     6     0.05088       0.09252       CD2W
     7     0.08678       0.04064       AGE
     8  -0.0006124     0.0002822       AG2
     9  0.00007004      0.002249       CD2
    10     -0.2213        0.2408       PATH(2).YRGP(2)
    11     -0.2265        0.2620       PATH(2).YRGP(3)
    12    -0.08016        0.4802       PATH(3).YRGP(2)
    13    -0.06820        0.6487       PATH(3).YRGP(3)
    14    0.006860       0.02582       PATH(2).CD2W
    15    -0.01852       0.04546       PATH(3).CD2W
    16    -0.01703       0.02557       YRGP(2).CD2W
    17    -0.06432       0.03176       YRGP(3).CD2W
    18   -0.006641       0.01124       PATH(2).AGE
    19    0.002405       0.01527       PATH(3).AGE
    20    -0.01107       0.01605       YRGP(2).AGE
    21    -0.01589       0.01719       YRGP(3).AGE
    22   -0.001494      0.001167       CD2W.AGE
scale parameter 0.1670
```

```
$plot %rs %fv $
       |                                           %
       |
    1. +
       |                                     %%   %      %
       |                      %         %      %      %%
       |                  %                %  %    %  %% %
       |                       %    %   %  % 2%%%%%% %2%2    %  %
       |                                    %%% %  2 %  % %   %%   % %
    0. +   %%            %     %2   2 %%%%3%  42%33%%%2 3%2%%%  2 %
       |                     2        %   %2 2%3% %2   %  2  %    %
       |                            %        %3%3 %2 % %%%      2%
       |                     %         % 2 %      4  %%%2
       |                                   %   %    % %
       |                                     %     %
   -1. +
        +-------------+-------------+-------------+-------------+------
       3.5           4.0           4.5           5.0           5.5
```

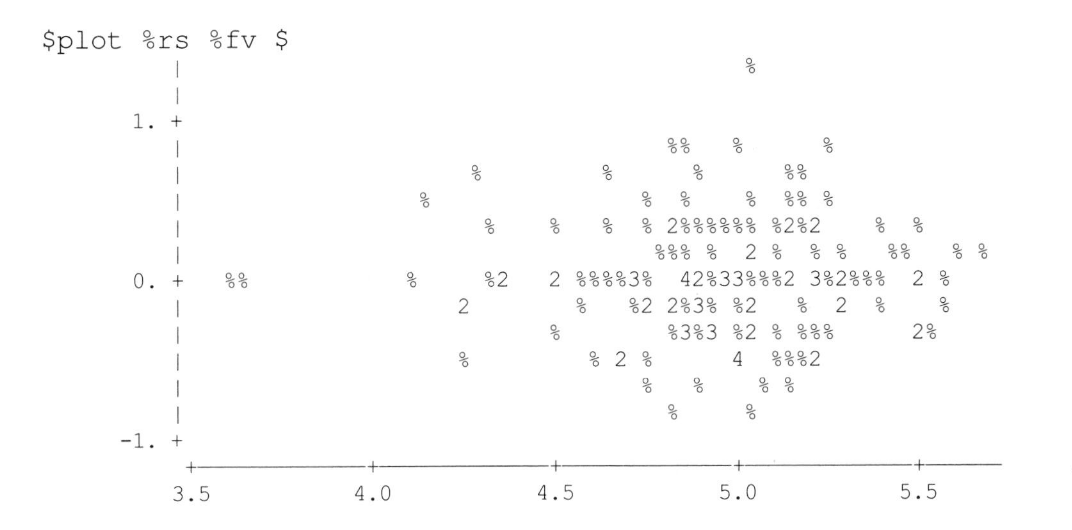

The plot of residuals against fitted values is somewhat improved although not entirely. Let us investigate the Normality on this scale.

```
$use wrep $
       |                                                               R
       |
    1. +
       |                                                     R RR R
       |                                                 RR2R
       |                                              222R
       |                                        244333R
       |                                     2542
    0. +                               2565543
       |                           45543
       |                     R33444
       |               2R22232
       |          RR RR
       |   R    R
   -1. +
        +----------+----------+----------+----------+----------+--------
       -3.        -2.        -1.         0.         1.         2.
```

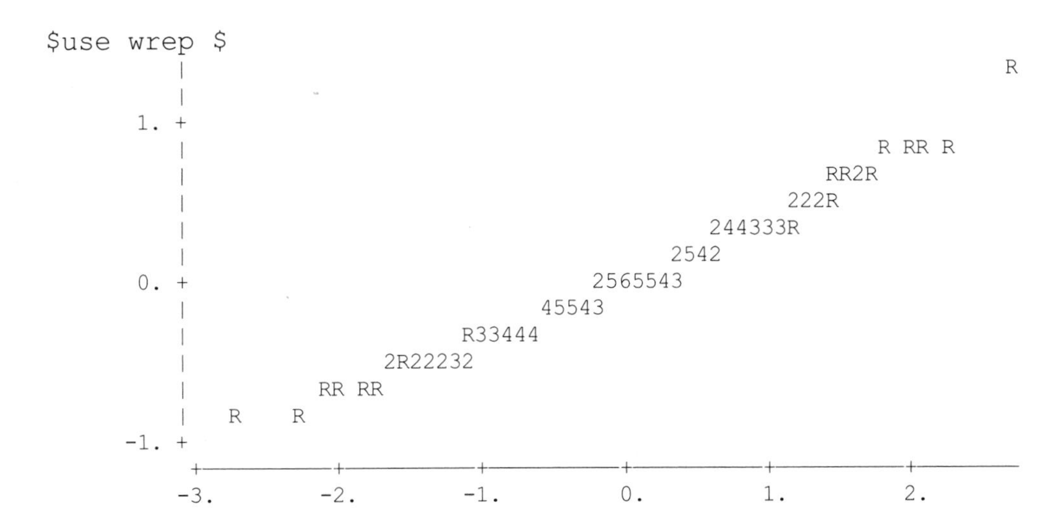

There is still a slight curve, but the plot is very much better.

To construct the full analysis of covariance from a sequence of fits it is necessary to store the residual sums of squares and degrees of freedom for each model to obtain the appropriate components of the F test calculations. The scalar variables to store these are declared by

```
$number s2 SS1 SS2 SS3 SS4 SS5 SS6 SS7 df1 df2 df3 df4 df5
df6 df7 $
```

First store the results from the full model above.

```
$cal SS1=%dv : df1=%df : s2=%sc $
```

We can now start to interpret the fitted parameters and consider ways in which the model can be simplified without significantly worsening the fit. A few of the interaction terms are greater than their s.e.s, but not by much. Let us start by dropping the interaction terms, including those between the factors and covariates, so the relationships with the covariates are modelled as a set of parallel quadratic surfaces with intercepts depending on the level of the two factors `path` and `yrgp`.

```
$fit path+yrgp + cd2w+cd2 + age+ag2 + age.cd2w $dis e $
   deviance =    21.050
residual df =   130       from 140 observations
          estimate          s.e.     parameter
     1       4.573        0.5994      1
     2     0.04607       0.08105      PATH(2)
     3    -0.02306        0.1394      PATH(3)
     4     -0.3756       0.09934      YRGP(2)
     5     -0.5590        0.1086      YRGP(3)
     6    -0.05964       0.05351      CD2W
     7    0.002777      0.001821      CD2
     8     0.05578       0.02061      AGE
     9  -0.0005002     0.0002143      AG2
    10  -0.0008495     0.0008179      CD2W.AGE
scale parameter 0.1619

$cal SS2=%dv : df2=%df $
```

The increase in deviance is of the order of 1.3 on 12 d.f., that is, an estimate of the residual variance of 1.3/12 = 0.1083 which, compared with the residual variance from the previous comprehensive model of 0.1670, gives a variance ratio F test of 0.65 on 12 and 118 d.f. For significance at the 5 per cent level this has to be greater than

```
$cal %fd(0.95,12,118) $
            1.835
```

Since this is much larger than 0.65 the observed F–value is not significant. This indicates that taken as a group these interaction terms are no further from zero than could easily occur by chance. However it is possible that some of the interaction terms may be reflecting real effects and if part of the exercise is to generate hypotheses it might be worth investigating them in more detail. For example the `yrgp(3).cd2w` coefficient is more than twice its s.e. which may indicate that the relationship between cyclosporinD and creatinine levels has changed over time. However this is rather speculative and since twelve parameters have been inspected this way the probability of at least one of them exceeding their s.e.s to this degree will be quite high. For the current purposes of a general assessment of the effects of these variables on creatinine it is probably reasonable to simplify the model by dropping the two– factor interactions and parameters allowing for non-parallelism in the covariate relationships. The model without them appears to give an adequate fit and it may be possible to simplify further.

In the reduced model, of the three second–order covariate terms, only the quadratic `age` term is more than twice its s.e. The coefficients of `cd2` and `age.cd2w` are not apparently significant and can probably be omitted without making the fit significantly worse.

```
$fit -cd2 - age.cd2w $dis e $
   deviance =   21.630 (change =  +0.5801)
residual df =  132      (change =  +2       ) from 140 observations
          estimate        s.e.      parameter
     1       4.809       0.3492      1
     2     0.03721      0.08106      PATH(2)
     3    0.006162       0.1384      PATH(3)
     4     -0.4277      0.09582      YRGP(2)
     5     -0.5930       0.1071      YRGP(3)
     6    -0.03952     0.009297      CD2W
     7     0.04276      0.01569      AGE
     8  -0.0004385    0.0002039      AG2
scale parameter 0.1639
```

Store the residual sum of squares and the degrees of freedom.

```
$cal SS3=%dv : df3=%df $
```

The F test p value is

```
$cal 1-%fp((SS3-SS2)/(df3-df2)/s2,df3-df2,df1) $
         0.1806
```

There is no evidence that these terms were necessary in the model. This model can therefore be considered as a reference model to be used as the baseline in this particular analysis. All the remaining terms should be tested by comparing the fit of this model with and without them thus avoiding assumptions about the effects of the other terms.

The fit of the model can be illustrated by obtaining the values predicted from the model for a chosen range of Cyclosporin D values, for example,

```
$ass Cyc_Dp=0,1...25 $
$pred   (s=-1)   age=50   ag2=2500   yrgp=1   cd2w=Cyc_Dp   $cal
cr1_=%exp(%pfv) $
$pred   (s=-1)   age=50   ag2=2500   yrgp=2   cd2w=Cyc_Dp   $cal
cr2_=%exp(%pfv) $
$pred   (s=-1)   age=50   ag2=2500   yrgp=3   cd2w=Cyc_Dp   $cal
cr3_=%exp(%pfv) $
```

```
$graph (t='Fig. 12.17 Creatinine by CyclosporinD and Year Tx'
        vla='Creatinine' hla='CyclosporinD'
        ys=#yscale xs=#xscale        pause=no)
        cr2w;cr1_;cr2_;cr3_  cd2w;Cyc_Dp;Cyc_Dp;Cyc_Dp
        1;6;2;10;16;14  yrgp;;; $
```

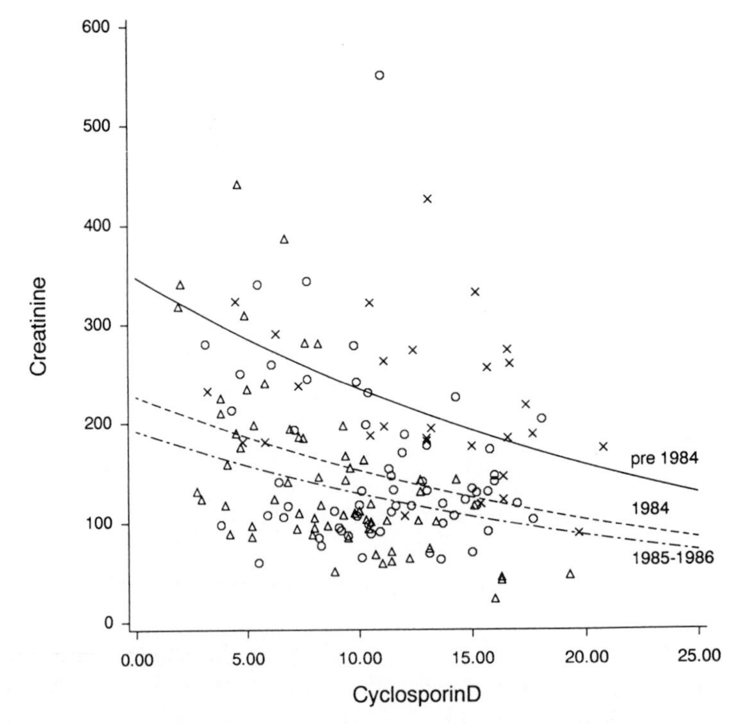

Then a sequence to label the curves representing the fitted model for each of the three time periods using suitable co-ordinates.

```
$ass xtxt=22,22,22 : ytxt=60,110,160 $
$gtext (m=p l=r) ytxt xtxt '1985-1986','1984','pre 1984' $
```

From this baseline model it appears that the difference between the pathology groups is far from significant. On the other hand the effects of year group, age (as a quadratic term), and cyclosporinD are all independently significant since their effects are several times their s.e.s. However the factor effect parametrizations only provide the differences between each level and the first. It is necessary to test for heterogeneity over all levels by F tests from comparisons of the models with and without each factor and it is useful to do this for the covariates as well. First drop the age terms:

```
$fit -age-ag2 $cal SS4=%dv : df4=%df $
   deviance =   23.791 (change =   +2.161)
residual df =  134      (change =   +2     ) from 140 observations
```

Then replace the age terms and drop the cyclosporinD term and so on through all the terms to be tested.

```
$fit +age+ag2 -cd2w $cal SS5=%dv : df5=%df $
   deviance =   24.592 (change =  +0.8007)
residual df =  133      (change =  -1      ) from 140 observations
```

```
$fit +cd2w -yrgp $cal SS6=%dv : df6=%df $
   deviance =   26.796 (change =   +2.204)
residual df =  134      (change =   +1     ) from 140 observations
```

```
$fit +yrgp-path $cal SS7=%dv : df7=%df $
   deviance =   21.667 (change =   -5.129)
residual df =  134      (change =    0     ) from 140 observations
```

The analysis of covariance is then obtained by

```
$use anco3125 $
```

Model	Residual SS	df	Mean square	Variance Ratio(F)
1) All two variable terms	19.71	118	0.17	
2) Main effects +quadratic terms	21.05	130		
(2-1) Due to Interns	1.34	12	0.11	0.67 (p= 0.7769)
3) -(CD2+AGE.CD2W) The baseline model	21.63	132		
(3-2) Due to (CD2+AGE.CD2W)	0.58	2	0.29	1.74 (p= 0.1806)
4) - AGE-AG2	23.79	134		
(4-3) Due to AGE+AG2	2.16	2	1.08	6.47 (p= 0.0022)
5) - CD2W	24.59	133		
(5-3) Due to CD2W	2.96	1	2.96	17.73 (p= 0.0000)
6) - YRGP	26.80	134		
(6-3) Due to YRGP	5.17	2	2.58	15.46 (p= 0.0000)
7) - PATH	21.67	134		
(7-3) Due to PATH	0.04	2	0.02	0.11 (p= 0.8971)

The analysis shows that cyclosporinD has a small, but highly significant, negative effect on creatinine levels. Since the rejection process, that cyclosporinD is used to suppress, generates creatinine this may well be evidence of it doing its job. The age relationship with a positive linear term and a negative quadratic term appears to indicate that the creatinine levels increase with age, but less and less rapidly. Pathology does not seem to have any effect on two week creatinine levels, but the year of transplant effect is very pronounced. Even after allowing for the effects of age and cyclosporinD dosages (which have on average decreased considerably over the period) there has been a systematic and highly significant decrease in the levels over the three periods.

12.1.2.6 A nested design with random effects

During a study of respiratory infections, crowding and family transmission dynamics a large number of throat swabs were taken at regular intervals from each member of families entered in the study. The families had to include at least three children including one pre-school child.

The swabs were taken from the father, mother, and three children including the pre-school child. The outcome variable is the count of swabs found positive to pneumococcus. The total number of swabs taken is not given so although the outcome variable looks as if it should have been a 'measured' proportion and probably analysed using Normal assumptions after transformation as available it is effectively a Poisson variable which has been assumed Normal. The data should more properly be analysed assuming the Poisson distribution, but it is a useful example to illustrate the analysis of a nested design matching a published analysis (Armitage and Berry, 1987). The design is partly crossed and partly nested. There are six replicate families in each of the three crowding categories and the individuals are nested within each family. The individual members of each family represent the five family positions: father, mother, first child (the oldest), second child, and third (pre-school) child.

The numbers of swabs positive to pneumococcus by crowding and family position are shown in the following table.

		Family Position				
				Child		
Crowding category	Family Number	Father	Mother	1	2	3
1. Overcrowded	1	5	7	6	25	19
	2	11	8	11	33	35
	3	3	12	19	6	21
	4	3	19	12	17	17
	5	10	9	15	11	17
	6	9	0	6	9	5
2. Crowded	7	11	7	7	15	13
	8	10	5	8	13	17
	9	5	4	3	18	10
	10	1	9	4	16	8
	11	5	5	10	16	20
	12	7	3	13	17	18
3. Uncrowded	13	6	3	5	7	3
	14	9	6	6	14	10
	15	2	2	6	15	8
	16	0	2	10	16	21
	17	3	2	0	3	14
	18	6	2	4	7	20

The data are input by

```
$un 90 $data psw $read
5 7 6 25 19
11 8 11 33 35
3 12 19 6 21
3 19 12 17 17
10 9 15 11 17
9 0 6 9 5
11 7 7 15 13
10 5 8 13 17
5 4 3 18 10
1 9 4 16 8
5 5 10 16 20
7 3 13 17 18
6 3 5 7 3
9 6 6 14 10
2 2 6 15 8
0 2 10 16 21
3 2 0 3 14
6 2 4 7 20
```

It is necessary to calculate variables for crowding category, family position, and family number. The last runs from 1 to 18, but since for the analysis the families only need to be numbered uniquely from 1 to 6 within each of the crowding categories a family number variable, within crowding category, has been created (`fam`). For complete and ordered tabular data such as this the directive `GFACTOR` can be used to generate factor variables with appropriate codes. The factors whose levels change least rapidly are given first.

```
$gfactor cr 3 fam 6 fpos 5 $
```

There are six different families within each crowding category so the family variable with values of 1 to 6 is nested within the three-category crowding variable. The families are effectively six replicates within each crowding category. The mean differences observed between the three crowding categories need to be compared with what might arise by chance as a result of the variation causing these replicates to differ.

In addition, because the families are represented by individuals from each of 5 position categories, there are in effect 6 x 5 replicate observations. However the variation between all the 30 'replicates' is partly due to systematic differences between the position categories. This needs to be removed, before the estimated effects of crowding are tested, to increase the sensitivity if nothing else.

The effects of position in the family need to be estimated within each family and then pooled over families so that family differences are kept from influencing the estimates. The possibility of the position effects being influenced by the level of crowding is investigated by fitting models with and without a `crowding.position` interaction. These effects are compared with the chance effects that might occur due to the variation about the model allowing for systematic family and position differences. In the Normal assumptions model this means the within-family residual variance.

This within-family variance is in fact the variation which would be explained by `family.position` interaction parameters. To use it as an estimate of the residual variance these have to be assumed zero and the variation they would explain taken as representative of the true underlying residual variance. Parameters for these interaction effects would actually saturate the model for this design and dataset. A more extensive design would be required for estimates of these interactions and the residual variance to be obtained.

The analysis is not simple so we will work through the model fitting sequence with some interpretation and then present the results in an analysis of variance tabulation equivalent to that used by Armitage and Berry(1981).

```
$yvar psw $
```

First fit a model with effects for the family differences nested six at a time within the crowding categories together with effects for the position differences within the families assumed the same for families within a crowding category, but differing between them.
This gives the residual variation about the pattern of changes with position within families assumed parallel for families within each of the crowding categories.

```
$fit cr/fam+cr.fpos $        $ass SS=%dv     : df=%df $
   deviance =  1516.7
residual df =     60
```

Storing the deviances and degrees of freedom from each of the models allows us to construct an analysis of deviance table using a sequence of models to address each of the relevant questions in turn. The scale parameter is the estimate of the within-families residual variance.

A check of assumptions using the residuals is obtained by

```
$sort rsor %rs $
$cal nq=%nd((%gl(%nu,1)-0.5)/%nu) $
$plot rsor nq $
```

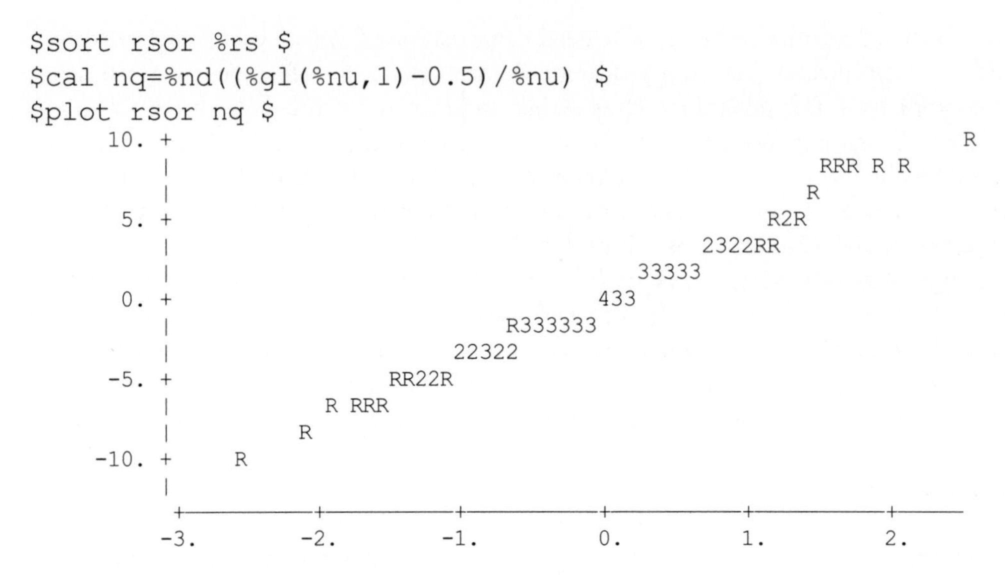

The Normal plot certainly looks sufficiently straight to justify an assumption of an underlying Normal error distribution. Let us look at the mean – variance relationship by plotting the residuals against the fitted values.

```
$plot %rs %fv 'r' $
```

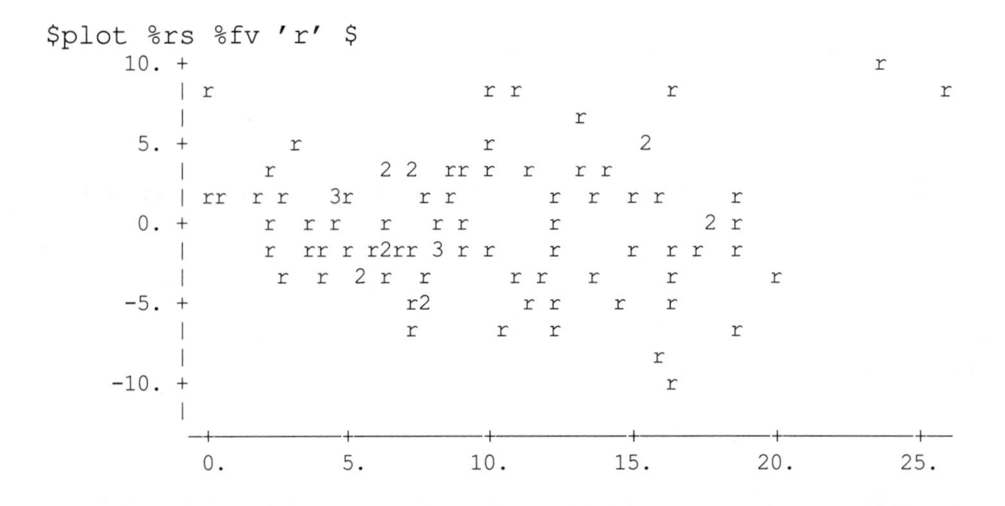

There does not seem to be any relation between the residuals and the fitted values in either their magnitude or their dispersion. However the constant variance assumption can be tested further by obtaining the mean fitted values and the standard deviations of the residuals within the `crowding` by `position` table cells and plotting one against the other.

```
$tab the %fv mean for cr;fpos into mean by crow;stat $
$tab the %rs devn for cr;fpos into rdev $
$tpr (s=1) rdev;mean crow;stat $
```

CROW	STAT	1	2	3	4	5
1	RDEV	5.245	5.738	5.698	7.699	5.493
	MEAN	6.833	9.167	11.500	16.833	19.000
2	RDEV	2.897	3.369	2.261	2.794	3.259
	MEAN	6.500	5.500	7.500	15.833	14.333
3	RDEV	3.975	2.285	2.077	3.935	6.074
	MEAN	4.333	2.833	5.167	10.333	12.667

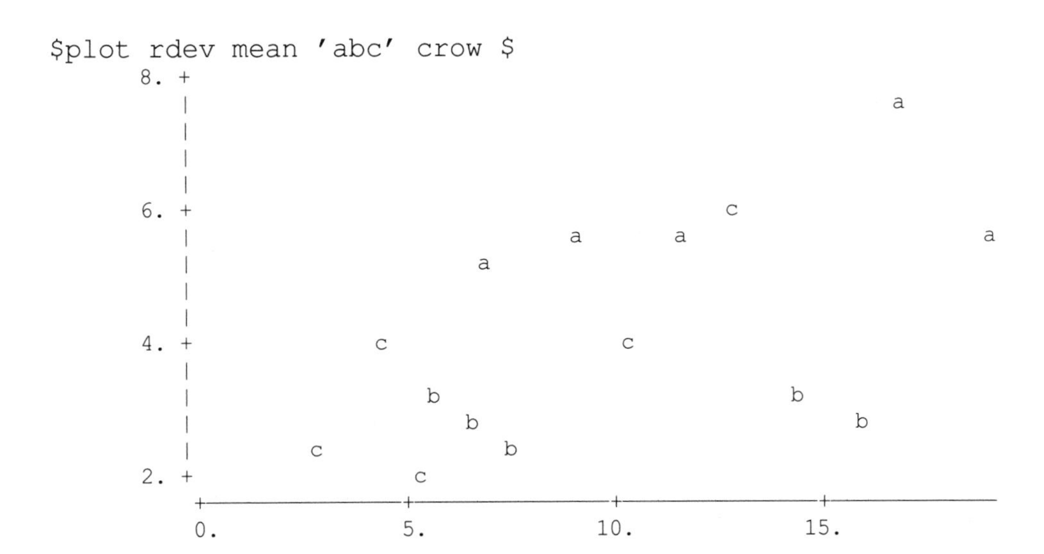

Within each crowding category there does not seem to be a relationship between the variation about the model and the magnitude of the mean although there are considerable differences between the mean positive swab count across the position variable. However there is some evidence that the variance changes systematically over the crowding categories. It is possible that the data should be transformed or analysed using the Poisson error distribution. However the assumptions are probably safe enough for us to complete and interpret the current analysis.

We now fit the model omitting the interaction term so the resulting increase in residual variation can be compared with the residual variance estimated from the first model. This model provides the best estimates of the position effects because they are adjusted for systematic differences between families.

```
$fit cr/fam+fpos        $dis u $ass SS=SS,%dv : df=df,%df $
   deviance =   1589.1
residual df =      68
          estimate          s.e.      parameter
     1       8.400         2.390       1
     2      -1.800         3.057       CR(2)
     3      -7.600         3.057       CR(3)
     4    -0.05556         1.611       FPOS(2)
     5       2.167         1.611       FPOS(3)
     6       8.444         1.611       FPOS(4)
     7       9.444         1.611       FPOS(5)
     8       7.200         3.057       CR(1).FAM(2)
     9     -0.2000         3.057       CR(1).FAM(3)
    10       1.200         3.057       CR(1).FAM(4)
    11   2.220e-15         3.057       CR(1).FAM(5)
    12      -6.600         3.057       CR(1).FAM(6)
    13  -1.006e-15         3.057       CR(2).FAM(2)
    14      -2.600         3.057       CR(2).FAM(3)
    15      -3.000         3.057       CR(2).FAM(4)
    16      0.6000         3.057       CR(2).FAM(5)
    17       1.000         3.057       CR(2).FAM(6)
    18       4.200         3.057       CR(3).FAM(2)
    19       1.800         3.057       CR(3).FAM(3)
    20       5.000         3.057       CR(3).FAM(4)
    21     -0.4000         3.057       CR(3).FAM(5)
    22       3.000         3.057       CR(3).FAM(6)
scale parameter 23.37
```

The magnitudes of the effects of position which are parametrized as excesses over the father's mean level of the positive swab count appear to be quite large and significant judging by their s.e.s. To test the factor as a whole we need the deviance from the model with FPOS omitted

```
$fit cr/fam           $        $ass SS=SS,%dv : df=df,%df $
   deviance =  3122.8
residual df =     72
```

A substantial increase in the deviance results from omitting these four parameters which, compared with the within-families residual in the analysis of variance below, confirms the impression obtained when the estimates above are compared with their standard errors.

The effect of crowding has to be compared with the residual variation including differences between families since these are only a sample of a hypothetical population of such families. To obtain an estimate of the appropriate between-family variance, including the within-family variation, but not that due to differences between the position categories, we fit a model with the crowding and position factors, but omitting the family factor. This is the model from which the crowding factor effects are estimated adjusted for `fpos` effects, but not the within-crowding category family differences. It assumes the `cr.fpos` interaction is zero and combines the within-family residual variation and that which might have been due to the `cr.fpos` interaction with the between-family variation to get an estimate of the residual variance with the maximum degrees of freedom for calculating the standard errors.

```
$fit cr+fpos            $dis u $ass SS=SS,%dv : df=df,%df $
   deviance =  2264.7
residual df =      83
         estimate          s.e.       parameter
     1      8.667         1.457        1
     2     -2.733         1.349        CR(2)
     3     -5.600         1.349        CR(3)
     4   -0.05556         1.741        FPOS(2)
     5      2.167         1.741        FPOS(3)
     6      8.444         1.741        FPOS(4)
     7      9.444         1.741        FPOS(5)
scale parameter 27.29
```

There is a reasonably clear tendency for the mean positive swab count to decrease with a decrease in crowding although only the `cr(3)` parameter, representing the 'uncrowded' minus the 'very crowded' difference actually exceeds twice its s.e. The difference between the residual sum of squares from this model and that from the equivalent model including `/fam` gives an estimate of the between-family variance. This is independent of the within-family variance obtained earlier which can be used to compare the within and between variances. Together with the increase in the residual sum of squares resulting from dropping the `cr` factor from this model it can also be used to obtain an overall (F) test of the differences between the three categories of crowding.

```
$fit    fpos    $
   deviance =  2735.2
residual df =      85
$ass SS=SS,%dv : df=df,%df $
```

This model completes the sequence of model fitting necessary to address all the questions implicit in the design and dataset. It simply remains to assemble the analysis of variance table which has been done using the specially written macro `ANCO3126`.

```
$use ANCO3126 $
            Model          Residual SS    df    Mean square      Variance Ratio(F)
  --------------------------------------------------------------------------------
    1) CR/FAM+CR.FPOS        1516.73      60       25.28
    2) CR/FAM    +FPOS       1589.13      68
  -----------------------------------------------
  (2-1)Due to CR.FPOS          72.40       8        9.05                    0.36
                                                                  (p=   0.9384)
  --------------------------------------------------------------------------------
    3) CR/FAM
       (i.e. -FPOS)          3122.80      72
  -----------------------------------------------
  (3-2)Due to FPOS           1533.67       4      383.42                   15.17
                                                                  (p=   0.0000)
  --------------------------------------------------------------------------------
    4) CR + FPOS
       (ie -/FAM)            2264.73      83
  -----------------------------------------------
  (4-2)Due to /FAM            675.60      15       45.04                    1.78
                                                                  (p=   0.0557)
  --------------------------------------------------------------------------------
    5)   1 + FPOS
       (ie - CR)             2735.22      85
  -----------------------------------------------
  (5-4)Due to CR              470.49       2      235.24                    5.22
                                                                  (p=   0.0073)
  --------------------------------------------------------------------------------
  --------------------------------------------------------------------------------
```

The analysis shows that, allowing for a crowding effect, a family position effect, and an interaction between them, the residual variance within families is 25.28. Using this the F test for the 'crowding.family position' interaction (0.36) is far from significant. This justifies a test of 'family position' as a main effect which gives an F test on 4 and 60 d.f. of 15.17 which is highly significant ($p < 0.0001$). Crowding must be tested by comparing groups of different families. Consequently the variation between families has to be included in the error variance. The appropriate estimate of variance is obtained by dropping 'family' from the second model with family 'nested' within the 'crowding' category. The mean square is then the between-family residual variance which is made up from the two components of variance within and between families. The variance ratio test here indicates whether between-family differences add significantly to the variation. It appears that they nearly double the variance (variance ratio = 1.78) although the increase does not quite reach significance ($p < 0.06$). The 'crowding' effect is then tested with an F test of the `cr` mean square against this residual variance. The result is an F-value, on 2 and 15 d.f., of 5.22 which is also highly significant ($p = 0.007$).

Since shifts on the 'numbers of swabs positive' do not have a very obvious interpretation and the precise definition of the crowding variable is not given there are no very obvious quantitative results that would be useful. This analysis is restricted to testing the hypotheses that:

Considering the outcome measure 'number of swabs positive' in general:

1.	Differences between different members of the family do not change with crowding.	It appears they do not.
2.	There are no real differences between different members of the family.	It appears there are pronounced differences.
3.	The level of 'number of swabs positive' does not systematically increase with crowding.	It appears that it does quite markedly.

It should be noted that if the design is unbalanced because values are missing this approach needs modification to give the correct estimate of the between-families variance. The modifications required are not simple, but they are described fully by Zambello de Pinho (1989).

12.1.2.7 Grouped data: the analysis of means

This data arose from a 'community intervention study' designed to help the control of hypertension in the community. One community was exposed to a number of health promotion campaigns and encouraged to use community and hospital clinics for diagnosing, monitoring, and advising on diet and other treatments for hypertension. The other community was observed, but not approached except for the purpose of obtaining measurements from representative samples of the residents. Samples were obtained in such a way that the age distributions were very similar. This analysis of mean systolic blood pressures shows how a complete analysis of variance may be constructed by fitting the appropriate models to grouped data available as a table of frequencies (n), means and standard deviations (sd).

Mean systolic blood pressure by sex and treatment group

		Control community	Intervention community
Males	n	200	223
	mean	158.5	150.1
	sd	24.2	21.0
Females	n	191	283
	mean	167.4	154.7
	sd	27.5	21.5

The frequencies, means, and standard deviations are input by

```
$units 4 $data n sbpm sd $read
 200 158.5 24.2
 223 150.1 21.0
 191 167.4 27.5
 283 154.7 21.5
```

Factor variables for the sex and treatment group factors are generated by

```
$gfactor  sex 2 ci 2  $
```

The variances of the means, assuming the populations of individual blood pressures to have a common variance v estimated with the pooled variance, are v/n. Each mean should be given a weight inversely proportional to its variance. Since v is constant the appropriate weights are the n's.

```
$weight n $
```

To perform the complete analysis requires the residual sum of squares and the pooled (that is, residual) variance for all the groups. These are obtained by

```
$number  SS1 df1 s2 $
$cal SS1=%cu((n-1)*sd*sd) : df1=%cu(n)-%cu(n/=0) : s2=SS1/df1 $

$pr 'the residual SS is' SS1 'with' df1 'df' : 'so the pooled
variance is' s2 $

 the residual SS is      488486. with   893.0 df
 so the pooled variance is   547.0
```

Then using the `SCALE` directive the pooled variance can be put into the fitting process so the s.e.s are exactly what would have been obtained if the individual systolic blood pressure values had been used in the analysis.

```
$scale s2 $
$yvar sbpm $
$fit sex*ci $dis e $
scaled deviance =  0.
    residual df =  0
           estimate          s.e.      parameter
     1        158.5        1.654        1
```

```
      2        8.900         2.366       SEX(2)
      3       -8.400         2.278       CI(2)
      4       -4.300         3.160       SEX(2).CI(2)
scale parameter 547.0
```

Note that because the means only appear as single observations in the analysis the model with interactions is saturated and fits exactly with no residuals and hence zero deviance. As a consequence the residual sum of squares and degrees of freedom are missing and for the analysis of variance below they have to be obtained from the pooled variance calculations above. However because the scale parameter has been set equal to the pooled variance the standard errors of the parameter estimates are correct. Thus the interaction term can be tested using the estimated `sex(2).ci(2)` effect divided by its s.e.

```
$number t $
$ext %pe %se $
$cal t=%pe(4)/%se(4) $
```

The parameter estimate and its se are

```
$print %pe(4)'   '%se(4) $
  -4.300     3.160
```

and the t test for interaction is

```
$print  t' with 'df1' df'$
  -1.361 with    893.0 df
```

Since the estimate is only slightly larger than its s.e. and the t test is well below 2 (1.96) there is little evidence of interaction.

The appropriate sum of squares for an analysis of variance is obtained by increasing the deviance, obtained from the model fitted to the grouped data, by the within groups residual sum of squares and degrees of freedom. This is most conveniently achieved using the `BASELINE` directive to declare that all deviances and degrees of freedom are to be increased by the baseline values `SS1` and `df1`.

The scale setting must also be cancelled, using the `SCALE` directive with no items, so the deviances from the models are the appropriate sums of squares. The main effects model, assuming that there is no interaction (with uncorrected standard errors) is then obtained by

```
$baseline SS1 df1 $scale $
$fit sex+ci $dis e $
   deviance =  489499. with baseline adjustment.
residual df =      894
          estimate        s.e.     parameter
     1       159.7        42.16       1
```

```
    2        6.489        46.91       SEX(2)
    3       -10.63        47.23       CI(2)
scale parameter 489499.
$number  SS2 df2 $
$cal SS2=%dv : df2=%df $
```

The control/intervention factor `ci` is then dropped to identify the sum of squares change in the residual sum of squares associated with it and to obtain the F test

```
$fit sex  $
   deviance =  514319. with baseline adjustment.
residual df =     895

$number  SS3 df3 $
$cal SS3=%dv : df3=%df $
```

Finally the `sex` factor is dropped from the model and the appropriate F test for it obtained

```
$fit ci $
   deviance =  498864. with baseline adjustment.
residual df =     895

$number  SS4 df4 $
$cal SS4=%dv : df4=%df $
```

The analysis of variance table is then

```
$use anov3127 $
           Model     Residual SS   df   Mean square    Variance Ratio(F)
 ______________________________________________________________________
   1)with interaction 488486.38  893        547.02
   2) Main effects    489499.34  894
 ______________________________________
(2-1)Due to intrns       1012.97    1       1012.97                 1.85
                                                       ( p =    0.1739 )
 ______________________________________________________________________
   3) SEX
       (that is - CI) 514318.88  895
 ______________________________________
(3-2)Due to  CI         24819.53    1      24819.53                45.37
                                                       ( p =    0.0000 )
 ______________________________________________________________________
   4) CI
       (ie - SEX)      498864.13  895
 ______________________________________
(4-2)Due to  SEX         9364.78    1       9364.78                17.12
                                                       ( p =    0.0000 )
 ______________________________________________________________________
 ______________________________________________________________________
```

The analysis of variance confirms that there is little evidence of an interaction, but definite and highly significant sex and treatment effects. This means that the main effects model is the simplest that adequately represents the data. To obtain confidence intervals for these effects it is necessary to fit the main effects model again with the scale parameter set to the value of the pooled variance and the baseline `%dv` and `%df` reset to 0.

```
$scale s2 $baseline $
$fit sex+ci $dis e $
scaled deviance =  1.8518
    residual df =  1
          estimate          s.e.       parameter
     1       159.7         1.409        1
     2       6.489         1.568        SEX(2)
     3      -10.63         1.579        CI(2)
scale parameter 547.0
```

The effects `sex(2)` and intervention `ci(2)` with 95 per cent confidence limits are then obtained by first specifying some scalar variables and then `EXTRACT`ing the appropriate parameter estimates and standard errors. There are sufficient degrees of freedom to permit the use of the Normal approximation to the t distribution so the formula

```
estimate +/- 1.96s.e.
```

may be used to calculate the 95 per cent confidence limits

```
$number sxlo sxup cilo ciup $
$ext %pe %se $
$cal  sxlo=%pe(2)-1.96*%se(2)  : sxup=%pe(2)+1.96*%se(2)
    : cilo=%pe(3)-1.96*%se(3)  : ciup=%pe(3)+1.96*%se(3) $
```

The `sex` and `ci` effects with their s.e.s and 95 per cent confidence limits are

```
$print : %pe(2)'   s.e.= '%se(2)' and 95% CLs 'sxlo' to 'sxup $
   6.489   s.e.=    1.568 and 95% CLs    3.415 to    9.563
```

and

```
$print   %pe(3)'   s.e.= '%se(3)' and 95% CLs 'cilo' to 'ciup $
  -10.63   s.e.=    1.579 and 95% CLs   -13.73 to   -7.540
```

It appears from these estimates that females such as these have systematically higher systolic blood pressures than males and although the difference may be as little as 3.5mm/Hg it may be as much as 9.6mm/Hg. Any values within this range are consistent with these results.

The treatment effect suggests that the intervention has had some beneficial effect. The intervention community sample has a mean blood pressure significantly lower than that of the control community and although the true reduction for individuals subject to such interventions may be no more than 7.5mm/Hg the results are consistent with a beneficial effect as much as 13.7mm/Hg.

12.1.3 Repeated measurements and bioassay

12.1.3.1 Paired measurements

This example uses data from an investigation of methods for teaching reading used a paired comparison design (Clarke and Cooke, 1983). Pairs of children were selected to be as alike as possible with respect to age, general ability, and social factors. One member of each pair was selected at random to be taught using traditional methods (T), the other was taught using a new (N) technique. After one year they were tested and their reading skills scored. The analysis is required to assess whether there is any systematic difference between the reading skills of the two groups that might indicate that one or other of the methods was more effective. (Source: Clarke,G.M., Cooke,D.; 1983, p166)

This example is intended to illustrate how modelling may be used for paired data. The simple paired t test is easily performed without modelling, but the general approach illustrated easily extends to much more complex situations where modelling is essential, for example, paired comparisons within a multi–way classification where pair differences may vary systematically from category to category.

```
$units 16  $data T_score  N_score  $read
37 41
51 49
46 50
46 46
55 53
38 42
43 48
53 55
50 49
41 45
47 52
58 60
32 35
40 39
41 42
48 51
```

The simplest way to compare the two sets of scores is to display their distributions on the same plot. This is most easily achieved by concatenating the two score variables and plotting them against an indicator variable taking the value 1 for the T scores and 2 for the N scores, that is,

```
$ass score=T_score,N_score $
```

Then to reset the standard length of variables to be the length of the new `score` variable.

```
$number len $
$cal len=%len(score) $slen len $
```

The teaching method indicator variable is created by

```
$cal TN=%gl(2,16) $factor TN 2 $

$plot (x=0,3 y=20,60) score TN 'TN' TN  $
```

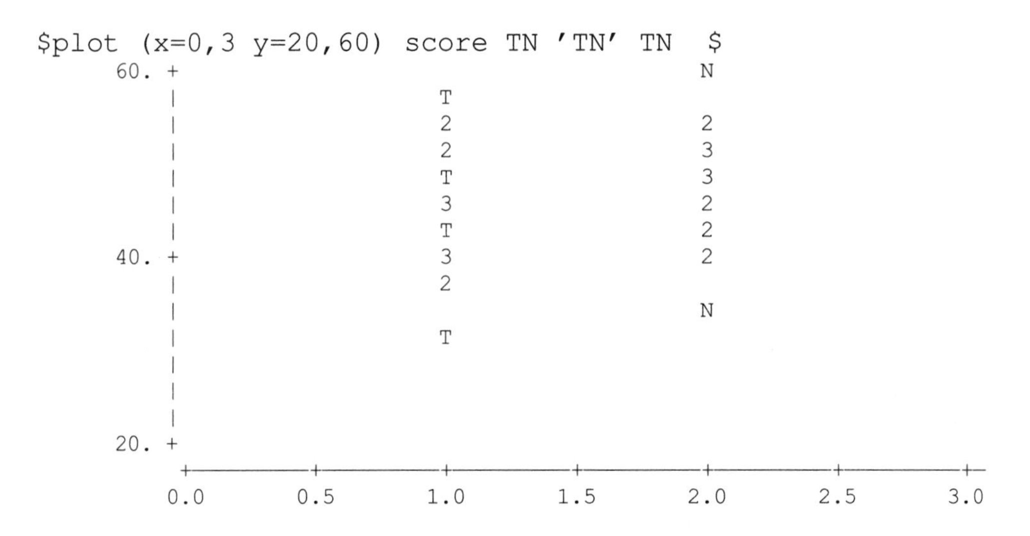

A graphical equivalent of this with the paired points joined illustrates the direction and magnitude of the differences between the paired values even more effectively.

```
$cal pair=%gl(16,1) $fac pair 16 $
$var 16 pen1 $cal pen1=15 $
$graph (t='Fig. 12.18 Paired reading scores by method T and
N'
     v='Score' h='Teaching method 1. T; 2. N' x=0,3 y=20,80)
        score  TN  pen1 pair $
```

Fig. 12.18 Paired reading scores by method T and N

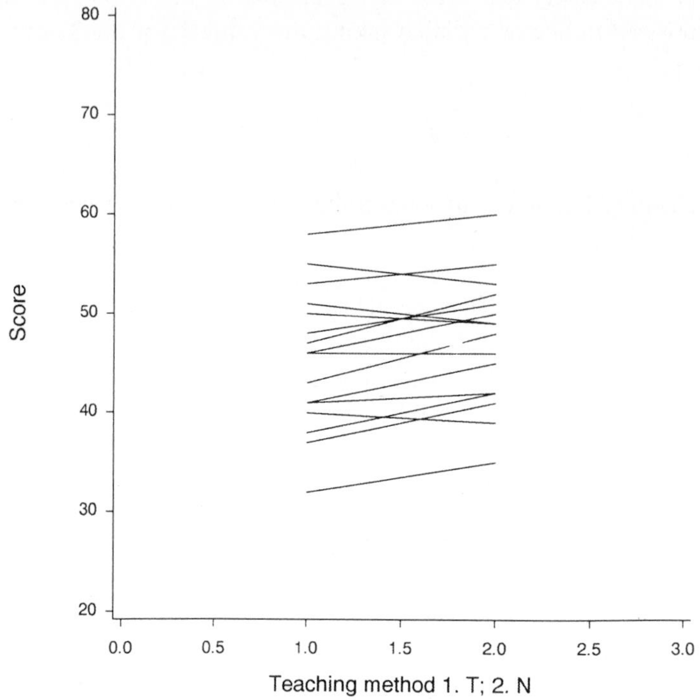

It seems that there is a general tendency for the lines to slope upwards. This indicates that the N scores are generally higher than their T score pair. However the visual impression is strongly influenced by the highest and lowest pairs so it would be unwise to regard this plot as conclusive.

The standard paired t test approach for paired data uses the differences as a single sample from a population assumed Normal and obtains an estimate of the mean of the population of differences. This is then tested, generally against zero representing the null hypothesis of no treatment difference, or used to obtain confidence limits on the estimated 'true' difference. The GLIM approach of modelling the systematic patterns in the data as shown in the plot is somewhat different, but it is to be preferred because it is easily generalized to much more complicated situations. Nonetheless the standard paired t test is easily performed. First calculate the differences

```
$cal NT_diffs=N_score - T_score $
```

The histogram shows how these are distributed and if there were no difference in the effectiveness of the two treatments this would be centred on zero. We can either use the `HISTOGRAM` directive

```
$hist NT_diffs $
[-2.00, 0.00) 4 NNNN
[ 0.00, 2.00) 2 NN
[ 2.00, 4.00) 4 NNNN
[ 4.00, 6.00] 6 NNNNNN
```

or we can use a macro to produce a high–resolution equivalent after providing appropriate title and label strings in macro form and specifying the intervals to be used.

```
$mac title Fig. 12.19 Histogram of Score Differences N - T
$endm $
$mac xlab  N - T differences $end $
$mac ylab  Frequency/interval $end $

$ass ix=-10,-9...10 $

$use ghist NT_diffs ix $
```

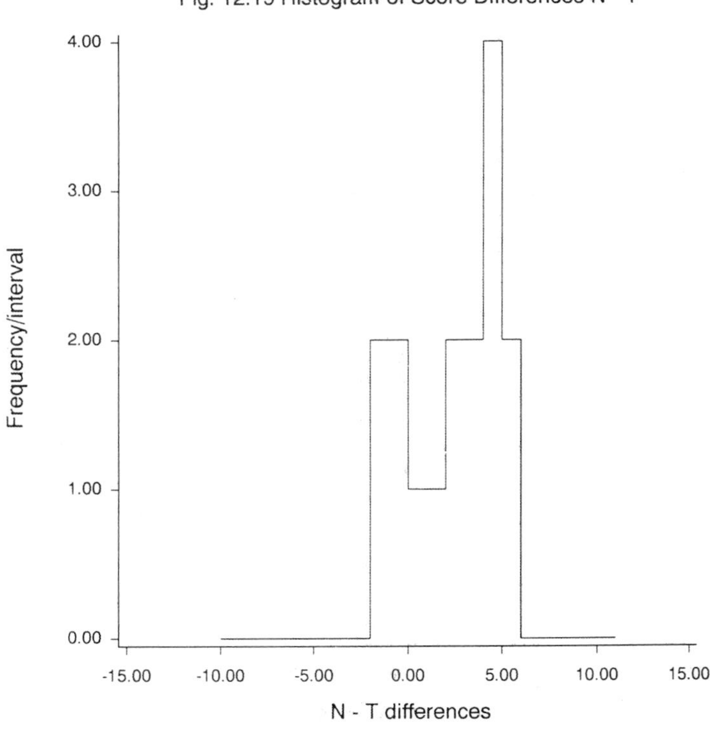

This shows that the sample distribution is centred to the right of zero confirming the impression from the earlier plot. First consider the standard approach to paired t test calculations. This is conveniently done with the TABULATE directive using scalar variables to hold the results

```
$number dse tval td tp dlo dhi $
$tab the NT_diffs mean;dev;weight into NT_dm;NT_ds;dn $
```

This calculates and stores the mean, standard deviation, and frequency of the differences used in the calculations. The standard error, t–value, p–value, and the 95 per cent confidence limits on the population mean difference are obtained and displayed with

```
$cal dse=NT_ds/%sqrt(dn)  : td=NT_dm/dse
    : tp=2*(1-%tp(%abs(td),dn-1))  : tval=%td(0.975,dn-1)
    : dlo=NT_dm - tval*dse : dhi=NT_dm + tval*dse $
```

The mean, sd, s.e. and n are

```
$print NT_dm NT_ds dse dn $
   1.938    2.462  0.6156   16.00
```

The t value is

```
$print td' with  (p='tp')' $
   3.148 with  (p=  0.0066)
```

The 95 per cent confidence limits are

```
$print dlo ' to ' dhi $
  0.6255 to    3.250
```

Clearly the difference is significantly different from zero and it indicates that the new teaching method is an improvement.

The modelling approach achieves the same effect as working with the differences by including a pair effect to allow for the varying heights of the different pairs, that is, lines in the earlier plot, using the factor ‘pair’. The ELIMINATE directive makes it possible to do this even when there are a large number of pairs. Essentially the eliminated factor effects are estimated in an efficient and space saving way, but not displayed.

The model is set up and fitted by declaring the y variable, that the factor pair is to be eliminated and fitting TN as the explicit part of the model.

```
$yvar score $
$eliminate pair $
$fit TN $dis e $
```

```
   deviance =  45.469
residual df =  15
          estimate        s.e.     parameter
     1       1.938      0.6156      TN(2)
scale parameter 3.031
    eliminated term: PAIR
```

The estimated parameter may be used to duplicate the paired t test performed above.

```
$extract %pe %se $
$cal td=%pe(1)/%se(1)  : tp=2*(1-%tp(%abs(td),%df))
:  dlo= %pe(1)  - tval*%se(1)  : dhi= %pe(1)  + tval*%se(1)  $
```

The mean, s.d., s.e. and d.f. are

```
$print %pe(1) %se(1) %df $
   1.938  0.6156   15.00
```

The t–value is

```
$print td' with  (p='tp')' $
   3.148 with  (p=  0.0066)
```

The 95 per cent confidence limits are

```
$print dlo ' to ' dhi $
  0.6255 to    3.250
```

The results are identical to those obtained above. This illustrates that modelling techniques may be used with paired data and, with certain assumptions of independence, for a variety of repeated measures problems.

12.1.3.2 Repeated measurements

These data arose during a trial of high–pressure oxygen for the treatment of multiple sclerosis (Wiles, 1986). It had been claimed that regular periods in a pressure chamber breathing oxygen under pressure alleviated symptoms and produced or maintained remission in patients suffering from multiple sclerosis. This trial was mounted to investigate that claim. Patients were assessed prior to treatment and at two and four weeks post entry to the study after weekly sessions in the pressure chamber. The active treatment group were exposed to oxygen at two atmospheres pressure for 90 minutes on each of 20 successive working days. The control group were exposed for the same time and frequency to a normal air mixture at a small increase in pressure (1.1 atmospheres). (Source: Wiles, C.M. *et al.*; 1986)

The variables are:

`sex`	1 male 2 female
`d_cat`	disease category: 1 progressive 2 relapse/progressive 3 static
`t_grp`	treatment group: 1 oxygen 2 control
`age`	in years calculated from DOB to date of entry to trial

time taken to walk 50 metres in seconds†

`t501`	immediately pre–treatment
`t502`	2 weeks post–treatment
`t503`	4 weeks post–treatment

† Times greater than 10 minutes were recorded as 601 seconds

There were a number of outcome (response) measures considered, but this analysis concentrates on ambulatory function, that is, the time taken to walk 50 metres.

The analysis needs to assess whether the pattern of change, in the response variable over the three assessments, differs between the two treatment groups. The average difference between the two treatment groups will include whatever initial differences there were in the sample which need to be excluded from the comparison. The technique is to estimate and test an interaction between the treatment group factor and a factor representing the occasions at which the subjects were assessed.

We read the data by

```
$data 84 sex d_cat t_grp age t501 t502 t503
$echo $trans -e $read
```

The `ECHO` and `TRANSCRIPT` directives are to suppress the output of the data (which was stored in the same file as these directives) to the screen and the transcript file.

Treating the observations at each assessment as separate units means there are 3 units for every one of the 84 subjects. This gives a total of 252 observations. The response variable has to be expanded, for example, by using `ASSIGN` to concatenate the vector of responses at each of the three occasions.

On the other hand the sex, age, disease category, and treatment group are the same for each assessment. Variables representing those factors can be incorporated in the model, if required, with an index identifying the appropriate subject. First it is necessary to create a subject index variable and a factor variable to identify the occasions.

Occasions could be treated as points on a continuous time scale, but they are not necessarily known very precisely and by using a factor it is possible to allow for any sort of non-linearity without having to specify it functionally.

It is also necessary to have a subject identifying factor so subject differences may be estimated and removed as a source of variation before the occasion to occasion differences, and their interactions with the other factors, are estimated. Then, on the assumption that there is no complex and strong correlation structure within the repeated measure sequences, the patterns of progress (trends) in subjects from one treatment group may be compared with those from the other.

```
$number len1=84 len3=252 $
```

```
$var len3 subj  occ   $
$cal subj=%gl(84,1)  : occ=%gl(3,84) $
$factor occ 3 subj 84 $
```

The response variable is constructed and declared by

```
$assign time50=t501,t502,t503 $
```

The other factors are

```
$fac sex 2 d_cat 3 t_grp 2 $
```

The unknown y values coded 999 and those values obtained from subject 30 that is 49, 285 and 29, which are so discrepant a transcription error is suspected, are weighted out by

```
$cal wt=(time50/=999)&(subj/=30) $weight wt $
```

The range of y values at each occasion is checked by

```
$plot (x=0,4) time50/wt occ $
```

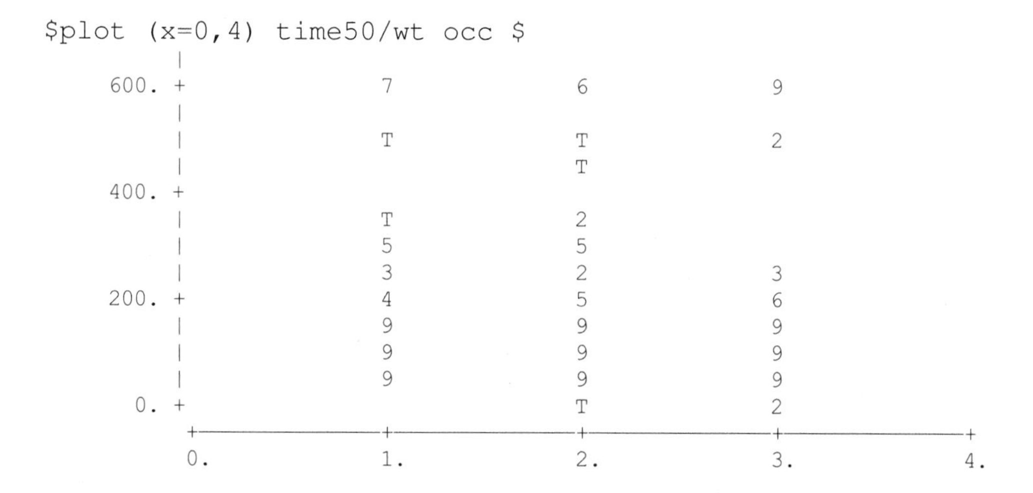

Let us look at the low values

```
$plot ( x=0,4 y=0,100 ) time50/wt occ $
```

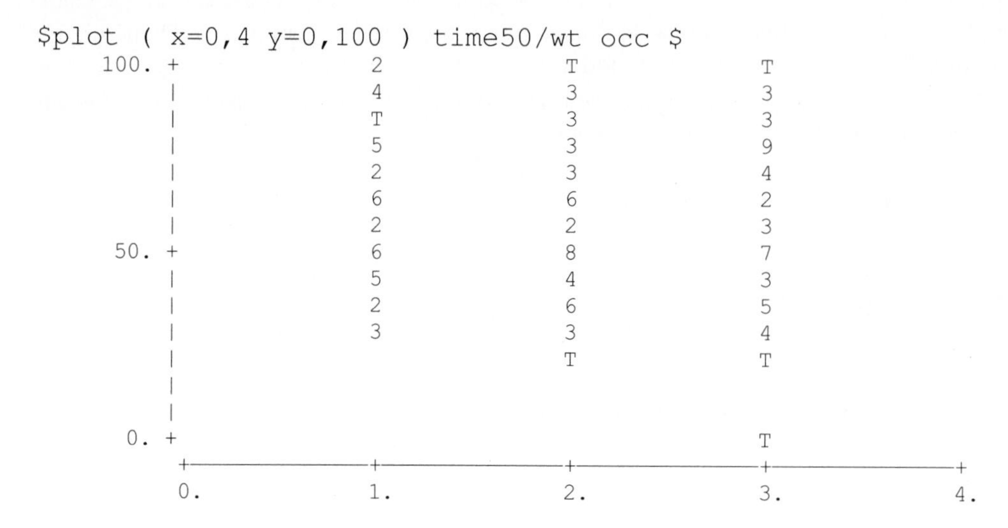

This shows a suspiciously low value for one subject at the third occasion. A plot of the original measurements at the third occasion against a variable indexing the 84 observations and a LOOK at the appropriate range of units allows us to identify the subject and the value.

```
$cal indx=%ind(t503) $
$plot (y=0,50) t503 indx $
```

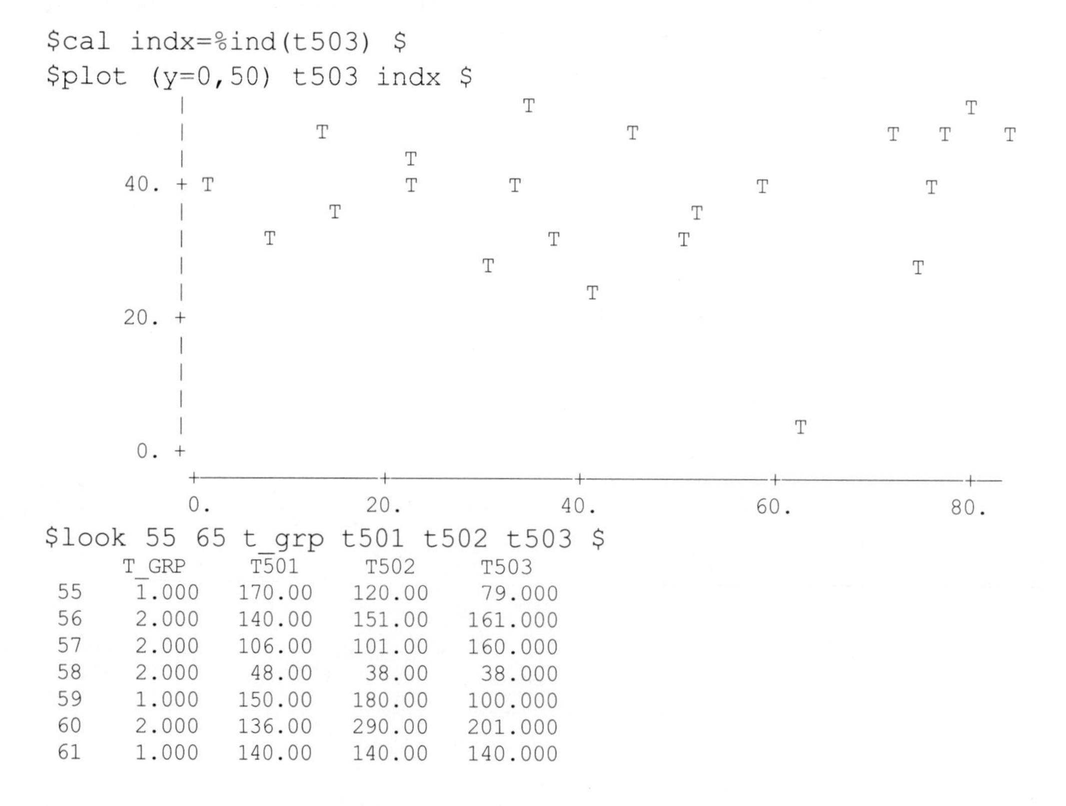


```
$look 55 65 t_grp t501 t502 t503 $
```

	T_GRP	T501	T502	T503
55	1.000	170.00	120.00	79.000
56	2.000	140.00	151.00	161.000
57	2.000	106.00	101.00	160.000
58	2.000	48.00	38.00	38.000
59	1.000	150.00	180.00	100.000
60	2.000	136.00	290.00	201.000
61	1.000	140.00	140.00	140.000

```
62    1.000   220.00   140.00     3.000
63    1.000   300.00   295.00   520.000
64    2.000    65.00    50.00    78.000
65    1.000    65.00    85.00    80.000
```

A value of 3 seconds has been recorded for subject 62 which is obviously an error. The weight variable is adjusted to exclude this reading for this subject by

```
$cal wt=wt*(time50 > 10)  $
```

The actual patterns in the data can be seen with a high–resolution plot of the response against occasion, separately for every subject, with the patterns for the two treatment groups identified with different pens. Note that the pens have to be specified explicitly for every subject line.

This means we need to use the factor `subj` to identify the pen to be used and 84 pens have to be supplied which are one style for subjects in treatment group 1 and another for patients in treatment group 2. This is achieved by

```
$assign ps=14,12 $
$cal   pens=ps(t_grp) $
```

To show the patterns for the two groups side by side it is necessary to create a new occasion variable with 1, 2 and 3, changed to 4, 5 and 6, for the second treatment group.

Fig. 12.20 Time to walk 50 metres by Occasion

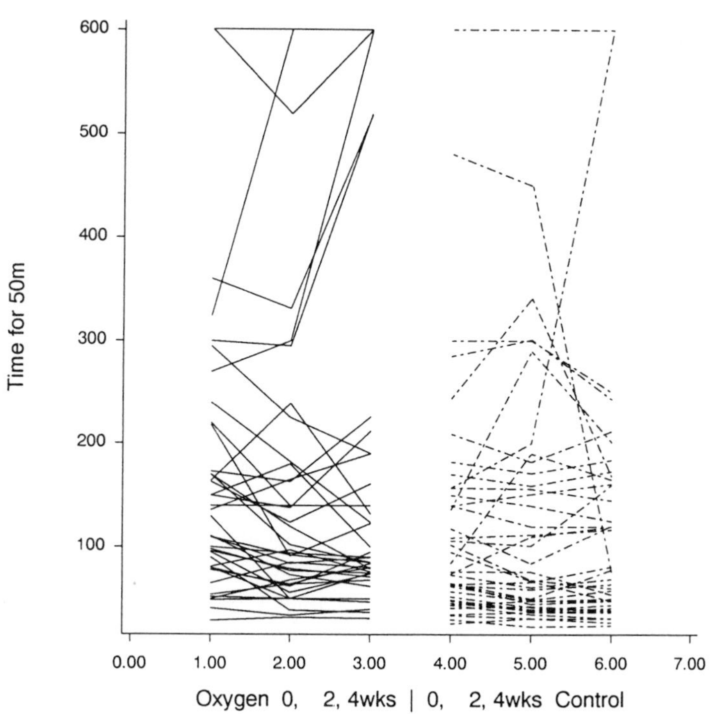

```
$cal occ2=occ+3*(t_grp(subj)==2) $

$graph (title='Fig. 12.20 Time to walk 50 metres by Occasion'
        v='Time for 50m'
        h='Oxygen 0,    2, 4wks | 0,    2, 4wks Control'
         x=0,7,1 )   time50/wt  occ2  pens  subj  $
```

The individual lines representing the change in the time variable over the three assessments are very compressed except for a few rather erratic lines at the higher values. There are no clear patterns or evidence of a difference between the two treatment groups. However part of the compression is due to the majority having covered the 50 metres in relatively short periods of time with a few taking much longer. These include those failing to complete the distance in 10 minutes given a time of 601 seconds. Working with the rate in metres/second at which the subjects covered the distance, that is 50/(time taken) which is a reciprocal transformation, may usefully counteract the compression. The censored values from those taking over 10 minutes will take the value 50/601 which will be a reasonable approximation to their true values which must be between that and zero.

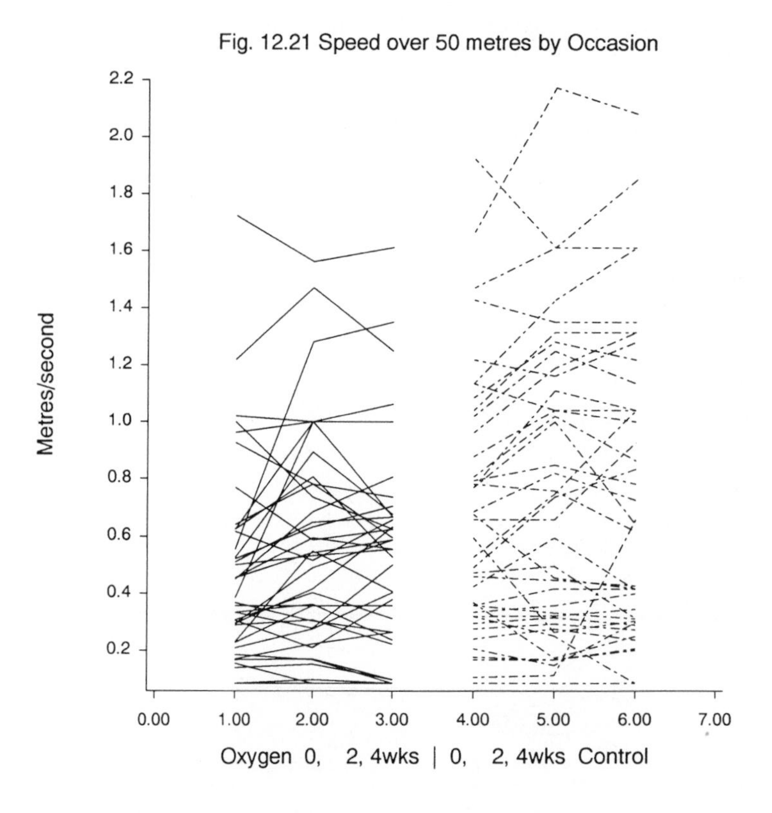

```
$cal rate50=50/time50 $

$graph (title='Fig. 12.21 Speed over 50 metres by Occasion'
        v='Metres/second'
        h='Oxygen  0,    2, 4wks  |  0,    2, 4wks  Control'
         x=0,7,1 )    rate50/wt  occ2  pens  subj  $
```

The subject lines are somewhat more spread out and patterns should be easier to distinguish. Nonetheless there are still no very obvious patterns in either treatment group nor evidence of systematically different trends in the two groups. However this can only be assessed properly by a comparison of models fitted to the data. As for paired data in the previous example the differences between the individuals, in this case represented by sets of repeated observations, must be allowed for in the model with a factor, identifying the individual subjects, which can be fitted but suppressed from the output using `ELIMINATE`. We declare the y variable and the factor to be eliminated by

```
$yvar rate50 $
$eliminate subj $
```

The sex, age and treatment group main effects are confounded with subject differences so we have no information on those unless we treat the subjects as replicates. However our interest is in whether the within–subject occasion to occasion differences differ from one treatment group to the other.

This means we need to test the `t_grp.occ` interaction. The effect of interest is an interaction. Therefore to check whether the response to treatment differs between sub populations of patients it is necessary to investigate the three way interactions between the `t_grp.occ` and factors representing the sex, age and disease category.

Terms not involving `occ` will all represent differences between one group of subjects and another so will be equivalent to, that is, aliased with, eliminated subject effects. This leads to a large number of aliased parameters in the model so instead of `DISPLAY E` it is convenient to use `DISPLAY U` which suppresses them.

Interactions of factors other than the treatment group factor, `t_grp`, with `occ` represent different patterns of response over time in different sub-groups. Some of these are hardly surprising. For example it is to be expected that the way clinical status changes over time is likely to differ according to disease category. However these effects will not necessarily interfere with the estimation of a single treatment effect applicable to all groups. As long as they are in the model the treatment effects are estimated making the appropriate adjustments for their effects.

```
$fit (sex(subj)+age(subj)+d_cat(subj)+t_grp(subj))**2*occ
$dis u $
   deviance =     2.1999
residual df =   134       from 247 observations
         estimate          s.e.      parameter
    6      0.2052        0.2635      OCC(2)
    7      0.3037        0.2686      OCC(3)
   17    -0.06269        0.2417      SEX(SUBJ)(2).OCC(2)
   18     -0.1948        0.2375      SEX(SUBJ)(2).OCC(3)
   19   -0.003329      0.005082      OCC(2).AGE(SUBJ)
   20   -0.005998      0.005226      OCC(3).AGE(SUBJ)
   21    -0.03785        0.2331      D_CAT(SUBJ)(2).OCC(2)
   22    -0.09672        0.2414      D_CAT(SUBJ)(2).OCC(3)
   23     -0.1634        0.2874      D_CAT(SUBJ)(3).OCC(2)
   24     -0.3161        0.2948      D_CAT(SUBJ)(3).OCC(3)
   25     0.01026        0.2213      T_GRP(SUBJ)(2).OCC(2)
   26     -0.1152        0.2197      T_GRP(SUBJ)(2).OCC(3)
   27   2.562e-06      0.004645      SEX(SUBJ)(2).OCC(2).AGE(SUBJ)
   28    0.002783      0.004638      SEX(SUBJ)(2).OCC(3).AGE(SUBJ)
   29     0.04945        0.1006      SEX(SUBJ)(2).D_CAT(SUBJ)(2).OCC(2)
   30      0.1948        0.1046      SEX(SUBJ)(2).D_CAT(SUBJ)(2).OCC(3)
   31      0.1840        0.1265      SEX(SUBJ)(2).D_CAT(SUBJ)(3).OCC(2)
   32     0.09975        0.1293      SEX(SUBJ)(2).D_CAT(SUBJ)(3).OCC(3)
   33    0.001711      0.004576      D_CAT(SUBJ)(2).OCC(2).AGE(SUBJ)
   34    0.001228      0.004773      D_CAT(SUBJ)(2).OCC(3).AGE(SUBJ)
   35    0.004249      0.006187      D_CAT(SUBJ)(3).OCC(2).AGE(SUBJ)
   36    0.008970      0.006401      D_CAT(SUBJ)(3).OCC(3).AGE(SUBJ)
   37     0.03805       0.08901      SEX(SUBJ)(2).T_GRP(SUBJ)(2).OCC(2)
   38     0.04389       0.09013      SEX(SUBJ)(2).T_GRP(SUBJ)(2).OCC(3)
   39  0.00006650      0.004347      T_GRP(SUBJ)(2).OCC(2).AGE(SUBJ)
   40    0.002444      0.004349      T_GRP(SUBJ)(2).OCC(3).AGE(SUBJ)
   41    -0.05817       0.09842      D_CAT(SUBJ)(2).T_GRP(SUBJ)(2).OCC(2)
   42     0.02550       0.09932      D_CAT(SUBJ)(2).T_GRP(SUBJ)(2).OCC(3)
   43     -0.1719        0.1154      D_CAT(SUBJ)(3).T_GRP(SUBJ)(2).OCC(2)
   44     -0.1012        0.1169      D_CAT(SUBJ)(3).T_GRP(SUBJ)(2).OCC(3)
scale parameter 0.01642
   eliminated term: SUBJ
```

None of the three factor interactions are more than twice their s.e.s., but they must be tested as a group. To do this we store the deviance and degrees of freedom to use in the appropriate model comparison test.

```
$number SS1 df1 s2 $
$cal SS1=%dv : df1=%df : s2=%sc $
```

The analysis is assuming that the *y* variable has a Normal distribution and this must be investigated. We will use a macro, previously input, for obtaining Normal (quantile/quantile) plots when observations have been weighted out.

```
$use wrep $
```

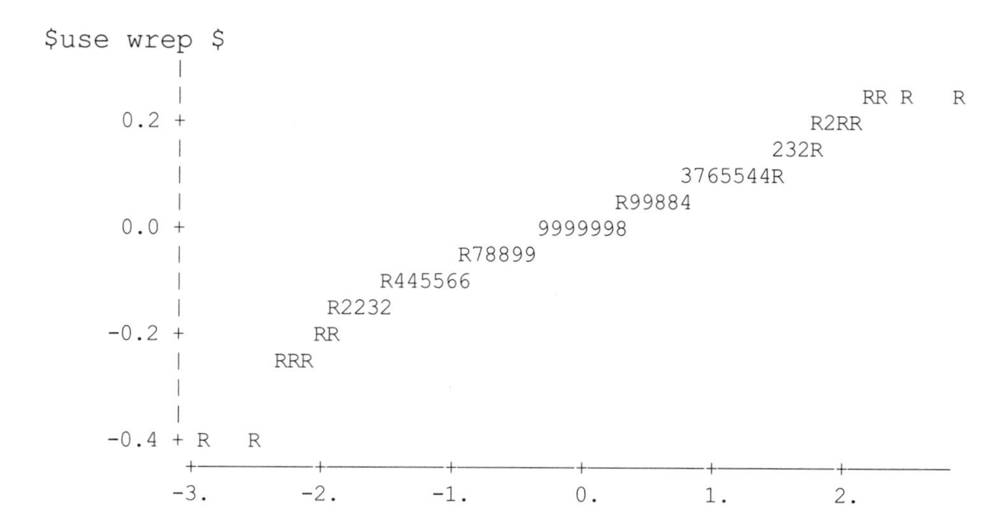

A slightly wavy line, but straight enough to justify the assumption of Normality.

The previous model allows a different pattern of change over occasions for every `factor.t_grp` combination. Ideally the differences will be no more than chance and we can concentrate on the difference between the pattern of change in one treatment group and that in the other assuming that within a particular treatment group the true underlying population patterns can be assumed parallel for all the factor combinations. The appropriate model is

```
$fit (sex+age+d_cat+t_grp)(subj)*occ $dis u $
   deviance =     2.4223
residual df =  152       from 247 observations
          estimate          s.e.      parameter
     6     0.08481        0.1032      OCC(2)
     7    -0.01841        0.1031      OCC(3)
     8     0.02722       0.04094      SEX(SUBJ)(2).OCC(2)
     9     0.08436       0.04094      SEX(SUBJ)(2).OCC(3)
    10  -0.0009458      0.001981      OCC(2).AGE(SUBJ)
    11  -0.0002874      0.001988      OCC(3).AGE(SUBJ)
    12     0.02674       0.04771      D_CAT(SUBJ)(2).OCC(2)
    13     0.06353       0.04803      D_CAT(SUBJ)(2).OCC(3)
    14     0.04355       0.05546      D_CAT(SUBJ)(3).OCC(2)
    15     0.06670       0.05596      D_CAT(SUBJ)(3).OCC(3)
    16    -0.02639       0.04006      T_GRP(SUBJ)(2).OCC(2)
    17    0.003040       0.04004      T_GRP(SUBJ)(2).OCC(3)
scale parameter 0.01594
    eliminated term: SUBJ
```

Note the last two parameters which represent the treatment effect are small compared to their s.e.s. However before we test or interpret these we should complete the analysis testing for higher order interactions.

The change in deviance, indicating how necessary the three factor interaction terms were for the fit of the model, is small. To perform the appropriate F test assign the results from this model to appropriate scalar variables (these are also needed for the analysis of variance table later).

```
$number SS2 df2 F21 df21 pF $
$cal SS2=%dv : df2=%df $
```

```
$cal df21=df2-df1 : F21 = ((SS2-SS1)/df21)/s2  :
 pF=(1-%fp(F21,df21,df1)) $
```

The F test of the three factor interactions involving treatment and occasion is

```
$print : 'F ='F21' with 'df21' and 'df1' degrees of freedom
(p='pF')' $
F =  0.7525 with    18.00 and    134.0 degrees of freedom (p=  0.7511)
```

This is far from significant so these interactions may be ignored. We can return to considering the `t_grp` effects which we already noted are small compared to their s.e.s. This indicates quite strongly that there is not much evidence of a treatment effect. Nonetheless they must be tested together by fitting the model without them.

```
$fit (sex+age+d_cat)(subj)*occ $
   deviance =     2.4327
residual df =  154        from 247 observations
$number SS3 df3 $
$cal SS3=%dv : df3=%df $
```

A relatively small change in deviance which confirms that they are not necessary for an adequate model. There is therefore no evidence of a treatment effect. The analysis of variance is obtained with a macro specific to this example by

```
$use anco3132 $
          Model     Residual SS   df   Mean square    Variance Ratio(F)
 ---------------------------------------------------------------------
   1)All 3 factor intrn
     terms with OCC       2.1999  134         0.0164
   2)All 2 factor intrn
     terms with OCC       2.4223  152
 --------------------------------------
 (2-1)Due to 3 factor
     interactn terms      0.2224   18         0.0124             0.7525
                                                          (p=  0.7511)
 ---------------------------------------------------------------------
   3)  - T_GRP.OCC        2.4327  154
 --------------------------------------
 (3-2)Due to T_GRP.OCC    0.0104    2         0.0052             0.3156
                                                          (p=  0.7299)
 ---------------------------------------------------------------------
 ---------------------------------------------------------------------
```

None of the interactions were significant and neither was the treatment effect. There is no evidence from this trial with this outcome measure that the treatment was of any benefit whatever.

12.1.3.3 Bioassay

This data arose from a quality control assay of penicillin with a concentration of 5 units/ml against an 'unknown' obtained from an accurate dilution of a standard (Wardlaw, 1985). This is a technique for assessing the quality and precision of assays and the true relative potency of the unknown is 3 units/ml.

In bioassays, to assess the potency of some preparation, portions of a standard and the unknown to be tested are diluted to produce samples of differing concentrations. These are then applied to the biological material which responds to the preparation in the required way so their potency can be measured and compared. The dose–response relationships are then estimated by fitting a linear model – generally in the form of parallel straight lines against the logarithm of the dose – and the factor by which the concentration of the unknown would have to be multiplied to generate the same response as the standard – known as the relative potency – is estimated.

In this assay samples of the two preparations are diluted twice by a factor of 3 to give concentrations of 1, 1/3, and 1/9 of the original concentrations. Four replicate observations are obtained for the response at each concentration. The response is the diameter of zones cleared in cultures of bacteria by a standard quantity of the preparation in a given time.

The results are

Concentrations	Standard				'Unknown'			
1 in 9	77	75	76	73	73	71	73	67
1 in 3	92	94	90	91	84	85	86	89
1 in 1	110	102	106	106	100	104	97	100

```
$units 24 $read z_dia
 77  75  76  73        73  71  73  67
 92  94  90  91        84  85  86  89
110 102 106 106       100 104  97 100
```

Then to calculate variables to indicate the preparation and dose

```
$cal prepn=%gl(2,4) : dose=1/(3**(3-%gl(3,8))) $
$factor prepn 2 $
```

Since the majority of biological dose–response relationships are proportional in nature, that is, multiplying the dose by a factor will cause the response to change additively by a constant, it is usual to work with the logarithms of the dose. Plots of the relationships are then generally linearized except at the extremes where the dose is so low there is no effect or so high that the effect is a maximum and cannot increase. This is seen in the plot of the mean responses at each dose

```
$tabulate the z_dia mean for prepn;dose into z_mean by
prep_;dose_ $
```

Then plot the mean against the tabulation dose variable `dose_`

```
$plot (x=0,1.5 y=40,120) z_mean dose_ 'su' prep_ $
```

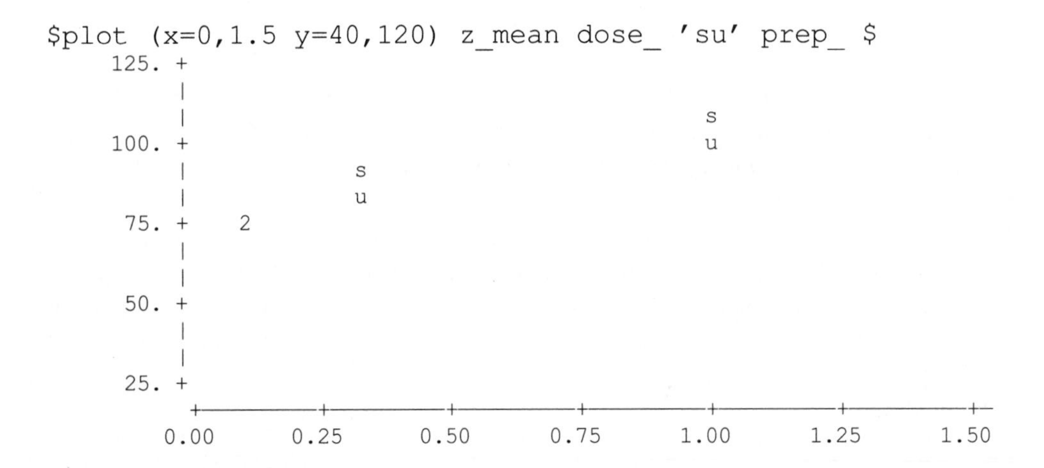

Check the effect of a logarithmic transformation on the dose scale by taking logs of the tabulation dose variable.

```
$cal log_d_=%log(dose_) $
$plot (x=-3,1 y=40,120) z_mean log_d_ 'su' prep_ $
```

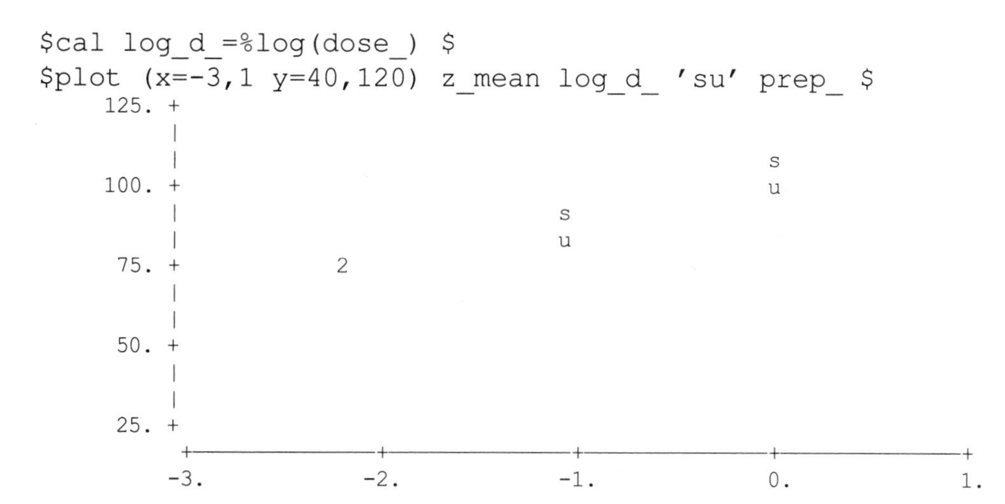

The transformation appears to linearize the relationship almost completely. For the analysis we also need to transform the original dose variable.

```
$cal log_dose= %log(dose) $
```

Now an equivalent high–resolution plot showing the individual points and the means joined by lines can be obtained by

```
$graph
    (title='Fig. 12.22  Mean zone diameter against Log-dose'
    xscale=-3,1,1 yscale=40,120,20)
    z_dia;z_mean log_dose;log_d_   4;2;15;14  prepn;prep_ $
```

The analysis needs to test assumptions that the relationships are linear and parallel and that an assumption of normality is justified. If these assumptions are met then the parallel line model can be used to estimate the relative potency with appropriate confidence limits.
The first step is to fit a model where the relationships are represented by two different curved lines. The simplest way is to fit quadratic curves using `log_dose` squared.

```
$cal log_d2=log_dose**2 $
```

If the nesting operator is used the coefficients for the two quadratic curves can be seen separately and an immediate assessment of the curvature in both can be made by inspection of the quadratic term coefficients and their s.e.s.

Fig. 12.22 Mean zone diameter against Log-dose

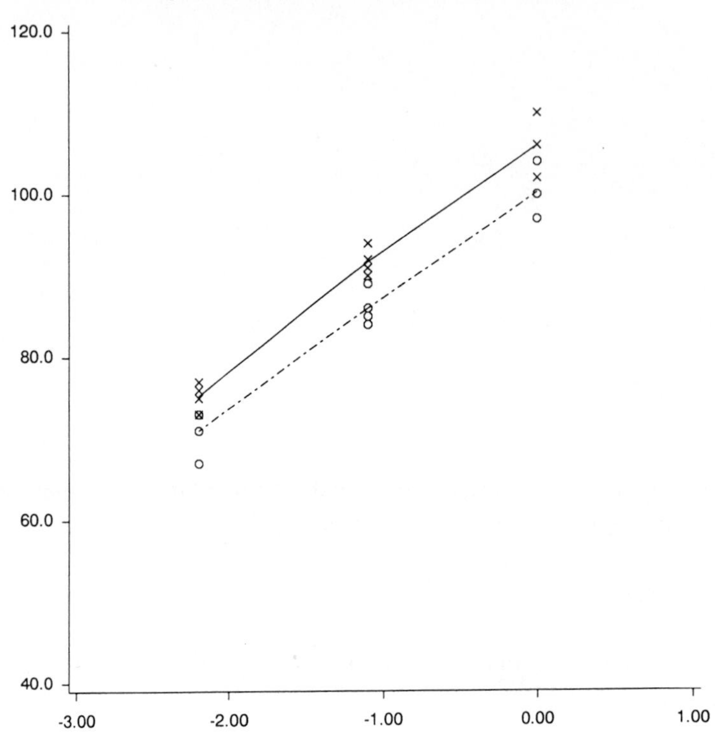


```
$yvar z_dia $
$fit prepn/(log_dose+log_d2) $dis e $
   deviance =  112.25
residual df =    18
           estimate        s.e.      parameter
     1        106.0       1.249       1
     2       -5.750       1.766       PREPN(2)
     3        11.95       2.898       PREPN(1).LOG_DOSE
     4        12.63       2.898       PREPN(2).LOG_DOSE
     5      -0.9321       1.267       PREPN(1).LOG_D2
     6      -0.3107       1.267       PREPN(2).LOG_D2
scale parameter 6.236
```

It appears that the quadratic terms, which are considerably less than their s.e.s, are not needed. However before proceeding further the assumptions of constant variance and Normality which are needed must be tested.

```
$plot %rs %fv 'r' $
```

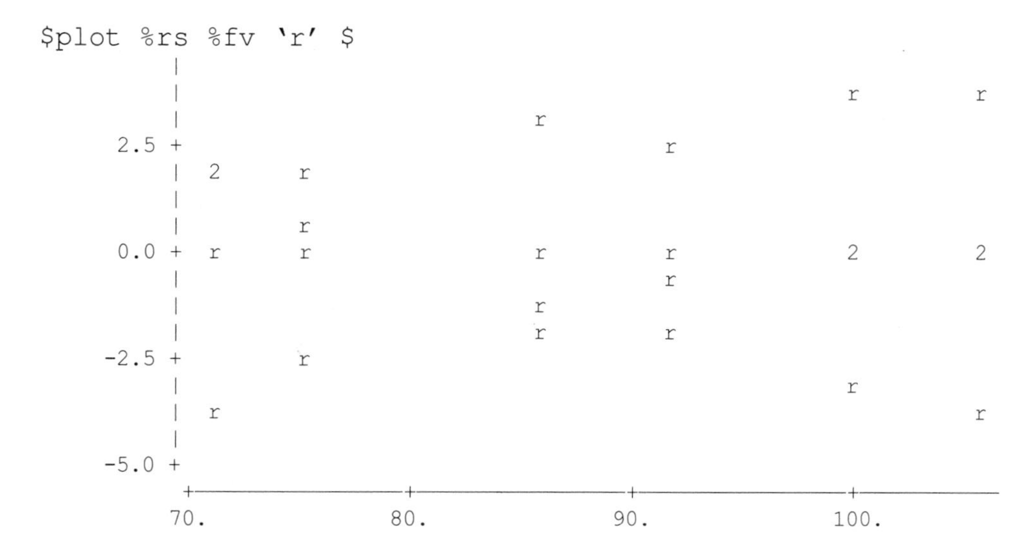

It seems that the variation in the residuals is relatively constant over the range of fitted values. That assumption appears quite justified. Check the normality with a stored macro.

```
$use crep $
```

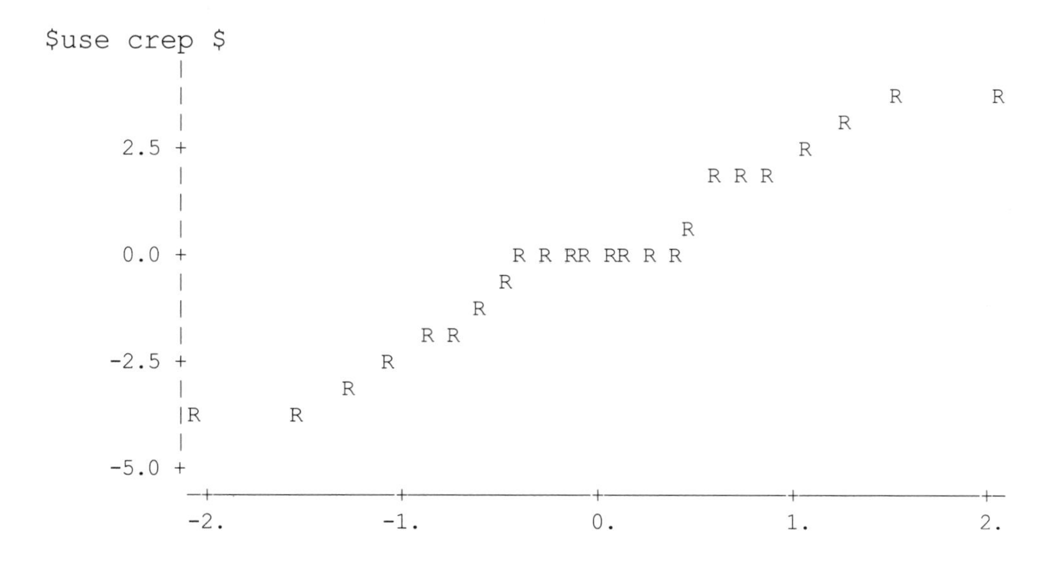

There is a bit of a 'hiatus' in the middle where the plot appears to be horizontal for a short stretch. This is due to an unexpectedly high number of zero residuals which may or may not have an explanation. However for current purposes the plot is adequately Normal.

The next step is to fit the straight line model without the quadratic terms, but not assuming parallelism. However for the analysis of covariance the residual sum of squares (SS) and degrees of freedom now need to be stored.

```
$number SS1=%dv df1=%df s2=%sc $
```

Then fit the two straight lines and store the residual sum of squares (SS) and d.f. from that

```
$fit prepn/log_dose $dis e $
   deviance =  116.00
residual df =    20
          estimate         s.e.      parameter
     1       106.4        1.099       1
     2      -6.000        1.555       PREPN(2)
     3       13.99       0.7750       PREPN(1).LOG_DOSE
     4       13.31       0.7750       PREPN(2).LOG_DOSE
scale parameter 5.800
$number SS2=%dv df2=%df $
```

The slopes are almost identical so it appears that a parallel line model will be appropriate. In fact the deviance has increased by 3.75 on 2 degrees of freedom, giving a mean square of 1.87 which is less than the residual variation which confirms the impression that the quadratic terms are not needed (see the analysis of covariance table below).

```
$fit prepn+log_dose $dis e $
   deviance =  118.25
residual df =    21
          estimate         s.e.      parameter
     1       106.0       0.9062       1
     2      -5.250       0.9688       PREPN(2)
     3       13.65       0.5400       LOG_DOSE
scale parameter 5.631
$number SS3=%dv df3=%df $
```

The residual sum of squares (SS) has increased by 2.25 on 1 d.f. which gives a mean square smaller than the residual variance from any of the models, so it is clearly no greater than could easily occur by chance. This is usually demonstrated more formally with an analysis of covariance as follows.

```
$use anco3133 $
```

Model	Residual SS	df	Mean square	Variance Ratio(F)
1) Different quadratics	112.25	18	6.24	
2) Different st. lines	116.00	20		
(2-1)Due to Curvatr	3.75	2	1.87	0.30 (p= 0.7440)
3) Parallel lines	118.25	21		
(3-2)Due to Non-parallelism	2.25	1	2.25	0.36 (p= 0.5555)

The parallel line model is clearly adequate and the effect of the unknown preparation can be described in terms of the standard after a constant shift on the `log(dose)` scale. This is equivalent to saying that the effect of the unknown at dose d is the same as the effect of the standard at dose ρd. The constant ρ is known as the 'relative potency'.

On the log dose scale this means that the effects are the same at log d and $\log(\rho d) = \log \rho + \log d$. The value of log ρ is estimated by the horizontal separation of the two parallel lines. Only if a parallel line model is a good fit can an 'unknown' be described in terms of a constant relative potency to the standard. If the unknown response line were above that of the standard then the vertical separation would be the intercept for the unknown (a_u) minus that of the standard line (a_s). The horizontal separation of the two lines is log ρ and so

$$(a_u - a_s)/\log(\rho) = b \quad \text{and therefore} \quad \log(\rho) = (a_u - a_s)/b$$

In terms of the model parameters this is

`prepn(2)`/log ρ = `log_dose` that is, log ρ = `prepn(2)/log_dose`.

If the unknown is less potent than the standard then the difference between the intercepts, log ρ, will be negative and ρ will be less than 1.

This can be understood more easily with the aid of a diagram which we can draw with GLIM by

```
$ass diagx=0,10,0,10,3,7,7,7 : diagy=0,5,2,7,3.5,3.5,3.5,5.5
   : line=1,1,2,2,3,3,4,4 : pend=10,10,11,11 $factor line 4 $
$graph  (t='Fig.  12.23   Relative  potency  as  a  horizontal
shift'
        v='Diameter' h='Log(dose)' x=-2,12,2 y=-2,8 pause=no)
        diagy diagx pend line $
$ass xtx=3.2,7.1    : ytx=3.2,4.8 $
$gtext (m=p l=r)  ytx xtx 'log(rho)','au - as' $
```

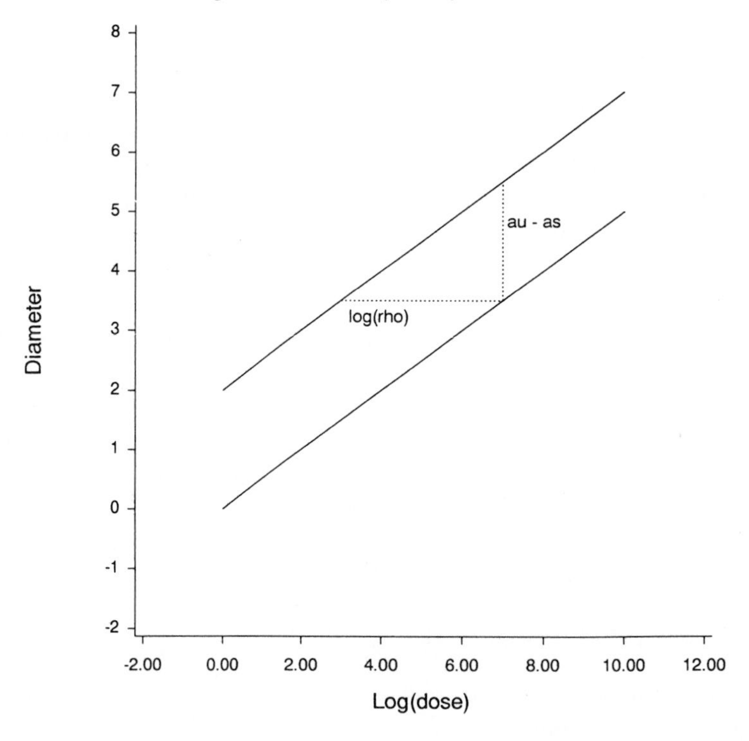

The estimate of log ρ is the appropriate maximum likelihood estimate, but obtaining confidence limits is not all that simple. The usual way is to employ an approximation for obtaining them known as Fieller's theorem. However with some ingenuity the model can be parameterised in terms of log ρ to give a non-linear model. This can be solved iteratively to give the exact confidence limits.

The results using Fieller's theorem are obtained with a macro as follows

```
$mac      !fieller
!
!  This macro uses Fiellers theorem to calculate
! the variance of a ratio and the confidence limits
! It is first necessary to extract %pe and %vc
! the parameters required are all scalar variables
!  1     N = the numerator
!  2     D = the denominator
!  3    VN = the variance of N
!  4    VD = the variance of D
!  5   CND = the covariance of N  and D
!  6    TC = the appropriate centile of the t distribution
(arg=N,D,VN,VD,CND,TC   loc=m,g,r,s,lo,hi)  fieller
$number m g r s lo hi $
$cal m=N/D : g=(TC/D)**2*VD : r=m-g*CND/VD
: s=TC/D*%sqrt(VN-2*m*CND+m*m*VD-g*(VN-CND*CND/VD))
: lo=(r-s)/(1-g)  : hi=(r+s)/(1-g)
$pr ' Estimated ratio = 'm
: 'Lower CL  'lo '  Upper CL  'hi $
$cal %m=m : %l=lo : %u=hi $
$endmac $

$extract %pe %vc $
$number num denom var_num var_den covar t_val $
$cal   num=%pe(2)   :   denom=%pe(3)      :   var_num=%vc(3)    :
var_den=%vc(5)    :  covar=%vc(4)  : t_val=%td(0.975,df3) $

$use fieller num denom var_num var_den covar t_val $

           Estimated ratio =  -0.3845
          Lower CL   -0.5579  Upper CL   -0.2260
```

This gives us an estimate of the log(relative potency) with 95 per cent confidence limits. Taking antilogs the relative potency estimate is therefore

```
$cal %l=%exp(%l)  : %m=%exp(%m)  : %u=%exp(%u)  $

$pr   %m ' with  95% CLs ' %l ' and ' %u $
  0.6808 with  95% CLs   0.5724 and   0.7977
```

The alternative approach uses the re-parameterisation of the model based on the fact that the relationships of the two preparations with a known relative potency rho i.e. k=log(rho) on the log(dose) scale are:

Preparation 1: `z_dia` = $a + b$ log(dose)
Preparation 2: `z_dia` = $a + b$ (log(dose) + k)

Thus for given k the X-variable is calculated appropriately, adding k for preparation 2, and the model $1 + X$ fitted to obtain the residual sum of squares for the given value of k. The value of k is then varied over a suitably wide range to determine 95 per cent confidence limits using the test region

(SS(k) - `SS3`)/`%sc` < F(0.95,1,`%df`)

SS(k), `%sc` and `%df` are the residual sum of squares, variance and degrees of freedom from the models with k fixed. SS(MLE) (which is `SS3` in this analysis) is the residual sum of squares from the model fitted to obtain the original maximum likelihood estimate of k. This is most easily done using a macro

```
$mac  loop !
$output $                              ! to suppress output
$cal xvar=log_dose+k*(prepn==2) $ ! calculate the X for
                                       ! prepn=2
$fit xvar $                            ! fit the fixed k model
$output 6 $                            ! restore the output
$ca diff=(%dv-SS3)/%sc  $              ! calculate the F statistic
$ass dd=dd,diff : kk=kk,k $            ! store the F and k values
$ca %d=0.001*(diff>4 & diff<5)         ! calculate increment for k
                                       ! smaller
      +0.01*(diff<=4 ? diff>=5) $ ! in range of interest
$ca k=k+%d $ca %z=(k<0) $              ! increment k and set %z
                                       ! (=0 to stop)
!                                      NB k is known to be < 0 .
$end $
```

Now declare `k` and `diff` as scalar variables, set flag `%z` non-zero and initialise `k` at a value certain to be less than the lower 95 per cent confidence limit. Then set the macro to execute repeatedly, until the flag `%z` becomes 0, with the `WHILE` directive.

```
$num k diff $cal %z=1 : k=-0.60 $echo off $trans -0 $
$while %z loop $echo on $trans +0 $
```

Obtain the 95th centile of the appropriate F distribution and generate x and y vectors of coordinates for two points with which to draw a horizontal line at this F–value.

```
$cal %f=%fd(0.95,1,%df) $ass k2=-0.60,0.0 : ff=%f,%f $

$graph ( t='Fig. 12.24 F values by log(relative potency)'
        pause=no)
        dd,ff  kk,k2  10,12 $
$gtext (m=p l=r) 4.8/1.1 -0.2 'F 0.95 on 1 and 21 df' $
```

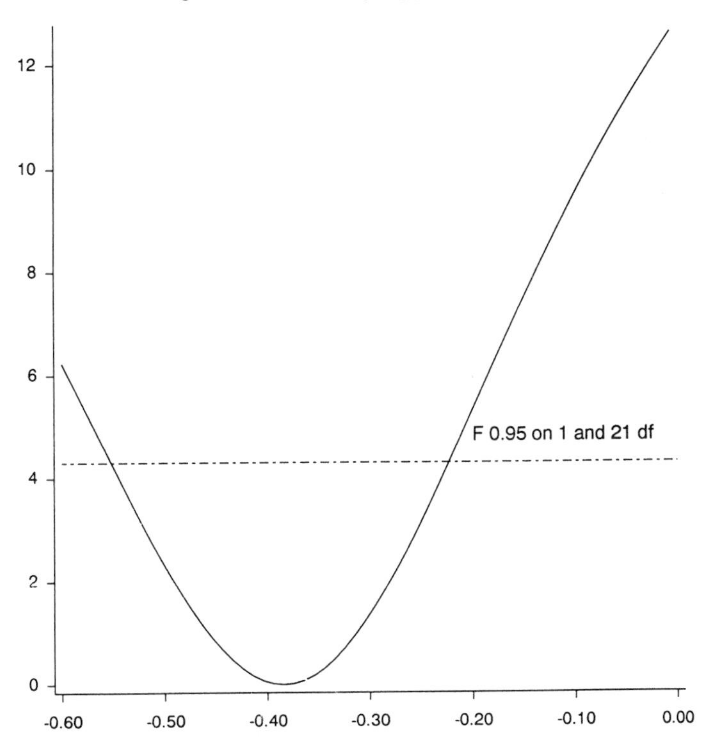

To derive the abscissae (k values) of the intersections requires identifying the upper and lower F–values either side of the intersections. This is done by creating a new vector of the F–values shifted one position to the left as they are seen on the graph.

```
$cal %l=%len(dd) : i=%ind(dd) : i=i+(i<%l)
  :  dd2=dd(i) : kk2=kk(i) $
```

Then calculate a 0/1 vector flagging the intervals containing the 95th centile F–value, put the upper and lower bounds of these intervals into new vectors using `PICK` and interpolate to get the estimates of the 95 per cent confidence limits on k.

```
$cal int=(%f>=dd)&(%f<dd2)?(%f>=dd2)&(%f<dd) $
$pick  dint1;dint2;klim1;klim2;il  dd;dd2;kk;kk2;i int $

$look  dint1 dint2 klim1 klim2 il $
     DINT1    DINT2    KLIM1    KLIM2     IL
 1   4.332    4.292  -0.5530  -0.5520  13.00
 2   4.301    4.344  -0.2230  -0.2220  55.00

$cal klim=klim1+(klim2-klim1)*(%f-dint1)/(dint2-dint1) $
$loo klim $
      KLIM
 1  -0.5522
 2  -0.2230
```

It appears that the Fieller's theorem approach is reasonably accurate. The limits calculated this way are a little closer to zero, but the differences are of the order of 1 per cent only.

Returning to the results of the analysis and the conclusions to be drawn from them. The experiment was set up with the relative potency set at 60 per cent. The assay estimated the relative potency as 68 per cent with limits of 57 per cent to 80 per cent. The limits contain the correct value, but only just.

Since the standard had a concentration of 5 units/ml the assay has estimated the unknown to be 5×0.6807 or 3.4 when it should have been 3.0 units/ml with 95 per cent confidence limits of 2.86 to 3.99. These contain the correct value of 3.0 units/ml, but it represents a 13.3 per cent error in the assay which clearly should provoke a review of technique, technician, or both.

12.2 Dichotomous *y* variables

12.2.1 Continuous and categorical *x* variables: logistic regression

12.2.1.1 Dose–response curves for dichotomous responses

This data arose from a test of an insecticide and is quoted in Finney (1971).

Dose of rotenone (mg/l) dose	Number of insects n	Number dead r
0	49	0
2.6	50	6
3.8	48	16
5.1	46	24
7.7	49	42
10.2	50	44

It is common for dose–response relationships to be such that differing doses need to be increased in proportion to produce the same increase in response. This means that a steady or straight line increase in the response is associated with a constant proportional increase in dose, that is, constant increases in log(dose). For this reason it is usual to work with log(dose) as the explanatory variable. One effect of this is that zero doses become a problem (since $\log 0 = -\infty$).

The simplest solution is to treat the zero as if it were a very small, but non-zero dose by adding a small constant to give a finite, large negative value, on the log(dose) scale. The effect on the model fitting is negligible. This is because the model is sigmoid or S shaped and largely determined by the linear part in the middle. Data points in the 'tails' with low response at low doses and close to 100% response at high doses have very little influence.

The data are input by

```
$data 6 dose n r $read
 0    49  0
 2.6 50   6
 3.8 48 16
 5.1 46 24
 7.7 49 42
10.2 50 44
```

The first thing to do is to examine how the proportion responding relates to dose.

```
$cal p=r/n $
$tpr (s=1) r;n;p dose $
```

DOSE	0.	3.	4.	5.	8.	10.
R	0.000	6.000	16.000	24.000	42.000	44.000
N	49.00	50.00	48.00	46.00	49.00	50.00
P	0.0000	0.1200	0.3333	0.5217	0.8571	0.8800

```
$plot p dose $
```

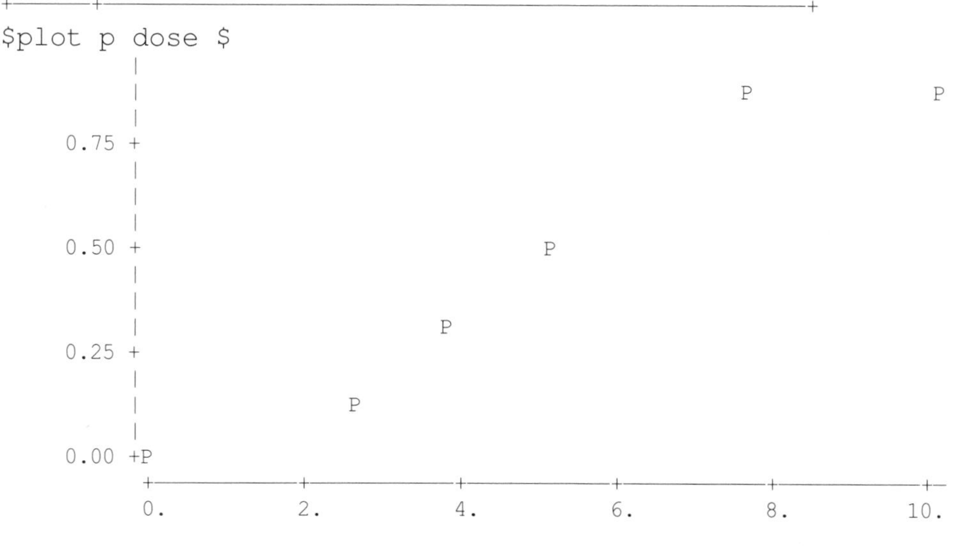

The proportion responding increases markedly with dose with a slight 'S' shape to the curve. This is because the curve is in effect estimating the cumulative 'tolerance' distribution. Now use the log(dose) scale with the zero dose increased to 0.01:

```
$cal logdose=%log(dose+0.01*(dose==0)) $
```

Then plot the observed proportion responding against `logdose`:

```
$plot (x=0,3 y=0,1) p  logdose  'o'  $
```

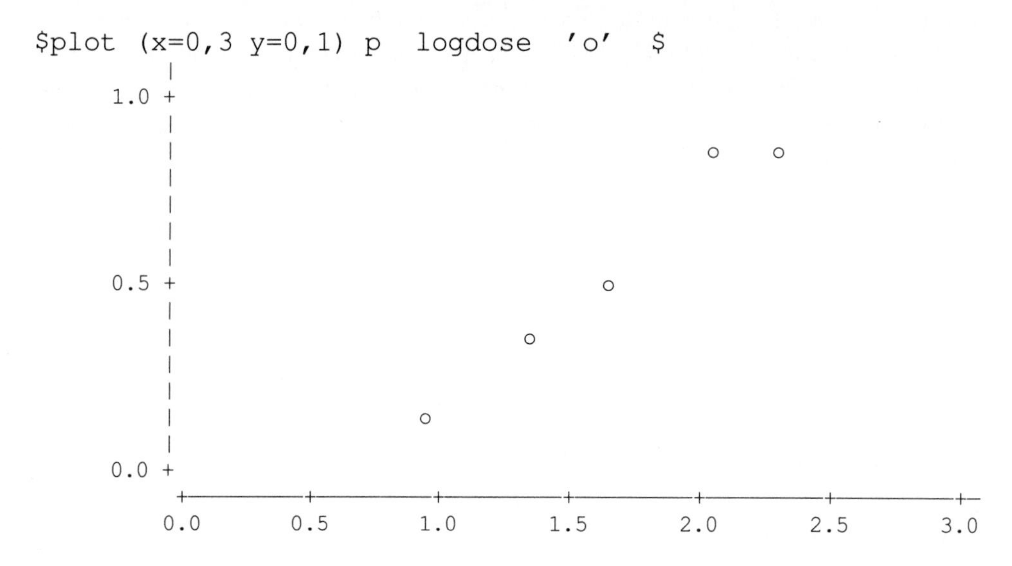

A slightly more 'S' shaped curve. If we transform the proportions to the logistic scale it should become more or less straight. The zero value is again a problem when taking the log solely for the purpose of the plot it is convenient to treat the `0/49` as if it were `0.25/49` (say).

```
$cal p=p+(p==0)*(0.25/49)  : lp=%log(p/(1-p)) $
$plot (x=0,3) lp  logdose  'o'  $
     2. +                                                   o    o
        I
        I
     0. +                                              o
        I                                        o
        I
    -2. +                                 o
        I
        I
    -4. +
        I
        I
    -6. +
        +-----------+-----------+-----------+-----------+-----------+-----------+-
       0.0         0.5         1.0         1.5         2.0         2.5         3.0
```

This is reasonably straight so it appears that the logistic scale may well be appropriate. On the assumption that, at each dose, the probability of death for each insect is constant and the events, that is, deaths are independent of each other, the number of deaths is a binomially distributed variable.

The above plots suggest that the relationship of the underlying probability with `logdose` may be adequately modelled by the logistic function. That means we should specify the error distribution of the y variable (that is, the number responding) as binomial and the link function as logistic. In fact the logistic link is the default for the binomial.

So to fit a model describing the dose response relationship we need to identify the number responding as the y variable. Then we must use the `ERROR` directive to declare the response distribution to be binomial with the numbers at risk, that is, the binomial denominators of the proportions, as a second parameter. Finally the `LINK` directive is used to specify the logistic link function (option `G`). Note that, since the logistic link is the default for Binomial errors this last step could have been omitted.

```
$yvar r $error b n $link g $
```

To quantify the form of the dose–response relationship fit a straight line regression model for the average increase in the proportion responding on the logistic scale per unit increase in `logdose`. It is also useful to store the deviance and degrees of freedom.

```
$fit logdose $dis e $
scaled deviance =  1.4241 at cycle 5
    residual df =  4
          estimate        s.e.      parameter
    1       -4.887      0.6429       1
    2        3.104      0.3877       LOGDOSE
scale parameter 1.000
$number SS1=%dv df1=%df $
```

Note that because the binomial distribution has a known mean–variance relationship the dispersion parameter, equivalent to the residual variance in the Normal case, is known. In these cases GLIM identifies the scale parameter as equal to 1. Since the deviance is measured from the value that would be obtained from a saturated model fitting the data exactly it is termed the scaled deviance.

It is possible that a simple model, such as that above, might be insufficient to describe the relationship. However if that were the case the deviance would be greater than expected. Since, for binomial data, the scaled deviance can be assumed, as a rough approximation, to come from a chi-squared distribution its expected value, if the model is adequate, will be equal to the degrees of freedom. This means that an inadequate model would be indicated by the deviance exceeding its degrees of freedom. This is clearly not the case here. The strong relationship observed in the plots of the observed proportions is confirmed by the `logdose` coefficient which is many times its s.e. and highly significant.

Note that the ratio of an estimate with its s.e. from non-Normal models is not an observation from a t distribution. As an approximation it may be taken as an observation from a standardised Normal distribution. However in small samples such tests can be seriously misleading. The approximation improves with sample size, but parameters in these models should properly be tested by removing them from the model and examining the resulting change in the scaled deviance. This is done as follows:

```
$fit -logdose $

scaled deviance =  163.74 (change =   +162.3) at cycle 3
    residual df =    5    (change =     +1   )

$number SS2=%dv SS21 df2=%df $
$cal SS21=SS2-SS1 : %p=1-%chp(SS21,df2-df1) $
$print  'Chi squared value 'SS21' on 1 df giving  p ='%p $

Chi squared value    162.3 on 1 df giving  p =  0.
```

The χ^2 value on 1 d.f. is highly significant. Clearly the response is strongly related to the dose. However before drawing any firm conclusions we must check whether any of the points have an undue influence on the model fitted. To do this we must first re-fit the full model, extract the leverage values and then plot them against an index variable representing their order in the dataset.

```
$fit logdose $
scaled deviance =  1.4241 at cycle 5
    residual df =  4
$ext %lv $
$cal order=%index(r) $
$plot (y=0,1) %lv order '1' $
         |
     1.0 +
         |
         |
         |
         |
         |
         |
     0.5 +                  1
         |                                                   1            1
         |                           1
         |                                       1
         |
         |
         |
     0.0 +1
         +------------------+------------------+------------------+------------------+------------------+
         1.                 2.                 3.                 4.                 5.                 6.
```

The data points all have much the same influence on the fit except for the modified zero dose point which we can see has almost no influence. Now to see how well the model fits the data we plot the data and fitted model together. First obtain predicted responses for a closely spaced sequence of log(dose) values.

```
$ass ld=0,.2...3  $
$predict (s=-1) logdose=ld $
$cal ppL=%pfv $
$plot (x=0,3 ) ppL;p ld;logdose 'eo' $
```

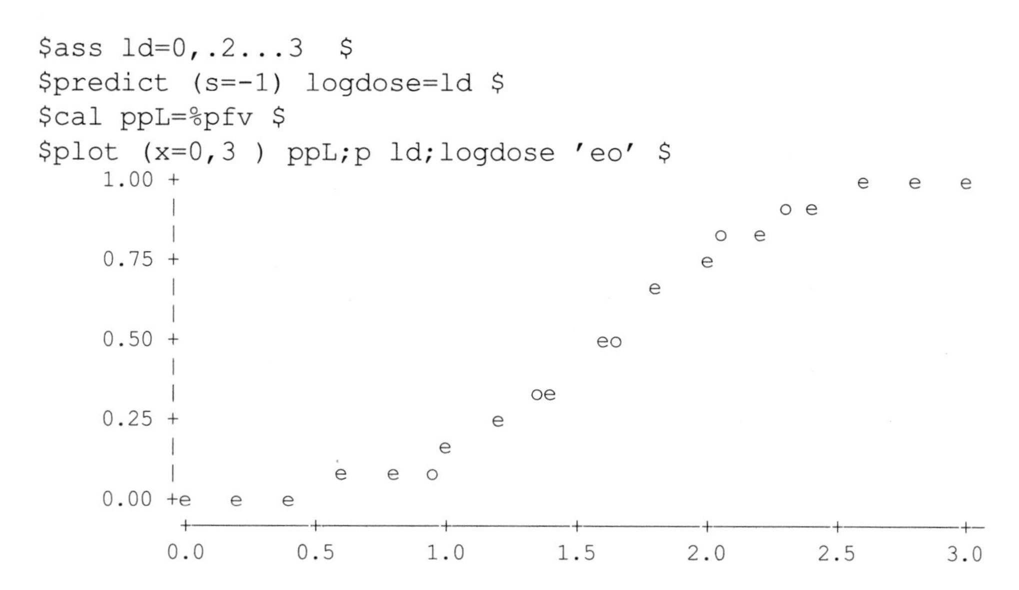

Fig. 12.25 Data and Fitted Logistic Curve

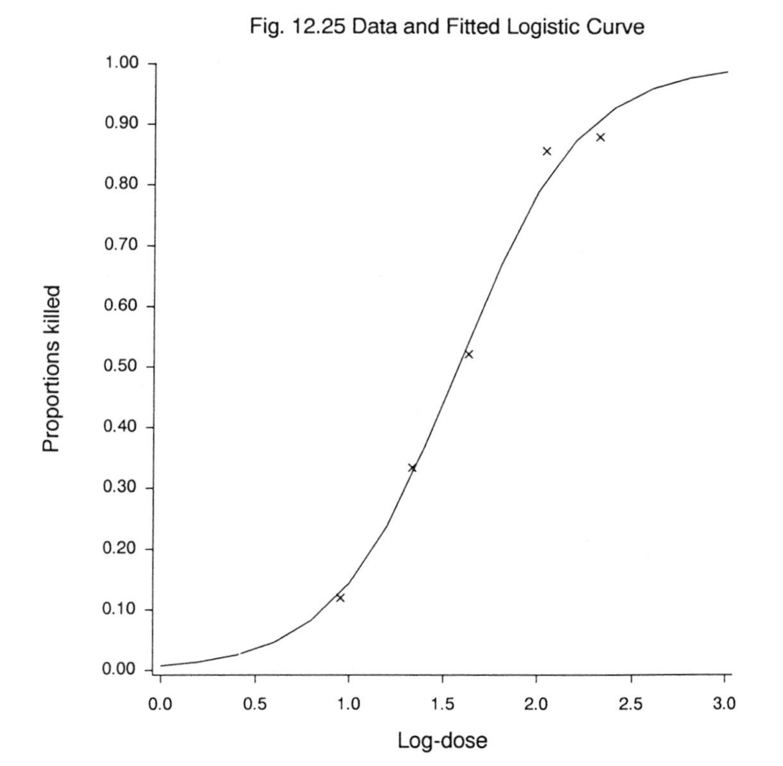

Then as a graph:

```
$gra (t='Fig. 12.25 Data and Fitted Logistic Curve'
   vla='Proportions killed' hla='Log-dose' x=0,3 )
p;ppL logdose;ld 1;14 $
```

The model fits the observed proportions very well and gives the complete estimated dose–response relationship. This could be used as a calibration curve, but a simpler summary measure is needed. The `logdose` regression coefficient is a summary measure which gives the estimated log of the ratio of the odds of an individual responding if exposed to a `logdose` of $X+1$ to those of an individual responding if exposed to X. However this is not a particularly easy to grasp or useful summary statistic to describe a dose–response relationship. A more useful statistic which is commonly used is the ED50. This is the dose required to provoke a response in 50 per cent of such individuals. It is in fact the estimated median tolerance. In this case the response is the death of an insect. In such toxicological experiments the ED50 is sometimes termed the LD50 standing for Lethal Dose 50.

To obtain the LD50 from the fitted relationship is relatively straightforward. The fitted function is of the form

$$p = 1/[1 + \exp\{-(a+bx)\}]$$

This means that on the logistic scale

$$y = \log(p/(1-p)) = a + bx$$

so LD50, the value of x, representing log(dose), that gives $p = 0.5$, is obtained from

$$a + b\,\mathrm{LD50} = \log(0.5/0.5) = 0.$$

Thus we require

$$\mathrm{LD50} = -a/b.$$

This can be calculated quite simply from the parameter estimates

```
$ext %pe %vc $cal %l=(-%pe(1)/%pe(2)) : %d=%exp(%l) $
$pr 'On the log dose scale ED50 is '%l
    '  On the absolute dose scale '%d $
On the log dose scale ED50 is     1.575  On the absolute dose scale     4.829
```

Obtaining confidence limits for a ratio of estimates is not so simple. One method is to use Fieller's theorem which is a rather complicated function of the variances and covariance of the numerator and denominator of the ratio estimate. Another method is to use likelihood– based confidence intervals as described in Aitkin et al. (1989). Here we shall use Fieller's theorem with a macro.

```
$mac    ! fieller
! This macro uses Fiellers theorem to calculate
!            the variance of a ratio a/b
!     and    the P% confidence limits.
! The arguments required are
!   a  = the numerator
!   b  = the denominator
!   va = the variance of a
!   vb = the variance of b
!   cab= the covariance of a and b
!   P  = the percentage CLs required
!   df = the degrees of freedom for the t distribution
!        (0 if the Normal distribution is appropriate)
(arg=a,b,va,vb,cab,P,df local=r)  fieller $
$number t2P_1 m g r s lower upper $
$cal t2P_1=%if(df==0,1.96,%td(1-(1-P/100)/2,df)) $
$cal m=a/b : g=(t2P_1/b)**2*vb : r=m-g*cab/vb !
: s=t2P_1/b*%sqrt(va-2*m*cab+m*m*vb-g*(va-cab*cab/vb))!
: lower=(r-s)/(1-g) : upper=(r+s)/(1-g)!
$pr ' Estimated ratio = 'm !
: 'Lower CL  'lower '  Upper CL  'upper $
$endmac $
```

To use the macro we must first assign to scalar variables the intercept and slope from the model, their variances and covariance, the percentage confidence required and the degrees of freedom. The latter should be as obtained from the model if the t distribution is to be used, but set to 0 if, as here, the Normal distribution is more appropriate.

```
$number a b c d e f P95 df $
$cal a=-%pe(1)  : b=%pe(2)
   : c=%vc(1)   : d=%vc(3)
```

and because we have changed the sign of the numerator, the first parameter, we must also change the sign of the covariance:

```
$cal e=-%vc(2)
   : P95=95 : df=0 $
$use fieller a  b c d e P95 df $
 Estimated ratio =     1.575
Lower CL      1.468  Upper CL      1.677
```

These are the estimated log(LD50) with approximate 95 per cent confidence limits. To convert them back to the original `dose` scale we need to use the exponential function, `%exp`, to obtain antilogs.

```
$cal %exp(lower) : %exp(m) : %exp(upper) $
          4.341
          4.829
          5.350
```

So the estimated median lethal dose is 4.83 mg/l with 95 per cent CLs about 10% above and below this figure.

If we believe that there is an underlying log-Normal tolerance distribution then we could use the `PROBIT` link, that is,

```
$link P $
$fit logdose $dis e $
scaled deviance =  1.7390 at cycle 5
    residual df =  4
          estimate        s.e.     parameter
     1      -2.887      0.3510      1
     2       1.830      0.2087      LOGDOSE
scale parameter 1.000
```

The fitted values as before are

```
$ass ld=0,.2...3 $
$pred (s=-1) logdose=ld $
$cal ppP=%pfv $plot (x=0,3) ppP;p ld;logdose 'eo' $
     1.00 +                                                   e   e   e
          |                                             o e
          |                                        o  e
     0.75 +                                           e
          |                                       e
          |
     0.50 +                                  eo
          |
          |                             oe
     0.25 +                         e
          |                     e
          |                  e  o
     0.00 +e   e   e   e
          +---------+---------+---------+---------+---------+---------+-
          0.0       0.5       1.0       1.5       2.0       2.5       3.0

$extract %pe %vc $
$number a b c d e f P95 df $
$cal a=-%pe(1) : b=%pe(2)
```

```
    : c=%vc(1)   : d=%vc(3)  : e=-%vc(2)
    : P95=95 : df=0 $
$use fieller a  b c d e P95 df $
 Estimated ratio =     1.578
Lower CL     1.473  Upper CL     1.678
```

These are the estimated log(LD50) with approximate 95 per cent confidence limits. On the original dose scale they are

```
$cal %exp(lower) : %exp(m) : %exp(upper) $
         4.364
         4.845
         5.354
```

So the estimated median lethal dose is 4.85 mg/l and the 95 per cent confidence limits are very close indeed to those obtained using the logistic link.

The overall contrast between the logistic link results and those using the probit are shown by a plot of the two sets of fitted proportions

```
$plot (x=0,3) ppP;ppL ld 'PL' $
```

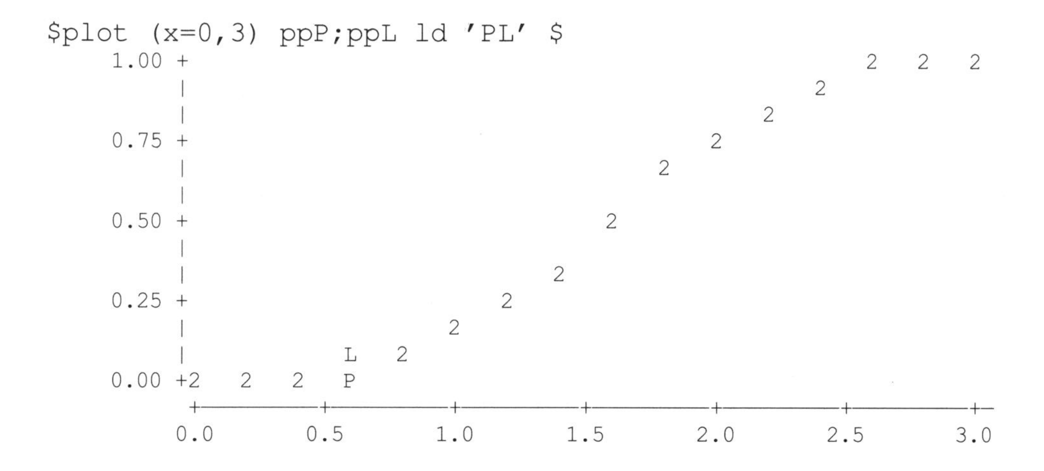

They are almost identical and even with a high–resolution graph the fitted curves are difficult to distinguish.

```
$graph (t='Fig. 12.26 A comparison of Probit and Logistic'
  vla='Proportion positive' hla='Logdose' x=0,3 )
ppP;ppL ld 10;12 $
```

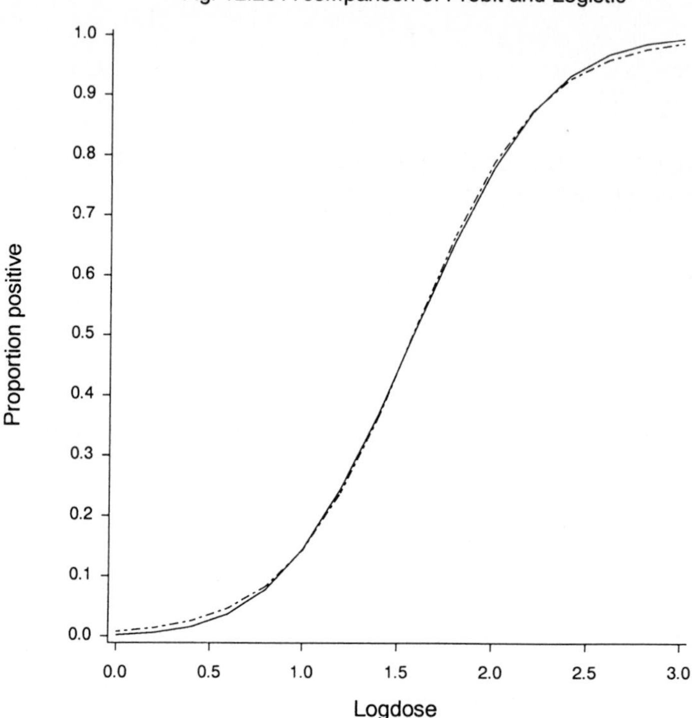

Fig. 12.26 A comparison of Probit and Logistic

The proportional responses from the two estimated curves and their differences are

```
$cal pdiff=(ppP-ppL) $
$look ppP ppL pdiff $
          PPP        PPL        PDIFF
  1   0.001942   0.007488   -0.005546
  2   0.005843   0.013841   -0.007998
  3   0.015559   0.025444   -0.009885
  4   0.036759   0.046319   -0.009560
  5   0.077274   0.082862   -0.005588
  6   0.145097   0.143888    0.001210
  7   0.244550   0.238185    0.006365
  8   0.372293   0.367734    0.004558
  9   0.516020   0.519680   -0.003661
 10   0.657673   0.668070   -0.010398
 11   0.779964   0.789212   -0.009247
 12   0.872445   0.874450   -0.002005
 13   0.933707   0.928349    0.005358
 14   0.969253   0.960163    0.009090
 15   0.987320   0.978183    0.009137
 16   0.995364   0.988153    0.007211
```

It is clear that in the middle of the range, which is the region of most practical interest, the two links will give almost the same results. The probit and logistic functions are very similar and differ mainly in the tails.

12.2.1.2 Two way classifications

The following table gives some data from a study of infant respiratory disease, namely the proportions of children developing bronchitis or pneumonia in their first year of life by sex and type of feeding.

	Bottle only	Breast +supplement	Breast only
Boys	77/458 (0.17)	19/147 (0.13)	47/494 (0.10)
Girls	48/384 (0.13)	16/127 (0.13)	31/464 (0.07)

The questions for the analysis are whether the risk of bronchitis and pneumonia is affected by the type of feeding. Subsidiary questions are whether the risk is the same for both sexes and whether differences between the different feeding groups, if they exist, are the same forboth sexes.

The data, in the form of the numerators (`ill`) and the denominators (`narisk`), are input by

```
$units 6  $data  ill  narisk  $read
77  458  19  147  47  494
48  384  16  127  31  464
```

Factors to identify the sex and feeding categories are obtained using the `GFACTOR` directive

```
$gfactor sex 2 food 3 $
```

and the proportions with bronchitis or pneumonia within each category by

```
$cal p=ill/narisk $
```

Tabulated, this gives

```
$tprint (s=1) ill;narisk;P  sex;food $
```

SEX		FOOD 1	2	3
1	ILL	77.00	19.00	47.00
	NARISK	458.0	147.0	494.0
	P	0.16812	0.12925	0.09514
2	ILL	48.00	16.00	31.00
	NARISK	384.0	127.0	464.0
	P	0.12500	0.12598	0.06681

A plot is useful to see the form of the relationships

```
$plot   (x=0,4) p  food  'bg'  sex  $
```

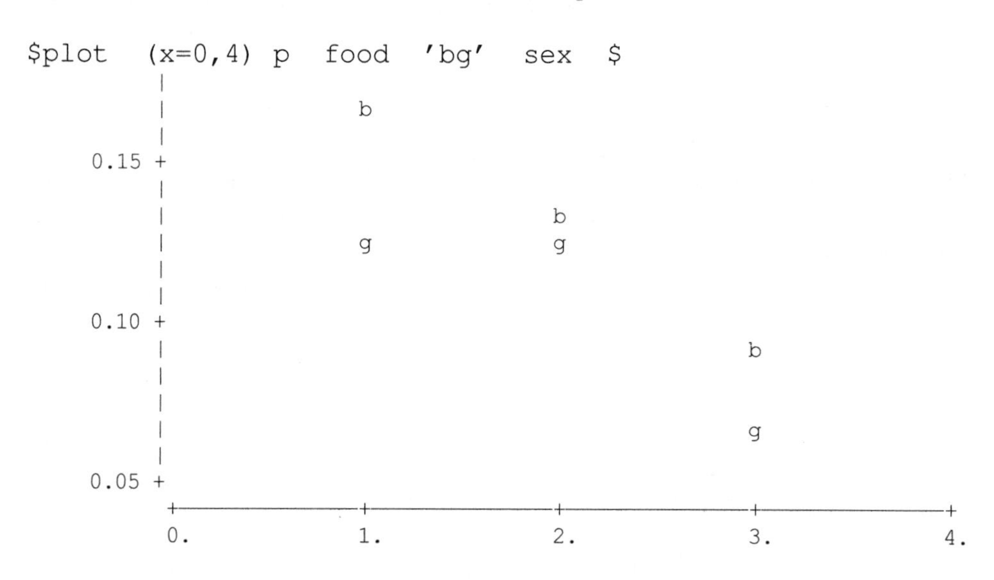


```
$cal wtm=(sex==1)  : wtf=(sex==2) $

$graph  (t='Fig.  12.27  Proportion  ill  by  Food  and  Sex'
        vla='Proportion'
        hla='Food: Bottle-Both-Breast'xsc=0,4,1 ysc=0,0.3,.1)
        p/wtm;p/wtm;p/wtf;p/wtf food 6;10;8;12 $
```

The proportion with respiratory disease appears to decline with breastfeeding and girls seem to have lower proportions than the boys. We assess these relationships by modelling the proportion ill, as the parameter of the binomial distributions followed by the numbers ill, in terms of the effects of the `sex` and `food` factors. With an appropriate sequence of models we can test assumptions and deduce estimates of the effects, of the two factors, on risk.

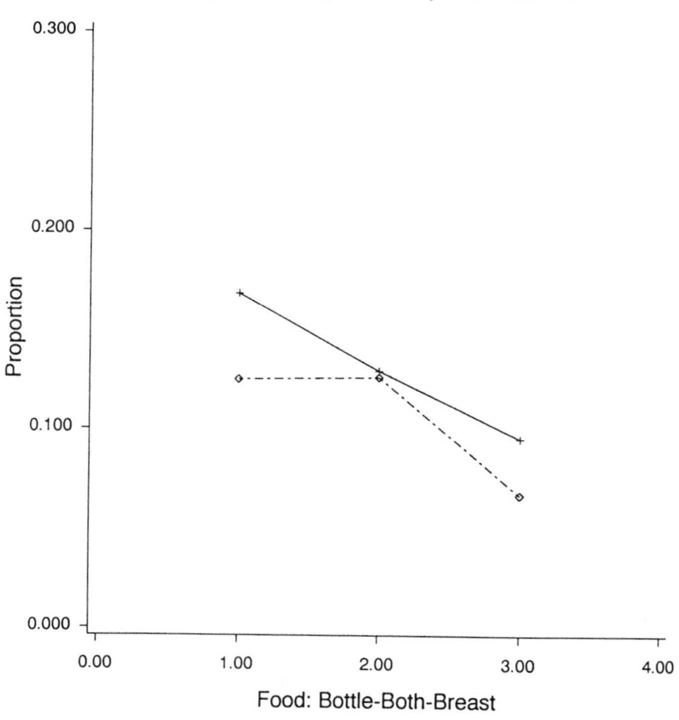


```
$yvar  ill  $
$error  B  narisk  $
```

We will use the logistic link which is the default so no `LINK` directive is required.

Now the first question is whether the effects of breast feeding, if any, are the same in both sexes. We need to test the interaction term `sex.food`. We can fit a model with this term and the main effects of `sex` and `food`, but there is a slight problem.

```
$fit sex*food $dis e $
scaled deviance =  0. at cycle 3
    residual df =  0
          estimate        s.e.      parameter
     1      -1.599      0.1249       1
     2     -0.3469      0.1985       SEX(2)
     3     -0.3086      0.2758       FOOD(2)
     4     -0.6534      0.1978       FOOD(3)
     5      0.3176      0.4140       SEX(2).FOOD(2)
     6    -0.03742      0.3123       SEX(2).FOOD(3)
scale parameter 1.000
```

We see that there are as many parameters as there were observations. This means there are no degrees of freedom left with which to assess the goodness of fit of the model. However the estimates and their standard errors can be usefully interpreted. For example the `food(3)` parameter is an estimate, on the logistic (or log(odds)) scale, of the difference in risk of bronchitis or pneumonia associated with breast milk compared to bottled milk in the boys. Using r for the numbers ill and n for the numbers at risk, that is, the denominators, this estimate is

$$\log(r_{13}/n_{13}/((n_{13}-r_{13})/n_{13})) - \log(r_{11}/n_{11}/((n_{11}-r_{11})/n_{11}))$$

where the first suffix indicates sex and the second food. The standard error (s.e.) is

$$\sqrt{[\ (1/r_{11}+1/(n_{11}-r_{11})+1/r_{13}+1/(n_{13}-r_{13}))\ ]}$$

The interaction terms are a little more complicated, but follow the same lines. If they are more than twice their s.e.s then food effects may have to be estimated separately for each sex and the above s.e.s can be used to obtain approximate confidence limits.

However to test whether any term involving more than one parameter is needed in the model, for example the interaction term here, the proper test is to compare the goodness of fit of the models with and without it. If the parameters are needed then the model without them will fit badly and the deviance will be increased by a significant amount.

In the case of non-Normal error distributions with the model fitted by maximising the likelihood and no least squares analogy the variance ratio F test cannot be used and instead the difference between the deviances is used as a likelihood ratio chi-squared test statistic. If the population parameters are zero, that is, the null hypothesis is true, this statistic has an approximate chi-squared distribution with the degrees of freedom equal to the number of parameters dropped from the model and a mean equal to its degrees of freedom. If the null hypothesis is not true then there will generally be an increase in deviance which will give the statistic a non-central chi-squared distribution.

Fortunately even with moderately small samples the approximation is quite close and it is usually safe to use such differences between deviances as chi-squared tests.

The next step in the analysis is to fit the model without the interaction term

```
$fit  sex+food  $dis  e  $
scaled deviance =  0.72192 at cycle 3
    residual df =  2
          estimate          s.e.      parameter
     1      -1.613        0.1124       1
     2     -0.3126        0.1410       SEX(2)
     3     -0.1725        0.2056       FOOD(2)
     4     -0.6693        0.1530       FOOD(3)
scale parameter 1.000
```

Since the deviance for the first model was zero the deviance difference chi-squared test is 0.72 on 2 d.f. We can assume, therefore, that there is no interaction between sex and food and accept this simpler model as an adequate representation of the data.

For an overall test of the differences between feeding groups drop the food factor

```
$fit  -food  $dis  e  $
scaled deviance =  20.899 (change =   +20.18) at cycle 3
    residual df =   4      (change =    +2    )
          estimate          s.e.      parameter
     1      -1.900       0.08965       1
     2      -0.3261       0.1403       SEX(2)
scale parameter 1.000
```

The deviance has increased to 20.899 and 4 d.f. and the increase is 20.18 on 2 d.f. which is much larger than the 5 per cent significance level for chi-squared on 2 d.f. which is 5.99. (`%chd(0.95,2)`) This is strong evidence that, on average, the differences between the food categories represent real effects and from the observed proportions it appears that breast milk is beneficial. The next step is to consider the difference between the sexes. We need to compare this with the baseline model `sex+food` to avoid assuming the food effect zero when testing the sex effect. Because of this it is not possible to specify the model by subtracting the term of interest from the previous model. We must either fit the baseline model again or, more simply, calculate the deviance differences directly from the appropriate models.

```
$fit  food  $dis e $
scaled deviance =  5.6990 at cycle 3
    residual df =  3
          estimate          s.e.      parameter
     1      -1.747       0.09693       1
     2      -0.1744       0.2053       FOOD(2)
     3      -0.6765       0.1528       FOOD(3)
scale parameter 1.000
```

The deviance is increased from that of the `sex+food` model. The chi-squared test is

$$\text{chi-squared} = 5.70 - 0.72 = 4.98 \text{ on } 3 - 2 = 1 \text{ d.f.}$$

Since this is considerably greater than 3.84, the 5 per cent significance level for 1 d.f. (`%chd(0.95,1)`), it suggests that the sex differences observed were not simply chance effects.

Since there is only one parameter for the sex factor an alternative approach can be used which involves calculating how many s.e.s the sex parameter, in the baseline sex+food model, is from zero. On the assumption that the estimate is an observation from a Normal distribution this gives a standardised Normal deviate which should be compared with +/-1.96 for a 95 per cent significance test. It is not a t test because there is no residual variance to estimate (the variance of the Binomial distribution is a known function of its mean). The value of the test statistic is

$$z = -0.3126/0.1410 = -2.22.$$

From this analysis there appears to be an effect on risk of both sex and feeding. For a complete analysis it is necessary to describe these effects in more detail.

```
$fit  sex+food  $dis e $
scaled deviance =  0.72192 at cycle 3
    residual df =  2
          estimate         s.e.      parameter
     1      -1.613        0.1124       1
     2     -0.3126        0.1410       SEX(2)
     3     -0.1725        0.2056       FOOD(2)
     4     -0.6693        0.1530       FOOD(3)
scale parameter 1.000
```

The fitted proportions can be calculated and the model seen with

```
$cal pf=%fv/%bd $
```

```
$graph  ( t='Fig. 12.28   Observed Data and Fitted Model'
        vla='Proportion'  hla='Food:  Bottle-Both-Breast'xsc=
        0,5,1 ysc=0,0.3,.1)
        p;pf food 6,8,10,10  sex;sex $
```

We see, from this model that children receiving some breast milk (food category 2) were slightly lower, on the log(odds) scale, than those receiving no breast milk (food category 1). The parameter food(2) represents the estimated expected distances, on this scale, between the children in the two categories. Because this parameter is the difference between two logs

$$\log(\text{odds } 1) - \log(\text{odds } 2) = \log(\text{odds } 1/\text{odds } 2).$$

The exponential of food(2) will give the odds ratio indicating how much the risk is reduced from category 1 to 2 and the exponential of food(3) the odds ratio for the

Fig. 12.28 Observed Data and Fitted Model

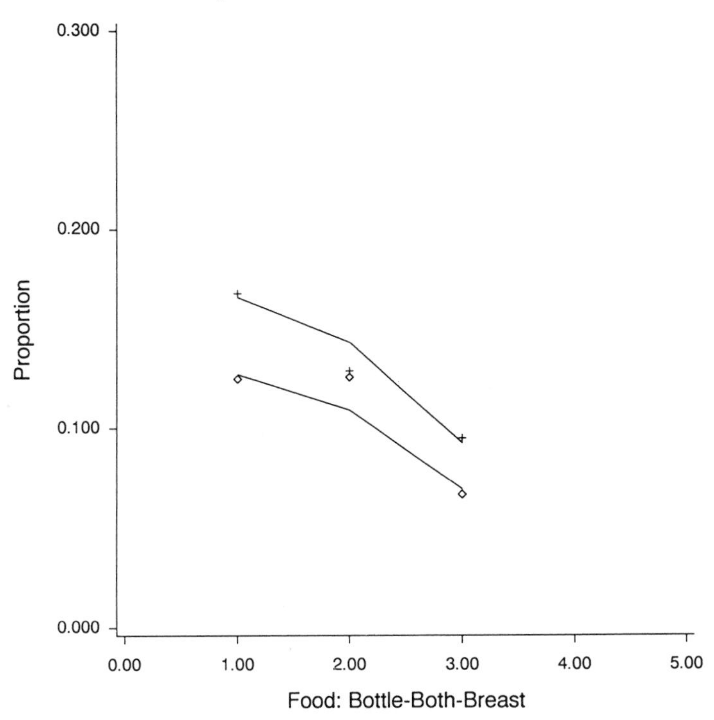

change in risk from the no-breast-milk category to the breast-milk-only category. Using the standard errors we can obtain approximate confidence limits for the estimates on the log(odds) scale and exponentiating will give us the appropriate intervals on the odds ratio scale. It is first necessary to make the parameter estimates and their standard errors available using EXTRACT. The odds ratio and confidence intervals are then obtained for all parameter estimates and printed as required.

```
$ext %pe  %se $
$cal odds=%exp(%pe) : low=%exp(%pe-1.96*%se) :
high=%exp(%pe+1.96*%se) $
```

The estimated Odds Ratios of respiratory disease with 95 per cent confidence limits are then

```
$pr 'for Some breast milk compared to None' odds(3)
     ' CLs'low(3)' to'high(3)
   ; 'for All  breast milk compared to None' odds(4)
     ' CLs'low(4)' to'high(4)
   ; 'for Girls             compared to Boys' odds(2)
     ' CLs'low(2)' to'high(2) $
```

```
for Some breast milk compared to None  0.8415 CLs  0.5624 to   1.259
for All  breast milk compared to None  0.5121 CLs  0.3794 to  0.6911
for Girls            compared to Boys  0.7316 CLs  0.5549 to  0.9645
```

For small risks the odds ratio (OR) is a good approximation to the relative risk. This is because

$$\textit{Odds ratio} = \frac{p_2/(1\text{-}p_2)}{p_1/(1\text{-}p_1)} \quad \text{(group2 versus group1)}$$

$$= \frac{p_2}{p_1} \; \frac{1\text{-}p_1}{1\text{-}p_2}$$

and if the probabilities are close to zero the second term is close to 1. This means that the odds ratio (OR) is approximately equal to the ratio of the two risks p_2/p_1, which is known as the relative risk (RR).

The risks estimated from this model can be obtained and compared with the observed proportions by

```
$tprint pf;p sex;food $
   FOOD      1          2          3
 SEX
   1 PF  0.16621  0.14366  0.09262
      P  0.16812  0.12925  0.09514
   2 PF  0.12728  0.10931  0.06949
      P  0.12500  0.12598  0.06681
```

For risks of the order of 10–12 per cent as here the odds ratio is not very good approximation to the relative risk and it is about 10–12 per cent further from the value 1, obtained when the risks are equal, than the relative risk. Nonetheless ignoring this for simplicity we see that:

(i) The risk for children getting some breast milk is reduced to about 84 per cent of the risk for children getting no breast milk.

(ii) The risk for children getting all breast milk is reduced to about 51 per cent

and

(iii) The risk for girls is about 73 per cent that of the boys.

12.2.1.3 A cross classification with covariates

These data arose from the Health and Lifestyle Survey (HALS) mounted by the Health Promotion Research Trust in 1986 (Cox 1987). They are the frequencies of male individuals reporting some form of 'heart trouble' together with the numbers at risk in a five-way classification according to the factors given below with the numbers of categories used and their coding

Smoking:	1. No;	2. Ex-smokers;	3. Current smokers
Alcohol:	1. Non-drinker;	2. Drinks	
Age:	1. <40;	2. 40–59;	3. 60+ years
Social class:	1. Non-manual;	2. Manual	
Family history:	1. No;	2. Yes	

The structure in the dataset is indicated to clarify the derivation of the factors representing the variables by which the sample was classified

```
$data 72 r n $read
!      Manual                   Non-manual
!No history History No history History !smoking drinking  age
   3   81    0   36   2   110   1   26 !   No      No    < 40
   1   75    3   31   2    53   0   15 !   No      No   40-59
   8   44    0   11  11    53   1   12 !   No      No     60+
   0  140    2   34   1   165   0   40 !   No     Yes    < 40
   2   41    0   19   2    54   1   16 !   No     Yes   40-59
   4   21    0    7   3    17   1    1 !   No     Yes     60+
   0   41    0    7   0    34   0   13 ! Ex-S      No    < 40
   3   50    1   28  10    90   3   39 ! Ex-S      No   40-59
  20   85    6   21  27   151  11   36 ! Ex-S      No     60+
   3   77    1   21   1    72   1   15 ! Ex-S     Yes    < 40
   3   68    3   41   5    72   6   41 ! Ex-S     Yes   40-59
  12   65    5   16  27    93   7   22 ! Ex-S     Yes     60+
   0   30    0   16   1    78   0   22 ! Curr      No    < 40
   3   26    2   15   4    55   4   30 ! Curr      No   40-59
   3   23    1    6  13    96   6   12 ! Curr      No     60+
   3   96    0   24   2   225   2   58 ! Curr     Yes    < 40
   2   54    3   34  13   142   7   44 ! Curr     Yes   40-59
   3   25    1    5  14    68   3    9 ! Curr     Yes     60+
```

Since the data were generated from a table with the dimensions ordered as above the factor variables can be generated with the `GFACTOR` directive.

```
$gfactor 72 sm 3 alc 2 age 3 scg 2 fht 2 $
```

We know that age will have a very definite effect, but the effects of the other factors are more contentious. A useful first step is to look at the apparent relationships allowing for age by plotting the prevalences (`r/n`) against `age` within the categories of each of these variables separately. This will give an idea of the sort of relationships present.

```
$cal p=r/n $
$tab the r total for age;sm into r_ by age_;sm_ $
$tab the n total for age;sm into n_  $
$cal p_=r_/n_ $
$plot (x=.5,3.5) p_ age_  'nos' sm_ $
```

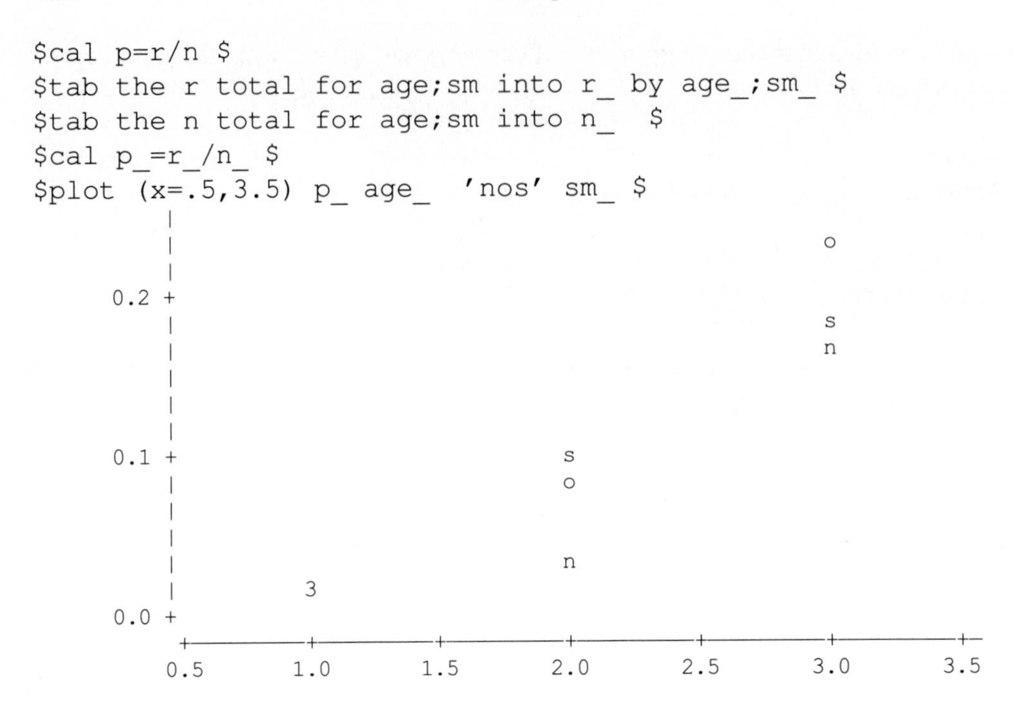


```
$del p_ r_ n_ age_ sm_ $
$tab the r total for age;alc into r_ by age_;alc_ $
$tab the n total for age;alc into n_  $
$cal p_=r_/n_ $
$plot (x=.5,3.5) p_ age_  'nd' alc_ $
```

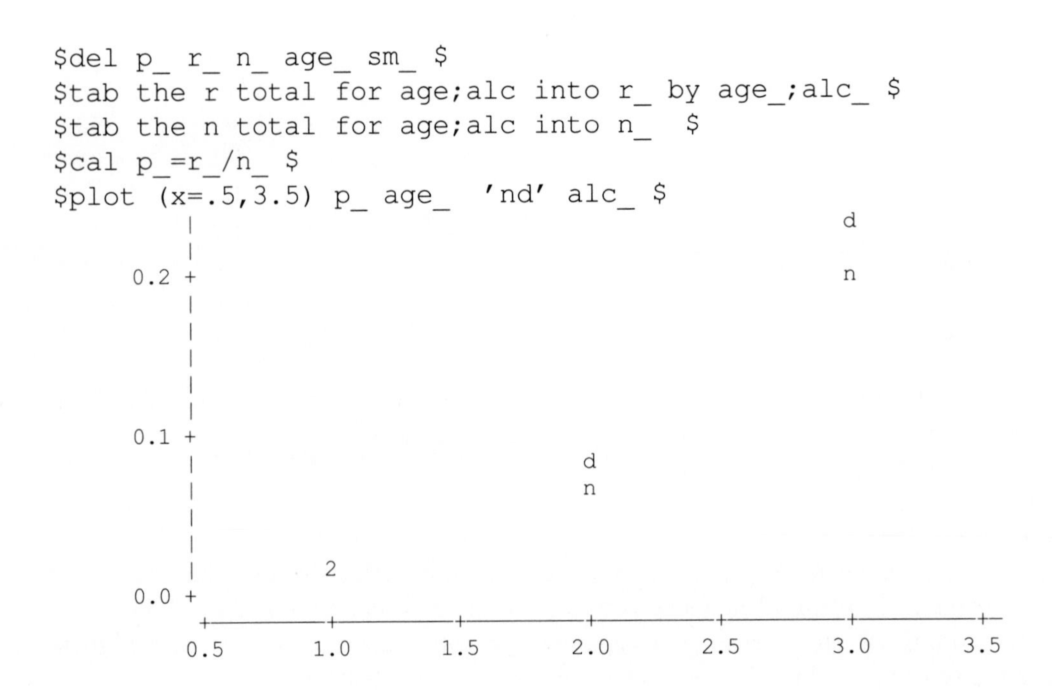

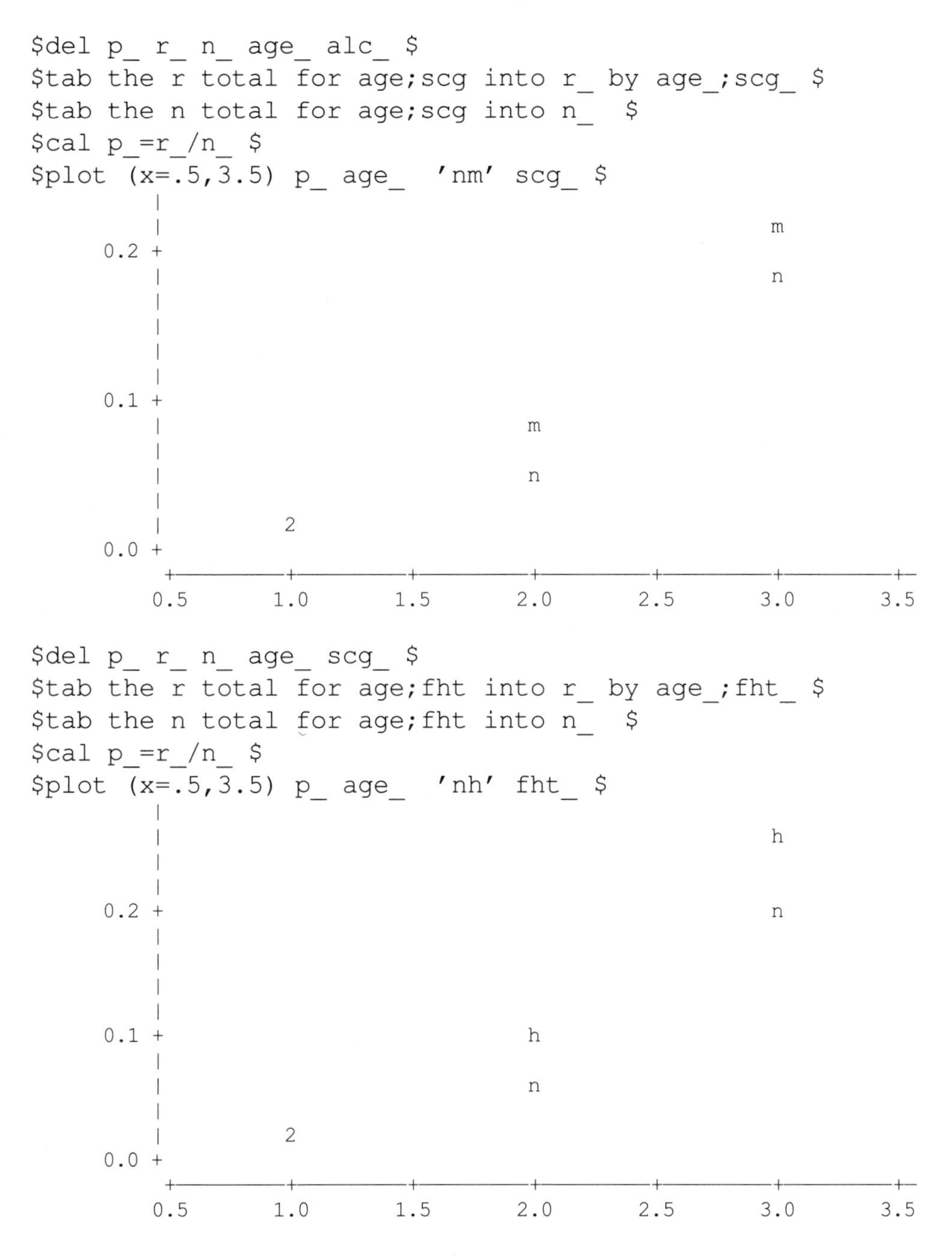

```
$del p_ r_ n_ age_ alc_ $
$tab the r total for age;scg into r_ by age_;scg_ $
$tab the n total for age;scg into n_  $
$cal p_=r_/n_ $
$plot (x=.5,3.5) p_ age_  'nm' scg_ $
     |
     |                                            m
0.2 +
     |                                            n
     |
     |
     |
     |
0.1 +
     |                           m
     |
     |                           n
     |
     |          2
0.0 +
    +---------+---------+---------+---------+---------+---------+-
   0.5       1.0       1.5       2.0       2.5       3.0       3.5

$del p_ r_ n_ age_ scg_ $
$tab the r total for age;fht into r_ by age_;fht_ $
$tab the n total for age;fht into n_  $
$cal p_=r_/n_ $
$plot (x=.5,3.5) p_ age_  'nh' fht_ $
     |
     |                                            h
     |
     |
0.2 +                                             n
     |
     |
     |
     |
0.1 +                            h
     |
     |                           n
     |
     |          2
0.0 +
    +---------+---------+---------+---------+---------+---------+-
   0.5       1.0       1.5       2.0       2.5       3.0       3.5
```

There is a very steep rise in prevalence with age and it appears that smokers, drinkers, those in the manual social class category, and those with a family history all have elevated risks of 'heart trouble' allowing for age. How they interrelate with each other, however, cannot easily be seen. We need to fit models to identify the form and magnitude of the separate and combined effects of these factors on the risk.

The number of men with 'heart trouble' has a binomial distribution with the numbers at risk as the denominators and the logistic link is a sensible way to start. We can investigate whether this gives a good enough fit with the standardized residuals. There are five factors so there is a very wide range of models we could fit. We need to have a strategy for determining an appropriate sequence of model fitting to answer the questions the analysis needs to address. We are not, with observational epidemiological data such as this, trying to identify a 'correct' model. Where possible we want to identify what simple general statements can be made about the effects on risk of the various factors allowing for confounding and after testing and taking account of any interactions there might be.

Since the effects of any or all of the factors may interact it is wise to investigate this possibility immediately. However with five factors there is a large number of possible interactions involving all the factors in combinations of 2, 3, 4 and 5. A reasonable strategy is to fit a model with all the two–factor interactions and only if they prove too large to be ignored should the more complex interactions be considered. The logic of this is that if there is, for example, a marked interaction with social class then it is probably best, from the point of view of simplifying interpretation, to analyse the data from the two social class groups separately. That would immediately reduce the number of interactions to be considered and the process can be repeated. Of course it will reduce the power of each analysis by reducing the degrees of freedom. Ideally the full model should be re-parametrized so the different effects in the two social class groups are estimated directly.

Returning to the analysis we fit the two-way interaction model with

```
$yvar r $error b n $
$fit (sm+alc+scg+fht+age)**2 $dis e $
scaled deviance =  51.137 at cycle 4
    residual df =  45
          estimate          s.e.      parameter
     1      -3.862        0.5095       1
     2     0.04721        0.6290       SM(2)
     3     -0.4016        0.6510       SM(3)
     4     -0.2319        0.5247       ALC(2)
     5     -0.7369        0.5136       SCG(2)
     6     -0.1772        0.5840       FHT(2)
     7      0.7070        0.5946       AGE(2)
     8       2.325        0.5402       AGE(3)
     9      0.3875        0.3987       SM(2).ALC(2)
    10      0.2929        0.4271       SM(3).ALC(2)
    11     -0.1198        0.3727       SM(2).SCG(2)
    12     0.01714        0.4206       SM(3).SCG(2)
    13      0.3878        0.2834       ALC(2).SCG(2)
    14      0.4946        0.4551       SM(2).FHT(2)
    15      0.8170        0.4861       SM(3).FHT(2)
    16      0.1537        0.3057       ALC(2).FHT(2)
    17      0.2453        0.3128       SCG(2).FHT(2)
    18      0.5212        0.6613       SM(2).AGE(2)
    19      0.1675        0.6114       SM(2).AGE(3)
    20      0.9689        0.6399       SM(3).AGE(2)
```

```
    21       0.1629      0.6076        SM(3).AGE(3)
    22      -0.4297      0.5419        ALC(2).AGE(2)
    23      -0.1510      0.5198        ALC(2).AGE(3)
    24       0.9321      0.5116        SCG(2).AGE(2)
    25       0.7891      0.4901        SCG(2).AGE(3)
    26      -0.2372      0.5365        FHT(2).AGE(2)
    27      -0.1318      0.5333        FHT(2).AGE(3)
scale parameter 1.000
```

The deviance is a little more than its degrees of freedom, but not enough to cast doubt on the distributional assumptions or to suggest that the model is inadequate to represent this data. None of the individual interaction terms exceeds twice its s.e. and there is no evidence of trends or patterns among them. Finally we should look at the standardized residuals. Plotting them against an index variable representing the ordered cells of the table from which the data was obtained will show up patterns if the model is fitting consistently less well for certain combinations of factor levels.

```
$cal indx=%cu(1) $
$plot %rs indx 'r' $
```

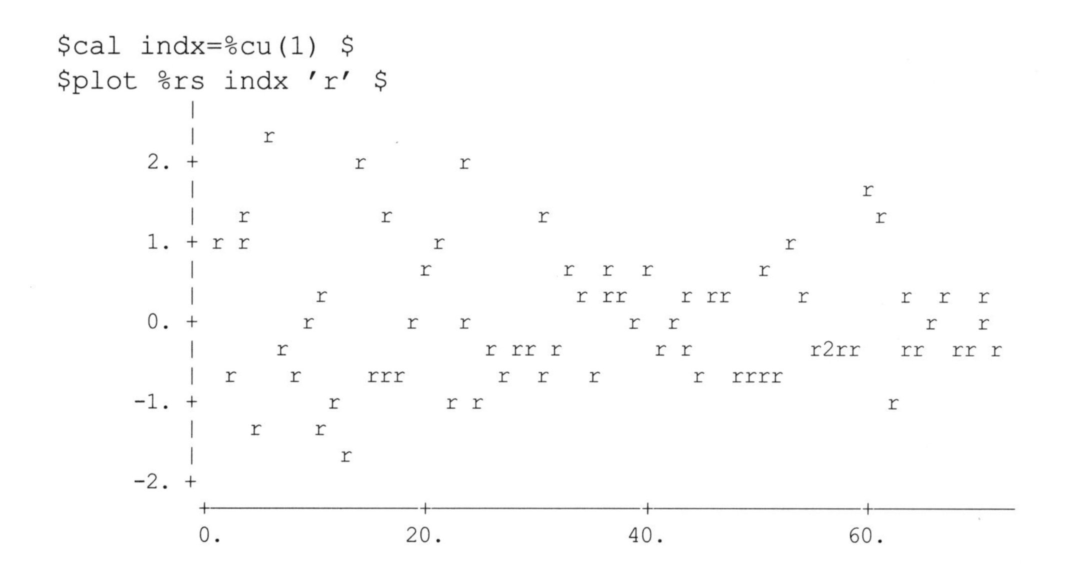

The residuals appear to be rather more widely spread for the early observations, but over the whole plot the spread is more or less what should be expected for a variable with zero mean and unit variance. A plot of the standardized residuals against the values of the linear predictor, not the fitted values because they are functions of the denominators as well as the fitted proportions, should give a similar plot if the binomial distribution is appropriate.

```
$plot %rs %lp 'r' $
        |
        |               r
     2. +         r                           r
        |                                          r
        |    r  r   r            r
     1. +   r       r            r      r
        |r            r  r r                   r
        |               r  r       r r  r    rr r 2r
     0. +         r        r      r r        rr r
        |    r    r     r     2r  r 2 r  rr2 r  r      r
        |  r r  2 3   r    r     r           rr         r
    -1. +           r   r             r     r
        |                  r            r
        |         r
    -2. +
         +---------+----------+----------+----------+-------
        -5.       -4.        -3.        -2.        -1.
```

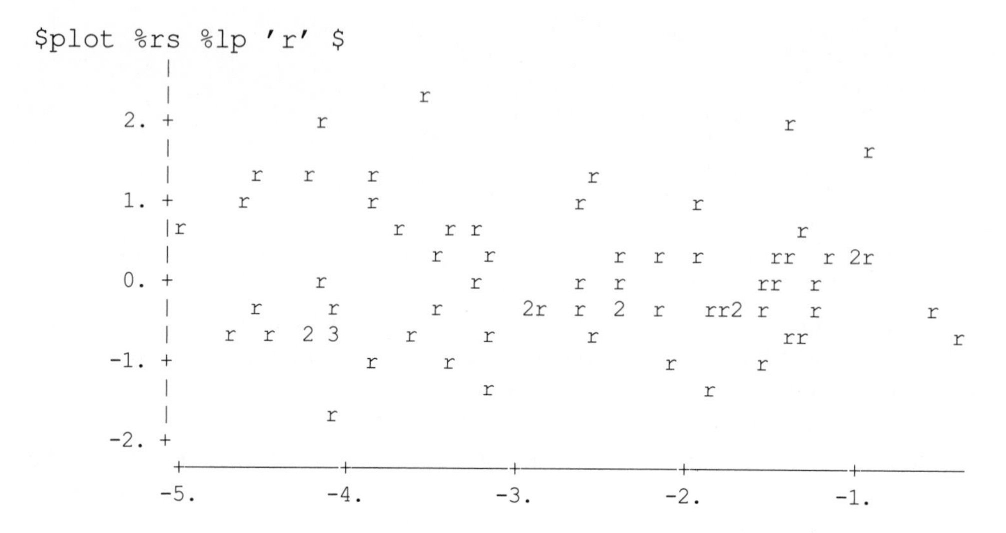

There appears to be little change in the spread of the residuals with increasing linear predictor values. The pattern is quite consistent with the residuals arising as samples from a distribution with mean zero and unit variance. The binomial assumption is adequately justified.

The next step is to test the interaction terms 'en masse'. This is done by fitting a model omitting them. The resulting increase in deviance gives a likelihood ratio chi-squared test of all the interactions at once with its degrees of freedom equal to the number of parameters dropped from the model. To calculate the chi-squared and compile an analysis of deviance (ANODE), equivalent to the analyses of variance/covariance covered in the sections dealing with continuous *y*–variables [12.1], it is useful to store the deviances and degrees of freedom. We will need to store them from several models.

```
$number D1 D2 D21 D3 D32 D4 D5 D6 D7 df1 df2 df21 df3 df32
df4 df5 df6 df7 $
$cal D1=%dv : df1=%df $
```

Now the model without interactions.

```
$fit sm+alc+scg+fht+age  $dis e $
scaled deviance =  68.977 at cycle 4
    residual df =  64
          estimate        s.e.      parameter
     1      -4.698      0.2668       1
     2      0.4916      0.1808       SM(2)
     3      0.3304      0.1943       SM(3)
     4      0.1367      0.1318       ALC(2)
     5      0.2105      0.1358       SCG(2)
     6      0.4135      0.1463       FHT(2)
```

```
     7       1.490       0.2422       AGE(2)
     8       2.722       0.2345       AGE(3)
scale parameter 1.000
```

The deviance has increased quite a lot, but so have the degrees of freedom. The likelihood ratio chi-squared is

```
$cal D2=%dv : df2=%df $
$cal D21=D2-D1 : df21=df2-df1 $
```

So the Chi-squared test of all two factor interactions is

```
$print 'Chi-sqd ='*r D21,6,2' df ='*i df21,2 $
Chi-sqd = 17.84 df =19
```

The probability of chi-squared values as large as this on 19 d.f. is obtained using the chi-squared cumulative distribution function `%chp(X,df)` (with six significant figures and four decimal places) by

```
$acc 6 4 $
$cal 1-%chp(D21,df21) $
         0.5332
```

It seems that the interactions are not significant and can be ignored. It remains to test the factors individually by removing them from this baseline model and obtaining chi-squared test statistics from the resulting increase in deviance. This should be done with all the other factors present in the baseline and reduced models to avoid unnecessary assumptions that their effects are zero, which may not be the case even if the estimated effects do not reach significance. We do this by fitting models with each factor removed in turn and replaced as the next is removed

```
$fit -age $
scaled deviance =  283.2672 (change =   +214.290) at cycle 4
    residual df =   66      (change =     +2    )
$cal D3=%dv : df3=%df : D32=D3-D2 : df32=df3-df2 $
```

Removing 'age' from the model has increased the deviance markedly while the degrees of freedom have only increased by 2 (that is, the number of parameters dropped from the model). The chi-squared test is clearly significant and the probability of values as high as this occurring by chance is extremely small

```
$cal 1-%chp(D32,df32) $
          0.
```

The remaining tests have been done within the formal analysis of deviance table below. Note, however, that because the models are not now 'nested' within one another the changes of deviance output by GLIM do not give the appropriate likelihood ratio chi-squared tests. Those have to be calculated specifically as in the table. The sequence of models required is:

```
$fit +age-fht $cal D4=%dv : df4=%df $
scaled deviance =  76.6825 (change =   -206.585) at cycle 4
    residual df =  65      (change =     -1     )
$fit +fht-scg $cal D5=%dv : df5=%df $
scaled deviance =  71.4096 (change =    -5.2728) at cycle 4
    residual df =  65      (change =     0      )
$fit +scg-alc $cal D6=%dv : df6=%df $
scaled deviance =  70.0528 (change =    -1.3569) at cycle 4
    residual df =  65      (change =     0      )
$fit +alc-sm  $cal D7=%dv : df7=%df $
scaled deviance =  76.7747 (change =    +6.7220) at cycle 4
    residual df =  66      (change =    +1      )

$use anod3213 $
```

	Model	Deviance	df	
1)	All 2f intrns	51.14	45	
2)	Main Effects	68.98	64	
(2-1)	Due to intrns	17.84	19	p =0.5332
3)	- AGE	283.27	66	
(3-2)	Due to AGE	214.29	2	p =0.0000
4)	+AGE-FHT	76.68	65	
(4-2)	Due to FHT	7.71	1	p =0.0055
5)	+FHT-SCG	71.41	65	
(5-2)	Due to SCG	2.43	1	p =0.1188
6)	+SCG-ALC	70.05	65	
(6-2)	Due to ALC	1.08	1	p =0.2996
7)	+ALC-SM	76.77	66	
(7-2)	Due to SM	7.80	2	p =0.0203

It is reasonably clear that the interactions can be assumed negligible with p=0.53. Of the other factors age is highly significant with $p<0.0001$ and, allowing for age and all the other factors, family history and smoking are also significant (p=0.0055 and p=0.02 respectively). Social class and drinking alcohol on the other hand do not appear to be important (p=0.12 and p=0.30 respectively). Their marginal effect seen in the plots must arise as a result of confounding due to associations with other factors, such as the well known association between smoking and the consumption of alcohol. We can check this by

```
$tabulate (s=1) the n total for sm;alc by smoking;alcohol $
          +------------------------+
  ALCOHOL |      1          2      |
  SMOKING |                        |
+---------+------------------------+
|       1 |  547.000    555.000    |
|       2 |  595.000    603.000    |
|       3 |  409.000    784.000    |
+---------+------------------------+
```

There is a marked association with the smokers tending to be drinkers (alcohol category 2) as well. Among the current smokers (smoking category 3) drinkers are in a two to one majority whereas the non- and ex-smokers (categories 1 and 2) appear to be split about 50/50 between drinkers and non-drinkers.

To complete the analysis we need to obtain the estimated odds ratios indicating how the risks for those in one category of a factor differ from those in the other categories. Since each parameter represents a difference on the log(odds) scale for individuals in that category of the factor from those in the reference category for that factor the parameters are the logs of odds ratios. Taking antilogs we obtain the required odds ratios (OR).

If we use the s.e.s to obtain confidence limits for the estimated parameters their antilogs provide confidence limits for the odds ratios. The estimates we want are those from the baseline model so we need to fit it again. We also need to consider which of the factor levels will be the most useful and easy to interpret reference categories. In fact the default reference categories (that is, category 1) are probably the most appropriate for all five factors. How other smoking groups compare with non-smokers is the most easily grasped comparison, the effect of increasing age is probably best seen by using the youngest category as a reference, and the other factors are all dichotomies so whichever category is taken as the reference it is simple to deduce the effect of reversing it. However to illustrate the facility we will use the regular smokers as the reference category so we are estimating the benefits of not smoking rather than the folly of smoking.

```
$fac sm 3(3) $
$fit sm+alc+scg+fht+age  $dis e $
scaled deviance =  68.9768 at cycle 4
```

```
 residual df =  64
          estimate          s.e.      parameter
     1     -4.3674        0.2744       1
     2     -0.3304        0.1943       SM(1)
     3      0.1612        0.1487       SM(2)
     4      0.1367        0.1318       ALC(2)
     5      0.2105        0.1358       SCG(2)
     6      0.4135        0.1463       FHT(2)
     7      1.4900        0.2422       AGE(2)
     8      2.7225        0.2345       AGE(3)
scale parameter 1.0000
```

Before proceeding it is wise to investigate the relative influence of the individual data points in case the estimates have been biased by some unrepresentative and unduly influential point. The leverage values for this will also provide a warning if the model is fitting some subset of the dataset inadequately.

```
$extract %lv %cd $
$plot (y=0,1)  %lv indx 'i' $
```

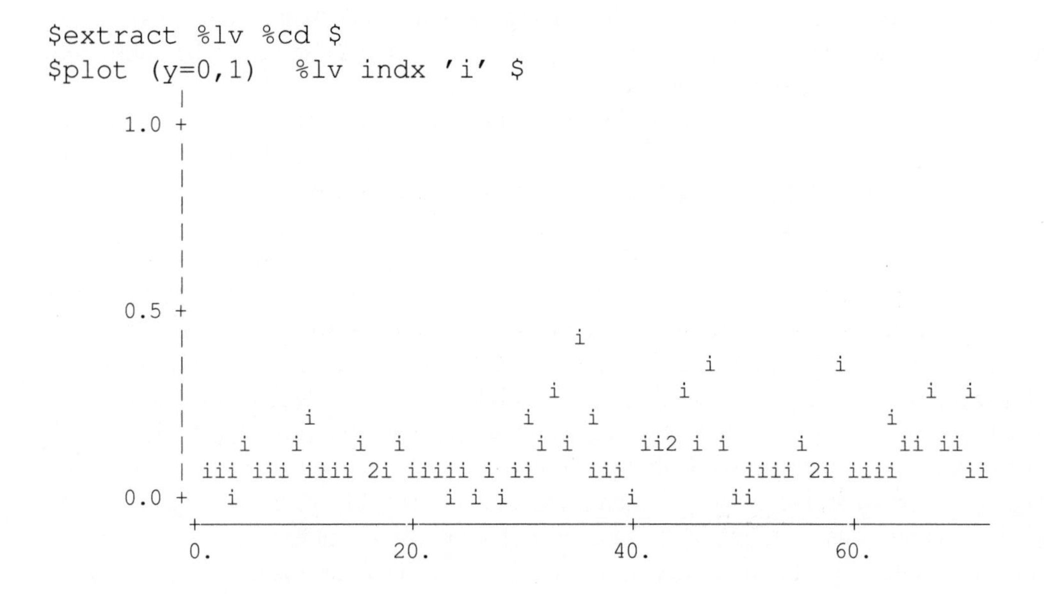

There are a large number of relatively uninfluential points with a scatter of rather more influential points reasonably evenly spread throughout the dataset. It looks reasonably acceptable. We can proceed to investigate the estimated parameters. It is convenient to use a macro to calculate the odds ratios together with their approximate 95 per cent confidence limits.

```
$use orcs $

95% CLs and the Odds Ratio for each factor level compared
with the base level 1 (for covariates the OR is for X+1 .v. X)
together with the fitted proportion for 1 + each factor level
in turn with the increase that results.
 Low 95% CL  Odds Ratio   High 95% CL

        L_          OR_         U_
 1   1.0000      1.0000      1.000
 2   0.4910      0.7186      1.052
 3   0.8780      1.1750      1.572
 4   0.8855      1.1465      1.484
 5   0.9458      1.2343      1.611
 6   1.1352      1.5121      2.014
 7   2.7603      4.4372      7.133
 8   9.6103     15.2181     24.098
```

From these results we can see that the effect of alcohol (parameter 4) could be, consistent with these data, equivalent to an odds ratio as high as 1.48, that is, moderately harmful, or as low as 0.89, that is, moderately beneficial. However the evidence is that it probably has no real effect. The social class effect (parameter 5) also failed to reach significance, but the estimated odds ratio (manual to non-manual) is 1.23 which is consistent with a true value as high as 1.61 with the lower confidence limit only just below 1 (0.95). It is possible that there is an independent effect which is not consistent or pronounced enough to be detected with this sample.

There is a clear and dramatic increase with age (parameters 7 and 8) and also, although not so dramatic, with family history (parameter 6; OR = 1.51). Finally, although the analysis of deviance shows that the smoking categories (parameters 2 and 3) differ significantly, that does not tell us exactly how. With smokers as the reference category we see that in fact neither of the odds ratios are significant. Both confidence intervals contain 1. Nonetheless the non-smokers have an odds ratio, compared to the regular smokers of 0.72. This is roughly equivalent to a risk three quarters that of the smokers. The ex-smokers appear to have a slightly higher risk than smokers, but far from significantly so.

It appears that the pattern is for the non-smokers to have the lowest risk with smokers and ex-smokers both considerably higher, but only the difference between the non- and ex-smokers reaching significance. This probably indicates a classic problem in using associations in cross-sectional data to investigate causal relationships. The most likely explanation of a raised prevalence of 'heart trouble' in the ex-smokers is that there will have been considerable self-selection among smokers diagnosed with heart disease into the ex-smoking category. This could be investigated further by considering various degrees of self-selection. Of course the never smoked/ever smoked comparison (which addresses a slightly different, but still interesting question) would not be biased in this way.

12.2.1.4 A non–linear dose–response problem with overdispersion

These data arose from an epidemiological study of toxoplasmosis. The values obtained are the proportions of subjects found positive in those tested for 'toxoplasmosis' in 34 towns and cities in El Salvador. The purpose of the analysis is to assess the association between the risk of infection and rainfall. The data, originally from Efron (1986) are analysed by Firth (1990).

```
$units 34 $data rainf ppos n $read
1735 .5    4 1936 .3    10 2000 .2    5 1973 .3    10 1750 1.0 2
1800 .6    5 1750 .25    8 2077 .368 19 1920 .5    6 1800 .8   10
2050 .292 24 1830 .0     1 1650 .5   30 2200 .182 22 2000  .0 1
1770 .545 11 1920 .0     1 1770 .611 54 2240 .444  9 1620 .278 18
1756 .167 12 1650 .0     1 2250 .727 11 1796 .532 77 1890 .471 51
1871 .438 16 2063 .561 82 2100 .692 13 1918 .535 43 1834 .707 75
1780 .615 13 1900 .3    10 1976 .167  6 2292 .622 37
```

Plot the data to assess whether there is any relationship between the proportions positive and rainfall.

```
$plot (y=0,1.1) ppos rainf 'x' $
```

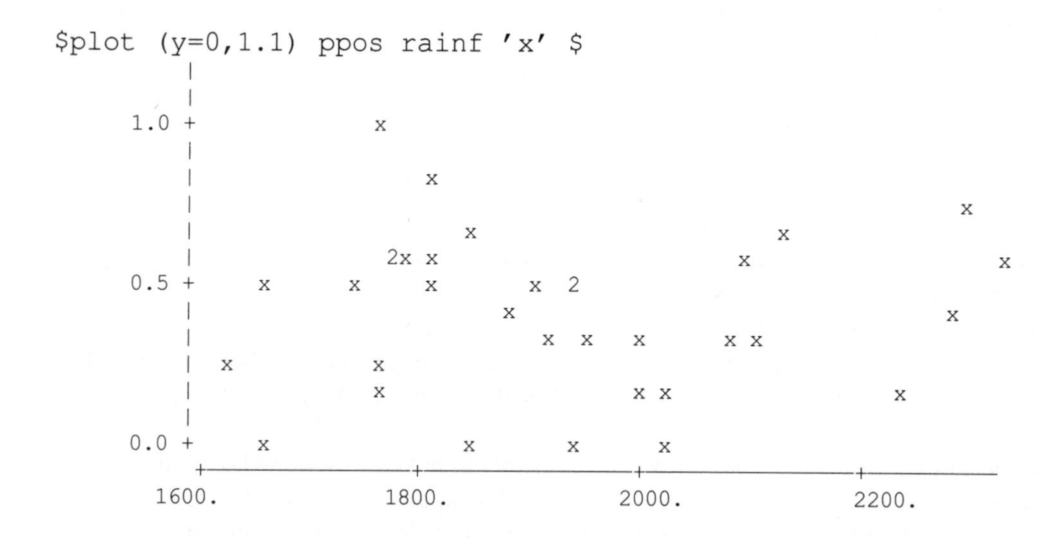

There is no very obvious relationship, but of course the relative importance of the points, determined by the denominators, varies considerably so a relationship might well be difficult to see. One approach is to use a facility in the GRAPH directive to vary the size of the plotting symbol according to a weight. Scaling the weights proportional to the denominators so that the largest weight is only just over 4 gives symbols of a reasonable size.

```
$cal wtg=n/20 $
$graph (title='Fig. 12.29 Observed proportions weighted for n'
        hlabe='Rainfall' vlab='Proportion positive' )
        ppos/wtg  rainf 5 $
```

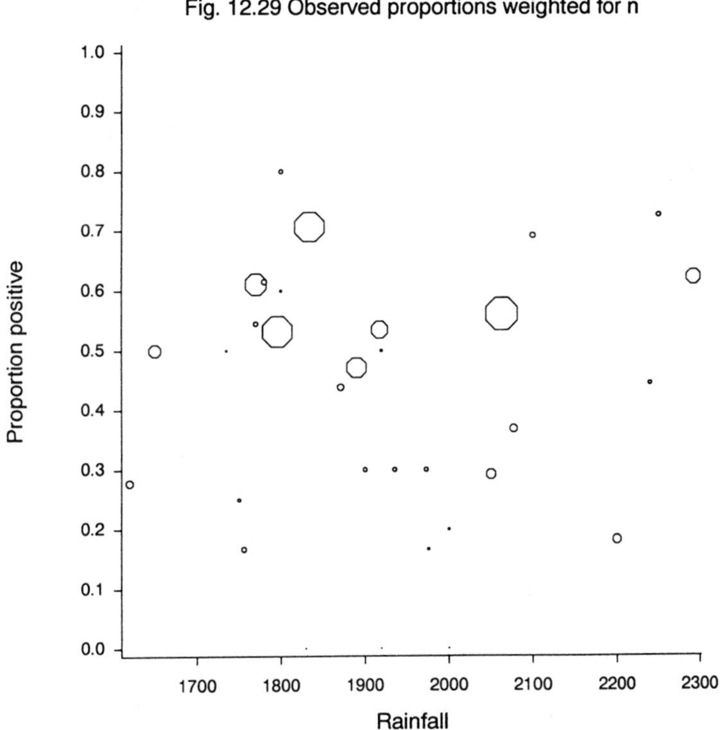

The points are very scattered. If there is any sort of dose–response relationship it may be easier to see on the logistic scale.

```
$cal logitp=%log(ppos/(1-ppos)) $

$graph (title='Fig. 12.30 Logit(proportions) weighted for n'
        hlabe='Rainfall' vlab='Proportion positive' )
        logitp/wtg rainf 5  $
```

Fig. 12.30 Logit(proportions) weighted for n

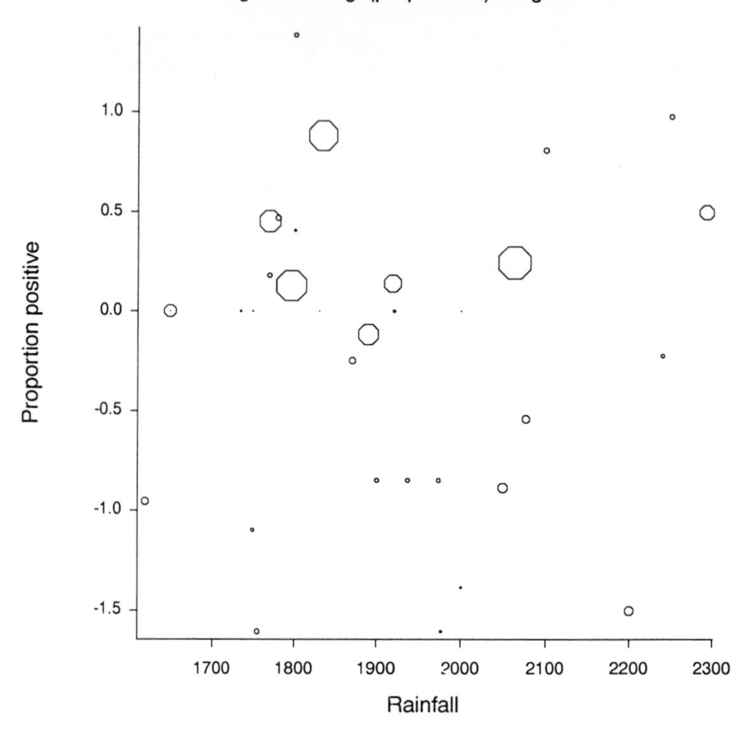

It is not much help, there is still no obvious relationship. However it could just be that there is a tendency for the risk to increase with rainfall in some non-linear manner. We can investigate this by fitting a sequence of polynomial models to identify the simplest that gives an adequate fit.

First we calculate the number positive to the test for toxoplasmosis from the given proportions and denominators and then we scale the rainfall figures to units of 1000. The latter is not necessary but it makes the fitted parameter values slightly easier to digest.

```
$cal npos=%tr(ppos*n+0.5) : rainf=rainf/1000 $
```

It is now necessary to set the y variable, binomial error distribution, and the binomial denominator. The efficient strategy is then to fit a moderately high–order orthogonal polynomial, such as the fifth, from which we can identify the highest–order coefficient that is significant and hence the highest–order polynomial required to model any trend present.

```
$yvar npos $error b n $
$fit rainf<5> $dis e $
scaled deviance =  61.196 at cycle 3
    residual df =  28
```

```
        estimate        s.e.      parameter
   1     0.02505     0.07709       1
   2     -0.2422      0.4861       RAINF<1>
   3     -0.2345      0.4902       RAINF<2>
   4       1.462      0.4317       RAINF<3>
   5     -0.2382      0.4750       RAINF<4>
   6      0.5155      0.4623       RAINF<5>
scale parameter 1.000
```

The fourth and fifth coefficients are considerably less than their s.e.s, but the third is more than three times its s.e. That indicates there is a trend that requires a cubic curve to model it. Fitting the cubic model we see that the deviance has changed little.

```
$fit rainf<3> $dis e $
scaled deviance =  62.635 at cycle 3
    residual df =  30
        estimate        s.e.      parameter
   1     0.02427     0.07693       1
   2    -0.08606      0.4587       RAINF<1>
   3     -0.1927      0.4674       RAINF<2>
   4       1.379      0.4115       RAINF<3>
scale parameter 1.000
```

The fourth and fifth powers added very little to the explanatory power of the model. The coefficients for the cubic polynomial and lower powers have changed little from the fifth order model. In the classic least squares case they would not change, but in the iterative fits the weighting deflects the polynomials from complete orthogonality.

From the coefficients we may deduce that the cubic model is adequate but the deviance is considerably more than its degrees of freedom. This could be due to the model being inadequate, but the data does not appear to warrant a more complex model. In this case the overdispersion suggests that there is more than binomial variation. A simple and generally adequate way to deal with that is to use a scale parameter correction.

First look at the predicted value cubic curve with 95 per cent confidence limits from the model as it is without correcting for overdispersion.

```
$ass rainp=1.5,1.55...2.5 $
$predict (s=-1) rainf=rainp $
$cal predcv=%pfv : lo=1/(1+%exp(-(%plp-1.96*%sqr(%pvl))))
                 : hi=1/(1+%exp(-(%plp+1.96*%sqr(%pvl)))) $
```

```
$plot (y=0,1,0.5 x=1.500,2.500,.500)
     ppos;predcv;lo;hi rainf;rainp;rainp;rainp 'xe^v' $
```

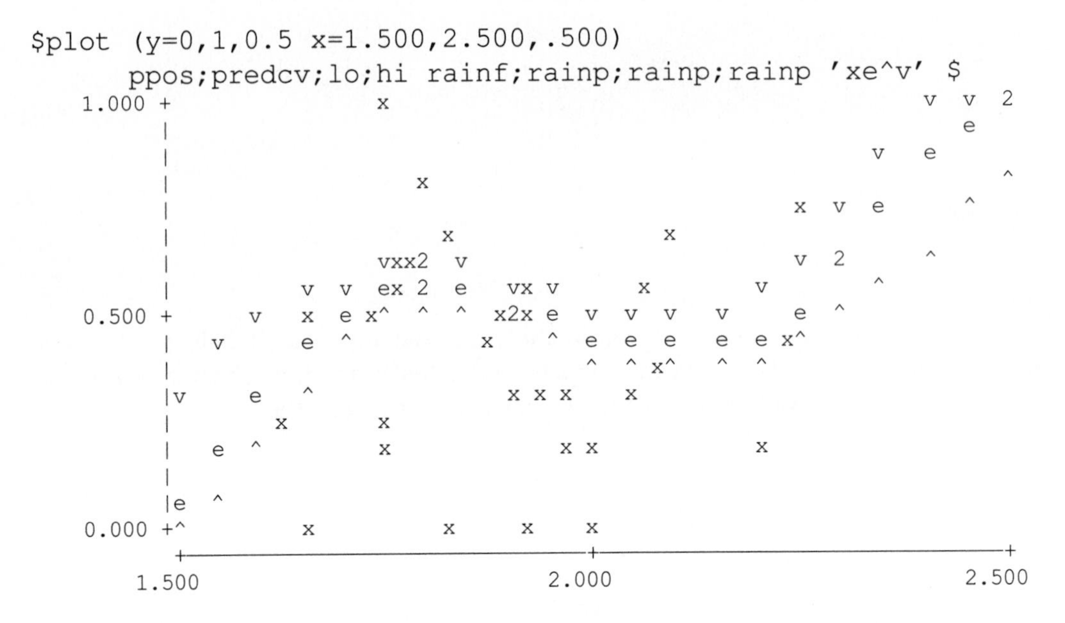

A high–resolution equivalent can be obtained, plus titles and labels, with

```
$mac title Fig. 12.31 Fitted Cubic $end $
$mac xlabel Rainfall $end $
$mac ylabel Propn +ve to test $end $
$mac xscale 1.500,2.500,.500 $end $
$mac yscale 0,1,0.5  $end $

$gra (t=title h=xlabel v=ylabel x=#xscale y=#yscale)
       ppos;predcv;lo;hi rainf;rainp;rainp;rainp 1;10,13,13 $
```

Now repeat the exercise with a correction for the overdispersion. The simplest correction (Hinkley et al 1990) is to use as a scale parameter the Pearson chi-squared value (the sum of the squared Pearson residuals) divided by the degrees of freedom.

```
$pr ;'Pearson''s Chi-square'%X2' with'%df' degrees of freedom'
$
Pearson"s Chi-square   58.21 with   30.00 degrees of freedom
$number scale_up $cal scale_up=%X2/%df $
```

Set the scale parameter and re-fit the model.

```
$scale scale_up $
$fit rainf<3> $dis e $
scaled deviance =  32.279 at cycle 3
```

```
    residual df =  30
          estimate          s.e.      parameter
     1     0.02427        0.1072       1
     2    -0.08606        0.6390       RAINF<1>
     3     -0.1927        0.6511       RAINF<2>
     4       1.379        0.5732       RAINF<3>
scale parameter 1.940
```

Fig. 12.31 Fitted Cubic

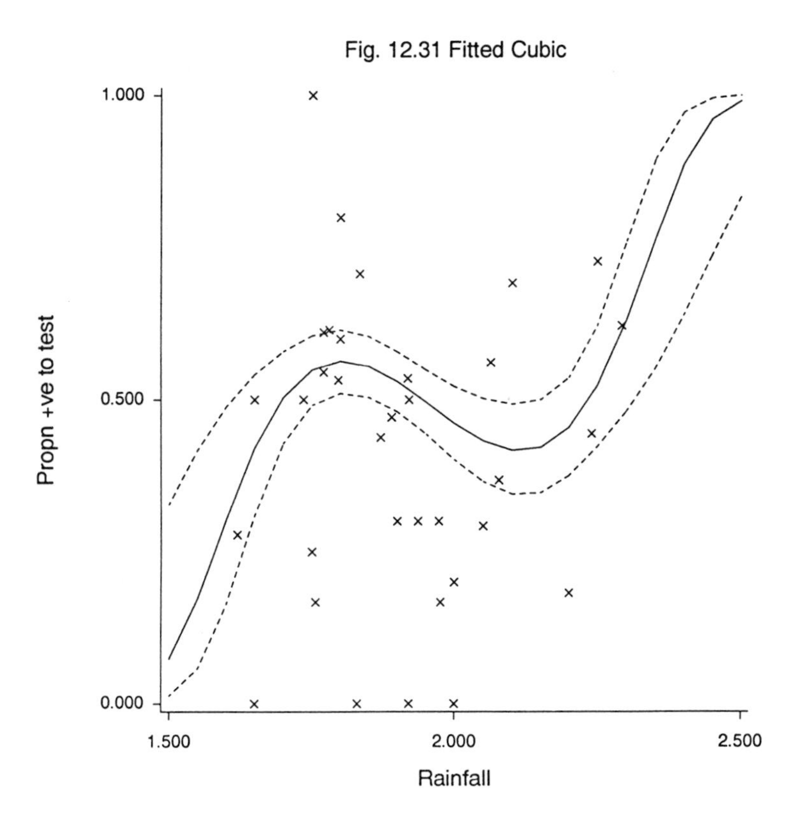

Notice that the scaled deviance is now much closer to the degrees of freedom. Re-calculate the predicted value from the cubic curve with 95 per cent confidence limits

```
$predict (s=-1) rainf=rainp $
$cal lo=1/(1+%exp(-(%plp-1.96*%sqr(%pvl))))
                   : hi=1/(1+%exp(-(%plp+1.96*%sqr(%pvl)))) $
```

and repeat the plot.

```
$plot (y=0,1,0.5 x=1.500,2.500,.500)
     ppos;predcv;lo;hi rainf;rainp;rainp;rainp 'xe^v' $
```

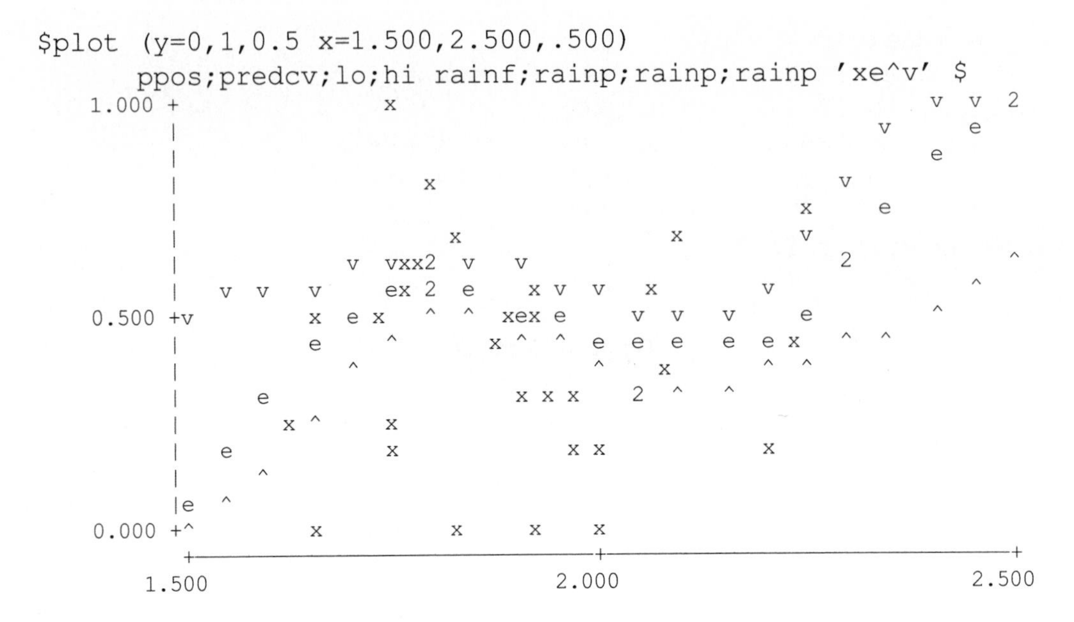

We see that the regions are much wider.

Although the limits are somewhat wider the general conclusions of the analysis are relatively unchanged. However the relationship is rather unusual. If rainfall is a risk factor it would seem likely that it would have some relatively standard dose–response relationship with risk. This would generally mean a straight line relationship on the logistic scale. We should investigate whether there are any unduly influential observations causing the 'cubic curvature'. This is done using an index plot of the Cook's distances and leverage values obtained from the corrected fit. The index is simply a variable indicating position in the dataset.

```
$extract %lv %cd $
$cal index=%ind(%cd) $
$plot %cd index 'c' $
```

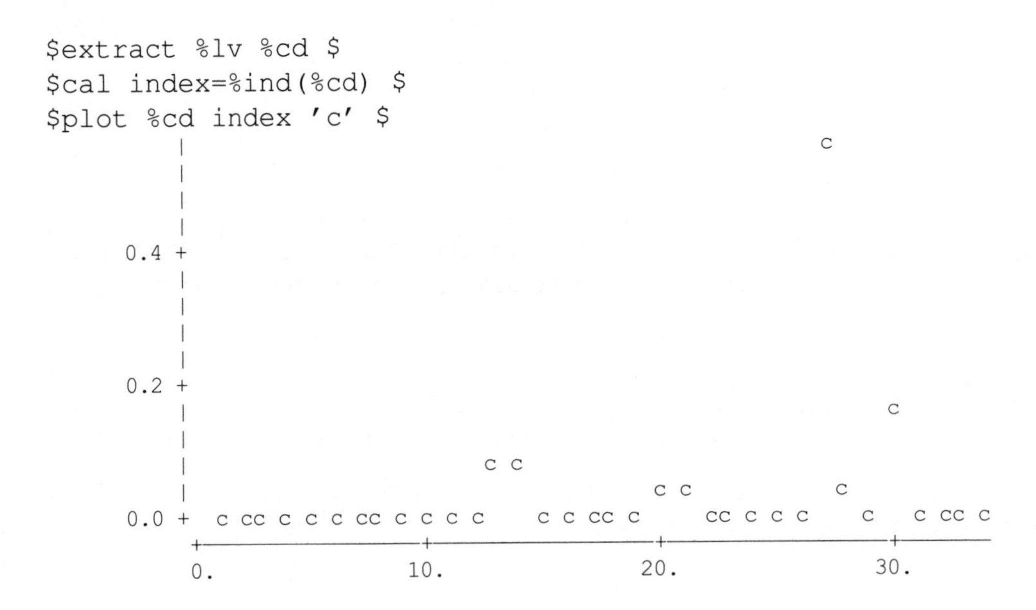

```
$plot %lv index 'i' $
```

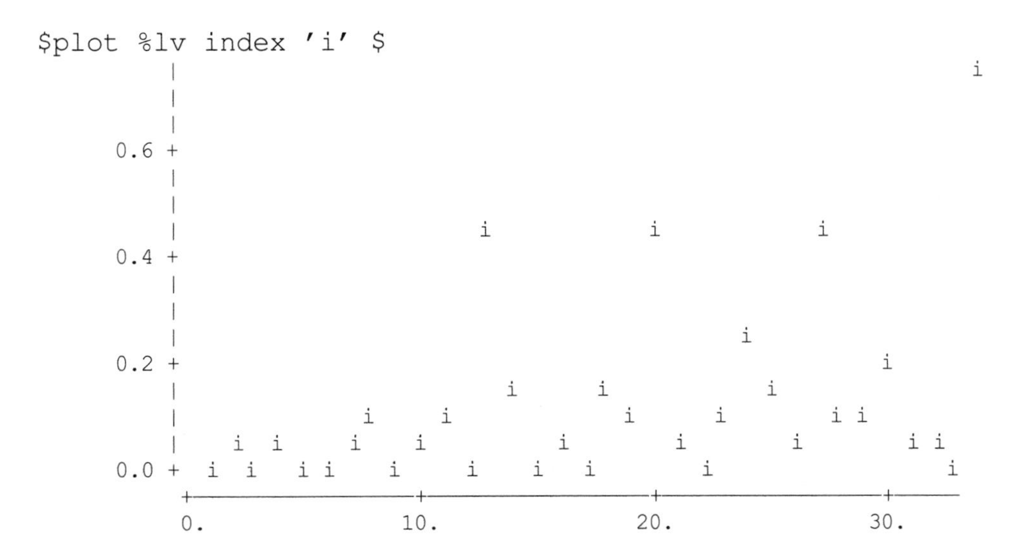

The 27th observation has the largest Cook's distance and the 34th the largest leverage. Both the values stand out from those from the other observations. To assess the effect of omitting the obvservation with the largest Cook's distance reset the scale parameter to the default, weight out the 27th value and re-fit the model.

```
$scale 1 $
$cal wt=(index/=27) $weight wt $fit . $dis e $
scaled deviance =  51.969 (change =    +19.69) at cycle 3
    residual df =  29      (change =     -1   ) from 33 observations
          estimate         s.e.      parameter
     1    -0.09431      0.08606       1
     2     -0.6969       0.5023       RAINF<1>
     3      0.1727       0.4854       RAINF<2>
     4       2.151       0.4827       RAINF<3>
scale parameter 1.000
$extr %vl %wt %cd $
```

Use wt to exclude the Cook's distance from the excluded observation and plot them again.

```
$plot %cd/wt index 'c' $
```

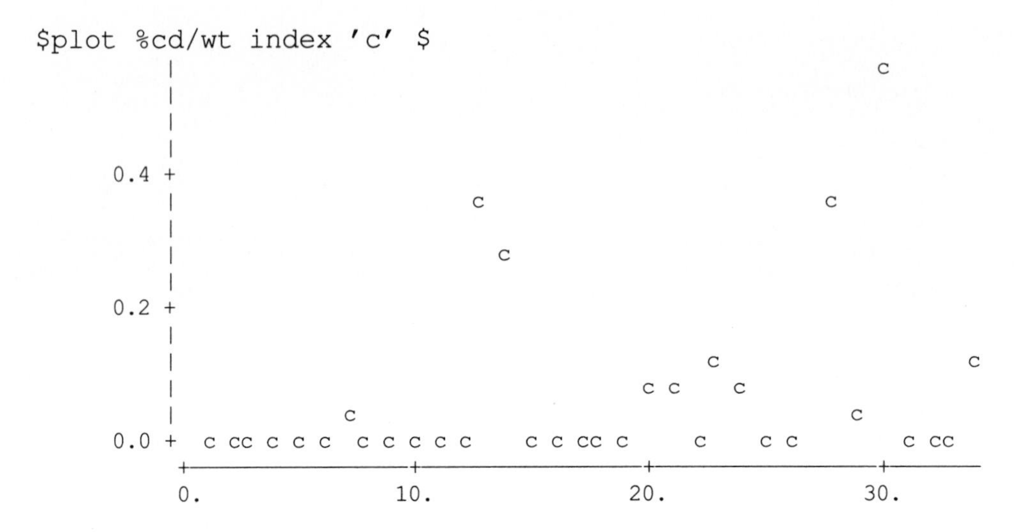


```
$plot %lv/wt index 'i' $
```

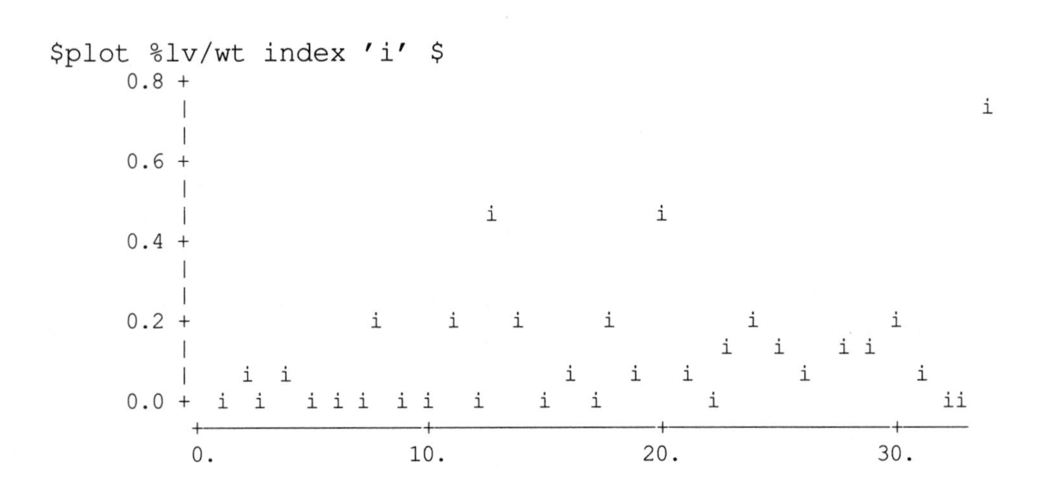

A rather smaller range of values for Cook's distances, but still one point with rather a large leverage. Obtain predicted values over the rainfall range from this model and plot the data with the two curves for comparison and the excluded point with a special symbol controlled by a two–level factor identifying included and excluded points.

```
$cal excfac=1+wt $fac excfac 2 $
$predict (s=-1) rainf=rainp $
```

```
$plot (y=0,1,0.5 x=1.500,2.500,.500)
    ppos;predcv;%pfv rainf;rainp;rainp 'oxem' excfac;; $
```

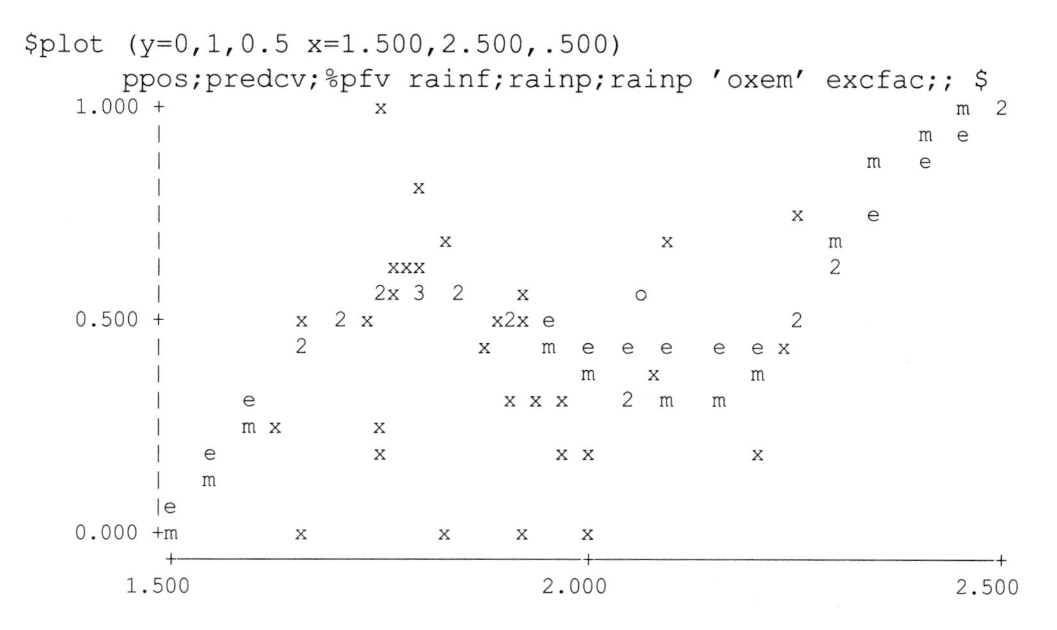

Some change in the positioning of the cubic curve, but not much change in shape. Repeat the process to assess the effect of omitting the observation with the largest leverage by weighting out the 34th value and re-fitting the model.

```
$cal wt=(index/=34)  $fit . $dis e $
scaled deviance =  62.572 (change =   +10.60) at cycle 3
    residual df =  29      (change =      0   ) from 33 observations
          estimate         s.e.      parameter
     1     0.01680      0.08247      1
     2     -0.1877       0.6121      RAINF<1>
     3     -0.2993       0.6317      RAINF<2>
     4       1.294       0.5333      RAINF<3>
scale parameter 1.000
$extr %vl %wt %cd $
```

Use `wt` to suppress the point from the excluded observation and plot them again.

```
$plot %cd/wt index 'c' $
```

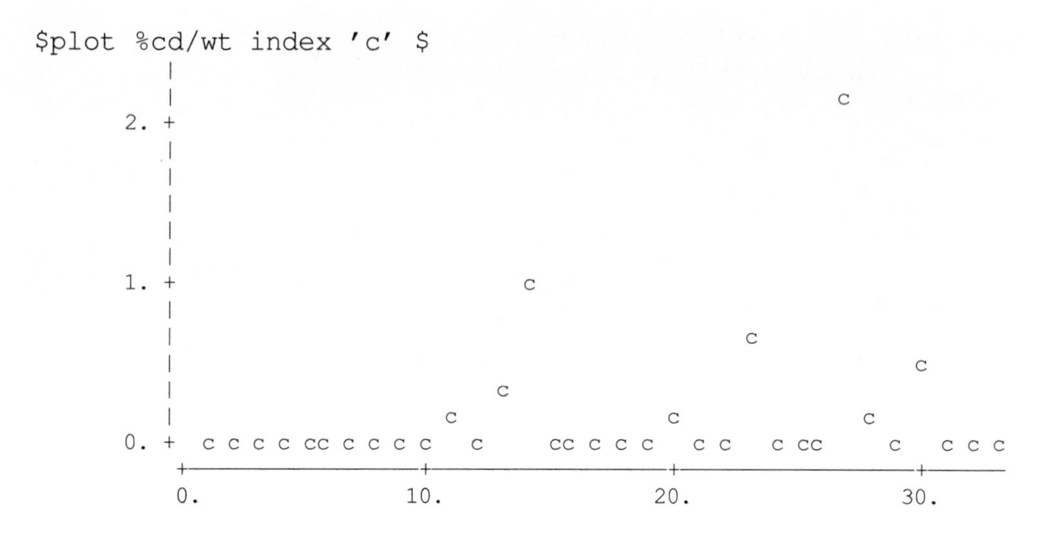


```
$plot %lv/wt index 'i' $
```

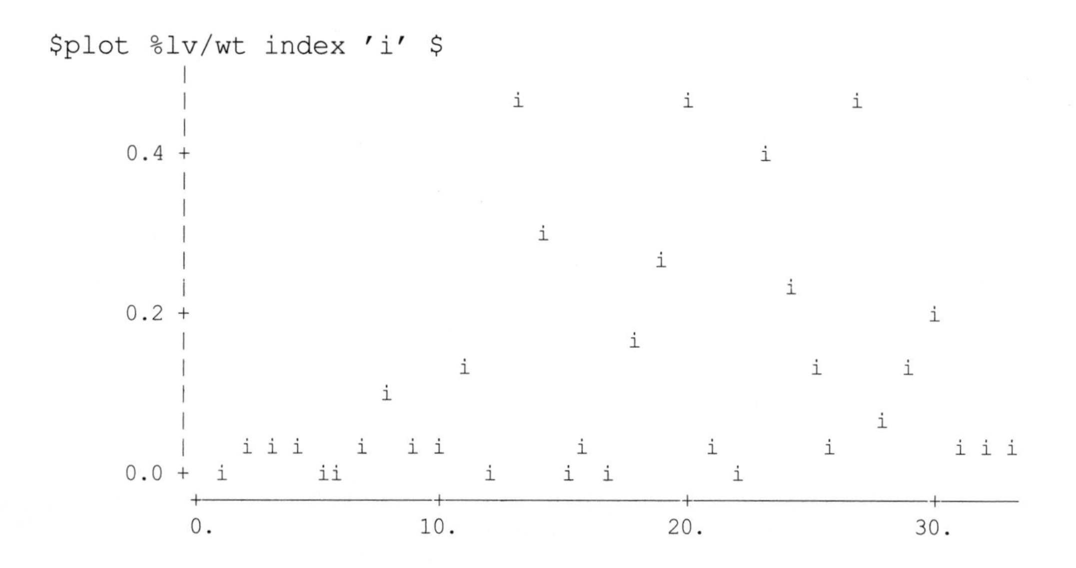

This time the Cook's distances are amplified with the 27th very extreme, but the range of influence values is reduced. As before it is instructive to obtain predicted values over the rainfall range and plot the data with this and the original curve for comparison. Again the excluded point is identified with a special symbol.

```
$cal excfac=1+wt $fac excfac 2 $
$predict (s=-1) rainf=rainp $
```

```
$plot (y=0,1,0.5 x=1.500,2.500,.500)
      ppos;predcv;%pfv rainf;rainp;rainp 'oxem' excfac;; $
```

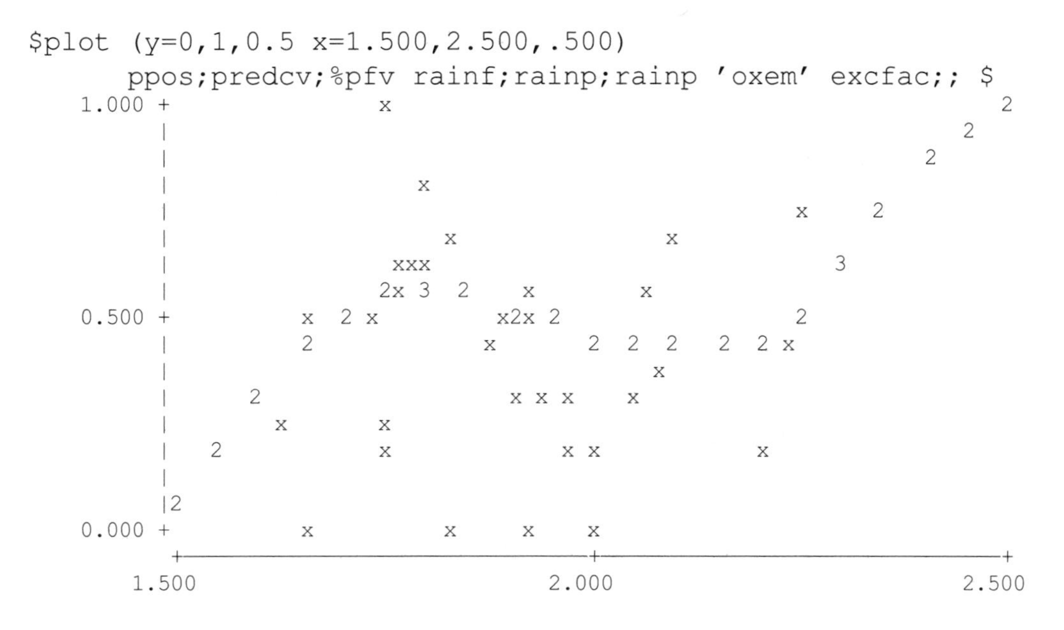

There is relatively little effect on the shape of the model. Finally omit both observations and observe the effect.

```
$cal wt=(index/=27)&(index/=34)   $fit . $dis e $
scaled deviance =  51.832 (change =    -10.74) at cycle 3
    residual df =  28      (change =     -1    ) from 32 observations
          estimate          s.e.      parameter
     1    -0.08576       0.08916        1
     2     -0.5571        0.6287        RAINF<1>
     3      0.3415        0.6668        RAINF<2>
     4       2.297        0.6248        RAINF<3>
scale parameter 1.000
$extr %vl %wt %cd $
```

Use `wt` to supppress the excluded observations and plot them again.

```
$plot %cd/wt index 'c' $
```

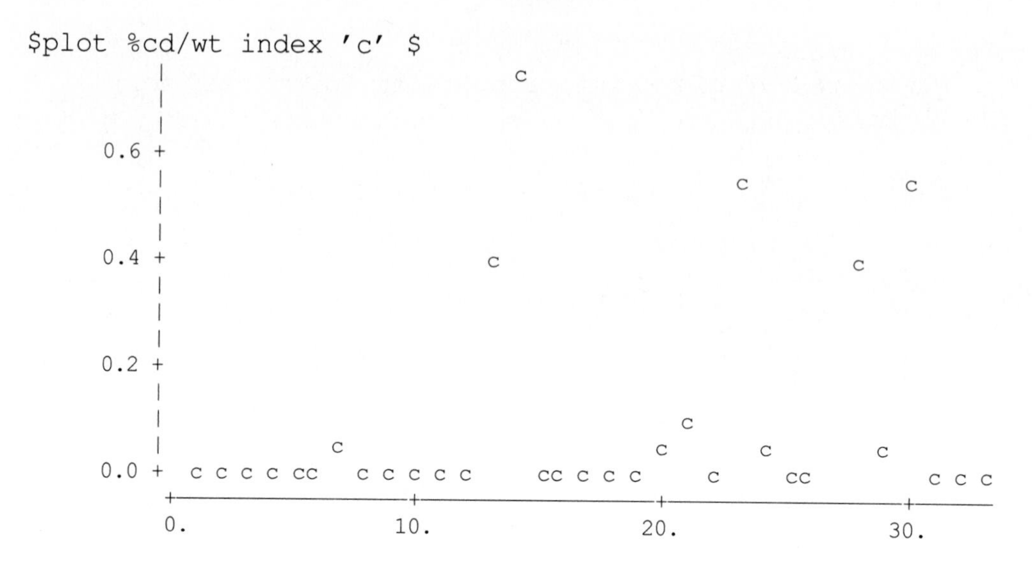


```
$plot %lv/wt index 'i' $
```

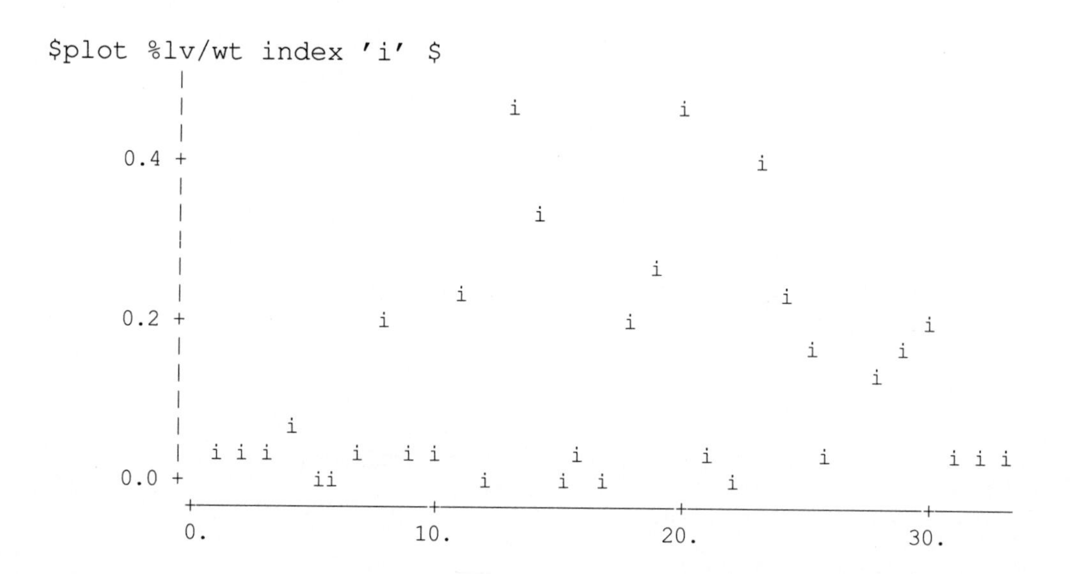

Both the 'influence' plots show a smaller spread of values. The fitted curves and excluded points are seen with the predicted value plots as before.

```
$cal excfac=1+wt $fac excfac 2 $
$predict (s=-1) rainf=rainp $
```

```
$plot (y=0,1,0.5 x=1.500,2.500,.500)
    ppos;predcv;%pfv rainf;rainp;rainp 'oxem' excfac;; $
```

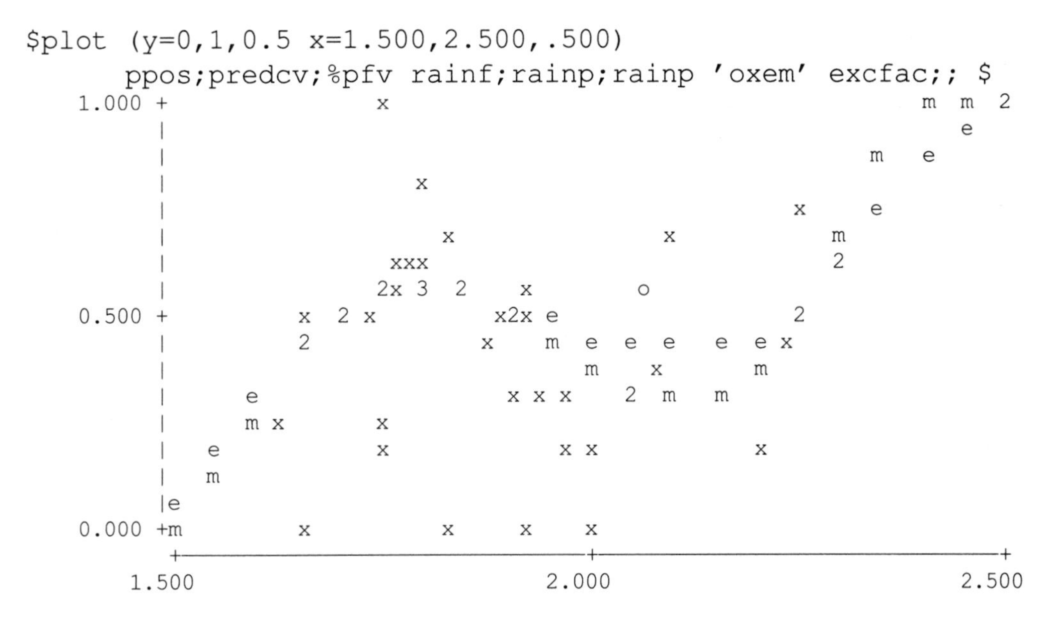

A graph indicating the excluded points with the before (solid) and after (dotted) curves is obtained by

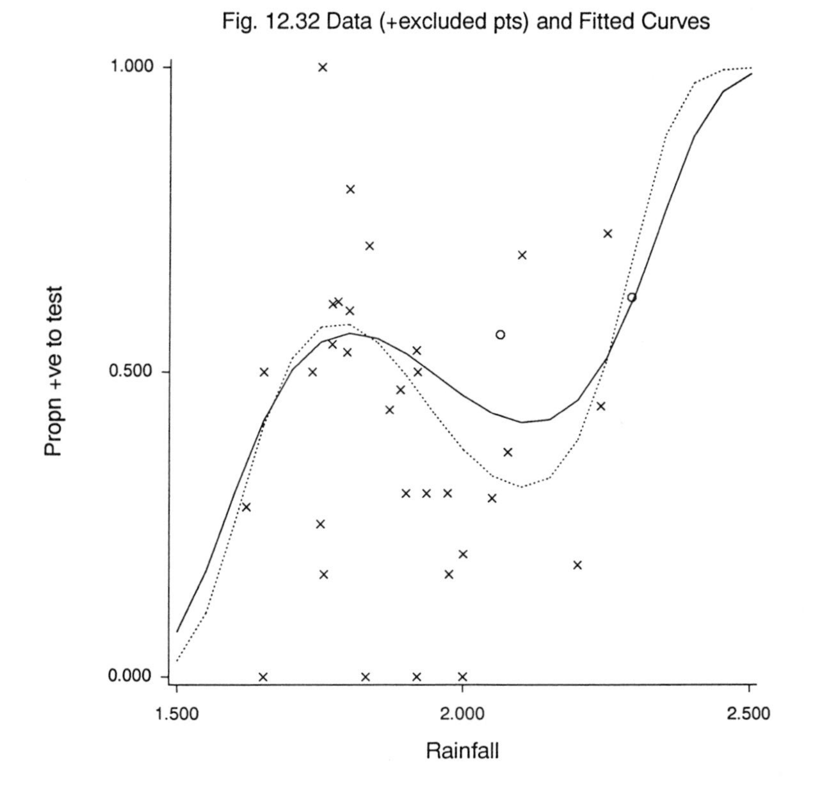

Fig. 12.32 Data (+excluded pts) and Fitted Curves

```
$gra (t='Fig. 12.32 Data (+excluded pts) and Fitted Curves'
      h=xlabel v=ylabel x=#xscale y=#yscale)
       ppos;predcv;%pfv rainf;rainp;rainp 5;1;10;11 excfac;; $
```

It seems that the shape of the curve was not determined by one or two 'rogue' points. There is definitely an odd structure to the data. If there is an underlying dose–response structure it appears to result from some sort of bimodal mixture of 'tolerance' distributions. This could mean that there are effectively two groups with different tolerance distributions, one in the lower part of the rainfall distribution and one in the upper part. On the other hand the influence of the data points is, to quite an extent determined by the number tested, that is, the binomial denominator. Since there is no information on the testing rate or the factors which influence it this dataset raises more questions than it answers. Obviously the correct explanation can only be established with further information.

12.2.1.5 Ungrouped data(y = 0 or 1): A study of vaccine efficacy

These data were obtained by observing a binary or dichotomous response in each of a sample of individuals. Data of this form are sometimes referred to as binary data. They arose from an investigation of an outbreak of a flu-like illness in a boarding school. The opportunity was taken to use this as an observational study of vaccine efficacy. (Source: Personal communication)

The information available for each individual included their vaccination history for 1989 and 1990, their school house, sex, and age.

The general question is: What does this outbreak tell us about the efficacy of flu vaccine and the benefits of repeating it at yearly intervals?

The main measure of the efficacy of a vaccine is how it affects the 'attack rate' (AR). That is the ratio of those who get the disease to the number at risk. The standard measure of efficacy is the proportional decrease in attack rate that a vaccine achieves. That is:

$$\frac{\text{AR without vaccine} - \text{AR with vaccine}}{\text{AR without vaccine}} = 1 - \text{RR}$$

where RR is the relative risk of attack with and without vaccination. The analysis needs to identify what can be deduced from these data on the efficacy of vaccination generally and for different vaccination histories. The need is to model the variation in numbers attacked to obtain estimates of the effect on AR of the various factors, in particular of vaccination, adjusting for other factors.

The variables are

1. Serial number;
2. Ill: 1=yes, 2=no;
3. sex: 1=M, 2=F;
4. Age (years);
5. House 1, ..., 8;
6. Vaccinated in 1990: 1=yes, 2=no;
7. Vaccinated in 1989: 1=yes, 2=no.

```
$data 470  ser_no ill sex age house vac90 vac89 $
```

We don't need the serial number. Declaring a list of the variables we do wish to input from which `ser_no` is excluded will allow us to read the appropriate subset of variables in the `DATA` list.

```
$list inp_vars=ill;sex;age;house;vac90;vac89 $
$echo off $read inp_vars
   59 2 2 12   1 2 2   58 2 2 12   1 1 2   57 1 2 12   1 1 2
   55 2 2 12   1 1 2   53 2 2 11   1 1 1   52 1 2 12   1 1 2
   51 1 2 10   1 1 2   50 1 2 12   1 2 2   49 2 2 11   1 1 2
   48 2 2 11   1 2 1   47 2 2 11   1 1 2   46 2 2 11   1 1 2
   45 2 2 11   1 1 2   44 2 2 11   1 1 2   43 2 2 11   1 1 2
      . . . . . . . . . . . . . . . . . . . . . . . . . .
      . . . . . . . . . . . . . . . . . . . . . . . . . .
```

The `house` and vaccination status variables are qualitative variables representing groups to be compared and therefore must be declared as factors.

```
$fac sex 2 house 8 vac90 2 vac89 2 $
```

The variable indicating whether the child was ill or not will be the outcome or *y* variable in the analysis and needs to be:

0. not ill, currently 2;
or 1. ill currently 1.

It is recoded by

```
$cal ill=2-ill $
```

There is one individual with unknown vaccination status. It is necessary to create a weight variable equal to 1 for each individual except the one with unknown vaccination status. The vaccination variables are coded awkwardly with 2. for no and 1. for yes and it is best to reverse this so model parameters represent the effect of vaccination.

```
$cal     wt=(vac89/=-1)&(vac90/=-1)     :     vac89=3-vac89     :
vac90=3-vac90$          :          vac89=%if(wt,vac89,1)          :
vac90=%if(wt,vac90,1) $
```

The houses are numbered 1 to 8, but four are all boys and four all girls. It is better to have them coded 1 to 4 within each sex category and then take the houses as nested within those categories. Thus

For Boys: recode houses 2,4,5,8 to 1,2,3,4
Girls: recode houses 1,3,6,7 to 1,2,3,4

This is conveniently achieved with the GROUP directive. First set–cut points on the 'house number' scale and identify code values for each interval defined by those cut-points. Values equal to a cut-point are taken to be in the upper interval. If 'below' and 'above' the range are to be considered as intervals then this must be indicated by an asterisk before and after the 'cut-point' vector in the GROUP directive.

```
$ass i=1,2...8 : val=1,1,2,2,3,3,4,4 $
```

The 'house' numbers are then recoded to the new values in hous4 which must then be declared as a factor with the appropriate number of levels, in this case 4. The GROUP directive does this automatically.

```
$group hous4=house intervals i * values val $
```

To look at the relationship between illness and vaccination status in the two years 1989 and 1990 the attack rates for the various vaccination categories are tabulated and plotted by

```
$tab the ill mean for vac89;vac90 with wt using n_ into p_ by
   vac8;vac9 $
$tpr p_;n_ vac8;vac9 $
    VAC9      1        2
 VAC8
    1 P_  0.4242  0.2290
      N_   99.00  131.00
    2 P_  0.2453  0.2634
      N_   53.00  186.00
$plot (x=0.5,2.5 y=0,0.6) p_ vac8 'ny' vac9 $
$del p_ n_ vac8 vac9$
    0.6 +
        |
        |
        |
        |           n
    0.4 +
        |
        |
        |                                         y
        |           y                             n
    0.2 +
        |
        |
        |
        |
    0.0 +
         +----------+----------+----------+----------+
        0.5        1.0        1.5        2.0        2.5
```

It seems reasonably clear that, ignoring the possible effects of other factors, any vaccine, last year, this year, or both, reduces the attack rate from about 40 per cent to about 25 per cent. It doesn't appear to matter if the vaccination is a year old and little appears to be gained by repeat vaccination.

However we should be careful not to jump to conclusions based on this marginal table. We need to assess this quantitatively and obtain estimates with confidence limits of the benefits of the various vaccination sequences adjusted for any possible effects due to other factors. These will arise as systematic differences in the 'attack rates' in the different school houses. These in turn may reflect the effects of sex, age, and contact with 'day' boys. To do this we must fit an appropriate sequence of models describing the variation in the dichotomous response variable `ill`.

The standard approach for dichotomous outcomes as here is to assume a binomial `ERROR` distribution for `ill` with the default logistic `LINK`. This is the classic logistic regression analysis in which dichotomous response variables are represented by models linear on the log(odds) scale. This will lead to estimates of odds ratios rather than the relative risks needed for estimates of efficacy, but it is the most efficient approach for testing hypotheses on the effects of the various factors. We will use this approach now and return to the estimation of efficacies later.

It is important to note that because the data are obtained from ungrouped individuals every y–value will be 0 or 1 and the denominator or number at risk will be 1. This causes no difficulty. A number at risk variable which has to be a vector of 1's needs to be calculated and identified in the binomial `ERROR` declaration. The deviances and degrees of freedom from the models tend to have rather large values and the residual plots from the 0/1 valued variable are rather uninformative, but otherwise the analysis proceeds exactly as for the grouped case where the numerators and denominators are potentially, and usually much greater than 1.

The model comparison results can be shown to be exactly the same as would be obtained from an equivalent grouped analysis when there are no measured covariates in the model as is the case here.

We proceed by declaring `ill` as the y variable, the `ERROR` as binomial with denominator `n` and `wt` as a weight to exclude the individual with unknown vaccination status.

```
$yvar ill $err b n $cal n=1 $
$weight wt $
```

Use the nesting operator to fit `hous4` effects within `sex` and fit a model with all possible two–factor interactions.

```
$fit (sex/hous4)*(vac89*vac90) $dis e $
scaled deviance =  500.73 at cycle 6
    residual df =  437     from 469 observations
          estimate          s.e.      parameter
     1   3.093e-16          1.000      1
     2      0.6061          1.121      SEX(2)
     3      -7.565          18.88      VAC89(2)
     4      -2.197          1.247      VAC90(2)
     5     -0.6931          1.173      SEX(1).HOUS4(2)
     6      -1.041          1.107      SEX(1).HOUS4(3)
     7     -0.6931          1.173      SEX(1).HOUS4(4)
     8     0.08701         0.8704      SEX(2).HOUS4(2)
     9      -1.012         0.8211      SEX(2).HOUS4(3)
    10     -0.9426         0.7749      SEX(2).HOUS4(4)
    11       6.554          18.91      SEX(2).VAC89(2)
    12       1.149          1.413      SEX(2).VAC90(2)
    13       8.782          18.91      VAC89(2).VAC90(2)
    14       7.565          18.93      SEX(1).HOUS4(2).VAC89(2)
    15       1.041          20.87      SEX(1).HOUS4(3).VAC89(2)
    16       7.411          18.90      SEX(1).HOUS4(4).VAC89(2)
    17     -0.3747          1.530      SEX(2).HOUS4(2).VAC89(2)
    18      0.7239          1.503      SEX(2).HOUS4(3).VAC89(2)
    19      -6.217          18.89      SEX(2).HOUS4(4).VAC89(2)
    20      0.9445          1.520      SEX(1).HOUS4(2).VAC90(2)
    21       1.293          1.710      SEX(1).HOUS4(3).VAC90(2)
    22       2.603          1.586      SEX(1).HOUS4(4).VAC90(2)
    23     -0.6568          1.132      SEX(2).HOUS4(2).VAC90(2)
    24      0.4418          1.094      SEX(2).HOUS4(3).VAC90(2)
    25     0.06268          1.049      SEX(2).HOUS4(4).VAC90(2)
    26      -8.245          18.95      SEX(2).VAC89(2).VAC90(2)
    27      -8.883          18.97      SEX(1).HOUS4(2).VAC89(2).VAC90(2)
    28      -3.202          20.95      SEX(1).HOUS4(3).VAC89(2).VAC90(2)
    29      -8.946          18.95      SEX(1).HOUS4(4).VAC89(2).VAC90(2)
    30       2.004          1.832      SEX(2).HOUS4(2).VAC89(2).VAC90(2)
    31     -0.5878          1.820      SEX(2).HOUS4(3).VAC89(2).VAC90(2)
    32       7.266          18.92      SEX(2).HOUS4(4).VAC89(2).VAC90(2)
scale parameter 1.000
```

The deviance is larger than the degrees of freedom, but it can be shown that the deviance from 0/1 'binomial' data is uninformative. There are no obviously significant interaction terms, but it is difficult to judge from such a complex model. The best approach is to test all the interactions together by fitting the model without them. However the `vac89.vac90` interaction is of specific interest so it should be retained.

```
$fit sex/hous4+vac89+vac90+vac89.vac90 $dis e $
scaled deviance =  515.36 at cycle 3
    residual df =  458     from 469 observations
          estimate          s.e.      parameter
     1     -0.6016         0.4785      1
     2       1.062         0.5043      SEX(2)
     3      -1.029         0.4068      VAC89(2)
     4      -1.076         0.3128      VAC90(2)
```

```
       5     -0.2989       0.5343       SEX(1).HOUS4(2)
       6     -0.7578       0.5860       SEX(1).HOUS4(3)
       7      0.6705       0.5292       SEX(1).HOUS4(4)
       8      0.1917       0.3842       SEX(2).HOUS4(2)
       9     -0.7433       0.3982       SEX(2).HOUS4(3)
      10     -0.6049       0.3955       SEX(2).HOUS4(4)
      11       1.266       0.4908       VAC89(2).VAC90(2)
scale parameter 1.000
```

The increase in deviance of 14.63 with 21 d.f. is an approximate chi-squared test of the interactions omitted. Since it is less than its degrees of freedom it is nowhere near significance – the *p*–value is

```
$cal 1-%chp(14.63,21) $
       0.8410
```

It appears that apart from the `vac89.vac90` interaction the others can be ignored.

The validity of such models is generally investigated by inspecting the leverage values –to check for unduly influential single or groups of observations – and by inspecting how the standardized residuals relate to the fitted values; – there should be no tendency for the variance to change with the fitted values (means) since the standardization will have removed that expected for the binomial distribution. However the fitted values involve the binomial denominator which may not always be constant. For this reason the residuals are plotted against the fitted proportions or the linear predictors which are the log(odds) of the fitted proportions. These plots also make it possible to check for systematic patterns which may indicate a model which is inappropriate for a number of other reasons.

```
$ext %lv $cal indx=%cu(1) $
$plot %lv indx 'i' $
```

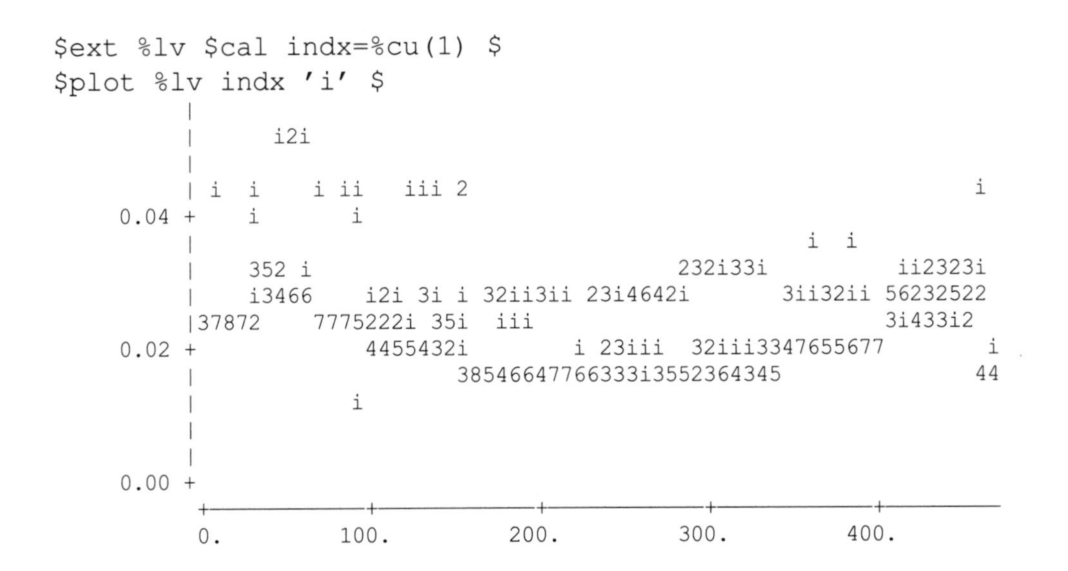

Unfortunately as this plot shows in ungrouped data analysis when every individual is either 0 or 1 the usual diagnostic plots are almost impossible to interpret.

```
$plot %rs %lp 'r' $
```

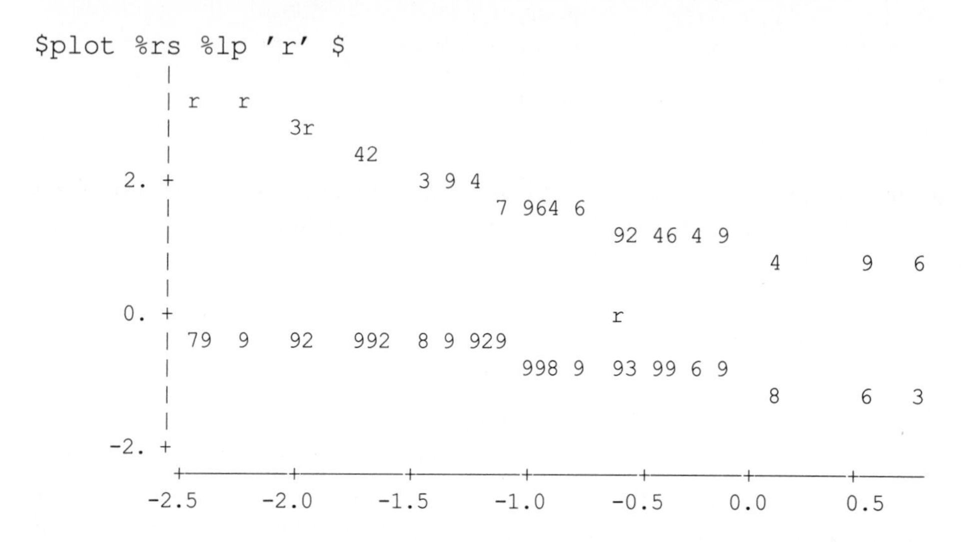

The residual plot against the values of the linear predictor is equally difficult to interpret because at each fitted value only two residual values occur: those for the 0's and those for the 1's. The data needs to be grouped to permit model checking with these methods. In cases where a possibly unique covariate value for each individual is involved grouping is not possible, but here where all the explanatory variables are factors it is relatively simple and generally advisable.

The first step is to aggregate the individual observations into grouped counts according to the multi-way classification represented by the factors of interest using `TABULATE`. This aggregation compresses the data so the number of units is reduced from the number of individuals to the number of cells in the classification. At the same time the appropriate reduced length explanatory variables have to be created which is most easily done with the `BY` keyword in the `TABULATE` directive. First create the number at risk in each cell for the binomial denominator. The weight variable to exclude the individual whose vaccination status was unknown is identified within the `TABULATE` directive by using the `WITH` keyword.

```
$tab for sex;hous4;vac90;vac89 with wt
     using nar_ by sex_;hous4_;vac90_;vac89_ $
```

The exercise then needs to be repeated to get the numbers ill for each cell in `ni_`. The tabulation is restricted to the ill individuals by using a modified weight variable set to 0 for those not ill.

```
$cal wti=wt*(ill==1) $
$tab for sex;hous4;vac90;vac89 with wti using ni_  $
```

The compressed data in tabular form is

```
$acc 4 0 $tpr ni_;nar_ sex_;hous4_;vac90_;vac89_ $acc 4 4 $
          VAC90_   1          2
          VAC89_   1    2     1    2
 SEX_ HOUS4_
    1      1 NI_   2.   0.    2.   3.
             NAR_  4.   2.   20.  11.
           2 NI_   4.   1.    3.   4.
             NAR_ 12.   3.   24.  35.
           3 NI_   6.   0.    1.   1.
             NAR_ 23.   9.    8.  19.
           4 NI_   4.   6.    3.   6.
             NAR_ 12.  20.    7.  17.
    2      1 NI_  11.   2.    9.   4.
             NAR_ 17.   5.   23.  14.
           2 NI_   6.   2.    4.  15.
             NAR_  9.   6.   15.  28.
           3 NI_   4.   2.    4.   7.
             NAR_ 10.   6.   15.  34.
           4 NI_   5.   0.    4.   9.
             NAR_ 12.   2.   19.  28.
```

It is convenient to inspect the way attack rates vary by sex, house, and vaccination status with a plot. The attack rates are calculated as follows

```
$cal att_rate=ni_/nar_ $
```

A factor to identify the four vaccination patterns is calculated by

```
$cal vacc4=2*(vac90_-1)+vac89_ $fac vacc4 4 $
```

The attack rates by sex, house and vaccination are

```
$tpr att_rate sex_;hous4_/vacc4 $
       VACC4     1        2        3        4
 SEX_ HOUS4_
    1      1  0.5000  0.0000  0.1000  0.2727
           2  0.3333  0.3333  0.1250  0.1143
           3  0.2609  0.0000  0.1250  0.0526
           4  0.3333  0.3000  0.4286  0.3529
    2      1  0.6471  0.4000  0.3913  0.2857
           2  0.6667  0.3333  0.2667  0.5357
           3  0.4000  0.3333  0.2667  0.2059
           4  0.4167  0.0000  0.2105  0.3214
```

and plots against vaccination status identifying houses are obtained separately for the males and females by

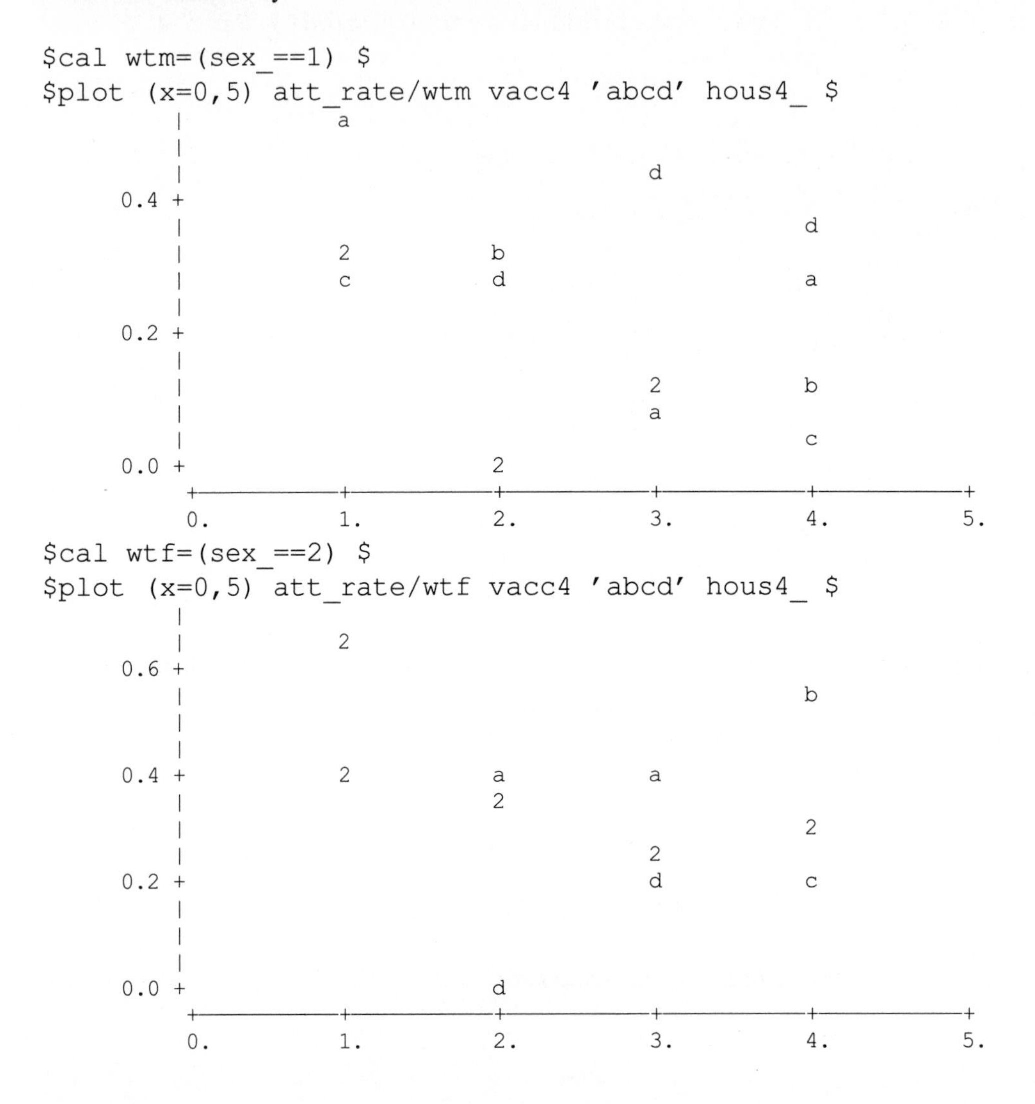

```
$cal wtm=(sex_==1) $
$plot (x=0,5) att_rate/wtm vacc4 'abcd' hous4_ $
         |         a
         |
         |                                  d
   0.4 + 
         |                                          d
         |         2          b
         |         c          d                     a
         |
   0.2 +
         |
         |                              2           b
         |                              a
         |                                          c
   0.0 +                    2
         +---------+---------+---------+---------+---------+
         0.        1.        2.        3.        4.        5.
$cal wtf=(sex_==2) $
$plot (x=0,5) att_rate/wtf vacc4 'abcd' hous4_ $
         |
         |         2
   0.6 +
         |                                          b
         |
         |
   0.4 +         2          a          a
         |                    2
         |                                          2
         |                              2
   0.2 +                              d           c
         |
         |
         |
   0.0 +                    d
         +---------+---------+---------+---------+---------+
         0.        1.        2.        3.        4.        5.
```

Equivalent high–resolution graphs separately for the two sexes and one overall title can be obtained with

```
$layout 0,0 1,1 $
$gstyle 1 1 1 s 1 $
$gstyle 2 1 2 s 2 $
$gstyle 3 1 3 s 3 $
$gstyle 4 1 4 s 4 $
$ass pens=1,2,3,4 : gy_=0,0,0 : gx_=1,1,1 $
```

```
$gra (t='Fig. 12.33 Attack Rates by Vaccination Status'
     style=-1 pause=no)
     gy_/gy_ gx_  $

$layout 0,0 .48,.8 $
$graph (t='- Boys -' h='Vaccination Status' pause=no
       v='Attack Rates' ref=no  x=0,5,1 y=0,1.1,.2 )
       att_rate/wtm   vacc4  pens   hous4_  $

$layout .52,0 1,.8 $
$graph (t='- Girls -' h='Vaccination Status'
                          ref=no  x=0,5,1 y=0,1.1,.2 )
       att_rate/wtf   vacc4  pens   hous4_  $
```

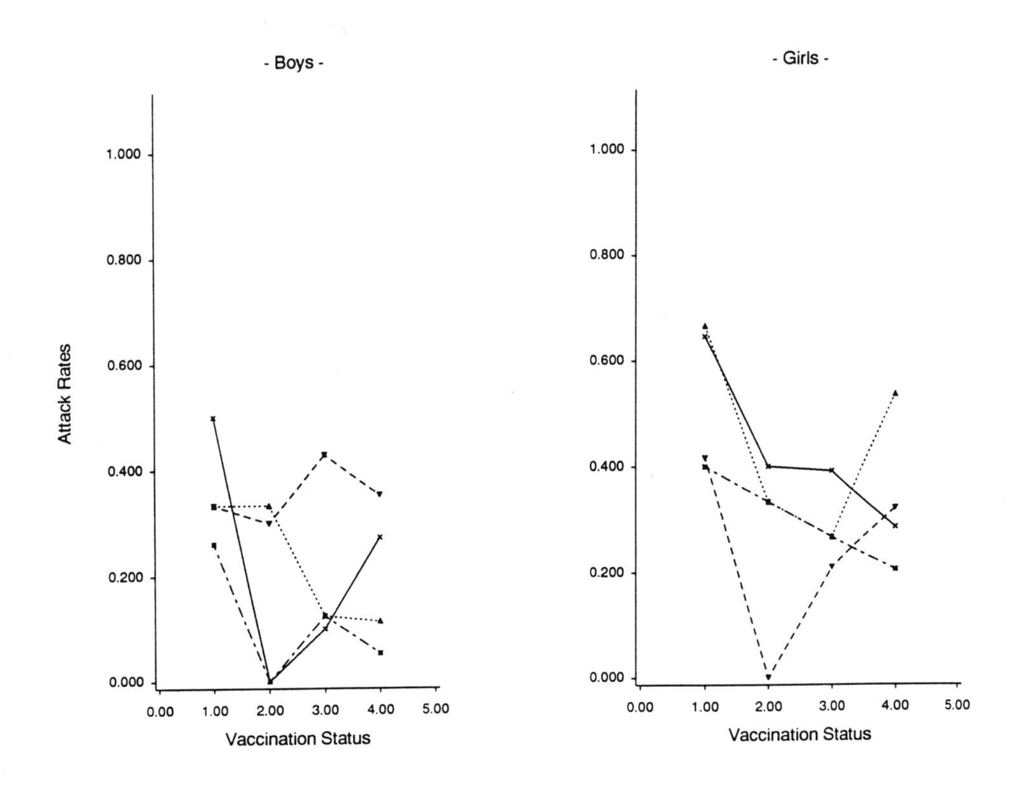

To fit the model it is now necessary to cancel the earlier weight declaration, and, just in case, create a new weight variable to exclude any cells with zero numbers at risk. The new *y* variable has to be declared and the binomial error to identify the new denominator variable.

```
$number n_units $cal n_units=%len(nar_) : wt_gpd=(nar_/=0) $
$un n_units $
$weight wt_gpd $
$yvar ni_ $error b nar_ $
```

```
$fit sex_/hous4_+vac89_+vac90_ +vac89_.vac90_ $dis e $
scaled deviance =  14.644 at cycle 3
    residual df =  21
          estimate         s.e.      parameter
     1     -0.6016       0.4784       1
     2       1.062       0.5042       SEX_(2)
     3      -1.029       0.4070       VAC89_(2)
     4      -1.076       0.3129       VAC90_(2)
     5     -0.2989       0.5342       SEX_(1).HOUS4_(2)
     6     -0.7578       0.5866       SEX_(1).HOUS4_(3)
     7      0.6705       0.5292       SEX_(1).HOUS4_(4)
     8      0.1917       0.3843       SEX_(2).HOUS4_(2)
     9     -0.7433       0.3983       SEX_(2).HOUS4_(3)
    10     -0.6049       0.3956       SEX_(2).HOUS4_(4)
    11       1.266       0.4910       VAC89_(2).VAC90_(2)
scale parameter 1.000
```

There is no sign of overdispersion now since the deviance is actually less than its degrees of freedom. But note that the parameter estimates are exactly as they were before with, apart from rounding errors, the same standard errors. Checks on the influence of the data points, on the mean–variance relationship and general fit of the model are made as before.

```
$ext %lv $cal indx_gpd=%ind(%lv) $plot %lv indx_gpd $
```

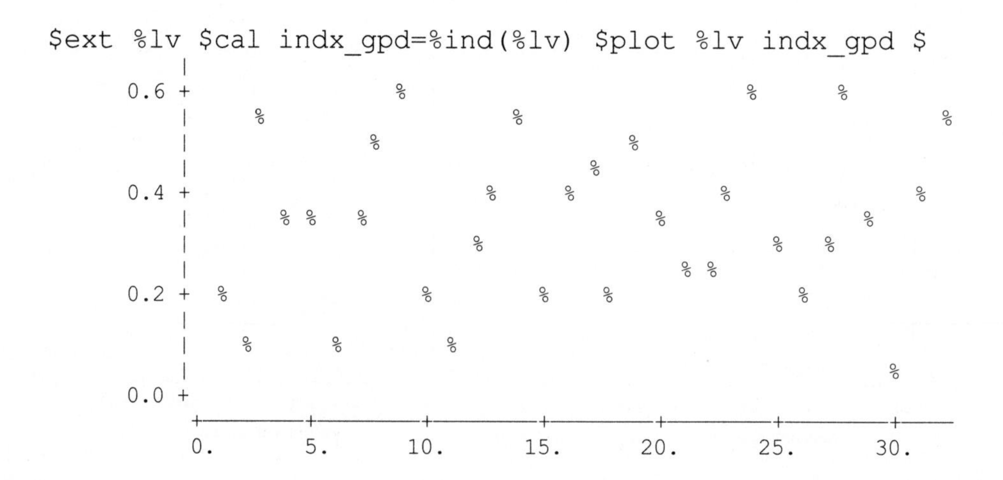

And now the influence is evenly spread among the grouped observations.

```
$plot %rs %lp 'r' $
```

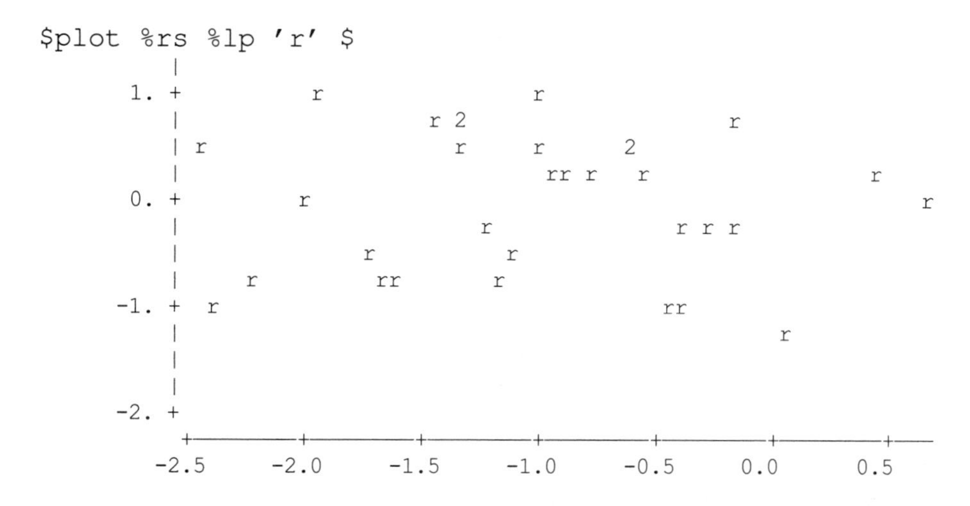

The residuals appear to have a uniform spread with no suggestion of any lack of fit.

The parameters of the model are differences on the log(odds) scale from the odds of being 'attacked', that is, the logs of odds ratios. Consequently they are not directly interpretable in terms of attack rates. Nonetheless it can be deduced that the `vac89_.vac90_` interaction is reflecting the pattern in the earlier plot of attack rates. Among those not vaccinated in 1990 the attack rate is markedly decreased from those not vaccinated in 1989 to those who were. Among those who were vaccinated in 1990 there was little difference between those who were and were not in 1989. The interaction reflects the fact that the effect of vaccination in 1990 cannot possibly be independent of whether they were vaccinated in 1989 or not, unless the 1989 vaccination has no effect at all. Sex and house appear to have effects, but these are not really of interest. They reflect variation in attack rates which needs to be taken into account in the estimation of vaccination effects. Equally it is not p–values that matter, but the magnitudes of the vaccine effects and confidence intervals that are required.

To get the fullest benefit from these results it is necessary to identify the useful questions that can be addressed by comparisons of the odds ratios in groups with the various different vaccination histories. Because of the interaction it is not possible to make simple additive generalizations about the effects of vaccination in the two years. It is necessary to consider separately the four groups, that is, never vaccinated, vaccinated in 1989 only, vaccinated in 1990 only, and vaccinated both years.

The interesting practical questions are:

1. How effective is one vaccination in the previous year?
   This is the (89 yes/90 no) versus (89 no/90 no) comparison.

2. How effective is recent vaccination, this year?
   This is the (89 no/90 yes) versus (89 no/90 no) comparison.

3. How effective are two vaccines, one in each year?
   This is the (89 yes/90 yes) versus (89 no/90 no) comparison.

In addition to these questions it would be interesting to know whether a recent vaccination, that is, in 1990, is more effective than one a year earlier and it may be of interest to know whether two vaccinations are of more benefit than either. These questions could be addressed using the current model and quantitative estimates with confidence limits obtained for the relevant effects. However, although parameters 3 and 4 are the appropriate measures for answering questions 1 and 2, the other questions can only be addressed using slightly obscure combinations of these and the interaction parameter. It is more sensible to re-fit the model with a different parametrization. This is done most conveniently using `vacc4` the single factor with four levels describing all four vaccination patterns created earlier for the plots. The levels are:

1. not vaccinated either year;
2. vaccinated 1989 only;
3. vaccinated 1990 only;
4. vaccinated both years.

With 1 as the reference category this will produce a specific parameter for the comparison of 'vaccination in both years' with 'no vaccination', that is, the estimate required for question 3 above. Changing the reference category will produce parameters specific for the other questions on how different vaccination sequences compare, or they can be obtained as simple differences of the parameters obtained with the default reference category (level 1). The appropriate model is fitted by

```
$fit sex_/hous4_+vacc4 $dis e $
scaled deviance =  14.644 at cycle 3
    residual df =  21
          estimate          s.e.     parameter
    1      -0.6016        0.4784      1
    2        1.062        0.5042      SEX_(2)
    3       -1.029        0.4070      VACC4(2)
    4       -1.076        0.3129      VACC4(3)
    5      -0.8385        0.2837      VACC4(4)
    6      -0.2989        0.5342      SEX_(1).HOUS4_(2)
    7      -0.7578        0.5866      SEX_(1).HOUS4_(3)
    8       0.6705        0.5292      SEX_(1).HOUS4_(4)
    9       0.1917        0.3843      SEX_(2).HOUS4_(2)
   10      -0.7433        0.3983      SEX_(2).HOUS4_(3)
   11      -0.6049        0.3956      SEX_(2).HOUS4_(4)
scale parameter 1.000
```

Because it is the same model with a different parametrization the scaled deviance is the same as before.

The estimates and confidence limits for the odds ratios required to answer the questions above are then obtained by:

1. Vaccination in 1989 compared to 'no vaccination':

```
$number es se  lo hi eff eflo efhi $
$extract %pe %se $
$cal es=%exp(%pe(3))
   : lo=%exp(%pe(3)-1.96*%se(3)): hi=%exp(%pe(3)+1.96*%se(3)) $
$pr ;;'Odds Ratio'es' with 95% CLs'lo' to'hi $
```

```
Odds Ratio  0.3574 with 95% CLs  0.1610 to  0.7936
```

The attack rate is reduced by the equivalent of an odds ratio of 0.3574 with 95 per cent confidence limits which come nowhere near including 1. The effect is clearly significant.

2. Vaccination in 1990 compared to 'no vaccination':

```
$cal es=%exp(%pe(4))
   : lo=%exp(%pe(4)-1.96*%se(4)): hi=%exp(%pe(4)+1.96*%se(4))
$pr ;;'Odds Ratio'es' with 95% CLs'lo' to'hi $
```

```
Odds Ratio  0.3411 with 95% CLs  0.1847 to  0.6299
```

Very similar results to those for vaccination in 1989 with slightly narrower confidence limits.

3. Vaccination both years compared to `no vaccination':

```
$cal es=%exp(%pe(5))
   : lo=%exp(%pe(5)-1.96*%se(5)): hi=%exp(%pe(5)+1.96*%se(5))
$pr ;;'Odds Ratio'es' with 95% CLs'lo' to'hi $
```

```
Odds Ratio  0.4324 with 95% CLs  0.2480 to  0.7539
```

Again the same sort of pattern although the odds ratio is not quite as small so the effect appears to be slightly less than for single vaccinations.

From these results it is reasonably clear that any pattern of vaccination is better than none, but that there is not much difference between the effects of the different vaccination sequences. Making level 2 the reference category and re-fitting allows a simple comparison of 1990 versus 1989 and of both versus 1989 alone.

```
$fac vacc4 4(2) $fit . $dis e $
scaled deviance =  14.644 (change =     0.) at cycle 3
    residual df =  21      (change =     0 )
          estimate         s.e.      parameter
     1      -1.630        0.5520      1
     2       1.062        0.5042      SEX_(2)
     3       1.029        0.4070      VACC4(1)
     4     -0.0467        0.4141      VACC4(3)
     5      0.1903        0.3880      VACC4(4)
     6     -0.2989        0.5342      SEX_(1).HOUS4_(2)
     7     -0.7578        0.5866      SEX_(1).HOUS4_(3)
     8      0.6705        0.5292      SEX_(1).HOUS4_(4)
     9      0.1917        0.3843      SEX_(2).HOUS4_(2)
    10     -0.7433        0.3983      SEX_(2).HOUS4_(3)
    11     -0.6049        0.3956      SEX_(2).HOUS4_(4)
scale parameter 1.000
```

4. Vaccination in 1990 only compared to vaccination in 1989 only

```
$extract %pe %se $
$cal es=%exp(%pe(4))
   : lo=%exp(%pe(4)-1.96*%se(4)): hi=%exp(%pe(4)+1.96*%se(4)) $
$pr ;;'Odds Ratio'es' with 95% CLs'lo' to'hi $

Odds Ratio  0.9543 with 95% CLs  0.4239 to   2.149
```

The estimate is very close to 1 so there is no evidence that the two vaccinations have any different effect.

5. Vaccination in both 1989 and 1990 compared to vaccination in 1989 only

```
$cal es=%exp(%pe(5))
   : lo=%exp(%pe(5)-1.96*%se(5)): hi=%exp(%pe(5)+1.96*%se(5)) $
$pr ;;'Odds Ratio'es' with 95% CLs'lo' to'hi $

Odds Ratio   1.210 with 95% CLs  0.5654 to   2.588
```

Again no evidence of any difference.

The value greater than 1 suggests that vaccination in both years is worse than only in 1989. However the confidence limits are very wide and it is far from significant and well within what could occur by chance.

Finally

6. Vaccination in both 1989 and 1990 compared to vaccination in 1990 only. A parameter for this is not directly available from this model. It could be obtained by changing the reference category and again re-fitting the model, but it is also possible to use the difference between parameters 4 and 5. The only complication is that the standard error, although it can be `DISPLAY`ed using option `S`, is not directly available for calculations. However if that is what is required the s.e. can be obtained from the appropriate parameter estimate variances and covariances in the covariance matrix. The lower triangle of this can be extracted into the system vector `%vc`. The elements of this, with the variances on the diagonal and the variances and covariance needed here in parentheses, are:

```
par 1    1
par 2    2  3
par 3    4  5  6
par 4    7  8  9 (10)
par 5   11 12 13 (14) (15)

  . . . . . .
```

```
$extract %vc $cal se=%sqrt(%vc(10)-2*%vc(14)+%vc(15)) $
```

So the standard error of the difference between parameters 4 and 5 is

```
$pr 's.e. ='se $
s.e. =  0.2809
```

The standard errors of the differences obtained by using the `DISPLAY` directive with option `S` are

```
$display s $
standard errors of parameter estimate differences
  1       0.000
  2       0.9653      0.000
  3       0.8437      0.6508      0.000
  4       0.8686      0.6380      0.3129      0.000
  5       0.8384      0.6260      0.2837      0.2809      0.000
  6       0.9767      0.4203      0.6798      0.6623      0.6690      0.000
```

```
 7        1.018      0.4779      0.7300      0.6797      0.6896      0.5081
 8        1.013      0.4082      0.6210      0.5882      0.6009      0.4443
 9       0.6784      0.7404      0.5423      0.5501      0.5559      0.6554
10       0.6841      0.7472      0.5615      0.5594      0.5709      0.6623
11       0.6714      0.7460      0.5743      0.5751      0.5768      0.6619
              1           2           3           4           5           6
 7        0.000
 8       0.4920       0.000
 9       0.7024      0.6506       0.000
10       0.7074      0.6581      0.3921       0.000
11       0.7071      0.6610      0.3928      0.4056       0.000
              7           8           9          10          11
```

The standard error we require is in row 5 and column 4 of this triangular matrix which we can see confirms that the calculations are correct. Using this s.e. the required test of the difference in the effect of vaccination in 1989 and 1990 against vaccination in 1990 alone is

```
$cal es=%pe(5)-%pe(4)
   : lo=%exp(es-1.96*se): hi=%exp(es+1.96*se)  :   es=%exp(es) $

$pr ;;'Odds Ratio'es' with 95% CLs'lo' to'hi $
Odds Ratio    1.268 with 95% CLs  0.7309 to   2.198
```

Again no evidence of a genuine difference in efficacy.

This more or less covers all the useful questions that need be asked on the effects of vaccination although if it can be assumed that all three vaccination sequences were equally effective then the best estimate of the effect would be obtained by comparing the three groups together against the non-vaccinated group. We will do this below in the sequence to estimate efficacy. The standard definition of 'vaccine efficacy' is such that it cannot readily be estimated from logistic regression using models linear on the log(odds) scale. The simplest solution to this problem is to use a different `LINK` function and work with models linear on the log scale.

The standard logistic regression analysis, as above, provides odds ratios as estimates of the effect of being in one vaccination group rather than another. This is usually the most sensible general measure for comparing risks or attack rates because it is generally more likely that the odds ratio, rather than the relative risk (RR), will remain constant as attack rates change. However for the efficacy we need (1 – RR) and RR, the relative risk, is not readily obtained from models linear on the logistic scale. The simplest solution is to change the `LINK`. Although the default link for the binomial response distribution is the logistic function it is perfectly possible and reasonable to use a log link if appropriate to a particular problem as here and it can be shown to give a model which fits the data adequately.

Using this approach involves assuming that the various effects, including those of vaccination, are to alter the risks of infection, and thus the attack rates, by a constant proportion. This is the same as saying that the models, for the risk per person, are linear on the log scale. Differences on this scale, which is what the parameters then represent, are the logs of risk ratios, that is, log(RR), and are modelled as constant at all risk levels. As a result vaccine efficacy, in terms of relative risks, can be estimated directly.

```
$factor vacc4 4(1) $
$yvar ni_ $err B nar_ $link L $
$fit sex_/hous4_+vacc4 $dis e $
scaled deviance =  15.760 at cycle 4
    residual df =  21
          estimate          s.e.      parameter
     1      -1.149        0.3518       1
     2       0.6962       0.3628       SEX_(2)
     3      -0.5834       0.2635       VACC4(2)
     4      -0.6376       0.1916       VACC4(3)
     5      -0.4515       0.1618       VACC4(4)
     6      -0.2409       0.4235       SEX_(1).HOUS4_(2)
     7      -0.5441       0.4679       SEX_(1).HOUS4_(3)
     8       0.4249       0.3889       SEX_(1).HOUS4_(4)
     9       0.1055       0.1930       SEX_(2).HOUS4_(2)
    10      -0.4767       0.2518       SEX_(2).HOUS4_(3)
    11      -0.3713       0.2399       SEX_(2).HOUS4_(4)
scale parameter 1.000
```

As for the logistic link fit there is no sign of overdispersion with the deviance less than its degrees of freedom. The same checks are needed for the data points, the mean–variance relationship and the general fit of the model.

```
$ext %lv $cal indx_gpd=%ind(%lv) $plot %lv indx_gpd $
```

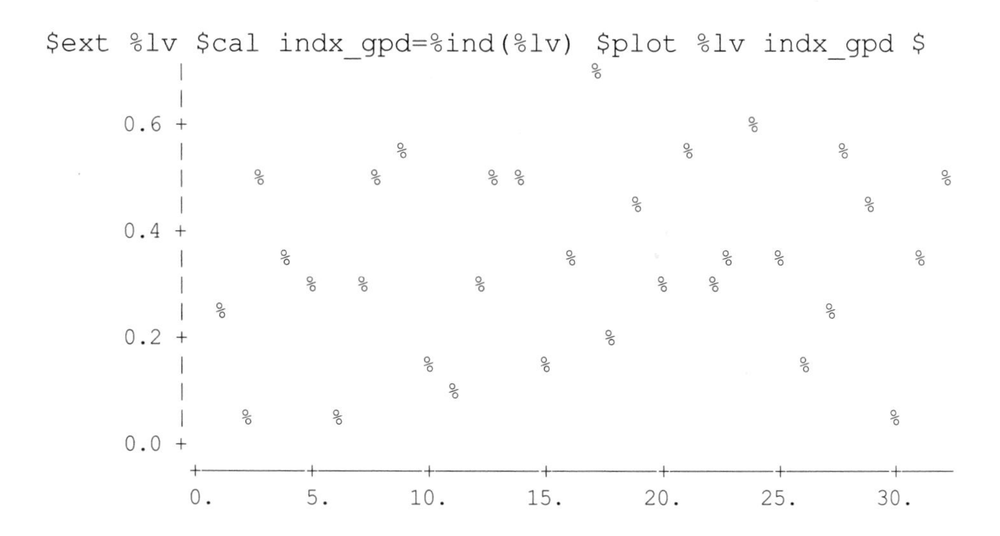

This is as well behaved as for the binomial model.

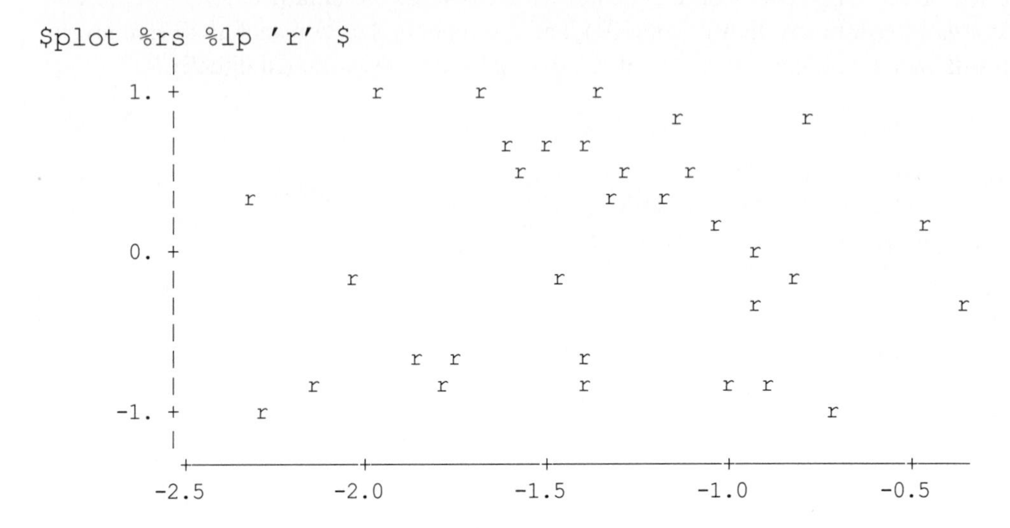

These residuals are more or less uniformly distributed about zero which indicates that this model is appropriate.

The questions are the same as before and because the model parametrization is unaltered the same functions of the parameters give the relative risks where before they gave the odds ratios.

The estimates and confidence limits for the relative risks required to answer the questions are obtained by:

1. Vaccination in 1989 compared to 'no vaccination':

```
$number eff eflo efhi $
$extract %pe %se $
$cal es=%exp(%pe(3))
   : lo=%exp(%pe(3)-1.96*%se(3)) :  hi=%exp(%pe(3)+1.96*%se(3))
   : eff=(1-es)*100 : eflo=(1-hi)*100 : efhi=(1-lo)*100 $
```

The relative risk estimate with 95 per cent confidence limits is

```
$pr 'RR ='es' with 95% CLs'lo'  to'hi $
RR =  0.5580 with 95% CLs  0.3329  to  0.9354
```

The estimate of efficacy with 95 per cent confidence limits is therefore

```
$pr ' %Efficacy ='eff' with 95% CLs'eflo'  to'efhi $
 %Efficacy =   44.20 with 95% CLs   6.465  to   66.71
```

The relative risk confidence interval does not quite include 1 so this analysis indicates, as before, that the effect of vaccination in 1989 just reaches significance. The efficacy is estimated as 44.2 per cent and the results are not inconsistent with a figure as high as 66.7 per cent or as low as 6.5 per cent. Now assign these and the other results to vectors for use in a later graph.

```
$ass cls2=eflo,efhi : cls=cls2 : effs=1,eff $
```

2. Vaccination in 1990 compared to 'no vaccination':

```
$cal es=%exp(%pe(4))
   : lo=%exp(%pe(4)-1.96*%se(4)) :  hi=%exp(%pe(4)+1.96*%se(4))
   : eff=(1-es)*100 : eflo=(1-hi)*100 : efhi=(1-lo)*100 $
```

The relative risk estimate with 95 per cent confidence limits is

```
$pr 'RR ='es' with 95% CLs'lo'  to'hi $
RR =  0.5286 with 95% CLs  0.3631  to  0.7695
```

The estimate of efficacy with 95 per cent confidence limits is therefore

```
$pr ' %Efficacy ='eff' with 95% CLs'eflo'  to'efhi $
 %Efficacy =    47.14 with 95% CLs    23.05  to    63.69
```

```
$ass cls3=eflo,efhi : cls=cls,cls3 : effs=effs,eff $
```

The efficacy is similar, but the limits are rather narrower. They are far from including zero so vaccination in 1990 is certainly protective.

3. Vaccination both years compared to 'no vaccination':

```
$cal es=%exp(%pe(5))
   : lo=%exp(%pe(5)-1.96*%se(5)) :  hi=%exp(%pe(5)+1.96*%se(5))
   : eff=(1-es)*100 : eflo=(1-hi)*100 : efhi=(1-lo)*100 $
```

The relative risk estimate with 95 per cent confidence limits is

```
$pr 'RR ='es' with 95% CLs'lo'  to'hi $
RR =  0.6367 with 95% CLs  0.4637  to  0.8743
```

The estimate of efficacy with 95 per cent confidence limits is therefore

```
$pr ' %Efficacy ='eff' with 95% CLs'eflo'  to'efhi $
 %Efficacy =    36.33 with 95% CLs    12.57  to    53.63
```

```
$ass cls4=eflo,efhi : cls=cls,cls4 : effs=effs,eff $
```

For vaccination in both years the estimated efficacy is slightly lower, but the difference is very small compared to the width of the confidence interval. The limits again exclude 1 so it reaches significance.

From these results it is reasonably clear that any pattern of vaccination is better than none, but that there is not much difference between the effects of the different vaccination sequences.

Finally to obtain an overall estimate the efficacy of any vaccination against none we need a two–level factor

1. not vaccinated; 2. vaccinated 1989, 1990 or both.

```
$cal any_vacc=1+((vac89_==2)?(vac90_==2)) $fac any_vacc 2 $
```

```
$fit sex_/hous4_+any_vacc $dis e $
scaled deviance =  16.719 at cycle 4
    residual df =  23
          estimate        s.e.      parameter
     1      -1.183       0.3521      1
     2       0.7210      0.3638      SEX_(2)
     3      -0.5363      0.1398      ANY_VACC(2)
     4      -0.1979      0.4235      SEX_(1).HOUS4_(2)
     5      -0.5032      0.4667      SEX_(1).HOUS4_(3)
     6       0.4582      0.3833      SEX_(1).HOUS4_(4)
     7       0.1150      0.1921      SEX_(2).HOUS4_(2)
     8      -0.4475      0.2493      SEX_(2).HOUS4_(3)
     9      -0.3586      0.2391      SEX_(2).HOUS4_(4)
scale parameter 1.000
```

Thus the estimate of efficacy for any vaccination sequence with 95 per cent confidence limits is

```
$extract %pe %se $
$cal es=%exp(%pe(3))
   : lo=%exp(%pe(3)-1.96*%se(3)) :  hi=%exp(%pe(3)+1.96*%se(3))
   : eff=(1-es)*100 : eflo=(1-hi)*100 : efhi=(1-lo)*100 $
```

The relative risk estimate with 95 per cent confidence limits is

```
$pr 'RR ='es' with 95% CLs'lo'  to'hi $
RR =  0.5849 with 95% CLs  0.4447  to  0.7694
```

The estimate of efficacy with 95 per cent confidence limits is therefore

```
$pr ' %Efficacy ='eff' with 95% CLs'eflo'  to'efhi $
 %Efficacy =   41.51 with 95% CLs   23.06  to   55.53
```

```
$ass cls5=eflo,efhi : cls=cls,cls5 : effs=effs,eff $
```

A plot of the efficacy estimates with their 95 per cent confidence limits is obtained as follows

```
$ass v_status=1,2...5 : v2_=2,2 : v3_=3,3 : v4_=4,4 : v5_=5,5
      : v_cls=v2_,v3_,v4_,v5_ $
$plot (x=0,6) cls;effs v_cls;v_status '+e' $
       |
       |                   +         +
  60.  +
       |                                       +         +
       |
       |                   e         e
  40.  +                                                 e
       |                                       e
       |
       |                             +                   +
  20.  +
       |                                       +
       |
       |                   +
   0.  +         e
       +---------+---------+---------+---------+---------+---------+-
      0.0       1.0       2.0       3.0       4.0       5.0       6.0
```

The high–resolution equivalent is

```
$layout 0,0 1,1 $
$ass cl_set=1,1,2,2,3,3,4,4 $fac cl_set 4 $

$graph (t='Fig. 12.34 Efficacies and 95% CLs' vla='Efficacy'
        p=no y=0,120,20
        h='No Vacc    1989      1990     89&90     Any' )
        cls;cls2;cls3;cls4;cls5;effs
v_cls;v2_;v3_;v4_;v5_;v_status
        6;10;14;15;16;10  $
$gtext (m=p l=r) 120/1.2 2 'By Vaccination status' $
```

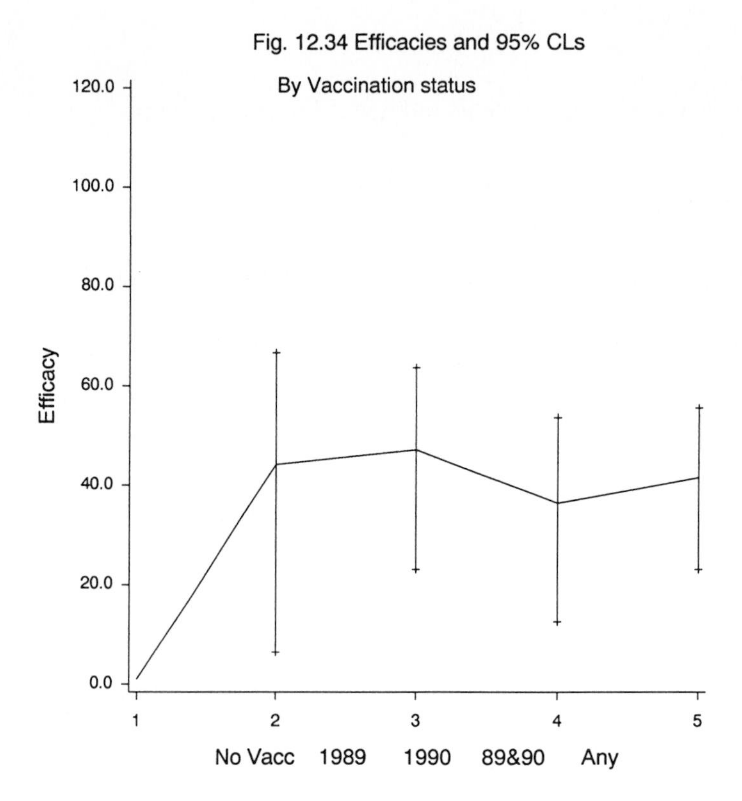

The overall efficacy of any vaccine versus none is 41.5 per cent with confidence limits of 23.1 per cent to 55.5 per cent. Pooling the three groups has narrowed the confidence interval, but the estimate of efficacy is somewhat disappointing. In theory these vaccines should, for the right strains of the flu virus, be about 90 to 95 per cent effective. However without data on the antibodies in the patients blood serum it is extremely difficult, especially in a school, to avoid false positive diagnoses. As a result the attack rate in the vaccinated group may be inflated which makes the efficacy looks much worse. These results must be treated with caution as estimates of efficacy, but they certainly provide evidence that the benefits of repeating vaccinations each year in teenage children are small if they exist at all.

12.2.2 Case–control study data

12.2.2.1 Unmatched case–control study: grouped data

This is an analysis of cases and unmatched controls from a study of oesophageal cancer. The data are grouped frequencies – cases and controls – within a three–way table according to age (six levels), alcohol, and tobacco consumption (four levels each). There are therefore 6×4×4 = 96 cells to be read as units.

The age groups are
1. 25–34 years
2. 35–44
3. 45–54
4. 55–64
5. 65–74
6. 75+

Tobacco consumption
1. 0–9 gms/day
2. 10–19
3. 20–29
4. 30+

Alcohol consumption
1. 0–39gms/day
2. 40–79
3. 80–119
4. 120+

(Source: Breslow and Day, 1980).

The essence of this form of analysis is to take the cases/(cases + controls) as if they were proportions from a cohort study. Although direct estimates of prevalence or incidence cannot be obtained from this type of sample odds ratios can. As long as the ratio of the sampling fractions for the cases and the controls may be assumed constant with respect to the explanatory variables then the odds ratios can be modelled using a binomial `ERROR` and the logistic `LINK`.

```
$units 96 $
$data ncas ncon $read
 0 40  0 10  0  6 0  5
 0 27  0  7  0  4 0  7
 0  2  0  1  0  0 0  2
 0  1  1  0  0  1 0  2
 0 60  1 13  0  7 0  8
 0 35  3 20  1 13 0  8
 0 11  0  6  0  2 0  1
 2  1  0  3  2  2 0  0
 1 45  0 18  0 10 0  4
 6 32  4 17  5 10 5  2
 3 13  6  8  1  4 2  2
 4  0  3  1  2  1 4  0
 2 47  3 19  3  9 4  2
 9 31  6 15  4 13 3  3
```

```
 9  9  8  7  3  3 4  0
 5  5  6  1  2  1 5  1
 5 43  4 10  2  5 0  2
17 17  3  7  5  4 0  0
 6  7  4  8  2  1 1  0
 3  1  1  1  1  0 1  0
 1 17  2  4  0  0 1  2
 2  3  1  2  0  3 1  0
 1  0  1  0  0  0 0  0
 2  0  1  0  0  0 0  0
```

First generate the age, alcohol, and tobacco factors using the `GFACTOR` directive. The factors to be created must be specified with their levels in increasing order of the rapidity with which their values should change.

```
$gfactor age 6 alc 4 tob 4 $
```

Then calculate the case/control 'binomial denominator' and weight vector values to indicate units with zero denominators to be excluded.

```
$cal bd = ncas+ncon $
$cal w = bd>0 $weight w $
```

Identify the number of cases variable as the *y*–variate and the `ERROR` as binomial with the appropriate denominator (by default GLIM will use the logistic `LINK`).

```
$error b bd $
$yvar ncas  $
```

First fit a model allowing all two–factor interactions to check whether it is safe to assume they are negligible and also use it to check the influence of the individual data points and the model assumptions.

```
$fit (age+tob+alc)**2 $dis e $
scaled deviance =  30.826 at cycle 10
    residual df =  37     from 88 observations
         estimate          s.e.     parameter
     1     -30.24         82.85      1
     2      24.54         82.85      AGE(2)
     3      24.77         82.85      AGE(3)
```

```
     4       27.23         82.85       AGE(4)
     5       28.13         82.85       AGE(5)
     6       27.54         82.85       AGE(6)
     7       19.05         64.74       TOB(2)
     8       1.048         86.23       TOB(3)
     9      0.5800         77.00       TOB(4)
    10       1.728         72.59       ALC(2)
    11       2.725         89.89       ALC(3)
    12       21.30         63.69       ALC(4)
    13      -16.44         64.74       AGE(2).TOB(2)
    14       1.122         86.23       AGE(2).TOB(3)
    15      -8.218         110.0       AGE(2).TOB(4)
    16      -17.40         64.74       AGE(3).TOB(2)
    17      0.4868         86.23       AGE(3).TOB(3)
    18       2.116         77.00       AGE(3).TOB(4)
    19      -17.38         64.74       AGE(4).TOB(2)
    20      0.4702         86.23       AGE(4).TOB(3)
    21       2.378         77.00       AGE(4).TOB(4)
    22      -18.30         64.74       AGE(5).TOB(2)
    23      0.4166         86.23       AGE(5).TOB(3)
    24      0.6454         77.01       AGE(5).TOB(4)
    25      -17.34         64.75       AGE(6).TOB(2)
    26      -11.63         175.6       AGE(6).TOB(3)
    27       1.798         77.01       AGE(6).TOB(4)
    28      0.3713         72.60       AGE(2).ALC(2)
    29      -10.23         113.4       AGE(2).ALC(3)
    30      -16.48         63.70       AGE(2).ALC(4)
    31       2.252         72.59       AGE(3).ALC(2)
    32       1.466         89.90       AGE(3).ALC(3)
    33      -14.11         63.70       AGE(3).ALC(4)
    34     -0.1263         72.59       AGE(4).ALC(2)
    35      0.1333         89.89       AGE(4).ALC(3)
    36      -17.68         63.69       AGE(4).ALC(4)
    37      0.3131         72.59       AGE(5).ALC(2)
    38     -0.7037         89.89       AGE(5).ALC(3)
    39      -17.72         63.69       AGE(5).ALC(4)
    40      0.4625         72.59       AGE(6).ALC(2)
    41       11.27         164.4       AGE(6).ALC(3)
    42      -6.781         140.3       AGE(6).ALC(4)
    43      -1.342        0.6550       TOB(2).ALC(2)
    44      -1.127        0.7151       TOB(2).ALC(3)
    45      -1.878        0.9306       TOB(2).ALC(4)
    46      -1.062        0.7750       TOB(3).ALC(2)
    47      -1.286        0.9265       TOB(3).ALC(3)
    48      -1.548         1.143       TOB(3).ALC(4)
    49     -0.8033         1.057       TOB(4).ALC(2)
    50     -0.6028         1.260       TOB(4).ALC(3)
    51      -1.379         1.425       TOB(4).ALC(4)
scale parameter 1.000
```

It is convenient to store the deviance and degrees of freedom in scalars for later model comparisons. It is also wise to investigate the influence of the individual data points by calculating the leverages and plotting them against an index of their position in the dataset.

```
$number D0=%dv df0=%df $
$extract %lv $
$cal index=%ind(%yv) $
$plot %lv/w index 'i' $
```

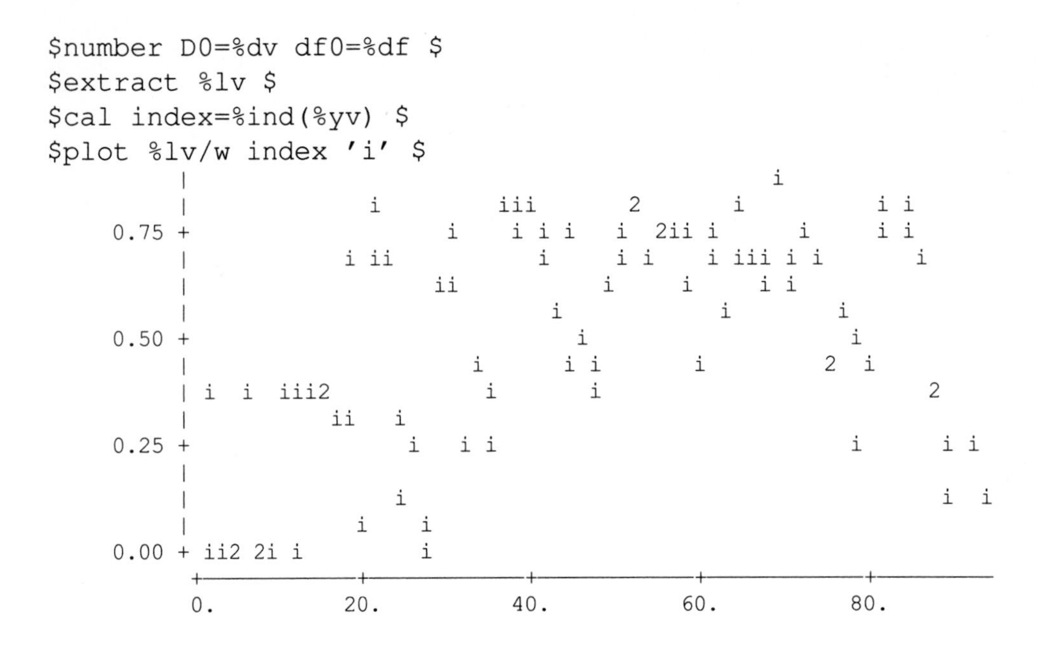

There is a slightly uneven spread of influence through the data, but there are no excessively influential points.

```
$plot %rs %lp 'r' $
```

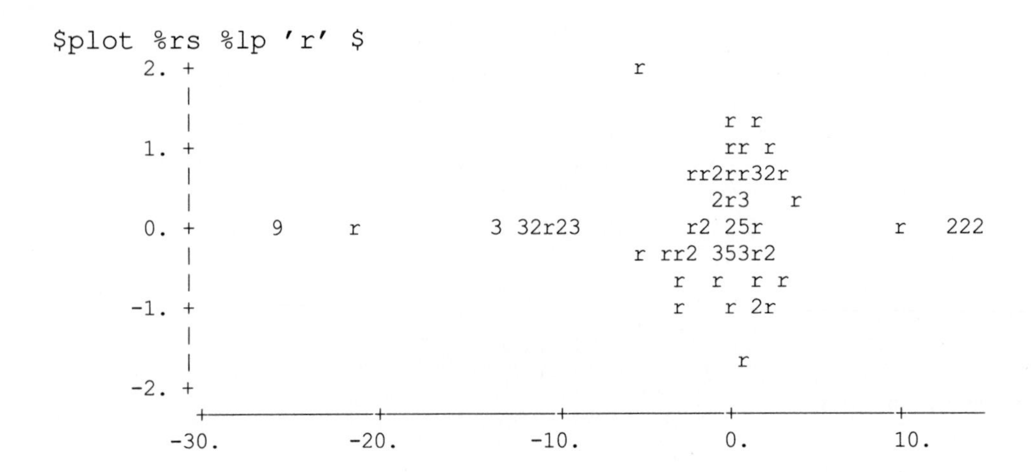

The plot looks messy, but it does not suggest that the mean-variance relationship is inappropriate for the binomial distribution. There is no evidence of systematic deviations from the model although there are a moderate number of points with large linear predictor values.

Returning to the model to assess whether it might be simplified we see that the interaction terms have large standard errors. However they cannot be assumed negligible without testing. It is necessary to fit the model without them.

```
$fit age+tob+alc $dis e $
scaled deviance =  82.337 at cycle 5
    residual df =  76      from 88 observations
          estimate          s.e.      parameter
     1      -6.895         1.084       1
     2       1.981         1.102       AGE(2)
     3       3.776         1.066       AGE(3)
     4       4.335         1.063       AGE(4)
     5       4.896         1.075       AGE(5)
     6       4.827         1.120       AGE(6)
     7      0.4381        0.2283       TOB(2)
     8      0.5126        0.2730       TOB(3)
     9       1.641        0.3441       TOB(4)
    10       1.435        0.2501       ALC(2)
    11       1.981        0.2848       ALC(3)
    12       3.603        0.3850       ALC(4)
scale parameter 1.000
$number D1=%dv df1=%df D10 df10 $
$cal D10=D1-D0 : df10=df1-df0 $
$pr 'Deviance difference ='D10' with df ='df10 $
Deviance difference =    51.51 with df =    39.00
```

The likelihood ratio chi-squared test of all the two-factor interactions taken together is $(82.34 - 30.83) = 51.51$ on $(76 - 37) = 39$ d.f. and the p-value is obtained (with, at most, 6 significant figures and 4 decimal places) by

```
$acc 6 4 $
$cal 1-%chp(51.51,39) $
           0.0866
```

This is not significant at the nominal 5 per cent level, but it is not far from significance. We should look at the interactions separately.

```
$fit (age+tob+alc).(age+tob+alc) - age.tob $dis e $
scaled deviance =  50.3110 at cycle 9
    residual df =  52       from 88 observations
            estimate            s.e.       parameter
     1      -13.4494          41.1359      1
     2        8.0860          41.1470      AGE(2)
     3        8.1989          41.1470      AGE(3)
     4       10.6910          41.1359      AGE(4)
     5       11.1133          41.1363      AGE(5)
     6       10.9183          41.1387      AGE(6)
     7        1.3321           0.5004      TOB(2)
     8        1.4625           0.6278      TOB(3)
     9        2.2401           0.6781      TOB(4)
    10        0.8433          56.2774      ALC(2)
    11        1.6583          67.9587      ALC(3)
    12       11.6310          41.1559      ALC(4)
    13        1.1908          56.2873      AGE(2).ALC(2)
    14       -8.2385          81.5719      AGE(2).ALC(3)
    15       -6.7324          41.1703      AGE(2).ALC(4)
    16        3.0094          56.2858      AGE(3).ALC(2)
    17        2.4419          67.9651      AGE(3).ALC(3)
    18       -4.7392          41.1709      AGE(3).ALC(4)
    19        0.6247          56.2776      AGE(4).ALC(2)
    20        1.0795          67.9586      AGE(4).ALC(3)
    21       -8.2757          41.1550      AGE(4).ALC(4)
    22        1.2866          56.2782      AGE(5).ALC(2)
    23        0.2868          67.9594      AGE(5).ALC(3)
    24       -8.3221          41.1620      AGE(5).ALC(4)
    25        0.7182          56.2826      AGE(6).ALC(2)
    26       11.3148          108.228      AGE(6).ALC(3)
    27        1.6210          86.3641      AGE(6).ALC(4)
    28       -1.1364           0.6171      TOB(2).ALC(2)
    29       -1.0801           0.6708      TOB(2).ALC(3)
    30       -1.2162           0.8860      TOB(2).ALC(4)
    31       -1.1347           0.7354      TOB(3).ALC(2)
    32       -1.2687           0.8938      TOB(3).ALC(3)
    33       -1.4016           1.0451      TOB(3).ALC(4)
    34       -0.6787           0.8654      TOB(4).ALC(2)
    35       -0.3439           1.1176      TOB(4).ALC(3)
    36       -1.3683           1.1633      TOB(4).ALC(4)
scale parameter 1.0000
```

The likelihood ratio chi-squared value, degrees of freedom and p-value are then obtained as follows

```
$number D2=%dv df2=%df D20 df20 p20 $
$cal D20=D2-D0 : df20=df2-df0 : p20=1-%chp(D20,df20) $
$pr ' Chi_sq ='D20' on'df20' df with p='p20 $
 Chi_sq =   19.4848 on    15.0000 df with p=  0.192601
```

This interaction is not significant. Replace it and test the `age.alc` interaction.

```
$fit +age.tob-age.alc $
scaled deviance =  56.8082 (change =   +6.4972) at cycle 9
    residual df =  52      (change =    0      ) from 88 observations
$number D3=%dv df3=%df D30 df30 p30 $
```

As before the likelihood ratio chi-squared value, degrees of freedom and p-value are

```
$cal D30=D3-D0 : df30=df3-df0 : p30=1-%chp(D30,df30) $
$pr ' Chi_sq ='D30' on'df30' df with p='p30 $
 Chi_sq =   25.9819 on    15.0000 df with p=  0.038213
```

This interaction does reach significance. There is an effect of alcohol which alters according to the age of the subjects.

```
$fit +age.alc-tob.alc $
scaled deviance =  37.5368 (change =   -19.2714) at cycle 10
    residual df =  46      (change =    -6      ) from 88 observations
$number D4=%dv df4=%df D40 df40 p40 $
$cal D40=D4-D0 : df40=df4-df0 : p40=1-%chp(D40,df40) $
```

Again the likelihood ratio chi-squared value, degrees of freedom and p-value are

```
$pr ' Chi_sq ='D40' on'df40' df with p='p40 $
 Chi_sq =   6.7106 on    9.0000 df with p=  0.667224
```

There is no evidence of an interaction between the effects of tobacco and alcohol. However the age/alcohol consumption term does appear to be reflecting a genuine interaction pattern in the data. Fit the model with this interaction omitting the other two.

```
$fit age+tob+alc+age.alc $dis e $
scaled deviance =  56.2576 at cycle 9
```

```
residual df =   61        from 88 observations
          estimate          s.e.       parameter
 1        -12.8771        39.4201       1
 2          7.9772        39.4328       AGE(2)
 3          8.1740        39.4328       AGE(3)
 4         10.6526        39.4212       AGE(4)
 5         10.9570        39.4213       AGE(5)
 6         10.7380        39.4238       AGE(6)
 7          0.4507         0.2316       TOB(2)
 8          0.4979         0.2775       TOB(3)
 9          1.6888         0.3522       TOB(4)
10          0.1338        55.5071       ALC(2)
11          1.1870        67.8630       ALC(3)
12         10.4821        39.4371       ALC(4)
13          1.3064        55.5188       AGE(2).ALC(2)
14         -8.3058        81.8627       AGE(2).ALC(3)
15         -6.3275        39.4557       AGE(2).ALC(4)
16          3.0529        55.5170       AGE(3).ALC(2)
17          2.2739        67.8715       AGE(3).ALC(3)
18         -4.4399        39.4576       AGE(3).ALC(4)
19          0.6837        55.5086       AGE(4).ALC(2)
20          0.9116        67.8645       AGE(4).ALC(3)
21         -7.9281        39.4408       AGE(4).ALC(4)
22          1.5026        55.5088       AGE(5).ALC(2)
23          0.2296        67.8649       AGE(5).ALC(3)
24         -7.7782        39.4472       AGE(5).ALC(4)
25          0.9122        55.5136       AGE(6).ALC(2)
26         11.2988        107.733       AGE(6).ALC(3)
27          2.2444        84.1677       AGE(6).ALC(4)
scale parameter 1.0000
$number D5=%dv df5=%df D50 df50 p50 $
$cal D50=D5-D0 : df50=df5-df0 : p50=1-%chp(D50,df50) $
```

This gives a likelihood ratio chi-squared value and *p*-value testing the omitted interactions as follows:

```
$pr ' Chi_sq ='D50' on'df50' df with p='p50 $
 Chi_sq =    25.4314 on    24.0000 df with p=  0.382632
```

Clearly the interaction terms involving tobacco consumption are not essential. Whatever the effect of tobacco consumption it is the same, on average, in each of the age and alcohol consumption categories.

Before considering the precise form of the effect of tobacco it is wise to investigate the form of the age/alcohol interaction a little further. As a first step obtain plots of the fitted values against age. These should clarify how the relationship changes from one alcohol group to another. The weight vector `wtg` is used to restrict the plots to the first tobacco group (the pattern will be the same for each) and the symbols are determined by alcohol group factor.

```
$cal wtg1=(tob==1) $plot %lp/wtg1 age 'abcd' alc $
```

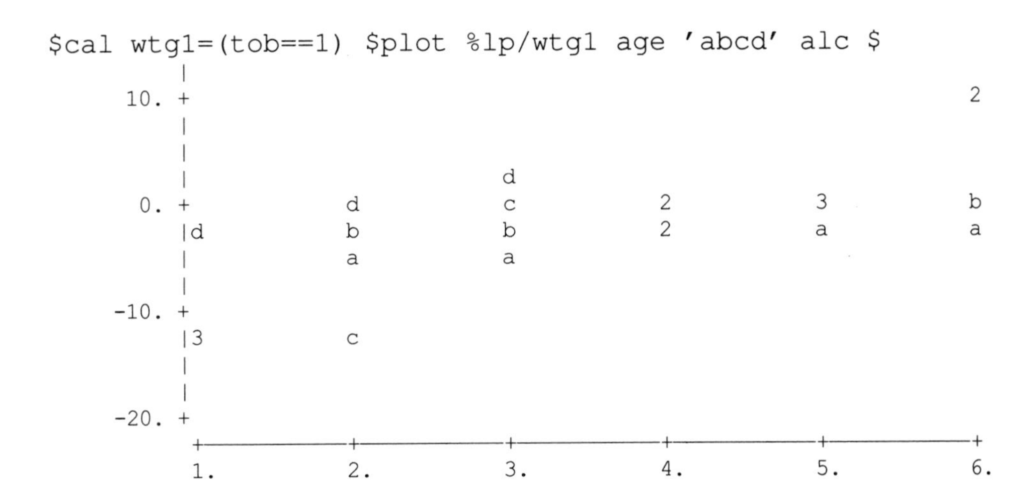

This is easier to see with a graph using the mean ages for each category as the *x*-variable and defining four pen styles with `GSTYLE` to indicate the alcohol categories. To label the lines it is necessary to obtain the coordinates of the last point on each to position an appropriate string with the `GTEXT` directive. This is done by using `PREDICT` to derive and store in `%plp`, the linear predictor values for age category 6 (`page`), tobacco consumption category 1 (`ptab`) and alcohol categories 1,2,3 and 4 (`palc`).

```
$cal agex=(age-1)*10+25 $
$gstyle 1 c 1 1 1 : 2 c 1 1 2 : 3 c 1 1 3 : 4 c 1 1 4 $
$ass pens=1,2,3,4 $

$ass page=6,6,6,6 : ptob=1,1,1,1 : palc=1,2,3,4 $
$predict age=page tob=ptob alc=palc $
prediction for the current model with AGE=PAGE, TOB=PTOB, ALC=PALC
Binomial denominator=1.0000
```

Because alcohol categories 3 and 4 have very similar linear predictor values in age category 6 it is necessary to adjust the values of `%plp` to separate them vertically before using `GTEXT`.

```
$cal %plp=%plp+0.5*%sgn(%plp(4)-%plp(3))*((palc==4)-(palc==3))
   : page=(page-1)*10+25 $
```

```
$graph ( t='Fig. 12.35 Log(odds) by Age and Alcohol (tob=1)'
         v='Log Odds (the Linear predictors)' pause=no
         h='Age categories' xscale=15,85,10 yscale=-20,20,10)
         %lp/wtg1  agex  pens alc $
```

```
$gtext (m=p l=r) %plp/1.2 page ' Alc = 1',' Alc = 2',' Alc = 3',
       ' Alc = 4' $
```

Fig. 12.35 Log(odds) by Age and Alcohol (tob=1)

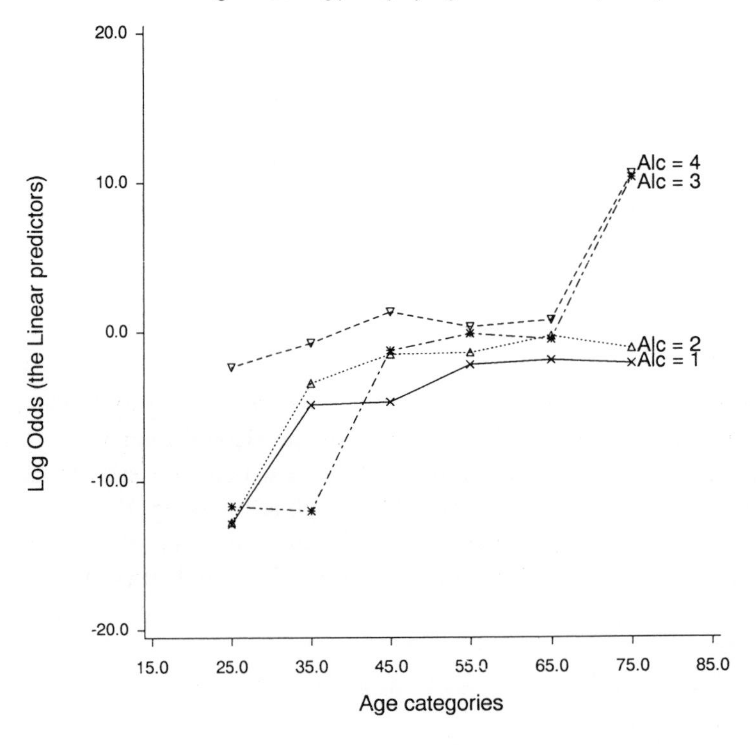

The patterns are not very clear. They indicate that the effect of age is steadily increasing over the first five age groups, but with the two higher alcohol consumption groups starting higher and increasing less so they converge. In the last age group the two higher alcohol groups increase markedly above the other two which continue the pattern of levelling with increasing age.

There is no very obvious explanation for this, but it is possible that there are some complex biases due to deaths from other causes making the older age groups unrepresentative in unpredictable ways. It could also be that the effect of alcohol is cumulative and it is total consumption that matters. This means that the effect will depend on average rates and the length of time involved. If so it considerably complicates the interpretation of age and current consumption effects. More information than is available in this dataset is needed to pursue this further.

Although the alcohol and age effects appear to be complex it is still possible to make simple generalisations about the effects of the other factors. On the assumption that there is not a three factor interaction between alcohol, age and tobacco consumption the effects of smoking may be summarised by constant odds ratios between the tobacco consumption categories within each of the age and alcohol categories using the tobacco consumption main effect parameters in this model.

The estimated effects of tobacco consumption represent the log(odds ratios) for each consumption category compared to the lowest. In the model these show a tendency to increase systematically with consumption. The form of this trend can be investigated using orthogonal polynomials. We do this and fit a model allowing for a quadratic trend by

```
$fit age*alc+tob<2> $dis e $
scaled deviance =  58.8827 at cycle 9
    residual df =  62       from 88 observations
          estimate          s.e.      parameter
    1     -12.2089        39.5524       1
    2       7.9914        39.5652       AGE(2)
    3       8.1642        39.5652       AGE(3)
    4      10.6407        39.5536       AGE(4)
    5      10.9296        39.5538       AGE(5)
    6      10.7793        39.5562       AGE(6)
    7       0.1452        56.0168       ALC(2)
    8       1.2814        68.6156       ALC(3)
    9      10.5386        39.5692       ALC(4)
   10       0.4931         0.1088       TOB<1>
   11       0.1244         0.1082       TOB<2>
   12       1.2860        56.0283       AGE(2).ALC(2)
   13      -8.4171        82.7005       AGE(2).ALC(3)
   14      -6.4893        39.5877       AGE(2).ALC(4)
   15       3.0271        56.0266       AGE(3).ALC(2)
   16       2.2038        68.6240       AGE(3).ALC(3)
   17      -4.4977        39.5896       AGE(3).ALC(4)
   18       0.6484        56.0183       AGE(4).ALC(2)
   19       0.8303        68.6171       AGE(4).ALC(3)
   20      -7.9654        39.5729       AGE(4).ALC(4)
```

```
    21        1.4648        56.0185        AGE(5).ALC(2)
    22        0.1777        68.6175        AGE(5).ALC(3)
    23       -7.8232        39.5793        AGE(5).ALC(4)
    24        0.8122        56.0232        AGE(6).ALC(2)
    25       11.2074        108.595        AGE(6).ALC(3)
    26        2.1705        84.9921        AGE(6).ALC(4)
scale parameter 1.0000
```

The deviance has increased by 2.62 on 1 d.f from the previous model with tobacco consumption represented by a four level factor. This model fits just as well.

```
$fit -tob<2> + tob<1> $dis e $
scaled deviance =  60.2147 (change =   +1.3320) at cycle 9
    residual df =  63      (change =   +1       ) from 88 observations
           estimate          s.e.      parameter
     1     -12.1977        39.8824      1
     2       8.0157        39.8951      AGE(2)
     3       8.1454        39.8951      AGE(3)
     4      10.6175        39.8836      AGE(4)
     5      10.9072        39.8837      AGE(5)
     6      10.7983        39.8862      AGE(6)
     7       0.1661        56.6795      ALC(2)
     8       1.3721        69.5458      ALC(3)
     9      10.6029        39.8987      ALC(4)
    10       0.4386         0.0982      TOB<1>
    11       1.2251        56.6908      AGE(2).ALC(2)
    12      -8.5609        83.5126      AGE(2).ALC(3)
    13      -6.6648        39.9168      AGE(2).ALC(4)
    14       2.9932        56.6891      AGE(3).ALC(2)
    15       2.0913        69.5540      AGE(3).ALC(3)
    16      -4.5735        39.9188      AGE(3).ALC(4)
    17       0.6138        56.6810      AGE(4).ALC(2)
    18       0.7165        69.5472      AGE(4).ALC(3)
    19      -8.0232        39.9023      AGE(4).ALC(4)
    20       1.4374        56.6812      AGE(5).ALC(2)
    21       0.0482        69.5476      AGE(5).ALC(3)
    22      -7.8859        39.9087      AGE(5).ALC(4)
    23       0.7301        56.6857      AGE(6).ALC(2)
    24      11.0417        108.965      AGE(6).ALC(3)
    25       2.0542        84.6980      AGE(6).ALC(4)
scale parameter 1.0000
```

The quadratic term does not appear to be needed since the increase in deviance from omitting it is 1.33 on 1 d.f. (cf. 3.84). In fact referring back to the model with tobacco consumption included as a four level factor the change in deviance from that to the current one with the effect of tobacco modelled as a straight line trend – which is a general test of non-linearity – is (60.22–56.26) = 3.96 on (63–61) = 2 d.f. which is also some way from significance with a *p*-value of 0.14

```
$cal 1-%chp(3.96,2) $
        0.1381
```

It appears that the effect of tobacco can be represented as a linear trend on the log(odds) scale. The `tob<1>` coefficient (parameter 10) is over four times its standard error so it is clearly significant. The odds ratio equivalent to the increase in 'risk' to be expected for one unit increase on the `tob` scale and its approximate 95 per cent confidence limits are obtained by

```
$number OR ORlo ORhi $
$extract %pe %se $
$cal OR=%exp(%pe(10))
   : ORlo=%exp(%pe(10)-1.96*%se(10))
   : ORhi=%exp(%pe(10)+1.96*%se(10)) $
```

The odds ratio with approximate 95 per cent confidence limits is

```
$pr ' OR ='OR' with 95% CLs'ORlo' to'ORhi $
 OR =    1.5505 with 95% CLs    1.2790 to    1.8796
```

One unit increase on the coded tobacco consumption scale used in the analysis is equivalent to 40 gms/day, that is, about 10 cigarettes per day. The risks involved are small enough for the odds ratio to be a reasonable approximation for relative risk. On this assumption, and with the proviso that they should not be taken to apply to consumptions outside the range of those in these data, the results can be expressed in terms of increased risk. They imply that for any combination of age and drinking habits a smoker with a consumption of $C+10$ cigarettes a day has a risk of oesophageal cancer which exceeds that of a smoker with a consumption of C per day, by at least 28 per cent, probably by as much as 55 per cent and possibly by as much as 88 per cent.

The effect of alcohol is probably much greater, but in this dataset the effects of alcohol and age are inextricably confused. Estimates of the effects for particular age groups could be obtained, but they would be based on rather small numbers and would have very wide confidence intervals as a consequence.

12.2.2.2 Matched case-control data

These data come from the Los Angeles case-control study of endometrial cancer to assess the magnitudes and relative importance of a number of risk factors. Cases of endometrial cancer, all female, were matched by age, with four controls. The data is described and analysed in chapter 7 of Breslow and Day (1980).

The variables are:

`CACT`	case status: 0. Control, 1. Case;
`SET`	identifier for each case/control set;
`AGE`	age in years at last birthday;
`GALL`	gall bladder disease: 0. no, 1. yes;
`HYPER`	hypertension (high blood pressure): 0. no, 1. yes;
`OBESE`	obesity: 0. no, 1. yes, 9. unknown;
`EST`	history of any oestrogen use: 0. no, 1. yes;
`EDOS`	dosage of oestrogen 0. None or single doses equivalent to < 0.1 mg/day, 1. 0.1 – 0.299 mg/day, 2. 0.3 – 0.625 mg/day, 3. 0.626+ mg/day, 9. Unknown;
`EDUR`	duration of oestrogen use in months (single doses are coded 0 and 99 = not known)
`ODRG`	other non-oestrogen drug: 0. no, 1. yes.

There are 63 cases, each matched to four controls to give 63 matched sets. There are thus 315 units in total. The analysis is required to assess the relative risks of endometrial cancer for those exposed and not exposed to these risk factors. The actual population risks are small so the odds ratios are reasonable approximations for the relative risks.

The appropriate analysis is to fit conditional logistic regression models giving estimates of the log(odds ratios). This is achieved by treating the control–case status as a 0/1 y variable with a Poisson distribution and performing the analysis in such a way as to keep the fitted totals for each matched case–control set the same as those observed. The most efficient way to do this is by constructing a factor (`set`) with as many levels as there are sets and using `ELIMINATE` to ensure that the main effects of that factor are included in all models fitted.

The data for a subset of variables are input with

```
$data 315 cact set age gall hyper obese est edos edur odrg
$read cact set     gall hyper           edos edur odrg
       1    1  74    0    0     1    1    3   96    1
       0    1  75    0    0     9    0    0    0    0
```

```
0    1  74    0    0    9    0    0    0    0
0    1  74    0    0    9    0    0    0    0
0    1  75    0    0    1    1    1   48    1
1    2  67    0    0    0    1    3   96    1
0    2  67    0    0    0    1    3    5    0
0    2  67    0    1    1    0    0    0    1
0    2  67    0    0    0    1    2   53    0
0    2  68    0    0    0    1    2   45    1
1    3  76    0    1    1    1    1    9    1
.    .  ..    .    .    .    .    .    .    .
.    .  ..    .    .    .    .    .    .    .
```

This analysis does not attempt to match that published in Breslow and Day (1980) because the intention is to illustrate an analysis investigating the effects of the various factors allowing for confounding and possible interactions in a sequence of model fits all using the same observations. This means that subjects lacking data on any variable are excluded from all analyses and where this involves the 'case' or all 'controls' the whole 'case–control' set is lost. Consequently the numbers available for these analyses and the results obtained are not necessarily the same as those published.

First we must calculate a weight vector for excluding the units with missing data for the two variables dose (`edos`) and duration (`edur`). All of the other variables have complete data.

```
$cal     wt = (edos/=9)&(edur/=99) $
```

It is useful to have the case/control make-up of the sets available for analysis. This can be obtained by

```
$cal control=1-cact $
$tab the cact    total with wt for set into castot by setno $
$tab the control total with wt for set into contot $
$tab for castot;contot using freq by cases;controls $
$tprint freq cases;controls $
 CONTROLS   3.000   4.000
    CASES
       0.   4.000   5.000
       1.   7.000  47.000
```

Out of the 63 sets of one case to four controls missing data has resulted in the loss of nine cases which means that the sets can contribute nothing to the analysis. In seven sets, one control has been lost, but there are still three controls remaining in the set so it contains useful information. All the individuals in case–control sets where there are no cases must be weighted out to get the correct degrees of freedom although it is not essential because the estimates and model comparisons are unaffected.

```
$cal w_set=(castot(set) > 0) $
$cal wt=wt*w_set $
```

To avoid the problem of determining the functional form of the relationship between the y variable and the explanatory variable 'duration of oestrogen use' used as a covariate it is better to create a variable grouping the subjects into duration categories and use it as a factor. This allows for any form of non-linearity in the relationship at the cost of ignoring the differences in duration among subjects in the same category. A reasonable strategy is to group the duration variable measured in months (`edur`) into five categories

1. `edur` < 1 ... mainly zeroes
2. 1≤= `edur` <12;
3. 12≤= `edur` <48;
4. 48≤= `edur` <96;
5. `edur` ≥= 96 months.

This is done relatively easily using the directive `GROUP` which, for a given set of intervals, creates a factor variable with values equal to the interval number

```
$ass idur=1,12,48,96 $
$group dur=edur intervals * idur * $
```

The variables `gall, hyper, odrg` and `dose` need to be declared as factors. Since they use the code 0 for the lowest category and GLIM expects factor variables to have positive integer values starting at 1 they need to be adjusted by adding 1.

In addition because the individuals with 0 or very low doses, that is, those in dose category 1, are all in one duration category the effects of duration for them will be aliased. To avoid this unduly complicating the model interpretations it is useful to set the `dose`=2 level to be the reference category.

Finally, for a conditional logistic regression, the 63 matched sets need to be equated with the levels of a factor (`set`). This is so their fitted total frequencies can be kept equal to those observed by including the main effects of the factor set in all models.

```
$cal gall=gall+1 : hyper=hyper+1 : odrg=odrg+1 : dose=edos+1 $
$fac gall 2 hyper 2  dose 4(2) dur 5 odrg 2 set 63 $
```

Age was used in the matching so it will be confounded with the `set` factor, but it is still possible to fit interactions between the other risk factors and age to assess whether their effects change with age. However this leads to a very large and relatively unstable model so, although there is in fact some evidence that the effects of some factors do change with age, we will not pursue it here.

An efficient strategy is to fit a model with all the two factor interactions and then models omitting them, separately or in groups using the parameter estimates as guide, to determine whether they can be safely assumed negligible. If they can the effects of the various risk factors can then be assessed using the main effects model as a baseline.

First it is necessary to specify the appropriate conditional logistic model by declaring the case–control indicator variable to be the y variable with a Poisson response distribution and ELIMINATE the factor set. It is also necessary to declare wt as the weight variable.

```
$yvar cact  $err p   $
$eliminate set   $
$weight wt $
```

It is useful for complex models with limited data when the fitting process may not converge quickly to use the CYCLE directive setting the maximum number of cycles at some value above the default of 10, e.g. 20, with a request for the deviance to be printed every 2 cycles (say). This allows the progress of the fitting process to be monitored.

```
$cycle 20 2 $
```

The model with all the two factor interactions is fitted by

```
$fit (gall+hyper+dose+dur+odrg)**2  $dis e $
scaled deviance =  100.76 at cycle 2
scaled deviance =   81.005 at cycle 4
scaled deviance =   78.406 at cycle 6
scaled deviance =   78.237 at cycle 8
scaled deviance =   78.215 at cycle 10
    residual df =  173      from 263 observations
          estimate          s.e.      parameter
    1        4.846         3.813        GALL(2)
    2       -4.069         3.534        HYPER(2)
    3       -14.93         15.65        DOSE(1)
    4       0.1195         2.896        DOSE(3)
    5      -0.3998         3.122        DOSE(4)
    6       -12.46         62.75        DUR(2)
    7       -4.522         3.673        DUR(3)
    8        3.436         4.159        DUR(4)
    9        0.000       aliased        DUR(5)
```

```
    10      -4.246       3.415     ODRG(2)
    11      -2.827       1.733     GALL(2).HYPER(2)
    12      -2.294       3.202     GALL(2).DOSE(1)
    13      -6.614       3.650     GALL(2).DOSE(3)
    14      -2.662       3.474     GALL(2).DOSE(4)
    15       2.499       2.056     HYPER(2).DOSE(1)
    16       1.128       1.544     HYPER(2).DOSE(3)
    17      -2.723       2.117     HYPER(2).DOSE(4)
    18      -9.777       127.3     GALL(2).DUR(2)
    19       2.179       3.293     GALL(2).DUR(3)
    20       3.164       3.418     GALL(2).DUR(4)
    21       0.000     aliased     GALL(2).DUR(5)
    22       3.169       2.186     HYPER(2).DUR(2)
    23       1.874       2.003     HYPER(2).DUR(3)
    24       2.284       1.876     HYPER(2).DUR(4)
    25       0.000     aliased     HYPER(2).DUR(5)
    26       0.000     aliased     DOSE(1).DUR(2)
    27       0.000     aliased     DOSE(1).DUR(3)
    28       0.000     aliased     DOSE(1).DUR(4)
    29       0.000     aliased     DOSE(1).DUR(5)
    30      -1.285       2.012     DOSE(3).DUR(2)
    31      -1.346       1.800     DOSE(3).DUR(3)
    32      -5.210       2.309     DOSE(3).DUR(4)
    33       0.000     aliased     DOSE(3).DUR(5)
    34      -3.356       2.251     DOSE(4).DUR(2)
    35       3.593       3.018     DOSE(4).DUR(3)
    36      -1.754       2.818     DOSE(4).DUR(4)
    37       0.000     aliased     DOSE(4).DUR(5)
    38       1.607       2.604     GALL(2).ODRG(2)
    39       1.976       2.896     HYPER(2).ODRG(2)
    40       12.93       15.55     DOSE(1).ODRG(2)
    41       1.368       2.761     DOSE(3).ODRG(2)
    42       4.811       3.442     DOSE(4).ODRG(2)
    43       11.06       62.73     DUR(2).ODRG(2)
    44       3.290       3.454     DUR(3).ODRG(2)
    45      -1.884       3.932     DUR(4).ODRG(2)
    46       0.000     aliased     DUR(5).ODRG(2)
scale parameter 1.000
   eliminated term: SET
```

There is only one interaction parameter estimate more than twice its s.e. `DOSE(3).DUR(4)`. But not only is this test only approximate it is a comparison against the reference category and tells us nothing about differences between the estimates. S.e.s of differences can be obtained with the `DISPLAY S` option, but the sets of interactions representing each interaction term need to be tested properly with model comparisons. First it is necessary to store the deviance and the degrees of freedom for this model. Then to test all the interactions as a group we fit the model without interactions that is the main effects model

```
$number D1 df1 $cal D1=%dv : df1=%df $
$fit (gall+hyper+dose+dur+odrg) $dis e $
scaled deviance =  127.69 at cycle 2
scaled deviance =  121.07 at cycle 4
scaled deviance =  121.05 at cycle 6
    residual df =  200     from 263 observations
          estimate          s.e.     parameter
     1       1.491        0.5092       GALL(2)
     2     -0.4488        0.3968       HYPER(2)
     3      -1.747        0.6550       DOSE(1)
     4     -0.1162        0.5497       DOSE(3)
     5      0.9507        0.5782       DOSE(4)
     6     -0.8925        0.6579       DUR(2)
     7     -0.2524        0.5758       DUR(3)
     8      0.1721        0.6047       DUR(4)
     9       0.000       aliased       DUR(5)
    10       1.189        0.5901       ODRG(2)
scale parameter 1.000
    eliminated term: SET
```

When several model comparisons have to be made using deviance differences as a likelihood ratio chi-squared test and p values calculated it is useful to construct a macro to avoid having to type in the same directives repeatedly. An appropriate macro (`chitst`) is

```
$mac (arg=Devp,dfp loc=dfcp,Devcp,Chi,p) chitst $
!
! This macro performs a chi-squared test comparing the deviance
!of the current model with the deviance (argument1: Devp) and
!the degrees of freedom (argument2: dfp) from a previous model.
!
$number dfcp Devcp Chi p $
$cal dfcp=%df-dfp : Chi=(%dv-Devp) : p=1-%chp(Chi,dfcp) $
$pr 'Chi-squared ='Chi' with 'dfcp' degrees of freedom and p ='p
$
$endm $
$use chitst D1 df1 $
Chi-squared =    42.84 with    27.00 degrees of freedom and p =  0.0272
```

The p value of 0.0272 is considerably less than 0.05 and so this is significant evidence, at the 5 per cent level, that not all the interactions can be assumed negligible. We now need to identify which of them are needed. Looking at the parameter estimates in the model with the interactions it is difficult to judge. The only safe way is to fit models with each interaction term removed in turn. The `chitst` macro can be used to obtain the appropriate test comparing each model with the model including them all. First reset `CYCLE` to suppress unecessary output.

```
$cycle $
$fit (gall+hyper+dose+dur+odrg)**2 - gall.hyper $
$use chitst D1 df1 $
scaled deviance =   81.161 at cycle 10
    residual df =  174      from 263 observations
Chi-squared =   2.946 with    1.000 degrees of freedom and p =  0.0861
```

```
$fit (gall+hyper+dose+dur+odrg)**2 - gall.dose $
$use chitst D1 df1 $
scaled deviance =   82.654 at cycle 10
    residual df =  175      from 263 observations
Chi-squared =   4.439 with    2.000 degrees of freedom and p =  0.1087
```

```
$fit (gall+hyper+dose+dur+odrg)**2 - gall.dur $
$use chitst D1 df1 $
scaled deviance =   80.572 at cycle 10
    residual df =  176      from 263 observations
Chi-squared =   2.357 with    3.000 degrees of freedom and p =  0.5017
```

```
$fit (gall+hyper+dose+dur+odrg)**2 - gall.odrg $
$use chitst D1 df1 $
scaled deviance =   78.535 at cycle 10
    residual df =  174      from 263 observations
Chi-squared =  0.3198 with    1.000 degrees of freedom and p =  0.5717
```

```
$fit (gall+hyper+dose+dur+odrg)**2 - hyper.dose $
$use chitst D1 df1 $
scaled deviance =    81.434 at cycle 10
    residual df =  175      from 263 observations
Chi-squared =   3.219 with    2.000 degrees of freedom and p =  0.2000


$fit (gall+hyper+dose+dur+odrg)**2 - hyper.dur $
$use chitst D1 df1 $
scaled deviance =    81.323 at cycle 10
    residual df =  176      from 263 observations
Chi-squared =   3.108 with    3.000 degrees of freedom and p =  0.3753


$fit (gall+hyper+dose+dur+odrg)**2 - hyper.odrg $
$use chitst D1 df1 $
scaled deviance =    78.706 at cycle 10
    residual df =  174      from 263 observations
Chi-squared =  0.4906 with    1.000 degrees of freedom and p =  0.4837


$fit (gall+hyper+dose+dur+odrg)**2 - dose.dur $
$use chitst D1 df1 $
scaled deviance =    92.996 at cycle 10
    residual df =  179      from 263 observations
Chi-squared =   14.78 with    6.000 degrees of freedom and p =  0.0220


$fit (gall+hyper+dose+dur+odrg)**2 - dose.odrg $
$use chitst D1 df1 $
scaled deviance =    80.320 at cycle 10
    residual df =  175      from 263 observations
Chi-squared =   2.105 with    2.000 degrees of freedom and p =  0.3491


$fit (gall+hyper+dose+dur+odrg)**2 - dur.odrg $
$use chitst D1 df1 $
scaled deviance =    80.926 at cycle 10
    residual df =  176      from 263 observations
Chi-squared =   2.711 with    3.000 degrees of freedom and p =  0.4383
```

Individually only the `dose.dur` interactions reaches significance. None of the other p-values were actually smaller than 0.05 although `gall.hyper` and `gall.dose` gave values less than 0.1. In practice they should all be investigated, but to keep the example as simple as possible we will assume that the interactions involving `gall` can be safely ignored, and restrict the remaining investigation of the interactions to that between dose and duration of oestrogen use. First we should fit the model with `dose.dur` as the sole interaction.

```
$fit (gall+hyper+dose+dur+odrg)+dose.dur $dis e $
scaled deviance =  113.39 at cycle 6
    residual df =  194     from 263 observations
          estimate          s.e.       parameter
     1       1.543        0.5339        GALL(2)
     2     -0.5823        0.4312        HYPER(2)
     3      -1.689        0.8423        DOSE(1)
     4      0.1869         1.064        DOSE(3)
     5      0.9993         1.091        DOSE(4)
     6     -0.5294         1.068        DUR(2)
     7     -0.5676        0.9741        DUR(3)
     8      0.6823         1.176        DUR(4)
     9       0.000       aliased        DUR(5)
    10       1.326        0.6319        ODRG(2)
    11       0.000       aliased        DOSE(1).DUR(2)
    12       0.000       aliased        DOSE(1).DUR(3)
    13       0.000       aliased        DOSE(1).DUR(4)
    14       0.000       aliased        DOSE(1).DUR(5)
    15     0.09286         1.550        DOSE(3).DUR(2)
    16    -0.02174         1.341        DOSE(3).DUR(3)
    17      -1.777         1.592        DOSE(3).DUR(4)
    18       0.000       aliased        DOSE(3).DUR(5)
    19      -1.861         1.666        DOSE(4).DUR(2)
    20       1.861         1.679        DOSE(4).DUR(3)
    21      0.1999         1.922        DOSE(4).DUR(4)
    22       0.000       aliased        DOSE(4).DUR(5)
scale parameter 1.000
    eliminated term: SET
$number D2 df2 $cal D2=%dv : df2=%df $
```

The best way to investigate the pattern of the relationships represented by the dose and duration interaction is to plot a graph of the values predicted from the model. We use `PREDICT` to obtain fitted values for the full range of durations, that is 1,2,3,4 for each of the 4 doses as follows

```
$var 20 pdur pdose $cal pdur=%gl(5,1) : pdose=%gl(4,5) $
$fac pdose 4 $
$ass pens=1,11,12,13 $


$pred dur=pdur dose=pdose $
prediction for the current model with GALL=1, HYPER=1, DOSE=PDOSE, DUR=PDUR,
ODRG=1, SET=1
```

The fitted values are held in `%pfv` and the linear predictors on the log(odds) scale in `%plp`. We will plot the latter since they generally give a clearer picture, after ascertaining the range of values and identifying a reasonable range for the vertical scale. Because the `dose=1` subjects all fell in the lowest duration category there is no information on what the effects of other durations would have been at this dose. The model assumes that they would have followed the same pattern as the reference dose group (2), but we will suppress them from the graph to simplify the picture.

```
$cal pwt=(pdose/=1)?((pdose==1)&(pdur==1)) $
$number ymin ymax midpt $
$tab the %plp small into ymin : the %plp large into ymax $
$cal midpt=(ymin+ymax)/2
: ymin=(ymin-midpt)*1.3+midpt : ymax=(ymax-midpt)*1.3+midpt $

$graph  (t='Fig. 12.36 The form of the dose.dur interactions'
        v='Predicted Log(Odds)' h='Duration category' p=no
        x=0,6 y=ymin,ymax) %plp/pwt pdur pens pdose $
$ass indx=5,10,15,20 : xtx=5,5,5,5 $cal ytx=%plp(indx) $
$gtext (m=p l=r) ytx xtx ' Dose 1',' Dose 2',' Dose 3',' Dose
4'$
```

The differences between the duration categories differ quite markedly with dose category. There is a suggestion that the effects increase with duration in dose category 4 (the largest) up to duration category 4 and then declines. In the two lower dose categories there does not seem to be much of a trend. The pattern is not susceptible to any more simpler description and further interpretation of these effects and what they tell us of how oestrogen affects the risk of endometrial cancer requires medical expertise.

The effects of the other three risk factors `gall`, `hyper` and `odrg` appear rather less complicated and are somewhat easier to interpret. The parameters representing the effects of `gall` and `odrg` are more than twice their s.e.s. They should be tested by model comparison, but it is instructive to obtain the estimated relative risks (ORs) with approximate 95 per cent confidence limits for these effects and that of `hyper`. They are 1,2 and 10 in the parameter list so the sequence required is

Fig. 12.36 The form of the dose.dur interactions

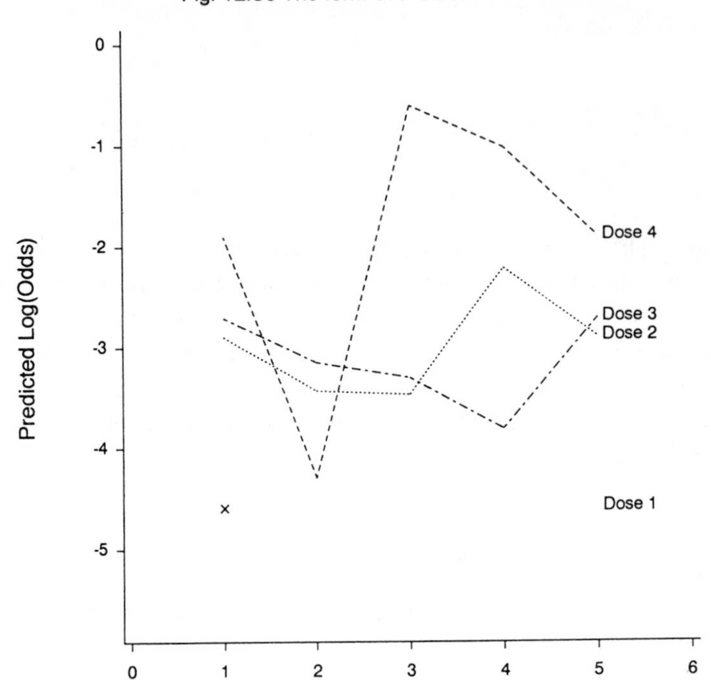


```
$extract %pe %se $ass indx=1,2,10 $var 3 lo OR hi $
$cal lo=%exp(%pe(indx)-1.96*%se(indx))  : OR=%exp(%pe(indx))
  :  hi=%exp(%pe(indx)+1.96*%se(indx)) $
$look lo OR hi $
       LO        OR        HI
 1   1.6427    4.6775    13.319
 2   0.2400    0.5586     1.301
 3   1.0915    3.7660    12.994
```

As should be expected the two parameters exceeding twice their s.e.s have confidence intervals which do not include 1. To test them formally we need to fit a sequence of models omitting each in turn and comparing their deviances and degrees of freedom with those from this model, `D2` and `df2`, using the macro `chitst`.

```
$fit (      hyper+dose*dur+odrg) $use chitst D2 df2 $
scaled deviance =  122.16 at cycle 5
    residual df =  195    from 263 observations
Chi-squared =   8.774 with    1.000 degrees of freedom and p =  0.0031
```

```
$fit (gall      +dose*dur+odrg) $use chitst D2 df2 $
scaled deviance =  115.27 at cycle 6
    residual df =  195     from 263 observations
Chi-squared =    1.880 with    1.000 degrees of freedom and p =  0.1703


$fit (gall+hyper+dose*dur      ) $use chitst D2 df2 $
scaled deviance =  118.53 at cycle 6
    residual df =  195     from 263 observations
Chi-squared =    5.139 with    1.000 degrees of freedom and p =  0.0234
```

The small p values for `gall` and `odrg` confirm that there is strong evidence that exposure to these risk factors, previous gall bladder disease and the use of drugs other than oestrogen, are strongly associated with the risk of endometrial cancer. Hypertension, on the other hand does not seem to be associated with this risk.

This is a very complex dataset. There are a number of factors associated with the risk of endometrial cancer, but there is some evidence that they interact with each other. There is reasonably strong evidence of an effect of oestrogen, but the form of the effect is not clear. There is evidence that previous gall bladder disease increases the risk and there is also evidence that those reporting taking drugs other than oestrogen are more at risk. Establishing precisely the form of the oestrogen effect and which, if any, of these relationships are in any way causal requires information beyond that available from this study.

12.3 Counted *y* variables, categorical responses and survival data

12.3.3 Poisson and multinomial *y* variables

12.3.1.1 Regression with a Poisson y *variable*

These data arose from an investigation of various media to be used for storing micro-organisms in a deep freeze. The results are bacterial concentrations (counts in a fixed area) measured at the initial freezing (–70°C) and then at 1, 2, 6, and 12 months afterwards.

Time (months)	0	1	2	6	12
Bacterial Concentration	31	26	19	15	20

(Source: Ludlam *et al.* 1989)

The aim is to obtain a model from which fractional recovery rates at specified times after freezing can be predicted.

The practical question is: How long can these organisms be kept at –70°C in this medium and still be usefully recovered?

We need to determine an appropriate model for the way concentration changes over time and use it to estimate recovery rates and a summary measure of how long bacteria stored in this way will, on average, last.

This type of counted variate frequently follows a Poisson distribution, but we cannot necessarily assume it. We must also determine the form of the relationship with time. We start by reading the five concentrations and using ASSIGN to input the times

```
$units 5 $data bac $read
  31 26 19 15 20
$assign mnth=0,1,2,6,12 $
```

A plot shows the form of the observed relationship.

```
$plot (y=0,40) bac mnth $
```

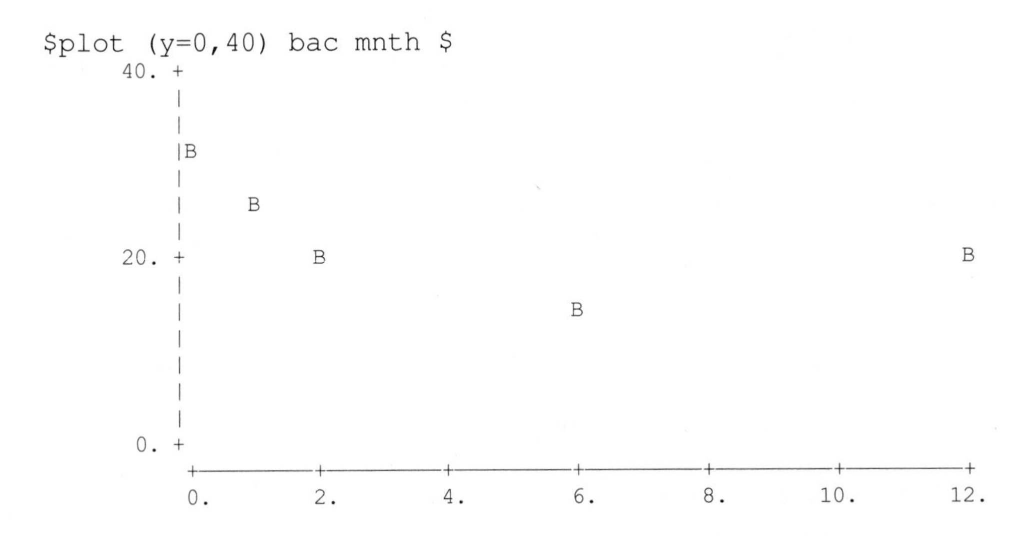

We might have expected some sort of exponential decay curve. The data do not appear to support that largely because the last value is actually an increase over the two preceding it. Even so the true concentrations could still be following such a decay curve with the observed values deviating from it, as here, due to sampling and measurement errors. If that were the case then the logarithm of the concentration would have a straight line relationship with time. We need to investigate the curvature and the error distribution.

The simplest test of whether the apparent curvature is more than chance is achieved by fitting a model quadratic in the time variable. First we need a variable for time squared. While assessing what ERROR is appropriate we will assume Normality and use the residuals to assess whether the Poisson distribution might be more appropriate.

```
$cal mnth2=mnth*mnth $
$yvar bac $
$fit mnth+mnth2 $dis e $
   deviance =   11.883
residual df =    2
          estimate        s.e.      parameter
     1      29.80        1.883       1
     2     -4.616        1.009       MNTH
     3     0.3186      0.08049       MNTH2
scale parameter 5.942
```

The quadratic term coefficient is many times its s.e. so it appears to be necessary. However before interpreting the model we must investigate whether the assumption of Normality is justified.

```
$cal nqs=%nd((%cu(1)-0.5)/%nu)
$sort sres %rs $
$plot sres nqs $
```

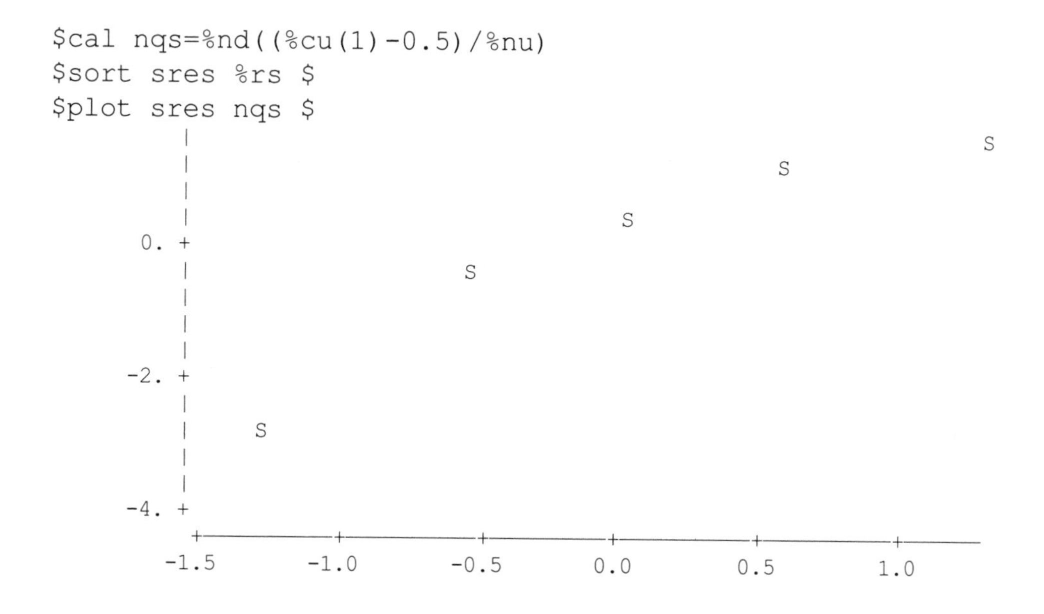

The plotted curve deviates markedly from a straight line. The underlying response distribution does not appear to be Normal.

A look at how the residuals relate to the fitted values should help clarify the mean variance relationship of the `bac` variable. In the Poisson case the variance equals the mean so the residual scatter should increase with the fitted values, that is, the estimated means.

```
$plot %rs %fv 'r' $
```

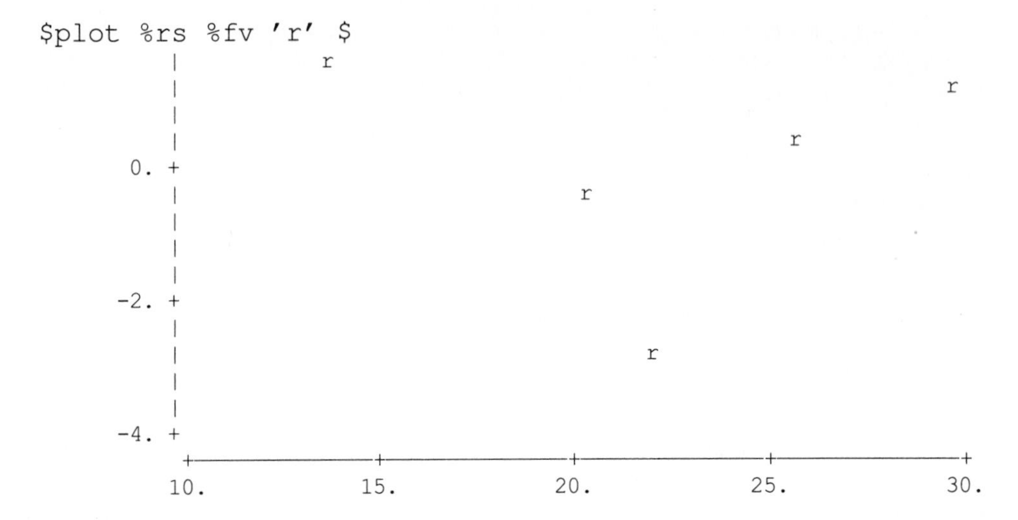

Because there are only five points patterns are difficult to see. It appears a bit rough. Nonetheless it is consistent with systematically increasing variability in the residuals as the fitted mean values increase although it would need more data to be really convincing.

We could attempt to find a transformation to Normality, but if we are prepared to assume a Poisson error it is not necessary. With GLIM we can specify the Poisson error distribution directly. In addition it is usual to represent Poisson means by models linear on the log scale, that is, we use a log link function. This means that if it is some sort of exponential decay curve we can model the means on the log scale which is appropriate while modelling the error distribution about the means as Poisson on the original scale.

To check whether the relationship is still curved on the log scale let us plot it.

```
$cal lbac=%log(bac) $
$plot lbac mnth 'x' $
```

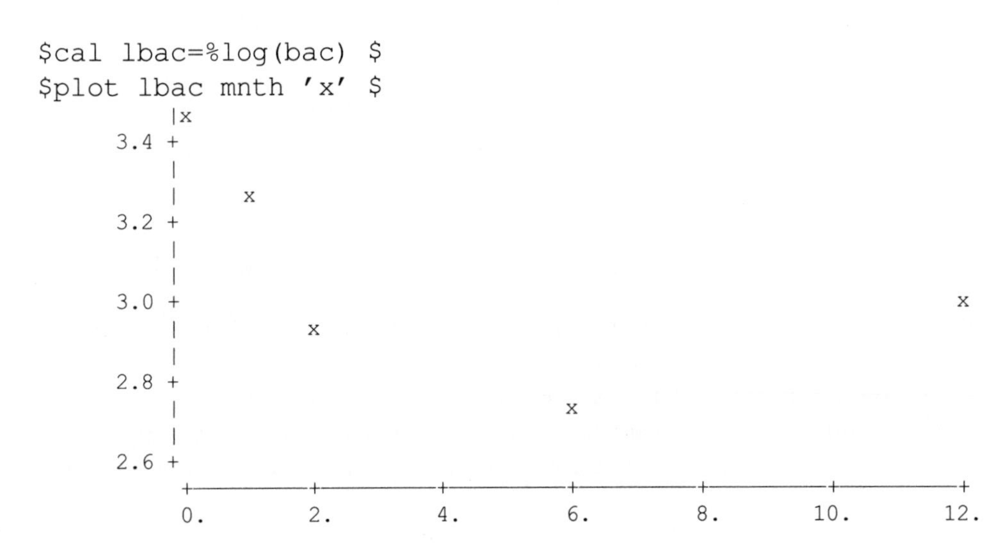

There is still considerable curvature so it is wise to start with a quadratic. We declare the ERROR as Poisson, use the default link LOG and fit the model

$$\log(\text{bacterial count}) = b_0 + b_1 \text{ month} + b_2 \text{ month}^2$$

by

```
$error P $
$fit mnth+mnth2 $dis e $
scaled deviance =  0.27930 at cycle 3
    residual df =  2
          estimate          s.e.      parameter
     1        3.424       0.1490       1
     2      -0.2214       0.09562      MNTH
     3      0.01553     0.007731       MNTH2
scale parameter 1.000
```

Both the deviance and the sum of the squares of the standardized residuals, that is, Pearson's chi-squared statistic, should be about the same as the degrees of freedom and provide a quick test of how well the model fits. Small values indicate very good fits. Large values approaching twice the d.f. indicate bad fits or overdispersion.

```
$pr %x2 $
  0.2745
```

The deviance and the Pearson chi-square are actually smaller than the degrees of freedom. The model fits very closely to the data and there is no evidence here of variation beyond what would be expected in Poisson data or that the assumption of a Poisson error distribution is inappropriate. We will test whether the quadratic team is needed in the model by comparing the fit with a model omitting it. First it is necessary to test whether the data shows the relationship between the mean and variance appropriate for the Poisson distribution. To do this we look at the standardized residuals.

Taking the variance as equal to the mean the residuals from the 'true' model would have variances equal to the fitted values. Thus dividing by the square root of the fitted values gives 'standardised' residuals which, if the distribution is Poisson, should have unit variance and give a moderately straight horizontal plot. The residuals GLIM provides, %rs, are automatically standardized in this way.

```
$plot (y=-2,2) %rs %fv 's' $
```

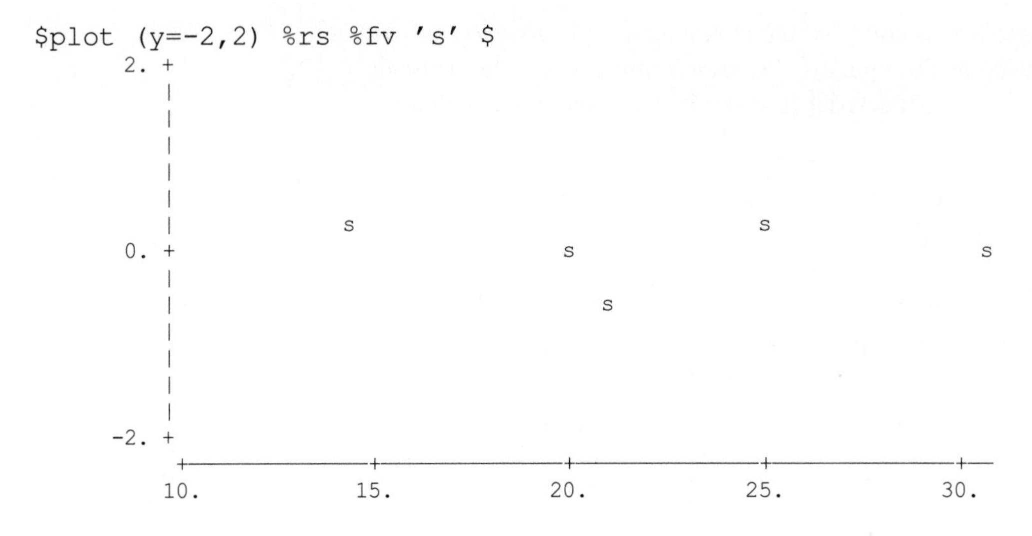

To see how closely the model fits the observed data points plot the fitted and observed values against month:

```
$plot bac %fv mnth 'oe' $
```

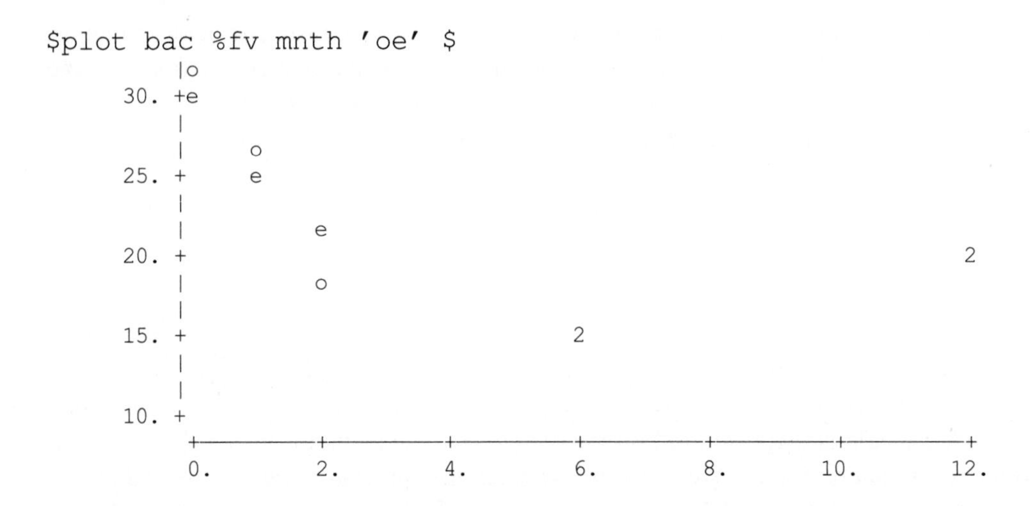

With only five points it would be difficult to see even if there were any sort of a trend, but the first plot looks relatively horizontal so the Poisson error assumption seems appropriate. The second plot seems to show the fitted values describing the curved relationship very well, but a three parameter polynomial is almost bound to fit a sequence of five points reasonably well. In addition, although measurement errors might lead to an increase in count, the true relationship should not increase. Because the quadratic increases after a certain time it cannot be the correct model. Although the coefficient is larger than its s.e. we should test it with a model comparison. First we need to store the deviance and degrees of freedom from the quadratic model.

```
$number D0=%dv df0=%df $
$fit mnth $dis e $
scaled deviance =  4.5249 at cycle 3
    residual df =  3
          estimate        s.e.     parameter
     1      3.240       0.1259      1
     2    -0.03613     0.02329      MNTH
scale parameter 1.000
```

The difference in deviance gives an approximate Chi-squared test on 1 d.f. for the quadratic term and the function `%chp(chi,df)` may be used to obtain the p-value for the test of significance.

```
$number D1=%dv df1=%df D10 df10 p10 $
$cal D10=D1-D0 : df10=df1-df0 : p10=1-%chp(D10,df10) $
$pr 'Chi-sqd ='D10' on 'df10' df with p-value ='p10 $
Chi-sqd =    4.246 on    1.000 df with p-value =  0.0394
```

Since this is less than 0.05 it is significant at the 5 per cent level and the curvature in the relationship is more than could easily occur by chance. Nonetheless we know that a quadratic relationship is not appropriate. It is possible that a simpler model may be achieved by transforming the time axis. Since the effect of time on the survival of the bacteria might well be multiplicative there is some argument for using a logarithmic scale on the time axis. However the zero origin must be moved by adding a constant to avoid the problem of the value of minus infinity at log(zero).

The transformation needed is log(month+constant), but there is a problem of determining what the constant should be. A useful strategy is to choose it to minimise the deviance from a suitable model. The simplest model to interpret will be a straight line relationship with the transformed explanatory variable. The 'best' value for the constant can be found by fitting the model

```
1 + log(mnth+c)
```

over a suitable range of values for `c`. This is most conveniently done by using a macro repeatedly. An appropriate macro is

```
$mac (arg=i,l,c,x,f,dd loc=lm) loop
$output $                    ! to suppress output during the iteration
$cal lm=%log(x+c(i)) $       ! re-calculate the log with the new origin
$fit lm $                    ! fit the  model
$output 6 $                  ! restore the output
$ass dd=dd,%dv $             ! store the deviance
$ca i=i+1 : f=(i < l+1) $ ! increment i and set flag f (=1 to stop)
$endmac                      ! where l is length of vector'c'
```

Now assign values to the vector `const` with length `%l`, set counter `i` to 1 and flag `flg` non-zero. Then set the macro to execute repeatedly, until the flag becomes 0, with the `WHILE` directive.

```
$assign const=0.01,0.02,...0.2,0.3,...1.0 $
$cal %l=%len(const) $
$num i=1 flg=1 $
$arg loop i %l const mnth flg devs $while flg loop $
```

The deviances obtained can now be plotted against the values of `const` to identify where it is smallest.

```
$graph ( t='Fig. 12.37 Deviance against Constant in log(x+const)')
        devs  const  10 $
```

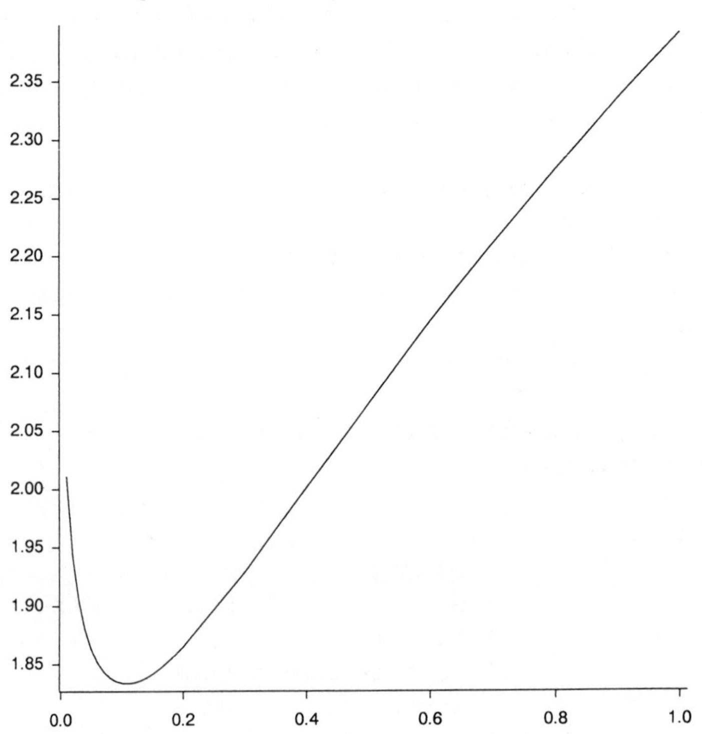

A clear indication that there is an unambiguous minimum around `const=0.1`. The values of the minimum deviance and the corresponding constant are obtained by sorting the two vectors with the permutation which puts the deviances into ascending size order

```
$sort sdevs;sconst devs;const devs $
$pr 'Deviance ='sdevs(1)' Constant ='sconst(1) $
Deviance =    1.834 Constant =  0.1100
$cal %z=sconst(1) $
```

This value of the constant can now be used for the log transformation and the models fitted explicitly. First of all it is wise to fit a model with a quadratic term to test whether there is still some curvature to explain.

```
$cal lm=%log(mnth+%z) : lgm2=lm*lm $

$fit lm+lgm2 $dis e $
scaled deviance =  1.7863 at cycle 3
    residual df =  2
          estimate         s.e.     parameter
     1       3.127       0.1554      1
     2     -0.1282      0.05517      LM
     3    0.008756      0.04027      LGM2
scale parameter 1.000
```

The coefficient of the quadratic term is now not significant so it suggests that we can simplify the model further. First save the deviance and degrees of freedom for a more formal test and then fit the model without the quadratic term.

```
$number D0=%dv  df0=%df $
$fit lm $dis e $
scaled deviance =  1.8336 at cycle 3
    residual df =  3
          estimate         s.e.     parameter
     1       3.153      0.09577      1
     2     -0.1289      0.05554      LM
scale parameter 1.000
```

The quadratic term can now be tested with a likelihood ratio chi-squared test by calculating the difference and the *p*-value from the chi-squared distribution with 1 degree of freedom.

```
$cal %x=%dv - D0 : %p=(1-%chp(%x,1)) $
$pr 'Chi-sqd ='%x' on 1 df with p-value ='%p $
Chi-sqd =  0.0473 on 1 df with p-value =  0.8278
```

It is nowhere near the 3.84 needed for significance for 1 degree of freedom. It appears that with the time axis transformed in this way a linear trend is adequate to describe the data. This model is the one to use for estimating the 'half-life' and the proportion surviving at any given time. However we must test the fit of this simplified model before using it for any sort of prediction

```
$plot (y=-2,2) %rs %fv 's' $
```

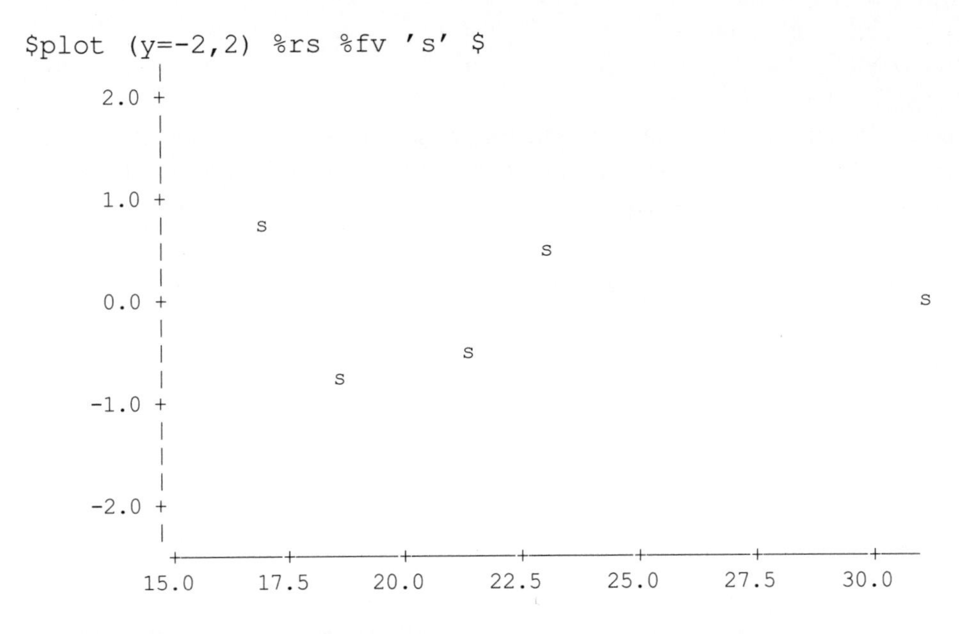

These are more or less horizontal and the spread is relatively constant. It appears that they justify the Poisson assumption.

The observed values and the fitted curve are obtained by

```
$plot bac %fv  mnth 'x.' $
```

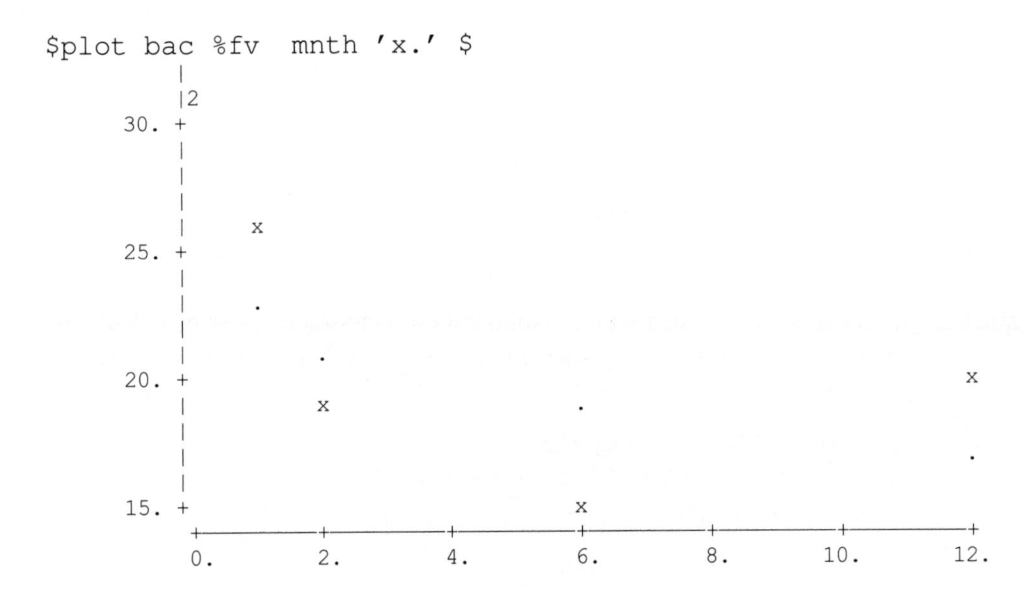

Using PREDICT to obtain predicted values (%pfv) over the whole time period to get a smooth curve the high–resolution equivalent is obtained as follows:

```
$ass time_p=0,0.002,...0.01,0.02,...0.1,0.2,...1,2,...36 $
$cal lm_p=%log(time_p+%z) $
$predict lm=lm_p $
prediction for the current model with LM=LM_P

$graph (t='Fig. 12.38 Data and fitted curve'
       v='Fitted log(counts)'
       h='Time in months' xscale=0,18,6 yscale=0,40,10 )
       bac;%pfv mnth;time_p 1;10 $
```

It appears to be a reasonably good fit. Now to investigate the influence of the individual points we again use the leverage values.

```
$extract %lv $
```

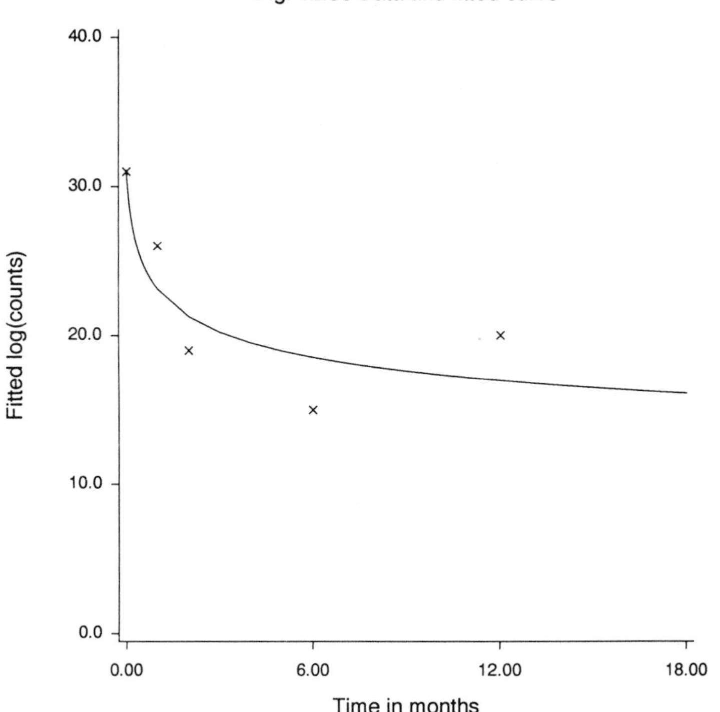

```
$cal indx=%ind(%lv) $
$plot %lv indx 'x' $
```

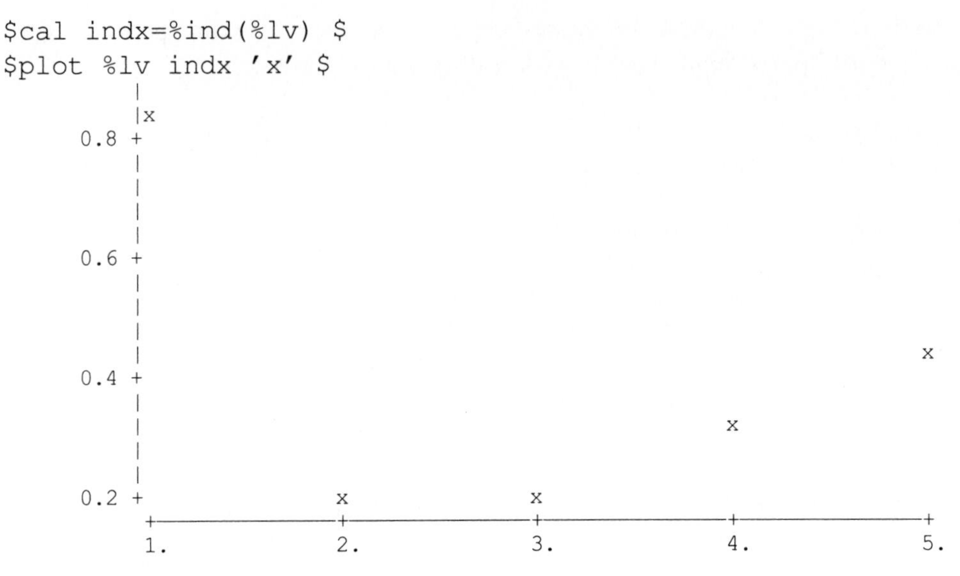

The observations at the extremes of the covariate range have the most influence. The range of values is slightly less so that means the influence is more evenly shared. This is much as should be expected from a model involving a single covariate.

This appears to be an equally adequate model compared with the earlier quadratic one and simpler at the cost of a slightly larger deviance.

The simplest way to obtain confidence limits for the line is by using the results from the `PREDICT` directive above. This generates not only the predicted fitted values for the specified model variable settings (`%pfv`), but also the predicted linear predictors and their variances (`%plp` and `%pvl`).

```
$cal low=%plp - 1.96*%sqr(%pvl): high=%plp + 1.96*%sqr(%pvl)  $
$cal low=%exp(low) : high=%exp(high) $
$plot bac low %pfv high mnth;time_p;time_p;time_p 'x^.v' $
         |v
         |6
     40. +4
         |4
         |
         |v
         |32
     30. +93
         |3v2
         | 2xv
         | 42  v v vv v vv v vv v
         |9^ .                   v vv v vv v vv v v vv v vv v vv v v vv
     20. + 52x . .         x
         |   ^     .. . . .. . .
         |     ^ ^  x        . . . .. . .. . .. . . .. . . .. . .
         |         ^^ ^ ^^                                          . . . ..
         |               ^ ^^ ^ ^ ^^ ^ ^^ ^ ^
     10. +                                   ^ ^ ^ ^^ ^ ^^ ^ ^^ ^ ^ ^^
         |
         +---------------+-----------------+-----------------+-----------
         0.             10.               20.               30.
```

A high–resolution plot of this can be obtained with

```
$macro xlab time in months since freezing  $end $
$macro ylab Concentration of bacteria $end $
$macro titl Fig. 12.39 Recoverable bacterial concns by time
$end $
$gra ( t=titl h=xlab v=ylab x=0,18,3 y=0,80,10)
      bac;%pfv;low;high mnth;time_p;time_p;time_p 1,10,12,12 $
```

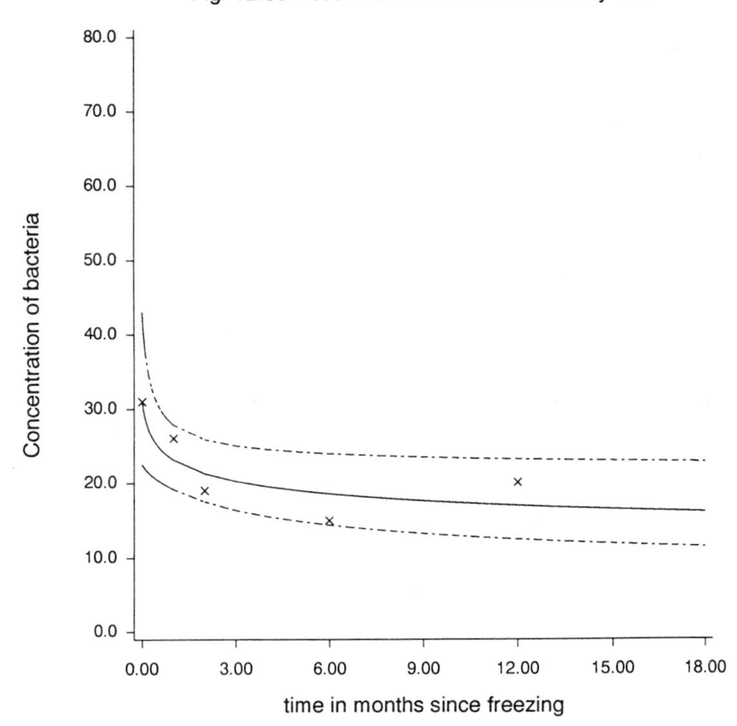

This shows us the fitted relationship on the original scale and could be used for prediction of various sorts. However the two measures of real interest are:

1. the half life of bacteria stored this way;

2. the proportion recoverable at 12 months (say).

1. Half-life

The derivation of a half-life in this context is slightly complicated by the transformation of the *x*-axis so on the time axis the half-life is not constant, but depends on the starting point. However taking 'freezing' as the origin we get the most relevant quantity. We start with the model for the expected y (bac) on the transformed axis `lm=log(mnth+%z)`. We want the distance lm_1 to lm_2 such that

$$E(y_1)=2\ E(y_2) \text{ that is, } \mu_1=2\mu_2.$$

Now with the log link and the linear model we have

$$\eta_i = \log(\mu_i) = a + b.\mathrm{lm}_i$$

so we need

$$\exp(a+b.lm_1) = 2\exp(a+b.lm_2)$$

that is, $\exp(a + b.lm_1 - a - b.lm_2) = 2$

so $$\exp(b.(lm_1-lm_2)) = 2$$

and $$lm_1-lm_2 = \log(2)/b$$

Taking lm_1 to be log(%z), the equivalent of month zero we get the point on the `lm` scale at which the concentration is predicted to have halved as lm_2=`log(%z) - log(2)/`b. For confidence limit calculations we have to use approximate s.e.s for the reciprocal of b. We obtain these from the variance of $1/b$ which is $\mathrm{var}(b)/b^4$ so the calculations in GLIM are

```
$extract %pe %se $
$number lm2 lcl ucl selm lg2 half loha hiha $
$cal lg2=%log(2)  : lm2=%log(%z)  - lg2/%pe(2)
 :  selm=lg2*%se(2)/(%pe(2)*%pe(2))
 :   lcl=lm2-1.96*selm : ucl=lm2+1.96*selm
 :  half=%exp(lm2)-%z : loha=%exp(lcl)-%z : hiha=%exp(ucl)-%z $
```

So the estimated time from freezing for the concentration to halve with approximate 95 per cent confidence limits is

```
$pr 'Half-life  ='half' months with 95% CLs ='loha'  to 'hiha ;$
Half-life  =   23.72 months with 95% CLs =  0.1437  to    2238.
```

This is clearly a very imprecise estimate. The combination of a flat curve and wide confidence limits means that a very large range of values on the time axis are consistent with a 50 per cent decrease in concentration from freezing.

2. Proportion Recoverable

The estimated proportion recoverable (`prec`) at a particular time, 12 months say, is

$$E(y_{12})/E(y_0) = \mu_{12}/\mu_0 = A.\exp(b.lm_{12})/(A.\exp(b.lm_0))$$

that is, `prec` $= \exp(b.(lm_{12} - lm_0))$

Therefore all we need is an estimate of $b.(lm_{12}-lm_0)$ with confidence limits and then take antilogs: that is,

```
$number prec lprc lopr hipr lm51 $
$cal lm51=lm(5)-lm(1)   : lprc=%pe(2)*lm51
: lopr=lprc-1.96*lm51*%se(2)  : hipr=lprc+1.96*lm51*%se(2)
: lopr=%exp(lopr)  : prec=%exp(lprc)  : hipr=%exp(hipr) $
```

The estimated proportion recoverable at 12 months with approximate 95 per cent confidence limits is

```
$pr ' Proportion ='prec ' with 95% CLs =' lopr'  to'hipr;$
Proportion =  0.5456 with 95% CLs =  0.3270  to  0.9101
```

This is also a relatively imprecise estimate, but it does indicate that at least a third of samples stored this way should be recoverable after 12 months.

The analysis shows how a model may be identified, fitted and used with this type of data. However the data were a bit skimpy. If anything important depended on the estimates the originators should be advised that the results should be confirmed with more comprehensive data obtained under carefully controlled experimental conditions.

12.3.1.2 Two way contingency tables with a response factor

These data arose from a multi-centre clinical trial of an improved treatment for Hodgkin's disease. The outcome of treatment after a fixed period of time was described by a classification according to remission status. The categories were: 'None/died', 'Partial' and 'Complete'. The distribution of patients according to outcome by treatment group was

	Remission category		
Treatment	None/died	Partial	Complete
A	16	19	15
B	11	4	31

This is actually a three-category outcome problem (trinomial). Ideally one would want to model the variation in how individuals fell on the remission category scale. However this is not straightforward in GLIM.

Effectively the analysis needs to assess how the distribution among the three outcome categories differs between the two treatment groups. If one treatment were markedly better than the other the relevant distribution would be shifted to the right with greater proportions in the partial and complete remission categories. However there are many ways in which the response distributions could differ in shape.

The analysis is most easily done using the log-linear modelling approach for contingency tables which uses what has become known as the Poisson 'trick'. This is achieved in modelling multinomial frequency data by treating them as Poisson variables and constraining the fitted values to have the same totals as those observed. As a result of this the *y*-variate used for the modelling is not the categorical outcome, but the frequencies in the various categories. This means that the analysis is modelling the variation in shape of the frequency polygons of the distributions and not individual shifts from one outcome category to another. One consequence of this is that vertical differences between one distribution's frequencies and another, that is, the main effects in a model, simply reflect differences in the sample sizes. They are a consequence of the design and convey no information on how the explanatory variables distinguishing the different distributions affect outcome. The interesting differences are those of shape reflecting how the individuals are dispersed along the categorical outcome scale. These are interaction terms in the model.

The data are input and appropriate factor variables generated by

```
$units 6 $data freq $read
 16           19          15
 11            4          31

$gfactor trt 2 remc 3 $
```

The two distributions are displayed with

```
$plot (x=0,4) freq  remc 'ab'  trt $
```

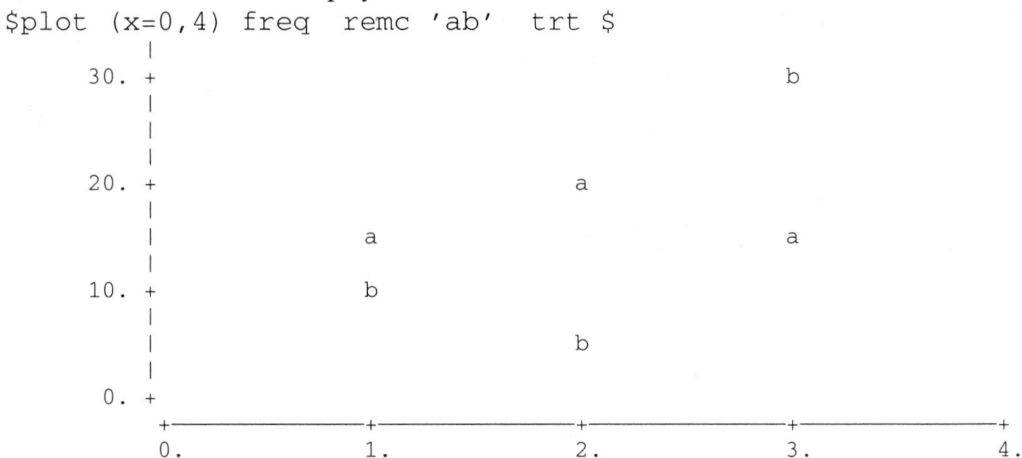

The distribution for the A treatment group (a) seems relatively uniform with roughly a third in each remission category, while that of the B treatment patients seem to be proportionately more in the 'complete' remission category. The pattern is reasonably clear, but try a logarithmic transformation of the frequencies since that is the most commonly used scale when the response variable has a Poisson distribution.

```
$cal logf=%log(freq) $
$plot (x=0,4) logf  remc 'ab'  trt $
```

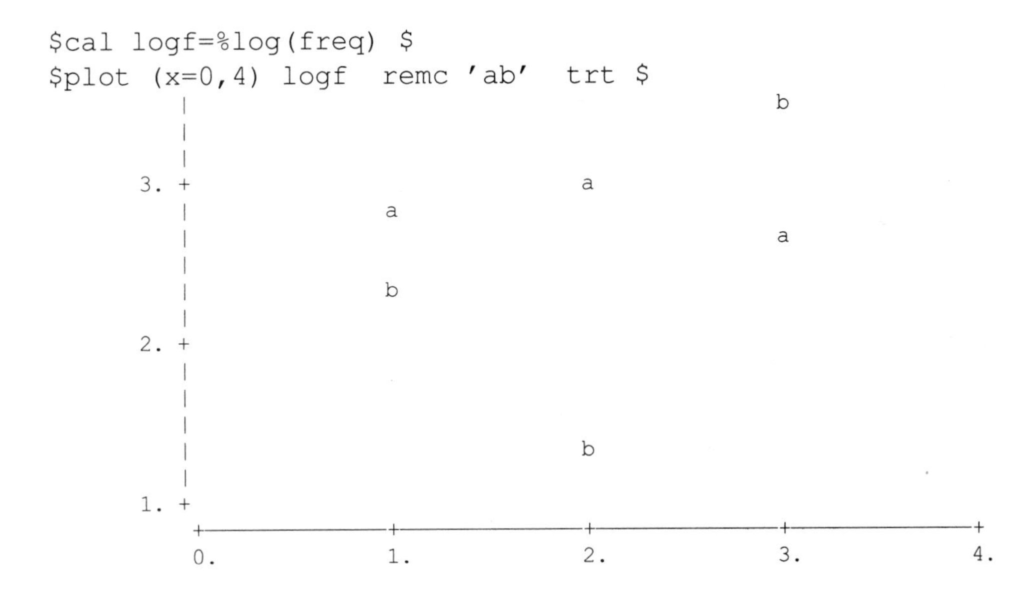

A plot almost identical in shape. Declare the y variable and Poisson response variable distribution and keep the default log link by

```
$yvar freq $error P $
```

and fit the saturated model with the treatment and remission factors and their interaction

```
$fit trt*remc $dis e $
scaled deviance =  0. at cycle 2
    residual df =  0
          estimate         s.e.      parameter
     1       2.773       0.2500       1
     2     -0.3747       0.3917       TRT(2)
     3      0.1719       0.3393       REMC(2)
     4    -0.06454       0.3594       REMC(3)
     5      -1.183       0.6753       TRT(2).REMC(2)
     6       1.101       0.5023       TRT(2).REMC(3)
scale parameter 1.000
```

Six parameters use all the information and the model fits exactly with no deviations from the data and hence zero deviance. This is a saturated model. However the parameters modelling the non-parallelism are large, of opposite sign and both around twice their s.e.s. We need to see the effect on the deviance of omitting them. First let us look at the observed and fitted values:

```
$tpr (s=1) freq;%fv trt;remc $
               +------------------------------------------+
       REMC    |      1            2             3        |
   TRT         |                                          |
+--------------+------------------------------------------+
|   1 FREQ     |   16.000       19.000        15.000      |
|       %FV    |   16.000       19.000        15.000      |
---------------------------------------------------------------
|   2 FREQ     |   11.000        4.000        31.000      |
|       %FV    |   11.000        4.000        31.000      |
+--------------+------------------------------------------+
```

The fitted frequencies are the same as observed. We need an unsaturated model if we are to assess the goodness (or badness) of fit. This will mean that the fitted frequencies differ from those observed, but their totals must remain equal to those observed because the analysis is conditional on the totals remaining fixed. This is achieved by keeping all the single factors in the model (that is, for the main effect parameters) and all the interactions not involving the outcome variable factor.

Now fit the model equivalent to assuming the 'frequency polygons' are parallel on the log(frequency) scale

```
$fit trt+remc $dis e $
scaled deviance =   17.079 at cycle 4
    residual df =   2
          estimate          s.e.      parameter
    1        2.644        0.2159       1
    2     -0.08338        0.2043       TRT(2)
    3      -0.1603        0.2837       REMC(2)
    4       0.5328        0.2424       REMC(3)
scale parameter 1.000
```

The deviance is a measure of how non-parallel the log(frequency) polygons actually are and can be used as an approximate chi-squared test of whether the population outcome distributions for these two treatments differ in shape. The probability of a chi-squared value on 2 degrees of freedom that large is

```
$cal 1-%chp(%dv,%df) $
      0.0001956
```

which is very small and therefore strong evidence that the two outcome distributions differ in shape. The simple parallel model does not fit the data adequately. To illustrate this it is useful to tabulate the observed and fitted frequencies with their totals

```
$tpr (s=1) freq;%fv trt;remc $
            +---------------------------------+
       REMC |     1         2         3      |
  TRT       |                                 |
+-----------+---------------------------------+
|    1 FREQ |  16.000    19.000    15.000    |
|       %FV |   14.06     11.98     23.96    |
+---------------------------------------------+
|    2 FREQ |  11.000     4.000    31.000    |
|       %FV |   12.94     11.02     22.04    |
+-----------+---------------------------------+
```

By forcing parallelism on the log(frequency) scale the fitted values have been forced away from those observed, but the totals are the same.

```
$tab (s=1) the %fv totals for trt by trt_ $
         +-------------------+
    TRT_ |   1         2     |
+--------+-------------------+
| TOTAL  | 50.00     46.00   |
+--------+-------------------+
$tab (s=1) the %fv totals for remc by rem_ $
         +-----------------------------+
    REM_ |   1         2         3     |
+--------+-----------------------------+
| TOTAL  | 27.00     23.00     46.00   |
+--------+-----------------------------+
```

The model and data may be displayed graphically with

```
$graph (t='Fig. 12.40 Saturated and Parallel Models'
        v='Frequencies' h='Remission Category'
        xscale=0,4.5,1 yscale=0,40,10 p=no )
        freq;freq;%fv remc 1,4,11,11,10,15  trt;trt;trt $
$ass gtx=3.1,3.1,3.1 : gty=15,31,23 $
$gtext (m=p l=r p=no) gty gtx
      'Treatment A','Treatment B','"Parallel" model' $
$ass gtx=1,2,3 : gty=1,1,1 $
$gtext (m=p l=c ) gty/1.1 gtx 'NONE','PARTIAL','COMPLETE' $
```

Because the parallel model is inadequate to describe the variation observed we cannot reject the saturated non-parallel model. This implies that there are more cases in the ‘complete remission’ category and less in the ‘partial remission’ and ‘no remission’ categories than would be expected by chance. This indicates that treatment B is somewhat better than A overall.

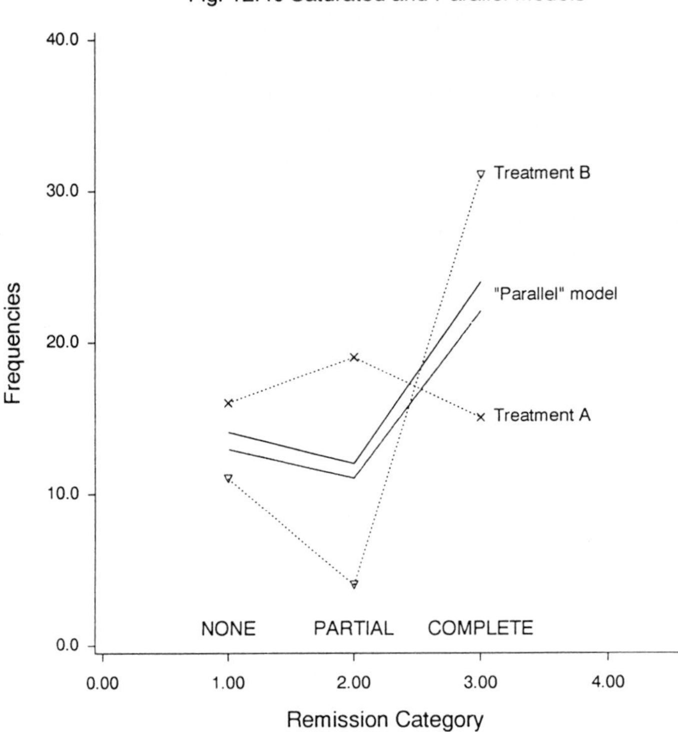

12.3.1.3 General cross classifications of frequencies with a response factor

This data was obtained from the MRC Derbyshire Smoking Study 1974–1984 (Swan et al 1991) in which a cohort of more than 6000 children were followed-up through their secondary schooling and subsequently as young adults. The analysis here is to determine how much the smoking behaviour of young adult males is associated with parental smoking when they were young children (at age 11–12 years). The parent questionnaires and those to young adults in 1984 were administered by post. The parental response in 1974 was 86%, but that for the young adults in 1984 only reached 60%. As a result data on the individual's smoking habits as young adults and that of their parents in 1974 were only available for 3387 subjects of whom 1684 were male.

The distribution of 1684 male subjects according to their smoking category when young adults by sex and parental smoking 10 years earlier is given in the table.

Current Smoking Behaviour

Parental smoking	Non-smoker	Ex-smoker	Occasional smoker	Regular smoker
Neither	279	79	72	110
Mother	109	29	10	64
Father	192	48	32	131
Both	252	65	32	180

The interesting questions are:

How much influence does parental smoking during early childhood have on adult smoking behaviour?

and

Do the effects of mother's and father's smoking differ?

We need to quantify and test how much the distribution among the current smoking categories (`s84`) differs according to the parental smoking factors.

First we input the 16 frequencies and calculate factor variables to represent the smoking categories of the mothers (`ms`), the fathers (`fs`), both parents together (`ps`) and of the subject's themselves in 1984 (`s84`).

```
$units 16 $data f $read
   279       79          72          110
   109       29          10           64
   192       48          32          131
   252       65          32          180

$cal s84=%gl(4,1) : fs=%gl(2,8) : ms=%gl(2,4) : ps=%gl(4,4) $
$factor fs 2 ms 2 ps 4 s84 4 $
```

The distribution of all the subjects among the smoking categories is obtained by

```
$tab the f total for s84 into s84t by s84_ $
$tprint s84t s84_ $
 S84_     1      2      3      4
 S84T  832.0  221.0  146.0  485.0
```

and plotted as a frequency polygon is

```
$plot (x=0,5) s84t s84_ 'T' $
```

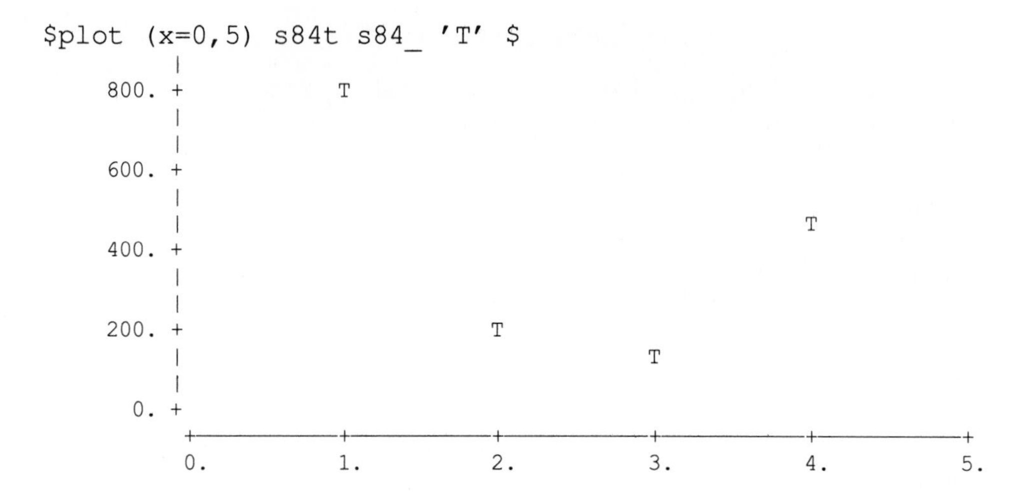

To see how the distributions change over the four parental smoking categories requires a plot of the frequencies against smoking category with the parental smoking categories identified by different symbols (`n` neither; `m` mother; `f` father; and `b` both).

```
$plot (x=0,5) f s84 'nmfb' ps $
```

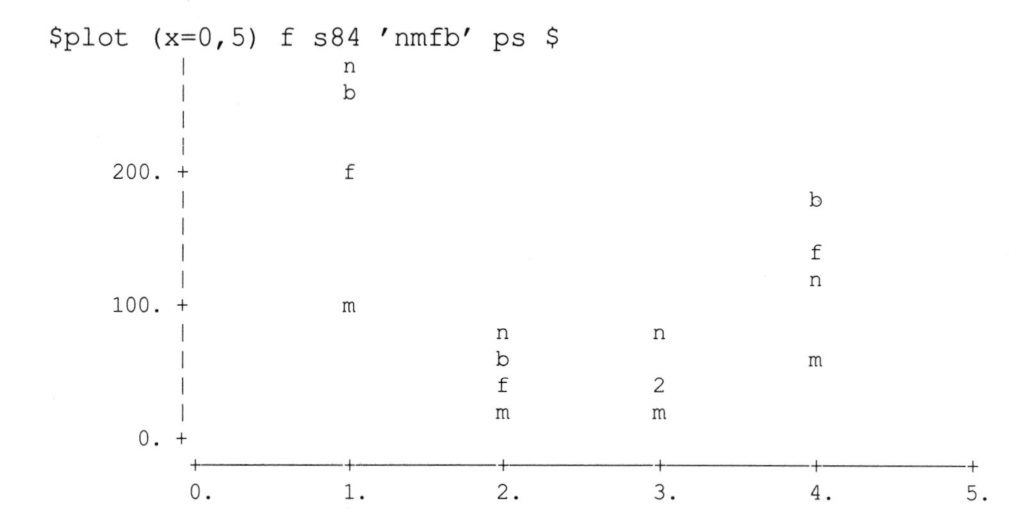

The graphical equivalent of this is obtained with

```
$graph ( t='Fig. 12.41 Boy''s smoking by parental smoking'
         v='Frequency'
         h='Non-smoker Ex-smoker Occ-smoker Regular'
         p=no xscale=0,5,1 y=0,300,50 )
         f    s84  10;12;13;11 ps $
```

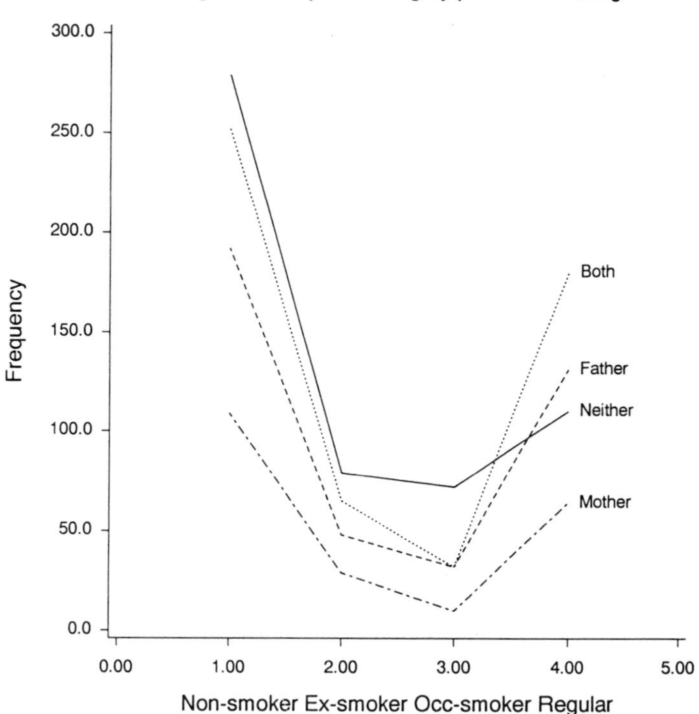


```
$ass gtx=4.1,4.1,4.1,4.1 : gty=110,64,131,180 $
$gtext (m=p l=r) gty gtx 'Neither','Mother','Father','Both' $
```

Using relative frequencies we can correct for the unequal numbers in the four parental smoking categories.

```
$tab the f total for ps into row_tot $
$cal pfreq=f/row_tot(ps) $

$graph ( t='Fig. 12.42 Boy''s smoking by parental smoking'
         v='Relative Frequency'
         h='Non-smoker Ex-smoker Occ-smoker Regular'
         p=no xscale=0,5,1 y=0,0.7,0.2 )
         pfreq  s84  10;12;13;11 ps $

$ass indx=1,2,3,4 $
$cal rf_gty=gty/row_tot(indx) $
$gte (m=p l=r) rf_gty gtx 'Neither','Mother','Father','Both' $
```

The general 'U' shape of the dis ͜utions is apparent, but how they compare is not at all easy to see. When they are converted to relative frequency distributions some features are

Fig. 12.42 Boy's smoking by parental smoking

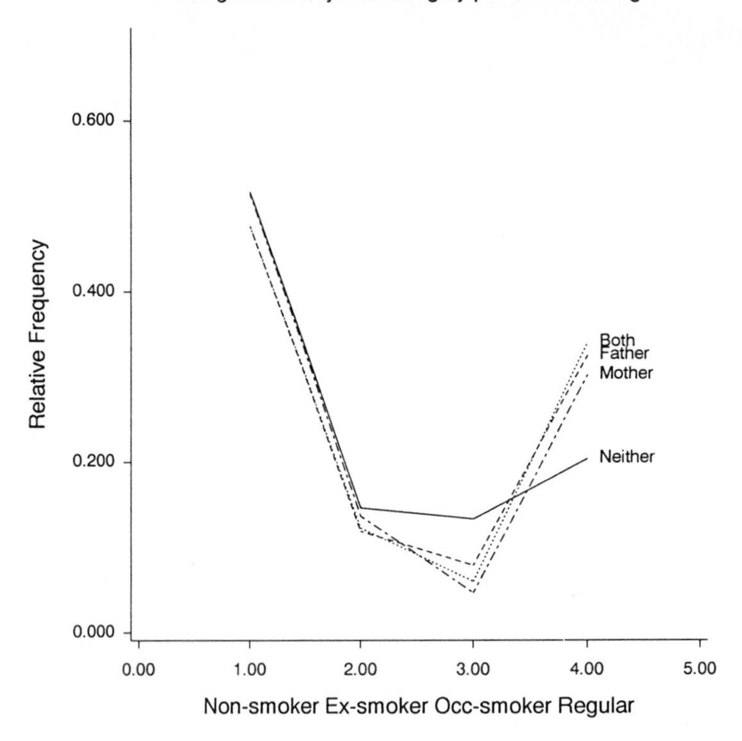

more clear, that is, the systematic difference between the boys without either parent smoking and the rest. However the other three lines fall so close together that differences in their orientation are obscured. The most useful approach is to use a logarithmic scale for the frequencies. This stops the very large values compressing the scale lower down so differences between the small frequencies are obscured. It is also the canonical link scale that GLIM assumes by default for Poisson distributed responses.

```
$cal logf=%log(f) $
$plot (x=0,5) logf s84 'nmfb' ps $
```

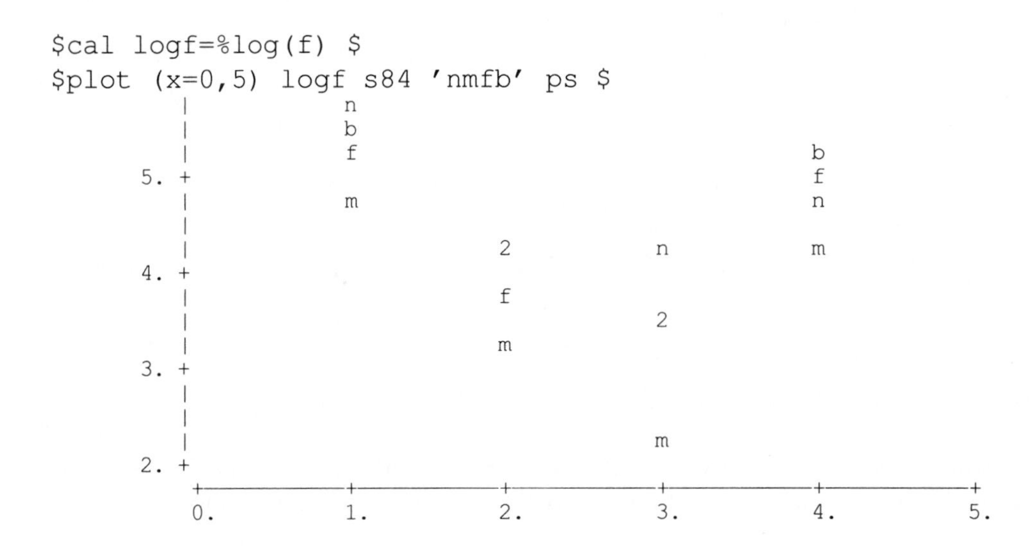

```
$graph ( t='Fig. 12.43 Boy''s smoking by parental smoking'
        v='Log(frequency)'
        h='Non-smoker Ex-smoker Occ-smoker Regular'
        p=no xscale=0,5,1 y=0,6,1 )
        logf  s84  10;12;13;11 ps $
```

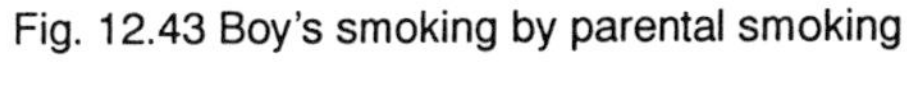

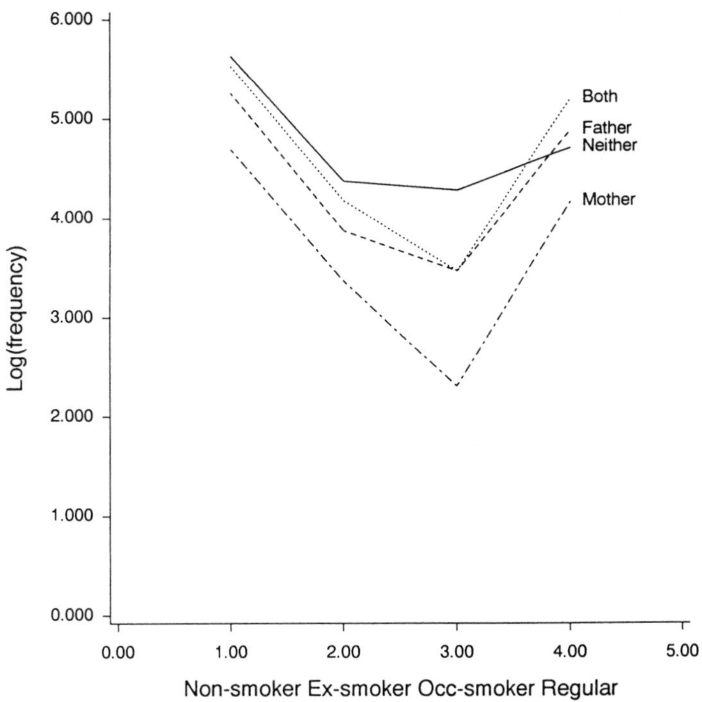


```
$cal lf_gty=%log(gty) $
$gte (m=p l=r) lf_gty gtx 'Neither','Mother','Father','Both' $
```

This scale gives a slightly clearer picture, but not much. However we can just about see that parental smoking appears to result in a shift of the distributions from the non-smoking to the smoking categories. To answer the questions above we need to describe this shift quantitatively with a model of some sort. The outcome or response scale is really the categorical `s84` scale. However it is simpler to model the frequency polygons, describe how they differ, and then interpret those differences in terms of shifts of the distributions over the 'current smoking categories' (`s84`).

```
$yvar f $error p $
```

We must fit all the possible terms involving `fs` and `ms` to ensure that the fitted totals in all the margins and cells of the `fs` by `ms` table equal those observed. This means that `fs*ms` must be in the model. The parameters involved are 'nuisance' parameters and of no interest in themselves and there are no other terms in the model involving `fs*ms` so it can be pre-fitted by the `ELIMINATE` directive.

The totals marginal to the subject's own smoking factor `s84` also need to be fixed so `s84` must be present. However only single terms may be eliminated and the sum of two terms such as $A*B + C$, is not a single term so `s84` cannot be eliminated as well as `fs*ms`. It has to be explicitly present in all the models.

The non-parallelism effects of interest are represented by the two-factor interactions between the independent variable factors (that is, `fs` and `ms`) and the dependent or outcome variable factor `s84`. The genuine interactions representing the effects due to one independent variable modifying the effects of another are represented by model terms involving three or more factors. Thus the model, using `ELIMINATE`, to include these and the three factor interaction which we need to test first is

```
$eliminate fs*ms $
$fit fs*ms*s84  $dis e $
scaled deviance =  0. at cycle 2
    residual df =  0
          estimate          s.e.      parameter
     1      -1.262        0.1274       S84(2)
     2      -1.355        0.1322       S84(3)
     3      -0.9307       0.1126       S84(4)
     4      -0.1245       0.2056       FS(2).S84(2)
     5      -0.4372       0.2322       FS(2).S84(3)
     6       0.5484       0.1597       FS(2).S84(4)
     7      -0.06229      0.2447       MS(2).S84(2)
     8      -1.034        0.3559       MS(2).S84(3)
     9       0.3983       0.1936       MS(2).S84(4)
    10       0.09354      0.3245       FS(2).MS(2).S84(2)
    11       0.7623       0.4453       FS(2).MS(2).S84(3)
    12      -0.3524       0.2446       FS(2).MS(2).S84(4)
scale parameter 1.000
    eliminated term: FS.MS

$number D1 D2 D3 D4 df1 df2 df3 df4 $
$cal D1=%dv : df1=%df $
```

This is a saturated model and so provides no information on goodness of fit. Even so the parameter estimates and standard errors can still be interpreted and we can see that the three-way interaction term parameters are large compared to their s.e.s. We need an overall test of them taken as a whole. This is achieved by fitting the model without them and using the change in deviance as a chi-squared test.

```
$fit                                   -fs.ms.s84 $dis e $
scaled deviance =  6.7587 (change =   +6.759) at cycle 3
    residual df =  3      (change =   +3     )
          estimate        s.e.       parameter
     1      -1.276      0.1178        S84(2)
     2      -1.421      0.1283        S84(3)
     3     -0.8532     0.09848        S84(4)
     4    -0.08669      0.1578        FS(2).S84(2)
     5     -0.2327      0.1877        FS(2).S84(3)
     6      0.3982      0.1222        FS(2).S84(4)
     7    -0.01032      0.1594        MS(2).S84(2)
     8     -0.5744      0.2029        MS(2).S84(3)
     9      0.1694      0.1196        MS(2).S84(4)
scale parameter 1.000
    eliminated term: FS.MS
$cal D2=%dv : df2=%df $
```

The deviance has increased by 6.76 as a result of dropping the 3 three-factor interaction parameters. Taken as an approximate (likelihood ratio) chi-squared on 3 d.f., it is not significant at the conventional 5 per cent level ($p = 0.08$). It would need to reach 7.82 for formal significance with $p < 0.05$. Nonetheless it cannot really be dismissed, but for the purposes of this analysis we will put it aside for the moment. A graph with this model and the observed frequencies on the log scale illustrates how well it fits.

```
$graph ( t='Fig. 12.44 Observed and Fitted log(frequencies)'
         v='Log(frequency)'
         h='Non-smoker Ex-smoker Occ-smoker Regular'
         p=no xscale=0,5,1 y=0,6,1 )
         logf;%lp  s84  1;2;3;4;10;12;13;11 ps;ps $
```

```
$gte (m=p l=r) lf_gty gtx 'Neither','Mother','Father','Both' $
```

To see the pattern of the interactions clearly the lines representing the model should all start from the same point. This is achieved by subtracting the appropriate `s84(1)` values from the *y* co-ordinates

```
$ass idx=1,1,1,1,5,5,5,5,9,9,9,9,13,13,13,13 $
$cal alp=%lp-%lp(idx) $
```

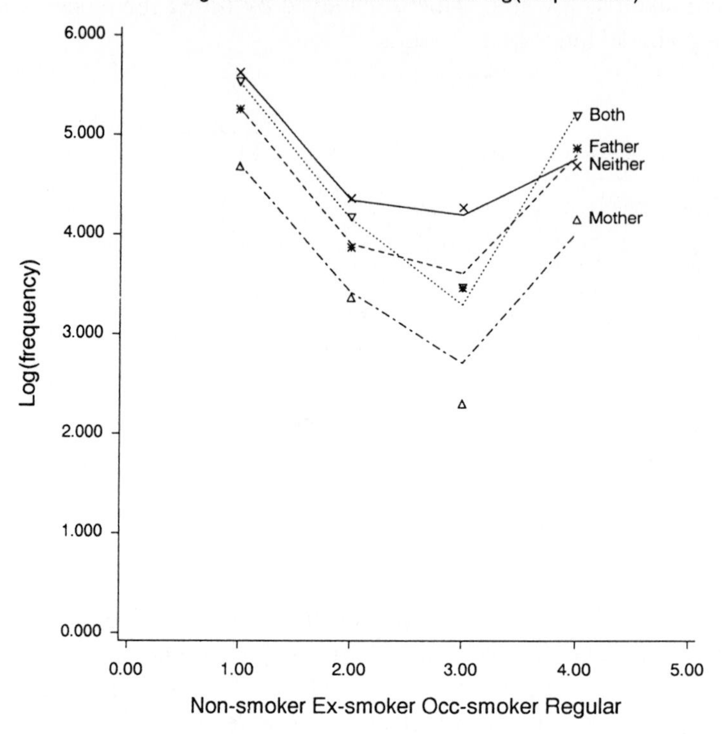


```
$graph    ( t='Fig. 12.45 Fitted model from the Non-smoking
          Baseline'
          v='Log(frequency) from Non-smoking category'
          h='Non-smoker Ex-smoker Occ-smoker Regular'
          p=no xscale=0,5,1 y=-3,1,1 )
          alp  s84  10;12;13;11 ps $

$ass gty=4,8,12,16 $cal gty=alp(gty) $
$gtext (m=p l=r) gty gtx 'Neither','Mother','Father','Both' $
```

The model and the graph indicate that a parent smoking tends to decrease the proportions in the ex- and occasional smoking categories, but increase the proportion smoking regularly. To test the effects of the parents separately we must drop each of them in turn from this 'baseline' model and use the changes in deviance as approximate chi-square tests of the null hypotheses that neither the mother's or the father's smoking has an effect on the children's eventual smoking behaviour as adults.

```
$fit                          -ms.s84  $dis e $
scaled deviance =  19.681 (change =    +12.92) at cycle 3
    residual df =   6      (change =     +3   )
          estimate          s.e.      parameter
```

```
     1        -1.279         0.1088          S84(2)
     2        -1.554         0.1214          S84(3)
     3       -0.8020        0.09123          S84(4)
     4      -0.08956         0.1514          FS(2).S84(2)
     5       -0.3827         0.1806          FS(2).S84(3)
     6        0.4459         0.1174          FS(2).S84(4)
scale parameter 1.000
    eliminated term: FS.MS
$cal D3=%dv : df3=%df $
```

Fig. 12.45 Fitted model from the Non-smoking Baseline

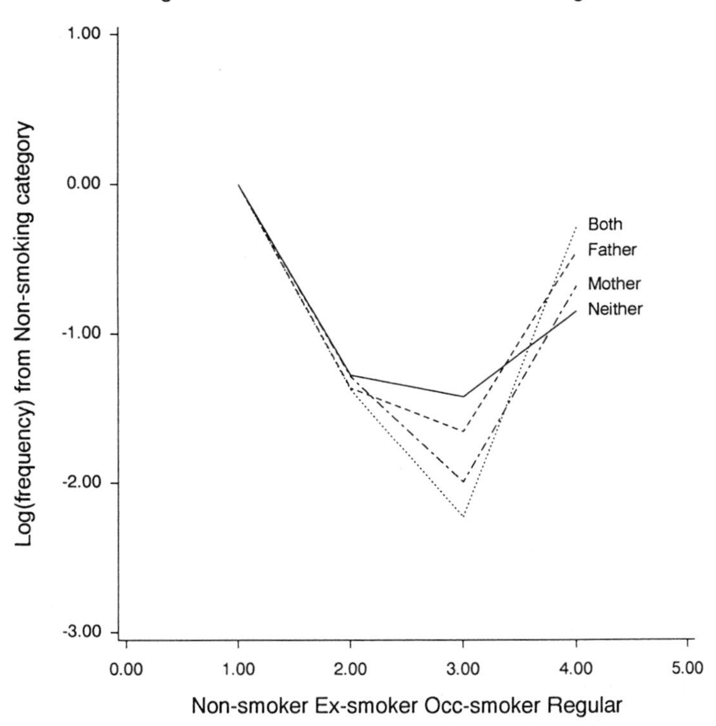

The change in deviance gives a chi-squared test of 19.68 – 6.76 = 12.92 on 3 d.f. for the effect of mother's smoking. The 95th centile of the Chi-squared distribution on 3 d.f. is 7.82 so this is highly significant. Clearly the mother's smoking when the children are 11 – 12 yrs old is strongly associated with their later smoking behaviour.

```
$fit                 -fs.s84   + ms.s84 $dis e $
scaled deviance =  23.300 (change =    +3.619) at cycle 3
    residual df =   6      (change =     0    )
          estimate          s.e.       parameter
     1      -1.311       0.09999        S84(2)
     2      -1.510        0.1083        S84(3)
```

```
     3       -0.6701       0.07919        S84(4)
     4      -0.03491        0.1530        MS(2).S84(2)
     5       -0.6407        0.1957        MS(2).S84(3)
     6        0.2784        0.1146        MS(2).S84(4)
scale parameter 1.000
    eliminated term: FS.MS
$cal D4=%dv : df4=%df $
```

The change for father's smoking is even more significant 23.3 – 6.76 = 16.54 on 3 d.f..
The full analysis can be summarized in the analysis of deviance table.

```
$use ANOD3313 $
```

Model	Deviance	df	
1) fs*ms*s84	0.00	0	
2) -fs.ms.s84	6.76	3	
(2-1) Due fs.ms.s84	6.76	3	
			p = 0.0800
3) - ms.s84	19.68	6	
(3-2) Due to ms.s84	12.92	3	
			p = 0.0048
4) +ms.s84-fs.s84	23.30	6	
(4-2) Due to fs.s84	16.54	3	
			p = 0.0009

It appears, unsurprisingly, that parental smoking during childhood has a strong association with later, adult smoking habits. There is also in this data a slight suggestion, from the nearly significant three-factor interaction, that the effects of parental smoking are not additive. The effects are more or less the same if either or both parents smoke.

In addition all interpretations of this analysis must be qualified by the fact that parents smoking when the children were 11 years old will in many cases continue smoking throughout the child's time at home. This means that the effect of parental smoking in the early years is confounded to an unknown degree with that at later stages. From the scientific point of view this is a serious limitation, we cannot deduce how important exposure at 11 years of age is in forming attitudes and influencing behaviour. However from a more pragmatic point of view parental smoking at one time may be taken as a proxy for all parental smoking. Consequently these data may certainly be used to assess what might be achieved in a health education sense if parents of 11 year old children were persuaded to stop smoking.

12.3.2 Survival data

12.3.2.1 Data on deaths for known person-years-at-risk

A study of occupational health produced the following mortality data from two factories. The deaths were grouped by age range and the person years at risk during those age ranges are given in parentheses.

	Factory	
Age range	1	2
50-59.9	7(4045)	7(3701)
60-69.9	27(3571)	37(3702)
70-79.9	30(1777)	35(1818)
80-89.9	8(381)	9(350)

The data are input with

```
$units 8 $data dths pyar $read
         7  4045     7  3701
        27  3571    37  3702
        30  1777    35  1818
         8   381     9   350
```

Calculate the death rates per person year at risk and age and factory factor variables.

```
$cal p=dths/pyar $
$gfactor agg 4 fac 2 $
```

Then plot against age with different symbols for the factories.

```
$plot (x=0,5) p agg 'AB' fac $
```

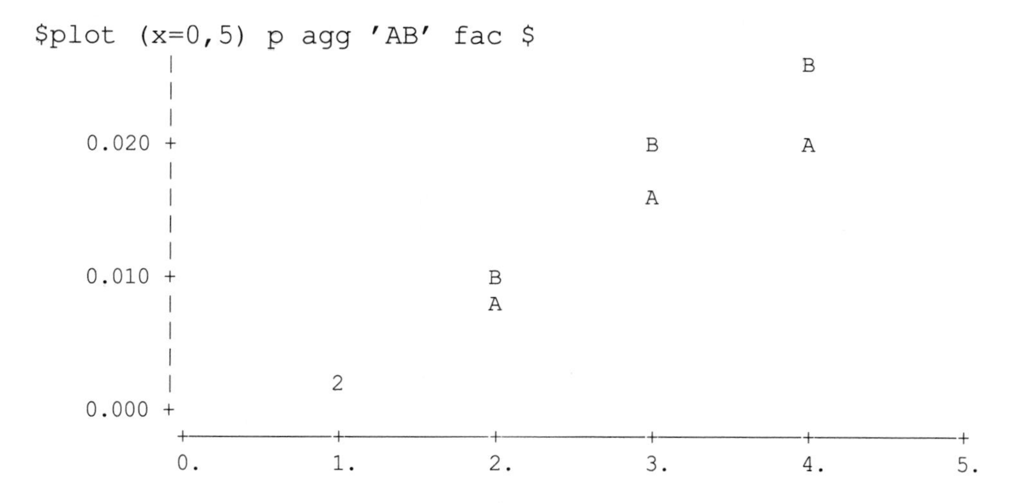

There is a steady increase with age in both factory groups and some evidence that factory 2 (B) workers have a worse mortality at each age range.

To fit a model to test this it is first necessary to declare the y variable. We then specify the Poisson error distribution and use the default log link. Finally it is necessary to calculate and specify an offset. This is achieved as follows

```
$yvar dths $error P $
$cal off=%log(pyar) $offset off $
```

Fitting a model treating the age as a factor, that is, allowing any form of non-linearity and allowing for non-parallelism between the two factories, that is, different age effects in each factory, is asking too many questions of the data. The number of parameters required is equal to the number of data points and the model fits perfectly. Nonetheless the parameter estimates can convey some information.

```
$fit agg*fac $dis e $
scaled deviance =  4.120e-18 at cycle 2
    residual df =       0
          estimate          s.e.      parameter
     1      -6.359        0.3780       1
     2       1.475        0.4241       AGG(2)
     3       2.278        0.4198       AGG(3)
     4       2.496        0.5175       AGG(4)
     5     0.08888        0.5345       FAC(2)
     6      0.1902        0.5914       AGG(2).FAC(2)
     7     0.04246        0.5896       AGG(3).FAC(2)
     8      0.1138        0.7224       AGG(4).FAC(2)
scale parameter 1.000
```

The deviance is zero, but the interaction parameter estimates give some guidance on whether they will be necessary to give a well-fitting model. They are all small compared to their s.e.s so they are unlikely to be significant. They are tested formally by fitting a model with the age effects the same in the two factories and the factory effect constant (on the log scale).

```
$fit agg+fac $dis e $
scaled deviance =  0.21568 at cycle 3
    residual df =  3
          estimate          s.e.      parameter
     1      -6.413        0.2800       1
     2       1.577        0.2951       AGG(2)
     3       2.298        0.2947       AGG(3)
     4       2.554        0.3609       AGG(4)
     5      0.1939        0.1590       FAC(2)
scale parameter 1.000
```

The deviance is very small so the model fits very well and, taken as a difference between the deviances for this and the previous model (zero), gives a likelihood ratio chi-squared test of the interaction terms which is clearly not significant.

The difference in the age trends between the factories appears to be no more than could easily occur by chance. However we have only tested a general form of non-parallelism. The apparent effect is one of gradual divergence. By fitting a trend using age as a continuous covariate we could focus the factory comparison down to a test of the difference between regression coefficients which is potentially a more sensitive test. We can most easily do this using factor polynomials and because the trends appear curved we will start with a quadratic. Using the nesting operator / allows us to see the coefficients of the two separate curves rather than the differences between them.

```
$fit fac/agg<2> $dis e $
scaled deviance =  0.55786 at cycle 3
    residual df =  2
          estimate         s.e.      parameter
     1      -4.797       0.1453       1
     2      0.1790       0.2006       FAC(2)
     3      0.8235       0.1446       FAC(1).AGG<1>
     4     -0.3134       0.1451       FAC(1).AGG<2>
     5      0.7982       0.1361       FAC(2).AGG<1>
     6     -0.3397       0.1382       FAC(2).AGG<2>
scale parameter 1.000
```

Apparently a very good fit, and the quadratic terms, `FAC(1).AGG<2>` and `FAC(2).AGG<2>`, certainly look as if they are necessary. However they are not very different. Try parallel quadratic curves:

```
$fit fac+agg<2> $dis e $
scaled deviance =  0.59167 at cycle 3
    residual df =  4
          estimate         s.e.      parameter
     1      -4.805       0.1320       1
     2      0.1941       0.1590       FAC(2)
     3      0.8101      0.09910       AGG<1>
     4     -0.3272       0.1000       AGG<2>
scale parameter 1.000
```

The change in deviance is trivial. One quadratic curve appears adequate to describe the data. We should test whether ‘factory’ matters at all:

```
$fit agg<2>  $dis e $
scaled deviance =  2.0908 at cycle 3
    residual df =  5
          estimate         s.e.      parameter
     1      -4.704       0.1001       1
     2      0.8099      0.09910       AGG<1>
     3     -0.3300       0.1000       AGG<2>
scale parameter 1.000
```

A change of 1.5 in the deviance for 1 d.f. (cf. 3.84). It appears that, despite the appearance of the plot, there is no factory effect beyond what could easily occur by chance. Finally for completeness we should test the need for the quadratic term in the parallel quadratic curves model

```
$fit fac+agg<1> $
scaled deviance =  12.712 at cycle 3
    residual df =   5
```

There is a large increase in deviance (12.1 on 1 d.f.) and, taken as a likelihood ratio chi–square, highly significant. The quadratic term is definitely significant and necessary. Incidentally it is worth noting that the square of the quadratic coefficient divided by its s.e. $[(-0.3272)/0.1]^2$ is 10.7 which is smaller than the chi-squared test. It appears that the standardized Normal deviate tests with the coefficients are relatively conservative in this case.

In conclusion it appears that there is a strong age effect, flattening off at the higher ages (the quadratic term has a negative coefficient) and no very strong evidence of a factory effect.

12.3.2.2 Cox's proportional hazards model for survival data

This data arose from a trial of 6-mercaptopurine on the duration of steroid induced remission in leukaemia patients quoted by Gehan (1965). Patients were allocated randomly to the active drug and placebo treatment groups. The response was whether a relapse occurred during the time (weeks) the patient was under observation in the study. The variables are:

`drug`	0. the active drug 1. the placebo treatment
`relapse`	0. no relapse during the observation period
	1. relapsed while under observation
`time`	time observed in weeks

```
$data 42 drug relapse time $
$read
0  0   6  0  1   6 0  1   6  0  1   6 0  1   7 0  0   9 0  1  10
0  0  10  0  0  11 0  1  13  0  1  16 0  0  17 0  0  19 0  0  20
0  1  22  0  1  23 0  0  25  0  0  32 0  0  32 0  0  34 0  0  35
1  1   1  1  1   1 1  1   2  1  1   2 1  1   3 1  1   4 1  1   4
1  1   5  1  1   5 1  1   8  1  1   8 1  1   8 1  1   8 1  1  11
1  1  11  1  1  12 1  1  12  1  1  15 1  1  17 1  1  22 1  1  23
```

Now use a macro to derive the lifetables and equivalent survival curves. The macro takes five arguments. The first three are variables specifying the time, the event that terminated observation and a 0/1 weight for restricting the analysis to a subset of individuals. We analyse the two treatment groups separately by using the weight variable `drg_wt` to specify which individuals to include. We start with the active drug group (code 0). The scalar parameter `occ_ints` is set to 0 indicating that intervals without an 'event' should be omitted from the lifetable and `risk_fr`, also a scalar, is set to 1 to indicate that individuals lost to observation should be considered at risk for the whole of the last interval in which they were observed. Several output variables are stored in new variables for use in plotting the survival curves together and for performing a 'log-rank' test comparing survival in the two groups.

```
$cal drg_wt=(drug==0) $
$number occ_ints=0 risk_fr=1 $
```

```
$use lfts time relapse drg_wt occ_ints risk_fr ti_d0 si_d0
ni_d0 nar0_wt $
  Lifetable with Estimated Survival Function (S) and se(S)
     Top of  No.at  Deaths Lost    Hazd          Est p
     Intvl   Risk                d/(n-fw)  (1-q)  Surv
       t       n       d      w     q        p    S(t)    se(S)
     TI_     NI_     DI_    WI_    QI_      PI_   SI_     SES_
  1   0.0    21.0    0.0    0.0   0.000    1.00   1.00    0.000
  2   1.0    21.0    0.0    0.0   0.000    1.00   1.00    0.000
  3   2.0    21.0    0.0    0.0   0.000    1.00   1.00    0.000
  4   3.0    21.0    0.0    0.0   0.000    1.00   1.00    0.000
  5   4.0    21.0    0.0    0.0   0.000    1.00   1.00    0.000
  6   5.0    21.0    0.0    0.0   0.000    1.00   1.00    0.000
  7   6.0    21.0    3.0    1.0   0.143    0.86   0.86    0.076
  8   7.0    17.0    1.0    0.0   0.059    0.94   0.81    0.087
  9   8.0    16.0    0.0    0.0   0.000    1.00   0.81    0.087
 10   9.0    16.0    0.0    1.0   0.000    1.00   0.81    0.087
 11  10.0    15.0    1.0    1.0   0.067    0.93   0.75    0.096
 12  11.0    13.0    0.0    1.0   0.000    1.00   0.75    0.096
 13  12.0    12.0    0.0    0.0   0.000    1.00   0.75    0.096
 14  13.0    12.0    1.0    0.0   0.083    0.92   0.69    0.107
 15  15.0    11.0    0.0    0.0   0.000    1.00   0.69    0.107
 16  16.0    11.0    1.0    0.0   0.091    0.91   0.63    0.114
 17  17.0    10.0    0.0    1.0   0.000    1.00   0.63    0.114
 18  19.0     9.0    0.0    1.0   0.000    1.00   0.63    0.114
 19  20.0     8.0    0.0    1.0   0.000    1.00   0.63    0.114
 20  22.0     7.0    1.0    0.0   0.143    0.86   0.54    0.128
 21  23.0     6.0    1.0    0.0   0.167    0.83   0.45    0.135
 22  25.0     5.0    0.0    1.0   0.000    1.00   0.45    0.135
 23  32.0     4.0    0.0    2.0   0.000    1.00   0.45    0.135
 24  34.0     2.0    0.0    1.0   0.000    1.00   0.45    0.135
 25  35.0     1.0    0.0    1.0   0.000    1.00   0.45    0.135
```

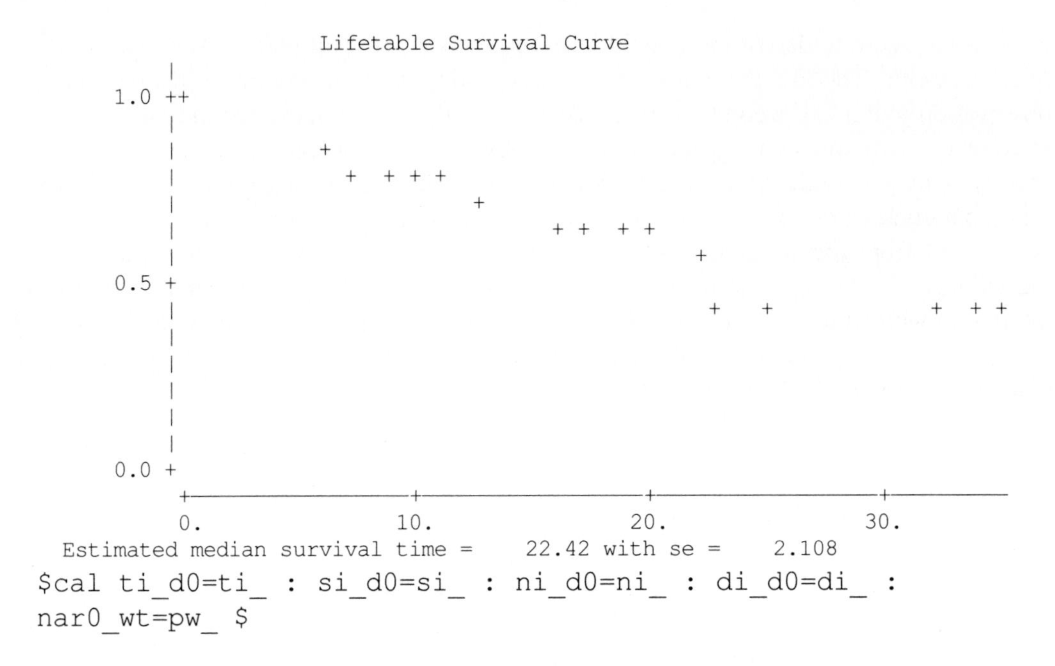


```
Estimated median survival time =    22.42 with se =    2.108
$cal ti_d0=ti_ : si_d0=si_ : ni_d0=ni_ : di_d0=di_ :
nar0_wt=pw_ $
```

Then to obtain the lifetable for the placebo group

```
$cal drg_wt=(drug==1) $
$use lfts $
```

Lifetable with Estimated Survival Function (S) and se(S)

	Top of Intvl t TI_	No.at Risk n NI_	Deaths d DI_	Lost w WI_	Hazd d/(n-fw) q QI_	(1-q) p PI_	Est p Surv S(t) SI_	se(S) SES_
1	0.0	21.0	0.0	0.	0.000	1.00	1.000	0.000
2	1.0	21.0	2.0	0.	0.095	0.90	0.905	0.064
3	2.0	19.0	2.0	0.	0.105	0.89	0.810	0.086
4	3.0	17.0	1.0	0.	0.059	0.94	0.762	0.093
5	4.0	16.0	2.0	0.	0.125	0.88	0.667	0.103
6	5.0	14.0	2.0	0.	0.143	0.86	0.571	0.108
7	6.0	12.0	0.0	0.	0.000	1.00	0.571	0.108
8	7.0	12.0	0.0	0.	0.000	1.00	0.571	0.108
9	8.0	12.0	4.0	0.	0.333	0.67	0.381	0.106
10	9.0	8.0	0.0	0.	0.000	1.00	0.381	0.106
11	10.0	8.0	0.0	0.	0.000	1.00	0.381	0.106
12	11.0	8.0	2.0	0.	0.250	0.75	0.286	0.099
13	12.0	6.0	2.0	0.	0.333	0.67	0.190	0.086
14	13.0	4.0	0.0	0.	0.000	1.00	0.190	0.086
15	15.0	4.0	1.0	0.	0.250	0.75	0.143	0.076
16	16.0	3.0	0.0	0.	0.000	1.00	0.143	0.076
17	17.0	3.0	1.0	0.	0.333	0.67	0.095	0.064
18	19.0	2.0	0.0	0.	0.000	1.00	0.095	0.064
19	20.0	2.0	0.0	0.	0.000	1.00	0.095	0.064
20	22.0	2.0	1.0	0.	0.500	0.50	0.048	0.046
21	23.0	1.0	1.0	0.	1.000	0.00	0.048	0.046
22	25.0	0.0	0.0	0.	0.000	1.00	0.048	0.046

```
23   32.0    0.0    0.0    0.   0.000   1.00   0.048   0.046
24   34.0    0.0    0.0    0.   0.000   1.00   0.048   0.046
25   35.0    0.0    0.0    0.   0.000   1.00   0.048   0.046
                     Lifetable Survival Curve
         |
    1.0 ++
         |  +
         |
         |    +  +
         |
         |         +
         |            +
    0.5 +
         |
         |                    +
         |                            +
         |                              +
         |                                      +
         |                                           +            +  +
    0.0 +
         +------------+------------+------------+------------+--------
         0.           5.           10.          15.          20.
  Estimated median survival time =     7.375 with se =     0.7578
$cal ti_d1=ti_ : si_d1=si_ : ni_d1=ni_ : di_d1=di_ : nar1_wt=pw_ $
$use del_ $
```

The output from these can be used to perform a log-rank test comparing the 'survival' in remission of the two groups.

```
$use lr2  ni_d0  di_d0  ni_d1  di_d1  $
The log-rank chi-squared statistic on 1 df is =    15.23 with p =  0.0001
```

This test shows the difference between the two groups to be highly significant with a p-value well below 1 per cent.

Both plots together

```
$plot si_d0/nar0_wt;si_d1/nar1_wt ti_d0;ti_d1 '.+' $
          |
   1.00 +2
          |  +
          |    +       .  .  .
   0.75 +      +             .  .
          |        +              .     .  .  .  .
          |          +
   0.50 +                                             .
          |              +                              .   .           .  .  .
          |
   0.25 +                      +
          |                      +    +
          |                          +        ++
   0.00 +
          +-----------------+-----------------+-----------------+-----------
          0.                10.               20.               30.
```

A higher resolution graph can be obtained with the aid of a macro steps to generate extra co-ordinates for the correctly 'stepped' plot, by

```
$use steps si_d0 ti_d0  s02 t02 $
$use steps si_d1 ti_d1  s12 t12 $

$mac title Fig. 12.46 Lifetable Survival Curves $end $
$mac xlab  Time since entry in days $end $
$mac ylab  Proportion Not Relapsed $end $
$ass pens=10,12 $

$graph (title=title h=xlab v=ylab y=0,1.4,.2 x=0,40,10 p=no)
        s02;s12 t02;t12 pens $
```

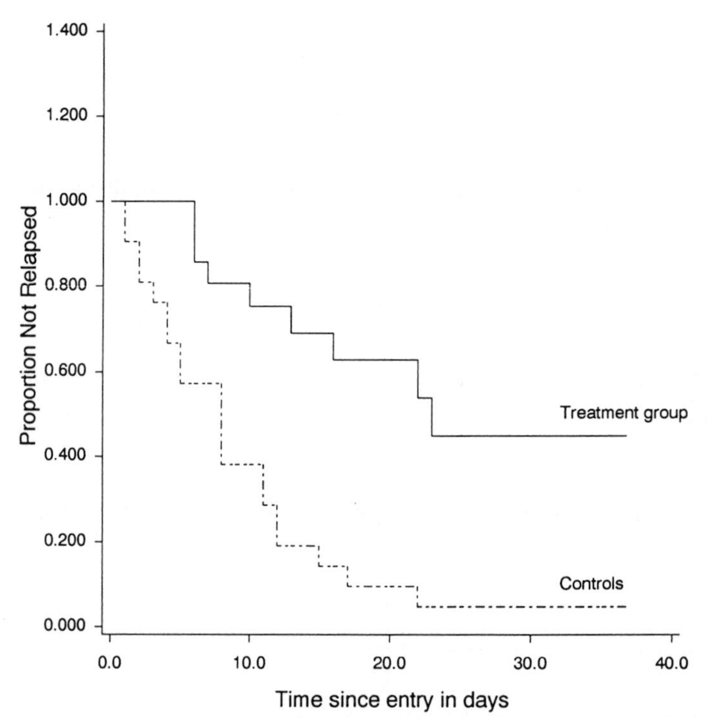


```
$cal %l=%len(s02)-3 : %y=s02(%l) + 0.05 : %x=t02(%l)
   : %l=%len(s12)-3 : %z=s12(%l) + 0.05 : %w=t12(%l) $
$ass gty=%y,%z : gtx=%x,%w $
$gtext (m=p l=r ) gty gtx 'Treatment group','Controls' $
```

For quantitative estimates of the effects of `drug` and the `drug.time` interaction it is necessary to fit models. Cox's proportional hazards model is fitted using a set of macros (Aitkin et al 1989) to identify intervals along the time scale which include all the events (relapses) of interest. The individual-based data is then expanded to a much larger set of 'person epochs at risk' within these intervals. Each individual may contribute a different time at risk for each epoch and these have to be calculated and summed. The number of relapses in interval i is then taken to have a Poisson distribution with a constant risk of relapse per person-unit-of-time-at-risk.

The time at risk and event indicator variables, which in this case are 'time' and 'relapse', need to be supplied to the macro `COXMODEL` as arguments. The macro then creates an appropriate y variable which indicates whether or not an event occurred for each person-epoch and an index variable ind_ to identify the values of the covariates which apply to all the epochs contributed by each individual. Finally the macros create a variable of times at risk for each person within each person-epoch which is converted to logarithms and used as an offset.

First, to simplify the interpretation later, reverse the treatment coding so the estimates obtained will be for the change in hazard (on the log scale) achieved by the active treatment:

```
$cal treat=(1-drug) $
```

Then invoke the appropriate macros with the time observed and event or not variables `time` and `relapse` as arguments and a vector `rlt` to hold the complete set of times at which the relapses occurred.

```
$use coxmodel time relapse rel_t $
There are 17 distinct death times.
```

There is a possibility that the difference between the effects of the two treatments changes over time. This will emerge as an interaction between the effect of treatment and time. Such an effect might result in the beneficial effect of a treatment, on the log(hazard) scale, changing from positive through zero to negative and back again. However on the assumption that a genuine effect of this nature would go through these changes, over time, in a relatively smooth manner the interaction effects can be represented by a relatively low degree polynomial function of the time at which the events occurred. Here we have used a cubic. Equating each level of the time_interval factor, `t_ints`, with the lower end time point to avoid the problem of an infinite upper end to the last interval the appropriately indexed time variable is obtained and the model with a 'cubic' interaction between time and the treatment effect on hazard fitted as follows

```
$ass r_pts=0,rel_t $
$cal r_pdx=r_pts(t_ints) $
```

```
$fit  treat(ind_)*r_pdx<3> $dis u $
scaled deviance =  200.10 at cycle 7
    residual df =  404
          estimate         s.e.      parameter
     1      -1.919       0.6694       TREAT(IND_)
     5       10.86        13.64       R_PDX<1>.TREAT(IND_)
     6      -13.53        11.80       R_PDX<2>.TREAT(IND_)
     7       12.17        10.44       R_PDX<3>.TREAT(IND_)
scale parameter 1.000
    eliminated term: T_INTS
$number D1=%dv  df1=%df  $
```

Removing the interaction term and remembering that the interval effects are in the model but `eliminated` from the displayed results we obtain the estimate of the treatment effect when it is assumed constant with the model:

```
$fit  treat(ind_)  $disp e $
scaled deviance =  202.37 at cycle 7
    residual df =  407
          estimate         s.e.      parameter
     1      -1.521       0.4102       TREAT(IND_)
scale parameter 1.000
    eliminated term: T_INTS
$number D2=%dv  df2=%df  $
```

A small increase in deviance, but less than the change in the degrees of freedom. There is therefore no evidence to suggest that the hazards are not proportional with a constant ratio as required for the Cox model. A plot of the log(hazard) against time since treatment for treatment group 1 can be obtained using a macro called `phaz` with the relapse times variable, `rel_t`, as the first parameter and a macro holding the plot title string as the second. Because the hazards have been assumed proportional the equivalent plots for other groups are simply multiples of this.

```
$mac title 'Fig. 12.47 The hazard function for treatment
group 1'
$end
$use phaz rel_t title $
prediction for the current model with TREAT(IND_)=0., T_INTS=BB
Offset=0.
```

Fig. 12.47 The hazard function for treatment group 1

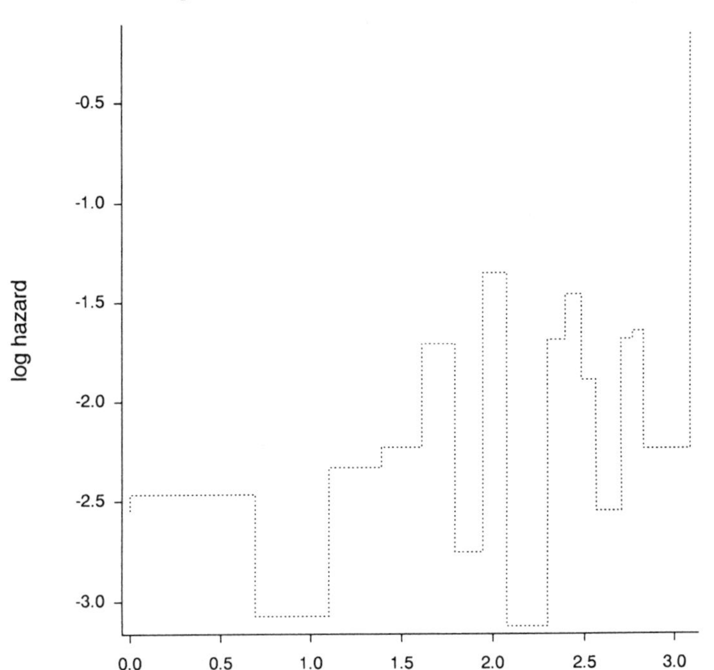

The plot is in the form of a step function because the hazards are only defined in those intervals where a death occurs.

The effect of `treat` is more than three times its s.e. so the evidence strongly indicates a beneficial effect of the treatment given to group 1 compared to the placebo given to group 2. The parametrization is such that the estimated coefficient is, on the log scale, the proportional change in risk associated with receiving the active treatment. Consequently the change in risk of relapse resulting from the use of the active treatment with approximate 95 per cent confidence limits is obtained by

```
$extract %pe %se $
$number  rel_rsk  rr_lo rr_hi $
$cal  rel_rsk=100*(1 - %exp(%pe(1)))
: rr_lo=100*(1-%exp(%pe(1)-1.96*%se(1)))
: rr_hi=100*(1-%exp(%pe(1)+1.96*%se(1))) $
```

The estimated percentage reduction in the risk of relapse is

```
$pri '   '*r rel_rsk,10,2' with 95% CLs    '*r rr_lo,10,2
     '   to'*r rr_hi,10,2 ;; $trans +e $
      78.16 with 95% CLs        90.23   to      51.20
```

Finally to obtain a model comparison test of the treatment effect a model assuming no difference is fitted.

```
$fit  $dis e $
scaled deviance =  217.79 at cycle 7
    residual df =  408
 - No parameters to display
scale parameter 1.000
    eliminated term: T_INTS
$number D3=%dv  df3=%df  $
```

The full analysis of deviance is obtained using a macro specific to this example by

```
$use anod3322 $
```

```
---------------------------------------------------------------
       Model                Deviance   df
---------------------------------------------------------------
   1) With treat.time<3>     200.10   404
   2) treat                  202.37   407
------------------------------------------
(2-1) Due to intrn             2.28     3
                                                p =   0.5170
---------------------------------------------------------------
   3) - treat                217.79   408
------------------------------------------
(3-2) Due to treat            15.41     1
                                                p =   0.0001
---------------------------------------------------------------
---------------------------------------------------------------
```

It can be seen that the interaction effect is far from significant, but the effect of the treatment is significant at the 1 per cent level at least.

12.4 Special problems

12.4.1 Linear

12.4.1.1 Dilution assay

In a dilution assay replicate samples of the test material at several dilutions are cultured and the number positive noted. This gives a binomial variable where the probability of a `success', that is, a positive culture, is a function of the number of bacteria per unit of the original material. It thus allows the original concentration to be estimated. In the microbiological literature this is often referred to as the 'most probable number' technique and sets of tables have been compiled to aid estimation for a range of designs. Inevitably these are restricted and the fact that the estimation can be modelled in GLIM provides a useful extension to the tables.

In this example the data arose as a sample from some contaminated process. One gram of the substance was put in suspension in 100 mls of water and then diluted 8, 16, 32, 64, and 128 times. Five replicates from each dilution were cultured and 3,2,2,0 and 0 were positive. The problem is to estimate the concentration of bacteria/gm in the original sample.

The theoretical argument is as follows. If λ is the number of bacteria per gm of the original substance and the concentration $z = 0.01/8, 0.01/16, ..., 0.01/128$ gm/ml then the number of bacteria per ml is λz. The probability of one 1ml sample proving fertile is

$$1 - \text{Probability of no bacteria} = 1 - p(0).$$

Assuming the Poisson distribution for the numbers of organisms arriving in the replicates at each of the dilutions the probability of no bacteria in a particular replicate is $\exp(-\lambda z)$ so the probability of a fertile sample is

$$1 - \exp\{-\lambda z\}$$

and therefore the number of fertile samples, r, in the n replicate samples at each dilution has a binomial distribution with mean

$$n(1 - \exp\{-\lambda z\})$$

The expected value of r/n is then

$$E(r/n) = \mu = (1 - \exp\{-\lambda z\})$$

Taking the complementary log log transformation of μ we get

$$\log(-\log\{1-\mu\}) = \log(-\log(\exp\{-\lambda z\})) = \log \lambda + \log z.$$

Since the complementary log log transformation is one of the links available for the binomial distribution specifying r as the y-variate with the binomial `ERROR`, this link will mean that the linear model represented by the linear predictor η is

$$\eta = \log(-\log\{1-\lambda\}) = \log \lambda + \log z.$$

This is thus a linear model for the unknown concentration λ on the log scale with $\log z$ in the model with a fixed coefficient of 1. This is known as an offset and can be incorporated in the analysis with the GLIM `OFFSET` directive. The model for the results of a single series is then simply the constant term obtained by fitting the parameter 1 and the exponential of the parameter estimate is then the required estimate of the concentration. The analysis is as follows:

```
$data 5 r d $read
  3   8
  2  16
  2  32
  0  64
  0 128
```

Specify the standard length of the variable vectors and create a denominator variable equal to 5 for all concentrations (it happens to be constant here but it is not necessary for this analytical approach).

```
$slen 5 $
$cal n=5 $
```

The concentration is the number of gm/ml (x) in the original sample divided by the dilution d = 8, 16, 32, 64, 128. Since the value of x here is 1/100, then z = 0.01/8, 0.01/16,... and so on.

Calculate the concentrations and the offset

```
$num x=0.01 $
$cal z = x/d : logz=%log(z)   $

$offset logz $
```

Inspect the results

```
$look n r z logz $
        N        R          Z         LOGZ
 1   5.000   3.000   0.00125000   -6.685
 2   5.000   2.000   0.00062500   -7.378
 3   5.000   2.000   0.00031250   -8.071
 4   5.000   0.000   0.00015625   -8.764
 5   5.000   0.000   0.00007812   -9.457
```

Then specify and fit the model, with `LINK C` to indicate the complementary log log link, by

```
$yvar r $error b n $link c $

$fit 1 $dis e $
scaled deviance =  2.6996 at cycle 3
    residual df =  4
          estimate       s.e.      parameter
    1        6.664      0.3852        1
scale parameter 1.000
```

The estimate and appropriate confidence limits are then obtained by

```
$extract %pe %se $
$number lower estimate upper $
$cal lower=%exp(%pe(1) - 1.96*%se(1)) :
$cal estimate=%exp(%pe(1)) :
$cal upper=%exp(%pe(1) + 1.96*%se(1)) $
```

The estimated number of bacteria per gm with 95 per cent confidence limits is then:

```
$print '     'estimate'     'lower' to' upper $

       783.704            368.367  to     1667.34
```

If the dilution series is one of several within a larger design the approach is easily extended to assess the effects of factors and covariates on the original concentrations.

12.4.1.2 Data with variance = mean squared – gamma errors

A field trial of insect control treatments produced counts of insects from two plots in each of 10 blocks for each of seven treatments (John and Quenouille, 1977). The data have a mean-variance relationship which, in the published analysis, led to the use of a logarithmic transformation which at the same time removed an interaction between treatments and blocks. However the mean-variance relationship suggests that the responses may follow a gamma distribution so the data have been analysed here on that assumption.

The treatments were applied to pairs of plots within each of 10 blocks as shown in the layout of the data read into GLIM.

```
$units 140 $data count $read

!                    Block
!  1   2   3   4   5   6   7   8   9  10  plot treat
!----------------------------------------!-----------
  32  38  27   7  13  14  26  25  22  30!  1     1
  18  40  39  12  19  26  30  19  18  28!  2     1
   6  23   8   4   3  18  26  27  17  19!  1     2
   9  14  20  13  15  14  15  19  19  10!  2     2
  10  21  25  10  13  20  33  48  28  27!  1     3
   4  21  26   4   9  14  30  18  27  18!  2     3
   2  17  11   3  10  10  26  13  22  17!  1     4
```

```
24  13  13  10   6  14  28  11  34   7!   2   4
13   2   5   1  18  10  33  23  20  34!   1   5
17  22  23   8  14  16  26  22  15  34!   2   5
13  10  21   4  10   8  17  15  13  16!   1   6
17   9  29   5  18  15  19  16  27  23!   2   6
37  58  28  11  24  44  30  44  56  45!   1   7
44  71  55  20  26  27  43  52  39  58!   2   7
```

Because the data are in a structured tabular form the treatment and block factor variables can be generated with the `GFACTOR` directive.

```
$gfactor treat 7 plot 2 block 10 $
```

As a first check of the underlying distribution it is useful to consider the 'plots' as replicates within the cells of the `treat.block` classification. The distribution can then be investigated using the residuals of these from their cell mean. The residuals are most easily obtained by fitting the full model `treat*block` with Normal assumptions. This fits each cell mean independently.

```
$yvar count $
$fit treat*block $
   deviance =  3346.0
residual df =     70
```

The distribution of the residuals is then inspected with a histogram and Normal plot

```
$histogram %rs 'r' $
[-15.00,-12.00)  2 rr
[-12.00, -9.00)  2 rr
[ -9.00, -6.00)  8 rrrrrrrr
[ -6.00, -3.00) 22 rrrrrrrrrrrrrrrrrrrrrr
[ -3.00,  0.00) 36 rrrrrrrrrrrrrrrrrrrrrrrrrrrrrrrrrrrr
[  0.00,  3.00) 30 rrrrrrrrrrrrrrrrrrrrrrrrrrrrrr
[  3.00,  6.00) 23 rrrrrrrrrrrrrrrrrrrrrrr
[  6.00,  9.00) 12 rrrrrrrrrrrr
[  9.00, 12.00)  3 rrr
[ 12.00, 15.00)  1 r
[ 15.00, 18.00]  1 r
```

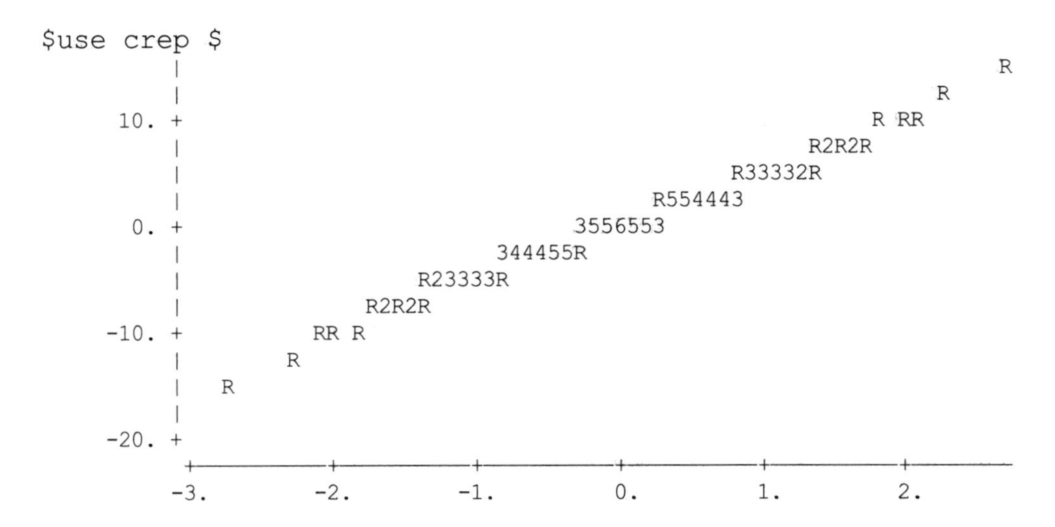

Both look very much as would be expected from Normal data. Look at the residuals plotted against the fitted values.

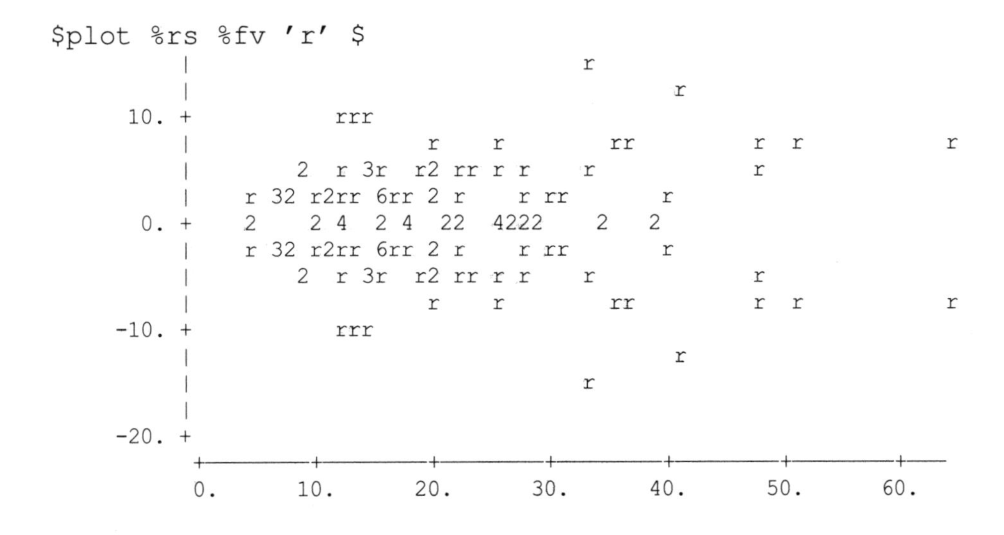

Note that because there was only a 'pair' of replicates in each cell this plot is symmetrical about the horizontal line through zero. The pattern is not very pronounced, but it can be seen that the vertical dispersion of the residuals appears to increase slightly as the fitted values increase. Let us use `TABULATE` to obtain the mean and standard deviations in each cell and inspect the relationship directly.

```
$tab the count mean;dev for treat;block into cell_mean;cell_sd
     by trt_;blk_ $
$plot cell_sd cell_mean 's' $
         |                           s
   20. +
         |                                   s
         |
         |              s
         |             s
         |               s
         |                                s           s
   10. +                   s     s
         |         s     s     s        s     s   s          s            s
         |            s        s
         |         s     2s  s   ss s                   s
         |    s 3    s s ss   s s       s          s
         |         s sss   5 s s       s
         |         s   2    2      ss s s      s
    0. +    s      s     s     ss  s s     s
         +---------+---------+---------+---------+---------+---------+--
         0.        10.       20.       30.       40.       50.       60.
```

Although there is a lot of scatter a definite tendency for the standard deviations to increase as the cell means increase can be seen. A clearer picture emerges if the between blocks variation is added to that within cells.

```
$tab the count mean;dev;weight for treat into t_mean;t_sd;t_n
     by treat_ $
$plot t_sd t_mean  's' $
```

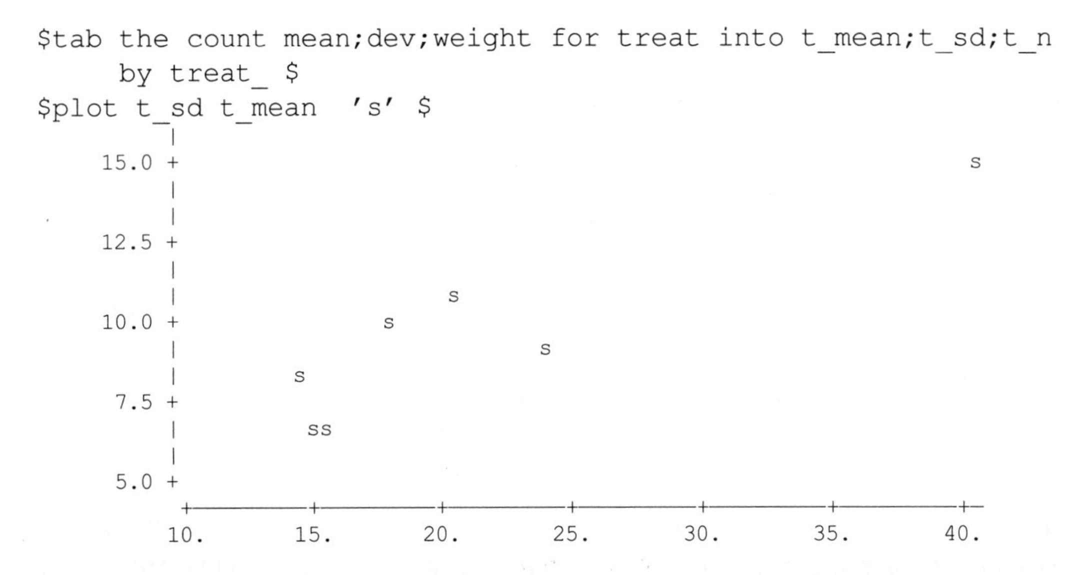

There is one extreme value with high standard deviation and mean, but there is also a reasonably clear trend in the other points. Since the points lie more or less on a straight line it suggests that the distribution from which they come has a standard deviation proportional to the mean. The gamma distribution has this property so we will assume that this response variable follows that distribution.

For the gamma `ERROR` specification the default link is the reciprocal function. However this causes complications with zero values and other aspects of the analysis. The identity link, although it can lead to negative fitted values, is generally adequate and somewhat simpler to handle so we will use that.

Remembering that the y variate is already declared we simply need to specify the gamma `ERROR` and identity `LINK` with

```
$error G $link I $
```

Note that the `LINK` directive must come after the `ERROR` or it will be reset to the default. Because the gamma, like the Normal distribution, has a free scale parameter it needs to be estimated and allowed for in the model comparison analyses. A suitable procedure for doing this is to estimate the scale parameter from a reasonable maximal model using the Pearson chi-square statistic divided by the degrees of freedom. This is then used to set the scale parameter. The maximal model is then re-fitted and the model fitting and comparison sequence proceeds as for the standard analyses of deviance used for binomial and Poisson data. The effect of scaling can be seen from the deviances and because there is a large number of parameters in this model and their estimates are not of immediate interest they have not been output.

The full sequence then, to estimate and test the treatment effects first testing for `block.treat` interactions and then estimating the main effects, is as follows.

```
$fit treat*block $
   deviance =  16.115 at cycle 3
residual df =  70
$ca %s=%X2/%df $
$scale %s $
$fit . $
scaled deviance =  91.333 (change =   +75.22) at cycle 3
    residual df =  70     (change =     0   )
```

```
$number D1 df1 $
$cal D1=%dv : df1=%df $
```

Now check the residuals.

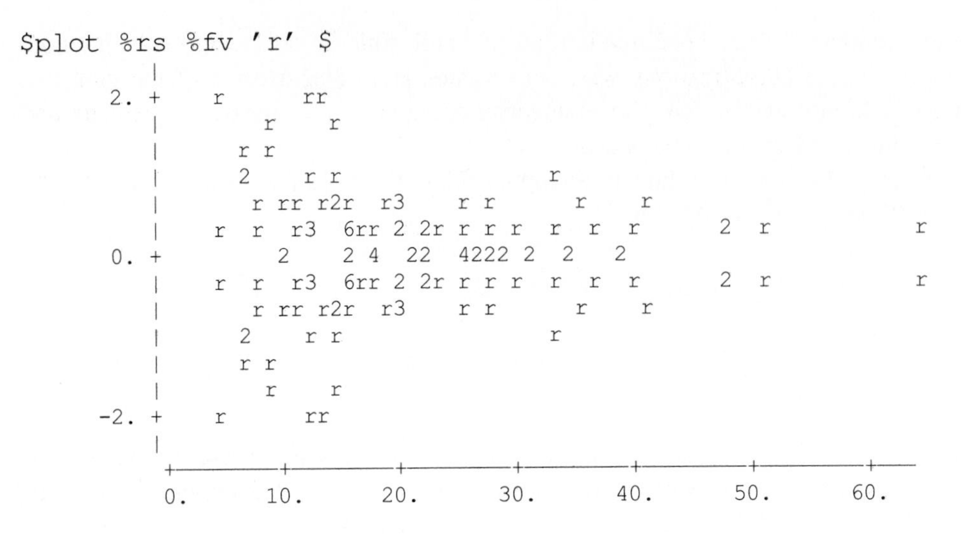

They now appear to have gone the other way and suggest a decrease in dispersion with the mean. It is possible that the gamma distribution is not entirely appropriate. Nonetheless continuing with the analysis we fit the main effects model and then models without each main effect in turn to construct the Analysis of Deviance table. Note that without the Normal assumptions the standard analysis of variance is not appropriate and model comparison tests have to be performed using the differences between the scaled deviances as likelihood ratio chi-squared tests.

```
$fit treat+block $dis e $
scaled deviance =  151.16 at cycle 7
    residual df =  124
           estimate          s.e.      parameter
     1        19.60         2.498       1
     2       -7.084         2.347       TREAT(2)
     3       -5.273         2.470       TREAT(3)
     4       -8.022         2.291       TREAT(4)
     5       -6.868         2.366       TREAT(5)
     6       -7.984         2.297       TREAT(6)
     7        15.49         4.099       TREAT(7)
     8        4.821         2.666       BLOCK(2)
     9        6.183         2.804       BLOCK(3)
    10       -6.450         1.752       BLOCK(4)
    11       -1.229         2.125       BLOCK(5)
    12        1.147         2.325       BLOCK(6)
    13        11.63         3.354       BLOCK(7)
    14        7.725         2.955       BLOCK(8)
    15        8.880         3.071       BLOCK(9)
    16        8.229         3.010       BLOCK(10)
scale parameter 0.1764
```

```
$number D2 df2 $
$cal D2=%dv : df2=%df $


$fit treat $
scaled deviance =  235.83 at cycle 3
    residual df =  133
$number D3 df3 $
$cal D3=%dv : df3=%df $


$fit block $
scaled deviance =  248.44 at cycle 3
    residual df =  130
$number D4 df4 $
$cal D4=%dv : df4=%df $
```

Then invoke a previously stored macro to complete the analysis of deviance with

```
$use anod3412 $
```

	Model	Deviance	df	
1)	All 2f intrns	91.33	70	
2)	Main Effects	151.16	124	
(2-1)	Due to intrns	59.82	54	p = 0.2726
3)	TREAT	235.83	133	
(3-2)	Due to BLOCK	84.67	9	p =1.8985e-14
4)	BLOCK	248.44	130	
(4-2)	Due to TREAT	97.28	6	p = 0.0000

There is no evidence of interaction and a chi-squared value of 59.82 on 54 degrees of freedom is more or less what one would expect by chance. The block and treatment effects are both highly significant ($\chi^2 = 84.67$ on 9 d.f. and 97.28 on 6 d.f.). These results are very similar to those obtained using a log transformation and a Normal error. Both approaches remove the significant interaction (or non-additivity) found if a Normal error is used without transformation.

It now remains to interpret the parameter estimates obtained from the main effects model. It could be argued that the scale parameter should be estimated now the simplest adequate model has been identified. However this is not generally necessary or appropriate. If the more comprehensive model used for estimating the scale parameter was not a better fit to the data the change in deviance will have been similar in magnitude to the change in degrees of freedom. This means that the numerator and denominator of the scale parameter estimate will have changed by similar amounts and the estimate will have changed little. This illustrates that as long as the model to be interpreted, in this case the main effects model, is not oversimplified the scale parameter estimate from a more complex model with adequate degrees of freedom will give appropriate standard errors for the parameter estimates.

The treatment means (from the `TREAT + BLOCK` main effects model) and their 95 per cent confidence limits may be obtained for block 1 (or any other block similarly) using `PREDICT`.

```
$fit treat+block $
scaled deviance =  151.16 at cycle 7
    residual df =  124
```

```
$ass treat_p=1,2,...7 : Block_p=1,1,1,1,1,1,1 $
$predict  treat=treat_p  block=block_p  $
prediction for the current model with TREAT=TREAT_P, BLOCK=BLOCK_P
```

Because the identity link was used the linear predictors and their variances can be used directly to obtain fitted values and confidence limits without the need to transform back to the original scale.

```
$cal lo=%plp - 1.96*%sqrt(%pvl) : hi=%plp + 1.96*%sqrt(%pvl) $
$gstyle 10 s 1 c 1 1 2 : 12 s 2 c 1 1 2 $
$ass tr2=treat_p,treat_p : cls=lo,hi
  :  pens=10,12,12,12,12,12,12,12 $
$fac tr2 7 $
```

High-resolution graphics

```
$ass cl_set=1,1,2,2,3,3,4,4 $fac cl_set 4 $
$graph (t='Fig. 12.48 Treatment means and 95% CLs for Block 1'
        v='Mean count' h='Treatment' xscale=0,8,1 yscale=0,60,20)
        %plp;cls      treat_p;tr2     pens ;tr2 $
```

Fig. 12.48 Treatment means and 95% CLs for Block 1

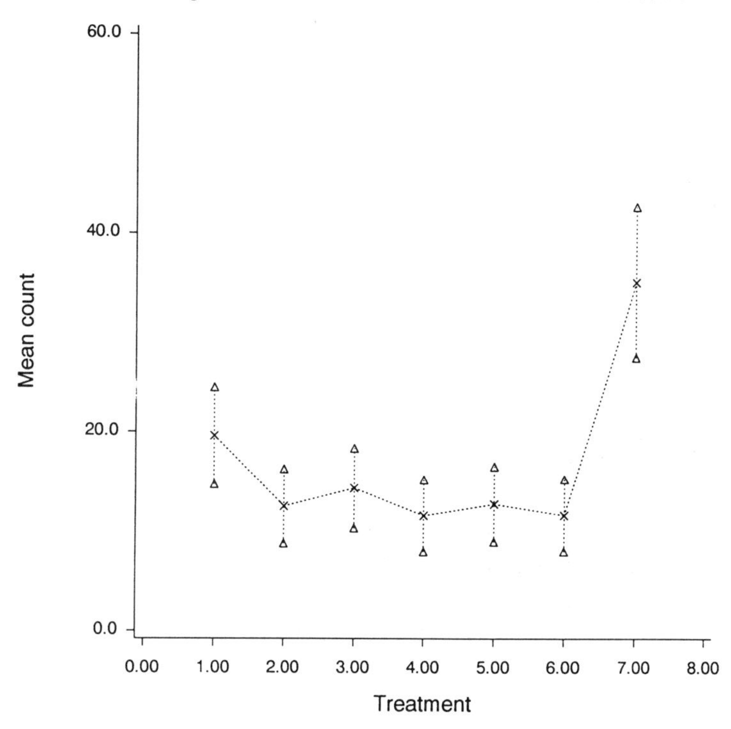

It appears that the treatment group differences are largely due to differences between treatment groups 1 and 7 differing from the rest with treatment 7 appearing to be by far the least effective.

12.4.2 Non-linear

12.4.2.1 Fitting a hyperbolic relationship

An assessment of the effect of haemolytic antiserum on the sensitivity of sheep erythrocytes to haemolysis (cell breakdown) induced by complement gave concentrations of complement required to produce 50 per cent haemolysis in the presence of different levels of antiserum. The data are given in Bliss (1970) and in example D4 of the GLIM3.77 manual (Payne, 1987). The need is for a model to describe the protective effect of antiserum which can also be used for prediction.

```
$units 9 $data antiserum complement $read
0.5 10.22 0.75 7.37 1.0 5.72 1.5 4.78 2.0 4.3
3.0 3.85 4.0 3.74 6.0 3.54 8.0 3.39
```

First the form of the relationship should be inspected graphically with

```
$plot antiserum complement'x' $
```

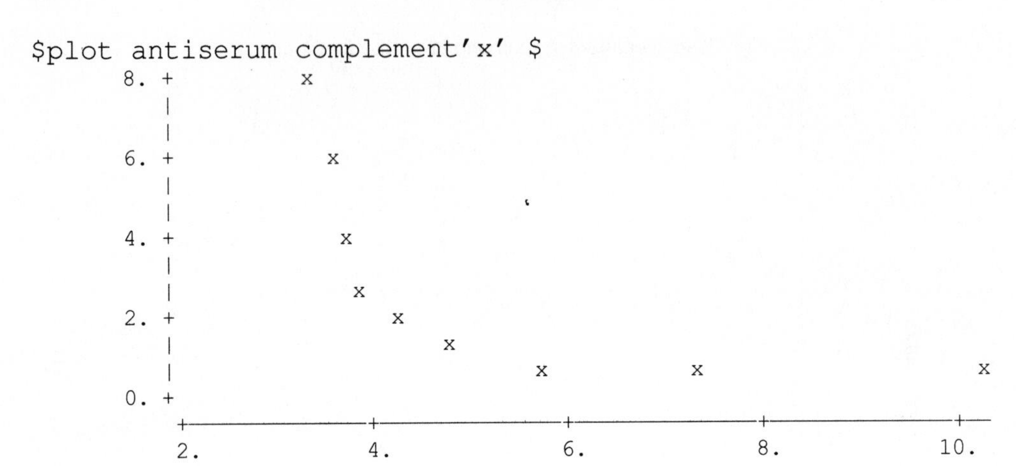

or, using high resolution graphics,

```
$graph (t='Fig. 12.49 Complemnt for 50% haemol. by antiserum'
       h='Anti-serum concn.' v='Complement level'
       y=0,11,2 x=0,10,2)
     complement;complement antiserum;antiserum  1,10 $
```

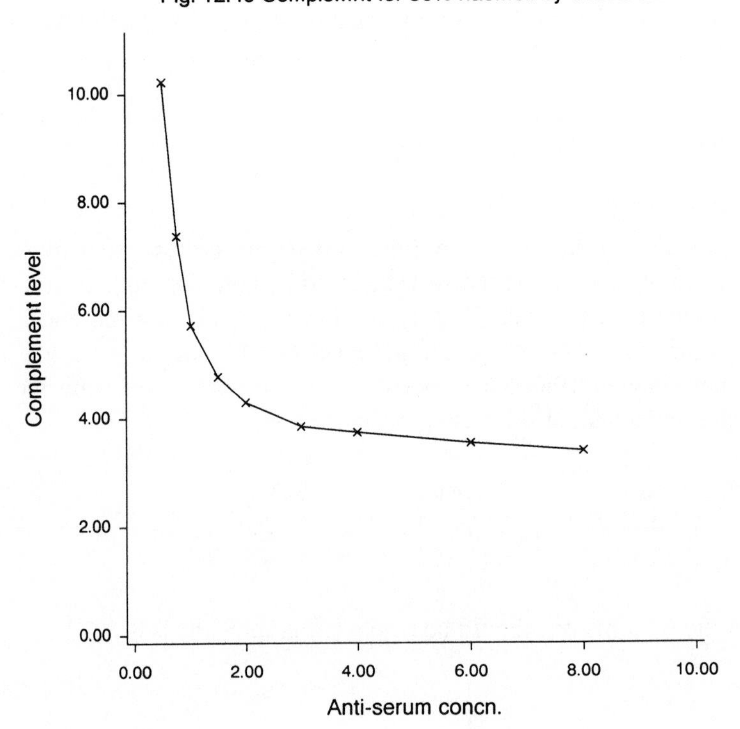

The relationship is strongly curved and appears hyperbolic. Such relationships can be modelled, but not straightforwardly. Check whether a quadratic model could be used. First specify the `YVARIATE` with

```
$yvar complement $
```

Then use orthogonal polynomials to specify a model `quadratic' in the antiserum variable and examine how well it fits using `PREDICT` to obtain fitted values for closely spaced x-values and hence smooth high-resolution plots of the resulting fitted curves.

```
$fit antiserum<2> $dis e $
   deviance =  10.537
residual df =    6
          estimate         s.e.     parameter
     1       5.212       0.4417      1
     2      -4.431        1.325      ANTISERU<1>
     3       3.313        1.325      ANTISERU<2>
scale parameter 1.756
$ass xfv=0,0.5...9 $
$predict (s=-1) antiserum=xfv $cal fv2=%plp $

$graph (title='Fig. 12.50 Observed data and Fitted Quadratic'
            h='Anti-serum concn.' v='Complement level'
            y=0,11,2 x=0,10,2)
            complement;fv2 antiserum;xfv 1,10 $
```

Clearly it is not going to be possible to find a well-fitting polynomial. The residuals from this model are unlikely to provide any useful information. Previous authors have suggested the gamma distribution as appropriate. Now consider the problem of fitting a hyperbola. If the relationship is hyperbolic in form then it can be modelled as

$$\mu = b_1 + \frac{b_2}{x - b_3}$$

This cannot be linearized. However the appropriate model can still be fitted by providing the iterative algorithm of GLIM with a set of covariates obtained as the derivatives of this function with respect to the b's, that is, the vector $\{\mathrm{d}\mu/\mathrm{d}b_i\}$. The mean of the y variable is related to x by the above function and that is used as the `OWN` predictor `%eta`. As a result of using this approach the formal `LINK` should be kept as the identity function (so that $\mu = \eta$). This means that the gamma `ERROR` default link must be overridden by a `LINK I` directive. The value of `%lp` is the scalar product $\{\mathrm{d}\eta/\mathrm{d}b_i\}.\{b_i\}$ which is obtained appropriately if the new covariates are correctly specified, but it needs to be initialised correctly. The derivatives of the model with respect to the parameters b_1,b_2 and b_3 to be used as covariates are

Fig. 12.50 Observed data and Fitted Quadratic

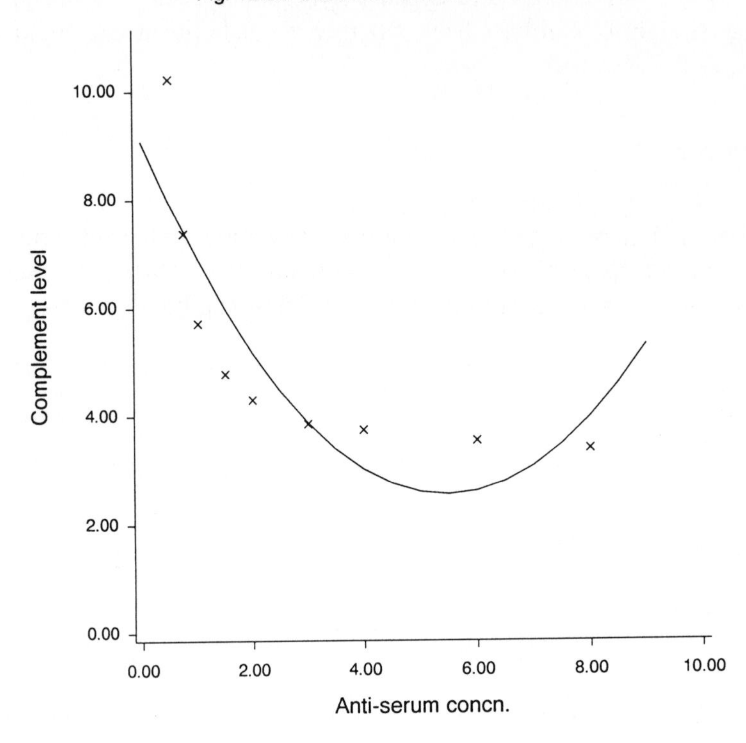

$$\frac{d\eta}{db_1} = 1 \qquad \text{for covariate X1}$$

which does not need to be supplied explicitly since it is automatically in the model if the constant term is included;

$$\frac{d\eta}{db_2} = \frac{1}{x - b_3} \qquad \text{for covariate X2}$$

and

$$\frac{d\eta}{db_3} = \frac{b_2}{(x - b_3)^2} \qquad \text{for covariate X3.}$$

The last two must be calculated and used in the model specification `X2+X3` and because they are needed during the initialisation as well as in a macro to be used in the fitting process it is convenient to calculate them in a macro. (Note that the parameters need to be obtained from `%pe`.)

```
$macro covars $
$cal X3=%pe(2)/((antiserum-%pe(3))**2) $
$cal X2=1/(antiserum-%pe(3)) $
$endmac $
```

Then the macro to calculate `%eta` and update the values of the working covariates must be provided. This is then defined as a model macro via the `METHOD` directive.

```
$macro Hyperb $
$pr 'calling Hyperb' $
$extract %pe $
$cal %eta=%pe(1)+%pe(2)/(antiserum-%pe(3)) $
$use covars $
$endmac $
```

The process must be initiated by a macro specifying starting values for the linear predictor `%lp` using some initial estimates of the `%pe(i)` values.

```
$macro first_lp
$pr 'calling first_lp' $
$cal %lp=%pe(1)+%pe(3)*X3+%pe(2)*X2 $
$endmac
```

The print statements in the above macros are not necessary, but inform the user each time they are called. This makes it easy to monitor the progress of the fitting process. To obtain fitted values for a smooth fitted curve cannot be done easily with `PREDICT` with this approach. It is necessary to use fictitious data points. Choosing antiserum values for the range of interest and supplementing the data variables appropriately is achieved by

```
$cal wt=1 $
$ass x_=0.2,0.25,...0.4,0.6...2,2.5,...10 $
$cal y_=x_ : y_=1 : wt_=y_*0 $

$ass wt=wt,wt_ : antiserum=antiserum,x_
  :  complement=complement,y_ $
```

and to create a weight variable to select the fictitious points

```
$cal wt2=0 $ass wt2=wt2,y_ $
```

Now to begin the modelling set the X variable to be 'antiserum'

```
$cal X=antiserum $
```

Then specify the gamma response distribution, the identity link, and
a weight vector with

```
$yvar complement $
$error G $link I $
$weight wt $
```

The linear predictor initialization is performed with

```
$ass %pe=0,8,0 $
$init first_lp $
```

Then the METHOD directive, with the existing method setting, is used to identify the %eta calculation and covariate updating macro:

```
$method * Hyperb $
```

```
$ass %pe=0,8,0 $
$use covars $
```

```
$fit X2+X3 $disp e $
calling first_lp
calling Hyperb
calling Hyperb
calling Hyperb
calling Hyperb
calling Hyperb
calling Hyperb
calling Hyperb
   deviance =  0.0028089 at cycle 7
residual df =  6          from 9 observations
          estimate        s.e.      parameter
     1       3.115      0.05789       1
     2       2.211       0.1497       X2
     3      0.1933      0.02394       X3
scale parameter 0.0004682
```

```
$graph (title='Fig. 12.51 Observed points and Fitted Hyperbola'
           h='Anti-serum concn.' v='Complement level'
           y=0,11,2 x=0,10,2)
     complement/wt;%fv/wt2 antiserum;X  1;10 $
```

Fig. 12.51 Observed points and Fitted Hyperbola

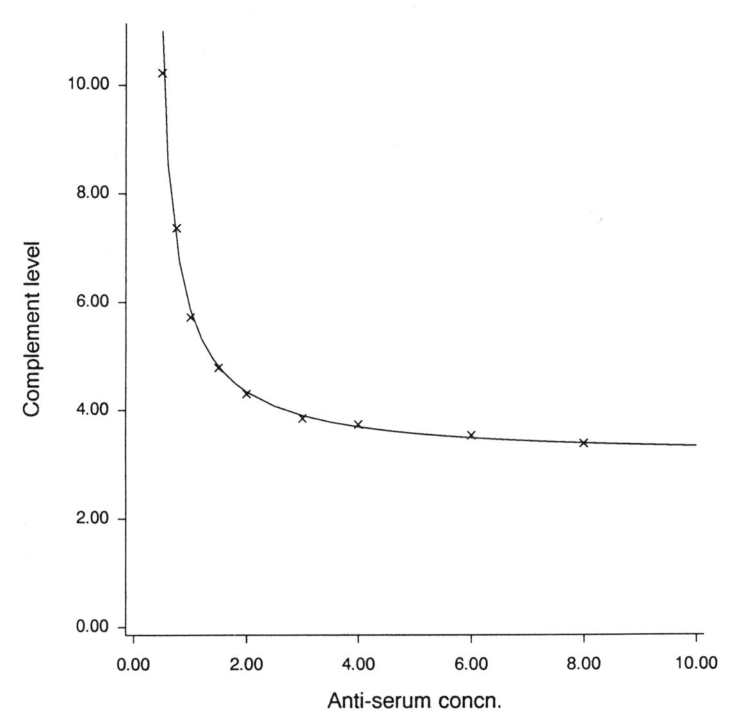

Check the residuals to see the evidence for or against the assumption of a gamma response distribution.

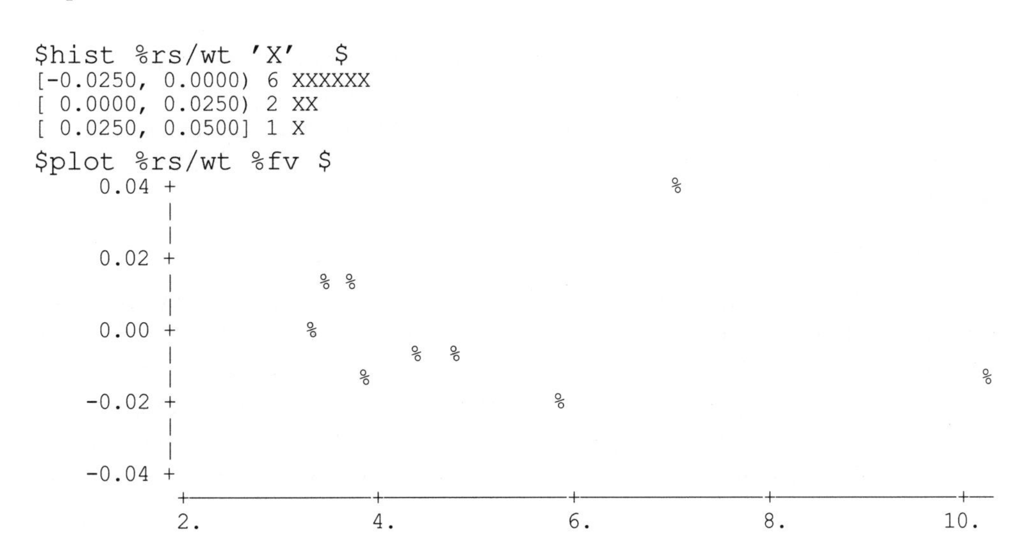

```
$hist %rs/wt 'X'  $
[-0.0250, 0.0000) 6 XXXXXX
[ 0.0000, 0.0250) 2 XX
[ 0.0250, 0.0500] 1 X
$plot %rs/wt %fv $
```

It is difficult to conclude much from nine residuals, but the histogram indicates the positively skew distribution one would expect from GAMMA data. At a stretch it could also be said that the dispersion in the plot of the residuals against the fitted values looks as if it might be increasing with the mean. It is certainly a well fitting model.

Finally we give a comparison of the results obtained here with those obtained in example D4 of the GLIM3.77 manual. Firstly the parametrization was different. In this example it is

$$\mu = a + \frac{b}{x - c}$$

whereas the GLIM3.77 manual example used

$$\mu = (a_3 + \frac{b_3}{d_3 + x})^{-1}$$

This means that the GLIM3.77 parameters in terms of those estimated here are

$$a_3 = 1/a \quad b_3 = -b/a^2 \text{ and } \quad d_3 = -c+b/a$$

Thus the equivalents of the GLIM3.77 parameters derived from the GLIM4 estimates are obtained as

```
$num a c b a3 b3 d3 $
$cal a=%pe(1) : c=%pe(3) : b=%pe(2) $
$cal a3=1/a : b3=-b/(a*a) : d3=-c+b/a $

$pr a3 b3 d3 $
  0.3210 -0.2279  0.5165
```

These are only slightly different to the estimates in the earlier manual which were 0.3212, -0.2288 and 0.52. This is probably due to the fact that the estimate of parameter d3 in the 3.77 analysis was found using a grid search and not by a full maximisation procedure. A major benefit of the GLIM4 approach is that standard errors are obtained without bias for all the parameters.

12.4.2.2 Log-logistic relationships for radio-immuno assay data

This is an example to illustrate the use of GLIM for a non-linear modelling problem arising from the widely used technique known as radio-immuno assay. The data, from an assay of thyrotrophin, are taken from Healy (1972). Radio-immuno assay is used to measure the quantity of a given biological substance in a sample by identifying the amount of a radio-active labelled anti-body removed from a reagent by subsamples of increasing concentration. In practice the actual response used is the amount of radio-active material remaining measured in counts per minute. The explanatory or x variable is the concentration of subsamples of the unkown obtained by successively diluting the original sample to give a suitable range of concentrations. The form of relationship that arises in practice is a sigmoid curve descending from an upper asymptote representing the zero concentration response to a lower asymptote representing an infinite concentration response generated by a reagent `blank' sample. An appropriate relationship for modelling this is the log-logistic function of the concentration x.

$$\mu = a + \frac{b}{1 + \exp\{-(c + d \log x)\}}$$

The readings were obtained as replicate pairs of 'counts per minute' for each concentration. To simplify data input enter the two replicates as if they were different variables. They can then be assigned to the same variable in an order that allows the concentrations to be entered only once and then duplicated by a similar assignment.

```
$units 12 $data cpm1 cpm2 $read
  37 5076
4789 4928
4979 4961
4769 5131
4270 4571
4462 3939
3374 3500
2793 2966
2482 3423
2111 2278
1676 1904
 686  700


$ass
cn=0.000001,0.1,0.5,1.0,2.0,5.0,10.0,16.0,20.0,50.0,100.0,100000
 : count=cpm1,cpm2 : concn=cn,cn $
$cal lconc=%log(concn) : lcn=%log(cn) $
$slen 24 $
```

It is also useful to identify the replicate pairs by an index variable

```
$cal pair=%gl(12,1) $
```

The mean response for each concentration can then be obtained and plotted against the original concentration variable of length 12.

```
$tabulate the count mean for pair into c_mean by pair_ $
$tprint c_mean pair_ $
  PAIR_    1.000    2.000    3.000    4.000    5.000    6.000    7.000    8.000
 C_MEAN  2556.5   4858.5   4970.0   4950.0   4420.5   4200.5   3437.0   2879.5
  PAIR_    9.000   10.000   11.000   12.000
 C_MEAN  2952.5   2194.5   1790.0    693.0
```

Although the extreme concentrations compress the plot the form of the relationship can be seen, on the log(concentration) scale, with

```
$plot (x=-14,12) count lconc 'x' $
```

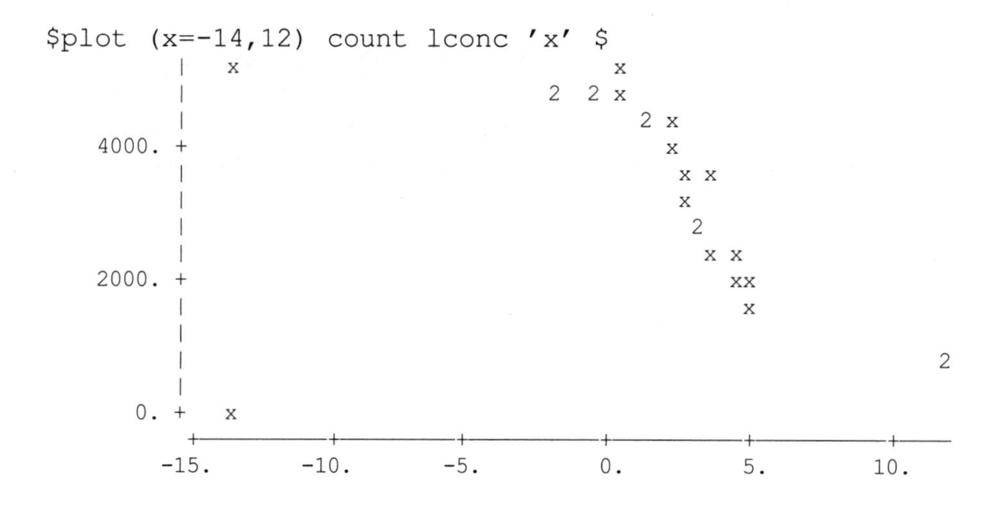

In practice some aberrant observations will occur and it is usual to screen the pairs by comparing the differences. In Healy's paper the large difference between the replicates for the first concentration lead to the exclusion of the complete pair. The methods used for detecting such points are discussed by Healy. In this analysis only the excessively low value, which is less than 100, will be excluded using a weight variable.

```
$cal wt=count>=100  $weight wt $
```

A high-resolution plot, using this weight variable to exclude the extreme point from the caculations of the means and restricted to the middle range of concentrations, is obtained by

```
$tabulate the count mean with wt for pair into c_mean by pair_$
$graph (title='Fig.12.52 Response as counts/min against Log(Concn)'
           h='Preparation log concn.' v='Counts per minute'
           y=0,6000,1000 x=-5,8,2)
     count;c_mean lconc;lcn  1,10 $
```

Fig. 12.52 Response as counts/min against Log(Concn)

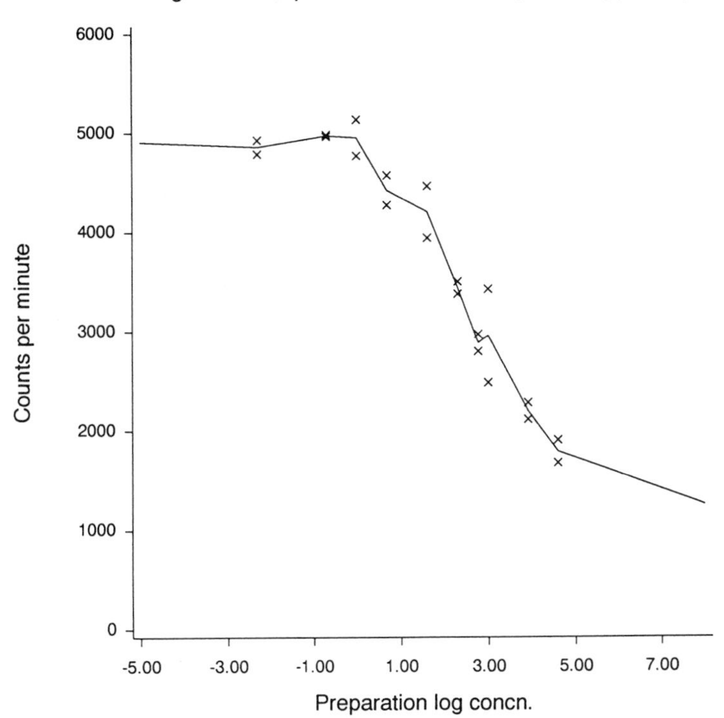

The counts are likely to follow a Poisson distribution, but they are so large that the usual assumption of Normality is probably justified in this case. However that may not always be true. This analysis assumes a Poisson response variable distribution, but a repeat analysis with Normal assumptions is shown for comparison.

The problem is to fit the non-linear log-logistic function above with the four parameters specified by b_1, b_2, b_3 and b_4

$$\mu = \eta = b_1 + \frac{b_2}{1 + \exp\{-(b_3 + b_4 \log x)\}}$$

Although this cannot be linearized the appropriate model can still be fitted using the `OWN` facility in GLIM. As in the previous example it is necessary to provide covariates which are the derivatives of this function with respect to the b's, that is, $\{d\eta/db_i\}$. The above function is the linear predictor of y and is used to calculate `%eta`. A macro to calculate `%eta` and the covariates is identified using the `METHOD` directive and called at each iteration. The `LINK` must be kept as the identity function which means that the Poisson default link must be over-ridden by a `LINK I` directive. With this approach `%lp`, which is not identical to `%eta`, is the scalar product $\{b_i\}.\{d\eta/db_i\}$. *Using this relationship* the values of `%lp` need to be initialized correctly in a separate macro identified by the `INIT` directive. The estimates of the parameters required during the iteration are held, as usual, in the system vector %pe. The derivatives of the model with respect to the parameters b_1,b_2 and b_3 to be used as covariates are

$$\frac{d\eta}{db_1} = 1 \qquad \text{for covariate X1}$$

This is automatically in the model if the constant term is included with the SET constant directive or by default;

$$\frac{d\eta}{db_2} = \frac{1}{1 + \exp\{ -(b_3 + b_4 \log x)\}} \qquad \text{for covariate X2}$$

$$\frac{d\eta}{db_3} = \frac{b_2 \exp\{ -(b_3 + b_4 \log x)\}}{(1 + \exp\{ -(b_3 + b_4 \log x)\})^2} \qquad \text{for covariate X3}$$

and

$$\frac{d\eta}{db_4} = \frac{b_2 \exp\{-(b_3 + b_4 \log x)\} \log x}{(1 + \exp\{ -(b_3 + b_4 \log x)\})^2} \qquad \text{for covariate X4}$$

The appropriate model specification is then X2+X3+X4. Because X1, X2 and X3 need to be re-calculated during the initialisation as well as in a macro called during each iteration it is convenient to calculate them in a macro.

```
$macro covars $
$cal X2 = 1/(1+%exp(-(%pe(3)+%pe(4)*%log(x))))
   :  X3 = %pe(2)*%exp(-(%pe(3)+%pe(4)*%log(x)))*X2**2
   :  X4 = %pe(2)*%log(x)*%exp(-(%pe(3)+%pe(4)*%log(x)))*X2**2
$endmac $
```

Then the macro to calculate %eta and update the values of the working covariates.

```
$macro loglo $
$pr 'calling Loglo' $
$extract %pe $
$cal %eta=%pe(1)+%pe(2)/(1 + %exp(-(%pe(3) + %pe(4)*%log(x)))) $
$use covars $
$endmac $
```

The process is initiated with a macro specifying starting values for the linear predictor %lp using some initial estimates of the %pe(i) values.

```
$macro first_lp
```

```
$pr 'calling first_lp' $
$cal %lp=%pe(1)+%pe(2)*X2+%pe(3)*X3+%pe(4)*X4 $
$endmac
```

Obtaining predicted values with their variances is not possible using the PREDICT directive with this approach. They can be obtained for the data points, but these do not necessarily cover an appropriate range of the explanatory variable values for obtaining smooth plots. The simplest solution is to add some fictitious data points with appropriate values to the data with zero weights. The fitting process will not use them, but they will still generate fitted values, linear predictor values, and the appropriate variances in the system vectors %fv, %lp, and %vl available directly or via the EXTRACT directive.

Choosing log concentrations for the range of interest and supplementing the data variables appropriately is achieved by

```
$ass x_=-3,-2.5,...8 $
$cal cx_=%exp(x_) $
```

Then variables y_ and wt_ with the same length as x_ and the values 1 and 0 can be created by

```
$cal y_=x_ : y_=1 : wt_=y_*0 $

$ass wt=wt,wt_ : count=count,y_
  : concn=concn,cx_ : lconc=lconc,x_ $
```

and to create a weight variable to select the fictitious points for the later plots:

```
$cal wt2=0 $ass wt2=wt2,y_ $
```

Then the Poisson response distribution and the identity link is specified with

```
$yvar count $
$error P $link I $
```

A variable X is created with the concentration values by

```
$cal X=concn $
```

The linear predictor initialization is performed by first setting the initial estimates for the parameters and then calling INIT. However such iterative processes are sometimes very sensitive to the choice of starting values. It is useful if some reasonable values can be identified. The first parameter, b_1, represents the lower asymptote which can be estimated by the mean of the pair with the 'infinite' concentration, that is pair 12, which is 693.

The upper asymptote is the sum of the first two parameters. Thus a starting value for the second parameter, b_2, can be obtained by estimating the upper asymptote using the one acceptable response at zero concentration and subtracting the estimate of the lower asymptote, that is, 5076 - 693 = 4383.

Deducing an initial estimate for the other two parameters is rather more difficult. However they can be calculated by assuming two data points lie on the 'true' model curve. These calculations require the solution of the simultaneous equations

$$b_3 + b_4 \log x_1 = \log(\frac{Y_1}{1-Y_1}) = Z_1$$

$$b_3 + b_4 \log x_2 = \log(\frac{Y_2}{1-Y_2}) = Z_2$$

where $Y_i = \frac{y_i - b_1}{b_2}$, that is,

$$b_4 = \frac{Z_1 - Z_2}{\log x_1 - \log x_2}$$

and $\quad b_3 = Z_1 - b_4 \log x_1$.

Using the 5th and 8th pair at concentrations 2.0 and 16.0 this gives

```
$number b1i b2i b3i b4i Y1 Y2 Z1 Z2 $
$cal b1i = 693 : b2i=5076-b1i
: Y1 = (c_mean(5)-b1i)/b2i : Z1 =  %log(Y1/(1-Y1))
: Y2 = (c_mean(8)-b1i)/b2i : Z2 =  %log(Y2/(1-Y2))
: b4i =  ( Z1 - Z2 )/(%log(cn(5)) - %log(cn(8)))
: b3i =    Z1 - b4i*%log(cn(5)) $
```

These are then assigned to the parameter estimate vector and the initialization achieved by

```
$ass %pe=b1i,b2i,b3i,b4i $
$init first_lp $
```

Then the `METHOD` directive, with the existing method setting, is used to identify the `%eta` calculation and covariate updating macro:

```
$method * loglo $
$use covars $
```

and the model fitted by

```
$fit X2+X3+X4 $disp e $
calling first_lp
calling Loglo
calling Loglo
calling Loglo
calling Loglo
scaled deviance =  346.25 at cycle 4
    residual df =   19     from 23 observations
          estimate          s.e.     parameter
     1       697.3         19.06       1
     2       4445.         43.69       X2
     3       2.281       0.07387       X3
     4     -0.7591       0.01928       X4
scale parameter 1.000
```

The deviance is far greater than the degrees of freedom so there is serious overdispersion. This should be allowed for by re-fitting the model with the scale parameter reset to a value calculated from the deviance divided by the degrees of freedom.

```
$cal %s=%dv/%df $scale %s $
$fit X2+X3+X4 $disp e $
calling first_lp
calling Loglo
calling Loglo
scaled deviance =  19.000 at cycle 2
    residual df =  19      from 23 observations
          estimate          s.e.     parameter
     1       697.3         81.37       1
     2       4445.         186.9       X2
     3       2.282        0.3148       X3
     4     -0.7593       0.08214       X4
scale parameter 18.22
```

```
$extract %pe $
```

```
$ext %vl $
$cal lo=%fv - 2*%sqrt(%vl) : hi=%fv + 2*%sqrt(%vl) $
```

```
$graph (title='Fig. 12.53 Fitted Curve and 95% CLs'
            h='Log concn' v='Counts/minute'
            y=0,6000,1000 x=-3,8,2)
     count/wt;%fv/wt2;lo/wt2;hi/wt2
        lconc;  lconc; lconc; lconc  1,10,11,11 $
```

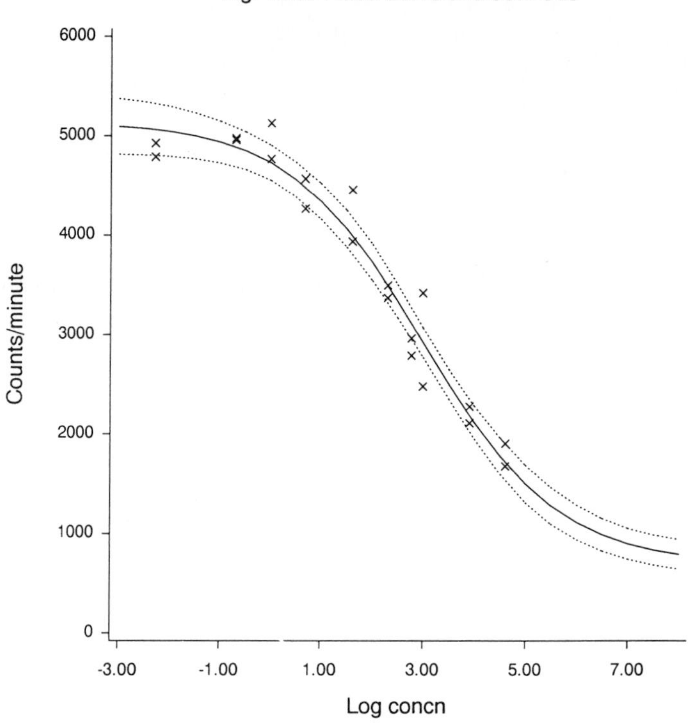

The fitted model may be used as a calibration curve for estimating concentrations from counts. Although it is not simple it is relatively straightforward to obtain such estimates with confidence limits from the parameter estimates and their covariance matrix. Comparing such a curve with a standard as in the usual bioassays requires that rather more complicated models are fitted.

Now repeat the exercise with Normal assumptions as in the published paper, first storing the fitted values from the Poisson fit.

```
$cal p_fv=%fv $
```

```
$error N $
```

The linear predictor initialization is again performed with

```
$ass %pe=b1i,b2i,b3i,b4i $
$init first_lp $
```

```
$use covars $
$fit X2+X3+X4 $disp e $
calling first_lp
calling Loglo
calling Loglo
calling Loglo
calling Loglo
   deviance =  1167556. at cycle 4
residual df =        19  from 23 observations
          estimate         s.e.      parameter
     1       742.6        173.2       1
     2       4359.        233.4       X2
     3       2.419       0.3233       X3
     4     -0.8085      0.09910       X4
scale parameter 61450.
```

Then to compare the Poisson and Normal fit

```
$graph (title='Fig. 12.54 Comparison of Poisson and Normal Fits'
           h='Log concn' v='Counts/minute'
           y=0,6000,1000 x=-3,8,2)
     count/wt;%fv/wt2;p_fv/wt2 lconc;lconc;lconc  1,10,12 $
```

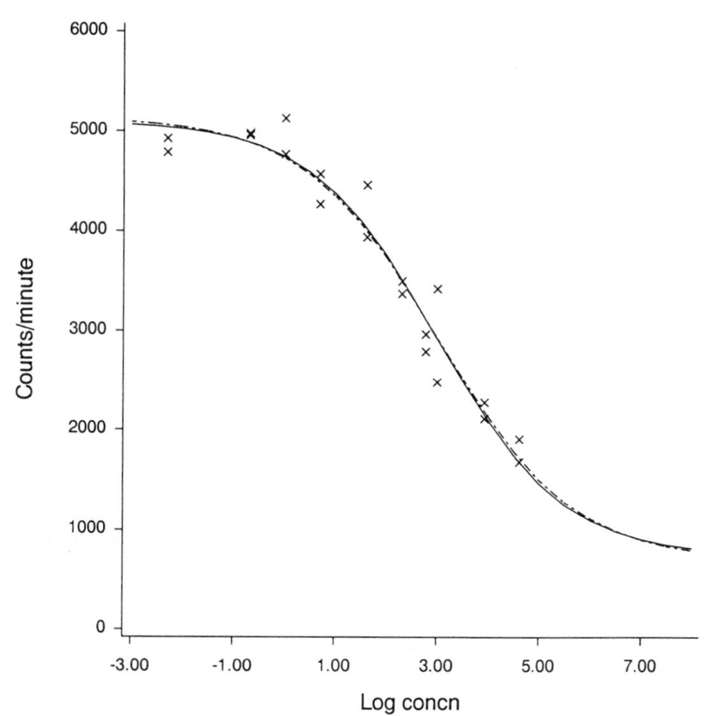

They appear almost indistinguishable. To see the curve fitted with Normal assumptions and 95 per cent confidence limits we use

```
$ext %vl $
$cal lo=%fv - 2*%sqrt(%vl) : hi=%fv + 2*%sqrt(%vl) $

$graph (title='Fig. 12.55  Fitted Curve and 95% Confidence
       Limits'
       h='Log concn' v='Counts/minute'
       y=0,6000,1000 x=-3,8,2)
       count/wt;%fv/wt2;lo/wt2;hi/wt2
       lconc;  lconc; lconc; lconc 1,10,11,11 $
```

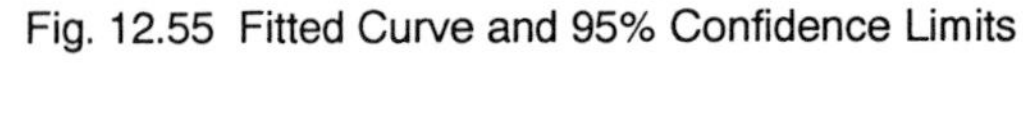

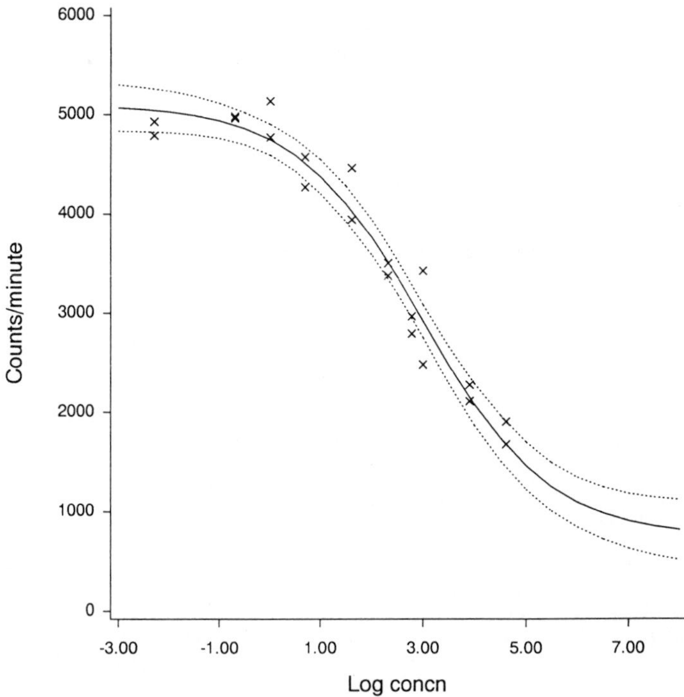

For estimating concentrations the Normal assumption seems reasonably justified for counts of this magnitude. There is some difference in that the confidence intervals obtained with Poisson assumptions are slightly wider. However over the middle range of concentrations the differences are very slight and unlikely to be of practical importance.

12.4.3 Graphics: keys and interactive macros

12.4.3.1 Keys and Labelling for Model and Data

This example is provided to illustrate how, with convenient macros, keys and labels can be positioned and drawn on a graph.

Consider the simple linear regression example [12.1.1.1]

```
$var 12 x y $read x y
      8     59
      6     58
     11     56
     22     53
     14     50
     17     45
     18     43
     24     42
     19     39
     23     38
     26     30
     40     27
```

The linear regression model is fitted and displayed with the data points by

```
$yvar y $
$fit 1+x $dis e $
   deviance =  273.84
residual df =    10
          estimate        s.e.     parameter
    1        64.25       3.603       1
    2       -1.013      0.1722       X
scale parameter 27.38
```

```
$ass xf=0,5...50 $
$predict (s=-1) x=xf $
$cal yf=%pfv $
```

Set up one macro to hold the title and axis labels and two macros to hold the x and y scales. This makes it easy to use them repeatedly with the text substitution facility, using `#macro_name`, for inserting the macro contents as an appropriate 'string' in other directives.

```
$mac title
t='Fig. 12.56 Data points with Fitted Straight Line'
v='y - axis' h='x - axis'
$endmac $

$mac xscale 0,50,10 $endmac $
$mac yscale 0,80,10 $endmac $
$ass pens=1,12 $

$gra ( #title x=#xscale y=#yscale )
        y;yf x;xf  pens $
```

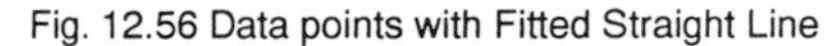

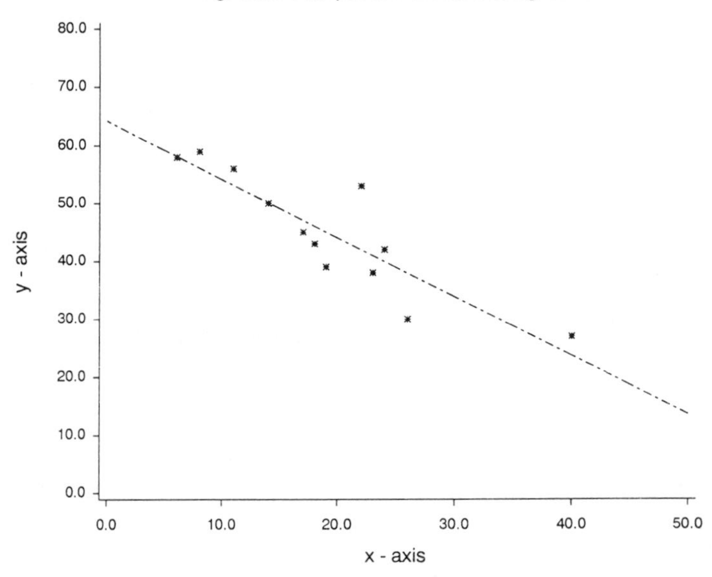

For presentation purposes a 'key' and text labelling can be added, with the GTEXT directive. However positioning the labelling and a 'key' or 'legend' box is not all that simple. In practice it is most easily done using appropriate *x,y* co-ordinates, on the same scales as the data, determined by superimposing a 'grid' on the graph. This can be achieved with the following macro.

```
$mac   ! grid
!
! This macro superimposes a grid of lines on the current graph
!which can be used to determine the positioning of text strings
!and keys to be added to the graph. The required vertical and
```

```
!horizontal grid lines are defined by their end-point
!coordinates at each x and y scale point respectively.
! The macro requires as arguments:
!
!  title  - a macro holding a string with the title and axis
!           label settings for the graph on which the grid
!           is to be plotted
!  xscale - macros holding the 'min,max,interval' strings
!  yscale   equivalent to those used to set the graph
!           directive xlimit and ylimit options e.g.
!           'xscale' for '0,50,10'
!  pen    - a scalar variable identifying an appropriate
!           pen for drawing the grid lines.
!           (dotted white lines are usually best)
!
(arg=title,xscale,yscale,pen
 local=ix,ax,bx,hx,iy,ay,by,hy,lv,lh,ixiy,
       xsc,ysc,xv,yv,fv,xh,yh,fh,xend,yend,pens )  grid $
!
$del xsc ysc xv yv fv xh yh fh xend yend pens $
$number ix ax bx hx iy ay by hy lv lh ixiy $
$ass xsc=#xscale : ysc=#yscale $
!
$cal ax=xsc(1) :  bx=xsc(2) : hx=xsc(3)
   : ay=ysc(1) :  by=ysc(2) : hy=ysc(3) $
!
$cal ix=%tr((bx-ax)/hx + 0.5)+1 : iy=%tr((by-ay)/hy + 0.5)+1 $
$cal lv=ix*2 : lh=iy*2 $
$var lv xv yv fv yend : lh xh yh fh xend $
!
$cal fv=%gl(ix,2) : fh=%gl(iy,2) $fac fv ix : fh iy $
$cal  yend=%gl(2,1) : xend=%gl(2,1)
 :   xv=(fv-1)*hx + ax : yv=%if(yend==1,ay,by)
 :   yh=(fh-1)*hy + ay : xh=%if(xend==1,ax,bx) $
!
$cal ixiy=ix+iy $
$var ixiy pens $cal pens=pen $
!
$gra ( ref=no #title y=#yscale x=#xscale) yv;yh xv;xh  pens
fv;fh $
!
$endmac $
```

The title is modified slightly with

```
$mac title
t='Fig. 12.57 Data points with Fitted Straight Line'
v='y - axis' h='x - axis'
$endmac $

$number pen=11 $
$use grid title xscale yscale pen $
```

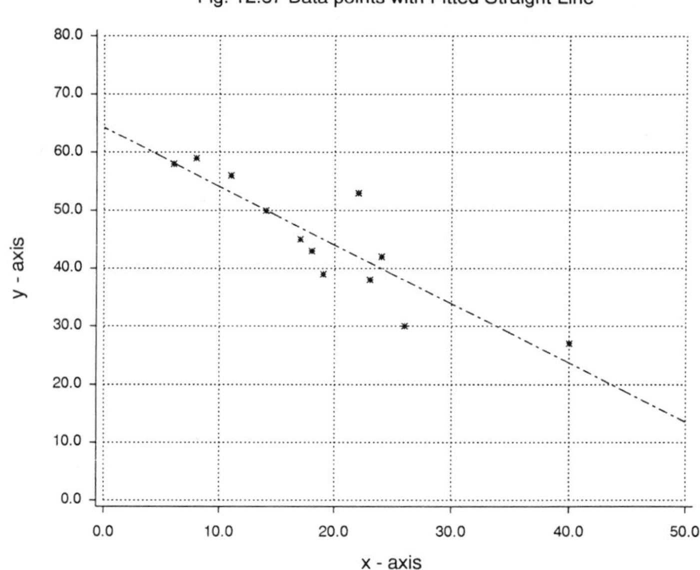

The 'key' needs a labelled data point and a segment of line drawn with the same pen as that used to represent the fitted model. And to ensure that the 'key' is clearly distinguished from the graph it is convenient to enclose it in a box.

A reasonable position for the box is extending from 30 to 50 on the x scale and from 55 to 70 on the y scale. An appropriate position for the 'key' title, `Key:`, is one unit down and along from the top left hand corner (31,69). The line segment is reasonably positioned one x unit in from the left hand edge of the box and 4 units up (roughly a quarter) from the bottom edge with a length of 6 units, roughly a third of the box length, (31,59) to (37,59). A good position for the data point is 3 units up from the middle of the line segment (34,62).

The `Data` and `Fitted Model` labels are best starting one unit on the x scale further to the right of where the line segment ends at the appropriate heights (38,63) and (38,60). A label for the fitted model is reasonably positioned to start at (22,45).

First create the data for plotting the 'key' point and line segment

```
$number pxkey=34 pykey=62 $
$ass xkey=31,37 : ykey=59,59 $
```

Then the coordinate vectors for the labelling.

```
$ass ytx=69,62,59,22  : xtx=31,38,38,45 $
```

Finally a macro to draw the box

```
$macro  ! box
! This draws a box on an existing graph given the coordinates
!of the bottom left and top right corners. The title and axis
!labels need to be supplied again as do the x and y scales
!to keep the absolute co-ordinate positions fixed when the
!box is plotted on an existing graph.
!
!  The parameters required are:
!
!  xbl,ybl,  - scalars holding the coordinates of the
!  xtr,ytr     bottom left and top right corners of the box
!
!  title     - a macro holding the current graph titles
!              as a string
!  xscale,   - macros holding the x and y plotting ranges
!  yscale      as strings in the form 'min,max,interval'
!
!  pen,col   - scalars indicating the pen number and colour
!              to be used (the pen chosen will be re-defined).
!
(arg=x1,y1,x2,y2,Tmac,xs,ys,pen,col  loc=x,y)  box $
$ass x=x1,x1,x2,x2,x1 : y=y1,y2,y2,y1,y1 $
$gstyle pen c col 1 1 $
$graph ( ref=no pause=no #Tmac xs=#xs ys=#ys)  y x pen $
$endmac $
```

This macro is used with the bottom left and top right corner coordinates of the box, as determined above from the grid, as arguments. These need to be stored in scalar variables. In addition the pens are required twice, once for the real data and once for the 'key' data.

```
$number xbl=30 ybl=55 xtr=50 ytr=70 white=1 pen=16 $
$ass pens=pens,pens $
```

The final plot, with a further slight modification to the title, needs a call to `GRAPH`, `GTEXT` and the `box` macro.

```
$mac title
t='Fig. 12.58 Data points with Fitted Straight Line'
v='y - axis' h='x - axis'
$endmac $

$gra ( #title x=#xscale y=#yscale pause=no)
        y;yf;pykey;ykey x;xf;pxkey;xkey  pens   $
$gt ( m=p l=r p=no)
       ytx xtx 'Key:','Data','Fitted model','y = a + bx' $

$use box xbl ybl xtr ytr title xscale yscale pen white $
```

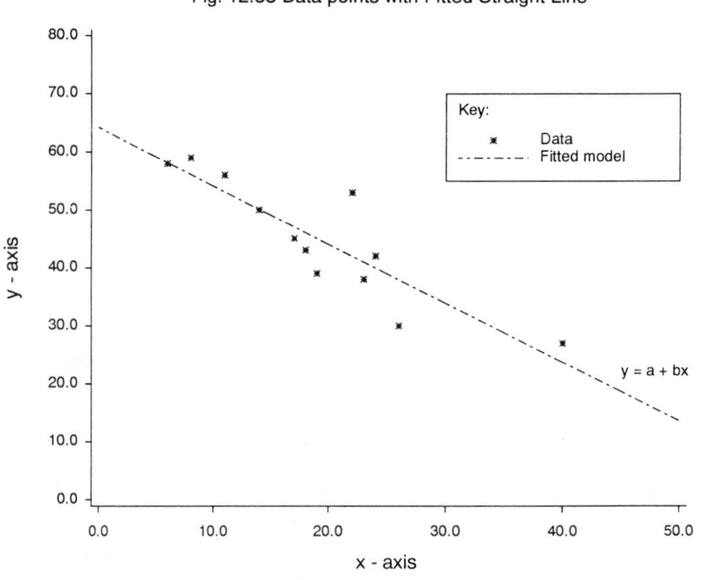

A hard copy version can be obtained by re-plotting the graph after setting the graphics device appropriately. For example 'PostScript' output is obtained by

```
$set device='post' $
```

```
$gra ( #title x=#xscale y=#yscale pause=no)
        y;yf;pykey;ykey x;xf;pxkey;xkey  pens    $
$gt  ( m=p l=r p=no)
       ytx xtx 'Key:','Data','Fitted model','y = a + bx' $

$use box xbl ybl xtr ytr title xscale yscale pen white $
```

This produces a file GLIMPS.PLT which can be printed on any printer with a 'Postscript' capability.

12.4.3.2 An interactive investigation of the F distribution

This example is designed to illustrate a further use of the graphics facilities and how interactive macros may be constructed with the GET directive. The example shows how the shape of the F distribution changes according to the degrees of freedom. It is assumed that there are four pens numbered 1 to 4 already specified with suitable and distinguishable line types, but the macro can be used for up to 16 plots if 16 pens, numbered 1 to 16, are available.

The main macro, Fcs, takes two arguments. The first is to specify the values of F, on the abscissa, for which the density function is to be plotted. The second is for specifying a device to be used for a hard-copy version of the graph if it is required. When invoked the macro Fcs calls a subsidiary macro Fc_ to plot the individual curves for selected degrees of freedom until the user indicates that the plot is complete and no further curves are required.

```
$macro    ! Fcs
!  This macro produces the probability density curves of selected
!  F distributions for the range of abscissa (F) values specified
!  in the first argument Fvals which must be a vector
!  containing the range of F values to be used. The degrees
!  of freedom for each curve are requested from within the macro.
!  The arguments are:
!    Fvals - an array specifying the x values at which the
!            F distribution function is to be evaluated
!    copy  - a macro specifying the device for hard copy
!            by a string enlosed in quotes e.g. 'post'
!
( arg=Fvals,copy local=nc,cd)   Fcs   $
$del df_1 df_2 $
$num df1 df2 cd nc $cal cd=1 : nc=1 $
$args fc_ Fvals nc cd copy df_1 df_2 $
$while cd Fc_  $
$tidy Fc_ $
$endmac $ ! Fcs
```

```
$mac  ! Fc_
!  This is an auxiliary macro called by Fcs to plot separate
!  F distribution curves for selected degrees of freedom input
!  using the GET directive.
(args=Fvals,nc,cd,copy,df_1,df_2
 loc=P,df1,df2,a,b,ab,xkey1,xkey2,ykey,
  ytx,xtx,xmax,xmin,kx,ky) Fc_ $
! set up incrementing y coordinates for the key to the graph
$number xkey1 xkey2  ykey ytx xtx $
$tab the fvals large,small into xmax,xmin $
$cal xkey1=xmax - .2*(xmax-xmin) : xkey2=xmax - .1*(xmax-xmin)
   : ykey=1.3-(nc-1)*0.08 $
$ass kx=xkey1,xkey2 : ky=ykey,ykey $
$del P $number df1 df2 a b ab $
$get ( prompt='df1,df2' type=macro ) df12m $
$ass dfs=#df12m $cal df1=dfs(1) : df2=dfs(2) $
!
$get ( ty=macro pr='refresh [no or yes] ') ref $
!
$cal a=df1/2 : b=df2/2 : ab=a+b $
$cal P=%exp(%lga(ab)-%lga(a)-%lga(b))
       *df1**a*df2**b*Fvals**(a-1)*(df2+df1*Fvals)**(-ab)
$graph
     (title='Fig. 12.59 F distribution curves' hlabel='F values'
      vlabel='Probability density' refresh=#ref copy=#copy
      ylimit=0,1.5,.1 pause=no ) P,ky Fvals,kx nc,nc $
!
$cal ytx=ykey+.04 : xtx=xkey2+.05  $
$ass ky=1.46,ytx : kx=xkey1,xtx $
$gtext (m=p l=r ) ky/1.4 kx ' F(df1,df2)',df12m $
$get (prompt='Another curve? [1 for no 2 for yes]') cd $
$cal cd=(cd/=1) : nc=nc+cd $
$endmac $ ! Fc_
```

For hard-copy output to be generated a macro must be specified to identify a hard-copy device. In this example we will use 'Postcript' output which will generate a file which can be used with a 'PostScript' printer to obtain paper output or incorporated into a document with a word processing package that accepts 'PostScript' files. The macro required is

```
$mac post 'post' $e $
```

In this example the array `F` is used to hold the array of abscissa (F) values which have been chosen more closely spaced near zero to make the plotted curves reasonably smooth.

```
$ass F=0.001,.1,.2,...,4 $
```

The `Fcs` macro is then called with the array of `F` values as the first parameter and the macro specifying the hard-copy device as the second. As shown below the user will then be prompted for the necessary input via the `GET` directive. The pair of values for the degrees of freedom are requested in the form of a string, as required for an `ASSIGN` directive, and then stored as the contents of a macro. This allows their values to be used in calculations and to be supplied to `GTEXT` for use in the key to the graph. For each curve the macro enquires whether the plot should be refreshed. For the first curve the answer should be `yes`, but subsequently, unless earlier curves are to be deleted, the answer should be `no`. Finally after each curve is plotted and labelled the user is asked whether another curve is required with the options 1 for no and 2 for yes. If the answer is no GLIM exits from the macro.

```
$use fcs F post $
df1,df2? 1,10
refresh [no or yes] ? y
Another curve? [1 for no 2 for yes]? 2
df1,df2? 2,10
refresh [no or yes] ? n
Another curve? [1 for no 2 for yes]? 2
df1,df2? 5,10
refresh [no or yes] ? n
Another curve? [1 for no 2 for yes]? 2
df1,df2? 10,10
refresh [no or yes] ? n
Another curve? [1 for no 2 for yes]? 1
```

See Fig. 12.59 overleaf. If at any point the user wishes to recall the last plot then using the directive `GTEXT` to plot nothing at a legal pair of coordinates will do it. For example:

```
$gtext 0 0 '' $
```

Finally the macro can be invoked again for another set of curves or the user can exit from the program and use the `PostScript` file to get a hard-copy of the plot or plots.

12.5 Computing Strategy

Finally it is worth making a few points on the computing stategy for analysing data with an interactive package such as GLIM. The package can be run and used simply as a

Fig. 12.59 F distribution curves

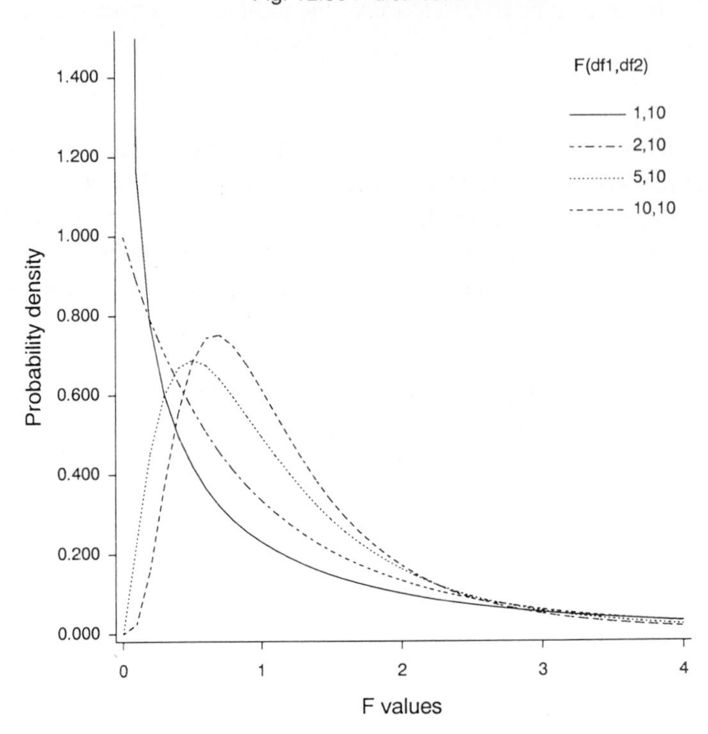

calculator if wished. Data can be generated or simulated, complex calculations may be performed, and high-quality graphical output produced. Small datasets may be read in directly, descriptive analysis performed and models fitted for inferential analyses. All this may be done interactively with the user providing all input via the keyboard. However there is a limit to what can sensibly be achieved interactively. Large datasets should certainly not be input directly to GLIM. They should be prepared, that is, checked, 'cleaned', and edited, with software designed for that purpose (GLIM has only limited data editing facilities) and input from a computer file. Equally, performing a complex sequence of calculations with a specific objective or performing an analysis, even with a small dataset, keeping to a consistent overall strategy is not easy to do interactively. In addition some parameter of the calculations or a part of the data might be changed so that the exercise needs to be repeated. GLIM will take directives from a file and it is often most efficient to perform a GLIM analysis by preparing a file of GLIM directives, for example, to read the data, to calculate derived variables and weights for variables with missing value codes, to initiate the analysis, and then return control to the user at the keyboard. This can be organized very conveniently by starting in the interactive mode, reading part of a large data file, performing some of the calculations needed and then exiting from GLIM. The journal file will then have a record of the keyboard input which can be edited into a GLIM batch job, starting with `$echo on $` and finishing with `$echo off $return $`, using a convenient text or screen editor or word processor which allows ASCII text output. This approach can also be used to develop macros. There is also a macro editor in GLIM which can be used during an interactive session and the final working macro may then be extracted from the journal file using an editor. For the more adventurous user the `PAUSE` directive may be used to exit from GLIM temporarily to identify a filename, perform some system operation, or even edit the batch file GLIM is executing. In the last case the file should be closed and the workspace cleared before it is re-input, `$new (close=yes) $` allows the user to do this. This general approach makes it very easy to develop complete sequences of complex analyses ready for when the data collection part of a study is complete. It also makes it easy to ensure that there is a precise record of the analysis and results in one complete whole. This approach is also a convenient, if not essential, technique for developing complicated hardcopy graphics and complex macros.

12.6 Conclusion

The examples have a heavy medical bias due to the author's particular background. However this should not be much of a limitation for users if their problem is one of explaining variation in a univariate response variable by modelling the systematic part of this variation as a linear function of the effects of one or more explanatory variables. The general approach is the same in any application area although the vocabulary may differ, for example, in medical statistics observed proportions are often taken to be estimates of 'risk' of disease or death and two risks are compared using ratios, that is, 'relative risks'. However the proportions could just as well be estimating the 'chance' of recovery or probability of voting for a particular political party and the general analytical approach would, apart from terminology, be much the same.

Many of the examples used macros that were not displayed. They are listed in the last section. A number of these are relatively trivial and users should easily be able to construct their equivalent. Some are more complex but rather specific such as the analysis of variance and deviance table calculations and output. These are intended to provide interested users with a 'template'. Finally there are several more difficult macros necessary for particular techniques such as those required for the piecewise exponential approach for fitting Cox's regression models for survival data. These are largely based on macros already available in the GLIM 3.77 macro library and will be available in the GLIM4 version.

As indicated in the introduction these examples are not, and could not possibly be, exhaustive. GLIM is such a flexible package that there is almost no limit to the examples that might prove useful. However those given here cover the use of most of the facilities in GLIM and should prove a reasonably comprehensive guide on how to use GLIM for a large proportion of the problems met in practice.

12.7 Macro listings

```
!____________________________________________________________________________
!
$subfile ANCO3123 $echo $
$mac ANCO3123 $
!
! This is a macro to obtain the various components of the Analysis of
!Co-variance table and print it. This is given in full here to illustrate,
!in particular, the use of the PRINT directive to obtain a relatively
!complicated output. Once constructed it is relatively easy, using one
!such macro as a template, to produce them for other analyses.
!
$trans -v $
$number p21 p32 p42  $
$cal p21=1-%fp(vr21,df21,df1) : p32=1-%fp(vr32,df32,df2)
   : p42=1-%fp(vr42,df42,df2) $
!
$pr '            Model     Residual SS    df   Mean square     Variance Ratio(F)'
  :  '  ______________________________________________________________________'
  :  '    1)  STATE*AGE     '*r RSS1,10,2 ' ' *i df1,4 '      '*r s2,10,2
  :  '    2)  STATE+AGE     '*r RSS2,10,2 ' ' *i df2,4
  :  '  ____________________________________'
  :  '(2-1)Due to STATE.AGE'*r SS21,10,2' '*i df21,4'     '*r MS21,10,2!
                                                      '          '*r VR21,10,2
  :  '                                                          ( p = 'p21' )'
  :  '  ______________________________________________________________________'
  :  '    3)  STATE'
  :  '        (i.e. -AGE)   '*r RSS3,10,2' '*i df3,4
  :  '  ____________________________________'
  :  '  (3-2)Due to AGE     '*r SS32,10,2' '*i df32,4'     '*r MS32,10,2!
                                                      '          '*r VR32,10,2
  :  '                                                          ( p = 'p32' )'
  :  '  ______________________________________________________________________'
  :  '     4) AGE'
  :  '        (ie -STATE)   '*r RSS4,10,2' '*i df4,4
  :  '  ____________________________________'
  :  '  (4-2)Due to STATE  '*r SS42,10,2' '*i df42,4'     '*r MS42,10,2!
                                                      '          '*r VR42,10,2
  :  '                                                          ( p = 'p42' )'
  :  '  ______________________________________________________________________'
  :  '  ______________________________________________________________________'
$
$trans +v $
$endmac  $
$echo $return $ ! ANCO3123
!
!____________________________________________________________________________
!
$subfile ANCO3124 $echo $
$mac ANCO3124 $
!
$trans -v $
$number SS21 SS32 SS42 df21 df32 df42 MS21 MS32 MS42
        vr21 vr32 vr42 p21 p32 p42 $
```

```
$cal  SS21=SS2-SS1 : df21=df2-df1 : MS21=SS21/df21 : vr21=MS21/s1
    : SS32=SS3-SS2 : df32=df3-df2 : MS32=SS32/df32 : vr32=MS32/s1
    : SS42=SS4-SS2 : df42=df4-df2 : MS42=SS42/df42 : vr42=MS42/s1
    : p21=1-%fp(vr21,df21,df1) : p32=1-%fp(vr32,df32,df2)
    : p42=1-%fp(vr42,df42,df2) $
$pr '           Model    Residual SS   df   Mean square    Variance Ratio(F)'
 :  '_____________________________________________________________________'
 :  '   1) B12*A_bio      '*r  SS1,10,2 ' ' *i df1,4 '     '*r s1,10,4
 :  '   2) B12+A_bio      '*r  SS2,10,2 ' ' *i df2,4
 :  '_____________________________________'
 :  '(2-1)Due to B12.A_bio'*r SS21,10,2 ' ' *i df21,4'    '*r MS21,10,2!
                                                     '        '*r VR21,10,2
 :  '                                                       ( p = 'p21' )'
 :  '_____________________________________________________________________'
 :  '   3)   B12'
 :  '       (i.e. -A_bio) '*r  SS3,10,2' '*i df3,4
 :  '_____________________________________'
 :  '(3-2)Due to A_bio    '*r SS32,10,2' '*i df32,4'    '*r MS32,10,2!
                                                     '        '*r VR32,10,2
 :  '                                                       ( p = 'p32' )'
 :  '_____________________________________________________________________'
 :  '   4)  A_bio'
 :  '       (ie - B12)    '*r  SS4,10,2' '*i df4,4
 :  '_____________________________________'
 :  '(4-2)Due to  B12     '*r SS42,10,2' '*i df42,4'    '*r MS42,10,2!
                                                     '        '*r VR42,10,2
 :  '                                                       ( p = 'p42' )'
 :  '_____________________________________________________________________'
 :  '_____________________________________________________________________'
$
$trans +v $
$endmac  $
$echo $return $ ! ANCO3124
!
!_____________________________________________________________________
!
$subfile ANCO3125 $echo $
$mac ANCO3125 $
!
$trans -v $
$num   df21 df32 df43 df53 df63 df73
       SS21 SS32 SS43 SS53 SS63 SS73 MS21 MS32 MS43 MS53 MS63 MS73
       vr21 vr32 vr43 vr53 vr63 vr73 p21 p32 p43 p53 p63 p73 $
$cal
     SS21=SS2-SS1 : SS32=SS3-SS2 : SS43=SS4-SS3 : SS53=SS5-SS3
  :  SS63=SS6-SS3 : SS73=SS7-SS3 : df21=df2-df1 : df32=df3-df2
  :  df43=df4-df3 : df53=df5-df3 : df63=df6-df3 : df73=df7-df3
  :  MS21=SS21/df21 : MS32=SS32/df32 : MS43=SS43/df43 : MS53=SS53/df53
  :  MS63=SS63/df63 : MS73=SS73/df73 : vr21=MS21/s2 : vr32=MS32/s2
  :  vr43=MS43/s2 : vr53=MS53/s2 : vr63=MS63/s2 : vr73=MS73/s2
  :  p21=1-%fp(vr21,df21,df1) : p32=1-%fp(vr32,df32,df1)
  :  p43=1-%fp(vr43,df43,df1) : p53=1-%fp(vr53,df53,df1)
  :  p63=1-%fp(vr63,df63,df1) : p73=1-%fp(vr73,df73,df1) $
!
$pr '           Model    Residual SS   df   Mean square    Variance Ratio(F)'
```

```
 :  '  _____________________________________________________________________'
 :  '   1)All two variable'
 :  '     terms          '*r  SS1,10,2 ' ' *i df1,4 '     '*r s2,10,2
 :  '   2) Main effects   '
 :  '     +quadratic terms'*r  SS2,10,2 ' ' *i df2,4
 :  '  _____________________________________'
 :  '(2-1)Due to Interns  '*r SS21,10,2' '*i df21,4'     '*r MS21,10,2!
                                                '              '*r VR21,10,2
 :  '                                                          (p='p21')'
 :  '  _____________________________________________________________________'
 :  '   3) -(CD2+AGE.CD2W)'
 :  '   The baseline model'*r  SS3,10,2' '*i df3,4
 :  '  _____________________________________'
 :  '(3-2)Due to'
 :  '      (CD2+AGE.CD2W) '*r SS32,10,2' '*i df32,4'     '*r MS32,10,2!
                                                '              '*r VR32,10,2
 :  '                                                          (p='p32')'
 :  '  _____________________________________________________________________'
 :  '   4) - AGE-AG2      '*r  SS4,10,2' '*i df4,4
 :  '  _____________________________________'
 :  '(4-3)Due to AGE+AG2  '*r SS43,10,2' '*i df43,4'     '*r MS43,10,2!
                                                '              '*r VR43,10,2
 :  '                                                          (p='p43')'
 :  '  _____________________________________________________________________'
 :  '   5) - CD2W         '*r  SS5,10,2' '*i df5,4
 :  '  _____________________________________'
 :  '(5-3)Due to CD2W     '*r SS53,10,2' '*i df53,4'     '*r MS53,10,2!
                                                '              '*r VR53,10,2
 :  '                                                          (p='p53')'
 :  '  _____________________________________________________________________'
 :  '   6) - YRGP         '*r  SS6,10,2' '*i df6,4
 :  '  _____________________________________'
 :  '(6-3)Due to  YRGP    '*r SS63,10,2' '*i df63,4'     '*r MS63,10,2!
                                                '              '*r VR63,10,2
 :  '                                                          (p='p63')'
 :  '  _____________________________________________________________________'
 :  '   7) - PATH         '*r  SS7,10,2' '*i df7,4
 :  '  _____________________________________'
 :  '(7-3)Due to  PATH    '*r SS73,10,2' '*i df73,4'     '*r MS73,10,2!
                                                '              '*r VR73,10,2
 :  '                                                          (p='p73')'
 :  '  _____________________________________________________________________'
 :  '  _____________________________________________________________________'
$
$trans +v $
$endmac  $! ANCO3125
$echo $return $
!
!_______________________________________________________________________________
!
$subfile ANCO3126 $echo $
$mac    ANCO3126 $
$trans -v $
!
!  Macro to calculate components of Nested Design Analysis of Variance
```

```
! and output the Anova table. Parameters are
!       SS - a vector of Sums of Squares from the sequence of models
!       df - a vector of the degrees of freedom from the models
!
$number SS1 SS2 SS21 SS3 SS32 SS4 SS42 SS5 SS54
      df1 df2 df21 df3 df32 df4 df42 df5 df54
      s2w s2b MS21 MS32 MS42 MS54 VR21 VR32 VR42 VR54
      p21 p32 p42 p54 $

$cal SS1=SS(1) : SS2=SS(2) : SS3=SS(3) : SS4=SS(4) : SS5=SS(5)
   : SS21=SS2-SS1 : SS32=SS3-SS2 : SS42=SS4-SS2 : SS54=SS5-SS4
   : df1=df(1) : df2=df(2) : df3=df(3) : df4=df(4) : df5=df(5)
   : df21=df2-df1 : df32=df3-df2 : df42=df4-df2 : df54=df5-df4
   : s2w=SS1/df1
   : MS21=SS21/df21 : MS32=SS32/df32 : MS42=s2b=SS42/df42 : MS54=SS54/df54
   : VR21=MS21/s2w : VR32=MS32/s2w : VR42=s2b/s2w : VR54=MS54/s2b $
$cal p21=1-%fp(vr21,df21,df1) : p32=1-%fp(vr32,df32,df2)
   : p42=1-%fp(vr42,df42,df2) : p54=1-%fp(vr54,df54,df4) $

$pr '          Model      Residual SS   df   Mean square    Variance Ratio(F)'
  : '  ___________________________________________________________________________'
  : '     1) CR/FAM+CR.FPOS '*r SS1,10,2 ' ' *i df1,4 '     '*r s2w,10,2
  : '     2) CR/FAM   +FPOS '*r SS2,10,2 ' ' *i df2,4
  : '  _________________________________________'
  : '   (2-1)Due to CR.FPOS '*r SS21,10,2' '*i df21,4'     '*r MS21,10,2!
                                                    '        '*r VR21,10,2
  : '                                                          (p='p21')'
  : '  ___________________________________________________________________________'
  : '     3) CR/FAM'
  : '        (i.e. -FPOS)   '*r SS3,10,2' '*i df3,4
  : '  _________________________________________'
  : '   (3-2)Due to FPOS    '*r SS32,10,2' '*i df32,4'     '*r MS32,10,2!
                                                    '        '*r VR32,10,2
  : '                                                          (p='p32')'
  : '  ___________________________________________________________________________'
  : '     4) CR + FPOS'
  : '        (ie -/FAM)     '*r SS4,10,2' '*i df4,4
  : '  _________________________________________'
  : '   (4-2)Due to /FAM    '*r SS42,10,2' '*i df42,4'     '*r MS42,10,2!
                                                    '        '*r VR42,10,2
  : '                                                          (p='p42')'
  : '  ___________________________________________________________________________'
  : '     5)  1 + FPOS'
  : '        (ie - CR)      '*r SS5,10,2' '*i df5,4
  : '  _________________________________________'
  : '   (5-4)Due to CR      '*r SS54,10,2' '*i df54,4'     '*r MS54,10,2!
                                                    '        '*r VR54,10,2
 : '                                                          (p='p54')'
  : '  ___________________________________________________________________________'
  : '  ___________________________________________________________________________'
$
$trans +v $
$endmac  $ ! ANCO3126
$echo $return $
!
```

```
!______________________________________________________________________________
!
$subfile ANOV3127 $echo $
$mac ANOV3127 $
$trans -v $
$number SS21 SS32 SS42 MS21 MS32 MS42
        df21 df32 df42 vr21 vr32 vr42 p21 p32 p42 $
$cal
  SS21=SS2-SS1   : df21=df2-df1 : MS21=SS21/df21
: vr21=MS21/s2   : p21=1-%fp(vr21,df21,df1)
: SS32=SS3-SS2   : df32=df3-df2 : MS32=SS32/df32
: vr32=MS32/s2   : p32=1-%fp(vr32,df32,df1)
: SS42=SS4-SS2   : df42=df4-df2 : MS42=SS42/df42
: vr42=MS42/s2   : p42=1-%fp(vr42,df42,df1) $
$pr '           Model    Residual SS   df   Mean square    Variance Ratio(F)'
 :  '   ___________________________________________________________________'
 :  '   1)with interaction'*r  SS1,10,2 ' ' *i df1,4 '     '*r s2,10,2
 :  '   2) Main effects   '*r  SS2,10,2 ' ' *i df2,4
 :  '   ___________________________________'
 :  '(2-1)Due to intrns   '*r SS21,10,2 ' ' *i df21,4'    '*r MS21,10,2!
                                                   '           '*r VR21,10,2
 :  '                                                       ( p = 'p21' )'
 :  '   ___________________________________________________________________'
 :  '   3) SEX'
 :  '       (i.e. - CI)   '*r  SS3,10,2' '*i df3,4
 :  '   ___________________________________'
 :  '(3-2)Due to  CI      '*r SS32,10,2' '*i df32,4'    '*r MS32,10,2!
                                                   '           '*r VR32,10,2
 :  '                                                       ( p = 'p32' )'
 :  '   ___________________________________________________________________'
 :  '   4) CI'
 :  '       (ie - SEX)    '*r  SS4,10,2' '*i df4,4
 :  '   ___________________________________'
 :  '(4-2)Due to  SEX     '*r SS42,10,2' '*i df42,4'    '*r MS42,10,2!
                                                   '           '*r VR42,10,2
 :  '                                                       ( p = 'p42' )'
 :  '   ___________________________________________________________________'
 :  '   ___________________________________________________________________'
$trans +v $
$end $ !ANOV3127
$echo $return $
!
!______________________________________________________________________________
!
$subfile ANCO3132 $echo $
$mac ANCO3132 $
!
$trans -v $
$num   df21 df32 df43 df54
       SS21 SS32 SS43 SS54 MS21 MS32 MS43 MS54
       vr21 vr32 vr43 vr54 p21 p32 p43 p54 $
$cal
     SS21=SS2-SS1 : SS32=SS3-SS2 : df21=df2-df1 : df32=df3-df2
  :  MS21=SS21/df21 : MS32=SS32/df32 : vr21=MS21/s2 : vr32=MS32/s2
  :  p21=1-%fp(vr21,df21,df1) : p32=1-%fp(vr32,df32,df1)
```

```
!
$pr '            Model     Residual SS    df    Mean square     Variance Ratio(F)'
  :  '  ____________________________________________________________________________'
  :  '   1)All 3 factor intrn'
  :  '     terms with OCC  '*r  SS1,10,4 ' ' *i df1,4 '     '*r s2,10,4
  :  '   2)All 2 factor intrn'
  :  '     terms with OCC  '*r  SS2,10,4 ' ' *i df2,4
  :  '  ____________________________________________'
  :  '(2-1)Due to 3 factor '
  :  '     interactn terms '*r SS21,10,4' '*i df21,4'      '*r MS21,10,4!
                                                  '             '*r VR21,10,4
  :  '                                                          (p='p21')'
  :  '  ____________________________________________________________________________'
  :  '   3) - T_GRP.OCC     '*r  SS3,10,4' '*i df3,4
  :  '  ____________________________________________'
  :  '(3-2)Due to T_GRP.OCC'*r SS32,10,4' '*i df32,4'      '*r MS32,10,4!
                                                  '             '*r VR32,10,4
  :  '                                                          (p='p32')'
  :  '  ____________________________________________________________________________'
  :  '  ____________________________________________________________________________'
$pr : $
$trans +v $
$endmac  $! ANCO3132
$echo $return $
!
!____________________________________________________________________________________
!
$subfile ANCO3133 $echo $
$mac ANCO3133 $
!
$trans -v $
$num   df21 df32
       SS21 SS32 MS21 MS32
       vr21 vr32 p21 p32  $
$cal
     SS21=SS2-SS1 : SS32=SS3-SS2
   : df21=df2-df1 : df32=df3-df2
   :  MS21=SS21/df21 : MS32=SS32/df32
   :  vr21=MS21/s2    : vr32=MS32/s2
   :  p21=1-%fp(vr21,df21,df1) : p32=1-%fp(vr32,df32,df1)
!
$pr '            Model     Residual SS    df    Mean square     Variance Ratio(F)'
  :  '  ____________________________________________________________________________'
  :  '   1) Different'
  :  '      quadratics      '*r  SS1,10,2 ' ' *i df1,4 '     '*r s2,10,2
  :  '   2) Different st.   '
  :  '      lines           '*r  SS2,10,2 ' ' *i df2,4
  :  '  ____________________________________________'
  :  '(2-1)Due to Curvatr  '*r SS21,10,2' '*i df21,4'      '*r MS21,10,2!
                                                  '             '*r VR21,10,2
  :  '                                                          (p='p21')'
  :  '  ____________________________________________________________________________'
  :  '   3) Parallel lines'
  :  '                      '*r  SS3,10,2' '*i df3,4
  :  '  ____________________________________________'
```

```
 :  '(3-2)Due to'
 :  '     Non-parallelism  '*r SS32,10,2' '*i df32,4'     '*r MS32,10,2!
                                                       '             '*r VR32,10,2
 :  '                                                                 (p='p32')'
 :  '  ____________________________________________________________________________'
 :  '  ____________________________________________________________________________'
$
$trans +v $
$endmac  $ ! ANCO3133
$echo $return $
!
!_______________________________________________________________________________
!
$subfile ANOD3213 $echo $
$mac ANOD3213 $
!
! This is a macro to obtain the various components of the analysis of
!deviance table and print it. This is given in full here to illustrate,
!in particular, the use of the PRINT directive to obtain a relatively
!complicated output. Once constructed it is relatively easy, using one
!such macro as a template, to produce them for other analyses.
!
$number D21 D32 D42 D52 D62 D72 df21 df32 df42 df52 df62 df72
        p21 p32 p42 p52 p62 p72 $
$cal D21=D2-D1 : D32=D3-D2 : D42=D4-D2: D52=D5-D2 : D62=D6-D2 : D72=D7-D2
 : df21=df2-df1 : df32=df3-df2 : df42=df4-df2 : df52=df5-df2 : df62=df6-df2
 : df72=df7-df2 : p21=1-%chp(D21,df21) : p32=1-%chp(D32,df32)
 : p42=1-%chp(D42,df42) : p52=1-%chp(D52,df52) : p62=1-%chp(D62,df62)
 : p72=1-%chp(D72,df72) $
!
$trans -v  $
$pr '   _______________________________________________________________'
 :  '         Model           Deviance   df '
 :  '   _______________________________________________________________'
 :  '      1) All 2f intrns '*r D1,10,2 ' ' *i df1,4
 :  '      2) Main Effects  '*r D2,10,2 ' ' *i df2,4
 :  '   ________________________________'
 :  '   (2-1) Due to intrns '*r D21,10,2' ' *i df21,4
 :  '                                                  p ='*r p21,6,4
 :  '   _______________________________________________________________'
 :  '      3) - AGE         '*r D3,10,2 ' ' *i df3,4
 :  '   ________________________________'
 :  '   (3-2) Due to AGE    '*r D32,10,2' ' *i df32,4
 :  '                                                  p ='*r p32,6,4
 :  '   _______________________________________________________________'
 :  '      4) +AGE-FHT      '*r D4,10,2 ' ' *i df4,4
 :  '   ________________________________'
 :  '   (4-2) Due to FHT    '*r D42,10,2' ' *i df42,4
 :  '                                                  p ='*r p42,6,4
 :  '   _______________________________________________________________'
 :  '      5) +FHT-SCG      '*r D5,10,2 ' ' *i df5,4
 :  '   ________________________________'
 :  '   (5-2) Due to SCG    '*r D52,10,2' ' *i df52,4
 :  '                                                  p ='*r p52,6,4
 :  '   _______________________________________________________________'
```

```
  :  '         6) +SCG-ALC      '*r D6,10,2 ' ' *i df6,4
  :  '                                              '
  :  '      ____________________________________________
  :  '      (6-2) Due to ALC      '*r D62,10,2' ' *i df62,4
  :  '                                                        p ='*r p62,6,4
  :  '      _________________________________________________________________'
  :  '         7) +ALC-SM        '*r D7,10,2 ' ' *i df7,4
  :  '                                              '
  :  '      ____________________________________________
  :  '      (7-2) Due to SM       '*r D72,10,2' ' *i df72,4
  :  '                                                        p ='*r p72,6,4
  :  '      _________________________________________________________________'
  :  '      _________________________________________________________________'
$
$trans +v $
$endmac  $! ANOD3213
$echo $return $
!
!_______________________________________________________________________
!
$subfile ANOD3313 $echo $
$mac ANOD3313 $
!
$number D21 D32 D42 df21 df32 df42 p21 p32 p42 $
$cal D21=D2-D1 : D32=D3-D2 : D42=D4-D2 : df21=df2-df1 : df32=df3-df2
 : df42=df4-df2 : p21=1-%chp(D21,df21) : p32=1-%chp(D32,df32)
 : p42=1-%chp(D42,df42)  $
!
$trans -v  $
$pr '    __________________________________________________________________'
  :  '          Model               Deviance   df '
  :  '    __________________________________________________________________'
  :  '      1) fs*ms*s84         '*r D1,10,2 ' ' *i df1,4
  :  '      2)  -fs.ms.s84       '*r D2,10,2 ' ' *i df2,4
  :  '                                              '
  :  '    ____________________________________________
  :  '    (2-1) Due fs.ms.s84    '*r D21,10,2' ' *i df21,4
  :  '                                                        p ='*r p21,8,4
  :  '    __________________________________________________________________'
  :  '      3) - ms.s84          '*r D3,10,2 ' ' *i df3,4
  :  '                                              '
  :  '    ____________________________________________
  :  '    (3-2) Due to ms.s84    '*r D32,10,2' ' *i df32,4
  :  '                                                        p ='*r p32,8,4
  :  '    __________________________________________________________________'
  :  '      4) +ms.s84-fs.s84    '*r D4,10,2 ' ' *i df4,4
  :  '                                              '
  :  '    ____________________________________________
  :  '    (4-2) Due to fs.s84    '*r D42,10,2' ' *i df42,4
  :  '                                                        p ='*r p42,10,4
  :  '    __________________________________________________________________'
  :  '    __________________________________________________________________'
$
$trans +v $
$endmac  $!ANOD3313
$echo $return $
!
!_______________________________________________________________________
!
$subfile ANOD3322 $echo $
```

```
$mac  ANOD3322 $
!
$number D21 D32 D43 df21 df32 p21 p32   $
$cal D21=D2-D1 : D32=D3-D2
 : df21=df2-df1 : df32=df3-df2
 : p21=1-%chp(D21,df21) : p32=1-%chp(D32,df32) $
!
$trans -v  $
$pr '      ________________________________________________________________'
 :  '          Model                  Deviance   df '
 :  '      ________________________________________________________________'
 :  '        1) With treat.time<3>'*r D1,10,2 ' ' *i df1,4
 :  '        2) treat              '*r D2,10,2 ' ' *i df2,4
 :  '      ____________________________________'
 :  '     (2-1) Due to intrn       '*r D21,10,2' ' *i df21,4
 :  '                                                          p ='p21
 :  '      ________________________________________________________________'
 :  '        3) - treat            '*r D3,10,2 ' ' *i df3,4
 :  '      ____________________________________'
 :  '     (3-2) Due to treat       '*r D32,10,2' ' *i df32,4
 :  '                                                          p ='p32
 :  '      ________________________________________________________________'
 :  '      ________________________________________________________________'
$
$trans +v $
$endmac  $ ! ANOD3322
$echo $return $
!
!_____________________________________________________________________________
!
$subfile anod3412 $echo $
$mac ANOD3412 $
!
$number D21 D32 D42 df21 df32 df42
        p21 p32 p42  $
$cal D21=D2-D1 : df21=df2-df1 : p21=1-%chp(D21,df21) $
$cal D32=D3-D2 : df32=df3-df2 : p32=1-%chp(D32,df32) $
$cal D42=D4-D2 : df42=df4-df2 : p42=1-%chp(D42,df42) $
$cal p21=1-%chp(D21,df21) : p32=1-%chp(D32,df32) : p42=1-%chp(D42,df42) $
!
$trans -v  $
$pr '       ________________________________________________________________'
 :  '          Model             Deviance   df '
 :  '       ________________________________________________________________'
 :  '         1) All 2f intrns '*r D1,10,2 ' ' *i df1,4
 :  '         2) Main Effects  '*r D2,10,2 ' ' *i df2,4
 :  '       _________________________________'
 :  '      (2-1) Due to intrns '*r D21,10,2' ' *i df21,4'    p ='*r p21,8,4
 :  '       ________________________________________________________________'
 :  '         3)  TREAT        '*r D3,10,2 ' ' *i df3,4
 :  '       _________________________________'
 :  '      (3-2) Due to BLOCK  '*r D32,10,2' ' *i df32,4'    p ='*r p32,8,4
 :  '       ________________________________________________________________'
 :  '         4)  BLOCK        '*r D4,10,2 ' ' *i df4,4
 :  '       _________________________________'
```

```
 :  '      (4-2) Due to TREAT  '*r D42,10,2' ' *i df42,4'    p ='*r p42,8,4
 :  '      _________________________________________________________________'
 :  '      _________________________________________________________________'
$
$trans +v $
$endmac  $ ! ANOD3412
$echo $return $
!
!__________________________________________________________________________
!
$subfile chitst $echo $
!
$mac (arg=Devp,dfp loc=dfcp,Devcp,Chi,p) chitst $
!
! This macro performs a chi-squared test comparing the deviance of
!the current model with the deviance (argument1: Devp) and the degrees
!of freedom (argument2: dfp) from a previous model.
!
$number dfcp Devcp Chi p $
$cal dfcp=%df-dfp : Chi=(%dv-Devp) : p=1-%chp(Chi,dfcp) $
$pr 'Chi-squared ='Chi' with 'dfcp' degrees of freedom and p ='p $
$endm $ ! chitst
$echo $return $
!
!__________________________________________________________________________
!
$subfile cox $echo $
!
!   macros to fit the piecewise exponential survival model
!
! ... taking each individual as contributing a number of person
!years/mths/weeks at risk in the set at risk when a death occurred....
!
!                        M. Green and B. Francis Centre for Applied Statistics
!                        University of Lancaster, U.K.    Sept 1991
!
!     macro        parameters
!     COXMODEL     %1   survival time
!                  %2   censor indicator  0 - censored
!                                         1 - uncensored
!                  %3   cut points for piecewise exponential (output)
!
!        sets up the environmment for fitting the piecewise exponential
!        model
!
!        CUTP  with the same arguments
!              finds the values of the uncensored survival times
!
!     INIB
!        INIB initialise the fitting procedure
!
!
!     PHAZ         %1   cut points
!        PHAZ  produces a plot of log hazard versus log survival time
!
```

```
!————————————————————————————————————————
!
$m (arg=time,censor,ddt) coxmodel
! NB could test if arguments set and GET if necessary
$use CUTP TIME CENSOR DDT $! uses 'time' in study var and 0/1 cens/dead var
                          ! to identify time intervals and risk sets
$ca %z1=%len(DDT)$
! ddt from CUTP holds the set of distinct time pts where deaths occurred
$pr 'There are ' *i %z1 ' distinct death times.'$
$use inib time censor ddt $
$e $
!
!————————————————————————————————————————
$macro (loc=temp) CUTP !
!  macro to find ordered observed death times
!    %1 -  survival times variable
!    %2 -  censor indicator
!    %3 -  ordered death times (cut-points) (output)
$ca %z1=%len(%1) $var %z1 temp $group temp=%1 $ ! factor with 1 level for each
                                                 ! distinct survival time
$sort temp $ca %z3=temp(%z1) $! highest level to %z3
$var %z3 val_ f_ $warn off $
$tab for %1 with %2 by val_ using f_ $warn on $! freq of 'deaths' in f_ for
                                                ! time points in val_

$cal %z3=%cu(f_/=0) $var %z3 %3 $ca %3=0! create o/p var with length
                                        ! = no of intvls with a death
$cal ind_=(f_/=0) : ind_=%cu(ind_)*ind_ : %3(ind_)=val_ $
!
! ind_ indexes the intervals with a death - o/p var = time pts for deaths
!
$del val_ f_ ind_ temp$!
$endmac
!
!————————————————————————————————————————
$macro INIT !
$del ni_ cni_ temni_ censor_ logtime_ gap_ ind_ tmp_ t_ints $
$group NI_=%1 intervals * %3 * $
!
! use 'time' values to make factor variable equivalent to %1 - the time in
!study variable with codes 1 to 1st cutpt; 2 there to 2nd; 3 to 3rd etc
!
$cal temni_=ni_ $
$cal NI_=NI_-(%1==%3(NI_-1)) : CNI_=%cu(NI_) $
!
! values equal to cut-points moved to lower interval category then accumulated
!!
!   expand censor variable -
!
$cal %z9=%cu(NI_) $var %z9 CENSOR_ LOGTIME_ $!
$cal CENSOR_=0 : CENSOR_(CNI_)=%2 $!
!   generate exposure time
! death times in %3, intervals between them in gap_
$cal GAP_=%3-%3(%cu(1)-1) $var %z9 IND_ $
$cal IND_=1 : IND_(CNI_)=1-NI_ : IND_=%cu(IND_) $!
```

```
$cal TMP_=GAP_(IND_) : TMP_(CNI_)=%1-%3(NI_-1) $!
$cal LOGTIME_=%LOG(TMP_) $!
!   generate block factor
$cal %z1=0 : %z1=%if((NI_>%z1),NI_,%z1) $!
$cal IND_(CNI_)=NI_ $assign t_ints=IND_ $!
$factor t_ints %z1 $DEL IND_ TMP_ GAP_$
$endmac
!
!-----------------------------------------
$macro INIB ! enter with args 'time' 'censor' 'ddt' and pass to INIT
$use INIT %1 %2 %3 $!
!   declare model
$yvar CENSOR_ $error P $link L $OFFSET LOGTIME_$ELIM t_ints $ !
$method * * DEVADJ$
$var %nu IND_$
$cal IND_=0 : IND_(CNI_-NI_+1)=1 : IND_=%cu(IND_) $DEL CNI_ NI_ $!
$endmac
!
$macro (arg=evntm,title loc=BB,W,PLP,LT) PHAZ  $
$del bb w plp lt$
$ca %z1=%len(evntm)$
$ca %z2=2*%z1 $
$gfac %z2 BB %z1 $var %z2 W $ca W=1 : W(1)=W(%z2)=0$
$predict T_INTS=BB$
$ca lt=%log(evntm(BB))$
$ass lt=0,lt : w=W,0: plp=%plp,0$
$pick plp,lt  w$
$gra (t=title      v='log hazard' h='log survival time' ) plp lt 11 $
$del bb w plp lt$
$e  $
!
$m devadj !
!
$ca %dv=%cu(%os*%yv)*2 -2*%cu(%yv*%log(%fv)-%fv)!
$e
!
!
!-----------------------------------------
!    END of piecewise macros...
!-----------------------------------------
$echo $return $ ! Cox-macs
!
!-----------------------------------------
!
$subfile crep $echo  $
!
$mac       ! crep $
!  This macro will obtain normal plots
!  of residuals from fits with continuous
!  y variables assuming all units were included in the fit
(local=r,n ) crep $!
$del %re r n $!
$sort r %rs $!
$cal n=%nd((%gl(%nin,1)-.5)/%nin) $!
$plot r n $!
```

```
$endmac $
$echo $return $ ! crep
!
!————————————————————————————————————————
!
$subfile ghist $echo $
!
$macro    !  ghist
!
! Macro to take y variable and cutpoints and plot an appropriate histogram
(arg=y,i  local=y_g,f_) ghist
!
$del y_g f_ $
$group y_g=y ints i * $
$tab for y_g using f_  $
!
$use histogram f_ i $
$endm $
!
$macro   ! histogram
!
! Macro to plot high resolution histograms given frequencies
!and the values specifying the lower end of the class interval.
!All intervals are assumed to be of the same length.
!
(arg=f,xlo local=i2,fex,xex,sf_,lenx,lenx2,
     f_max,f_int,x_min,x_max,x_int,tem)      histogram $
!
$del i2 fex xex sf_ $
$number lenx lenx2 f_max f_int x_min x_max x_int tem $
$sort sf_ f $
$cal lenx=%len(xlo) : lenx2=2*lenx $
$cal f_max=sf_(lenx) $
$cal f_max=%tr(f_max*1.1) : f_int=%if(f_max>5,%tr(f_max/5),1) $
$var lenx2 i2 fex xex $
$cal x_int=xlo(2)-xlo(1) : x_min=xlo(1)-5*x_int
   : x_max=xlo(lenx)+5*x_int
$cal i2=%gl(lenx,2) : xex=xlo(i2)
  : fex=f(i2) : tem=xex(lenx2)+x_int $
$ass fex=0,fex,0 : xex=xex,tem,tem $
!
$graph (title=title h=xlab v=ylab
        y=0,f_max,f_int x=x_min,x_max,x_int)
        fex xex 15 $
!
$endmac $! ghist
$echo $return $
!_____________________________________________________________________________
$subfile lfts $echo $
!
$mac               ! lfts
!                                                        AVS 12.12.89
!                                                            21. 4.91
!
! This macro obtains a lifetable and the associated Survival curve
```

```
!from a vector of survival times, in units of the intervals of
!interest, and a status vector indicating whether the value is
!exact or right censored. The estimated probabilities of survival
!to the end of the intervals is given in si_ and ti_ holds the
!upper ends of the intervals.
!
! time    is the vector of survival times
!         (rounded up to the next whole unit)
! event   is the status variable: 0 censored 1 died
! wt      is a 0/1 weight variable indicating which
!         cases to include (the 1s)
! ints_in is a scalar indicating the number of time
!         intervals to be used.
!             0 to omit those within which no deaths
!               or withdrawals occurred
!             1 for all intervals up to the maximum
!               time that occurs
!             k for all intervals up to the maximum
!               of (k,intervals occurring)
!
! r_frac  indicates for how much, of the intervals,
!         withdrawals should be considered at risk :-
!             0   for not at all
!             0.5 for half the interval
!                 (Standard Actuarial approach)
!             1   for all the interval  (Kaplan-Meier)
!                 (any value between 0 and 1 may be used)
!
(arg=time,event,wt,ints_in,r_frac )  lfts $
$delete  na_ ni_ nj_ ta_ ti_ da_ di_  !
  wa_ wi_ wd_ pi_ qi_ ses_ si_  pw_ %re $!
$cal wd_=1-event  $!
$cal ev_w=event*wt : wd_w=wd_*wt $
$tab the wt   tot for time  into na_ by ta_ !
 :   the ev_w tot for time  into da_        !
 :   the wd_w tot for time  into wa_       $!
!
$cal %z1=%cu(ta_==ta_) : %z9=%if(ints_in>1,2,ints_in+1 ) $!
$arg lt2_ ints_in r_frac $switch %z9 lt1_ lt2_ $!
!
$cal %z4=1-%if(r_frac<0?r_frac>1,0.5,r_frac)  $!
$cal pw_=ni_/=0  !
 : ni_=%cu(ni_) : %z3=ni_(%z2) : ni_=%z3-ni_ : ni_(%z2)=%z3 $ !
$sort ni_ ni_ 2 $ !
$ass ni_=%z3,ni_ : ti_=0,ti_ : di_=0,di_ : wi_=0,wi_ : pw_=1,pw_ $ !
$cal nj_=(ni_-wi_*%z4) : qi_=di_/nj_ : pi_=1-qi_ !
 : si_=%exp(%cu(%log(pi_))) : ses_=si_*%sqrt(%cu(qi_/(nj_*pi_))) $!

$cal %z6=ti_(2)==0 $while %z6 warn $

$pr ::'    Lifetable with Estimated Survival Function (S) and se(S)'!
    ::'      Top of  No.at  Deaths Lost   Hazd          Est p'  !
     :'      Intvl   Risk               d/(n-fw) (1-q)  Surv'!
     :'        t       n       d     w     q        p    S(t)    se(S)':$!
$acc 2 $look ti_ ni_ di_ wi_ qi_ pi_ si_ ses_  $acc 4 $ !
```

```
!
$pr :: '                        Lifetable Survival Curve' :: $!
$plot ( y=0,1 ) si_/pw_ ti_ '+' $del %re $!
$cal %z6=(si_(%z2+1) < 0.5 ) + 1 $!
$arg med_ ti_ si_ $switch %z6 nmd_ med_ $!
!
$endmac $ ! lfts
!
$mac warn $
$pr ::
    '      *******   WARNING   *******':
    ' The times observed for each individual should be given':
    ' as the next whole unit above them ie rounded up.':
    ' You do not appear to have done that.' $cal %z6=0 $
$endmac $
!
$mac lt1_ $!  Auxiliary macro for lfts
$cal %z2=%z1 :ni_=na_ :di_=da_ :wi_=wa_ :ti_=ta_ $!
$endmac $
!
$mac lt2_ $!  Auxiliary macro for lfts
$cal  %z2=%if(%1 <= ta_(%z1)+1,ta_(%z1)+1,%1) $!
$var %z2 ni_ di_ wi_ ti_ $!
$cal ni_=di_=wi_=0 :  ta_=ta_+1 : ni_(ta_)=na_ : di_(ta_)=da_ !
 : wi_(ta_)=wa_ : ti_=%gl(%z2,1)  $!
$endmac $
!
$mac med_ $
!  Auxiliary macro for lfts for median estimation
!  to calculate the estimated median survival time and se
!
!   %1 is the vector holding the upper ends of the time intervals
!   %2 is the vector holding the estimated survivor function S(t)
!
$cal %z1=%cu(%2>=0.5) : %z2=%1(%z1+1)-%1(%z1) !
 : %z3=(%2(%z1)-%2(%z1+1))/%z2 : %z4=%1(%z1)+(%2(%z1)-0.5)/%z3 !
 : %z5=%sqrt(1/(4*ni_(%z1)*%z3*%z3)) $
$pr : '  Estimated median survival time = ' %z4 ' with se = '%z5 :$!
$endmac $ !
!
$mac nmd_ $
!  Auxiliary macro for lfts for situations when
!  median estimation is not possible.
!
$pr : ' Since the estimated survival function did not '!
 :   ' fall below 50% the median could not be estimated':$!
$endmac$!
!
$mac del_ $! to delete variables declared in lfts after use
$delete  na_ ni_ nj_ ta_ ti_ da_ di_  !
  wa_ wi_ wd_ pi_ qi_ ses_ si_  pw_ %re $$endmac $
!
 $mac (loc=Chi2,p) lr2 $!
!           Log-rank test for 2 groups
!
```

```
!    This uses the output from the lifetable macro called twice with
!    a weight variable to select the two groups of individuals to be
!    compared. The parameters required are
!
! %1  the numbers at risk in group 1 for the intervals when a death occurred
!     (the ni_ from the first call to lfts)
!
! %2  the numbers of deaths in group 1
!     (the di_ from the first call to lfts)
!
! %3  the numbers at risk in group 2 for the intervals when a death occurred
!     (the ni_ from the second call to lfts)
!
! %4  the numbers of deaths in group 2
!     (the di_ from the second call to lfts)
!
!  NB The time intervals must be the same for both groups, but this
!     will automatically be the case if lfts is called with %4=0
!
$del di_ ni_ $!
$num Chi2 p $
$cal di_=%2+%4 : ni_=%1+%3 : %z1=%cu(di_*%1/ni_) !
 : %z2=%cu(di_*%3/ni_)  : %z3=%cu(%2) : %z4=%cu(%4) !
 : Chi2=(%z1-%z3)*(%z1-%z3)/%z1 + (%z2-%z4)*(%z2-%z4)/%z2 !
 : p=1-%chp(Chi2,1) $!
$pr
:'The log-rank chi-squared statistic on 1 df is = '*4 Chi2' with p ='p $!
$endmac $
$macr   ! steps
! Macro to supplement vectors of co-ordinates (y,x) with extra points
!for the 'curve' to be plotted as a step function. The macro adds a point
!to extend the last step horizontally for 5% of the horizontal range
(arg=s,t,sel,tel
local=ss,ssx,tt,ttx,ii,jj,len_s,len_x,ss_last,tt_last) steps
!
$number len_ len_ ss_last tt_last $
$sort tt t : ss s t $
$cal len_=%len(ss) : ss_last=ss(len_) : tt_last=tt(len_) $
$ass ss=ss,ss_last : tt=tt,tt_last $
$cal ii=%ind(s) : jj=(ss(ii)/=ss(ii+1)) : len_=%cu(jj) : jj=jj*%cu(jj) $
$var len_ ssx ttx ix $cal ix(jj)=ii  $
$cal ssx=s(ix) : ttx=t(ix+1)*(0.9999)**%sgn(t(ix+1)) $
$ass sel=s,ssx : tel=t,ttx $sort sel;tel sel;tel tel;tel $
$cal len_=%len(sel) : ss_last=sel(len_)
 : tt_last=tel(len_)+(tel(len_)-tel(1))*0.05 $
$ass sel=sel,ss_last : tel=tel,tt_last $
$del ss ssx tt ttx ii jj ix len_ len_ ss_last tt_last $
$endm $! steps
$echo $return $ ! lfts
!
!_______________________________________________________________________
!
$subfile orcs $echo $
$macro orcs $
!
```

```
! Assuming a Fit with binomial error and logistic link has been!
! performed this macro produces an estimate of the odds ratio and
! 95% cls appropriate for each parameter $
!
$del  i_ se_ l_ or_ u_ j_ pr_ ex_ $
$var %pl i_ se_ l_ or_ u_ j_ pr_ ex_ $
$cal j_=%gl(%pl,1) : i_=j_*(j_+1)/2 $extr %pe %vc $
$cal se_=%sqrt(%vc(i_)) : l_=%exp(%pe-1.96*se_)*(j_/=1)+(j_==1) :
     or_=%exp(%pe)*(j_/=1)+(j_==1)!
 : u_=%exp(%pe + 1.96*se_)*(j_/=1)+(j_==1)!
 : pr_=1/(1+%exp(-%pe(j_)-(i_>1)*%pe(1))) : ex_=pr_-pr_(1) $
$print '95% CLs and the Odds Ratio for each factor level compared'!
:'with the base level 1'!
' (for covariates the OR is for X+1 .v. X)'!
:'together with the fitted proportion for 1 + each factor level'!
:'in turn with the increase that results.'!
   ::' Low 95% CL  Odds Ratio   High 95% CL '$
$ACC 4 $look l_  or_  u_   $ACC $
$del i_ se_ l_ or_ u_ j_ pr_ ex_ $
$endmac $! orcs
$echo $return $ ! orcs
!
!_____________________________________________________________________________
!
$subfile pens3116 $echo $
$macro pens3116 $!
!
!   This macro sets up four pen styles
!          1 yellow crosses
!          2 magenta triangles
!          3 dotted yellow line
!          4 solid magenta line
!
$gsty 1 s 1 c 4 1 0 :  2 s 2 c 7 1 0 :  3 s 0 c 4 1 2 :  4 s 0 c 8 1 1 $
$end $
$echo $return $
!
!_____________________________________________________________________________
!
$subfile wrep $echo $
$mac   ! wrep $
!   this macro will obtain normal plots
!   of residuals from weighted fits with continuous
!   y variables excluding the observations
!   with zero weights.
(local=wts,res,nqs,n )  wrep $
$del %re wts res nqs $
$cal wts=(%pw/=0) : %n=%cu(wts)
$var %n res nqs
$cal res(wts*%cu(wts))=%yv-%fv
$sort res
$cal nqs=%nd((%gl(%n,1)-.5)/%n)
$plot res nqs $!
$endmac
$echo $return $
```

```
!
!_______________________________________________________________________
!
$subfile xprd $echo $
$macro     ! xprd
! Macro to predict x with 95% CLs from a set of 'observed'
! y values assuming a straight line relationship
! between the y and x variables
!       1 the y variable
!       2 the x variable
!       3 is the weight variable if set
!       4 is the vector of y values for which predictions are required
!          (transformed beforehand if necessary)
!
(arg=y,x,w,yp local=yd_,b_,c_,wr,r_,y_,l_)  xprd $
$trans -o $
$del %re yd_ b_ c_ wr r_ xl_ xh_ x_ y_ l_ $       !
$number xmean Sxx ymean byx ttsig foura t_975 $
$weight w $yvar x $fit 1 $dis e $ext %pe $
$cal xmean=%pe(1)
 : Sxx=%dv $  !
$yvar y $  !
$fit 1 $dis e $ext %pe $cal ymean=%pe(1) : t_975=%td(0.975,%df) $ !
$fit x $dis e $ext %pe $cal yd_=yp-ymean : byx=%pe(2) : ttsig=t_975*t_975*%sc !
 : foura=ttsig/Sxx - byx*byx : b_=byx*yd_     !
 : c_=ttsig*(%df+3)/(%df+2)-yd_*yd_ : r_=%sqrt(b_*b_-foura*c_) !
 : xl_=(-b_ + r_)/foura+xmean : xh_=(-b_ - r_)/foura+xmean   !
 : x_=(yp-%pe(1))/byx $          !
$trans +o $
$print ;;'95% CLs using t = 't_975 ; '  for the range of x values '  !
'consistent with a set of "observed" y values '$    !
$look yp xl_ x_ xh_ $           !
$cal wr=r_/=0 $
$pr ; ' Plot of predicted "x" values with CLs against chosen "y" values' ;; $
$plot xl_/wr x_/wr xh_/wr yp '^.v'$       !
$endmac $! xprd
$echo $return $
!_______________________________________________________________________
!
```

III

The Reference Guide

Introduction

GLIM is an interactive system for statistical modelling. This Reference Guide provides a comprehensive definition of its language and facilities.

The general features of the language of GLIM 4 are described in the Language chapter [13]. A bottom-up approach to syntax is taken, whereby we firstly define the character set [13.1], then the tokens (such as identifiers and numbers) that can be formed with these characters [13.2], and finally the statements that can be formed using these tokens [13.3]. Then in [13.4] we give a set of substitution rules (for arguments, string insertions and so on) that extend the allowable forms of statements, while in [13.5] we collect together various restrictions on the form of statement allowed. In [13.6] we describe the general semantic rules on how GLIM interprets language statements and in [13.7] we describe how GLIM obtains statements and the order in which it deals with them.

It must be remembered that the Language chapter [13] deals only with general rules that apply to all statements but does not (except as examples) refer to any particular statements. The templates (known as directives) from which actual statements may be derived are defined in the Directives chapter [15].

In the Structures chapter [14] we give a complete classification of the set of data structures supported by the system, and we define the properties of those structures that are created and maintained by the system. The names given to data structures are discussed in [14.1]. The set of data structures includes scalars [14.2], vectors [14.3], pointers [14.4], macros [14.5], functions [14.6], arrays [14.7] and lists [14.8]. Internals (structures hidden from the user) are discussed in [14.9] and subfiles (named portions of external files) are covered in [14.10].

In the Directives chapter [15] we give a complete definition of the syntax and semantics of all directives; the sections in [15] are ordered alphabetically by directive name.

The Diagnostics chapter [16] contains a detailed description of the warning and error handling mechanisms within GLIM, and a list of all warning messages is provided.

Finally, [17] contains a description of the PASS facility, which allows users to link in their own FORTRAN subroutines to GLIM. Details are provided of the FORTRAN subroutines PASS, EPASS, and HPASS which need to be amended by the user.

Syntax Conventions

The following conventions are used in this Guide.

(1) Elements consisting of *italic* lower-case letters, digits, and underlines (for example, *integer*) denote objects defined in [13] of this Guide. In the syntax definitions of [15] they represent objects of a particular structure type, and may

have extra digits appended to denote separate (though not necessarily different) instances from that structure type. For example the construction

integer1 integer2

could be replaced by

3 4

or

91 91

In addition, an element followed by an '*s*' denotes unlimited repetition of the element so that, for example,

integers

could be replaced by one or more integer values.

(2) Elements appearing in `mono` typeface are fragments of GLIM text. In the syntax definitions of [15] they stand for themselves and must appear in GLIM programs as they are shown in this Guide (but subject to any other conditions as given in [13] and [14]). For example,

`rows =` *integer*

could be used in a GLIM program as

```
rows = 16
```

or

```
rows = %a
```

or

```
rows = NROWS
```

(if the identifier `NROWS` had been previously defined as a scalar).

(3) Square brackets in ordinary typeface are used for two purposes:

(a) In the text they enclose section numbers, so that '[15.5]' refers to Section 15.5 of this Guide.

(b) In the syntax definitions they are used as metasymbols to denote optional appearance of their contents; thus

`$skip` [*integer*]

could be replaced by

`$skip` 2

or by just

`$skip`

In addition an 's' following the right bracket denotes optional unlimited appearance of the contents so that, for example,

[*integer*] *s*

could be replaced by zero or more integer values.

Note that square brackets are also part of the GLIM character set and may thus appear in GLIM program text. However, when used for this purpose they will, in line with (2) above, appear in `mono` typeface.

(4) Curly brackets (in ordinary typeface) appear in the syntax definitions of [15], and denote a forced choice from alternatives. For example

$$\texttt{\$set} \left\{ \begin{array}{c} \texttt{batch} \\ \texttt{constant} \\ \texttt{device} = \textit{keyword} \\ \texttt{interactive} \\ \texttt{noconstant} \end{array} \right\}$$

could be used (to give two examples) as

```
$set batch
```

or

```
$set noconstant
```

(5) Such constructions will be freely mixed so that, for example,

$$\left\{ \begin{array}{c} \texttt{[a[b]]} \\ \texttt{c} \end{array} \right\}$$

could be replaced by '`a b`', by '`a`', or by nothing (choosing from the first alternative) or by `c` (choosing the second alternative).

(6) **Bold** is used where terms are being defined.

(7) ***Bold italic*** is used for emphasis.

13 The language

This chapter defines the general features of the GLIM language.

13.1 The character set

At its lowest level, a GLIM session consists of the input to GLIM of a sequence of *characters* from a given *character_set*, together with the output from GLIM of another sequence of *characters* from the same set. The GLIM *character_set* consists of

letters: A B C D E F G H I J K L M N O P Q R S T U V W X Y Z

a b c d e f g h i j k l m n o p q r s t u v w x y z

digits: 0 1 2 3 4 5 6 7 8 9

ordinary operators: . , + - * / ( ) =

special_characters:

directive_symbol	$	(dollar)
repetition_symbol	:	(colon)
system_symbol	%	(per cent)
substitution_symbol	#	(hash)
end_of_line_symbol	!	(shriek)
string_symbol	'	(quote)
and_symbol	&	(ampersand)
or_symbol	?	(query)
open_angle_symbol	<	(less than)
close_angle_symbol	>	(greater than)
join_symbol	_	(underline)
open_square_symbol	[	(left-hand bracket)
close_square_symbol	]	(right-hand bracket)
dimension_symbol	;	(semi-colon)
invalid_character_symbol	@	(at)
line_symbol			(modulus)

space
newline

Lower-case *letters* are everywhere equivalent to the corresponding upper-case *letters*; *letters* in *macros* and *strings* [13.2.3] are stored as they are input; the equivalence is used if they are interpreted but not otherwise.

The representation (and usual nomenclature) of the *special_characters*, as given above, is used throughout the Guide, but it must be remembered that the representations used on a particular implementation may be different from these. (See the `ENVIRONMENT` *directive* [15.24] for details of how to get GLIM to display the representations used on the local implementation).

A *space* is the blank character.

Newline does not have an external representation but is generated on input by pressing the 'return' key on a keyboard, and characterised on output by the start of a new line.

13.2 Tokens

Formally, the input sequence of *characters* is interpreted as a sequence of *tokens*, each *token* consisting of one or more adjacent *characters*. Two types of *token* are recognized: *directive_names* and *items*. An *item* will be either an *identifier*, a *value*, a *keyword*, an *operator*, or a *separator*.

13.2.1 Directive_names

Directive_names introduce a sequence of *items* and specify a particular operation. Each *directive_name* consists of a *directive_symbol* followed by one or more *letters*, *digits*, and *underlines*, (for example, `$data, $fit`). Only 96 such *directive_names* are currently supported by GLIM; this set is defined in [15]. Only the first four *characters* of a *directive_name* are significant, so that *directive_names* which are identical up to four *characters* denote the same *directive* (for example, `$env, $environment` are interchangeable). In addition, some *directive_names* have synonyms that are formed by omitting the fourth, and possibly third, *characters*; the minimum number of *characters* that identifies the *directive_name* is indicated by underlining in the syntax subsections of [15], (for example, `$fit` may also be written as `$fi` or `$f` [15.31]).

13.2.2 Identifiers

Identifiers stand for objects within GLIM. They are of two kinds: those designated by the system (*system_identifiers*) and those designated by the user (*user_identifiers*). The former consist of the *system_symbol* followed by a specified number of *letters*, *digits*, or *underlines*, (for example, `%a, %z1, %log`). The latter consist of a *letter*, optionally followed by *letters*, *digits*, or *underlines*, (for example, `a, y, ,z1, item_cost, response`). Only the first eight *characters* of a *user_identifier* are significant so that *user_identifiers* which are identical up to eight *characters* denote the same object, (for example, response, response1, response_vector are interchangeable). Note also that since upper- and lower-case *letters* are equivalent, the *identifiers* `response`, `RESPONSE`, and `ResPONSE` are also interchangeable. Only the first four *characters* of a system identifier are significant.

13.2.3 Values

Values are items manipulated by GLIM and intended to represent real-world objects. There are three kinds of value: *integers*, *reals* ,and *strings*.

(1) An *integer* represents a non-negative whole number; it consists of one or more digits with the usual mathematical meaning. Note that by definition it is unsigned. Its value may not exceed that of the largest integer representable on the local implementation, (see [15.24] for details of how to display this value). For example,

```
0, 21, 1349, 043
```

are all integers.

(2) A *real* represents an ordinary real number. It is defined as:

$$[\left\{ \begin{array}{c} + \\ - \end{array} \right\}] [\textit{ integer }] [\, . \,] [\textit{ integer }] [\left\{ \begin{array}{c} \mathtt{e} \\ \mathtt{d} \end{array} \right\}] [\textit{space }]s [\left\{ \begin{array}{c} + \\ - \end{array} \right\}] \textit{ integer }]$$

where at least one of the first two integers must be present, and the value of the third integer must be such that the value of the resulting real does not exceed the value of the largest real number representable on the local implementation. For example,

```
2.5, 55, -6.34e-2, 4E, .37
```

are all real.

(3) A *string* represents a sequence of *characters* from the GLIM character set. It is defined as

' [*character*]s '

where *character* can be any character from the character set except the *end_of_line_symbol*. This restriction is imposed as the *end_of_line_symbol* is always interpreted by GLIM as the end of the line, and the symbol is thus never stored [13.6.3]. A string symbol may be included in the string by representing it with two adjacent string symbols. Thus

''''

refers to a string containing the single character '. Note that " is not an acceptable alternative to '' .

13.2.4 Keywords

A *keyword* is the name of an option. Each context defines its own set of options and the *keywords* used to denote them. In some contexts a single *letter* is used, in others a *letter* followed by *letters, digits,* or *underlines*, in others an asterisk * followed by *letters, digits,* or *underlines*. Details of which form is allowed in each context is given in [15].

13.2.5 Operators

An *operator* is either an *ordinary_operator* or a *special_character*.

13.2.6 Separators

A *separator* may be of two types:

(1) A standard separator is either a *space* or a *newline.*

(2) A list *separator* is a *comma*, or *dimension_symbol*, optionally prefixed and postfixed by one or more standard *separators*.

See [13.5.1] for further information on when the two types of separator are used.

13.3 Further language features

The *items* are the building blocks from which *phrases*, *lists*, *option lists*, *file specifiers*, and *statements* are constructed.

13.3.1 Phrases

A *phrase* is a keyword, optionally followed by one or more *identifiers*, *operators, values,* and *keywords.* For example, the following are instances of *phrases:*

`on`	(where `on` is a keyword);
`the X mean`	(where `the` and `mean` are keywords, and `X` is anidentifier);
`colour 6`	(where `colour` is a keyword and `6` is a value).
`interval * CUT`	(where `interval` is a keyword, `*` is an operator, and `CUT` is an identifier).

13.3.2 Option lists

An *option list* is used to modify the operation of a particular instance of a *directive*, and has the formal syntax:

([*keyword* = *item* [, *item*]*s*] *s*)

The first character only of the *keyword* is significant. The *item* is usually a *value* or an *identifier*, but may be another *keyword.* If more than one item follows a keyword, then the *items* must be separated by a comma. Typical instances of option lists are

```
( device='post')
(direction=ascending)
(x=10,50,10  y=0,10,1 title='test plot')
```

13.3.3 Lists

A *list* is either a sequence of *identifiers* or a sequence of *values*. The *identifiers* or *values* in a list are known as *elements*. Each *element* is separated from the next *element* by the *list_separator*. Examples of *lists* are:

```
A,B,C,D
1,2,3,4
```

13.3.4 File specifiers

A *file_specifier* is used within GLIM to specify either the name of an external file or a channel number or both, and has the formal syntax:

$$\left\{ \begin{array}{c} '\textit{filename}\,' \\ \textit{integer} \\ \textit{integer} \;=\; '\textit{filename}\,' \end{array} \right\}$$

where *filename* is the name of an external file, and *integer* represents the internal channel number to be used by GLIM. See [13.6.1] for further information on channels and files.

13.3.5 Statements

A *statement* is a *directive_name* optionally followed by an option list and then by a (possibly empty) sequence of *items*. Each *statement* is interpreted by GLIM as an instruction to perform some action (as specified by the *directive_name*), the sequence of *items* supplying further information to the system on how the action is to be performed.

The different instructions recognized by GLIM are termed *directives*, of which there are 96, each such *directive* being known by its *directive_name*; for example, the `FIT` *directive* is the *directive* specified by `$fit`. The sections of [15] define the syntax (that is, the allowable sequences of *items*) and the semantics (that is, what actions GLIM will then perform) for each of these *directives*. In this sense, a *statement* is then an instance of a *directive*, formed according to the prescribed syntax of that *directive* and having an effect as defined by the semantics of the *directive*.

A *session* represents a complete set of *statements*, and is defined as

[*job* [`$newjob` *job*]*s*] `$stop`

where *job* is defined as

[*statement*]*s*

and *statement* is any *statement* except a NEWJOB *statement* or a STOP *statement*. A *job* usually represents the portion of a *session* devoted to the analysis of one dataset [see 15.54].

13.4 Substitution rules

The following rules extend the allowable forms of *statement* beyond those permitted by the syntax definitions of [15].

13.4.1 Scalars

A *scalar* [see 14.2] may replace a *real* or *integer*; in the latter case the value of the *scalar* is rounded to the nearest integer. For example, the SLENGTH *directive* [15.77] has the syntax:

$slength *integer*

but

```
$slength  %n
```

is a valid SLENGTH *statement*, provided the rounded value of %n satisfies the restrictions on *integer* imposed by the SLENGTH *directive*. Exceptions to this rule for certain directives are described in [15].

13.4.2 Macro substitution

The occurrence of the form:

[*space*]*s* *macro*

implies replacement of the form by the contents of *macro* [14.5], causing the *macro* contents to be treated as part of the input stream. Such replacement is effectively an invocation of *macro* and follows the usual rules for invocations [given in 13.7.3]. If the *macro* has no contents then the form is effectively ignored. The form may not immediately follow the MACRO or SUBFILE *directive_names*.

Example

```
$macro  FUN  %log  $e
$calc   Y = #FUN(X) $
```

is interpreted as

```
$calc  Y = %log(X) $
```

13.4.3 Formal arguments

Within a *macro*, any identifier may be replaced by a *formal_argument* or a *keyword_argument*. A *formal_argument* is defined as either

% *digit*

or

% *ordinary_scalar*

or

% *system_scalar*

where the (rounded) value of the *scalar* must lie between 1 and 9 inclusive when interpretation occurs, and *digit* is not `0`. When the *macro* is invoked, and if the value of (respectively) *digit* or *ordinary_scalar* or *system_scalar* is *i*, or if the *keyword_argument* is the *i*th *identifier* in the *macro's keyword* list, then the *formal_argument* or *keyword_argument* will be replaced by the *i*th *actual_argument* of that *macro*, as set in the latest `ARGUMENT` or `USE` *statement* [see 15.3 and 15.90]; the *i*th *actual_argument* must have been defined at the time the substitution is attempted.

No substitution occurs within an `ARGUMENT` *statement* or within the argument specification section of a `USE` *statement*; in these contexts *formal_arguments* have a different interpretation [13.7.2].

Example

```
$macro (ARG=FIRST,SECOND,THIRD) M
$print THIRD %2 %%a $
$endmac
$calc  %a=1  $
$use   M  Z  Y  X  $
```

has the same effect as

```
$print  X  Y  Z $
```

13.5 Restrictions on statement forms

The following restrictions apply to the syntax of all *statements*, and are in addition to those explicitly specified in the syntax specfication of individual *directives* [15].

13.5.1 Separation of tokens

The following sequences of *token* pairs must be separated by at least one *standard_separator:*

directive_name	and *user_identifier*
directive_name	and *real*
directive_name	and *integer*

The following sequences of *token* pairs must be separated by at least one *standard_separator* or a *list_separator,* depending on the context:

identifier	and *user_identifier*
identifier	and *real*
identifier	and *integer*
real	and *identifier*
real	and *real*
real	and *integer*
integer	and *real*
integer	and *integer*

The context is defined by the syntactic definition of the *directive* given in [15]. Note that adjacent *reals* occurring in fixed-format contexts in data input [15.15] do not need a *separator* between them.

13.5.2 Inaccessible characters

All *characters* to the right of an *end_of_line_symbol,* and on the same line, are ignored.

All *characters* following a `RETURN`, `FINISH`, `NEWJOB`, or `STOP` *statement*, and on the same line, are ignored.

13.6 Interpreting statements

13.6.1 Channels

GLIM communicates with the user's screen, keyboard, and files via 'channels'. Each channel is denoted by a unique positive integer called the channel number. Each channel is used for a single purpose, such as for primary input, dumping, and so on. There must be some device or file, where GLIM can either obtain input (an input channel) or write output (an output channel), assigned to the channel. At the start of a *session* GLIM recognizes certain channels (and knows their channel numbers), and it is possible that the operating system has connected devices/files to them. These channels are:

primary input	where GLIM expects to get its first statement after reading an initialisation file if present;
primary output	where, by default, GLIM will send its output;
primary dump/restore	where, by default, GLIM will dump and restore [13.7.6];
transcript	where GLIM writes transcript output [13.6.4];

macro library	where macro library input [13.7.5] can be obtained;
journal	where GLIM writes a copy of the primary input;
initialisation input	if implemented, where GLIM expects to get its initial statements.

The channel numbers used and the names of the devices/files are system dependent; the former can be inspected by option `C` of the `ENVIRONMENT` *directive* [15.24]; these channels cannot be used by the user for other purposes. Any other channel number (subject to any limitations imposed by the operating system) can be used to connect to other devices/files. These connections (termed secondary channels) can either be made:

(a) before the GLIM session (details will depend on the operating system); or

(b) by the user using a file_specifier in the `INPUT`, `OUTPUT`, `DUMP`, `RESTORE`, `REWIND`, `DINPUT`, or `REINPUT` directives. If the file specifier does not contain a channel number, then GLIM will automatically assign the next available channel to the specified file. If a channel number is used that is not preconnected and no filename is given, GLIM will prompt the user for the name of a device/file to connect to the channel. If the filename and channel number specified are found to be inconsistent (for example, the file may already be open on a different channel), then an error is generated.

GLIM does not note which secondary channels are used for which purpose. Hence it cannot prevent a user from writing to a secondary input channel, or inputting from a secondary output channel with possibly disastrous consequences. Primary channels are protected from such misuse.

Each channel has associated with it a record length, which specifies to the program how many characters per line it can expect to read from or write to the associated device/file. The values used will be system dependent, but must reflect the capabilities of the actual device/file. Output channels also have a page height, which is used solely to set the size of plots and, for the primary output channel, for pagination purposes [13.6.4].

13.6.2 Mode of operation

GLIM is designed as an interactive system, that is, it interacts dynamically with the user via a keyboard and screen. There are many advantages to the user from this type of arrangement. It is also possible, however, to run GLIM in batch mode, when it gets its *statements* from an input file and writes its results to an output file. For the remainder of this guide we assume that GLIM is being used in interactive mode, and mention only in passing the differences that occur in batch mode.

A *job* [13.7.8] is processed a *statement* at a time. However the *statements* that constitute a *job* need not all originate from the same source: some may be entered from the keyboard, some may be input from files that are external to GLIM, and some may be stored internally as *macros* [14.5]. In this section we discuss how an individual *statement* is processed regardless of its source, while in [13.7] we describe how to direct GLIM to obtain *statements* from a new source, and how GLIM keeps track of such redirection.

A GLIM *statement* begins with a *directive_name* and terminates at the *directive_symbol* that begins the next *statement*. In practice, *statements* are often followed by a terminating *directive_symbol*, whose purpose is simply to mark the end of the previous *statement* and initiate execution.

Example

```
$calc  X = 0  $
```

the final `$` being the terminating *directive_symbol*.

Each *statement* is read, interpreted, and executed to completion before reading of the next *statement* is attempted.

The GLIM interpreter reads, interprets, and executes as much of each *statement* as it can, as it proceeds through the *statement*. For some *directives* (such as the `DISPLAY` *directive* [15.16], which produces output from a fit) the system can execute part of the *statement* before the whole of the *statement* has been read. For other *directives* (such as the `FIT` *directive*) the whole *statement* must be read and interpreted before any execution is possible. Thus for the former the results of the execution (such as printed output) may be seen before the terminating *directive_symbol* has been encountered. For the latter, no execution can occur until the terminating *directive_symbol* is encountered. If in doubt, always end a *statement* with a *directive_symbol* on the same line.

13.6.3 Input sources

At the start of each *job* GLIM expects to get *statements* from the primary input channel and will print its response on the primary output channel. In interactive mode the keyboard and screen serve these purposes; in batch mode two separate files are used.

GLIM can also be directed to read *statements* from secondary input channels; these channels are usually connected to files containing GLIM statements. How this is achieved is described in [13.7.4].

However, before reading the first directive from the primary input channel, GLIM will first check the filestore to see if an initialization file is present. The name of this initialization file is system dependent. If this file exists, then it is treated as a secondary input file, and input is read from this file before reading the first directive from the primary input channel.

When reading from the keyboard GLIM indicates that it is expecting the user to enter *statements* by issuing a prompt to the screen; that is, it prints a few characters as an invitation to the user. Usually the character '`?`' followed by a *space* is printed, but when expecting items for a particular *directive* the '`?`' is preceded by the *directive_name*. For example if, in response to the '?' prompt the user enters '`$calc %a=5 $print %a`', GLIM will respond with '`$PRI?`' to indicate that the PRINT *statement* has not been terminated, and can be continued.

If a *character* that is not part of the GLIM *character_set* is input, it is interpreted as an *invalid_character* and appears as such in any fault messages [16.2], echoing [15.18], and so on; the *invalid_character* has no other use in the language. Note that some *characters* generated by keyboards (such as control characters) are not printed on the screen but are still received by GLIM; they will appear on output as an *invalid_character.*

Characters on an input line following the first *end_of_line_symbol* are ignored by GLIM, and may be used as in-line comments.

13.6.4 Output

GLIM can be directed to write its output to secondary output channels. Again, these channels will usually be connected to files, and such files are often submitted for printing at the end of a *session*. To give flexibility to the user, output from the program is classified into seven types:

input	reproduction of primary input *statements* [13.6.3];
verification	reproduction of executing macros [13.7.3];
warnings	information messages [16.1];
faults	fault/interrupt messages [16.2] and [16.3];
help	diagnostic messages [16.2];
ordinary	primary ordinary output from directives;
echoing	reproduction of secondary input and output statements.

The `OUTPUT` *directive* [15.59] sets the current output channel, that is, the channel to which ordinary output is sent; it can also be used to switch off ordinary output until further notice. Echoing, verification, and warnings are only produced if ordinary output is switched on, and their output is then sent to the current output channel. When ordinary output is switched on echoing, verification, and warnings may be independently switched off (or back on again) through the `ECHO`, `VERIFY`, and `WARN` *directives.* [15.18], [15.92] and [15.93]. Fault and interrupt messages cannot be switched off; they are always sent to the primary output channel. Diagnostic messages can be switched off and on by the `BRIEF` *directive* [15.7]; they are always sent to the primary output channel. All settings remain in force until reset by another appropriate *statement*. They are reset to default values at the start of a *job*.

A record of the *session* will be preserved in the 'transcript file' in interactive mode. This file can receive all seven types of output. By default, the transcript file will reproduce the output to the primary output channel. Use of the `BRIEF`, `ECHO`, `OUTPUT`, `VERIFY`, and `WARN` *directives* will therefore automatically affect the record of the session stored in the transcript file. It is possible to specify, through the `TRANSCRIPT` *directive* [15.88], which type of output is to be sent; any subset of the seven types can be selected, independently of whether such output is also being sent to its usual destination. (For example, warning *statements* can be reproduced on the transcript file, whether or not the `WARN` facility is switched on.) Each line written to the transcript file is prefixed by five characters

space [*type*] *space*

where *type* is `i, e, v, w, f, h,` or `o`, denoting the output type of the line, and standing for the types given above. In interactive mode, input, warnings, faults, help, and ordinary output but not echoing and verification are sent to the transcript file by default at the start of each *job*. In batch mode no output is sent to the transcript file, by default. The transcript file is connected to its own channel, which cannot therefore be used for other purposes. The name of the transcript file is system dependent.

In interactive mode, output to the primary output channel can be 'paged', that is, GLIM will pause every time a screen-full of information has been sent (including lines input by the user), and will not continue until the user presses the 'return' key. This facility is controlled by the `PAGE` *directive* [15.61]. By default the facility is switched off at the start of each *job*. Paging is automatically enabled in the `MANUAL` *directive*. The facility is not available in batch mode.

Additionally in interactive mode, a copy of input from the primary input channel is written to the journal channel. All statements are copied, except those issued by the user when in the macro debugger [13.7.3]. The journal facility can be turned on and off by the `JOURNAL` *directive*.

The accuracy to which *reals* are printed is controlled by the `ACCURACY` *directive* [15.1]; its setting determines the accuracy of all *numbers* subsequently printed for ordinary output (see above), but has no effect on the other five types of output. Each GLIM *directive* determines the number of significant figures in the *numbers* that it prints and their format (including whether fixed or floating point); the `ACCURACY` *directive* can be used to increase or decrease this number of significant figures. Each *directive* ensures that within each set of related *reals* that it prints, all *reals* have the same format and the smallest *real* is printed to at least *n* significant figures, where *n* is 4 by default; the `ACCURACY` *directive* can change *n* to any value between 1 and 9. For example the statement

```
$look  X  Y  Z  $
```

where X has the values 1.23456, 12.3456, 123.456 would print the `X` values as `1.235`, `12.346`, and `123.456`, ensuring that the smallest *real* has 4 significant figures and all *reals* have the same number of decimal places; the formats for `Y` and `Z` would be similarly determined, but independently of `X` and of each other. But,

```
$accuracy  2
$look  X  Y  Z  $
```

would print the values of X as 1.2, 12.3, and 123.5, and similarly for Y and Z.

It is also possible, given the current accuracy setting, to further restrict the number of decimal places output; this is controlled by the second setting to the ACCURACY *directive*.

13.6.5 Workspace

All data structures [14] are stored in the workspace. The size of this workspace is system dependent, but may be inspected via option U of the ENVIRONMENT *directive* [15.24]. The size is given in terms of cells, each cell being capable of storing a number (typically 1 or 2) of *reals* or *characters*; this storage capacity per cell is system dependent.

As new data structures are defined and given *values* so the amount of available workspace decreases. When insufficient workspace is available to perform the operation implied by the current *statement*, GLIM will abandon the *statement* and print a fault message [16.2]. This does not necessarily mean that the workspace is full, only that there is not enough space to complete the current operation.

Used workspace may be recovered by deleting redundant structures (through the DELETE and TIDY *directives*). Since GLIM locates data structures in the workspace in a manner that optimizes the available space, it is not necessary to know where individual structures are stored in the workspace, but only the total amounts of used and available workspace.

GLIM also stores certain internal structures [14.9] in the workspace, where they compete for space with the other data structures. The space taken up by these structures may be inspected via option S of the ENVIRONMENT *directive*.

13.6.6 Retrieving system information

GLIM stores certain values that reflect its current operating state in user-accessible system structures. These structures and the values they contain are described in detail in [14]. The current values of most of the settings described in this section can be inspected by printing the appropriate system structure.

13.7 Obtaining statements

This section describes how GLIM obtains the program *statements*.

13.7.1 The program control stack

As described in [13.6], at the start of a *job* and after taking input from any initialization file GLIM will obtain its first *statement* from the primary input channel. Subsequent *statements* may, however, originate from other sources, that is, from other input channels or from *macros* [14.5].

Each time GLIM starts reading from a new source it notes its position within the old source so that it can return there when the last *statement* from the new source has been read. To keep track of this nesting of input sources GLIM maintains the program control stack (or simply the stack). The first level of this stack contains information on the original input source, that is, the primary input channel. When a new source is used, GLIM stores within the first stack level details of how far it has read along the current line, and then uses the second stack level to store information about the new source. It then starts reading from that source. If the end of this second source is reached with no other new source being requested, then GLIM will abandon the second stack level and, using the information stored in the first level, continue reading the line from the primary input channel. If, however, the second source contains a request to read from another, third, source then GLIM will note, within the second level, its current position within the second source, use the third level to store details of the third source, and then start reading from the third source. When the end of this source is reached GLIM returns to the previous level in the same manner as before.

There are 16 levels in the stack; thus a maximum of 16 sources may be nested at any one time. The level number of the current source (with level 1 being used by the primary input channel) is known as the current stack level; its value is also stored in the system scalar `%cl` [14.2.2].

Macros may call themselves [13.7.3], and thus the same *macro* may appear more than once in the stack at a given time. Input channels, with one exception, may appear only once in the stack at any given time; the only exception is that the primary input channel can appear any number of times (in addition to its appearance at level 1 and subject, of course, to the maximum size of the stack).

GLIM terminates a *macro* source when it reaches the end of the *macro* (unless invoked by `WHILE`; see [13.7.3]). Input sources are terminated when a `RETURN` or `FINISH` *statement* is encountered. Sources may also be terminated prematurely by a SKIP or `EXIT` *statement*; these *statements* request GLIM to jump back through the stack and continue executing at a lower level, abandoning intervening levels; see [15.76] and [15.26]. User faults, either unintentional or generated by the `FAULT` *statement* [15.29], may also cause premature termination; see [16.2].

For obvious reasons, a macro cannot be deleted as long as it appears in the current stack.

The stack is re-initialized at the start of each *job*. This implies that a `NEWJOB` *statement* terminates reading from all input sources in the current stack, the next input being obtained from the primary input channel as level 1 of the stack for the new *job*.

Option `P` of the `ENVIRONMENT` *directive* allows the current stack to be inspected.

13.7.2 Arguments

GLIM allows nine *arguments* per *macro*. A *formal_argument* or *keyword_argument* [13.4.3] is a place holder for some *identifier*; this *identifier* is termed the *actual_argument*, and the association is defined through an `ARGUMENT` [15.3], `GET` [15.33] or `USE` [15.90] *statement*. When GLIM encounters a *formal_argument* or *keyword_argument* within a macro it interprets it as a reference to the associated *identifier*; such an association must exist at that time.

Arguments exist on a per-*macro* basis. If a *macro* occurs more than once in the stack then all instances use the same *argument* associations.

A further mechanism exists, whereby *formal_argument -actual_argument* associations can be carried through several levels of the stack. If an *actual_argument* at some level is, in fact, a *formal_argument* then it is taken as a reference to the *formal_argument* (and its associated *actual_argument*) at the immediately preceding stack level. For example, if the *formal_argument* `%2` had been associated with the *identifier* `%3` then use of `%2` would refer to the third *argument* of the *macro* at the previous level. Thus

```
$macro FIRST $use SECOND * %3 $$endmac
$macro SECOND $print %2 $$endmac
$use  FIRST * *  X  $
```

would print the values of `X`. The facility is useful when an initial *macro* (such as `FIRST`) is used to collect all *arguments* from the user and then pass them on to other *macros* that it calls.

The *scalars* `%a1, %a2, ..., %a9` are set by GLIM to indicate whether corresponding *actual_arguments* have been set; see [14.2.2].

13.7.3 Macro invocations

A request to GLIM to read *statements* from a *macro* is termed a *macro_invocation*. The ways in which this can be done are described below.

The `USE` *directive* [15.90] performs a *macro_invocation*; it may also be used to set/reset actual *arguments* for the *macro*.

The `SWITCH` *directive* [15.83] requests GLIM to execute one *macro* out of a list of *macros*, on the basis of the value of a supplied *scalar*. It thus enables the user to program a choice of macros to execute. In its simplest form, with a single *macro*, it permits conditional execution of the *macro*.

The `WHILE` *directive* [15.95] requests GLIM to execute a *macro* repeatedly, while the value of a given *scalar* is non-zero. It thus provides a looping facility.

Macro_substitution [13.4.2] causes GLIM to read immediately from the named *macro*. As such, it is identical in its effect to a `USE` *statement* without the *actual_argument* settings. Syntactically it is different, though, since it can be used anywhere within a *directive*. Moreover, the inserted *macro* may contain another *directive_name*, as in

```
$macro BITS +2 $print $endmac
$calc X = Y # BITS X $
```

which is equivalent to

```
$calc X = Y + 2 $print X $
```

It should be carefully noted from the above that there is no requirement for *macros* to contain only sequences of ***complete*** *statements*. The beginnings and endings of *macro_substitutions*, and the endings of *macro_invocations* in general, are completely transparent to the GLIM interpreter, and for this reason the above example works as intended. But this convention also leads to one of the commonest programming problems with GLIM. Consider the following:

```
$macro M $print %a $calc %a = %a-1  $endmac
$calc  %a = 1 $while %a M $
```

Note that M is actually stored as

```
$print  %a  $calc  %a = %a-1
```

since ' $endmac ' is not part of the *macro*.

Now M will be executed twice, not once as may be intended. The reason is that when GLIM reaches the end of *macro* M for the first time it will have to check the value of %a (because of the WHILE *statement*) but, since the CALCULATE *statement* has not yet been executed (it cannot be executed until a *directive_symbol* is encountered), it finds %a still has the value 1, and therefore executes M again. On reading through M again it finds a *directive_symbol* (from '$print') so it executes the CALCULATE *statement* and when it reaches the end of M for the second time it finds %a to be zero and terminates M. It then finds the *directive_symbol* following the WHILE statement so completes the execution of the CALCULATE *statement*, leaving %a with the value -1. Replacing M with

```
$macro M $print %a $calc  %a = %a-1  $$endmac
```

solves the problem.

Thus if a *macro* is to be used as a sequence of complete *statements* then its final *statement* should ***always*** be terminated with a *directive_symbol* in addition to that contained in the $endmac *directive_name*.

Lines of currently executing *macros* can be printed as they are executed, through the VERIFY *directive* [15.92]. This can help with program debugging, as can the macro debugger, available through the DEBUG [15.13] and NEXT *directives* [15.55].

13.7.4 Input requests

A request to GLIM to read *statements* from an input channel is known as an *input_request*.

For each input channel that it reads from, GLIM maintains a current position, which indicates the ***next*** line to be read from that channel. (Strictly, this applies only to files, since the current position on a terminal is simply the next line entered by the user.) When the channel is first opened the current position is the first line in the file. Beyond the last line of the file as stored by the user there is considered to be an end-of-file line. Whenever GLIM reads a line from a file it also moves the current position to the succeeding line in the file. If GLIM, at the user's request, attempts to read the end-of-file line it reports a fault; to avoid this problem the `FINISH` *directive* [15.30] should be used as the last *directive* in the file.

The only explicit controls that the user has over the current position are:

(i) to move it to the first line of the file (through the `REWIND` [15.73] or `REINPUT` [15.70] directives);

(ii) through the subfile facility described below; and

(iii) through the `FINISH` directive to move the current position to the first line of the file.

The current position within a file is maintained between *input_requests*, so that a subsequent *input_request* will read from the current position as set by the previous *input_request* unless, of course, the current position was reset explicitly as described above.

The current position within each open input file is not affected by default when starting a new *job* [15.54], although an option exists to close all secondary input and output files (using the `CLOSE` *directive*).

The `INPUT` *directive* [15.42] requests GLIM to start reading from the current position on the specified channel. The `REINPUT` *directive* is similar, except that it first sets the current position to be the first line in the file. Both *directives* can also be used to supply *subfile_identifiers*. If a *subfile_identifier* is supplied then GLIM (after repositioning for the `REINPUT` *directive*) searches through the file, from the current position, for a matching `SUBFILE` *statement*; if such a match is not found by the end of the file it starts searching from the beginning of the file again; if a match is still not found by the end of the file the current position is left at the first line of the file and a fault is reported. A matching `SUBFILE` *statement* is one whose *identifier* matches that supplied with the `INPUT/REINPUT` *statement*. If a match is found the current position is set to the line on which the match was found, before reading begins. Note that although there may be several possible matches within the file, only the first found by the above method is relevant.

If more than one *subfile_identifier* is supplied with an `INPUT` or `REINPUT` *statement* the process as described above is performed for the first such *subfile_identifier*. When reading from this channel is terminated GLIM then performs the same process for the second *identifier*, except that if requested by a `REINPUT` *statement* the current position is not initially reset to the first line of the file but keeps its current value. Again, when reading is terminated GLIM performs the same process for the third *subfile_identifier* (if any), and so on.

The `DINPUT` *directive* [15.15] acts like the `READ` *directive* in so far as it reads *reals* for the *identifiers* named in the `DINPUT` *statement* or in the last `DATA` *statement*, but it differs in that it reads them from the current position within the specified channel. It continues to read from the channel until it has satisfied the `DATA` *statement*. It will leave the current position at the line following that containing the last *real* read.

The `SUSPEND` *directive* [15.82] causes input to be read from the primary input channel. It is useful, when teaching `GLIM`, to insert `SUSPEND` *statements* at appropriate points in input files so that the current values of variables, etc, can be inspected. `SUSPEND` is not available in batch mode.

GLIM terminates reading from an input channel that has been originated by an `INPUT`, `REINPUT`, or `SUSPEND` *statement*, when a `RETURN` or `FINISH` *statement* is encountered, or when a `SKIP` or `EXIT` *statement* causes the corresponding stack level to be abandoned [13.7.1]. Reading originated by a `DINPUT` *statement* terminates when all *reals* have been read.

The `ECHO` *statement* causes lines read from a secondary input channel to be printed on the current output channel as they are read.

13.7.5 The macro library

A library of predefined *macros* (known as the macro library) is supplied with the program as a text file. It is preconnected to GLIM and the channel number is stored in the scalar `%plc`. It contains numerous *subfiles*, each of which contain *macros* that can be used to help solve a particular class of problem. Fuller details of its contents will be found in its *subfile* `INFO` or by issuing the *statement*

```
$manual library$.
```

13.7.6 Dumping and restoring

At any given time the GLIM program state can be defined by its internal data values together with current positions in open files. The internal data values include data supplied by the user, the currently specified model, the stack, and other internal information. The `DUMP` and `RESTORE` *directives* [15.17], [15.71] allow the user to save and later recover this program state.

A `DUMP` *statement* causes GLIM to copy all its program data values to an external system file. Its use does not otherwise affect the current operating state of GLIM, so that further program *statements* may be entered and executed as normal. By default, GLIM uses the primary dump/restore channel [13.6.1], but a secondary channel may be used in its place. Writing starts at the current position within the file.

A `RESTORE` *statement* causes GLIM to read a previously dumped program state from an external system file. By default, GLIM will read from the primary dump/restore channel, but a secondary channel may be used in its place. Reading begins at the current position within the file. The file must contain at its current position a program state that was previously dumped by the same release of GLIM (that is, 4 at present) on the same machine range. Moreover, the size of certain internal arrays contained in the dumped program state must not exceed those of the restoring program; this will occur only if GLIM has been reconfigured between the dumping and the restoring in order to change the size of the workspace, the directory, and so on (as described in the User Note), or if the program state to be restored was dumped by another, different version of GLIM; if the same GLIM program is used for both dumping and restoring no such problems will occur.

A dump/restore file must not be used for text input/output, nor vice versa; GLIM enforces this restriction on primary channels but its enforcement on secondary channels is system-dependent. The only operation allowed on dump/restore channels, other than dumping and restoring, is to reset the current position to the start of the file through the `REWIND` *directive* [15.73]. At the start of a *session* a dump/restore file may contain zero or more program dumps. After the last program dump (or at the start of the file if it contains no program dumps) there is an end-of-file marker. At the start of a *session* the current position in the file is either the start of the first program dump if the file contains one, or otherwise on the end-of-file marker. A `DUMP` *statement* writes the program state to the current position in the file (overwriting any dump or end-of-file marker already present), then writes an end-of-file marker and leaves the current position on the end-of-file marker. Before a `RESTORE` *statement* the current position must point to a program dump, not the end-of-file marker; after the program dump pointed to has been read the current position is moved to the start of the next program dump in the file if there is one, or otherwise to the end-of-file marker. A `REWIND` *statement* moves the current position to the start of the first program dump in the file if there is one, or otherwise to the end-of-file marker.

Some implementations do not allow arbitrary sequences of `DUMP`, `RESTORE` and `REWIND` *statements* on the same file; however, sequences in which `DUMP` *statements* and `RESTORE` *statements* are separated by a `REWIND` *statement* are always valid.

After restoring a program dump the new program state is identical to that which existed at the time the dump was made, with the following exceptions:

(1) the local pseudo-random number generator retains the state it had prior to the restoration;

(2) the set of open channels, together with the current position in each channel and the system vector `%oc`, is unchanged;

(3) the graphics state is unchanged.

Note that exception (2) will cause problems if the dumped program contained an *input_request* in its stack from a channel that is not available in the restoring *session*. Similarly, even if the same channel is available when restoring the dump, unexpected results may still occur since the current position in the restoring *session* will not necessarily be the same as that in force at the time of the dump.

In addition, it must be remembered that those *statements* that follow the original DUMP *statement* on the same line, are also saved, and will be the first *statements* to be executed following the restore. This can be useful, as in:

```
$dump $print 'dumped by J.K.S 3/12/92'$
```

but

```
$dump  $stop
```

should not be used as this will terminate not only the dumping *session* but also the restoring *session*.

13.7.7 Pausing

The PAUSE *directive* [15.63] by default returns control temporarily to the operating system: subsequent input from the terminal is passed to and interpreted by the operating system, until an (implementation-dependent) instruction causes the operating system to return control to GLIM. When control returns, GLIM continues by executing the *statements* following the original PAUSE *statement*. The *directive* enables the user to perform those operating system functions (such as inspecting file directories, and so on) that cannot be performed from within the program. If implemented, it is also possible for the PAUSE *directive* to issue an operating system command from within GLIM, in which case control will stay with GLIM.

The PAUSE facility is not available on all implementations of GLIM. When it is available care must be taken with its use, as it may have unfortunate side-effects. For example, on some implementations, running another program while GLIM is pausing causes the GLIM process to be erased. On almost all implementations, editing a file connected to GLIM will corrupt the current position within the file as held by GLIM, leading to unpredictable consequences when GLIM next attempts to read from it. The CLOSE *directive* may be used to circumvent this problem. The User Note should be consulted for details of further restrictions.

The PAUSE *directive* is not available in batch mode.

13.7.8 Jobs and sessions

Jobs and *sessions* were defined syntactically in [13.3.5] as sequences of *statements* terminated by, respectively, a NEWJOB and a STOP *statement*. Here we describe the effect of these *statements* on program execution.

At the start of each *session* the primary channels are opened and the pseudo-random number generators (PRNG) are initialized. The GLIM initialization file is searched for, and if found, it is read and the statements in it are executed. Control then passes to the primary input channel.

At the start of each *job* the mode (interactive/batch) is reset to its default, all *internals* and *system_structures* are initialized as described in [14], the workspace is cleared, the stack is initialized at level 1 (the primary input channel), and input/output settings return to their default values. Optionally, all secondary input and output files are also closed. Note the following settings that do not return to default value at the start of a new *job* :

(a) the current position within each open channel is unchanged (unless the option `close=yes` appears in the option list for the NEWJOB directive);

(b) the seeds for the PRNGs are unchanged;

(c) the system scalar `%jn` [14.2.2] is incremented by one;

(d) the graphics state remains unchanged, including the style definition of all pens and the default device. Every open device is refreshed (that is, the screen is cleared or a new page is started), and the layout setting reverts to its default.

At the end of each *session* the workspace is cleared, all open channels are closed, and GLIM returns control to the operating system.

13.8 Data manipulation

This section provides a brief survey of GLIM data manipulation facilities. Full details of the *directives* mentioned will be found in [15].

- *Scalars* may be declared explicitly through the `NUMBER` *directive.*
- *Vectors* may be declared explicitly through the `VARIATE` and `FACTOR` *directives.*
- *Arrays* may be declared explicitly through the `ARRAY` *directive.*
- *Lists* of identifiers may be declared explicitly through the `LIST` *directive.*
- Values may be assigned to a set of *vectors* via the `READ` or `DINPUT` *directives*; the *vectors* are first named in a `DATA` *statement* and the format of values is specified through a `FORMAT` *statement.*
- Values may be assigned to an individual *vector* or *scalar* by the `ASSIGN` *directive.*
- *Factors* with a regular pattern may be generated by the `GFACTOR` *directive* or with the `%gl` function.
- The contents of a *macro* may be assigned by the `EDMAC` and `MACRO` *directives* and also by the `GET` and `PRINT` *directives.*

- *Vectors* and *scalars* can have their values changed with the CALCULATE *directive*, which permits general arithmetic expressions to be used. Individual values may be changed through the use of indexing in CALCULATE or via the EDIT *directive*.

- The SORT *directive* will produce permutations of the values of *vectors*, while the GROUP and MAP *directives* enable *vector* values to be recoded.

- The TABULATE *directive* provides a general-purpose tabulation facility. Sets of input *vectors* can be tabulated to output *vectors*, with or without the results being printed.

- The ENVIRONMENT *directive* allows the set of current structures (termed the directory) to be printed.

- The LOOK *directive* prints the values of *vectors* or *scalars* in parallel, while the PRINT *directive* gives complete control over the printing of combinations of *vectors, scalars, macros, lists,* and the names of *identifiers*. The PLOT *directive* produces a scatter plot of *vectors*, while the HISTOGRAM *directive* produces a histogram display. The TPRINT *directive* allows *vectors* to be printed in tabular form. The GRAPH *directive* produces high-quality graphical output; the style of the pens used is controlled with the GSTYLE *directive* and the position of the graph is controlled by the LAYOUT *directive*. Graphical text may be positioned with the GTEXT *directive*.

- A pseudo-random number generator is available through the %sr function in the CALCULATE *directive*. It uses a multiplicative congruential generator, so that the ith number generated, x_i, is given by

  $$x_i = ax_{i-1} + c \pmod{m}$$

  where $c = 1$, $m = 2^{35}$ and $a = 8\,404\,997$. It has a period of 2^{35}. The initial value of x is set by GLIM at the start of each session but may be reset through the SSEED *directive*. A local PRNG may also be provided through the %lr function (see the User Note for details) but if it is not, %lr will be a separate copy of the standard generator %sr.

- The DELETE and TIDY *directives* enable data structures to be removed from the system and their workspace to be recovered.

14 Structures

14.1 Identifiers

Various kinds of object are recognized by the system, and have names by which the user can refer to them; these names are known as *identifiers*. (See [13.2] for a discussion of the syntax of identifiers.) *Identifiers* refer to either data structures or *subfiles*.

There are eight types of data structure in GLIM:

scalar :	holds a single real [14.2];
vector :	holds a list of reals [14.3];
pointer :	holds the name of a vector [14.4];
macro :	holds program text [14.5];
function :	maps real numbers [14.6];
array :	a multi-dimensional vector [14.7];
list :	holds a list of identifiers [14.8];
internal :	holds GLIM's internal data [14.9].

The individual types are discussed in more detail in the subsections below.

When a data structure is created it is associated with an *identifier*, by which the user can refer to it. An *identifier* is said to be **defined** when it has been associated with a data structure; otherwise it is said to be **undefined.** A defined *identifier* has the same **type** as the associated data structure. An undefined *identifier* has **unknown** type. The *identifier* remains defined until the associated data structure is deleted, when it becomes undefined again. An *identifier* is declared as a *global identifier* by default, unless is is specifically declared as a *local identifier* to a macro via the `LOCAL` option to the `MACRO` *directive* [15.50].

Identifiers are stored in directories; global identifiers being stored in the main directory and local identifiers being stored in a local directory associated with their macro.

An *identifier* may normally be associated with at most a single data structure at any one time. However, if an identifier is declared as a local identifier to a macro, then that identifier can be the same as a global identifier and also the same as another local identifier in any other macro. In this case, the data structure associated with the *identifier* will depend on the context in which it is being used. If a macro is currently being executed, then the local directory to that macro, if it exists, will be searched first, followed by the global directory. If a macro is not currently being executed, only the global directory is searched.

The type of a data structure determines the kind of **values** that it will contain (that is, its contents). These are discussed below for each type of structure.

Identifiers can be used not only to refer to data structures but also to *subfiles* (that is, to parts of files); these are discussed in [14.10].

All *identifiers* are stored in the **directory**. The number of identifiers available depends on the machine type; this number can be discovered through the *statement* `$environment u$.` Sites with access to the source code can change this number as described in the User Note. Current usage of directory space can be inspected by using the `U` option to the `ENVIRONMENT` *directive* [15.24].

14.2 Scalars

Scalars may be categorized into three types: ordinary scalars, system scalars, and user scalars.

Ordinary scalars and *system scalars* are defined by the system at the beginning of a *job* and, with the exceptions noted below, are initially assigned the value zero. They may be assigned values at any time by the `ASSIGN`, `CALCULATE`, or `GET` *directives* [15.5], [15.8] and [15.33]. They may be substituted for *reals* or *integers* as appropriate at any time.

User scalars are defined by the user by means of the `NUMBER` *directive* [15.56]. When defined, they are by default assigned the value zero. They may be assigned values at any time by the `ASSIGN`, `CALCULATE`, `GET`, or `NUMBER` *directives*. In most circumstances, they may be substituted for reals and integers as appropriate; instances where this is not true are documented under the appropriate directives in the next chapter.

14.2.1 Ordinary scalars

Ordinary_scalars have *identifiers* of the form:

`%`*letter*

There are thus 26 of them with *identifiers* `%a, %b, ..., %z.`

14.2.2 System scalars

There are 74 *system_scalars*. They are assigned values by the system as given below. Such values are copies of values held by the system, so that values assigned to *system_scalars* by the user do not in general affect the values kept by the system.

(a) `%jn`

At the beginning of each *job* `%jn` is assigned a value equal to one plus the number of `NEWJOB` *statements* so far encountered.

(b) `%pi`

is assigned the value of the mathematical constant π at the beginning of each *job*.

(c) `%cl`

is assigned the current level of the program control stack [13.7.1] each time this changes. This value is 1 at the beginning of each *job*.

(d) `%a1, %a2,..., %a9, %at`

When executing a *macro*, `%a1` has the value 1 if the first *argument* of that *macro* has been set and 0 otherwise; similarly for `%a2` to `%a9` in respect of the second to ninth *arguments*. `%at` is assigned the sum of the values `%a1` to `%a9`. When input is being taken from an input channel they all have the value 0.

(e) `%z1, %z2,..., %z9`

These *scalars* are not assigned values by the system but are intended for use by library *macros* as temporary storage; they should not therefore be used for other purposes. See the User Guide, Appendix A.

(f) `%im`

has the value 1 if GLIM is operating in interactive mode, otherwise 0.

(g) `%s1 %s2 %s3`

are assigned the values of the three seeds for the standard pseudo-random number generator at the start of each *job* and are reassigned the current values of the random number seeds at the current point of the random number sequence whenever the random number generator is called or the seeds are changed.

(h) `%bri %ech %war %ver %pag %jou`

If, respectively, the `BRIEF`, `ECHO`, `WARN`, `VERIFY`, `PAGE`, or `JOURNAL` facility is switched on the corresponding *scalar* has the value 1, otherwise the value 0.

(i) `%pic %pil %cic %cil`

The first two hold the primary input channel number `%pic` and its record length `%pil` ; the second two hold the current input channel number `%cic` and its record length `%cil`.

(j) `%poc %pol %poh %coc %col %coh`

The first three hold the primary output channel number `%poc`, its record length `%pol`, and its height `%poh`; the second three hold the current output channel number `%coc`, its record length `%col`, and its height `%coh`.

(k) `%pdc %plc %jrc`

These hold the channel numbers for the primary dump channel `%pdc`, the primary macro library channel `%plc`, and the journal channel `%jrc`.

(l) `%acc %ndp`

`%acc` holds the number of significant figures, and `%ndp` holds the number of decimal places, as determined by the `ACCURACY` *directive* [15.1]. If the number of decimal places has not been set, then `%ndp` holds the value -1.

(m) `%tra`

has the value

$$l_i + 2l_v + 4l_w + 8l_f + 16l_h + 32l_o + 64l_e$$

where l_i has the value 1 if input is being transcribed and 0 otherwise, while l_v, l_w, l_f, l_h, l_o, and l_e are similarly defined for verification, warnings, faults, help messages, ordinary output, and echoing respectively.

(n) `%nu`

is assigned 0 at the start of a *job*, and the current number of units by the `YVARIATE` *directive* [15.96]. If no `YVARIATE` *directive* has been issued, it contains the standard length.

(o) `%err %lin`

are assigned 0 at the start of a *job*. Following an `ERROR` or `LINK` *statement* they hold codes for the current error and link settings as follows:

`%err`	Error	`%lin`	Link
1	Normal	1	identity
2	Poisson	2	logarithm
3	binomial	3	logit
4	gamma	4	reciprocal
5	inverse Gaussian	5	square root
		6	probit
		7	complementary-log-log
		8	exponent
		9	inverse quadratic
10	OWN error	10	OWN link

(p) `%yvf %bdf %pwf %osf %ivf %lof`

If, respectively, a *y*-variate `%yvf`, binomial denominator `%bdf`, prior weight `%pwf`, offset `%osf`, initial values `%ivf`, or load vector `%lof` has been defined the corresponding *scalar* has the value 1; otherwise it has the value 0.

(q) `%cyc %prt %cc %tol`

hold the current values of the four *items* for the `CYCLE/RECYCLE` *directives*, that is, the maximum number of cycles `%cyc`, the printing frequency `%prt`, the convergence criterion `%cc`, and the aliasing tolerance `%tol`.

(r) `%sc %scf`

If the scale parameter has been set (by a `SCALE` *statement* or by default) to a positive value, `%sc` is then set to this value, and again set to this value at the start of each fit. Otherwise it is first assigned the value 0, and then the value of the mean deviance after each cycle of a fit. `%scf` takes the value 1 if the scale parameter has been explicitly fixed by the user, and 0 otherwise.

(s) `%pl %ml`

`%pl` is assigned the number of parameters (excluding those intrinsically aliased) in the current model at the end of the first cycle of a fit. `%ml` is assigned the number of elements in the lower triangle of the (co)variance matrix of non-intrinsically aliased parameters in the current model after the first cycle of a fit.

(t) `%df %dv %x2`

are assigned the current values of (respectively) the degrees of freedom `%df`, the (scaled) deviance `%dv`, and Pearson's χ^2 statistic `%x2` after each cycle of a fit. Both `%dv` and `%x2` are calculated from the current fitted values. Where no model has been fitted, and at the start of each *job* `%df`, `%dv`, and `%x2` are assigned the value 0.

Note that the `OWN` weights macro can assign values to `%dv` to change the internal stored deviance.

(u) `%met`

This scalar holds the current setting of the algorithmic method, as set by the `METHOD` *directive*:

`GIVENS`	1
`GAUSS-JORDAN`	2

(v) `%blv` `%blf`

contain, respectively, the baseline deviance `%blv` and the baseline degrees of freedom `%blf`.

(w) `%sl`

contains the standard length of vectors as set by the `SLENGTH` *directive*. If no standard length has been set and a `YVARIATE` is in force, `%sl` contains the number of units as set by the `YVARIATE` *directive*. It contains 0 at the start of a job.

(x) `%nin`

contains the number of valid observations weighted into the fit.

(y) `%cv`

contains the current state of fitting.

-1	no model yet fitted;
0	model fitted and iterated to convergence;
1	model fitted but has not yet converged;
2	model fitted but failed because of loss of d.f.;
3	model fitted but failed because of divergence;
4	model failed because of invalid linear predictor or fitted value.

(z) `%itn`

after a fit, contains the number of cycles (or iterations) used in that fit. During a user-defined fit, `%itn` contains the current iteration number. It contains 0 at the start of a job until the first model is fitted.

14.3 Vectors

A *vector* is a list of *reals*. A *vector* has three attributes, its **length**, its **number of levels,** and its **reference category.**

The length of a *vector* is equal to the number of *reals* that it can hold. A *vector* is given its length (which must be positive) when it is defined; this length may change during the *vector's* existence. A standard length for *vectors* is set by the `SLENGTH` *directive*; certain *vectors* take this length automatically as explained below and in [15]. A *vector* need not be given values when it is given a length. However, when it is given values the number of these values must equal the declared length.

The number of levels of a *vector* is either a positive integer (when it is known as a *factor*) or zero (in which case it is known as a *variate*). A *vector* is given its number of levels when it is defined; this setting may change during its existence. In general, *variates* are intended to hold sequences of arbitrary real numbers, while *factors* are intended to hold elements from the set of integers 1, 2, ... up to the declared number of levels. Certain *directives* enforce this interpretation and will treat a *variate_vector* differently from a *factor_vector* (for example, the `FIT` *directive*), in particular expecting the values of a *factor* (after rounding) to be within the declared range of levels.

A *factor* must have a further attribute — the *reference category* of the *factor*. This is used in model fitting and is by default 1 (the first level of the factor) but may be set to any number between 1 and the number of the levels of the factor. The reference category of a *variate* has no meaning and is undefined. If a *factor* is redeclared as a *variate*, then information on the number of levels and the reference category is lost.

Vectors are defined either by the user (*user vectors*) or by the system (*system vectors)*.

14.3.1 User vectors

An undefined *identifier* becomes defined as a *user_vector_identifier* when used in the appropriate *directive*; the corresponding *vector* is created at the same time. The length, number of levels, and reference category are determined by the particular *directive* which defines it. *Variates* may be redefined as *factors*, and vice versa, through the `FACTOR` and `VARIATE` *directives*, the values being unaffected.

14.3.2 System vectors

System_vectors are created by the system as variates at various times during a *job*. *System_vectors* can be deleted, in which case they lose their values.

14.3.2.1 System vectors available after a fit

These system vectors exist after a fit, and can be used in all directives.

`%fv` assigned the *f*itted *v*alues by the `FIT` *directive*.

`%lp` For iterative fits, `%lp` is assigned the *l*inear *p*redictor by the `FIT` *directive*. For non-iterative fits, it points to `%fv` and shares its values and `%lp` does not therefore appear in the directory. Redeclaration of a non-iterative model deletes any current values and `%lp` again shares the same values as `%fv`.

`%eta` Usually, `%eta` points to `%lp`, and contains the same values. However, if the `OWN` predictor macro (set by the `METHOD` *directive*) is declared, then `%eta` is a distinct system vector which must be assigned values by the user. It needs to contain the value of η as a function of the current values of the parameter estimates and/or the linear predictor. Resetting of a model where this macro is no longer set deletes any current values and `%eta` again shares the same values as `%lp`.

`%rs` contains the Pearson residuals from the current fit. These are defined as

$$(\text{Observed - Fitted}) * \sqrt{\frac{\text{Prior weight}}{k * \text{Variance function}}}$$

where k is the scale parameter if set and $k = 1$ otherwise. The Modelling Guide [11.3.1] gives a definition of the variance function and its value for each of the standard error distributions.

The default length of `%fv`, `%lp`, `%eta`, and `%rs` is the number of units, that is, the length of the y-variate (or its index, if set), and the standard length if no y-variate is set. If neither a y-variate nor a standard length has been defined, then the above vectors are not defined. Subsequent `YVARIATE` *statements* will redefine the above system vectors with the new number of units deleting any current values.

If any of the above vectors are deleted, then use of the `DISPLAY` and `EXTRACT` *directives* from the current model is inhibited.

14.3.2.2 Extractable system vectors

These system vectors will contain useful quantities following the use of the `EXTRACT` *directive* after a fit. Unless otherwise stated, the length of each system vector is `%nu`, the number of units.

`%wt` For an iterative fit, it is assigned the iterative *w*eigh*ts*. For a non-iterative fit, it is assigned a vector of ones.

`%wv` When an iterative model has been fitted it is assigned the values of the *w*orking *v*ector by the `EXTRACT` *directive.* For a non-iterative fit, its values will correspond to those of the y-variate.

`%dr` During the fitting of a model with a standard link, it will contain the values of the link function derivative. During the fitting of a model with a user-defined link via the `LINK` *directive* it must be assigned the values of the derivative $\frac{d\eta}{d\mu}$ within the `OWN` link macro [15.45].

`%va` During the fitting of a model with a standard error distribution, it is assigned the values of the variance function. During the fitting of a model with a user-defined error (that is, the `OWN` option to the `ERROR` *directive*) it must be assigned the values of τ^2 (the variance function) by the user within the `OWN` error macro.

`%di` During the fitting of a model with a standard error distribution, it may be assigned the values of the *d*eviance *i*ncrement During the fitting of a model with a user-defined error via the `ERROR` *directive* it must be assigned the values of the *d*eviance *i*ncrement within the `OWN` error macro [15.45].

`%ft` is assigned the state of each observation in the fit:

0 observation is present in the fit;
1 observation is weighted out;
2 observation is weighted out and has an invalid binomial denominator or y-variate. Fitted values or fitted probabilities can be calculated.
3 observation is weighted out and has an invalid vector in the model formula (for example, a factor level out of range). Fitted values cannot be calculated.

`%lv` is assigned the leverage values from the current fit as defined in the Modelling Guide [11].

`%cd` is assigned the Cook's statistic from the current fit as defined in the Modelling Guide [11].

`%vc` is assigned the lower triangle of the (co)*v*ariance matrix of the current non-intrinsically aliased parameter estimates. The length of this *vector* is given in the system scalar `%ml`.

`%vl` is assigned the *v*ariances of the *l*inear predictors.

`%pe` is assigned the non-intrinsically aliased *p*arameter *e*stimates (the length is given in `%pl`).

`%se` is assigned the standard errors of the parameter estimates (the length is given in `%pl`).

`%al` is assigned a vector containing the extrinsic aliasing structure of the parameter estimates. If a parameter is extrinsically aliased, it contains the value 1.0, otherwise it contains the value 0.0. (the length is given in `%pl`) .

`%sb` when used in the `EXTRACT` *directive* with the syntax `%sb`(*model_term*), is assigned a vector of length `%pl` which indicates the position in the vectors `%pe`, `%se`, and `%al` of the model parameters contributing to the *model_term*. If a parameter contributes towards the set of estimates defined by *model_term*, then `%sb` contains 1.0, otherwise it contains 0.0.

14.3.2.3 High precision system vectors

These system vectors exist after a fit, but, as they are stored in high precision, they may only be accessed via the `CALCULATE` *directive.*

`%wvd` contains a high-precision version of the working vector z in the IRLS algorithm. Its length is `%nu`.

`%wtd` contains a high-precision version of the iterative weights in the IRLS algorithm. Its length is `%nu`.

`%tri` contains the lower triangle of the current working matrix; it will occupy $(n+1)(n+2)/2$ cells, where n is the number of (non-intrinsically aliased) parameters in the current model (n is stored in `%pl`). The lower triangle contains different quantities depending on the algorithmic method being used. When the method is Gauss–Jordan, then `%tri` usually contains the values of the working matrix, which is output by the `T` option to the `DISPLAY` *directive*. This is true except when `%tri` is accessed from within the `OWN` triangle macro, where `%tri` will contain instead the lower triangle of the weighted sums of squares and products matrix augmented by the working vector. When the method is Givens, then the working triangle contains quantities which are described in the Modelling Guide [11].

14.3.2.4 Predicted quantities

These system vectors will contain useful quantities following the use of the `PREDICT` *directive* after a fit. All have length equal to the number of predicted units [15.66].

`%plp` the *p*redicted *l*inear *p*redictor for the current model;

`%pfv` the *p*redicted *f*itted *v*alues for the current model;

`%pvl` the *p*redicted *v*ariance of the *l*inear predictor for the current model.

14.3.2.5 Other system vectors

`%re` the *re*strict vector, to which the user can assign values either to select units in plotting or to restrict the display of fitted values after a fit. Rules for definition and assignment are as for *user variates*.

(a) If `%re` has values when the `PLOT` *directive* or the `GRAPH` *directive* is used, then if there is no explicit weight vector for the *x-y* pair being plotted, the units corresponding to zero elements in `%re` will not be plotted. The length of `%re` must be the same as that of all valid *x-y* pairs in the directive.

(b) If `%re` has values when the `DISPLAY` *directive* is used with option `W`, then units corresponding to zero elements in `%re` will be omitted. The length of `%re` must be `%nu`.

`%oc` contains a list of all non-graphics *o*pen *c*hannel numbers in GLIM. It is defined automatically at the start of a job, and its length increases or decreases as files are opened and closed.

14.4 Pointers

A *pointer* is a system identifier which refers to another *vector* or *value*. For example, `%yv` is a pointer which refers to the current *y*-variate. The value of a *pointer* is therefore either a *vector_identifier*, a *real*, or is null. When used in place of a *vector_identifier* then its value must not be null and the *vector_identifier* or *value* pointed to will be substituted for the *pointer_identifier*; for example, if `%yv` points to the user *vector* `y` then

```
$print  %yv   $
```

is equivalent to

```
$print   y   $
```

There are eleven *pointers*, all of which are defined by the system at the start of a *job*.

`%yv` has the value null until the first `YVARIATE` *statement* when it points to the values of the declared *y*-variate. Each new `YVARIATE` *statement* causes `%yv` to point to the values of the new *y*-variate. It has the value null after a `NEWJOB` *statement* or when the declared *y*-variate is deleted or removed.

`%pw` points to the scalar value 1.0 until the first `WEIGHT` *statement* when it points to the values of the declared *p*rior *w*eight vector. Each new `YVARIATE` *statement* causes `%pw` to point to the values of the new prior weight vector. It has the scalar value 1.0 after a `NEWJOB` *statement* or when the declared prior weight vector is deleted or removed.

`%os` points to the scalar value 0.0 until the first `OFFSET` *statement* when it points to the values of the declared *o*ff*s*et. Each new `OFFSET` *statement* causes `%os` to point to the values of the new declared offset vector. It has the scalar value 0.0 after a `NEWJOB` *statement* or when the declared offset is deleted or removed.

`%iv` has the value null until the first `INITIAL` *statement* when it points to the values of the declared *i*nitial *v*alues variate. Each new `INITIAL` *statement* causes `%iv` to point to the values of the new *i*nitial *v*alues variate. It has the value null after a `NEWJOB` *statement* or when the declared *i*nitial *v*alues variate is deleted or removed.

`%lo` has the value null until the first `LOAD` *statement* when it points to the values of the declared *lo*ad variate. Each new `LOAD` *statement* causes `%lo` to point to the values of the new *lo*ad variate. It has the value null after a `NEWJOB` *statement* or when the declared *lo*ad variate is deleted or removed. This system pointer additionally becomes unset when the algorithmic method is changed to Givens via the `METHOD` *directive*.

%bd has the value null until the first ERROR *statement* with option B, when it points to the values of the declared *b*inomial *d*enominator. Each new ERROR *statement* with option B causes %bd to point to the values of the new binomial denominator. It has the value null after a NEWJOB *statement*, after each ERROR *statement* with option other than B, or when the declared binomial denominator is deleted or removed.

%yi has the value null until the first YVARIATE *statement* when it points to the index vector of the declared *y* variate, if it is set. Each new YVARIATE *statement* causes %yi to point to the index vector of the new *y*-variate, if an index is set. It has the value null after a NEWJOB *statement* or when the declared *y*-variate is deleted or removed or when a new YVARIATE statement is issued which does not specify an index vector.

%wi as for %yi, but with reference to the WEIGHT *statement* and the index of the declared prior weight.

%oi as for %yi, but with reference to the OFFSET *statement* and the index of the declared offset.

%ii as for %yi, but with reference to the INITIAL *statement* and the values of the declared initial values variate.

%bi has the value null until the first ERROR *statement* with option B, when it points to the index of the declared *b*inomial *d*enominator if such an index has been set. Each new ERROR *statement* with option B causes %bi to point to the index of the new binomial denominator. It has the value null after a NEWJOB *statement*, after each ERROR *statement* with option other than B, when the declared binomial denominator is deleted or removed, or when an ERROR *statement* with option B is given and no index vector has been specified.

14.5 Macros

A *macro* contains a sequence of GLIM *characters*; the sequence may be empty. Each *macro* has nine arguments [13.7.2].

The *characters* stored in a *macro* may be used either to provide output text (for example, through the PRINT *directive*), or as GLIM *statements* for subsequent interpretation (for example, through the USE *directive* or *macro_substitution* [13.4.2]), or both.

All *macros* are user defined. (*Macros* in the macro library [13.7.5] become defined when they are read in by the user.) Previously undefined *identifiers* become *macro_identifiers* when used in the appropriate *directive*. They are defined explicitly, and given values, by the MACRO, EDMAC, and GET *directives* and by the STORE option to the PRINT *directive* , and are defined as a side-effect by certain uses of the ERROR, LINK,

`LOAD, INITIAL,` and `METHOD` *directives*. A *macro* may not be deleted while it appears in the current program control stack [13.7.1].

A *macro* must not contain the character strings ' `$fin` ' or '`$ret`'.

A macro may have local identifiers, providing local storage which may only be accessed when the macro is in use. The local storage is defined when the macro is defined. The local identifiers initially have no type, and need to be defined before use. By default, local identifiers keep their type, length, and contents between calls of the macro unless they are deleted via the `TIDY` or `DELETE` *directives*.

Deletion of a local identifier will remove its contents, length, and type but not its existence.

14.6 Functions

Functions are supplied and defined by the system at the start of each *job*. They perform certain mathematical or logical operations. They can only be used within the `CALCULATE` *directive* or as actual *arguments* to be passed to the `CALCULATE` *directive*.

Each *function* has a fixed number of arguments. The arguments of a *function* follow the *function_identifier*, must be separated by commas (if there are two or more), and must be enclosed in parentheses (unless there is only one argument, when the parentheses are optional).

Example

```
%gt (5, 6)
%if (A, B, C)
%log X
```

The following is a complete list of *functions*, together with their arguments and resultant values. The symbols '*x*', '*y*' and '*z*' stand for the first, second and third arguments respectively.

14.6.1 Mathematical functions

Function	Number of arguments	Result	Notes
`%abs`	1	absolute value function $\lvert x \rvert$	
`%angle`	1	arcsin ($\sqrt{x}$)	(b) (c)
`%cos`	1	cosine function	(e) (n)
`%cu`	1	cumulative function	(a) (i)
`%day`	3	days since 13 September, 1752	
`%dig`	1	digamma function	
`%exp`	1	e^x	(a) (m)
`%gl`	2	generate levels from 1 to *x* in blocks of *y*	(l)
`%ind`	1	index function — index or unit of *x*	(f)
`%lga`	1	Log gamma function $\log_e(\Gamma(x))$	(o)

Function	Number of arguments	Result	Notes
`%log`	1	natural logarithm of x	(a) (b) (d)
`%lr`	1	local PRNG	(h)
`%sgn`	1	sign function	(q)
`%sin`	1	$\sin(x)$	(e)
`%sqrt`	1	$\sqrt{x}$	(b) (d)
`%sr`	1	standard PRNG	(h)
`%tr`	1	integer part of x	(j)
`%trg`	1	trigamma function	(p)

14.6.2 Statistical functions

14.6.2.1 Cumulative probabilities of selected distributions

Given d, these functions calculate $\int_K^d f(x)\,dx$ (continuous) or $\sum_K^d f(x)\,dx$ (discrete)

Function	Number of arguments	Result
`%bip`	3	Binomial distribution cumulative probability. K=0. First argument is the deviate, d, which is rounded to the nearest integer. Second argument is the binomial denominator. Third argument is the mean probability of the binomial distribution
`%btp`	3	Beta distribution probability integral (incomplete beta function). K=0. $f(x) = \frac{x^{a-1}(1-x)^{b-1}}{\mathrm{BETA}(a,b)}$ where $\mathrm{BETA}(a,b) = \frac{\Gamma(a)\Gamma(b)}{\Gamma(a+b)}$. First argument is the deviate d ($d \le 1$) Second argument is a. Third argument is b.
`%chp`	2	Chi-squared probability integral. K=0. First argument is the deviate d. Second argument is the degrees of freedom.
`%fp`	3	F distribution probability integral. K=0. First argument is the deviate d. Second argument is the numerator degrees of freedom. Third argument is the denominator degrees of freedom.
`%gp`	2	Gamma distribution probability integral. Also known as the incomplete gamma function. $f(x) = \frac{y^{a-1}\exp(-y/b)}{b^a\,\Gamma(a)}$ with b=1.

Function	Number of arguments	Result
		First argument is the deviate d. Second argument is a. If b not equal to 1, standardize (use d/b as the first argument).
`%np`	1	Normal probability function. $K = -\infty$. Argument is the deviate d.
`%pp`	2	Poisson distribution cumulative probability. First argument is the deviate, d, which is rounded to the nearest integer. Second argument is the mean of the Poisson distribution.
`%tp`	2	t distribution probability integral. $K = -\infty$. First argument is the deviate d. Second argument is the degrees of freedom.

14.6.2.2 *Deviates or percentage points of a distribution*

For these functions:

(continuous) given $p = \int_K^d f(x)\,\mathrm{d}x$, returns d;

or

(discrete) given $p_1 = \sum_{K=0}^{d-1} f(x)\,\mathrm{d}x$ and $p_2 = \sum_{K=0}^{d} f(x)\,\mathrm{d}x$ and $p_1 \leq p < p_2$, returns d if $d > 0$, else returns 0.

Function	Number of arguments	Result
`%bid`	3	Binomial distribution deviate. First argument is the probability p. Second argument is the binomial denominator (which is rounded to the nearest integer). Third argument is the mean probability of the binomial distribution.
`%btd`	3	Beta distribution deviate (inverse of the incomplete beta function). K=0. $f(x) = \frac{x^{a-1}(1-x)^{b-1}}{\mathrm{BETA}(a,b)}$ where $\mathrm{BETA}(a,b) = \frac{\Gamma(a)\Gamma(b)}{\Gamma(a+b)}$. First argument is the probability p. Second argument is a. Third argument is b.
`%chd`	2	Chi-squared deviate. K=0. First argument is the probability p. Second argument is the degrees of freedom.

Function	Number of arguments	Result
%fd	3	F distribution deviate. K=0. First argument is the probability p. Second argument is the degrees of freedom of the numerator. Third argument is the degrees of freedom of the denominator.
%gd	2	Gamma distribution deviate (inverse of the incomplete gamma function). First argument is the probability p. Second argument is a.
%nd	1	Normal distribution deviate. $K = -\infty$. First argument is the probability p. See note (b).
%pd	2	Poisson distribution deviate. First argument is the probability p. Second argument is the mean of the Poisson distribution.
%td	2	t distribution deviate. $K = -\infty$. First argument is the probability p. Second argument is the degrees of freedom.

14.6.3 Logical functions

Function	Number of arguments	Result
%eq	2	1 if x = y, 0 otherwise
%ge	2	1 if $x \geq$ y, 0 otherwise
%gt	2	1 if x > y, 0 otherwise
%if	3	if x is non-zero then y, otherwise z. See note (k).
%le	2	1 if $x \leq$ y, 0 otherwise
%lt	2	1 if x < y, 0 otherwise
%ne	2	1 if $x \neq$ y, 0 otherwise

14.6.4 Structure functions

Function	Number of arguments	Result	Notes
%dim	2	The length of the yth dimension of x	(r)(v)
%in	2	The position of the identifier x in the list y	(g) (r) (s)
%len	1	The length of the identifier	(r) (u)
%lev	1	The number of levels of the identifier	(r) (u)
%match	2	Matches the contents of the source macro x to the target macro or string y	(r) (t)
%ref	1	The reference category of the identifier	(r) (u)
%typ	1	The type of the identifier	(r) (u)

Notes

(a) Care must be taken by the user to ensure that argument value(s) are not so large (or small) as to produce floating-point overflow [16.4].

(b) If an argument is not within the valid range, as defined in these notes, then the result is zero and a warning message is printed.

(c) $0 \leq x \leq 1$; the result is in radians.

(d) x must be positive for `%log` and non-negative for `%sqrt`.

(e) x is in radians.

(f) returns i for the ith unit of the calculation.

(g) If the identifier is not present in the list, the value 0 is returned.

(h) If $x \leq 0$ then the result is a uniform pseudo-random number between 0 (inclusive) and 1 (exclusive); otherwise the result is a uniform pseudo-random number between 0 (inclusive) and $(x + 1)$ (exclusive) and rounded down to the nearest integer.

(i) The ith value is $\Sigma_{j=1}\, x_{j,}$ where x_j is the jth value of the *function* argument.

(j) The result is always rounded towards zero.

(k) Both y and z are evaluated, regardless of the value of x.

(l) The `%gl` *function* returns a *vector*, the ith value of which is defined as

$$\text{mod}(\ [\ (i - 1)/y\ ], x\ ) + 1$$

where 'mod' is the usual modulus function and '[ ]' denotes the integer part; the length of the *vector* is determined by the context [15.8].

(m) The valid range for the exponential function is $-\infty < x \leq K$, where K=log(max_real).

(n) The valid range for the sine and cosine functions is $-L \leq x \leq L$, where L is machine dependent.

(o) The valid range for the log gamma function is $0 < x \leq M$, where M is machine dependent. The algorithm used is Macleod (1989).

(p) The valid range for the trigamma function is $P < x \leq Q$, where P and Q are machine-dependent constants. The algorithm used is Schneider (1978) modified by Francis (1991).

(q) Result is 1 for positive or zero x, and -1 for negative x.

(r) x must be an identifier and cannot be an expression or value.

(s) y must be an identifier and cannot be an expression or value.

(t) The match function provides fuzzy matching between strings and macros. The task of this function is to compare the contents of a macro x (the source) to target text supplied by the second argument y, and to report the level of the matching detected. We refer to the contents of the two arguments as the source and target strings, whether they are stored in macros or given directly as strings. The matching is not exact, but instead obeys the following rules:

- All leading and trailing spaces, tabs, and new lines in the source and target strings are ignored.
- Any collection of adjacent spaces, tabs, and new lines in either the source or the target string is replaced by a single space.
- Matching is not case-sensitive.
- If, after the above adjustments, suppose the source string contains k characters and the target string contains l characters. Various levels of matching can occur, and the level of matching affects the value returned by the function in the following way:

Condition	Value returned by `%mat`
The source string is the same length as the target string (k=l) and the source string matches the target string	3
The source string is longer than the target string (k>j)and the target string matches the first j characters of the source string	2
The source string is shorter than the target string (k<l) and the source string matches the first k characters of the target string	1
No matching is possible.	0

(u) The real values returned by the monadic structure functions are as follows:

Identifier type	Value returned by			
	`%len`	`%typ`	`%lev`	`%ref`
Variate	Actual length if factor has values, otherwise 0	11	0	0
Factor	Actual length if variatehas values, otherwise 0	17	Actual number of levels	Actual reference category or 1 if not defined
Macro	Length given by `$env d$` directive	33	0	0
List	Number of elements in list	39	0	0
Array	Actual number of values	As for factor or variate	Actual number number of levels	As for factor or variate
Scalar	1	1	0	0
Existing identifier but unknown type	0	0	0	0
Identifier which does not exist	-1	-1	-1	-1

(v) For the dyadic structure function `%dim`, if *x* is an array, returns the length of the *y*th dimension. If *y*=0, `%dim` returns the number of dimensions.

Identifier type of *x*	Value returned by `%dim (x,y)`		
	with *y*=0	with *y*=1	with *y*>1
Variate	1	length of *x*	0
Factor	1	length of *x*	0
Macro	1	length of *x*	0
List	1	length of *x*	0
Array	Actual number of dimensions	Length of first dimension	Length of *y*th dimension or 0
Scalar	1	1	0
Existing identifier but unknown type	0	0	0
Identifier which does not exist	-1	-1	-1

(w) The date-to-day function. Converts a date into the number of days since 13 September 1752. Day, month, and year are all rounded to the nearest integer.

Year should be specified as a full four-digit integer; that is, 1991 rather than 91. Day is specified as an integer between 1 and 31; month is specified as an integer between 1 and 12. The function checks for invalid combinations of day, month, and year, so, for example, `%day(29,2,1988)` is valid but `%day(29,2,1987)` is not valid. A zero value is returned for invalid combinations. Note that 13 September 1752 is the date after the last adjustment of the calendar.

14.7 Arrays

An *array* is a *vector* with a dimensional structure. It has attributes as follows:

(a) the number of dimensions;
(b) the number of levels;
(c) the reference category;
(d) the length or size of each dimension;
(e) the total length.

The total length of an *array* is equal to the number of *reals* that it can hold. An *array* is given its number of dimensions and lengths of dimensions (which must be positive) when it is defined via the `ARRAY` *directive*; the product of the lengths of each dimension must equal the total length. The total length of an array may change during the *array's* existence. There is no standard length for *arrays*. An *array* need not be given values when it is defined. However, when it is given values, the number of these values must equal the declared length.

Variates and factors may be redefined as arrays:

(1) If a factor is redefined as an array, it keeps its attributes, including the number of levels and reference category.

(2) If a variate is redefined as an array, then the number of levels is set to zero and the reference category is undefined.

For a two-dimensional array (a matrix) the length of the first dimension is the number of rows and the length of the second dimension is the number of columns.

When a vector is defined as an array, the vector's values are accessed so that the first dimension varies slowest, the second dimension the next slowest, and so on. This means that the contents of a two-dimensional array will be stored row by row.

In most directives, the dimensionality of the array is ignored, and the array is treated as a variate or factor, according to whether the number of levels is zero or greater than zero.

Arrays have special meaning in the following contexts.

- Output structures produced by the `TABULATE` *directive* are defined as arrays.
- Arrays are recognized by the `TPRINT` *directive*, and the lengths of the dimensions of an array are used to provide the shape information.
- Arrays are recognized and treated differently in model formulae and in the `PREDICT` *directive*.
- The `PASS` *directive* recognizes arrays, and a special option to the `PASS` *directive* allows arrays to be transposed so that they are stored in a manner compatible with the FORTRAN language.

There is one system array:

`%dm` can be extracted following a valid `TERMS` statement or after or during a fit. It contains the *d*esign *m*atrix (model matrix) generated by the current model formula. It has two dimensions, and is of size `%nu` × `%pl`.

14.8 Lists

A *list* is a list of *identifiers*. A *list* has one attribute, its **length.** A list itself has an identifier, referred to as the *list_identifier*.

Lists are defined by the user through the `LIST` *directive*. There are no system lists.

The length of a *list* is equal to the number of *identifiers* that it can hold. A *list* is given its length (which must either be positive or zero) when it is defined; this length may change during the *list's* existence. A *list* is always given values when it is defined. These values are assigned in one of two ways.

(i) They may be declared by the user and may be defined or undefined *identifiers*. The length of a list may be declared explicitly, in which case the number of identifiers must equal the declared length. If an identifier on the right-hand side is itself a list, then that list is expanded, and its identifiers form part of the new list. For example:

```
$list 3 L = X,Y,Z$
```

sets up a list `L` of length 3 taking as its three values the identifiers `X`, `Y`, and `Z`.

```
$list L2=A,B,L,C
```

sets up a list of length 6 taking as its six values the identifiers
`A, B, X, Y, Z` and `C`.

Alternatively,

(ii) Values can be generated automatically by GLIM. Each identifier in the list is generated by appending a integer (the position in the list) to the *list_identifier*. The *list_identifier* is truncated if necessary to ensure a set of unique identifiers. For example:

```
$list 3 L $
```

sets up a list `L` of length 3 taking as its three values the identifiers `L1, L2,` and `L3.`
The generated identifiers may or may not already exist, but if they do exist, none of the identifiers can be a list.

The *integer*th element of a list may be accessed anywhere in a GLIM statement by referring to it as *list_identifier* [*integer*]. For example, if the list `L` contains `A,B,C,` and `D`, then `L[4]` refers to the identifier `D`.

A list may be modified by using the + and - operators in the `LIST` *directive*. The + operator allows an identifier or the contents of a list to be added to the end of a list, and the - operator allows an identifier (not a list) to be removed from a list. In the latter case GLIM searches backwards from the end of the list and removes the first element found which matches the specified identifier. All elements to the right of this element are shifted one place to the left.

A list may be null, in which case it contains no identifiers.

If an element of a list is deleted, the list itself is also deleted. Deletion of a list does not however cause the deletion of the identifiers in that list.

Every identifier in a list (but not the list identifier itself) may be deleted by using the `TIDY` *directive*.

14.9 Internals

For housekeeping purposes, the system maintains certain private data structures, known as *internals*. Since they are stored in workspace they occupy space that would otherwise be directly available to the user. The user has no access to these structures but the `S` option to the `ENVIRONMENT` *directive* prints details about their space usage.

14.10 Subfiles

Subfiles are portions of external files. Each *subfile* begins with a source line that has a `SUBFILE` *statement* as its first non-space *token* and ends at the next `RETURN` or `FINISH` *statement*. Note that *subfiles* do not have to be disjoint (that is, the current subfile does not need to be terminated by a `RETURN` statement before the next `SUBFILE` *statement* is encountered), nor does any part of a file have to belong to a *subfile* (that is, a portion of a file may belong to zero or more *subfiles*). For example, a file may be divided as

source lines — not in a *subfile*

`$subfile a`

source lines — in *subfile* `a`

`$subfile b`

source lines — in *subfiles* `a` and `b`

`$return`

source lines — not in a *subfile*

`$subfile c`

source lines — in *subfile* `c`

`$finish`

As shown in the above example, the *directive_name* `$subfile` is followed by an *identifier*, known as a *subfile_identifier*. This *identifier* is used by the system solely to locate the *subfile* within the file, following an `INPUT` or `REINPUT` *statement* [15.42] [15.70]. It is immaterial whether or not the *subfile_identifier* is also currently defined as a *structure_identifier*, since its use as a *subfile_identifier* is independent of its other uses. Moreover, more than one *subfile* may have the same *identifier*, though this will not usually be required in practice. *Subfile_identifiers* can be passed as *actual_arguments* to *macros*.

15 The directives

This chapter defines the syntax and semantics of each directive in GLIM, and in some cases gives examples of their use. It presupposes the terms and definitions given in [13] and [14].

The underlining within directive names in the syntax definitions indicates their minimum abbreviation [13.2].

15.1 The `ACCURACY` Directive

Syntax: `$accuracy` [*integer*1 [*integer*2]]

Semantics: Sets the accuracy of *reals* output by directives and sets the number of decimal places. If *integer*1 is $n1$, say, then each directive will ensure that, within each set of reals that it prints, the smallest real will be printed with $n1$ significant figures. If *integer*1 is 0 or is omitted, the default value 4 is used; if *integer*1 > 9, the value 9 is used.

If *integer*2 is present and is $n2$, say, then if the number of decimal places printed would have been d, and $n2<d$, the number of decimal places printed are further restricted to $n2$. If $n2 \geq d$ then *integer*2 has no effect.

Until this directive is used, the default values are in force. The current value of *integer*1 is stored in the system scalar `%acc` and that of *integer*2 in the system scalar `%ndp`.

Examples:

```
$accuracy 6  3$      ! sets future output to have
                     ! at least 6 significant
                     ! figures, with no more than
                     ! 3 decimal places
$ac 20$              ! sets 9 significant figures
```

15.2 The `ALIAS` directive

Syntax: `$alias` [$\left\{ \begin{matrix} \texttt{on} \\ \texttt{off} \end{matrix} \right\}$]

Semantics: Alters the state whereby rows for intrinsically aliased parameters are or are not excluded from the formation of the weighted SSP matrix or its equivalent in subsequent fits. If the setting is on, then intrinsically aliased parameters are excluded. If the setting is off, then intrinsic alias detection is disabled and it is left to the algorithmic method to detect both extrinsic and intrinsic aliasing. A declaration persists until the next `NEWJOB` or `ALIAS` *statement*. At the beginning of a job it is assumed that such rows will be excluded.

Example: `$alias on$    ! excludes intrinsically`
`                ! aliased parameters.`

15.3 The ARGUMENT directive

Syntax: `$argument` *macro items*

where *item* is *identifier* or `%`*digit* or `*` or `-`, and the number of *items* ≤ 9.

Semantics: An *identifier* may be a macro, list, scalar, vector, subfile, or function. An *identifier* may also be a formal argument (that is, `%`*digit*) or a keyword argument of another macro. It must not be an integer, a real, a constant, or an expression. Undefined identifiers may appear in an argument list, and their types will become defined when first used as actual arguments in the macro.

If the *i*th item is *identifier*, then this directive replaces the *i*th argument of *macro* by *identifier*, that is, future occurrences in *macro* of the string `%`*i* are to be replaced by *identifier*. If the *i*th *item* is `%`*digit*, then future occurrences in *macro* of `%`*i* will be replaced by the *digit*th argument of the macro, using the immediately preceding level of the program control stack at the time of replacement; a fault occurs if the system or an input channel is using this level. If the *i*th *item* is `*`, the *i*th argument remains unchanged.

If the *i*th argument is a minus sign then this indicates that the *i*th argument and all subsequent arguments (that is, from the (*i*+1)th to 9th arguments) will be unset.

Arguments may be set or unset at any level of the program control stack whether or not the macro is on the program control stack. If the macro is at the current level of the program control stack (that is, it is currently being executed), then the directive additionally recalculates the system scalars `%a1`,`%a2`,...,`%a9` and `%at`.

Examples:

```
$argument M  A  B   ! sets A as first and B as
                    ! second argument of M; %3
                    ! to %9 are unchanged.
$a  M * %log  %2    ! sets second argument to
                    ! %log, and third argument
                    ! to the second argument of
                    ! the macro that invokes M;
                    ! other arguments are
                    ! unchanged
$arg M  D,*,-       ! sets first argument to D,
                    ! keeps the second argument
                    ! set to %log and unsets
                    ! all other arguments.
```

15.4 The ARRAY directive

Syntax: $array *identifier integer* [, *integer*]*s*

Semantics: Defines the *identifier* to be an array of *k* dimensions, where *k* is the number of *integers* in the list following the identifier. The *i*th dimension of the array is defined to be of size *integer*i, where *integer*i is the *i*th *integer* in the list. For a two-dimensional array, the first dimension will represent the number of rows, the second dimension the number of columns. All *integers* must be positive and non-zero.

Arrays are recognized only in model formulae and by the PASS and TPRINT directives. They are also produced as output by the into phrase of the TABULATE directive. In all other directives, arrays will be treated as a variate or factor with length equal to the product of the dimensions.

If *identifier* exists and contains values, then it must be a variate or factor. The product of the *k integers* must equal the length of the identifier. If *identifier* does not exist, then it will be created as a variate array.

Example:

```
$assign F=1...20$   ! defines F as a variate of
                    ! length 20
$array  F 5,4       ! redefines F as a two-
                    ! dimensional 5x4 array
                    ! with 5 rows and 4 columns.
$calc F=F/2         ! divides each element of F
                    ! by 2
```

15.5 The ASSIGN directive

Syntax:

$$\text{\$assign } \mathit{vector}1 = \left\{ \begin{array}{c} \mathit{vector} \\ \mathit{real} \\ \mathit{assign\_sequence} \end{array} \right\} \left[[\,,\,] \left\{ \begin{array}{c} \mathit{vector} \\ \mathit{real} \\ \mathit{assign\_sequence} \end{array} \right\} \right] s$$

or

$assign *scalar* =*real*

where

assign_sequence is *real*1 [,] . . . [,] *real*3
or *real*1 [,] *real*2 [,] . . . [,] *real*3

Semantics: **First syntax.** Assigns to *vector*1 the concatenated list of values given by the right-hand side (RHS) of the expression. The length of *vector*1 will become the length of the concatenated values. If *vector*1 is undefined it will become defined as a variate with the given number of values. Each *vector* must have values, except in one special case described below.

Assign_sequences provide a method of specifying regular sequences. If an *assign_sequence*, which has precedence in the syntax, appears in the RHS of the expression, it is interpreted according to the following rules. *Real*1 and *real*3 define the starting value and finishing value of the sequence, with default *step* interval 1.0 if *real*3 > *real*1 and -1.0 if *real*3 < *real*1. If *real*2 is specified, then the *step* interval is taken to be the difference between *real*2 and *real*1. The sequence is defined as:

*real*1, *real*1 + *step*, *real*1 + 2**step* , ..., *real*1 + k**step*, *real*3

where $k \geq 0$ and *real*1 + *k***step* < *real*3 for positive *step*
*real*1 + *k***step* > *real*3 for negative *step*.

System vectors that take their length from the standard length or number of units cannot have their lengths changed by this directive. Following a model fit, if `ASSIGN` changes the length or contents of any identifier used in the current model, then display and extraction of model structures is inhibited. During a user-defined fit, `ASSIGN` may change the contents, but is not allowed to change the length of a vector involved in the model fitting process.

Normally, all *vectors* on the RHS of the `ASSIGN` directive must exist and have values. However, in the special case where a *vector* on the RHS is followed by further values, and *vector identifier* is *vector*1, that is, used as the destination vector on the left-hand side of the expression, then the *vector identifier* need not exist and/or have values. Any occurrence of *vector* on the RHS will in this special case be ignored and the vector length taken as 0.

Second syntax. Assigns to *scalar* the value of *real.*

Examples:

```
$assign X=1.9, 3, 6      ! X will have length 3
:       X = X, X, %a     ! X will now have length 7
$assign A=1,3,...,9      ! A contains 1,3,5,7,9
$assign B=2  ...  7      ! B contains 2 3 4 5 6 7
$assign C=20,...,80      ! C contains the values
                         ! 20,21,22,23, up to 80
$assign D=2 4... 11      ! D contains 2 4 6 8 10 11
$assign E=-5 ... 4       ! E contains -5 -4 -3 -2 -1
                         !        0 1 2 3 4
$assign F=0.1 ...0.9! F contains 0.1  0.9
$del N $
$assign N=N,1             ! N contains 1
```

15.6 The BASELINE directive

Syntax: `$baseline` [*real1* [*real2*]]

Semantics: Sets the baseline deviance and degrees of freedom, which provides for a constant to be added to either or both of the deviance and degrees of freedom before printing. The baseline deviance is added to the deviance produced by the fit, and the baseline degrees of freedom are added to the residual degrees of freedom (but not to the number of units weighted into the fit `%nin` and not to the total number of units `%nu`). The resultant deviance and degrees of freedom are printed out as specified by the `CYCLE` and `RECYCLE` directives during and after a fit and by the `D` option to the `DISPLAY` directive.

If a baseline setting is in operation, then the message '`with baseline adjustment`' is printed following the deviance. The current setting of the baseline directive may be obtained via the `M` and `D` options to the `DISPLAY` directive.

If *real*1 is present, then the baseline deviance is set to *real*1, otherwise it is set to the value zero. *Real*1 can be positive or negative.

If *real*2 is present, the baseline degrees of freedom is set to *real*2, otherwise it is set to zero. *Real*2 may be negative.

The setting remains in force until the next `NEWJOB` or `BASELINE` statement.

The default values for the baseline values at the start of a job are zero.

Examples:

```
$baseline 62.5$
    ! sets the baseline deviance to 62.5, and
    ! the baseline degrees of freedom to zero.
$baseline $
    ! Sets both baseline values to zero.
```

15.7 The BRIEF directive

Syntax: `$brief [ {on | off} ]`

Semantics: The `BRIEF` directive is used to control the level of information which is provided following a GLIM fault. Following a fault, a fault message is displayed and an optional help message depending on the setting of the `BRIEF` directive.

If the setting of the `BRIEF` directive is `on`, then the brief switch is set, and a fault message but no help information will be provided. If the setting is `off`, then both a fault and a help message are provided. Giving a new value to the BRIEF setting will also affect the appearance of help messages in the transcript file, if one exists. If the brief setting is turned on by the BRIEF directive, then the help output source records are included in the transcript output by default; if turned off, then they are excluded by default.

The BRIEF directive has no effect in *batch* mode.

Example:
```
$brief on$        ! asks for brief, not detailed
                  ! fault messages.
```

15.8 The CALCULATE directive

Syntax: `$calculate` *arithmetic_expression*

where *arithmetic_expression* is an expression of the form:

(1) simple operand: an unsigned *real*, *scalar* or *vector*. In addition, the *system internals* %wtd, %wvd and %tri are also simple operands.

(2) *monadic_operator operand*

where monadic_operator is – or /

(3) dyadic expression of the form:

*operand*1 *operator operand*2

where *operator* is one of

```
+  -  *  /  **   >  >=  <   <=  ==  /=
&  ?   or   =
```

(4) function_identifier (operand1[,operand]s)

where *function_identifier* can be either:

(i) arithmetic function

(ii) structure function, when *operand*1 must be an identifier and other *operands* are simple operands;

(iii) string function, when *operand*1 is a *macro identifier*, and other *operands* are *strings* or *macros;*

(5) indexed vector: a *vector identifier* followed by an *expression* enclosed in matching brackets.

Notes:

(a) Operands other than simple operands can be expressions.

(b) Functions with a single argument do not require the brackets.

Precedence rules The order in which operations are performed are determined by the following rules.

(a) Expressions within brackets are performed first. If brackets are nested the inner-most expression is performed first.

(b) Expressions with multiple operators of the form:

*operand*1 *operator*1 *operand*2 ... *operand*(*n*)

have the following rules:

(i) If any of the operators is an assignment the expression to the right of the assignment operator is evaluated first.

(ii) For any other operators the operator with the highest precedence (as defined in the table below) is performed first. If operators have equal precedence the left-most operator is performed first.

Precedence table

	**	/ *	- +	< <= > >= == /=	&	?	=	
high	1	2	3	4	5	6	7	low

Length rules Each expression has a length; operands within an expression may or may not define a length.

(1) Simple operands: *reals* and *scalars* do not define a length. If a *vector* has a length this defines the operand length. A *vector* without a previously defined length may only occur, as a simple operand, as the left-hand operand for the assignment operator, when it takes the length of the expression or, if the expression does not otherwise define a

length, it takes the standard length [3.1.2] which also becomes the *operand* length.

(2) Negation and not: the length of these operands is the length of the operand which follows the minus sign or /.

(3) A dyadic expression takes the length of its operands.

(4) (i) Arithmetic function: the length of an arithmetic function is the length of its argument(s).

(ii) and (iii) Structure or string function: As these return a scalar value, they do not define a length.

(5) Indexed vector: the length of this operand is the length of the index. The vector must have a length unless this operand is the left-hand operand of an assignment operator, in which case, if the vector has no length, it takes the length of the expression or, if the expression does not define a length, the standard length.

If an operand defines a length then all other operands defining a length must define the same length, and operands not defining a length take this length, which then becomes the length of the *arithmetic_expression*. If no operands define a length then the *arithmetic_expression*, and all operands within it, have length one.

Semantics: The `CALCULATE` directive evaluates the *arithmetic_expression*.

Mode of evaluation If the length of the *arithmetic_ expression* (as defined above in the length rules) is n ($n \geq 1$) then the *arithmetic_expression* has n values, produced by n evaluations, the ith evaluation ($1 \leq i \leq n$) producing the ith value.

Values of expressions

(1) Simple operand: the ith value of a real is the real. The ith value of a scalar is the value of the scalar at the start of the ith evaluation. The ith value of a vector is the value of the ith element of the vector at the start of the ith evaluation. Furthermore, if the operand is a scalar or vector and is the left-hand operand of an assignment *operator* then the ith value of the right-hand operand is assigned to the scalar if it is a scalar, or to the ith element of the vector if it is a vector.

(2) Negation: the ith value of this operand is the negation of the ith value of the operand following the minus sign.
Not: the ith value of the operand is 1 or 0, as the ith value of the operand following the / is zero or non-zero respectively.

(3) If the expression is a dyadic expression of the form

*operand*1 *operator* *operand*2

then the table below gives the value of the expression for each *operator*, where x and y stand for the ith values of *operand*1 and *operand*2, respectively:

Operator	Result	Notes
+	the sum $x + y$	(a)
-	the difference $x - y$	(a)
*	the product xy	(a)
/	the quotient x/y	(a) (b)
**	x raised to the power y	(a) (c)
>	1 if $x > y$, 0 otherwise	
>=	1 if $x \geq y$, 0 otherwise	
==	1 if $x = y$, 0 otherwise	
<	1 if $x < y$, 0 otherwise	
<=	1 if $x \leq y$. 0 otherwise	
/=	1 if $x \neq y$, 0 otherwise	
&	1 if x and y non zero, 0 otherwise	
?	1 if x or y non zero, 0 otherwise	
=	y	(d)

Notes:

(a) Care must be taken by the user to ensure that the values of x and y are not so large (or small) as to produce floating-point overflow [16.4].

(b) If y is zero then the result is zero and a warning message is printed.

(c) The definition holds when $x > 0$ for all y, or when $x < 0$ and y is integral; or when x equals 0 and y is a positive integer; otherwise the result is zero and a warning message is printed.

(d) As a side-effect, the value y is assigned to *operand*1 (which must be a scalar or vector identifier).

(4) (i) Arithmetic function: the *i*th value of a function is defined for each function in [14.6], where the value(s) of the argument(s) is/are taken as the *i*th value(s) of the argument expression(s).

(ii) and (iii) Structure or string function: The *i*th value of a structure or string function is the scalar value defined at the start of the first evaluation. All structure and string functions are evaluated once before the first evaluation, and the resulting scalar value is then used as in (1).

(5) Indexed vector: let the *i*th value of the index expression, when rounded be *j* (where *j* must lie between 0 and the length of the indexed vector, inclusively). If *j* is positive the *i*th value of the operand is the *j*th value of the indexed vector, while if $j = 0$ the *i*th value is zero. Furthermore, if this operand is the left-hand operand of an assignment operator then:

(i) if the indexed vector does not exist, or exists but does not have values, then all values of the indexed vector are initialized to zero before carrying out any evaluations. If the indexed vector exists and has values, no such initialization occurs.

(ii) if *j* is positive, the *j*th element of the indexed vector is assigned the *i*th value of the right-hand operand while, if $j = 0$, no action is taken and the result of the operand is discarded.

Precision Evaluation of all expressions use the highest precision available in the local implementation of GLIM. Assignments to vectors or scalars will use the standard precision used for storing these quantities.

Output of values If the *arithmetic_expression* does not contain an operator or if the last operation of the evaluation is not an assignment then the values of the *arithmetic_ expression* are printed, one to a line, on the current output channel, provided that channel has been set. The accuracy with which the values are printed can be altered by the `ACCURACY` directive.

Examples:

```
$calculate 192/3$               ! prints 64
$calculate V=A+B/2$
$factor F 3$
$calculate %n=%lev(F)$
$calculate %p=%chp(%dv,%df)
$calculate A=%typ(2)            ! generates an error
$calculate B=%len(2+6*A)        ! generates an error
```

```
$calculate Z=%exp(600)
      ! generates a warning on most systems and
      ! sets the values of Z to zero.
$slength 10$del x$
$calc x(2)=5$
      ! sets the 2nd element of X to 5 and all
      ! other elements to zero
$ca %a=(%in(A,L)==%len(L))
      ! %a is 1 if A is the last element in the
      ! list L, 0 otherwise.
```

15.9 The CLOSE directive

Syntax: `$close` *file_specifier*

Semantics: The directive provides a mechanism for closing files from within GLIM. This allows channels to be reused, and is helpful in environments where there is a system restriction on the total number of channels which can be opened.

The *file_specifier* will be closed. If the channel *integer* or file *filename* is not open, the directive will be ignored.

The primary input and output channels and the transcript channel and the journal channel cannot be closed. The CLOSE directive is also useful in the following specific circumstances:

a) Terminating a graphics output file, before a different file is opened on the same unit for further graphics output;

b) If the PAUSE directive is implemented, then an input file may be closed, edited by calling the system editor via the PAUSE directive, and reinputted to GLIM.

Examples:

```
$calc %a=16$
$close %a$
    ! will close channel 16, if open
$close 'data13'
    ! will close the file data13, if open
```

15.10 The COMMENT directive

Syntax: `$comment` *characters*

where *characters* must not include the *directive_symbol*.

Semantics: Introduces a comment, which ends at the next *directive_symbol*. The *characters* are ignored by the program, except that any *macro_substitutions* are executed [13.4.2].

Example: `$c  this is a comment$`

15.11 The CYCLE directive

Syntax: `$cycle` [*integer*1 [*integer*2 [*real*1 [*real*2]]]]

Semantics: Controls the algorithmic aspects of model fitting. If *integer*1 ≠ 0, then iteration continues until convergence or for *integer*1 cycles, whichever occurs first; if *integer*1 = 0 or is omitted then a value of 10 is used. If *integer*2 ≠ *0* then printing occurs every *integer*2 cycles and at the end; if *integer*2 = 0 or is omitted then printing occurs at the end of iterations only. *Real*1 is used to decide when convergence has occurred; it is initially set to 1.0 e-4; increasing/decreasing its value will tend to decrease/increase the number of cycles needed before convergence is declared; if omitted then the default setting is used. *Real*2 is used in the detection of extrinsic aliasing; it is initially set to 1.0 e-10; increasing/decreasing its value will tend to increase/decrease the number of parameters declared to be aliased; if omitted then the default setting will be used.

Use of this directive implies that either the initial values variate, if set with the INITIAL directive, or the observations are to be used as initial estimates of the fitted values when fitting standard models (compare RECYCLE). If the initial values variate is set to %fv, either explicitly or implicitly via the RECYCLE directive, then the CYCLE directive has the effect of unsetting the initial values variate, with the effect that the initial estimates of the fitted values will be calculated by default from the *y*-variate.

If the initial values variate is set to any other vector, or is indexed, or is unset, then the CYCLE directive has no effect on the initial values.

A CYCLE statement remains in force until the next CYCLE, RECYCLE, or NEWJOB statement.

The system scalars %cyc, %prt, %cc and %tol [14.2.2] hold the current values of these four settings. %itn holds the iteration number, and %cv the convergence state.

Examples:

```
$cycle   4  1
    ! sets a maximum of 4 cycles with printing
    ! each cycle; other 2 settings are
    ! unchanged
```

```
$cy %cyc %prt 1.e-6
    ! sets convergence criterion to 1.e-6;
    ! other settings are unchanged.
```

15.12 The DATA directive

Syntax: <u>$da</u>ta [*integer*] { *identifiers* / *list* }

Semantics: Defines the set of *identifiers* whose values will be read and optionally stored by the next READ statement(s). If a *list* is given, then the *list* provides a set of identifiers, which are then processed as though the *identifiers* themselves had been given. The *identifiers* must be either existing vectors or be undefined, the latter becoming defined as variates. The optional *integer* defines the length of undefined identifiers; if absent, undefined identifiers take the length of the first *identifier* which, if also undefined, takes the standard length. All vectors must have the same length. The set of identifiers defines the *data list*; subsequent READ or DINPUT statements will place values into all, or a subset, of identifiers in the data list.

A DATA statement remains in force until the next DATA or NEWJOB statement or until one of the *identifiers* is deleted. The current *data list* may be examined via the F option to the ENVIRONMENT directive.

Examples:

```
$data 20 A B C D$
    ! defines a data list to contain four
    ! identifiers: A, B,C and D
$delete A$
    ! data list is abolished.
```

15.13 The DEBUG directive

Syntax: <u>$de</u>bug [{ on / off }]

Semantics: The DEBUG directive is used to control the debugging of macros to the primary output channel. If the setting of the DEBUG switch is on, then debugging of macros to the primary output channel will be enabled. If the switch is off, then such debugging is disabled.

If debugging is enabled, then this will have no effect until the user starts to execute a macro, in which case GLIM will stop before each directive symbol to allow the user to examine the contents of data structures and to change values if necessary. A prompt will be given (<D>), and GLIM directives may then be issued by the user at the primary input channel. The user may proceed to the next directive symbol by using the NEXT directive;

if debugging is still enabled, then GLIM will again cease execution, and so on. Issuing the statement `$debug off$` will stop the debugging process and let macros execute normally.

When debugging is enabled and a macro is in the process of execution:

(1) Most GLIM directives are allowed but the following directives which change the level of the program control stack are not allowed: `DINPUT, INPUT, OUTPUT, WHILE, USE, SKIP, SWITCH, EXIT, SUSPEND, RETURN, REINPUT.`

(2) The macro editor cannot be used to change the contents of the currently invoked macro.

(3) Local identifiers to the currently invoked macro are available to the debugger. The `D` option to `ENVIRONMENT` gives appropriate names and lengths.

(4) The system scalars `%a1, ...,%a9,%at` relating to the currently invoked macro are available to the debugger.

(5) Debug output is treated as ordinary output and is copied to the transcript file unless the o transcript switch is off.

(6) Input directives issued while debugging are not copied to the journal file.

Examples:

```
$debug on       ! Enables echoing information to the
                ! screen and to the transcript file.
$use MAC1       ! start to use macro MAC1
                ! macro stops before first directive
$next           ! execute first directive in macro and
                ! stop before next directive
$next           ! etc...until
$debug off$
                ! Disables debugging and continues
                ! normal execution of macro.
```

15.14 The `DELETE` directive

Syntax: `$delete` *identifiers*

Semantics: Erases the values and attributes of *identifiers*, which must refer to macros, lists, scalars, or vectors.

If *identifier* is user-defined, then it is also removed from the directory and becomes undefined.

If *identifier* is a list, then the list loses its contents and becomes undefined; identifiers present in the list however do not lose their values.

If *identifier* is a macro, then the local identifiers to the macro, if any, will also be deleted; if one of its local identifiers is a macro, then any local identifiers to that macro will also be deleted, and so on. Deletion of a local identifier without the deletion of its macro will make the local identifier lose its values and type, but it is not removed from the local directory. A macro currently appearing in the program control stack cannot be deleted.

If *identifier* is system-defined then its values and attributes are lost. Its name is not deleted from GLIM's internal directory, but it will not appear in the output from the `D` option to the `ENVIRONMENT` directive. Deletion of a system or ordinary scalar has no effect. Deletion of an *identifier* which is undefined is ignored.

If an *identifier* appears in the data list, then the data list becomes undefined. If an *identifier* appears in a list, then the list is deleted. If an *identifier* appears as an actual argument to a macro, then that argument is unset. During the execution of a `FIT` statement for a user-defined model, no vectors, scalars, or macros associated with the model can be deleted. See also the `TIDY` directive.

Examples:

```
$delete X$          ! removes identifier X,
                    ! recovering its space
$list l=a,b,c$
$delete l$          ! removes identifier L
                    ! but not A,B or C
```

15.15 The `DINPUT` directive

Syntax: `$dinput` *file_specifier* [*integer*] [$\left\{ \begin{array}{c} \textit{list} \\ \textit{identifiers} \end{array} \right\}$]

Semantics: As for the `READ` directive, except that the values will be read from the next record of *file_specifier.* The record width is taken as *integer* (≥ 30 and ≤ 132) or, if it is omitted, as that of the primary input channel. The level of the program control stack is increased by one. When sufficient numbers have been read the level is decreased by one and the statement following the `DINPUT` statement is read.

Example:

```
$format free$
        ! no data list - free format.
$dinput 'ex.dat' A B C D$
```

```
! reads values from file ex.dat, storing
! them in A,B,C and D.
```

15.16 The DISPLAY directive

Syntax: `$display` *letters*

Semantics: Displays components of output from the previous fit in sequence given by *letters*. The valid options for *letter* are:

A As for option E but intrinsically aliased parameters are also listed, in the same manner as for extrinsically aliased parameters, but with the parameter number given as 0.

C The correlations of parameter estimates given by U.

D The (scaled) deviance and degrees of freedom. The change of deviance and degrees of freedom are also given if the model fitted by the most recent FIT directive started with an operator. Information on the baseline deviance and degrees of freedom is also provided.

E The parameter estimates and their standard errors, including extrinsically aliased parameters. The number of each parameter is also given, as is the (estimated) scale parameter, together with the eliminated model formula, if one has been set.

I Information on the algorithmic aspects of the current model. This comprises the current settings of the CYCLE, RECYCLE, and INITIAL directives, together with information on the model algorithm in use.

L The composition of the linear model formula as a sum of simple terms, and the current eliminated term if this facility is in use.

M The model specifications for the *y*-variate, error, link, prior weight (if any), offset (if any), together with information on the current setting of the SCALE directive followed by details on any macros defining a user-defined fit. This is followed by the output for option L.

O Lists the definitions of any orthogonal variate polynomials in the model. The definition is given in the form of a lower triangle containing for each orthogonal term, a set of coefficients in the untransformed variate.

P A listing of the current parameters in the model, in the order in which they will be fitted in GLIM (that is, the order in which option E will display them, and %pe will, if extracted, contain them).

R Parallel listing of *y*-variate, fitted values, and generalized residuals. For binomial error models the binomial denominator is also printed.

S The standard errors of differences of the parameter estimates listed in option U.

T Displays the working triangle.
If the method is Gauss–Jordan, contains the generalized inverse of the SSP matrix, printed as a lower triangle with rows ordered as parameters in option E and signs reversed, and a final line for the *y*-variate having the correct sign.
If the method is Givens, then the form of the working triangle is given in the Modelling Guide [11].

U As for option E but lines corresponding to aliased parameters are omitted.

V The variance–covariance matrix of the parameter estimates listed in option U.

W As for option R but if the vector %re has been assigned values then units corresponding to zero elements of %re are omitted.

The output from all options except I, L, M, and P is inhibited in the following circumstances:

(i) before the first fit;

(ii) by the use of YVARIATE, ERROR, LINK, SCALE, WEIGHT, OFFSET, LOAD, ALIAS, and BASELINE statements;

(iii) by any alteration to the values or the length of vectors or structures in the current model;

(iv) by the use of the TERMS or ELIMINATE directives to define a new model.

The output from all options except I, L, and M is inhibited in the following circumstances:

(i) by redefining the number of levels of a factor in the model;

(ii) by otherwise redefining the parametrization of the model (for example, by changing the order of an orthogonal polynomial by altering the value of a scalar);

(iii) by deleting one or more of the structures or vectors of the current model.

All options except `C`, `I`, `L`, `M`, `O`, `P`, and `T` assume the current value for the scale parameters as given by default or by the `SCALE` directive.

Examples:

```
$dis i $      ! displays information about the
              ! algorithm.
$dis er $     ! displays estimates and fitted values
              ! following a fit.
```

15.17 The DUMP directive

Syntax: `$dump` [*file specifier*]

Semantics: Dumps the current program state on the channel or filename specified by *file_specifier* or, if omitted, on the default dumping channel. The dump is placed after any previous dump unless a `REWIND` statement precedes its use. Dump files cannot be transferred across machine types.

Example: `$dump 'apr30-92.dmp' $`

15.18 The ECHO directive

Syntax: `$echo` [{ `on` / `off` }]

Semantics: Controls the state whereby lines from a secondary input channel are or are not printed on the current output channel. If no *phrase* is given, then the echoing state is reversed. If the echoing state is `off`, then no echoing takes place. The directive takes effect from the line following its occurrence. The echoing state is initially set to `off`. Lines read under fixed format are never echoed. A change in echoing state is automatically reflected in the output to the transcript file, with, for example, echoing output not being written to the transcript file if the echo state is `off`.

Example: `$echo on$`

15.19 The EDIT directive

Syntax: `$edit` [*integer*1 [*integer*2]] { *vectors* / *list* } *reals*

Semantics: If *integer*1 and *integer*2 are present, assigns *reals* to consecutive elements *integer*1 to *integer*2 of each *vector*. The number of *reals* must agree with the number of values implied by *integer*1 and *integer*2. If *integer*1 only is present, assigns *real* to element *integer*1 of each *vector*. If neither *integer*1 nor *integer*2 is present, assigns *reals* consecutively to elements 1 to *n* of each *vector*, where *n* is the length of the first *vector*.

If *list* is present, then all identifiers in the list must be *vectors*, which are then used as described above.

Examples:

```
$assign A=1,3,...9$
$calc B=2*A$
$edit 5 A B 23 $          ! makes the last
                          ! element of A and B
                          ! the value 23.
$var 6 C$
$edit C D 1 1 1 2 2 2$    ! assigns the same
                          ! values to C and D
```

15.20 The EDMAC directive

Syntax: `$edmac` *identifier*

Semantics: Initiates the macro line editor. If *identifier* exists, then it must be a macro, and the contents of *identifier* is used as input to the macro editor. If *identifier* does not exist, then *identifier* is created with no contents. The macro editor may thus be used for the creation of new macros as well as the editing of old macros.

The macro editor is a self-contained subsystem to GLIM, and editing commands have a distinct syntax. The macro editor is available from any input source and in either interactive or batch mode. All edit commands start on a new line. In interactive mode when input is from the primary input channel, the macro editor prompt * is displayed.

Line numbers are displayed at the beginning of each line. Any keyword arguments and local structures are enclosed in parentheses and located at the beginning of the macro, where the user has the ability to redefine, delete, or create new keyword arguments and local structures. The S command will interpret these lines as an option list, whereas the T command will treat them as text and store them as part of the macro contents.

A summary of the editing facilities follows.

- The **move** command moves the current line to the the line specified.
- The **relativemove** command moves the current line a relative distance forward or backward in the macro.
- The `N` command or 'return' key makes the **next** line the current line.
- The `P` command **prints** a range of lines.
- The `I` command **inserts** a line of text before the specified line.
- The `D` command **deletes** the specified lines of the macro.
- The `C` command **changes** the specified line, replacing it with the specified text.
- The `F` command **finds** text in a macro. There is an optional query search mode.
- The `R` command finds text in a macro and **replaces** it with new text. There is an optional query search and replace mode.
- The `S` command **saves** the changes to a named macro and exits from the macro editor.
- The `T` command **textsaves** the changes and exits from the macro editor.
- The `E` command **exits** from the macro editor and saves the edited macro under its default name.
- The `Q` command **quits** from the macro editor without saving any changes made to the macro.

A full description of the macro editor can be found in the User Guide [7.10]

15.21 The `ELIMINATE` directive

Syntax: `$eliminate` [*model_formula*]

Semantics: Specifies the eliminated model formula to be used in subsequent model fitting. The *model_formula* should reduce to a single term which would generate a set of mutually orthogonal columns in a model matrix. The *model_formula* must also not start with an updating operator, nor should it contain orthogonal polynomial terms. The eliminated *model_formula* is pre-fitted, and the resulting (scaled) deviance and degrees of freedom from any subsequent fits will be as though the eliminated *model_formula* had been fitted as an extra term in the terms model formula. The parameter estimates associated with the eliminated model formula are not displayed.

If *model_formula* is omitted, then any current eliminate model formula is abolished.

The eliminate setting remains in force until a new `ELIMINATE` or `NEWJOB` is issued, or until one of the identifiers in the model formula is deleted (this last action has the effect of abolishing the eliminate model formula).

Example: `$eliminate A.B $fit X+C$`

15.22 The `END` directive

Syntax: `$end`

Semantics: The `END` directive is obsolete in GLIM4, and has been replaced by the `NEWJOB` directive.

15.23 The `ENDMAC` directive

Syntax: `$endmac` (abbreviation: `$e`)

Semantics: Signifies the end of assignment of contents to a macro. See the `MACRO` directive.

15.24 The `ENVIRONMENT` directive

Syntax: `$environment` *letters* (abbreviation: `$env`)

Semantics: Gives information on the current state of the program. The following are valid options for *letter*:

`A` Gives information on *a*rguments to macros. This provides, for each macro, a list of the *keyword_arguments* available and the currently set *actual_arguments* If a macro has no *keyword_arguments* and no actual arguments set, then only the name of the macro is displayed.

C Lists the settings for input, output, and dump *c*hannels; for input, the primary, current, and macro library channels and their record widths are given; for output the primary output channel and the current channels for the various output types [13.6.1] are given, together with their record widths (and for ordinary output, the page height) and whether reproduced on the transcript file; the primary dump/restore channel is given. Finally, a list of the open secondary channels as recorded in the system vector `%oc` is given, together with their associated file names.

D Lists part of the current contents of the *d*irectory [14.1]; this includes currently defined system vectors that have values, and user-defined identifiers. The list will also include local identifiers of the current macro if this directive is called from the macro or from the debugger. In the '`type`' column, the type of the structure is given: '`undefined`' means that the name is known to GLIM (for example, an identifier in a list) but no type has yet been given to the identifier. In the '`levels`' column a positive entry indicates a factor. If an integer in parentheses follows the number of levels, then this indicates that a reference category has been explicitly defined for that factor. Entries in the '`length`' column indicate the length of the identifier as defined by the `%len` function in the `CALCULATE` directive [14.6.4] The '`space`' column indicates the number of workspace cells allocated to the identifier; zero indicates that an identifier has no values or contents. The '`space`' column is followed by two further messages if certain conditions hold:

(a) if the identifier has been declared as an array, then the declared length of each dimension is given (for example, `3 x 2 array` will appear against `XARR` if it has been declared by the statement `$array xarr 3,2$`);

(b) if a macro is in use or being debugged, then the message `local identifier` will appear against all local identifiers of the currently invoked macro.

The output is terminated by a message giving the total space usage of all local identifiers not relating to the current macro.

E Gives a description of the available *e*xternal facilities in the `PASS` directive [17], if implemented.

F Gives the identifiers in the current *data list* (as set by the DATA directive) if it is defined, and the current *f*ormat setting (as set by the FORMAT directive).

G Gives text information on the *g*raphical facilities The following information is produced:

- a list of valid devices, together with the GLIM device names, the channel number associated with the device, and an indication of whether the device is open;
- the current default graphical device;
- the current settings of the LAYOUT directive;
- an identification table which maps colours, line types and symbols to integers; these integers may be used in the GSTYLE directive to set pen styles;
- a list of the current pen styles; if no pen styles have been explicitly defined, the 16 default pen styles are displayed.

H Gives *h*igh-quality graphical output on the available colours, line types, and symbols, and the current graphical settings. The output is produced on the current graphical output device, as set via the SET directive or through the most recent use of the DEVICE option to the GRAPH or GTEXT directives. The output produced may differ from device to device.

I Gives information on the local *i*mplementation of GLIM: installation name, GLIM version number, local values for special symbols, and largest integer permissible in GLIM on the machine. Also gives the default names of the journal, transcript, and initialization files.

M Provides information on the *m*aximum and *m*inimum argument values to the GLIM functions which have a restricted domain.

P Lists the contents of the *p*rogram control stack [13.7.1] with two types of entry:

(a) input channel numbers and line lengths;

(b) macro identifiers, with an asterisk if invoked by the `WHILE` directive. Each entry is preceded by its stack level.

`R` Gives the current seed values for the standard pseudo-*r*andom number generator.

`S` Lists the allocation of *s*pace to internals [14.9]:

`constants`	storage of constants and scalars;
`user directory`	storage of user identifiers;
`entity table`	storage of the internal directory;
`model fitting`	storage associated with fitting;
`PCS (stack)`	program control structures.

`U` Lists the current *u*sage (and the total available) of the workspace, identifiers, vectors in the model, terms, and program control stack levels.

Example:
```
$env D U S $          ! displays information on the
                      ! directory, workspace usage and
                      ! system internals.
```

15.25 The `ERROR` directive

Syntax: `$error` *letter* [*identifier*1 [(*identifier*2)]]

Semantics: Declares the probability or error distribution to be used in subsequent fits. Valid options for *letter* are:

`B`	binomial;
`G`	gamma;
`I`	inverse Gaussian;
`N`	Normal;
`O`	`OWN` or user-defined probability distribution;
`P`	Poisson.

If *letter* = `G, I, N,` or `P`, there must be no *identifier*.

If *letter* = `B`, *identifier*1 will be expected to hold the values of the binomial denominator during a fit and *identifier*2 represents an optional index variate.

If there is no index variate, then *identifier*1 must be either a variate of length the number of units or previously undefined; if the latter, then it will become defined as a variate of length the number of units.

If indexing is being used, then *identifier*1 need not have the length of the number of units. *Identifier*2, if it is defined, must have the length of the number of units.

References to `%bd` will point to *identifier*1 and references to `%bi` will point to *identifier*2.

Deletion of the binomial denominator vector or its index leaves the error set as binomial but causes the binomial denominator or its index to become undefined. The `YVARIATE` directive may also unset the binomial denominator and its index, if set, where the new number of units differs from the length of the binomial denominator or its index, if set.

If *letter* = `O`, *identifier*1 must be present, and *identifier*2 with its parentheses must not be present. *Identifier*1 must be either an existing macro or previously undefined; if the latter, then it will become defined as a macro; it will be expected to hold GLIM statements which calculate values for the two system vectors `%va` (the variance function) and `%di` (the deviance increment) during a user-defined fit.

The declared `ERROR` option remains in force until redeclared or until the next `NEWJOB` statement. The default option is `N`. This directive resets the `LINK` and `SCALE` options to their default values for the declared error. There is no default link for an OWN error; the link is unchanged from the previous model if this option is specified. If there is no previous model, then the default link for a user-defined distribution is the identity link.

The system scalar `%err` contains the current setting of this directive.

Example:

```
$error B N              ! sets the error distribution to
                        ! be binomial with binomial
                        ! denominator vector N.The link
                        ! is set to the default link
                        ! (G=logit)
```

15.26 The `EXIT` directive

Syntax: `$exit` [*integer*]

Semantics: Exits unconditionally from a macro or input channel. If *integer* = $n > 0$, the current macro or input channel is left and the program control stack reduced by n. n must be less than the current control level, whose value is held in `%cl`.

If *integer* = 0 or is absent, the directive has no effect. See also the `SKIP` directive.

15.27 The `EXTRACT` directive

Syntax: `$extract` *extractable* [[`,`] *extractable*] *s*

where *extractable* is

```
{  %pe
   %vl
   %vc
   %va
   %di
   %dr
   %ft
   %wv
   %wt
   %se
   %al
   %dm
   %cd
   %lv
   %sb(term)  }
```

that is, a system *identifier*, optionally followed by a *term* in parentheses, and *term* is a model term present in the current model formula.

Semantics: Causes values to be extracted from internal structures relating to the current fit and assigned to the corresponding *identifiers*. . If the model has not been changed since the previous fit, the `EXTRACT` directive saves the specified quantities into their system *identifiers*. The system *identifiers* are available for use until the start of the next `FIT` directive, when they become deleted. Most of the system vectors are deleted immediately, although `%pe`, `%se,` and `%al` are not deleted until after the initialization macro, if it exists, has been called.

If identifier is `%sb` then a *term* must be specified, and a variate of length `%pl` identifying the position of that *term* in the parameter names will be produced in `%sb`.

User-defined fits may use the `EXTRACT` directive if the structure is available and is not in the process of being updated; the values obtained are then those for the current iteration. The initialisation macro may additionally use `%pe`, `%sb`, `%se,` and `%al` from the previous fit if they have been previously extracted.

See [14.3.2.2] for details of the values extracted.

Examples:

```
$extract %pe %se $          ! extracts the parameter
                            ! estimates and the
```

```
                                  ! standard errors into their
                                  ! system structure
                                  !
$fit 1+X*A$                       ! extracts an indicator
$extract %sb(X.A)$                ! vector with 1 where the
                                  ! equivalent element of %pe
                                  ! is a parameter belonging
                                  ! to the X.A term, and 0
                                  ! otherwise.
```

15.28 The FACTOR directive

Syntax: `$factor` [*integer*1] [*identifier integer*2 [(*integer*3)]]*s*

Semantics: Declares each of a set of identifiers to be a factor with *integer*2 levels and reference category *integer*3.

If *integer*2 > 0, the directive defines or redefines *identifier* to be a factor with *integer*2 levels. The optional *integer*3 which is enclosed in parentheses and which must satisfy the constraint 1 ≤ *integer*3 ≤ *integer*2, defines the reference category of the factor; if *integer*3 is omitted then the reference category is set or reset to 1 (the first category).

The optional *integer*1 defines the lengths of undefined *identifiers*; if omitted or zero, undefined *identifiers* take the length of the first *identifier* which, if undefined, takes the standard length. All *identifiers* must either be undefined or vectors of the same length.

If *integer*2 = 0 or is omitted, it defines *identifier* to be a variate.

Example:

```
$factor B 3 C 4(3)$
       ! defines two factors. B has 3 levels,
       ! C has 4 levels
       ! with reference category 3.
```

15.29 The FAULT directive

Syntax: `$fault` *scalar* { *string* / *macro* }

Semantics: The FAULT directive provides a facility to generate a GLIM fault using the standard error handling mechanism [16.2].

If *scalar* is zero, the directive has no effect.

If *scalar* is non-zero, then after the successful completion of the FAULT directive, the following fault message is displayed

```
** user generated error, at
```

The usual GLIM fault message is then produced.

If *scalar* is followed by a *string* or *macro*, then the *string* or the contents of *macro* provide an optional help message which may be printed following the fault message according to the setting of the `BRIEF` directive. The usual GLIM fault handling procedure is then followed.

Example:

```
$number A
$calc A= (%len(M)<=0) ? (%typ(M)/= 33)
$fault A 'macro M is not defined'$
```

15.30 The `FINISH` directive

Syntax: `$finish`

Semantics: Acts as an end-of-file marker when searching for subfiles [14.10]. It also has the same effect as the `RETURN` directive, although it cannot be used on the primary input channel. The directive name must be the first directive name on its line.

15.31 The `FIT` directive

Syntax: `$fit` [*model_formula*]

Semantics: Causes the model to be fitted as specified by *model_formula* and the most recent `TERMS`, `ELIMINATE`, `SET`, `ERROR`, `LINK`, `YVARIATE`, `WEIGHT`, `OFFSET`, `INITIAL`, `LOAD`, and `SCALE` statements (or their default options). Computational aspects are controlled by `CYCLE`, `RECYCLE`, `METHOD`, `BASELINE`, and `ALIAS` directives. The directive sets the values of various system scalars [14.2.2] and system vectors [14.3.2]. The following items may occur in *model_formula*:

operators	.	/	*	,	+	-	-/	*/
brackets	(	)						
angle brackets	<	>						
operands	*vectors*	*integers*	1					

The *model_formula* is stored by a `FIT` or `TERMS` statement and may be updated by beginning the model formula of the next `FIT` statement with an operator, which implies that the new model formula is appended to the old one. If the new model formula contains only an operator, the previous model formula will be unchanged. The absence of *model_formula* implies the null model (that is, by default, the constant term only or the most recent setting of the `SET` directive). If a vector or scalar in *model_formula* is subsequently deleted then *model_formula* is replaced by the null model for later updating. If a fault occurs in the specification of *model_formula*, then the previous *model_formula* remains unaltered.

See the detailed specification of model formulae in the User Guide [8.2].

15.32 The FORMAT directive

Syntax: `$format` [{ `free` / *items* }]

Semantics: Declares the format to be used in subsequent READ statements. The *items* must constitute a valid FORTRAN format statement, except that if the outer pair of parentheses is omitted these will be supplied by GLIM. If FREE is used, or if both FREE and *items* are omitted, then free format will be used for the next READ or DINPUT statement; otherwise fixed format using *items* will be used.

If fixed format is specified, then *items* are syntax-checked. F, E, and G field specifiers are allowed, together with X and T and the new line field specifier. The I field specifier will be translated into an F field specifier; for example, I2 will be translated to F2.0. All other field specifiers will generate a syntax error, and the format setting will revert to FREE. The total number of translated non-space characters in the format must not exceed 238.

The current format setting is available through the statement `$environment f$`

The directive remains in force until redeclared or until a NEWJOB directive, when the default free-format option is assumed.

Example:

```
$format (4f1.0,2f2.0)
        ! declares the format to be the specified
        ! fixed format
```

15.33 The GET directive

Syntax: `$get` [(*option list*)] *identifier*

The options available are:

`prompt` = { *string* / *macro* }

`type` = *name*

where *name* is `scalar` or `macro` or `identifier`, and *name* can be abbreviated to the first character.

Semantics: The `GET` directive provides a mechanism for macros to prompt the user for an *integer* or *real* value, for a text *string,* or for an *identifier.* The directive is usually used from within a macro, but this is not mandatory.

If the program is in interactive mode, then the prompt *string* defined by the `prompt` option is displayed on the primary output channel and either a *real*, a *string,* or an *identifier* (depending on the setting of the `type` option) will be expected to be read on the primary input channel.

If the program is in batch mode, no prompt string will be displayed.

If `prompt` is a *macro,* then the contents of the *macro* will be taken as the prompt string. If `prompt` is *string,* then that will be taken as the prompt string. If no `prompt` option is given, a default prompt will be displayed according to the `type`.

If `type` is `scalar`, then *identifier* must be a scalar identifier. If *identifier* is undefined, then *identifier* is created with type *scalar.* The user is subsequently prompted to input a scalar value. The default prompt is '`scalar?`'.

If `type` is `macro`, then *identifier* must be a macro if it exists; otherwise it will be created with type *macro.* The user is subsequently prompted to input a *string* which will define the contents of *identifier.*

The string may be input in one of two ways:

(a) If the quote symbol is the first non-space character found, then the string input follows the usual rules for string definition [13.2.3], and may extend over several lines of input.

(b) If a character other than a quote symbol is the first non-space character found, then the string is terminated by the end of the line. The length of the string is then limited by the width of the primary input channel less the number of characters in the prompt string.

The default prompt is '`text?`'

If `type` is `identifier`, then *identifier* must be defined and must be a *keyword_argument* or *formal_argument*. The user is subsequently prompted to input an identifier name which will set the *actual_argument*, and will thus be associated with the *keyword_argument* or *formal_argument*. This provides a mechanism both for dynamic assignment of arguments and reassignment of arguments within a macro. The default prompt is '`identifier?`'.

If no `type` option is present, the default type is taken to be `scalar`. The `GET` directive temporarily overrides any disablement of output to the primary output channel set by a previous `OUTPUT` directive.

Example:

```
$mac (loc=name,age) trivial
$get (type=macro prompt='What''s your name')
    name
$get (prompt='What about your age') age
$pr 'Your name is ' name ' and you are ' *i age
    ' years old.'$
$endm

? $use trivial
What's your name? Tom Payne
What about your age? 16
Your name is Tom Payne and you are 16 years old.
?
```

15.34 The GFACTOR directive

Syntax: `$gfactor` [*integer*1] [{ *identifier* | * } *integer*2 [(*integer*3)]]*s*

Semantics: The `GFACTOR` directive provides a quick and convenient method of generating one or more factors which have a regular pattern. *Integer*1 is the length of the resultant factors. If this value is not specified the length of the factors is taken to be the length of the first *identifier*, and if this *identifier* is also undefined, the length is taken to be the standard length. We refer to this length as *len*. All *identifiers* must be undefined or vectors with length *len*.

Each *identifier* will be defined as a factor, and set up with levels specified by the following *integer*2, which must be a divisor of *len*. The optional *integer*3 which is enclosed in parentheses and which must satisfy the constraint 1 ≤ *integer*3 ≤ *integer*2, defines the reference category of the factor; if *integer*3 is omitted then the reference category is set or reset to 1 (the first category), and values will be assigned in a regular fashion, with the first factor varying least frequently, and the last vector varying most frequently. Formally, the first factor contains values which would be produced by the GLIM function `%gl(`*integer*2 , *l*`)`, where *l* =*len*/*integer*2 , the *i*th factor contains values which would be produced by `%gl(`*k*, *l*`)`, where *k* is equal to *integer*2 for the *i*th factor, and *l* is equal to *len* divided by the product of the (*i*-1)th previous *integer*2 values. The product of the levels for all factors must not be greater than *len*.

The asterisk `*` may be used in place of the identifier to skip over sequences which are not required to be assigned to factor identifiers. Trailing asterisks and their associated factor levels may be omitted.

Example:

```
$gfac  8  x 2  * 2  y 2 $
         !will generate the values
         ! 1 1 1 1 2 2 2 2     for x
         ! 1 2 1 2 1 2 1 2     for y
```

15.35 The GRAPH directive

Syntax: `$graph` [(*option list*)] *y_list* *x_list* [{ *style_list* / * } [*factor_list*]]

where

y_list is { *y_vector* [/*wt_vector*] [, *y_vector* [/*wt_vector*]]s / *list1* [/*list2*] }

x_list is { *x_vector* [, *x_vector*]s / *list3* }

style_list is { *integer* [, *integer*]s) / *vector* }

factor_list is { *factor* [, [*factor*]]s / *list4* }

The options available in the *option list* are:

`device`	= *string*
`copy`	= *string*
`ylimit`	= *real*1 , *real*2 [, *real*3]
`xlimit`	= *real*1 , *real*2 [, *real*3]
`refresh`	= { `yes` / `no` }
`annotation`	= { `yes` / `no` }
`pause`	= { `yes` / `no` }
`style`	= *real*
`title`	= { *string* / *macro* }

`hlabel`	=	{ *string* / *macro* }
`vlabel`	=	{ *string* / *macro* }

Semantics: The `GRAPH` directive produces plots on a graphical output device. The syntax is similar to that used in the `PLOT` directive. The `GRAPH` directive causes a multiple scatter or line plot of the values of *y_vectors* in the *y_list* against the equivalent *x_vectors* in the *x_list* with scaling of both axes.

For each device, a plotting area is defined, which contains one or more plotting regions, each containing axes, axis annotation, titling, and points and lines. The width and height of the plotting area are equal, and depend on the particular graphical output device in use. By default, the whole of the plotting area on the output device will be taken as the current plotting region, but this can be changed using the `LAYOUT` directive.

We define a *polyline* to be the graphical representation of a plot of a *y_vector* against an *x_vector* on the graphical output device. The polyline may be:

(a) a sequence of lines of a defined line type which joins the *x-y* pairs of the *y_vector* and *x_vector*.

(b) a scatter plot which marks the points of the *x-y* pairs with a defined plotting symbol.

(c) a scatter line with a defined linetype and a defined plotting symbol, which both marks the points and joins the *x-y* pairs of the *y_vector* and *x_vector*.

A polyline also has a defined colour.
The number of *y_vectors* must equal the number of *x_vectors* with the *i*th *y_vector* in the *y_list* plotted against the *i*th *x_vector* in the *x_list* , except in the following cases:

(a) A single *y_vector* is allowed with multiple *x_vectors*. If there are *k* *x_vectors*, this has the effect of defining *k* superimposed polylines, with *y_vector* plotted against each *x_vector* in the *x_vector_list*.

(b) A single *x_vector* is allowed with multiple *y_vectors*. If there are *k* *y_vectors*, this has the effect of defining *k* superimposed polylines, with each *y_vector* in the *y_vector_list* plotted against the *x_vector*.

Any or all of the *y_vectors* may be associated with a *wt_vector* which defines its *associated* weight vector. Units corresponding to zero elements of the *wt_vector* will be omitted from the polyline associated with the *y_vector*; that is, if the polyline is a line, a continuous line will be drawn joining the points with non-zero weights. If the polyline is a scatter plot or a scatter line, then the values of the *wt_vector* will scale the plotting symbol from its default size for each unit in the polyline; a different scaling is possible for each unit.

If the system vector `%re` has been assigned values by the user, then for *y_vector-x_vector* pairs without an associated *wt_vector,* the weight vector for that pair is instead taken from the values of `%re`.

For each *y_vector-x_vector* pair, the *y_vector*, *x_vector,* and the *wt_vector* if present and `%re` if assigned values must all be of the same length. However, the lengths of *y_vector-x_vector* pairs may differ from each other.

A plot of a *y_vector* against an *x_vector* may produce more than one polyline if a *factor* is present in the *factor_list* for that pair. If a *factor* is present, then it must have the same length as the *y_vector* and *x_vector*. If the factor has *k* levels, then *k* polylines are defined. The ith polyline corresponds to points defined by units for which the factor has value i; if the polyline is a line or scatter line, then these points will be connected by a line.

The `device` option specifies a new default graphical output device, overriding the currently set default. The graphical output device specified will be used for the current graph and all subsequent graphs, until a new `SET` directive is issued or a new `device` option to `GRAPH` is specified.

The `copy` option specifies a device to which the graphical output will be sent in addition to the default graphical output device; this will usually be a plotter or printer.

The `refresh` option allows the drawing of multiple plots on the same page or screen. The default, `refresh=yes,` means that the graphical output device will be cleared before plotting. If `refresh=no,` no such clearing will take place. The `copy` option is not available with `refresh=no` on some implementations.

The `pause` option allows pausing to occur at the end of the `GRAPH` directive if output is being written to a screen-based device and if the mode is interactive. The default, `pause=yes,` means that pausing will take place, with an audible indication sent to the output device. The user should press the 'return' key to continue. If `pause=no,` no such pausing will take place.

The `annotation` option allows a user to omit axis annotation from a plot. This helps to prevent clutter in presenting many scatter plots on the same plotting area. If `annotation=no`, then no numbering will appear on either axis. Axis tick marks are still displayed and are calculated as if annotation was present. If `annotation=yes`, numbering will appear. The default is `annotation=yes` and this option will reduce the size of the plotting area available for the axes, annotation, and polylines.

If the `ylimit` option is used and if only the first two reals are supplied, then *real*1 is taken as the lower and *real*2 as the upper plotting limit for the *y*-(vertical) axis, and only those *y*-values that fall within these limits will be included in the graph; *real*1 must be strictly less than *real*2. If the option is omitted then GLIM chooses limits that will permit all *y*-values with non-zero weights from all *y*-vectors to be plotted. Printed scale labels at the tick marks on the *y*-axis will have 'neat' values falling within the range between *real*1 and *real*2, and printed scale labels will not necessarily be given at *real*1 and *real*2.

*Real*3 is optional and specifies a user-supplied step length which affects the printed scale labels at the tick marks. If *real*3 is specified, *real*1, as well as specifying the lower plotting limit, also specifies the first printed label. Subsequent printed labels are given at increments of *real*3 until *real*2 is reached or exceeded; the last scale label may be therefore be greater than *real*2. If the value of *real*3 is such that the number of scale labels specified exceeds a suitable maximum, then *real*3 is ignored, and a plot is produced as if *real*3 had not been specified.

The `xlimit` option has the same effect as the `ylimit` option (with the same restrictions), but on the *x*- (horizontal) axis.

The `style` option determines the appearance of the plot. If the *real*, when rounded, is zero or if the option is omitted then the plot is annotated on the left with a *y*-axis and *y*-values and at the bottom with an *x*-axis and *x*-values. If the rounded real is positive then in addition the whole plot is enclosed in a box, with axis tick marks. If the rounded *real* is negative then no axes or annotation are printed, only the plotted polylines.

The `title` option allows the user to specify a title, which can be specified either by a *string* or by a *macro*. The title will be centred above the scatter plot.

The `vlabel` option allows the user to specify a text label for the *y*-axis. The text label is rotated if the device allows or stacked vertically otherwise and is printed to the left of the *y*-axis.

The `hlabel` option allows the user to specify a text label for the *x*-axis. The text label is centred and printed below the scatter plot.

The `title`, `vlabel` and `hlabel` options will reduce the size of the plotting area available for the axes, annotation, and polylines.

If *style_list* is included then it defines the pen style of the polylines to be used for plotting points. The *k*th *integer* or the *k*th element of *vector* in *style_list* determines the pen style of the *k*th polyline. If no style_list is present or the symbol * is present, then the default pen style *k* is taken for the *k*th polyline.

If a *factor_list* is present, then a factor may or may not be present for the *i*th *x-y* pair. If the *i*th factor has l_i levels (l_i is taken to be 1 if the factor is omitted) and if there are *n* *y*-vectors (or 1 *y*-vector and *n* *x*-vectors), then the list must contain at least $\sum_{i=1}^{n} l_i$ integers. If the *m*th unit of the *k*th factor is *j* then the *m*th unit of the *k*th *y*-vector will be plotted using the style defined by the $(\sum_{i=1}^{k-1} l_i + j)$ th integer in *style_list*.

For example with

```
$graph  AVEC,BVEC,CVEC  XVEC  1,2,6,8,9,10
                FAC1,,FAC3 $
```

where `FAC1` has two levels and `FAC3` has three levels, then for `FAC1`(*i*)=1 and `FAC1`(*i*) =2 respectively the styles 1 and 2 will be used for the polylines `AVEC against XVEC` ; the style 6 will be used for the polyline `BVEC` against `XVEC` and for `FAC3`(*i*)=1, `FAC3`(*i*)=2, and `FAC3`(*i*)=3 respectively the styles 8, 9, and 10 will be used for the polylines `CVEC` against `XVEC`.

If no *factor_list* is present, the number of pen style integers should be equal to the number of *y_vectors* or the number of *x_vectors*, whichever is the greater. If the *factor_list* is present and starts with a comma (that is, there is no factor for the first *x-y* pair) and default pen styles are required, then the * for the *style_list* is mandatory.

*List*1, *list*2, *list*3 and *list*4 represent identifiers of type *list*. If *list*1 is present, then it should contain vector identifiers representing *y_vectors*. If *list*2 is present, then it should contain vector identifiers representing *wt_vectors*. If *list*3 is present, then it should contain vector identifiers representing *x_vectors*. *List*4, if present, should contain factor identifiers representing the factors in *factor_list*. If *list*4 is specified, a factor must be present for each element of the list.

Examples:

```
$yvariate y$fit x$
$gstyle 1 c 4 s 2 1 0      ! set up style 1 to be
```

```
                                  ! a coloured symbol
$gstyle 2 c 5 s 0 1 1             ! set up style 2 to be
                                  ! a coloured line
$graph  y,%fv x$                  ! a simple graph of
                                  ! observed and fitted
                                  ! values (assuming x is
                                  ! in ascending order)
$factor f 2$fit +f$
$gstyle 3 c 3 s 3 1 0             ! another coloured
                                  ! symbol for style 3
$gstyle 4 c 6 s 0 1 2             ! another coloured line
                                  ! style for style 4
$graph y,%fv x 1,3,2,4 f,f $
                                  ! plot different
                                  ! symbols for the two
                                  ! levels of f, and
                                  ! different fitted line
                                  ! styles.
$graph y1,y2 x * ;f               ! Plots y against x1
                                  ! with pen style 1 and
                                  ! y against x2 with
                                  ! pen styles 2 and 3
```

15.36 The GROUP directive

Syntax: `$group` [*vector*2 =] *vector*1 [*phrase*]*s*

The following *phrases* are available:

```
values      vector4
interval    [*] vector3 [*]
```

The *phrases* may appear in any order and if a *phrase* is repeated then its last instance will be used. Only the first letter of each *phrase-name* is significant.

*Vector*1, (and also *vector*3 and *vector*4 if specified) must have values.

Semantics: The GROUP directive creates a factor from a variate or factor, and optionally allows cut-points on the vector values to determine the groups. Formally, the values of *vector*1 are grouped on the basis of the transformation defined with *vector*3 as the input cut-points and *vector*4 as the output group levels, and stored in *vector*2.

*Vector*1 supplies the input values to be grouped and stored in *vector*2. If *vector*2 is absent then *vector*1 will be used in its place. *Vector*2 (or *vector*1 in its place) will be (re)defined as a factor of the same length as *vector*1 with *k* levels where *k* is the number of distinct values of *vector*4.

The `interval` *phrase* defines the group cut-points of the transformation. The values of *vector*3 must be in strictly ascending order. If *vector*3 has *n* elements then it defines *n*-1 intervals of the form:

$$[x_1, x_2),\ [x_2, x_3),\ \ldots,\ [x_{n-1}, x_n)$$

where square brackets denote inclusion, round brackets denote exclusion, and x_i is the *i*th value of *vector*3. If the left-hand `*` is present then an extra interval of the form

$$(-\infty, x_1)$$

is assumed; if the right-hand `*` is present then an extra interval of the form

$$[x_n, +\infty)$$

is assumed. In this manner the `interval` *phrase* defines *l* contiguous intervals, in an increasing sequence, where *l* is either *n* -1 or *n* or *n* +1; n must be greater than 1 unless a `*` is present. If the `interval` *phrase* is omitted then a *phrase* of the form

`interval` *vector* `*`

is assumed, where *vector* contains the distinct values of *vector*1 in ascending order.

The `values` *phrase* defines the output group categories, the *i*th interval mapping onto the *i*th value of *vector*4. The length of *vector*4 must equal *l*, the number of intervals. If the `values` *phrase* is omitted then a vector of length *l* with values 1, 2, ..., *l* is assumed.

If the *j*th value of *vector*1 lies in the *i*th interval (i =1, ..., *l*) then the *j*th element of *vector*2 is assigned the *i*th value of *vector*4. Each value of *vector*1 must lie in one interval.

Examples: To group a variate `X` into a factor `F`, with each level of `F` corresponding to a distinct value of `X`, use

```
$group  F = X
```

To store these values in X and implicitly redeclare it as a factor, use

```
$group X
```

To map values of X in the intervals 0–9, 10–19, 20–30 onto the values 1, 2, and 3 respectively, storing in GX, use

```
$assign   I = 0,10,20,30
$group   GX = X   intervals I
```

If X contains year values and we want to group 1945–1960 as level 1 of a new factor, 1962–1968 as level 2, and 1961 and 1969 as level 3 (defining GX as a three-level factor), use

```
$assign I=1945,1961,1962,1969,1970
$assign V=1,3,2,3
$group  GX=X   values V  intervals I
```

15.37 The GSTYLE directive

Syntax: $gstyle *integer phrases*

where the following phrases are available:

```
colour      integer
linetype    integer
symbol      integer
```

and at least one *phrase* must be present.

Semantics: Changes the definition of the style of the *integer*th pen.

There are 30 pen styles available in GLIM. Each pen style has associated with it a colour setting, a line type setting, and a symbol type setting. If *phrase* is present, and its associated *integer* is positive, then the *phrase* setting will be set to *integer.* The *phrases* may appear in any order and if a *phrase* is repeated then its last instance will be used. Only the first letter of each *phrase-name* is significant.

For any pen, the setting controlled by a particular *phrase* does not affect the other settings for which no *phrase* is present. If the *integer* is zero then the setting becomes zero; this means that if *phrase* is colour 0, then the colour is set to anti-background; if *phrase* is linetype 0, then no lines will be plotted; and if *phrase* is symbol 0, then no symbols will be plotted.

By default, GLIM sets up 16 default pen styles at the start of a session. The remaining 14 pen styles are initialized to zero settings on colour, line type, and symbol. All pen styles may be redefined. A pen style remains defined until a subsequent `GSTYLE` directive redefines it. The `NEWJOB` directive does not redefine the pen styles. A list of colours, line types, and symbols and their associated indices available on the current graphical output device is available via the `G` or `H` options to the `ENVIRONMENT` directive. The current definitions of the first 16 pens, (or more, if more have been defined by the user) are also provided.

Example:
```
$gstyle 1 colour 3 linetype 2$
       ! redefines pen 1 to have colour 3 and
       ! linetype 2, keeping the previously
       ! defined symbol setting.
```

15.38 The GTEXT directive

Syntax: `$gtext` [(*option list*)] *y_item* [/ *wt_item*] *x_item text_list*

where

y_item is $\left\{ \begin{array}{c} real \\ y\_vector \end{array} \right\}$

wt_item is $\left\{ \begin{array}{c} real \\ wt\_vector \end{array} \right\}$

x_item is $\left\{ \begin{array}{c} real \\ x\_vector \end{array} \right\}$

text_list is $\left\{ \begin{array}{c} string \\ macro \end{array} \right\}$ [, $\left\{ \begin{array}{c} string \\ macro \end{array} \right\}$]s

The options available in the *option list* are:

`device` = *string*

`copy` = *string*

`ylimit` = *real*1 , *real*2

`xlimit` = *real*1 , *real*2

`refresh` = $\left\{ \begin{array}{c} \texttt{yes} \\ \texttt{no} \end{array} \right\}$

`pause` = $\left\{ \begin{array}{c} \texttt{yes} \\ \texttt{no} \end{array} \right\}$

```
style         = real

orientation   = { horizontal }
                {  vertical  }

mode          = { previous }
                {  window  }

                {   left   }
                { centred  }
location      = {  right   }
                {  above   }
                {  below   }
```

Semantics: The `GTEXT` directive displays text on a graphical output device. The text may either be superimposed using the coordinates of the previous graph or the directive can define its own coordinate system and use the whole of the current plotting region. *Y_item* and *x_item* together specify a set of one or more coordinate pairs which define the location of the strings in *text_list*, and with the size of characters determined by the weights in *wt_list*.

If *y_item* or *wt_item* or *x_item* is *vector,* then the *vector* must exist and have values. All vectors must then have the same length, say *k*, which also defines the number of text strings to be displayed. If *reals* are used for all three items, then $k = 1$. If $k > 1$ and *y_item* or *x_item* or *wt_item* is *real* or *scalar,* then the *real* is treated as a constant *vector* of length *k*.

There must be exactly *k strings* or *macros* in the *text_list* . If *macro* is present, then the contents of *macro* are taken as the text string to be displayed. Strings are truncated to a maximum of 80 characters, and may be further truncated on display if part of the text string would fall outside the plotting area. New line characters are ignored, but spaces are displayed.

The *i*th string in the *text_list* is displayed at the coordinate (x(*i*) ,y(*i*)), where x(*i*) is either *real* or the *i*th element of the *x_vector*, and *y*(*i*) is either *real* or the *i*th element of the *y_vector* in the coordinate system defined by the `mode` option. If *wt_item* is present, then the *real* or the *i*th element of *wt_vector* defines a multiplicative scaling factor which changes the size of the *i*th string to be displayed. The scaling factor must be positive.

The `mode` option controls the coordinate system used by the `GTEXT` directive.

If `mode` is `previous,` then the coordinate system used by the previous `GRAPH` directive is used. The current plotting region is that defined by the last `LAYOUT` directive, and text will be displayed within the axes used by the previous `GRAPH` directive, using the *x* and *y* maxima and minima defined by that directive.

If mode is `window`, then the whole of the current plotting region is used. If there is no `xlimit` option in the *option list*, then the minimum *x*-coordinate is taken as 0.0 and the maximum as 1.0. If there is no `ylimit` option in the *option list*, then the minimum *y*-coordinate is taken as 0.0 and the maximum as 1.0.

If the `GRAPH` directive has not been previously used or has not actually produced a graph, or if the `LAYOUT` directive has altered the plotting area since the last graph, or if the default device has been changed via the `SET` directive, then the mode must be `window`.

The `device` option specifies a new default graphical output device, overriding the currently set default. The graphical output device specified will be used for the current graph and all subsequent graphs, until a new `SET` directive is issued or a new `device` option to `GRAPH` is specified. If the `device` option is used, the `mode` must be set to `window`. If `mode=previous`, the `device` option cannot be used and instead the device setting from the previous graph will be used.

The `copy` option specifies a device to which the graphical output will be sent in addition to the default graphical output device; this will usually be a plotter or printer. If the `copy` option is used, the `mode` must be set to `window`. If `mode=previous`, the `copy` option cannot be used and the copy device setting from the previous graph will be used.

The `refresh` option specifies whether a new page is to be started or the screen is to be cleared. The default, `refresh=no`, means that the graphical output device will not be cleared before plotting. If `refresh=yes`, clearing will take place. `Refresh=yes` is not available with `mode=previous`.

The `pause` option allows pausing to occur at the end of the `GTEXT` directive if output is being written to a screen-based device and if the mode is interactive. The default, `pause=yes`, means that pausing will take place, with an audible indication sent to the output device. The user should press the 'return' key to continue. If `pause=no`, no such pausing will take place.

If the `ylimit` option is used then *real*1 is taken as the lower and *real*2 as the upper plotting limit for the *y*-coordinate, and only those strings whose *y*-values fall within these limits will be displayed; *real*1 must be strictly less than *real*2. If the option is omitted then GLIM sets *real*1 to be 0.0 and *real*2 to be 1.0. The `ylimit` option may only be used if `mode=window`.

The `xlimit` option (with the same restrictions) has the same effect as the `ylimit` option, but on the *x*-coordinates.

The `style` option determines the appearance of the plot. If the *real*, when rounded, is zero or negative or if the option is omitted then no box is produced around the plotting region. If the rounded *real* is positive then the plotting region is enclosed in a box. The `style` option is only available when `mode=window`.

The `location` and `orientation` options control the position of the text string relative to the *x*,*y*-coordinate. If `orientation` is `horizontal` (the default), then the text string is written from left to right, parallel to the *x*-axis. If `orientation` is `vertical`, and if text rotation is supported on the plotting device, the text string is written from the bottom to the top, parallel to the *y*-axis and rotated; if rotation is not supported, then it is written from top to bottom parallel to the *y*-axis and unrotated. The `location` option specifies where the text string appears in relation to the plotting position. If `location` is `left`, then the text string will be displayed to the left of the coordinate, if `location` is `centre`, it will be centred on the co-ordinate, and if `location` is `right` or omitted, it will be displayed to the right of the coordinate. If `location` is `above`, the text string is centred, but is displayed above the coordinate. If `location` is `below`, the text string is centred, and is displayed below the coordinate. The orientation of the text string has precedence; so the location of the text string refers to the string ***after*** rotation. In addition, when `orientation` is `vertical` the `location` keywords will apply to the text string as if it had been rotated whether or not rotation is actually available for that device.

Examples:

```
$gtext (l=1) 50 100 'control group'
      ! plots "control group" to the left of
      ! (100,50) superimposing on the previous
      ! graph.
$gtext (l=c) 50/0.5 200 'treatment 1'$
      ! plots 'treatment 1' centred at (200,50)
      ! in half-size text.
$ass y=1,2,3$
$gtext y 3 'one','two','three'$
      ! plots "one" to the right of (3,1), "two"
      ! to the right of  (3,2) and "three" to the
      ! right of (3,3), superimposing on the
      ! previous graph.
```

15.39 The HELP directive

Syntax: `$help`

Semantics: Immediately following a GLIM fault and if the `BRIEF` setting is `ON` and the session is interactive, the `HELP` directive prints the help message which would have been printed if the `BRIEF` setting was `OFF`. Otherwise, the `HELP` directive prints a message referring users to the `MANUAL` directive.

15.40 The `HISTOGRAM` directive

Syntax: `$histogram` [`(`*option-list* `)`] [*y_vector* [`/`*weight*]]*s* [*string* [*factor*]]

where *y_vector* and *weight* are *vectors*

or

`$histogram` [`(`*option-list* `)`] *y_list* [`/`*wt_list*] [*string* [*factor*]]

where *y_list* and *wt_list* are *lists* of *vector identifiers*.

The options available are:

```
ylimit   = real1, real2
xlimit   = real1, real2
rows     = integer
cols     = integer
tails    = real1, real2
style    = real
```

In the first syntax, there must be at least one *y_vector* present. In the second syntax, the *y_list* must contain at least one vector identifier.

Semantics: The `HISTOGRAM` directive plots the joint histogram of the *y_vectors* (weighted by their *weight* vectors, if supplied), using the symbols supplied by the *string* and partitioned as implied by the levels of the *factor*. All *y_vectors* (and *weights* and the *factor* if present) must have values. If a *weight* is included then all its values must be non-negative and it must have the same length as its associated *y_vector*. If the *factor* is included then all *y_vectors* must have the same length, which must equal that of the *factor*.

In the second syntax, the identifiers in the *y_list* are treated as *y_vectors*; those in the *wt_list* are treated as *weights*, with association of *y_vector* to *weight* defined by the position in the lists. The length of the *y_list* must be equal to the length of the *wt_list*, if it is present.

The *y_vector* values are plotted on the vertical scale, the frequencies on the horizontal scale. If there are n vectors and if the factor has k levels (k is taken as 1 if the factor is omitted) then the full plot consists of $n \times k$ interleaved subplots. Each interval of the histogram will be plotted on a separate row but if the histogram contains $n \times k$ subplots each such row consists of $n \times k$ lines of output. The first k lines of a row correspond to the k plot lines obtained by partitioning the first vector into k subsets on the basis of the values of the factor, the first line corresponding to the subset of values of the first vector that lie in the interval and for which the factor has the value 1, and so on for the second up to the *k*th lines. Similarly the second *k* lines correspond to such a subset of the values of the second vector, and so on.

The *i*th subplot will use the *i*th character of the string if supplied, otherwise the initial character of the identifier of the corresponding vector. If supplied, the string must have at least $n \times k$ characters (excluding blanks and new lines, which are ignored) where *n* and *k* are defined above.

Each line of the plot corresponds to some subset of the values of some vector in some interval and consists of a number of such plotting symbols. The number of symbols plotted is based on f, the total frequency for the line, which is computed as the sum of the weight vector values (taken as 1 if the weight is omitted) for the given subset in the given interval. The number of symbols plotted is determined as

$$\text{maximum}\{0, \text{minimum}\ (f, f_{\text{max}}) - f_{\text{min}} + 1\} / m$$

rounded to the nearest integer, where f_{min} and f_{max} are the smallest and largest plotting frequencies (see the `xlimit` option) and *m* is the number of frequencies that a plotting symbol represents.

If the `ylimit` option is used then *real*1 and *real*2 are taken as the lower and upper limits (respectively) of the plot, and only values of the *y_vectors* falling within or on these limits are displayed; the first number must be strictly less than the second. Otherwise, GLIM chooses limits that enclose all the vector values.

If the `xlimit` option is used then *real*1 and *real*2, rounded to the nearest integer, give the smallest and largest (respectively) frequencies to be plotted; both frequencies must be positive and the smallest frequency must not be greater than the largest. Otherwise GLIM uses a lower limit of 1 and an upper limit equal to the largest frequency in the histogram.

If the `rows` option is used then the *real*, when rounded, specifies the number of *y*-intervals to be used; that is, it sets the number of rows (see above) to be used for the histogram, excluding any tail rows (see below). If the rounded *real* is zero then the default option is used. The rounded *real* must not be negative. If the `rows` option is not used (or if the rounded *real* is zero) then GLIM, by default, uses *m* rows, where *m* is either the square root of the largest subplot frequency, or the number of rows that will conveniently fit on a page of output, whichever is the smaller. There will be at least one row, even if this exceeds the page size.

If the `cols` option is used then the *real*, when rounded, specifies the maximum number of columns to be used for plotting frequencies; GLIM will use as many of these columns as possible to fit the largest plotted frequency on to a line of output with each column representing an integral number of observations. If the rounded *real* is zero or if the option is omitted GLIM computes the frequency scale on the basis of the number of columns on the current output record. The *real* must not be negative.

The `tails` option determines whether vector values lying below or above the main body of the histogram are plotted as single grouped intervals. If the `ylimit` option is not used then no tail intervals are printed and the `tails` option is ignored. Otherwise, if *real*1 when rounded, is positive then a lower-tail row is included in the histogram and that interval represents the y-values strictly less than the lower y-limit; if the rounded *real*1 is zero or negative or if the option is omitted, no such row is produced; similarly, *real*2 controls the inclusion of an interval for y-values greater than the upper y-limit.

The `style` option determines the layout of the plot. If the *real* when rounded is zero, or if the option is omitted, then the plot is annotated on the left with intervals and their frequencies and at the bottom with a frequency axis and frequency values. If the rounded *real* is positive then, in addition, a grid is drawn and the whole plot is enclosed in a box, with some extra annotation. If the rounded *real* is negative then no axes or annotation are printed, only the plot points.

Each printed interval consists of the lower limit and the upper limit enclosed in brackets; a left square bracket denotes inclusion of the left end-point in the interval, a left round bracket denotes its exclusion, and similarly for right brackets and end-points. Adjacent to the printed intervals are one, two, or three integer frequencies. If the lower frequency limit is greater than 1 then the first integer is

$$\text{minimum}(f, f_{\text{min}} - 1)$$

rounded to the nearest integer. The next integer printed is

$$\text{maximum}(0, \text{minimum}\ (f, f_{\text{max}}) - f_{\text{min}} + 1)$$

rounded to the nearest integer, but zero values are not printed if an initial integer is present.

If the upper frequency limit is less than the largest individual frequency in the plot then the final integer is

$$f - f_{\text{max}}$$

rounded to the nearest integer, but it is printed only if it is greater than zero.

Examples:
```
$histogram  Y                 ! histogram of vector Y
$histogram(style=-1)  Y  ! histogram of Y with
                          ! no annotation
$histogram(style=1)  Y   ! histogram of Y with
                          ! full annotation
$hist(ylim=0,40 rows=8 tails=1,1 style=1) Y
      ! restricts the plot to Y values between
      ! 0 and 40 in 8 rows, and shows the grouped
      ! tails as extra rows
$hist(y=8,40  r=16  x=41,300  t=1,1  s=1) Y
      ! restrict the plot to Y values between
      ! 8 and 40 and frequencies between 41 and
      ! 300 (but showing tails) 41 is used
      ! rather than 40 as the lower frequency
      ! limit to produce more attractive
      ! intervals
$calc W = %ind(Y) >4    $hist  Y/W
      ! omits the first four values of Y.
$hist  Y  '123'  A       ! plots histograms for the
                          ! three levels of A
```

15.41 The `INITIAL` directive

Syntax: `$initial` [*vector*1 [(*vector*2)] [*macro*]

Semantics: Declares the starting values for subsequent fits. *Vector*1 will be expected to hold the initial values for all future fits and *vector*2 represents an optional index variate.

If there is no index variate, then *vector*1 must be either a variate of length the number of units or previously undefined; if the latter, then it will become defined as a variate of length the number of units.

If indexing is being used, then *vector*1 need not have the length of the number of units. *Vector*2, if it is defined, must have the length of the number of units.

References to `%iv` will point to *vector*1 and references to `%ii` will point to *vector*2.

If *macro* is present, then this specifies the macro which defines the initial conditions of model fitting in a user-defined fit. The macro will be called after initial model fitting checks have been performed, but before the system vectors `%pe`, `%se`, `%sb`, and `%al` lose their contents. The macro must exist.

If *vector*1 is omitted, then initial values of μ will be calculated from the *y*-variate according to the error distribution declared at the time of the fit, and `%iv` and `%ii` become undefined. If *macro* is omitted, the name of the macro defining the initial conditions also becomes unset.

If a user-defined link is set, then any initial values variate or default setting of the variate is ignored, and initial values of the linear predictor vector will need to be provided via the system vector `%lp`. If the `RECYCLE` directive is subsequently used, then the initial values variate is set to `%fv`, and the initial values index becomes undefined. If an `OWN` predictor macro is set, than starting values will also need to be assigned to `%eta`. See the User Guide [9].

The setting remains in force until the next `INITIAL, RECYCLE,` or `NEWJOB` directive. If *vector*1 or *vector*2 is deleted, then this has the effect of unsetting the initial values variate. The `YVARIATE` directive will also unset the initial values variate and the initial values index (if set), when the new number of units differs from the length of the initial values vector or its index (if set).

Example:

```
$ini i$              ! initial values taken from
                     ! vector i
$ini i(IND)$         ! initial values taken from
                     ! vector i indexed by IND
$ini inimac$         ! No initial values vector.
                     ! Initial macro specified as
                     ! inimac.
```

15.42 The `INPUT` directive

Syntax: `$input` *file_specifier* [*integer*] [*subfile*]s

Semantics: Causes input to be read from *file_specifier* with record width *integer* (≥ 30 and ≤ 132). If *integer* is omitted, the record width is taken to be that of the primary input channel. The current program control stack is increased by one level. *File_specifier* must not already be in the program control stack unless it is the primary input channel.

If *subfiles* are absent, input is read from the record following the last record previously read on *file_specifier* or, if this is the first input of the session from this file, input is read from the first record on the file.

If *subfiles* are present then, starting from the present position and rewinding if necessary, records are scanned until a matching `SUBFILE` statement for the first subfile is found and input starts from the record after the matching identifier.

Input continues to be read from *file_specifier* until

(i) an `INPUT`, `REINPUT`, `DINPUT`, or `SUSPEND` statement or macro invocation is met, when input is diverted as specified therein, returning to *file_specifier* following its completion; or

(ii) the next `END`, `EXIT` or `SKIP` statement; or

(iii) the next `RETURN` or `FINISH` statement.

Following (iii), the next *subfile* (if any) is located and the appropriate input read as specified above or, if no *subfiles* remain, the program control stack is decreased by one level and the text following the original `INPUT` statement is read.

The order in which records are read can also be changed by the `REWIND` directive.

Example: `$input 'mymacro.dat'$`

If the file `mymacro.dat` has not already been opened, this directive will open the file `mymacro.dat` on the next available channel, append the channel number to the system vector `%oc`, and input the contents of the file into GLIM until a `RETURN` statement is encountered.

`$input 23$`

If channel 23 is not open, GLIM will prompt for a filename before opening the file and proceed as above.

15.43 The `JOURNAL` directive

Syntax: `$journal [ {on | off} ]` (the `$j` of `$journal` is underlined on the page; `on` and `off` are stacked inside braces)

Semantics: The `JOURNAL` directive is used to control whether a journal is stored for an interactive session.

A journal file contains a record of most statements input on the primary input channel during a GLIM session. The file may be replayed in a subsequent session to reproduce an analysis; errors in the interactive session may be edited out before replay with the local system editor. The journal file is also of use in the event of a system crash or communications failure. A journal file may be replayed:

(a) in batch mode, by using a suitable command line option to assign the file to the primary input, renaming the file if necessary. See the User Guide [4].

(b) in interactive mode, by reading the file into GLIM by using the `INPUT` directive. In this case, it will usually be necessary to replace the terminating `$stop` statement from the end of the file with a `$return` statement.

No journal file is kept if the user is in batch mode.

If the journal is `on`, then journal information will be written to the journal file, the file being opened if necessary. If the journal is `off`, then the journal file is kept open but is not written to. The journal file may not be closed.

All input on the primary input channel is written to the journal file except:

- commands issued to the GLIM debugger;
- 'return' keys pressed in response a pause caused by the `MANUAL` and `PAGE` directives and the `PAUSE` option to the `GRAPH` and `GTEXT` directives;
- Break-ins.

The journal channel number and state of the journal switch may be found by using the `C` option to the `ENVIRONMENT` directive. The system scalar `%jrc` also contains the journal channel number. The width of the journal channel at any point in a session is taken to be that of the primary input channel. The system scalar `%jou` provides information on whether the journal setting is on or off.

The default state is implementation dependent (usually on) and the name of the default journal file is also implementation dependent (it may be found for the local installation by using the `I` option to the `ENVIRONMENT` directive).

Example: `$jou on$       ! enables output to the journal file.`

15.44 The `LAYOUT` directive

Syntax: `$layout` [*real*1 [,]*real*2 *real*3 [,]*real*4]

Semantics: The `LAYOUT` directive defines that part of the plotting area (the current plotting region) which is to be occupied by subsequent graphs produced by the `GRAPH` and `GTEXT` directives. If no *reals* are present, and by default, the whole of the plotting area is defined as the current plotting region.

If *reals* are present, then *real*1 and *real*2 define the lower left corner of the current plotting region (as x and y coordinates respectively) and *real*3 and *real*4 define the upper right corner of the current plotting region (as x and y coordinates respectively). The x and y coordinates are given in plotting area coordinates, such that the lower left coordinate of the whole plotting area is defined as (0,0) and the upper right coordinate as (1,1). The *reals* must satisfy the constraints

$$0 \leq real1 < real3 \leq 1$$
$$0 \leq real2 < real4 \leq 1$$

Any scaling applies to the whole graph, including text strings, titles, axis annotation, and labelling. Note that symbols and lettering will change size and not distort.

The setting remains in force until altered by another `LAYOUT` directive or until a `NEWJOB` directive. A `NEWJOB` directive resets the `LAYOUT` setting to the default. The setting applies to all devices.

The current `LAYOUT` settings may be obtained via the `G` option to the `ENVIRONMENT` directive.

`LAYOUT` may be used with the `refresh =no` option to `GRAPH` to produce multiple plots on the screen.

This directive has no effect on the output produced by the `H` option to the `ENVIRONMENT` directive.

Examples:

```
$layout 0.5,0.5 1,1 $
      ! sets the current plotting region to be
      ! the top right quadrant of the plotting
      ! area
$graph y x$
      ! display a graph
$layout 0,0  1,0.4 $
      ! sets the current plotting region to be a
      ! rectangle at the base of the plotting
      ! area.
$gra(r=n) z x f$
      ! display another graph
```

15.45 The `LINK` directive

Syntax: `$link` *letter* [{ *real*1 [*real*2] | *macro* }]

Semantics: Declares the link function to be used in subsequent fits. Valid options for *letter* are

letter	Link function	Mathematical description
C	complementary log-log	$\eta = \log(-\log(1-\mu))$
E	exponent	$\eta = (\mu + real2)^{**}(real1)$
G	logit	$\eta = \log\{\mu/(1-\mu)\}$
I	identity	$\eta = \mu$
L	log	$\eta = \log(\mu)$
O	OWN or user-defined link	
P	probit	$\eta = \Phi^{-1}(\mu)$
Q	inverse quadratic	$\eta = 1/\mu^2$
R	reciprocal	$\eta = 1/\mu$
S	square root	$\eta = \sqrt{(\mu)}$

If *letter* is `C, G, I, L, P, Q, R,` or `S`, there must be no reals.

If *letter* is `E`, then the optional *real*1 and *real*2 respectively define the parameters δ and α in the exponential or power link $\eta = (\alpha+\mu)^{\delta}$. If *real*2 is omitted, then α is taken as 0.0; if both *real*1 and *real*2 are omitted then the default values of $\alpha = 0.0$ and $\delta = 1.0$ are taken, and the link reverts to the identity link.

If *letter* is `O` and *identifier* is defined, the *identifier* must be a macro. If it is undefined, it will be defined as a macro with no contents. Before a fit, the macro must contain statements which will define

(i) `%fv` as a function of the linear predictor `%lp;`

(ii) `%dr`, the vector containing the derivative of the link function.

When a user-defined link is set, the macro is also used by the `PREDICT` directive to convert from predicted values of the linear predictor to predicted fitted values.

A `LINK` statement remains in force until the next `NEWJOB`, `ERROR`, or `LINK` statement. Following an error declaration by the `ERROR` directive, or if the `LINK` directive is omitted, the default link declaration is as follows

Error	Link
B	G
G	R
I	Q
N	I
O	Unchanged
P	L

Examples:

```
$link Q          ! sets the link function to
                 ! be inverse quadratic.
$link O CUBIC    ! defines the macro CUBIC as the
                 ! macro which specifies the
                 ! values of %dr and %fv in a
                 ! subsequent fit.
```

15.46 The LIST directive

Syntax: `$list` *integer list_identifier*

or

`$list` [*integer*]*list_identifier* =*identifier* $\left\{ \begin{array}{c} , \\ + \\ - \end{array} \right\}$ *identifier*]*s*

Semantics: Defines the *list_identifier* to be of type *list,* with values specified below. *List_identifier* must either not already exist or be of type *list* or be of undefined type.

The first syntax generates a list of length *integer* with values which are identifiers of the form of *list_identifier digits.* The *i*th identifier in the list will have identifier *list_identifier i.* For example, `$list 5 VAR$` would create a list `VAR` which contains five identifiers: `VAR1`, `VAR2`, `VAR3`, `VAR4,` and `VAR5`. When generating identifiers, the *list_identifier* will be truncated if necessary to produce identifiers of at most 8 characters. If the *identifiers* exist, they may be of undefined type or of type *scalar*, *variate, factor,* or *macro,* but not of type *list,* and may or may not have values.

If an *identifier* does not exist, then the *identifier* will be created with undefined type.

The second syntax defines the *list_identifier* and takes as values the identifiers generated by the expression on the right-hand side of the assignment.

The expression is dealt with from left to right. An expression of the form

LHS_identifier *operator* *RHS_identifier*

where *operator* is + `,` or − is treated as follows.

If *LHS_identifier* is not a list, then it is treated as a list of length 1 with value *identifier.* The *LHS_identifier* is modified by the *operator* and produces a temporary list, which then becomes the left-hand side of the next *operator*, and so on, until the expression is exhausted.

If *operator* is + or , then:

(a) If *RHS_identifier* is not of type *list*, the *identifier* on the RHS is added to the end of the *list*. If the *identifier* exists, it may or may not have values. If *identifier* does not exist, then *identifier* will be created with unknown type.

(b) If *RHS_identifier* is of type *list*, all the *identifiers* in the list on the RHS are added to the end of the *list*.

If *operator* is - then the *RHS_identifier* is removed from the *list* and all identifiers following this position are moved one position up towards the beginning of the list. *RHS_identifier* must not be a *list*. If the *RHS_identifier* occurs more than once in the list, the position closest to the end of the list is deleted. If the *identifier* is not present in the list, the operator and its RHS are ignored.

When the expression is exhausted, then the temporary list is assigned to the *list_identifier*. The *list_identifier* is thus not updated until the end of the expression. If an optional *integer* has been specified as the length of the *list_identifier*, then the length of the temporary list must be *integer*.

Within a macro, *identifier* can be an *argument_identifier* or *formal_argument*, but each *identifier* must not be a function or subfile.

Examples:

```
$var A$fac B 2
$num C=1 $
$ass A=1...10
    :B= 1 1 2 2
$mac D
$ca L[C]=L[C]*4 $
$endm!
$list L=A,B,C,D$                ! defines the list L to
                                ! contain identifiers
                                ! A,B,C and D
$use D$                         ! execute the macro D.
$pr L[1]                        ! print the values of A
$pr L$                          ! prints the
                                ! identifiers in L
$list 3 MEASUREMENT$
      ! creates a list with 3
      ! identifiers MEASURE1,
      ! MEASURE2 and MEASURE3.
```

15.47 The LOAD directive

Syntax: `$load` [*vector*] [*macro*]

Semantics: This directive allows the weighted sums of squares and products matrix to be altered before solving.

If specified, the *vector*, if defined, needs to be of length `%pl`. If undefined, it will be defined as a variate of length `%pl`. If the algorithmic method is Givens, then the algorithm is changed to Gauss–Jordan and a warning message is displayed. The directive causes the *vector* to be added to the diagonal of the working triangle before inversion.

If a *macro* is specified, then the OWN triangle macro is defined to be *macro*. In the model fitting cycle, the *macro* is executed after the working triangle (or its Givens equivalent) has been formed but before the triangle is inverted. If both *vector* and *macro* are specified then the *macro* is executed before the *vector* is added to the diagonal of the triangle.

If a *vector* is not specified and a load vector is currently defined, then the load vector becomes unset. If a *macro* is not specified and an OWN triangle macro was previously defined, then the OWN triangle macro becomes unset.

Examples:
```
$var %pl T$ass T=2,2,2,0,0,0
$load T$
```

15.48 The LOOK directive

Syntax: `$look` [(*option-list*)] [*integer*1 [*integer*2]] $\left\{ \begin{matrix} \textit{vectors} \\ \textit{list} \end{matrix} \right\}$

or

`$look` [(*option-list*)] *scalars*

where *option-list* has one *option* :

`style` = *real*

Semantics: Displays in parallel, down the page, the consecutive values of the elements of *vectors* or *scalars*.

The *option* controls the format used. If the *option* is omitted, or if the *real* when rounded is zero, then the identifiers of the *vectors* will be printed as column headings to their values and a left-hand column of indices is given; if the rounded *real* is positive then, in addition, a border is printed around the table and between the columns; if the rounded *real* is negative then no borders, identifiers, or left-hand index vector are printed, only the values. When `LOOK`ing at *scalars* only the values are printed irrespective of the `style` setting.

If both *integer*1 and *integer*2 are present they specify, respectively, the first and last elements of the *vectors* to be printed. If *integer*1 only is present, *integer*2 is taken to be the same as *integer*1. If neither are present then *integer*1 is taken as 1 and *integer*2 as the length of the first *vector*. *Vectors* of varying length are allowed if the elements specified are within range for all of them. The number of significant digits is controlled by the current setting of the `ACCURACY` directive.

Only as many *vectors* or *scalars* are displayed as can fit on the current output record; if the list is truncated a warning is given.

If *list* is present, then all identifiers in the list must be *vectors*, which are then used as described above.

Example:

```
$look 2 4 a b c$          ! prints elements 2,3 and 4
                          ! of the vectors A,B and C
                          ! in parallel.
```

15.49 The `LSEED` directive

Syntax: `$lseed` [*integer*1 [*integer*2 [*integer*3]]]

Semantics: In the program as supplied, this directive acts like the `SSEED` directive but on the alternative, local pseudo-random number generator whose values are generated by the function `%lr`. If an installation has replaced this generator with a local version the syntax and semantics for this directive may also be locally defined.

15.50 The `MACRO` directive

Syntax: `$macro` [`(`*option_list*`)`] *macro_identifier space characters* `$endmac`

where *characters* may not contain the character sequences `$m` or `$ret` or `$fin` and where *option_list* has two options:

`arg` = *argument_identifier* [, *argument_identifier*] *s*
`loc` = *local_identifier* [, *local_identifier*] *s*

and *macro_identifier*, *argument_identifier*, and *local_identifier* are *identifiers* and the number of *argument_identifiers* is less than or equal to 9.

Semantics: Causes the *characters* to be named *macro_identifier* and stored as a macro for future use. If *macro_identifier* is already defined it must be a macro and must not appear in the current program control stack; if it already has contents these will be deleted, together with any argument settings. If previously undefined, *macro_identifier* is defined as a macro with undefined arguments. The *space* immediately following *macro_identifier* is not part of the macro contents; it can be replaced by a *newline* symbol. The *characters* may include *newline* symbols. For the first line only the *characters* after space and up to and including the first new line symbol (or `$e` if it occurs) are stored; for subsequent lines the *characters* up to and including the first *newline* symbol or `$e` (whichever occurs first) are stored. A line may contain solely a new line symbol. If a fault occurs during the reading of the *characters* the reading terminates and those *characters* already stored are erased.

Note that the `$` symbol in `$endmac` is not part of the macro contents. Some directives (such as `CALCULATE`) have a variable number of items and are thus not executed until a terminating `$` is encountered. Hence, if the macro ends with such a directive, a terminating `$` must be included in the macro if the directive is to be executed before the end of the macro's invocation. See [13.7.1].

Local_identifiers provide a mechanism for declaring *identifiers* which are local to the macro. *Argument_identifiers* or *keyword arguments* provide a way of assigning more meaningful names to *formal_arguments.*

Local_identifiers and *argument_identifiers* are entirely local to the macro. Access to any *local_identifier* or *argument_identifier* is only available while the *macro_identifier* is either on the current level of the program control stack or from within the debugger when the *macro_identifier* is on the previous level of the program control stack, that is, while the macro is being executed.

If an *identifier* is found during execution of the *macro_identifier*, then the list of *local_identifiers* and *argument_identifiers* is searched before the standard user directory. *Local_identifiers* and *argument_identifiers* can therefore have identical names to *identifiers* present in the user directory.

Any *local_identifier* or *argument_identifier* declared for a particular *macro_identifier* must be different from other *local_identifiers* and *argument_identifiers* for that *macro_identifier*, but can be the same as a *local_identifier* or *argument_identifier* for another *macro_identifier* or the same as any identifier in the user directory.

If the *i*th *argument_identifier* is *identifier*, then any occurrence of *identifier* detected within the macro during execution of the macro will be taken as referring to the *i*th argument of the macro.

If a *local_identifier* is declared for a particular macro, then the *local_identifier* is of unknown type and has no attributes or values until these are explicitly declared during execution of the macro by suitable directives. Once declared, the *local_identifier* keeps its attributes and values even if the macro is no longer on the program control stack. It therefore acts as any other GLIM identifier, but may only be accessed during execution of its macro. Deleting a *local_identifier* during the execution of its macro or from the debugger causes the *local_identifier* to lose its attributes and values. There is no limit to the number of *local_identifiers* except the global limits to the total number of identifiers set by the GLIM system as a whole.

Example:

```
$mac (loc=a,b) m
$number a=6 : b=3$          ! declare two local
                            ! identifiers to be scalars
$endmac
```

15.51 The MANUAL directive

Syntax: `$manual` [*keyword*1 [*keyword*2]]

where *keyword*1 is one of

directive_name
glossary_name
system_identifier
GLOSSARY
INFO
DIRECTIVES
LOCAL
LIBRARY

Semantics: If *keyword*1 is *directive_name* then information on the syntax and use of that directive will be displayed.

If *keyword*1 is *glossary_name* then information on the meaning of the term *glossary_name* will be displayed.

If *keyword*1 is *system_identifier* then if *system_identifier* is a *function*, information on the use of that function will be given, whereas if *system_identifier* is a *system_scalar*, *system_vector*, or *system_array* then information on the meaning and use of that system structure will be displayed.

If *keyword*1 is GLOSSARY, a list of *glossary_names* will be displayed.

If *keyword*1 is DIRECTIVES a list of *directive_names* will be displayed.

If *keyword*1 is LOCAL then information on the local features of the implementation of GLIM will be displayed.

If *keyword*1 is `INFO` or *keyword*1 is omitted, then the list of available options for *keyword*1 is displayed.

If *keyword*1 is `LIBRARY`, then if *keyword*2 is absent, information on the available macros in the macro library is displayed. If *keyword*2 is present, then it should contain the name of a *macro* in the macro library. Information on the use and output of that macro will be displayed.

The `MANUAL` directive temporarily overrides the setting of the `PAGE` directive, setting the pagination state to be on for the duration of the directive.

Examples:

```
$manual edmac$              ! produces information on
                            ! the EDMAC directive
$manual commands$           ! produces a list of valid
                            ! directives in GLIM
$man library qplot$         ! produces information on
                            ! the QPLOT macro in the
                            ! macro library.
```

15.52 The `MAP` directive

Syntax: `$map` [*vector*2 =] *vector*1 [*phrase*]*s*

The following *phrases* are available:

`values` *vector*4
`interval` [*] *vector*3 [*]

The *phrases* may appear in any order and if a *phrase* is repeated then its last instance will be used. Only the first letter of each *phrase_name* is significant. *Vector*1, *vector*3, and *vector*4 must have values.

Semantics: The values of *vector*1 are mapped on the basis of the mapping defined with *vector*3 as domain and *vector*4 as range, and stored in *vector*2.

*Vector*1 supplies the input values to be mapped and stored in *vector*2. If *vector*2 is absent then *vector*1 will be used in its place. *Vector*2 (or *vector*1 in its place) will be (re)defined as a variate of the same length as *vector*1.

The `interval` *phrase* defines the input values or domain of the mapping. The values of *vector*3 must be in strictly ascending order. If *vector*3 has *n* elements then it defines *n* -1 intervals of the form:

$$[x_1,x_2),\ [x_2,x_3),\ ...,\ [x_{n-1},x_n)$$

where square brackets denote inclusion, round brackets denote exclusion, and x_i is the *i*th value of *vector*3. If the left-hand * is present then an extra interval of the form

$(-\infty, x_1)$

is assumed; if the right-hand * is present then an extra interval of the form

$[x_n, +\infty)$

is assumed. In this manner the `interval` *phrase* defines *k* contiguous intervals, in an increasing sequence, where *k* is either *n*-1 or *n* or *n*+1 according to whether and which asterisks are present; *n* must be greater than 1 unless a * is present. If the `interval` *phrase* is omitted then a *phrase* of the form

`interval` *vector* *

is assumed, where *vector* contains the distinct values of *vector*1 in ascending order.

The `values` *phrase* defines the output values of the mapping, the *i*th interval mapping onto the *i*th value of *vector*4. The length of *vector*4 must equal *k*, the number of intervals. If the `values` *phrase* is omitted then a default vector is assumed; the `interval` *phrase* must not contain a point at minus infinity if the `values` *phrase* is omitted. If the `interval` *phrase* contains a point at plus infinity then the default vector for the `values` *phrase* is taken as the vector of length *n* containing the left end-points of the *n* intervals; if no point at infinity is included then the default vector is the vector of length *n* -1 containing the *n*-1 mid-points of the intervals.

If the *j*th value of *vector*1 lies in the *i*th interval (*i* =1, ..., *l*) then the *j*th element of *vector*2 is assigned the *i*th value of *vector*4. Each value of *vector*1 must lie in one interval.

Examples:

```
$assign  I=0,10,20,30
      !  maps values of X in the intervals
      ! 0-9,10-19,20-30 onto the values
      ! 5, 15 and 25 respectively, storing in GX
      !
$map     GX=X     intervals I
$assign  I=20,30  :  V=1.8, 3.2, 5.0
$map     GX=X    values V  intervals * I *$
```

```
! maps values of X in the intervals
! minus infinity-20, 20-30,
! 30-infinity, onto the values
! 1.8, 3.2, 5.0
```

15.53 The METHOD directive

Syntax: $method { *letter* | * } [{ *identifier1* | * } [*identifier2*]]

Semantics: Declares the algorithmic method to be used in subsequent fits. Valid options for *letter* are

G	Givens;
J	Gauss–Jordan.

If * is specified in place of *letter* then the algorithmic method is left unchanged.

The declared METHOD option remains in force until redeclared or changed automatically by GLIM via the LOAD or NEWJOB directives. The default option is G. The METHOD option changes the meaning of certain system vectors and scalars available after a fit. The system scalar %met contains the current setting of the method (1 = Givens, 2 = Gauss–Jordan).

If *identifier*1 or *identifier*2 are present, then they specify, respectively, the OWN predictor macro and the own weights macro. If *identifier*1 or *identifier*2 is defined, they must each be a macro. If either or both are undefined, then the appropriate macro or macros will be defined as having no contents. If *identifier*1 and/or *identifier*2 is not present, then the specified user-defined macro is unset.

An asterisk * may be used in place of *identifier*1, and has the effect of not altering that setting. If *identifier*1 is unset, it will remain unset; if set, if will remain defined.

If *option* is G and a load variate is set, then the load variate is unset and a warning is printed.

Examples:

```
$method J          ! sets the algorithmic method to
                   ! be Gauss-Jordan.
$method * MAC1     ! Sets the OWN predictor macro
                   ! to be MAC1. It leaves
                   ! the algorithmic
                   ! setting unchanged.
```

15.54 The NEWJOB directive

Syntax: `$newjob [(option_list)]`

where *option_list* has one option

close = { yes / no }

Semantics: Signifies the end of a job [13.7.8]. If the `close` option is set to `no` then the workspace is cleared, the model is reset to the default model (Normal error, identity link), and the various system settings are reset to their state at the start of the GLIM session. The initialization file, if it exists, is then opened, and input is read from this file. Any open graphics device is refreshed, that is, the graphics screen is cleared or a new page started, and the `LAYOUT` setting is set to its default.

If the `close` option is set to `yes` then additionally all secondary files are closed, and the system vector `%oc` is reinitialized. The journal file and transcript file are not closed, and neither are high-resolution graphics files. The default is `close=no`.

Note that with either option, the `NEWJOB` directive does not affect the graphics state except as described above; pen styles remain defined, and graphics devices and files remain open.

15.55 The NEXT directive

Syntax: `$next`

Semantics: If macro debugging is `on`, and a macro is in use but temporarily suspended by the debugging facility, the `NEXT` directive moves to the next directive symbol in the program control stack. If this directive symbol is also in a macro, and the debugging facility is still on, then GLIM will suspend execution, allowing the user to issue further GLIM directives to examine and reset the contents of variables or to issue a further `NEXT` or `DEBUG` directive. The `NEXT` directive must not occur in a macro.

Examples:

```
$debug on     ! Enable debugging
$use MAC1     ! start to use macro MAC1
              ! macro stops before first directive
$next         ! execute first directive in macro and
              ! stop before next directive symbol
$next         ! etc...until
$debug$       ! Disable debugging - continue normal
              ! execution of macro.
```

15.56 The NUMBER directive

Syntax: $number *identifier* [= *real*] [*identifier* [= *real*]]*s*

Semantics: The NUMBER directive allows the definition of scalars with user-defined names. In general, *identifier* may be used in all places where *ordinary scalars* or *system scalars* are used.

The exceptions are:

(a) as a scalar in the TRANSCRIPT directive;

(b) as a scalar in expressions of the form %*scalar* in macros;

(c) as a scalar in PRINT items of the form **scalar*.

If *identifier* does not exist, then it will be defined as a *scalar* with value set equal to 0.0. If a *real* is present, then the value of *real* will be assigned to *identifier*. If *identifier* exists it must be of type *scalar*, in which case the value of *real*, if present, will be assigned to *identifier*, otherwise the contents of *identifier* are left unchanged.

Examples:
```
$number old_dev=%dv old_df=%df$
$number maxiter=20$
```

15.57 The OFFSET directive

Syntax: $offset [*vector*1 [(*vector*2)]]

Semantics: Declares an a priori known component to be included in the linear predictor during fitting.

*Vector*1 will be expected to hold the offset values for all future fits and *vector*2 represents an optional index variate.

If there is no index variate, then *vector*1 must be either a variate of length the number of units or previously undefined; if the latter, then it will become defined as a variate of length the number of units.

If indexing is being used, then *vector*1 need not have the length of the number of units. *Vector*2, if it is defined, must have the length of the number of units.

References to %os will point to *vector*1 and references to %oi will point to *vector*2.

If *vector*1 is omitted, then the offset is unset, with no known component being included in the linear predictor. %os will point to an internal scalar containing the value zero, and %oi will become undefined.

The setting remains in force until the next `OFFSET` or `NEWJOB` directive or until *vector*1 or *vector*2 is deleted. The `YVARIATE` directive will also unset the offset variate and the offset index, if set, where the new number of units differs from the length of the offset variate or its index, if set.

Example:
```
$offset A(I)          ! declares the offset to have
                      ! values of A and index of I
```

15.58 The OPEN directive

Syntax: `$open [(`*option_list*`)]` *file_specifier*

where *option_list* is

```
status= { new }
        { old }
```

`status=old` specifies that the file must already exist; `status=new` specifies that the file must not already exist. The default is `status=old`.

Semantics: The directive provides a mechanism for opening files specified by *file_specifier* according to the `status` setting of the *option_list*. This allows channels to be opened ready for subsequent input or output. If `status` is set to `old` then the *file_specifier* must already exist; this is normally associated with input. If `status` is set to `new` then the *file_specifier* should not already exist; this is normally associated with output. Note however that GLIM has no restriction on reading and/or writing to secondary files once opened.

If the *file_specifier* is open already, the directive will be ignored. If the file exists and `status=new` is specified then an error will be generated in batch mode and in interactive mode users will be asked if they wish the file contents to be overwritten. If the file does not exist and `status=old` is specified an error message is generated.

Graphics output files may be opened; this provides a mechanism for changing the default name of a graphics output file. The correct channel number (as given by the `G` option to the `ENVIRONMENT` directive) must be used.

Examples:
```
$calc %a=16$
$open %a$             ! will open channel 16 for
                      ! reading, prompting for a file
                      ! name.
$open 'data13'        ! will open the file data13 on
                      ! the next available channel
```

15.59 The OUTPUT directive

Syntax: `$output` [*file_specifier* [*integer*1 [*integer*2]]]

Semantics: Makes *file_specifier* if present, the current output file and channel with record width *integer*1 and page height *integer*2. If *integer*1 or *integer*2 is omitted or is 0, the value(s) for the primary output channel will be used. The value of *integer*1 ($\geq$ 52 and $\leq$ 132) affects the output from all directives. The value of *integer*2 ($\geq$ 12) affects in particular the output from the PLOT and PAGE directives.

If *file_specifier* is omitted or if the channel of the *file_specifier* is 0, no output will be generated until another OUTPUT statement specifies a *file_specifier* or *channel*. Ordinary output is also not written to the transcript file by default. The output may be restored to the primary output channel by the statement `$output %poc$`; this also causes ordinary output to be written to the transcript file by default.

Graphics output cannot be suppressed or redirected by the OUTPUT directive.

Example:

```
$output 'results.dat'$              ! write the values
$look(style=-1) %yv %fv %lp$        ! of %yv, %fv and
$output %poc$                       ! %lp to the file
                                    ! results.dat,
                                    ! then restore
                                    ! output to the
                                    ! screen
```

15.60 The OWN directive

Semantics: The OWN directive of earlier versions of GLIM has been replaced by relevant options in the ERROR, INITIAL, LINK, LOAD, and METHOD directives. Use of the OWN directive now results in an error message.

15.61 The PAGE directive

Syntax: `$page` [{ on / off }]

Semantics: Alters the current pagination state. If pagination is on then GLIM pauses after every *n* lines of input/output on the primary channel, where *n* is the height of the primary output channel. The pause takes the form of a prompt to the user to press the 'return' key when inspection of these *n* lines is complete. If pagination is off then no such pauses occur. The directive is ignored in batch mode.The pagination state is overridden in the `MANUAL` directive, where it is always `on`.

Example: `$page on$`

15.62 The `PASS` directive

Syntax: `$pass` [(*option_list*)] { *keyword* / *integer* } [*vector_list* [*macro*]]

where *option_list* has one option:

`array=fortran`

vector_list is { *vector* / *scalar* } [, { *vector* / *scalar* }] *s*

and *keyword* is defined by the writer of the PASS routine.

Semantics: The `PASS` directive enables users to access GLIM data structures for reading and/or writing via their own Fortran routines, and permits them to implement, via these routines, data manipulation facilities, and so on, that are not available (or not easily available) via GLIM directives.

The `PASS` directive makes available to the user-supplied routine the values of the *vector_list* and *macro*; the value of the *integer* or the *keyword* is also made available and is provided to enable the user to distinguish between the different facilities that may be implemented within the user's routine. On return from the routine the values of the individual *vectors* and *scalars* in the *vector_list* and *macro* (which may have been changed by the routine) are again available to the user in GLIM (for example, through the `PRINT` directive). A fault indicator enables GLIM to treat faults and break-ins detected within the routine via the usual GLIM fault/help mechanism. If the option `array=fortran` is set, then any GLIM array passed will be transposed so that its storage is suitable for the FORTRAN language. The default is that no transposition takes place. Details of the `PASS` facilities available on the local installation may be found by issuing the statement `$environment e $`.Full details on implementing the `PASS` facility can be found in [17].

Examples:

```
$pass 3 A,B,C MAC$      ! pass the vectors A,B and C
                        ! and the macro MAC to the
                        ! fortran subroutine called
                        ! by option 3.
```

15.63 The PAUSE directive

Syntax: `$pause` [*string*]

Semantics: If *string* is absent, then control is returned temporarily to the operating system. The method of returning to the GLIM session is operating system dependent, and will be given in the User Note and by the statement `$man local$`.

If *string* is present, then the *string* is passed to the operating system for execution as an operating system command. When the operating system command terminates, control returns to GLIM automatically. On some operating systems it may be only possible to run commands which use a small amount of memory.

When control returns to GLIM, the next statement executed is the one following the PAUSE statement.

The PAUSE directive is implementation dependent, and may not be available on the local version of GLIM.

Examples:

```
$pause 'dir'                    ! VAX/DOS
$pause 'more glim.log'          ! UNIX
$pause 'format a:'              ! DOS
```

15.64 The PICK directive

Syntax: `$pick` [[*destination_list*] *source_list*] *key_vector*

where both *destination_list* and *source_list* are

$$\left\{ \begin{array}{c} \textit{vector} \;[\,,\textit{vector}\,]\text{s} \\ \textit{list} \end{array} \right\}$$

and *key_vector* is a vector.

Semantics: This directive is used to select or pick elements from a set of vectors and place them in new vectors, according to whether the values of a *key_vector* are non-zero. For all vectors in *source_list*, the procedure below is followed.

If an element of *key_vector* is non-zero, then for the *i*th *vector* in *source_list* the equivalent element is selected and copied to the next available element in the *i*th vector in *destination_list*.

If an element of *key_vector* is zero, then the equivalent elements in the vectors in *source_list* are not selected or copied. If no *destination_list* is present, then each vector in the *source_list* is overwritten.

If no *source_list* and no *destination_list* are present, then both the *source_list* and the *destination_list* are taken to be the *key_vector*, and the directive has the effect of removing all zero elements of *key_vector*.

All identifiers in the *source_list* must be vectors, which must already exist and have values, and have the same length as *key_vector*. All identifiers in the *destination_list* must, if they already exist, be vectors and have length the number of non-zero elements of *key_vector*. If any identifier in the *destination_list* does not exist, it will be created with the same type as its equivalent vector in the *source_list* and length the number of non-zero elements of the *key_vector*. *Key_vector* must contain at least one non-zero element.

Example:
```
$assign x  =1,2,3,4,5$
$assign key=0,5,0,0,10$
$pick x key$                    !x contains 2,5
$pick key$                      !key contains 5,10
```

15.65 The PLOT directive

Syntax: `$plot` [(*option_list*)] *y_list* *x_list* [*string* [*factor_list*]]

or

`$plot` [(*option_list*)] *y_vectors* *x_vector* [*string* [*factor*]]
(old GLIM 3.77 syntax)

where

y_list is { *y_vector* [/*wt_vector*] [, *y_vector* [/*wt_vector*]]*s* | *list*1 [/ *list*2] }

x_list is { *x_vector* [, *x_vector*]*s* | *list*3 }

factor_list is { *factor* [, [*factor*]]*s* | *list*4 }

and

y_vector, *wt_vector*, and *x_vector* are variates.

The options available in the *option_list* are:

```
ylimit  = real1  ,real2  [,real3 ]
xlimit  = real1  ,real2  [,real3 ]
rows    = real
cols    = real
style   = real
title   = { string
          { macro
hlabel  = { string
          { macro
vlabel  = { string
          { macro
```

Semantics: The PLOT directive causes a multiple scatter plot of the values of the *y*-vectors in the *y_list* against the equivalent *x*-vectors in the *x_list*, with scaling of both axes, and width and height set by *integer*2 and *integer*3 of the OUTPUT directive.

The second syntax is retained for compatibility with GLIM 3.77. If the second syntax is used, the *y-vectors* may be thought of as elements of a *y_list,* and the *x_vector* can be thought of as the only element of the *x_list.*

The number of *y_vectors* must equal the number of *x_vectors* except in the following cases.

(a) One *y_vector* is allowed with multiple *x_vectors*. If there are *k* *x_vectors*, this has the effect of defining *k* superimposed scatter plots, with *y_vector* plotted against each *x_vector* in the *x_list.*

(b) One *x_vector* is allowed with multiple *y_vectors.* If there are *k* *y_vectors*, this has the effect of defining *k* superimposed scatter plots, with each *y_vector* in the *y_list* plotted against the *x_vector.*

Some or all of the *y_vectors* may be associated with a *wt_vector*. Units corresponding to zero elements of the *wt_vector* will be omitted from the scatter plot associated with the *y_vector*.

For each scatter plot defined by a *y_vector-x_vector* pair, the *y_vector*, *x_vector*, the *wt_vector* if present, and the *factor* if present (and %re if it has values) must all be of the same length. There is no requirement that different *y_vector-x_vector* pairs must have the same length.

If %re has been assigned values by the user then units corresponding to zero elements of %re are omitted from all scatter plots which do not have an explicit *wt_vector*.

If the `ylimit` option is used and if only the first two *reals* are supplied, then *real*1 is taken as the lower and *real*2 as the upper plotting limit for the *y* (vertical) axis, and only those *y*-values that fall within these limits (or no more than half an interval outside these limits) will be included in the plot; *real*1 must be strictly less than *real*2. If the option is omitted then GLIM chooses limits that will permit all non-weighted out *y*-values from all *y_vectors* to be plotted; if `%re` has values then the scales are computed only from points for which `%re` has a non-zero value. Printed scale labels at the tick marks on the *y*-axis will have 'neat' values falling within the range between *real*1 and *real*2, and printed scale labels will not necessarily be given at *real*1 and *real*2.

*Real*3 is optional and specifies a user-supplied step length which affects the printed scale labels at the tick marks. If *real*3 is specified, *real*1, as well as specifying the lower plotting limit, also specifies the first printed label. Subsequent printed labels are given at increments of *real*3 until *real*2 is reached or exceeded; the last scale label may be therefore be greater than *real*2. If the value of *real*3 is such that the number of scale labels specified is greater than the number of rows of the scatter plot, or the resultant height of the plot would be above the allowed height as defined above, then *real*3 is ignored, and a plot is produced as if *real*3 had not been specified.

The `xlimit` option has the same effect as the `ylimit` option (with the same restrictions), but on the *x*-(horizontal) axis. If *real*3 is used, it is not possible to force tick marks at intervals closer than 9 columns apart, or to exceed the width of the plot as specified above.

If the `rows` option is used then the *real*, when rounded, specifies the number of rows to be used for the *y*-axis of the plot, except that at least two rows will always be used. If the rounded real is zero, or if the option is omitted, then GLIM chooses that number of rows that will fit conveniently on to a page of output. The rounded real must not be negative.

The `cols` option has the same effect as the `rows` option (with the same restrictions), but on the number of columns used for the *x*-axis, and at least 21 columns will always be used.

The `style` option determines the layout of the plot. If the *real*, when rounded, is zero or if the option is omitted then the plot is annotated on the left with a *y*-axis and *y*-values and at the bottom with an *x*-axis and *x*-values. If the rounded *real* is positive then in addition the whole plot is enclosed in a box, with some extra annotation. If the rounded *real* is negative then no axes or annotation are printed, only the plotted points.

The `title` option allows the user to specify a title, which can be specified either by a *string* or by the text contents of a *macro*. The title will be centred above the scatter plot.

The `vlabel` option allows the user to specify a text label for the *y*-axis. The text label is printed above the *y*-axis and below the title, if specified.

The `hlabel` option allows the user to specify a text label for the *x*-axis. The text label is centred and printed below the scatter plot.

The `title, vlabel,` and `hlabel` options will reduce the size of the scatter plot if the number of rows and number of columns are not specified.

If *string* is included then it defines the symbols to be used for plotting points. If no *string* is present, then the initial letter of the name of each *y_vector* in the *y_list* is used as the plotting symbol, or the initial letters of each *x_vector* in the *x_list* if there is one *y_vector* and multiple *x_vectors*.

If a *factor_list* is present, then a factor may or may not be present for the *i*th scatter plot. If the *i*th factor has l_i levels (l_i is taken to be 1 if the factor is omitted) and if there are *n* *y*-vectors (or 1 *y*-vector and *n* *x*-vectors), then the string must contain at least $\sum_{i=1}^{n} l_i$ characters (excluding blanks and new lines which will be ignored). If the *m*th unit of the *k*th factor is *j* then the *m*th unit of the *k*th *y_vector* will be plotted using the $\sum_{i=1}^{k-1} l_i + j$ th symbol in *string*.

For example with

```
$plot AVEC,BVEC,CVEC XVEC 'Aa B GCc'
         FAC1,,FAC3
```

where `FAC1` has two levels and `FAC3` has three levels, then for `FAC1`(*i*)=1 and `FAC1`(*i*) =2 respectively the symbols '`A`' and '`a`' will be used for the scatter plot `AVEC` against `XVEC` ; the symbol '`B`' will be used for the scatter plot `BVEC` against `XVEC` and for `FAC3`(*i*)=1, `FAC3`(*i*)=2, and `FAC3`(*i*)=3 respectively the symbols '`G`', '`C`', and '`c`' will be used for the scatter plot `CVEC` against `XVEC`.

If no *factor_list* is present, the number of plotting symbols should be equal to the number of *y_vectors* or the number of *x_vectors*, whichever is the greater.

*List*1, *list*2, *list*3, and *list*4 represent identifiers of type *list*. If *list*1 is present, then it should contain vector identifiers representing *y_vectors*. If *list*2 is present, then it should contain vector identifiers representing *wt_vectors*. If *list*3 is present, then it should contain vector identifiers representing *x_vectors*. *List*4, if present, should contain factor identifiers representing the factors in *factor_list*. If *list*4 is specified, a factor must be present for each element of the list.

Examples:

```
$plot (t='multiple plot') y1/w1,y2,y3 x1,x2,x3 $
      ! multiple plot of y1 vs x1 with weight w1,
```

```
        ! y2 vs x2 and y3 vs x3
        ! a title will be produced.
$plot (y=10,19,2) y1 x1,x2,x3 '+ox'$
        ! multiple plot of y1 vs x1 with symbol '+'
        ! y1 vs x2 with symbol 'o' and y1 vs x3
        ! with symbol 'x'. On the y-axis, user
        ! labels will be placed at 10,12,14,16,18
        ! and 20.
$plot y1 y2 x ' ab cd' f
        ! old GLIM377 syntax is still valid
```

15.66 The PREDICT directive

Syntax: `$predict` [(*option_list*)] [*item*] [[,] *item*] s

where the only *option* available in the *option_list* is

`style` = *real*

and where *item* is of the form *LHS* = *RHS* where

LHS is a variate, factor, or array, optionally indexed, corresponding to a model component in the current model formula, and

RHS is an identifier or real according to the specification below.

Semantics: For each *LHS* corresponding to a model component in the model formula, the associated values of the *RHS* identifier or real are used to construct vectors of predicted fitted values, predicted linear predictor, and predicted variance of the linear predictor from the current model. These are stored in system structures and optionally displayed according to the `style` option.

The **number of predicted units** is the number of predicted fitted values produced by a `PREDICT` statement. By default, the number of predicted units is 1. If an identifier is present on the *RHS* of any *item*, then the number of predicted units is taken to be the length of the first non-scalar identifier encountered in the list of *item*s. If this identifier is a two-dimensional array, then the length is taken to the number of rows of the array.

If the number of predicted units is greater than 1, all *RHS* identifiers which are not scalars must have the same length, (where the length of a two-dimensional array is the number of rows of that array) and all *RHS* reals or scalars are treated as constant vectors with values equal to *real* for all units.

For each predicted unit, the parameter estimates, variance–covariance matrix and other stored quantities of the current model are used to construct the predicted linear predictor, fitted values, and variance of the linear predictor, taking as input the values of the identifiers and reals specified on the *RHS* of an *item* corresponding to each of the specified model components.

The type of the model component in the model formula determines the *LHS* specification, and the valid syntax for the *RHS* of the *item*.

In the table below, `X` is a variate, `A` is a factor, `IND` is a variate or factor representing an index, and `Q` is a two-dimensional vector array. The valid values of *LHS* and *RHS* for various types of model component are as follows:

Model component	Value of *LHS*	Valid values of *RHS*
`X`	`X`	*real, vector*
`A`	`A`	*real, vector*
`Q`	`Q`	*vector, array*
`X(IND)`	`X or X(IND)`	*real, vector*
`A(IND)`	`A or A(IND)`	*real, vector*
`Q(IND)`	`Q or Q(IND)`	*vector, array*
X<*integer*>	`X`	*real, vector*
A<*integer*>	`A`	*real, vector*
<*integer*>	`%units`	*real, vector*
binomial denominator	`X` or `%bd`	*real, vector*
offset	`X` or `%os`	*real, vector*

An *item* may be specified which does not necessarily correspond to a model component in the ***current*** model, but must be of the correct type and length according to the rules above. This allows the same `PREDICT` statement to be stored in a macro and called following each of a number of fits, some of which will be submodels of a more complex model.

If no offset is set in the current model, then the *item* corresponding to the offset is ignored. If the `ERROR` setting is not binomial, the predict term corresponding to the binomial denominator is similarly ignored. Note however that the *RHS* of the predict *item*s must be of the correct length and type according to the rules above.

If a model component is present in the model formula, but there is no equivalent *item* present, then a default value is taken according to the rules below.

LHS	Default value of *RHS* if *item* missing
`X`	0.0
`A`	The reference category, if defined, otherwise 1
`Q`	A row vector of zeros
`X(IND)`	Values taken as `X` if present, otherwise 0.0
`A(IND)`	Values taken as `A` if present, otherwise the reference category, if defined, otherwise 1
`Q(IND)`	Values taken as `Q` if present, otherwise a row vector of zeros
`%units`	`%nu+1`
binomial denominator	1.0
offset	0.0

The output from `PREDICT` consists of the following.

(a) A set of three system vectors, of length the *number of predicted units*, as follows:

`%plp`	the predicted linear predictor;
`%pfv`	the predicted fitted values;
`%pvl`	the predicted variance of the linear predictor.

If one or more of `%plp`, `%pfv`, and `%pvl` already exist then their contents are deleted. `%plp`, `%pfv`, and `%pvl` are automatically deleted following the fit of a new model.

(b) If the *real* associated with the `style` option is zero or positive, a statement describing the *RHS* of the predict items for all model components in the current model (and the offset and binomial denominator if appropriate), including the default values assumed if no predict item exists for that model component.

(c) If the *real* associated with the `style` option is positive, the values of the system vectors `%plp`, `%pfv`, and `%pvl` are displayed.

Examples:

```
$fit A+X<2>$
$predict a=4 x=3$
      ! produces a single predicted unit
$assign agelist=20,25...60$
$fit (age+gender)(ind)
```

```
$predict age=agelist gender=2$
      ! produces 9 predicted units with constant
      ! gender group and for 9 different ages
$error b n$fit X+X(LAG)$
$predict x=3 x(lag)=5 $
      ! produces a single predicted unit with X
      ! and X(LAG) taking different values and
      ! with the binomial denominator taken as 1
      !(i.e. predicted probabilities)
```

15.67 The `PRINT` directive

Syntax: `$print` [(*option_list*)] [*item*]*s*

where the only *option* is `store=`*macro*

and where *item* is *real* or *string* or *identifier* or `*`*phrase* or `*`*integer* or `/` or `;` where *phrase* is one of

`margin` *real*1 [`,` *real*2 [`,` *real*3]]

`integer` { *real* | *vector* } [`,` *real*1 [`,` *real*2 [`,` *real*3]]]

`real` { *real* | *vector* } [`,` *real*1 [`,` *real*2 [`,` *real*3 [`,` *real*4]]]]

`line` { *string* | *macro* }

`text` { *string* | *macro* }

`name` *identifier* [`,` *real*1 [`,` *real*2]]

and only the first letter of each phrase is significant.

Semantics: Produces output of *items* on the current output channel or stores output in *macro*, with the layout format determined by the *phrases* and the destination by the *option_list*.

If the `store` option is absent, the `items` will be displayed on the current output channel.

If the `store` option is present, then *macro* specifies the name of a macro to which all *items* will be written. If *macro* exists, then it must not be present on the program control stack. At the end of the `PRINT` statement, the contents of *macro*, if it exists, will be replaced by text specified by *items*. If *macro* does not exist, it will be created.Thus, the contents of macros may be defined dynamically during a GLIM run, and passed as titles, and so on, to plots and graphs. *Items* are written to the macro line by line in the same format as that which would have been used if the items had been output to the current output channel. New line characters are stored in the macro, but a terminating new line character is not stored by default. Thus the `margin`, `text` and `line` phrases all may be used and have an appropriate effect. If a fault occurs during construction of the macro, and *macro* exists, then it will be left unchanged.

Each *item* either causes characters to be output or stored at the current print position or changes the output format of subsequent output. The current print position is initially set to the start of the next line of output; as characters are output it is moved to the end of the last character produced; when the current print position reaches the end of the line a new line character is generated.

An *identifier* must be a *macro*, *vector*, *list*, or *scalar* identifier; a *vector* must have values.

For a macro *identifier* or a *string* the first line of the macro or string is output at the current print position, succeeding lines being printed on new lines; if a line of a macro or *string* cannot be accommodated on the current line it will be broken as specified by the `margin` *phrase* (see below).

For *lists*, starting from the current print position, each identifier in the *list* is printed separated by commas.

Arrays are treated as *vector identifiers*, with their dimensionality ignored. Note that the `TPRINT` directive will print arrays in a tabular format.

For *vector identifiers*, *scalar identifiers*, and *reals* each value is printed at the current print position, provided it can be accommodated on the current line, otherwise printing starts at the beginning of the next line. Values are printed in FORTRAN F format with the number of significant digits n given in the last **integer item* if present, or by the current `ACCURACY` setting if not. Each value usually occupies $n + 4$ characters; however, if the integer part of the value has m digits, with $m > n$, then it will be printed to the nearest integer in $m + 4$ characters. Exact zeros appears as `0.`, that is, with the fractional zero digits suppressed. Fractional zero digits may be suppressed everywhere by using negative values for *integer*.

The *phrases* are defined as follows:

(1) `margin`

Sets the margins and word-break. When the `margin` *phrase* is encountered GLIM prints the current line buffer if it is not empty. If *real*1, when rounded, is non-negative it specifies the left margin, that is, the number of spaces to be inserted at the start of subsequent lines. The default setting is 0. If *real*2, when rounded, is positive it specifies the right margin, that is, the maximum number of characters per line. The default setting is one less than the length of a current output record. If *real*2, when rounded, is zero the right margin will be set to the default value.

If *real*3, when rounded, is non-negative it sets the word-break size (that is, when printing macros and strings, the largest word that will be moved to the next line if it cannot be accommodated on the current line); the default setting is 0.

If the right margin setting is greater than or equal to the current line width it will be truncated to one less than the current line width; if the left margin is greater than or equal to the right margin it will be truncated to one less than the right margin; if the word-break is greater than the difference between right and left margins it will be truncated to that difference.

If any setting is negative or is omitted then the current value is retained.

Example

```
*margin 8,60,10
```

would cause subsequent output to appear in columns 9 to 60 inclusive; words of length 11 or more will be broken if they cannot be accommodated on the current line, but smaller words would be moved to the next line.

(2) `integer`

The value(s) of the *real* or *vector,* rounded to the nearest integer, are printed as integer value(s) under the specified format starting at the current print position.

If the optional *real*1, when rounded, is positive then it specifies the field width for the integer value. If it is zero or negative or is omitted, then each integer value is printed in the minimum field width needed for the value without leading or trailing spaces.

If *real*2, when rounded, is non-negative then it specifies the minimum number of digits to appear in the printed value, with extra zeros on the left if necessary; if it is zero and the integer value is zero, no digits are printed. If it is negative or is omitted then a setting of 1 is assumed.

If *real*3, when rounded, is non-negative or is omitted then the integer value is right justified within its field; otherwise it is left justified.

If an integer value cannot be accommodated within its field then the field will be extended to the right. If the field, extended as necessary, cannot be accommodated on the current line, printing will begin at the start of the next line.

Examples

`*integer 321`	appears as	`321`
`*i 321,6,4,-1`	appears as	`0321`
`*i 1234.5,8`	appears as	`1235`
`*i 1234,2`	appears as	`1234`

(3) `real`

Prints the *real* or *vector* as real value(s) under the specified format, starting at the current print position.

If the optional *real*1, when rounded, is positive then it specifies the field width for the real value(s). If it is zero or negative or omitted, then each real value will be printed in the minimum field width needed for that value, without leading or trailing spaces.

If *real*2, when rounded, is non-negative then it sets the number of decimal places to appear in the real value; that is, the number of digits after the decimal point, whether in E or F format. If *real*2 is negative or is omitted, then GLIM will use enough decimal places to ensure that each real value is printed with the number of significant digits given by the current accuracy setting.

If *real*3, when rounded, is non-negative or is omitted then each real value is right justified within its field. Otherwise it is left justified.

If *real*4, when rounded, is zero then a real value will be printed in F format, provided that a field width has been set and the real value can be accommodated within this field or, if no field width is set, that the real value fits into a field of width $8 + n$, where n is the current accuracy setting; otherwise E format will be used. If *real*4 is negative then E format will be used unconditionally. If the *real* is positive then F format will be used unconditionally.

If a real value cannot be accommodated within its field then the field will be extended to the right. If the field, extended as necessary, cannot be accommodated on the current line, printing will begin at the start of the next line.

Examples

`*real 123.456`	appears as	`123.5`
`*r 123.456,8,2`	appears as	`  123.46`
`*r 12345.6,10,2,-1,-1`	appears as	`1.23e+04`

(4) `line`

Prints the *string* or *macro* on a line by line basis, starting at the current print position. Each line of the *string* or *macro*, except the first, starts on a new line, and lines of the *string* or *macro* that cannot be accommodated on the current output line are broken as specified in the word-break setting of the `margin` option.

(5) `text`

Prints the *string* or *macro* as a continuous stream of characters, starting at the current print position.*Newlines* are ignored, and the layout of the output is determined solely by the `margin` option settings.

(6) `name`

Where *identifier* refers to a currently defined scalar, vector, list, or macro, prints the *identifier* name under the given format, starting at the current print position.

If *real*1, when rounded, is positive then it specifies the field width for the *identifier* name. If it is zero or negative or is omitted, then the name is printed in a minimum field width, without leading or trailing spaces.

If *real*2, when rounded, is non-negative or is omitted, then the *identifier* name is right justified within its field. Otherwise it is left justified.

If an *identifier* name cannot be accommodated within its field then the field is extended to the right. If the field, extended as necessary, cannot be accommodated on the current line, printing will begin at the start of the next line.

Examples

`*name A`	appears as	`A`
`*n ABCD`	appears as	`ABCD`
`*n A,5`	appears as	`    A`
`*n A,5,-1`	appears as	`A`
`*n  ABCD,2`	appears as	`ABCD`

(7) `;`

The character `;` acts as a *newline* symbol, and causes the current buffer to be printed; if the buffer is empty a blank line is printed.

Example

```
$print 'line 1';; 'line 3'
```

prints as

```
line 1

line 3
```

(8) `/`

The *item* `/` will produce a new page if the output device is suitable. Otherwise it will have no effect.

(9) A null statement outputs a blank line.

Examples:

```
$print (sto=title)
    ' Run ' *i %a ' with kappa = ' *r %k $
$graph (t=title) y x$
!
$ca %n=%n+1
$print(store=model) model '+ var' *i %n $
$fit #model$          ! adds a new variate to the
                      ! model formula stored in macro
                      ! model.
```

15.68 The READ directive

Syntax: $\underline{\texttt{\$r}}$`ead` [*identifier_list*] *reals* for free format.

$\underline{\texttt{\$r}}$`ead` [*identifier_list*] `$` *newline* for fixed format.
data records

where *identifier_list* is $\left\{ \begin{array}{c} \textit{identifier} \ [[\texttt{,}]\ \textit{identifier}\ ]s) \\ \textit{list} \end{array} \right\}$

Semantics: If *identifier_list* is absent, initiates reading of the *reals* into the vectors specified by the most recent DATA statement.

If *identifier_list* is present, then each *identifier* (or every *identifier* in the *list*) must be of type *vector.* If no data list has been specified, then a temporary data list consisting of the *identifiers* in the *identifier_list* is set up for the duration of the directive. If a data list exists, then the *identifier_list* should contain a subset of the identifiers in the data list in the same order, and the effect is to read values for all identifiers in the data list, but to store values only for those *identifiers* in the *identifier_list*.

If fixed format has been specified by the FORMAT directive the values of the vectors must follow in a series of *data_records*, starting on the next line, and with each data row starting on a new line, in the format declared in the most recent FORMAT statement; the *reals* are not echoed, regardless of the setting of the statement. The *directive symbol* following the optional *identifiers*, and before the data values, is mandatory for fixed-format reading of the data.

If free format, values follow unit by unit with *reals* separated by spaces, tabs, or new lines and an optional comma.

A READ statement occurring in a macro must not use fixed format.

Examples:

```
$format free$
$data 3 A B C D! specifies 4 vectors of length 3
$read B D            ! Only store values of B and D.
1 2 3 4
5 6 7 8
9 1 2 3              ! A has no values. B contains
                     ! 2,6,1.   C has no values.
                     ! D contains 4,8,3
                     !
$newjob
$units 10
$read x              !temporary data list contains x.
12 3 4 5 6           ! read values
7 8 9 10
$var 5 y
$format (f3.0)
$read y$             !temporary data list contains y.
1.0                  ! read values
2.3
5.6
7.8
9.2
```

15.69 The RECYCLE directive

Syntax: $recycle [*integer*1 [*integer*2 [*real*1 [*real*2]]]]

Semantics: As for the CYCLE directive, except that the values of the vector %fv will be used as initial estimates of the fitted values in subsequent fits of standard models, cancelling any previous setting which was set explicitly through the INITIAL directive, or implicitly through the CYCLE directive.

For user-defined links this directive has the same effect as CYCLE. A RECYCLE statement remains in force until the next CYCLE, RECYCLE, INITIAL, or NEWJOB statement.

Example:
```
$recycle 10 1$      ! recycles from the previous
                    ! fitted values with up to 10
                    ! iterations and printing every
                    ! iteration.
```

15.70 The REINPUT directive

Syntax: $reinput *file_specifier* [*integer1*] [*subfile*]s

Semantics: As for the INPUT directive, except that reading starts at the first record of the file.

Example:
```
$reinput 'mymacro'$
```

15.71 The RESTORE directive

Syntax: $restore [*file_specifier*]

Semantics: Restores the previously dumped program from *file_specifier* or, if omitted, from the default dumping channel. The program continues from the statement following the statement that produced the dump. Note that GLIM cannot read dump files produced on a different machine type, or where the dump file was produced on a machine with a larger workspace size. GLIM4 cannot read GLIM 3.77 dump files.

Example:
```
$restore 'ct.dmp'$      ! restores a GLIM4 dump
                        ! from the file ct.dmp
```

15.72 The RETURN directive

Syntax: $return

Semantics: Causes control to be returned from the current channel to the immediately preceding level of the program control stack. Reading of further subfile identifiers continues under the control of the `INPUT` or `REINPUT` statement that caused input to be read from this channel, or control returns to the previous level of the program control stack if input was started by a `SUSPEND` statement. If no subfiles remain in the `INPUT/REINPUT` statements, the program control stack is decreased by one level. This directive must not occur at the first level of the stack or in macros.

15.73 The `REWIND` directive

Syntax: `$rewind` [*file_specifier*]

Semantics: Causes *file_specifier* (or, if omitted, the default dumping channel) to be rewound, that is, the next input from the channel will come from the first record on the channel. If *file_specifier* occurs elsewhere in the program control stack, reading of the current record of the file is completed before rewinding. The primary input channel, the primary output channel, the journal, and the transcript file cannot be rewound.

Example: `$rewind 'votedata'$`

15.74 The `SCALE` directive

Syntax: `$scale` [*real* [*keyword*]]

where *keyword* is { `deviance` / `chisquared` } and may be abbreviated to its first character.

Semantics: Sets the value of the scale parameter in model fits.

If present, *real* must be non-negative. If *real* > 0, then *real* will be used as the value of the scale parameter. If *real* = 0 or is omitted, then the scale parameter, when needed, will be estimated by a method specified by *keyword*. If this directive is omitted, following an `N`, `G`, `O` or `I error` declaration or following a user-defined error declaration, the default option of *real* = 0 is assumed; following a `B` or `P error` declaration, the default value of 1 is assumed. The declaration persists until the next `NEWJOB`, `ERROR`, or `SCALE` statement.

If *keyword* is `deviance`, then the scale parameter will be estimated by the mean deviance. If *keyword* is `chisquared`, then the scale parameter is estimated by the mean generalized Pearson chi-squared statistic. The estimation setting is not altered by changes in error distribution, and remains in force until the next `SCALE` or `NEWJOB` directive. The default setting is `deviance`.

The current value of the scale parameter, whether estimated or fixed, is stored in the system scalar `%sc`. The system scalar `%scf` contains 1 if the scale parameter is fixed, and 0 if it is to be estimated.

Example:
```
$scale 0.0 c$          ! specifies that the scale
                       ! parameter is to be
                       ! estimated by the mean
                       ! chi-squared statistic.
```

15.75 The `SET` directive

Syntax: `$set` *phrase*

where *phrase* is one of { `batch` / `interactive` / `device = ` *string* / `constant` / `noconstant` }

and where only the first letter of *phrase* is significant.

Semantics: If `batch` or `interactive` is specified, the `SET` directive defines the mode of execution of GLIM. If `batch` is used then GLIM acts as if operating in batch mode (that is, no prompts, only syntax checks following a fault, no `PAUSE`ing, no `SUSPEND`ing, no transcription); if `interactive` is used then GLIM operates in normal, interactive mode. The current mode is stored in the system scalar `%im` [14.2.2].

The `constant` and `noconstant` keywords determine whether or not a constant term (a variate of 1's) is included in a `TERMS` or `FIT` model formula by default. If `constant` is specified, then a constant is included in the model by default; if `noconstant` is specified, no constant term is included in the model by default. Note that neither option restricts the range of models which can be fitted: if `constant` is set then the constant term may be omitted by the addition of `-1` to the model formula; if `noconstant` is set, then the constant term is included by the addition of `+1` to the model formula. The default at the start of a job is `constant`; this is compatible with previous releases of GLIM.

The `device` keyword allows the default graphical output device to be set before any graphical work is done. *String* must contain the name of a valid graphical output device which is implementation dependent (a list of valid devices and the current default may be obtained via the `$environment g$` statement).

Examples: `$set device='t4014'$`
`$set noconstant$`

15.76 The SKIP directive

Syntax: `$skip` [*integer*]

Semantics: Exits conditionally from a macro called by the WHILE directive. Otherwise it behaves as the EXIT directive.

If *integer* = $n > 0$, the current macro or input channel is left and the program control stack is reduced by (n -1). If the new level corresponds to a macro invoked by a WHILE statement, the WHILE scalar is tested, and if non-zero the macro is re-entered. Otherwise the control stack is reduced by a further level, making the action equivalent to `$exit` *integer*. n must be less than the current control level, whose value is held in `%cl`.

If *integer* = 0 or is omitted, the directive has no effect.

Example: `$number A=2$skip A$`

15.77 The SLENGTH directive

Syntax: `$slength` *integer*

Semantics: The SLENGTH directive defines the standard length of vectors. If no SLENGTH directive has been issued, then the standard length is taken to be `%nu` (which is set by the YVARIATE directive), if that is defined, otherwise no standard length is defined. The SLENGTH setting remains in force until the next SLENGTH or NEWJOB statement. The current standard length is stored in the system scalar `%sl`. The UNITS directive is a synonym for the SLENGTH directive, but the latter is now recommended.

Example: `$slen 30$`

15.78 The SORT directive

Syntax: `$sort` [(*option_list*)] *destination_list* [{ *source_list* | * } [*key_list*]]

where

destination_list is { *list* | *vector* [, *vector*]s }

and

source_list and *key_list* are $\left\{ \begin{matrix} list \\ \left\{ \begin{matrix} vector \\ real \end{matrix} \right\} \left[, \left\{ \begin{matrix} vector \\ real \end{matrix} \right\} \right]_s \end{matrix} \right\}$

The following option is available:

`order` = { `ascending` / `descending` }

Semantics: In full form with three lists, applies to each vector in the *source_list* the permutation required to sort the vectors of the *key_list* and assigns the result to the corresponding vector of the *destination_list.* Vectors must have the same length and the number of vectors in the *destination_list* and *source_list* must be equal. The sort would put values into an order of increasing numerical value for option `order=ascending` and of decreasing numerical value for `order=descending`. The sort takes the vectors in the *key_list* one at a time in order so that ties in the values of the first key vector are broken by use of values of the second key vector and so on. Where *real* is used instead of a vector it is replaced by a vector with values generated in the following way. Each *real* is rounded (to, say, *r*) and its absolute value is taken mod *n*, (where *n* is the length of the vectors in the directive) giving, say, *m*.

If *r* is positive then a value *m* indicates a cyclic shift of the vector 1,...,*n* by *m*-1 places left (that is, to *m*, *m* + 1,...,*n*, 1, 2, ..., *m*-1); if *r* is negative then *m* indicates a right shift of *m*-1: the case *m*=0 is treated as +1 or -1 and therefore causes no shift.

Shortened forms:

*list*1 *list*2	is equivalent to	*list*1 *list*2 *list*2
*list*1	is equivalent to	*list*1 *list*1 *list*1
*list*1 * *list*2	is equivalent to	*list*1 *list*1 *list*2

Example:

```
$sort newx,newy,newkey x,y,key key$
       ! sorts the vectors x,y and key by the
       ! vector key, and stores the sorted values
       ! in new structures.
```

15.79 The SSEED directive

Syntax: $sseed [*integer*1 [*integer*2 [*integer*3]]]

Semantics: Resets the *i*th starting value (i =1, 2, 3) of the standard pseudo-random number generator with *integeri* if it is present;

$1 \leq integer1 \leq 4095,$
$1 \leq integer2 \leq 4095,$
$1 \leq integer3 \leq 2047.$

If the *i*th *integer* is absent or zero the *i*th starting value is unchanged. The next set of numbers generated by `%sr` starts from the values as set by this directive. The scalars `%s1`, `%s2`, and `%s3` contain the values of the random number seeds at the current point of the random number sequence, and are updated after every call to the random number generator. The starting values to the generator are given by the `R` option to the `ENVIRONMENT` directive.

Examples: `$sseed 3527 35 941 $`

15.80 The `STOP` directive

Syntax: `$stop`

Semantics: Ends a session and returns control to the operating system.

15.81 The `SUBFILE` directive

Syntax: `$subfile` *identifier*

Semantics: Denotes the beginning of the subfile *identifier*, which ends at the next `RETURN` or `FINISH` statement. A `SUBFILE` statement must stand first on its line. The *identifier* must be a subfile identifier. When a search for a subfile is not in progress the directive is ignored. The substitution symbol # must not appear between the directive name and *identifier.*

15.82 The `SUSPEND` directive

Syntax: `$suspend`

Semantics: Causes subsequent input to be read from the primary input channel. The action taken by the program is as for the `INPUT` directive, with *file_specifier* and *integer2* given the values for the primary input channel and no subfiles being assumed, the next input being taken from the next record of that channel. Control returns to the previous level of the program control stack through the `RETURN` directive. This directive is not available in batch mode.

15.83 The SWITCH directive

Syntax: `$switch` *scalar* $\left\{ \begin{matrix} \textit{macros} \\ \textit{list} \end{matrix} \right\}$

Semantics: Conditionally invokes a macro. If *scalar* = *i* ($1 \le i \le k$) after rounding to the nearest integer, and if *macros* are *macro*1, *macro*2, ..., *macrok,* and *macroi* has contents, then *macroi* is invoked. If $i < 1$ or $i > k$ or *macroi* has no contents, the directive is ignored.

If *list* is present, then all identifiers in the list must be *macros* which are then used as described above.

Example:

```
$number option=3$
$switch option opt1 opt2 opt3 opt4$
             ! will invoke opt3.
```

15.84 The TABULATE directive

Syntax: `$tabulate` [(*option_list)*] [*phrase*]*s*

The following *options* are available:

```
style  = real
cols   = real
```

The following *phrases* are available:

(a) input *phrases* :

`the` $\left\{ \begin{matrix} \textit{variate} \\ * \end{matrix} \right\}$ *statistic* [*real*] [*statistic* [*real*]]s

where *statistic* (of which the first character only is significant) is one of:

```
mean (or * )
total
variance
deviation
smallest
largest
fifty
percentile
interpolate
weight
```

and the *real*, in the range (0.0,100.0), must be present if and only if the *statistic* is `percentile` or `interpolate`

`with` { *variate* / * }

`count` { *variate* / * }

`for` { *vector* [, *vector*]*s* / *list* / * }

(b) Output phrases :

`into` { *variate* / `[ ]` / * } [, { *variate* / `[ ]` / * }] *s*

`using` { *variate* / `[ ]` / * }

`by` { *vector* [, *vector*]*s* / *scalar* [, *scalar*]*s* / *list* / * }

where `[ ]` is an *output_request*; note that this usage of square brackets should not be confused with their use elsewhere in the manual as metalanguage symbols and for enclosing section references.

Any selection of the *phrases* may appear provided at least one input *phrase* is used. The *phrases* may appear in any order and if a *phrase* is repeated then its last instance will be used. The use of a * following a *phrase_name* is equivalent to omitting the *phrase*, that is, is a null setting. Only the first letter of each *phrase_name* is significant.

Semantics:

Summary

For each combination of values of the `for` *vectors* the *statistic(s)* (weighted as specified by the `with` or `count` *variate*) are calculated from the corresponding values of the `the` *variate*. The output classification is stored in the `by` *vectors* or *scalars*, the resultant weights in the `using` *variate*, and the calculated statistic(s) values in the `into` *variate(s)*; in each case use of an *output_request* causes the corresponding values to be printed as a table instead, using the supplied *options*.

Using the phrases

The `for` *phrase* defines the classification of the input data, each *vector* in the list contributing a dimension to the table being formed. The number of levels for a dimension is defined as 'the effective number of levels' of the corresponding *vector*, where the effective number of levels of a factor is defined as its number of levels, and of a variate as its number of distinct values. If the `for` *phrase* is null the table has one cell and is of dimension 0. In all cases there may be zero, one, or more data points contributing to a given cell, and the ith values of the *vectors* determine to which cell the ith data point belongs.

The `by` *phrase* is used to store the classification of the output table; *vectors* will be given values such that they classify the output table in standard order, that is, completely, uniquely, and with the values of the right most *vector* varying fastest; *scalars* will be given the effective number of levels of the corresponding `for` *vectors*.

The `with` or `count` *phrase* supplies input weights; if omitted, a variate of 1's is assumed. For *statistics* `mean`, `total`, `variance`, and `deviation`, the `with` *variate* is treated as if its values were inversely proportional to the variances of the values of the `the` *variate*; for the remaining *statistics*, or if the `the` *phrase* is not used, it is treated as a frequency count. If a `count` phrase is used the weights are treated as frequency counts for all statistics. Zero weights always cause the corresponding data points to be omitted from the calculations.

The `using` *variate* holds the output weights, computed either as variance divisors of the `into` *variate* or as cell frequencies, as given for the `with` or `count` *phrase*.

The `the` *variate* supplies the values to be summarized, in the sense that, for each cell of the output table a single value of the given *statistic* is computed from all values in the *variate* that contribute to that cell in the table being formed.

The `into` variate(s) are used to store the computed values of the statistic(s), in the same order as specified in the `the` phrase.

Definition of calculations

The following formulae are used for computing the `into` and `using` values, for each *statistic* setting.

In each case y is a `the` *variate* value, w is a `with` or `count` *variate* value, the subscript i varies over all data values with non-zero weight contributing to a given cell, m is the number of such data values for the cell, *delta* is the Kronecker delta function, and p is a percentage. The functions percentile (p) and interpolate (p) are defined as follows.

Let

$$y_{(1)}, \ldots, y_{(m)}$$

be the ordered data points for a cell, and

$$w_{(1)}, \ldots, w_{(m)}$$

be their weights. Define

$$w_{(0)} = 0$$

$$W = w_{(0)} + w_{(1)} + \ldots + w_{(m)}$$

and

$$u_{(j)} = \frac{(w_{(0)} + w_{(1)} + \ldots + w_{(j)}) \times 100)}{W} \qquad j = 0,\ldots,m$$

and

$$v_{(j)} = v_{(j-1)} + \frac{0.5\, w_{(j)} \times 100}{W} \qquad j = 1,\ldots,m.$$

Then

$$\text{percentile}(p) = y_{(j)} \qquad \text{if} \quad u_{(j-1)} < p < u_{(j)}$$

$$\text{percentile}(p) = \frac{y_{(j+1)} - y_{(j)}}{2} \qquad \text{if} \quad p = u_{(j)}$$

and

$$\text{interpolate}(p) = y_{(1)} \qquad \text{if} \quad p \le v_{(1)}$$

$$\text{interpolate}(p) = y_{(j-1)} + (y_{(j)} - y_{(j-1)}) \; \frac{p - v_{(j-1)}}{v_{(j)} - v_{(j-1)}}$$

$$\text{if} \quad v_{(j-1)} < p \le v_{(j)}$$

interpolate(p) = $y_{(m)}$ if $v_{(m)} < p$.

For each statistic the `into` and `using` *variate* values are calculated as:

Statistic	`into` *variate*	`using` *variate*
`mean`	$y = \frac{\Sigma(w_i y_i)}{\Sigma(w_i)}$	$\Sigma(w_i)$
`total`	$\Sigma(w_i y_i)$	$\frac{1}{\Sigma(w_i)}$
`variance`	$s^2 = \frac{\Sigma(w_i (y_i - y)^2)}{N-1}$	N-1
`deviation`	$\sqrt{s^2}$	N-1
`smallest`	y_{min} = minimum$\{y_i\}$	$\Sigma(w_i$ delta$(y_i, y_{min}))$
`largest`	y_{max} = maximum$\{y_i\}$	$\Sigma(w_i$ delta$(y_i, y_{max}))$
`fifty`	percentile (50)	$\Sigma(w_i)$
`percentile`	percentile(p)	$\Sigma(w_i)$
`interpolate`	interpolate(p)	$\Sigma(w_i)$
`weight`	(see Note)	

where N=n for the `with` phrase and N=$\Sigma(w_i)$ for the `count` phrase.

Note: The values of the `into` variate for `weight` are the same as those for the `using` variate for the other specified statistic(s). If there is ambiguity, that is, different definitions for different statistics, it will be defined to be $\Sigma(w_i)$.

Restrictions

The *variates* and *vectors* appearing in the input *phrases* must all have values and be of the same length. Those in the output *phrases* need not be defined but if they are defined they must be of the correct length, as given below. An identifier must not appear in both an input and an output *phrase,* nor must it appear more than once in the output *phrases*.

If the `for` *phrase* is omitted the length of the output *variates* and *vectors* is 1. Otherwise the length is the product of the effective number of levels (as defined above) of each *vector* in the `for` *phrase*.

Undefined identifiers in the output *phrases* become defined to be of the same type as the corresponding *variates* and *vectors* in the input *phrases* and to be of the correct length. Each factor in the `by` *phrase* will have its number of levels reset to that of the corresponding factor in the `for` *phrase*.

The `by` *phrase* may be non-null only if the `for` *phrase* is non-null. If both a non-null `for` *phrase* and a non-null `by` *phrase* are used then they must both contain the same number of *vectors* or *scalars*, and if the `by` *phrase* contains *vectors* then each of those *vectors* must be of the same type (that is, variate or factor) as the corresponding vector in the `for` *phrase*.

If the `with` or `count` *phrase* is null then a *variate* of 1's is assumed. All values of the *variate* must be non-negative.

If the `the` *phrase* is null and the `into` *phrase* is non-null the `the` phrase will be taken to be `the * weight.`

Printing of results

If a null `into` *phrase* is used with a non-null `the` *phrase* then an output request is assumed for `into`. If both the `into` and `using` *phrases* are null then an *output_request* is assumed for `using`.

If the table is printed, then it is printed in the same format as by the `TPRINT` directive, and its layout may be controlled by the `style` and `cols` options as described for that directive. If both the `into` and the `using` *variate* are printed then the `into` *variate* is printed first.

If any cell in the output table receives no contributions from the input data then a warning message to that effect is printed.

Examples:

```
$tabulate the Y mean$
      ! find the overall mean of Y
$tabulate the Y mean   for GROUP
      ! find the mean of Y for each
      ! level of GROUP
$tabulate(style=1)  the Y mean  for GROUP
      ! add a border
$tabulate  the Y mean  with WEIGHT  for GROUP
      ! weighted group averages
$tabulate   the Y variance for GROUP
               into VAR    using V_WT
      ! store the group variances in VAR
      ! and the group totals in V_WT
$tabulate    the Y percentile 25   for  A;B;C
                into TABLE          by  AA;BB;CC
      ! store the cross-classifying factors
      ! in AA,BB and CC
```

```
      !----------------------
      ! Formation of contingency tables -
      ! there is no THE variate and the results
      ! are the weights of the various factor
      ! combinations:
$tabulate   for  A,B,C
               into ABC_TABLE  by  AA,BB,CC
      !
      !tables can be collapsed over dimensions:
      !
$tabulate   with ABC_TABLE    for  AA,BB
               using AB_TABLE    by  AAA,BBB
      !
      ! The classification vectors need not be
      ! factors.
$tabulate   for X    into FREQ    by VALUE
```

15.85 The TERMS directive

Syntax: `$terms` [*model_formula*]

Semantics: Sets the specified *model_ formula* to be the current model formula. This is useful in three situations.

(1) It allows a *model_formula* to be specified before using a macro (to fit models).

(2) It allows the *model_formula* to be syntax checked, and structures to be set up before a fit takes place. The system scalars `%pl` and `%ml` [14.2] are set following the `TERMS` statement, and the number of non-intrinsically aliased parameters in the fit can therefore be determined. The option `P` to the `DISPLAY` directive becomes available, and allows users to determine the order of fitting of the parameters in the model. This is useful for users who wish to use the `LOAD` directive.

(3) The system structures `%sb` and `%dm` may be extracted immediately following a `TERMS` directive.

To fit the current `TERMS` model formula, use the `FIT` directive:

`$fit` *operator*`$`

where *operator* is one of `+ - . *` or `/`

Example:
```
$terms A+B $         ! specifies the current model
$print %pl$          ! print the number of parameters
                     ! in the model
$display p$          ! display the order of parameters
                     ! in the model
$fit .$              ! fit the current models
```

15.86 The TIDY directive

Syntax: `$tidy` *identifiers*

Semantics: Provides a method of deleting all identifiers in a list, or reinitializing all local identifiers to a macro.

If *identifier* is of type *list*, then all identifiers in the list are deleted. The list itself is not deleted, but loses its contents, becoming a null list. The effect of deleting an identifier is defined to be as though the identifier was an argument to the DELETE directive.

If *identifier* is of type *macro*, then any *local_identifiers* to that *macro* become devalued, losing their values, type, and other attributes.
If *identifier* is of any other type, the *identifier* is ignored.

Example:
```
$list L=A,B,C$
$tidy L$
      ! has the same effect as
$delete A B C$list L=L-A-B-C$
```

15.87 The TPRINT directive

Syntax: `$tprint` [(*option_list*)] *list*1 [*list*2]

The following *options* are available:

`style` = *real*
`cols` = *real*

and *list*1 = { *array* [, *array*]s ; *vector* [, *vector*]s ; *list* }

and *list*2 = { *vector* [{ / , } *vector*]s ; *real* [{ / , } *real*]s ; *list* ; * }

and the use of `*` is equivalent to omitting *list*2

Semantics: The `TPRINT` directive prints the values of the *vectors* in *list*1 as the body of a table classified by the *vectors* in *list*2.

The `style` *option* determines the layout of the table. If the *real*, when rounded to the nearest integer, is zero or if the *option* is omitted, then the table is annotated with the names of the *vectors* in *list*2, (or with dimension numbers if *reals* are used in *list*2) and with the names of the *vectors* in *list*1 if there is more than one *vector* in that list; it is indexed by the distinct values of the *vectors* (or by the values implied by the *reals*) in *list*2. If the *real*, when rounded, is negative only the values of the *vectors* in *list*1 are printed, with no annotation. If the *real*, when rounded, is positive then the table is drawn as for `style=0` but borders are drawn around the table and around each of its dimensions.

The `cols` *option* sets the page width to be used for printing the table. If *real*, when rounded, is positive then it sets the maximum number of columns (that is, characters) per output line, except that values greater than the current line width are reduced to the current line width, and values less than 20 are treated as 20. If the rounded *real* is zero then the current line width is used. The rounded *real* may not be negative.

All *vectors* or *arrays* must be defined and have the same length. *List*2 may be omitted if *list*1 contains only *vectors,* or contains only *arrays* of the same shape. The *vectors* or *arrays* from *list*1 form the body of the table, multiple *vectors* or *arrays* being written in parallel down the page.

If *list*2 is *vectors* then each *vector* contributes a dimension to the table, the number of levels of the dimension being the effective number of levels of the *vector*. The elements of the vectors in *list*2 must be in regular order, but not necessarily standard order. Vectors are in **standard order** when they form a complete and unique classification with the values of the right-most vector varying fastest. A list of vectors which is a permutation of a list of vectors in standard order is said to be in **regular order**. Note that if any of the *vectors* in *list*2 is a variate, its distinct values must occur in numerical order for *list*2 to be in regular order.

`TPRINT` puts as many (column) dimensions across the page as the output width and the setting of the `column` option permits. This can be overwritten by the use of / as a separator in *list*2, and which can occur only once in *list*2. *Vectors* before the / will be used as row dimensions, and *vectors* following the / will be used as column dimensions if the page width permits.

If *reals* are used in *list*2 they specify the number of levels in each dimension and the values of the table are taken to be in standard order.

Examples:
```
$tprint  VAR  2,3,4
      ! print VAR as a 2x3x4 table
$factor A 2 B 3 C 4$
$calc A=%gl(2,12) : B=%gl(3,4) : C=%gl(4,1)
$tprint  VAR  A,B,C $
      ! an alternative method of printing
      ! the table, which also allows permutation
      ! of the cross-classifying factors
$tprint  VAR  B,A,C  :  VAR C,A,B
      ! Multiple tables can be printed in
      ! parallel if they have the same shape
$tprint  VAR1,VAR2  A,B,C
      !increasing annotation
$tprint(style=-1)  VAR1,VAR2    A,B,C
   :               VAR1,VAR2    A,B,C
   :   (style= 1)  VAR1,VAR2    A,B,C
```

15.88 The TRANSCRIPT directive

Syntax: transcript [(*option_list*)] $\left\{ \begin{array}{c} name\ [\ [,]\ name\ ]\ s \\ \left\{ \begin{array}{c} + \\ - \end{array} \right\} name\ [\ [,]\ \left\{ \begin{array}{c} + \\ - \end{array} \right\} name\ ]\ s \\ integer \end{array} \right\}$

where *name* is echo, fault, help, input, ordinary, verify, or warn but only the first character is significant.

The following option is available:

style = *real*

Semantics: Determines which input and output sources are written to the transcript file. The *names* stand for the following sources:

Tag	Name	Source	Value of name
[i]	input	Primary input	1
[v]	verify	Verification of macros	2
[w]	warn	Warning messages	4
[f]	fault	Fault messages	8
[h]	help	Help messages	16
[o]	ordinary	Ordinary output	32
[e]	echo	Echoing of secondary input and output	64

The `style` *option* determines the style of the transcript lines sent to the transcript file. If the *real*, when rounded to the nearest integer, is zero or if the *option* is omitted, then a transcript tag given in the above table (the initial letter of each *name* enclosed in square brackets) precedes each transcript line. If *integer*1 is negative then no such tag is produced.

Three forms of the `TRANSCRIPT` directive exist.

The first form sets the absolute setting. If a *name* is used, then subsequent input or output from the corresponding source will be sent to the transcript file. If no *name* is supplied, then no input or output is sent to the transcript file. At the start of a job in interactive mode, the *names* `input, warn, fault, help, and ordinary` are assumed. This reproduces the user's screen on most systems.

The second form allows the updating of a previous setting by the addition and/or subtraction of sources. If a *name* is preceded by a `+,` then the specified source is added to the contributing transcript sources. If a *name* is preceded by a `-,` then the specified source is deleted from the contributing transcript sources. Addition of a source which is already present (or deletion of a source which is not present) is ignored.

The third form allows an integer to be used in place of *name*s. The integer is defined by the sum of a set of source values; integer values for each source are given in the table above. The system scalar `%tra` contains the sum of the source values for the current transcript setting (and thus its values may be copied for later use). Note that for this directive, the integer must not be substituted by a *user scalar* but only by an *ordinary* or *system scalar*.

The `VERIFY`, `BRIEF`, `WARN`, `ECHO,` and `OUTPUT` directives act on the transcript setting as well as on the output to the current output file, that is, `$warn on$` would do an implicit `$tra +w$`

Examples:

```
$cal %o=%tra        ! save current transcript setting
$use mymacro        ! use a macro (which might change
                    ! the transcript settings)
$tra %o$            ! recover the old settings
$tra +e$            ! and add echoing if not already
                    ! present.
                    !
```

15.89 The `UNITS` directive

Syntax: `$units` *integer*

Semantics: This directive is a synonym for the `SLENGTH` directive. Note that the `UNITS` directive sets the value of the system scalar `%sl`. It has no effect on the system scalar `%nu`, which is set by the `YVARIATE` directive.

15.90 The USE directive

Syntax: `$use` *macro* [*items*]

where *item* is *identifier* or `%digit` or `*` or `-`, and the number of *items* ≤ 9.

Semantics: Sets the arguments for *macro* in the same fashion as for the `ARGUMENT` directive, and then invokes the *macro*; if the macro has no contents then the invocation has no effect.

Example: `$use mac t v$`

15.91 The VARIATE directive

Syntax: `$variate` [*integer*] $\left\{\begin{array}{c}\textit{identifiers}\\ \textit{list}\end{array}\right\}$

Semantics: Declares that *identifiers* (or *identifiers* in the *list)* are variates of length *integer* (> 0) or, if *integer* = 0 or is omitted, of standard length. The *identifiers* must be previously undefined or vectors of the length implied by *integer.*

Example: `$var 20 a b c$`

15.92 The VERIFY directive

Syntax: `$verify` [$\left\{\begin{array}{c}\texttt{on}\\ \texttt{off}\end{array}\right\}$]

Semantics: Controls the macro verification state. If macro verification is switched `on` then each line of an executing macro is written to the current output channel (and to the transcript channel, if set) before it is executed. If verification is switched `off` no such writing occurs. If verification is turned `on` by the `VERIFY` directive, then the verification output source records are included in the transcript output by default; if turned `off`, then they are excluded by default.

15.93 The WARN directive

Syntax: `$warn` [$\left\{\begin{array}{c}\texttt{on}\\ \texttt{off}\end{array}\right\}$]

Semantics: Controls the state whereby certain warnings are or are not printed. See [16] for the list of warnings that are affected by this directive. If warnings are turned `on` by the `WARN` directive, then the warning output source is included in the transcript output by default; if turned `off`, then they are excluded by default.

15.94 The `WEIGHT` directive

Syntax: `$weight` [*vector*1 [(*vector*2)]]

Semantics: Declares prior weights for subsequent fits.

*Vector*1 will be expected to hold the prior weight values for all future fits and *vector*2 represents an optional index variate.

If there is no index variate, then *vector*1 must be either a variate of length the number of units or previously undefined; if the latter, then it will become defined as a variate of length the number of units.

If indexing is being used, then *vector*1 need not have the length of the number of units. *Vector*2, if it is defined, must have the length of the number of units.

The variance of each unit will be divided by its prior weight during the fit. Zero prior weight values cause the unit to be excluded from the fit. Subsequent references to the system pointer `%pw` will refer to *vector*1 and to the system pointer `%wi` will refer to *vector*2.

If *vector*1 is omitted, then the prior weight is unset, with no adjustment of the variance of each unit. `%pw` will point to an internal scalar containing the value 1.0, and `%wi` will become undefined.

The setting remains in force until the next `WEIGHT` or `NEWJOB` directive or until *vector*1 or *vector*2 is deleted. The `YVARIATE` directive may also unset the prior weight variate and the prior weight index, if set, where the new number of units differs from the length of the prior weight variate or its index, if set.

Example:

```
$weight W(I)          ! declares the prior weight to
                      ! have values of A and index of I
```

15.95 The `WHILE` directive

Syntax: `$while` *scalar* *macro*

Semantics: Causes repeated invocation of *macro* for as long as the rounded value of *scalar* is non-zero. The test occurs before each (including the first) invocation of *macro*. If *macro* has no contents or if the rounded value of *scalar* is zero when the directive is first issued then the directive is ignored.

Example:

```
$macro mac
```

```
$pr I$ca I=I-1$
$endmac
!
$number I=3 $while I mac$      ! produces the output
                               ! 3 2 1
```

15.96 The YVARIATE directive

Syntax: $\underline{\$y}$variate *vector*1 [(*vector*2)]

Semantics: Declares the *y*-variate or response variate for subsequent fits.

*Vector*1 will be expected to hold the values of the *y*-variate for subsequent fits and *vector*2 represents an optional index variate.

If no index variate is present, then *vector*1, if previously defined, must be a variate. The directive will cause the number of units %nu to be redefined to be the length of *vector*1. If *vector*1 is undefined then it will be defined as a variate of length %nu; in this case %nu must be non-zero.

If indexing is present, then *vector*2, if defined, must be a variate or a factor; it will cause the number of units %nu to be redefined to be the length of *vector*2; if *vector*2 is undefined then it will be defined as a variate of length %nu; in this case %nu must be non-zero.

Subsequent references to %yv will point to *vector*1, and to %yi will point to *vector*2.

If the length of any of the offset, prior weight, binomial denominator, or initial values variate (or the index of these variates, if appropriate) is not equal to the new value of %nu, then the appropriate pointer is unset and a warning is displayed. The YVARIATE directive does not otherwise affect the model settings.

The setting remains in force until the next YVARIATE or NEWJOB directive or until *vector*1 or *vector*2 is deleted. The YVARIATE directive may also unset the prior weight variate and the prior weight index, if set, where the new number of units differs from the length of the prior weight variate or its index, if set.

Example:

```
$yvar Y$
$fit X1+X2$                ! fit model
$ass IND=1,3...%nu$        ! every second observation
$yvar Y(IND)
$fit (X1+X2)(IND)$         ! refit model
```

16 Faults and error handling

The sections in [15] describe the normal behaviour of GLIM, that is, when the *statements* presented are syntactically correct and specify valid operations. In this chapter we describe how GLIM handles exceptions. Such exceptions occur when GLIM detects a potentially dangerous occurrence ([16.1] on warnings), or is presented with a syntactically or semantically incorrect *statement* or encounters the `FAULT` *directive* ([16.2] on faults), or is interrupted by the user while executing a *statement* ([16.3] on break-ins), or is interrupted by the operating system when detecting a floating point error [16.4], or finds an inconsistency within its own data structures ([16.5] on system faults), or finds a FORTRAN inconsistency which GLIM cannot handle ([16.6] on FORTRAN level faults).

16.1 Warnings

A warning message is a line of output beginning with '`--`' They are for information purposes only, and are not fatal in either interactive or batch mode. They indicate that although the *directive* was syntactically and semantically correct, some occurrence was detected which was unusual or potentially dangerous. In general, the execution of the *directive* proceeds as normal. Occasionally, however, GLIM may take some action to protect the user from a potential problem.

Warnings are sent to the current output channel (provided there is one) and may be inhibited on the current output channel by the use of the `WARN` *directive* [15.93]. The `TRANSCRIPT` *directive* [15.88] can be used to control warning messages output to the transcript channel. The following is a list of warnings.

16.1.1 Warnings during a fit

Warnings are given as part of the output from a `FIT` *statement* [15.31] when a condition has occurred that indicates that the results may be unreliable.

16.1.1.1 `-- (change in df)`

A parameter has been aliased during the fit. Results will be incorrect.

16.1.1.2 `-- (iterations diverged)`

The numerical algorithm has detected divergence. Results may be meaningless.

16.1.1.3 `-- (no convergence yet)`

The numerical algorithm has not converged after the maximum number of iterations. Results will be only approximately correct. The statements

```
$recycle $fit .$
```

may be used for further iterations of the same model.

16.1.1.4 `-- unit(s) held at limit`

This message occurs with the logit, probit, and complementary log-log links. The linear predictor for one or more units is tending to plus or minus infinity. The fitted value is set to either 0 or the binomial denominator for the binomial distribution, and to either 0 or 1 for other distributions. This problem might possibly be caused by the model being over-parametrized. The (scaled) deviance will be slightly too large.

16.1.2 Warnings relating to model checking

These warnings will occur during model definition or in the early model checking carried out by the `FIT` *directive* [15.31].

16.1.2.1 `-- METHOD changed to Gauss-Jordan`

The `LOAD` *variate* [15.47] has been set when the algorithmic method has been set to Givens and GLIM does not allow this combination.[1] GLIM therefore changes the method automatically to Gauss–Jordan.

16.1.2.2 `-- LOAD variate unset`

The `METHOD` [15.53] has been set to Givens when a `LOAD` variate has been set and GLIM does not allow this combination.[1] .GLIM therefore unsets the `LOAD` variate.

16.1.2.3 `-- Null fit - LOAD variate ignored`

There are no parameters to be estimated (the model is null) but a `LOAD` variate is set. This is meaningless, and the `LOAD` variate is ignored for the duration of the current fit.

16.1.2.4 `-- prior weight unset`

`-- offset unset`

`-- initial values unset`

`-- binomial denominator unset`

By changing the *y*-variate, the number of units in the fit has changed, and the prior weight (offset, initial values, binomial denominator) vector is now of the wrong length. The setting is inappropriate and is unset.

16.1.3 Warnings related to model display and extraction

Use of the `DISPLAY` and `EXTRACT` *directives* ([15.16], [15.27]) is inhibited as soon as the definition of the current model is changed in some way by user action ([16.1.3.1] to

[1] Users who insist on a load variate with the Givens algorithmic method may write a `LOAD` macro to adjust the triangle.

[16.1.3.10]). This is to prevent users misinterpreting the extracted quantities or displayed output as being from the adapted model (which has not yet been fitted) rather than from the fitted model. GLIM employs some extensive checking to ensure that when the model is changed, possibly inadvertently by some action unconnected with model fitting or model definition, then this is detected. The following warning messages are possible.

16.1.3.1 `-- model changed`

This message occurs when model settings are changed (error, link, *y*-variate, offset, prior weight, binomial denominator, initial values, load, baseline, terms formula, eliminate formula, or a macro defining a user-defined fit).

16.1.3.2 `-- change to length of model vector`

A vector in the terms formula, eliminate formula, or set as a *y*-variate, weight, offset, binomial denominator, initial value, or load or an index to these, has had its length increased by an `ASSIGN` *statement* [15.5].

16.1.3.3 `-- change to parametrization of model`

The parametrization of the model has been changed by the redefinition of a factor reference category.

16.1.3.4 `-- change to number of parameters in model`

Either a variate in one of the model formulae has been redefined as a factor (or vice versa), or a scalar in the terms model formula representing the order of a polynomial term has been modified.

16.1.3.5 `-- change to data values affects model`

A vector in the terms formula, eliminate formula, or set as a *y*-variate, weight, offset, binomial denominator, initial value, or load or an index to these, has had its contents changed by a *directive*.

16.1.3.6 `-- change to macro affects model`

A macro has been modified by the `PRINT` or `EDMAC` *directives* ([15.67], [15.20]) or redefined by the `MACRO` or `GET` *directives* ([15.50], [15.33]), and that macro defines a part of the model fitting algorithm.

16.1.3.7 `-- deletion affects model`

This message occurs when model identifiers are deleted (error, link, *y*-variate, offset, prior weight, binomial denominator, initial values, load, an identifier in the terms formula, eliminate formula, or a macro defining a user-defined fit). Redefine and refit the model if required.

16.1.3.8 `-- no parameters in model`

There are no parameters in the model and either one or more of the `A`, `E`, `P`, or `U` options to the `DISPLAY` *directive* have been used or one or more of the identifiers `%pe`, `%se`, `%al`, `%vc`, `%dm`, or `%sb` were specified in the `EXTRACT` *directive*. The relevant `DISPLAY` option is ignored and the specified vector or vectors are not extracted.

16.1.3.9 `-- No orthogonal polynomials present in model`

The `O` option to the `DISPLAY` *directive* has been used to examine the definition of the orthogonal polynomials in the current model, but none are present. The option is ignored.

16.1.3.10 `-- number of units changed`

By changing the *y*-variate, the number of units in the fit has changed. Check the terms and eliminate model formulae if set for vectors of inappropriate length.

16.1.3.11 `-- identifier X needed in model fitting - unable to delete`

A user-defined fit is in progress, and one of the macros (or the user from the macro debugger) is attempting to delete identifier *X*. This identifier is used to define some part of the fit (*y*-variate, weight, part of the model formula, and so on) and so *X* cannot be deleted.

16.1.3.12 `-- macro X needed in model fitting - unable to delete`

A user-defined fit is in progress, and one of the macros (or the user from the macro debugger) is attempting to delete macro *X*. This macro defines one of the macros in the user-defined fit and so *X* cannot be deleted.

16.1.4 Other warnings

16.1.4.1 `-- new job begins`

This message occurs at the start of every job except the first (that is, after a `$NEWJOB` *statement*).

16.1.4.2 `-- execution ends - syntax checks continue`

This message occurs when GLIM is used in batch mode and a fault occurs.

16.1.4.3 `-- standard length re-initialized`

This occurs when the `SLENGTH` or `UNITS` *directives* ([15.77], [15.89]) are used to change the standard length.

16.1.4.4 `-- program dump completed`

This message follows the successful completion of a `DUMP` *directive* [15.17].

16.1.4.5 `-- program restarts from previous dump`

This message follows the successful completion of a `RESTORE` *directive* [15.71].

16.1.4.6 `-- pausing`

Control is returned to the operating system by a `PAUSE` *directive* [15.63].

16.1.4.7 `-- the pause directive is not implemented`

The `PAUSE` *directive* is ignored.

16.1.4.8 `-- floating-point error(s)`

This message occurs when floating-point error detection is installed, and a floating-point error occurs in the `CALCULATE` *directive* [15.8]. The result of the calculation is set to zero and the calculation proceeds.

16.1.4.9 `-- invalid function/operator argument(s)`

This occurs when invalid values are given in the `CALCULATE` *directive* for the `/` or `**` operators or for one of the function arguments. The calculation proceeds, assigning zero to the result of the operation or function. Valid ranges for some of the function arguments are given by the `M` option to the `ENVIRONMENT` *directive* [15.24].

16.1.4.10 `-- the table contains empty cell(s)`

This indicates that the table resulting from a `TABULATE` *directive* [15.84] has at least one cell that had insufficient data to compute the required statistic. The output weight(s) for the corresponding cell are set to zero.

16.1.4.11 `-- data list abolished`

This message occurs if a vector included in the most recent `DATA` *statement* [15.12] is deleted. The data list is lost.

16.1.4.12 `-- list X abolished`

This occurs if an identifier included in a list *X* is deleted. The list *X* is abolished.

16.1.4.13 `-- macro X in use - unable to delete`

Deletion of macro *X* has been requested. This macro, however, is being executed on the program control stack and so cannot be deleted.

16.1.4.14 `-- No manual entry for requested item`

There is no information on the requested topic within `MANUAL` [15.51]. Check the spelling of the item, and issue the statement `$man info$` for more information on how to use the `MANUAL` *directive*.

16.1.4.15 `-- list truncated`

This occurs if too many items are supplied for the `LOOK` *directive* [15.48]. Only as many items as can be accommodated by the width of the current output channel will be output, the remainder will be ignored.

16.1.4.16 `-- X not in list`

This message occurs if an attempt is made to modify a list using the `LIST` *directive* [15.46] with the - operator, and the item *X* to be subtracted from the list is not in the list.

16.1.4.17 `-- filename truncated`

The filename given exceeds the maximum length allowed in GLIM on the local system, and has been truncated.

16.2 Faults

A fault occurs when GLIM encounters a *statement* that is syntactically incorrect (that is, is not formed correctly) or semantically incorrect (that is, requests an operation that GLIM cannot perform). In addition, faults may be generated by the `FAULT` *directive.*

When a fault occurs, GLIM abandons execution of the statement and outputs a message of the form:

`**` *fault_message* `at [` *statement_context* `]`
`on level` *integer* `from` *stack_context*

but the second line is not printed if the program is currently at level 1 of the program control stack. The *fault_message* describes the detected fault (for example, '`no such directive name`'). The *statement_context* consists of those *characters* that were most recently read by GLIM when the fault was detected, and serves to indicate the *statement* that gave rise to the fault; up to nine *characters* are printed by GLIM, but only back as far as the beginning of the input line or the start of the *macro* containing the *characters.* The *integer* on the second line (if the line is printed) gives the current stack level where the fault was detected. The *stack_context* describes the source occupying that level; if it is an input channel source then *stack_context* has the form:

`channel` *integer*

where *integer* is the channel number of the source; if it is a *macro* source then *stack_context* has the form:

`macro` *identifier*

where *identifier* is the name of the *macro.* Note that the *statement_context* and *stack_context* printed by GLIM cannot be regarded as definitive guides to the origin of the fault: a fault that originated in one place may not be detected by the program until several more *characters* (including *newlines*) have been read, and since *statements* may extend over several sources [13.7.4] even the program control stack level may have changed.

The *fault_message* indicates the type of fault that has occurred. For example, if a macro MAC1 contained a statement

```
$tprint Y 4,8 $
```

when Y was a *vector* of length 24, use of MAC1 would result in

```
** shape inconsistent with length of vector
at [nt Y 4,8 $] on level 2 from macro MAC1
```

If BRIEF diagnostic messages are supressed [15.7] a further message of more detailed diagnostics may be given with possible remedies. If the BRIEF setting is on, this information may still be obtained if HELP is the next *directive* [15.39]. For example:

```
$help$
The table vector(s) have length 24, whereas the
shape specified by the integers implies a length
of 32. Re-enter the directive with the correct
shape.
```

Subsequent action of the program depends on the mode of use. In batch mode, reading and checking of the *directives* continues without execution; some subsequent faults may be spurious, being caused by a non-execution of preceding *directives*. In interactive mode, control returns to the most recent instance of the primary input channel in the program control stack, intermediate execution of macros and input from secondary channels is abandoned, and input is obtained from the primary input channel.

Wherever possible GLIM restores the program state that existed prior to execution of the offending statement. However, this may not be possible, for example with an index out of range in a CALCULATE *statement*. The result can be checked only by inspection of the vectors involved using the LOOK or PRINT *directives*.

16.3 Break-ins

On some implementations a soft break-in facility is available: pressing the break-in key at any point during a GLIM session causes GLIM to abandon the execution of the current *statement* and return to the terminal for more input. The actual key or keys used is implementation dependent and the User Note should be consulted for further details. The break-in facility is available in batch mode, but on many implementations it will not be possible to generate the break-in signal in that mode. When GLIM receives a break-in signal it behaves in precisely the same way as it would if it had detected an ordinary fault at that point, and follows the procedure described in [16.2] except that, instead of printing a fault (and diagnostic) message, it prints a message of the form:

```
** break-in, at [ statement_context ]
      on level integer from stack_context
```

where *statement_context*, *integer,* and *stack_context* are as previously defined. Except for the message printed, the only difference between break-ins and faults is that the former are generated by the user while the latter are detected by the program.

On some implementations the break-in key also serves to abort the program in the sense that if the key is repeatedly pressed (usually at least four times) without GLIM responding then the session will be abandoned.

16.4 Floating-point errors

On some implementations software floating-point error detection is available: if a floating-point error (for example, division by zero or a real number which is too large) occurs at any point during a GLIM session then GLIM will usually abandon the execution of the current *statement* and return to the terminal for more input. This does not happen when executing the `CALCULATE` *directive*; the effect of a floating-point error is to set the value to zero and to print a warning message. Floating-point error detection is available in batch mode.

When GLIM receives a floating-point error signal and the `CALCULATE` *directive* is not being executed, it behaves in precisely the same way as it would if it had detected an ordinary fault at that point, and follows the procedure described in [16.2] except that, instead of printing a fault message, it prints a message of the form:

```
** floating point error, at [statement_context ]
      on level integer from stack_context
```

where *statement_context*, *integer,* and *stack_context* are as previously defined. If not inhibited by the `BRIEF` *directive*, a diagnostic message giving the *directive* name in which the floating-point error occurred is then printed. Except for the message printed, there is no difference between break-ins and floating-point error detection.

16.5 GLIM system faults

GLIM maintains internal checks on the validity and consistency of its own data structures and operations. If an inconsistency or invalid status is detected it will print a system fault message, together with a system fault number (beginning at 900 and indicating the problem), and abandon the session. Such occurrences are believed to be very rare, but should one happen the user is asked to follow the fault reporting procedure given below [16.7].

16.6 FORTRAN-level faults

GLIM has no control over certain FORTRAN operations. In some cases an invalid operation will result in a FORTRAN execution error and the response of the system will be system dependent — a helpful message may or may not be given, the operating system may allow recovery from the error, or execution of GLIM may be prematurely terminated. On some systems the transcript file may be lost unless a specific command to the operating system has been given.

Listed below are the types of operation for which GLIM does not perform prior checks and the *directives* during whose execution they are most likely to occur. It should be emphasized that GLIM maintains extensive checks against possible faults but that detection of the faults indicated below is practicable only by the operating system. Hence, should an abnormal termination occur during a GLIM run the user is advised to check the list below for a possible cause before following the procedure indicated in [16.7].

Input/output channel numbers On some systems, there is only a partial check on the use of invalid channel numbers for secondary channels, beyond a check that the number is positive (`INPUT, REINPUT, DINPUT, OUTPUT, DUMP, RESTORE, REWIND, OPEN, CLOSE`) [13.6].

Format specification The format statement used in the `FORMAT` *directive* [15.32] is only partially checked for adherence to FORTRAN specifications. Any fault will appear when reading of data commences (`READ, DINPUT`).

Real number overflow If no floating-point error detection is in operation, attempts to produce larger real numbers than are permissible on the machine are not prevented (all *directives are affected*, but especially `CALCULATE, FIT,` and `DISPLAY`). Note all *integers* are checked for size by the program.

Non-printing characters Precautions must be taken that non-printing characters (such as carriage control characters) do not appear on input files, and so on, as the result of attempting to read such characters can be unpredictable (this affects all *directives*).

Rewinding The implementation of the rewind facility included in the program is not guaranteed. In particular, the system may impose restrictions on arbitrary sequences of dumping, restoring, and rewinding (`REWIND, REINPUT, DUMP, RESTORE` [13.6]).

Pausing The warnings given in [13.7.7] should be noted (`PAUSE`).

System restrictions Restrictions on resource allocation (for example, file space, file access privileges, cpu time, and so on) cannot be circumvented by GLIM, and actions that exceed any system-imposed limitations may be fatal (all *directives*).

16.7 Fault reporting

Should GLIM abandon a session with a system fault or otherwise terminate abnormally the user is requested to send details, together with all other relevant input and output, to the local distributor or to:

The GLIM Co-ordinator
NAG Central Office
Wilkinson House
Jordan Hill Road
Oxford OX2 8DR
UK

If GLIM was being used in interactive mode, then a copy of the transcript file and journal file, together with all input files used by the run, should be sent. Details of the version of GLIM4 and the type of machine being used should also be provided.

17 The PASS facility

17.1 Introduction

The PASS directive enables values stored in GLIM to be accessed and changed by an implementation-supplied subroutine.

The PASS directive has the syntax

$pass [(*option_list*)] { *keyword* / *integer* } [*vector_list* [*macro*]]

where *option_list* has one option:

array=fortran

vector_list is { *vector* / *scalar* } [, { *vector* / *scalar* }] s

keyword is defined by the writer of the PASS routine.

Execution of a PASS statement causes the *keyword* or *integer*, the values of the list of *vector(s)/scalar(s),* and the elements of the *macro* to be passed to the FORTRAN subroutine PASS. As distributed in the standard release of GLIM, the body of subroutine PASS includes some routines for illustrative purposes. It is anticipated, though, that the user may wish to supply their own body for PASS, which will perform certain operations on the input parameters that are not available (or not easily available) via the standard GLIM directives. This note explains how this may be achieved.

17.2 Scope of the PASS facility

The three items for the PASS directive are:

(1) An unsigned *integer* or a *keyword* which may be used to specify an option within the PASS subroutine; this enables more than one operation to be implemented, the particular operation required being selected by specifying the corresponding option. If a *keyword* is given it is passed to PASS as a character string of up to 8 characters (after all characters have been converted to upper case).

(2) A list of *vector(s)* and/or *scalar(s).* All *vectors* must have a defined length. If a *vector* has not yet been assigned values, all its elements will be set to zero. The

elements of the *vectors* and *scalars* are presented to the PASS subroutine in a ***single*** FORTRAN REAL array of suitable length, and the values may be changed by the PASS routines; the values stored in the array on exit from `PASS` become the new values of the *vectors* and *scalars*, and may be accessed in subsequent GLIM directives (for example, by `PRINT`ing them out). If this *list* is omitted then the FORTRAN array will have length 0.

(3) A *macro*, which must have values (for example, via the `MACRO` directive). The elements of the *macro* are presented to the `PASS` subroutine as a FORTRAN INTEGER array of the same length. Each element of the array is the integer representation of the corresponding *macro* character. For example, if the fifth character of the *macro* is 'H', then the following FORTRAN statement is .TRUE.

CHAR(CARRAY(5)).EQ.'H'

where CARRAY is the name of a FORTRAN INTEGER array. End-of-line symbols are stored as the value ICHAR ('!'). On return from PASS all characters in the array are checked by GLIM and any character not in the GLIM character set (such as, a control character) is replaced by the character '`@`'. The *macro* may be accessed in sub-sequent GLIM directives (for example, by `PRINT`ing it out).

The next section discusses implementation of the PASS subroutine. Section [17.4] discusses the HPASS subroutine used for printing diagnostic messages on PASS faults, while Section [17.5] discusses the use of the subroutine EPASS in printing information on the user's implementation of the PASS subroutine. Note that only the subroutines PASS, HPASS, and EPASS need to be modified to implement the `PASS` facility. Section [17.6] describes the GLIM routines that may be called from PASS, EPASS, and HPASS while Section [17.7] deals with some general considerations. Section [17.8] gives an illustrative example.

17.3 The PASS subroutine

The formal and actual arguments of PASS are supplied by GLIM. The user supplies the appropriate body of code for the required facility. The formal arguments are:

OPT	Integer	Input	Holds the (non-negative) value of the *integer* supplied by the user or a negative value if none was supplied.
NAME	Character*8	Input	Holds the *keyword,* if supplied (converted to upper case).

RARRAY	Real array (1:RLEN)	Input and/ or Output	Holds the values of the *vectors* and *scalars* supplied by the user; the array will be empty if no values are supplied by the user.
RLEN	Integer	Input	Holds the length (≥ 0) of RARRAY; zero if no *list* supplied.
RORV	Integer array(1:NS)	Input	Holds the position in RARRAY of the first (or only) element of each item in the *list* of *vector(s)* and *scalar(s).*
RNV	Integer array(1:NS)	Input	Holds the length of each *vector/scalar.*
NS	Integer	Input	The number of items in the *list* of *vector(s)/scalar(s).*
CARRAY	Integer array (1:CLEN)	Input and/ or Output	Holds the integer representation of the characters in the *macro* is supplied by the user; the array will be empty if no *macro* is supplied.
CLEN	Integer	Input	Holds the length (≥ 0) of CARRAY; that is, the length of the *macro*. Is zero if none was supplied.
RMV	Real	Input	RMV is not needed in GLIM but may be needed in future to hold the representation of a real missing value.
IFT	Integer	Input and Output	See below.
IFTA	Integer array (1:2)	Output	See below.

The body of PASS, as implemented by the user may access the actual arguments only as specified above, that is, 'input' only means the value may not be changed, 'output' only means the input value is undefined.

The values of the *vectors* and *scalars* specified by the user are assembled as contiguous portions of the single one-dimensional array, RARRAY. To access the values of an item (vector or scalar) the user needs to know which portion of RARRAY to use. This is provided by RORV, whose *i*th element is the position in RARRAY of the first element of the *i*th item in the *list* specified in the `PASS` statement. The length of the *i*th item is given by the *i*th value of RNV (scalars having length 1).

Since the internal storage of arrays is different for GLIM and FORTRAN it is necessary to request that GLIM makes the necessary conversions of its array structures if PASS routines are to be used that will define such items as FORTRAN multi-dimensional arrays. This is achieved by using the *option* `array=fortran`. Then any GLIM structure that has been declared as an `array` in GLIM will be automatically converted before and after the PASS routine is called.

All arrays may be accessed ***only*** within the bounds given; when RLEN, NS, or CLEN are zero then the corresponding array ***must not be accessed at all***. Any accessing outside the array bounds is expressly forbidden as it can cause unpredictable side-effects which may not become manifest until much later in the session. A possible protection against inadvertent, illegal accessing is to incorporate dynamic array bound checks on the PASS subroutine and any further subroutines it calls; the formal arguments RARRAY and CARRAY must then be declared of size RLEN and CLEN respectively, but the user must ensure that the cases when RLEN = 0, NS = 0, or CLEN = 0 are adequately handled.

The PASS subroutine may call further subroutines or functions, in particular those written to implement the PASS facility. The only GLIM routines that it may call are those given in [17.6]; no other GLIM routines should be called. If further routines are used then it is suggested that, to avoid conflict with existing GLIM routines, they begin with the letters PS (for example, PSXXXX).

No COMMON declarations that occur in GLIM proper may occur in PASS or any routine that PASS calls (except of course in GLIM routines called from PASS).

Care should be taken when using file input in the PASS subroutine for two reasons:

(1) There is the danger when reading files that are also read by GLIM, of unwittingly changing the position of the current record in the file.

(2) Use of fixed channel numbers, and so on, will decrease the portability of the PASS subroutine, an important consideration if the user's PASS facility should prove to be of general applicability. It is suggested that files be opened in GLIM, and the channel number passed to the PASS subroutine as one of the scalars.

Any output should be produced via the GLIM formatting routines described in [17.6] to ensure that transcript, pagination, and channels are correctly maintained.

Apart from these restrictions users may include in PASS whatever facilities they need. Because the PASS subroutine may need to perform several different functions the arguments OPT or NAME may be used to differentiate between these functions. If a *keyword* has been used then the equivalent value of OPT should be found using the character string NAME. Then the PASS subroutine becomes simply a list of calls to further subroutines, each one performing a specific function, together with a jump to the relevant subroutine on the basis of the value of OPT. In this manner it is easy to add further, separate facilities, and the porting of `PASS` facilities between sites is simplified. Because of this consideration, the argument OPT should not be used to distinguish options within a facility (this can de done via the values of RARRAY and CARRAY), but rather to distinguish different facilities.

A mechanism is provided, via the arguments IFT and IFTA, of informing GLIM of faults or break-ins occurring during the execution of PASS. (Such faults may be the detection of invalid input values, incompatible data and so on.) Before calling PASS, GLIM sets IFT to zero. If on return from PASS, IFT is still zero, GLIM assumes that PASS terminated successfully and continues executing further directives; in this case PASS must have performed its function satisfactorily and the values of the *vector* and *macro* must be as expected by the user.

If, however, such a satisfactory completion is not possible because of a fault, PASS must set IFT to any positive value (to inform GLIM that a fault has occurred) and store further information on the fault in IFTA — this aspect is discussed below in [17.4]. When GLIM finds a positive value in IFT it prints the fault message 'problems in PASS subroutine', prints a help message as described in [17.4], and continues as for any other GLIM fault, that is, it tidies up and prompts for more input. As GLIM has no control over the actions of the PASS subroutine it is ***very important*** that fault situations, particularly those that might result in an operating system error, are foreseen by the implementor, and are trapped and returned with fault codes.

A break-in mechanism and floating-point overflow detection may also be implemented in GLIM (see the User Note) whereby the terminal user can request, by hitting a break-in key, that the current directive be aborted without the operating system aborting the program itself. To inform GLIM that the break-in key has been hit the operating system causes the fault indicator to be set to a certain negative value, if that indicator was previously zero. For this reason it is important not to reset the value of IFT unless a fault occurs, nor to set it to a negative value. Instead, the value of IFT must be monitored regularly for a negative value, and when such a value is found then PASS must tidy up and return control to GLIM, that is, the same procedure as would follow the detection of a fault. The monitoring of IFT must be performed as frequently as is necessary to ensure that no appreciable delay occurs between the user hitting the break-in key and PASS returning control to GLIM. Monitoring for floating-point overflow should take place following lines of code where overflow is most likely.

Following a fault, break-in, or floating-point overflow, the values of the *vector* and *macro* are still available to the user, and it is desirable that 'reasonable' values, given the fault, should be stored before the return from PASS.

17.4 The HPASS subroutine

It a fault occurs in PASS then IFT must be set to a positive value, as described above. Further information on the fault can be stored in the array IFTA, which is of length 2. If, on return from PASS, IFT is positive then GLIM prints a fault message. It then calls subroutine HPASS to print a help message, if the HELP facility is switched on. The formal arguments to HPASS are

OPT	Integer	Input	Holds the non-negative value of the *integer* set by the user in the PASS directive, or a negative value if the *integer* was omitted.
IFTA	Integer array(1:2)	Input	Holds the values as set in PASS.

The HPASS subroutine must print messages which explain why the fault occurred, from the information supplied via the arguments. The messages are to be printed via the output routines described in [17.6].

As with the PASS subroutine, it will prove useful to construct HPASS as a list of subroutine calls, one for each of the functions performed by PASS, with a jump to the appropriate routine being made on the basis of the value of OPT. In this manner, it is possible for each function in PASS to set values in IFTA independently of the other functions.

17.5 The EPASS subroutine

This subroutine is called when the user requests option `E` in the `ENVIRONMENT` directive. The EPASS subroutine is intended to provide information to the user on the functions provided by PASS. It prints messages via the output routines described in [17.6]. It has no arguments.

As with PASS and HPASS, it will prove useful to construct EPASS as a list of subroutine calls, each subroutine providing information on one of the PASS functions. EPASS then goes through the list, for each one printing the option value, then calling the appropriate routine to print the information.

Enough information should be provided to inform the user of the purpose of each option, but naturally it may not be possible to provide full details on their use. This must be provided elsewhere; see [17.7].

17.6 The output routines

This section describes the GLIM output routines that may be called from the PASS routines:

PUTREL	NCLBUF
PUTINT	NEWPAG
PUTSTR	NEWLIN
PUTCHR	

The routines use a buffer to hold the line currently being built up and when the line is complete this buffer is printed. The buffer is a CHARACTER*135 array. Characters are inserted into this buffer only by means of the subroutine PUTCHR and the buffer is printed only through the subroutine NEWLIN. The other subroutines do not need to know anything about the structure of the buffer. If PUTCHR attempts to go beyond the end of the buffer, NEWLIN is called to start a new line. NEWLIN also keeps a count of the number of lines printed. The carriage control character is controlled by the subroutine NEWPAG; a call to this ensures that the next line is at the head of a new page.

The function NCLBUF indicates how much space would remain in the buffer after inserting an item of given length.

Numbers are inserted into the buffer by two subroutines, PUTINT and PUTREL, for integer and real values respectively. Parameters to these routines allow the user to specify the field width, the number of decimal places and the justification (left or right). Text is printed using the subroutine PUTSTR. Text is automatically broken on word boundaries.

17.6.1 Specification of the routines

INTEGER FUNCTION NCLBUF(IFWD)

Returns the number of characters that would be left on the current line if a field of width IFWD were inserted (with result < 0 if there is insufficient space for IFWD on the current line). Thus the number of characters remaining on the current line can be obtained by N = NCLBUF(0).

IFWD	Integer	Input	Field width to be checked.

SUBROUTINE NEWLIN(NLINES)

Prints new lines.

NLINES	Integer	Input	Number of lines to be printed. For NLINES zero or negative, the buffer is printed if it is non-empty (otherwise no lines).

SUBROUTINE NEWPAG

Sets the carriage control character to ensure that the next line will be printed at the start of a new page. The carriage control character is then reset automatically by NEWLIN.

No parameters.

SUBROUTINE PUTCHR(CHAR,NREP)

Puts NREP copies of a character into the buffer.

CHAR	Character	Input	Character to be put into the buffer.
NREP	Integer	Input	Number of copies of CHAR to be inserted (none if NREP is zero or negative).

SUBROUTINE PUTINT(INTVAL,IFWDTH,NDGT,RJUST,OVRFLG)

Converts an integer into the appropriate string of characters and then inserts them into the buffer.

INTVAL	Integer	Input	Integer value to be put into the buffer.
IFWDTH	Integer	Input	Field width into which INTVAL is to be placed. For IFWDTH zero or negative a minimum width is used (that is no spaces before or after INTVAL).
NDGT	Integer	Input	This specifies a minimum number of digits to be used to represent INTVAL (with initial zero digits if necessary).
RJUST	Logical	Input	Justification. If .TRUE. then INTVAL is right justified (that is, INTVAL is printed on the right of the field, preceded initially by spaces if necessary).

			If .FALSE. it is left justified (that is, INTVAL is printed on the left of the field, followed by spaces if necessary).
OVRFLG	Logical	Output	Overflow flag: .FALSE. if the integer was successfully inserted into the buffer. .TRUE. if the field width was too small for the integer; the field will then be filled with asterisks.

SUBROUTINE
PUTREL(RELVAL,IFWD,NDECPL,MMODE,RJUST,OVRFLG)

Converts a real number into a string of characters and then inserts them into the buffer.

RELVAL	Real	Input	Real number to be placed into buffer.
IFWD	Integer	Input	Field width into which RELVAL is to be placed. For IFWD zero or negative a minimum width is used (that is, no spaces before or after RELVAL).
NDECPL	Integer	Input	Number of decimal places to be used in the representation of RELVAL. For NDECPL negative, the number of decimal places is taken as zero.
MMODE	Integer	Input	Format If < 0 RELVAL is output in E format, that is, as a decimal number in the interval (-10.0, 10.0) followed by the appropriate exponent of 10. If = 0, F format, otherwise (if F format would cause the number to overflow the field) E format; this setting thus has the same effect as Fortran G format.

			If > 0 F format, the standard decimal representation, with NDECPL decimal places.
RJUST	Logical	Input	Justification If .TRUE. then RELVAL is right justified (i.e. RELVAL is printed on the right of the field, preceded initially by spaces if necessary). If .FALSE. it is left justified (that is, RELVAL is printed on the left of the field, followed by spaces if necessary).
OVRFLG	Logical	Output	Overflow flag: If .FALSE. the number was successfully inserted into the buffer. If .TRUE. the field width was too small for the number; the field will then be filled with asterisks.

SUBROUTINE PUTSTR(STRING,NCHAR)

Inserts a string of characters into the buffer.

STRING	Character*(*)	Input	Array of characters to be put into the buffer.
NCHAR	Integer	Input	The number of characters in a STRING. If NCHAR ≤ 0 no action is taken.

17.7 Further considerations

Documentation: If the facilities implemented in the PASS subroutine are to be made available to other users then it is essential that adequate documentation for them is provided, especially if a wider audience is envisaged (for example, through the GLIM Newsletter). As with all documentation it should cover the following points:

(1) the purpose of the facility;

(2) the input values required;

(3) the output values produced;

(4) the general method used;

(5) the fault conditions that can arise;

(6) references to further reading.

FORTRAN All code should be produced in standard FORTRAN 77, and any necessary deviations from the standard should be clearly documented. Such deviations, especially undocumented ones, may cause problems when code is ported to another machine.

Linking: To make the linking of the PASS routine to GLIM as efficient as possible (especially during testing when relinking may occur often) it may be helpful to construct GLIM (without the PASS facility) as a separate library module to be linked later with the latest PASS module. The details of this will depend on the operating system.

17.8 Example

The following is a portion of the PASS code supplied with the standard version of GLIM for illustrative purposes. This will be used to illustrate the principles involved in implementing a PASS routine. The particular example is that of providing a matrix multiplication routine using GLIM array structures. An example of its use within a macro is as follows:

```
$macro (args=A,B,C local=NRA,NCA,NCB) MMULT !
$num NRA NCA NCB $!
$cal NRA=%dim(A,1) : NCA=%dim(A,2) : NCB=%dim(B,2) $!
$array C NRA,NCB $!
$pass(a=f) %matmult A,B,C,NRA,NCA,NCB $!
$endmac

!  example usage

$array X 2;3 : Y 3;4 $
$ass X=1,0,1,1,1,0 : Y=1,2,3,4,2,3,4,5,3,4,5,6 $
$tprint X $
           1        2        3
```

```
    1  1.000  0.000  1.000
    2  1.000  1.000  0.000
$tprint Y $
          1      2      3      4
    1  1.000  2.000  3.000  4.000
    2  2.000  3.000  4.000  5.000
    3  3.000  4.000  5.000  6.000
$use MMULT X Y Z $
$tprint Z$
           1        2        3        4
    1   4.000    6.000    8.000   10.000
    2   3.000    5.000    7.000    9.000
```

Notes:

(1) Properties of the arrays (such as the number of levels of a dimension) are accessed at the GLIM level and passed to the routine via scalars.

(2) The *keyword* `%matmult` is passed to choose the particular routine. An integer value of 3 would have the same result as `%matmult` is the third keyword in the NAMES array.

(3) The lengths of the items passed in the *list* are identified by RNV.

(4) The portions of RARRAY for particular arrays are identified using RORV in routine PSMMCH and passed to routine PSMMLT.

(5) The final routine PSMMLT declares the structures as two-dimensional arrays. For this to work correctly the *option* `array=fortran` has been used.

```
C*************************************************************
C-- GLIM 4  PASS ROUTINES
C--------------------------------
C
      SUBROUTINE PASS(OPT,NAME,RARRAY,RLEN,RORV,RNV,NS,
     -                   CARRAY,CLEN,RMV,IFT,IFTA)
C
C     PASS routines  M.Green  Centre for Applied Statistics
C                             University of Lancaster
C
C        OPT    1 = run demonstration program
C               2 = fast search for subfile
C               3 = matrix multiplication
C--parameters
      INTEGER OPT,CLEN,RLEN,NS,IFT,IFTA(2)
      INTEGER RORV(NS),RNV(NS)
```

```
      REAL RARRAY(RLEN),RMV
      CHARACTER*8 NAME
      INTEGER CARRAY(CLEN)
C
C--local
      INTEGER MXN
C============================================================
C  if new keywords are added MXN should be increased
C============================================================
      PARAMETER (MXN=3)
      INTEGER I,NOP(MXN)
      CHARACTER*8 NAMES(MXN)
C=================================================================
C  if new keywords are added NOP should be extended and new
C  keywords added to NAMES list
C=================================================================
      DATA NOP/ 1,2,3 /
      DATA NAMES / '%RUNDEMO','%SUBSRCH','%MATMULT' /
C
C ------------------------------
C
      IF(IFT.NE.0) RETURN
      IF(OPT.EQ.0) THEN
C  search for keyword
        DO 100 I=1,MXN
        IF(NAME.EQ.NAMES(I)) GO TO 150
  100   CONTINUE
        GO TO 999
  150   OPT=NOP(I)
      ENDIF
C===============================================================
C  if new routines are added following test should be amended
C===============================================================
      IF(OPT.GT.3) GOTO 999
C===============================================================
C  if new routines are added computed GOTO should be extended
C===============================================================
      GO TO (1001,1002,1003),OPT
 1001 CALL ....
....
 1003 CALL PSMMCH(RARRAY,RLEN,RORV,RNV,NS,IFT,IFTA)
      GOTO 900
  999 IFT=1
      IFTA(1)=-1
  900 RETURN
      END
C***************************************************************
C-- GLIM 4  PASS ROUTINE
C------------------------------
C
      SUBROUTINE PSMMCH(RARRAY,RLEN,RORV,RNV,NS,IFT,IFTA)
C
C   performs checks on structures and calls matrix
C   multiplication routine
C
```

```
C—parameters
      INTEGER RLEN,NS,RORV(NS),RNV(NS),IFT,IFTA(2)
      REAL RARRAY(RLEN)
C
C—local
      INTEGER NRA,NCA,NCB
      INTEGER I,IA,IB,IC
C
C———————————————————————————————
C
      IF(NS.NE.6) GOTO 900
C  number of rows of A
      IF(RNV(4).EQ.1) THEN
        I=RORV(4)
        NRA=RARRAY(I)
      ELSE
        GOTO 900
      ENDIF
C  number of columns of A
      IF(RNV(5).EQ.1) THEN
        I=RORV(5)
        NCA=RARRAY(I)
      ELSE
        GOTO 900
      ENDIF
C  number of columns of B
      IF(RNV(6).EQ.1) THEN
        I=RORV(6)
        NCB=RARRAY(I)
      ELSE
        GOTO 900
      ENDIF
C  check sizes of arrays compatible
      IF(RNV(1).NE.(NRA*NCA)) GOTO 901
      IF(RNV(2).NE.(NCA*NCB)) GOTO 901
      IF(RNV(3).NE.(NRA*NCB)) GOTO 901
      IA=RORV(1)
      IB=RORV(2)
      IC=RORV(3)
      CALL PSMMLT(RARRAY(IA),RARRAY(IB),RARRAY(IC),NRA,NCA,NCB)
      RETURN
  900 IFT=1
      IFTA(1)=1
      RETURN
  901 IFT=1
      IFTA(1)=2
      RETURN
      END
C*****************************************************************
C— GLIM 4  PASS ROUTINE
C———————————————————————————————
C
      SUBROUTINE PSMMLT(A,B,C,NRA,NCA,NCB)
C
C  performs matrix multiplication C = AB
```

```
C
C—parameters
      INTEGER NRA,NCA,NCB
      REAL A(NRA,NCA),B(NCA,NCB),C(NRA,NCB)
C
C—local
      INTEGER I,J,K
      DOUBLE PRECISION VAL
C
C  ----------------------------------
C
      DO 100 I=1,NRA
      DO 200 J=1,NCB
      VAL=0.0
      DO 300 K=1,NCA
  300 VAL=VAL+DBLE(A(I,K))*DBLE(B(K,J))
C
C  check for break-in or overflow
C
      IF (IFT.LT.0) GOTO 900
  200 C(I,J)=SNGL(VAL)
  100 CONTINUE
      RETURN
C
C   check for overflow IFT=-10
C
  900 IF (IFT.EQ.-10) THEN
        IFT=1
        IFTA(1)=3
      ENDIF
      RETURN
      END
C***********************************************************************
C— GLIM 4  PASS ROUTINES
C----------------------------------
C
      SUBROUTINE HPASS(OPT,IFTA)
C
C   This routine is called by the HELP routine to print
C   information on a fault occurring in the PASS subroutine
C
C—parameters
      INTEGER OPT,IFTA(2)
C
C—subroutines
      EXTERNAL PUTSTR
C
C      ----------------------------------
C
      IF(IFTA(1).EQ.-1) THEN
        CALL PUTSTR('Invalid option number',21)
      ELSE
        GOTO (101,102,103),OPT
      ENDIF
      GO TO 1000
```

```
  101 ....
....
C  matrix multiplication
  103 GOTO (111,112,113),IFTA(1)
  111 CALL PUTSTR('Error in parameters',19)
      GO TO 1000
  112 CALL PUTSTR('Incompatible matrices',21)
      GOTO 1000
  113 CALL PUTSTR('Floating-point overflow in ',27)
      CALL PUTSTR('matrix multiplication',21)
 1000 CALL NEWLIN(0)
      RETURN
      END
```

Bibliography

Adena, M. and Wilson, S.R. (1982) *Generalized Linear Models in Epidemiological Research: Case Control Studies.* INSTAT, Sydney.

In addition to a description of case control analyses from a GLM viewpoint this includes worked GLIM examples and macros. However, the new facilities in GLIM4, particularly the `ELIMINATE` directive, will make it much simpler to fit the relevant models.

Agresti, A. (1984) *Analysis of Ordinal Categorical Data.* Wiley, New York.

A detailed treatment of models for ordinal categorical data with an emphasis on log-linear and logit models. Some of the simpler methods can be handled in the GLIM framework although no details are given in the book.

Agresti, A. (1990) *Categorical Data Analysis.* Wiley, New York.

A thorough modern treatment of categorical data analysis based logit and log-linear models. GLIM4 can be used to fit most of the models described in this book.

Aitkin, M.A., Anderson, D.A., Francis, B.J., and Hinde, J.P. (1989) *Statistical Modelling in GLIM.* Oxford University Press.

A detailed practical description of statistical modelling in GLIM. The presentation is through worked examples reflecting the authors' interactive approach to modelling. The material covered includes models for normal, binomial, multinomial, count, and survival data.

Andersen, E.B. (1990) *The Statistical Analysis of Categorical Data.* Springer-Verlag, Berlin.

This covers log-linear models for multi-way contingency tables, logit models for categorical responses, interaction models, correspondence analysis, and latent structure models.

Atkinson, A. (1985) *Plots, Transformations and Regression: An Introduction to Graphical Methods of Diagnostic Regression Analysis.* Oxford University Press.

A detailed treatment of regression diagnostics primarily for the normal regression model. A brief reference is made to similar techniques for generalized linear models and a chapter is devoted to goodness of link tests.

Bishop, Y.M.M., Fienberg, S.E., and Holland, P.W. (1975) *Discrete Multivariate Analysis: Theory and Practice.* M.I.T. Press, Cambridge, Mass.

A classic text on models for categorical data.

Collett, D. (1991) *Modelling Binary Data.* Chapman and Hall, London.

A practical and comprehensive text containing many useful examples on the statistical modelling of binary data. Contains chapters on model diagnostics, extra-variation and case-control studies. An appendix of GLIM macros is provided.

Cox, D.R. and Oakes, D. (1984) *Analysis of Survival Data.* Chapman and Hall, London.

A concise treatment of models for duration data with applications to both medical survival and industrial life-testing data. The presentation is fairly mathematical and although illustrated by examples few computational details are included.

Cox, D.R. and Snell, E.J. (1989) *The Analysis of Binary Data,* (2nd Ed). Chapman and Hall, London.

A detailed description of the analysis of binary response data using the logistic regression model with many illustrative examples. This includes a section on residuals and diagnostics which serves as a good introduction to this topic for non-normal models.

Crowder, M.J., Kimber, A.C., Smith, R.L., and Sweeting, T.J. (1991) *Statistical Analysis of Reliability Data*. Chapman and Hall, London.

This presents techniques suitable for reliability data with detailed analyses of several datasets. Many of the methods can be handled by GLIM either as standard models or through macros.

Decarli, A., Francis, B.J., Gilchrist, R., and Seeber, G.U.H. (Eds) (1989) *Statistical Modelling: Proceedings of the GLIM89 and the 4th International Workshop on Statistical Modelling*. Lecture Notes in Statistics, **57**. Springer-Verlag, Berlin.

A collection of papers on theoretical and applied aspects of generalized linear modelling from the third GLIM conference.

Dobson, A. (1990) *An Introduction to Generalized Linear Models*. Chapman and Hall, London.

This describes the GLM framework starting from regression and analysis of variance and moving on to logistic regression for binary data and log-linear models for counted data. The techniques are illustrated using examples together with GLIM analyses.

Draper, N. and Smith, H. (1981) *Applied Regression Analysis*. (2nd Ed). Wiley, New York.

A comprehensive description of the theory and application of the normal theory regression model.

Gilchrist, R. (Ed.) (1982) *GLIM 82: Proceedings of the International Conference on Generalised Linear Models*. Springer-Verlag, New York.

Papers from the first conference devoted to generalized linear models and GLIM.

Gilchrist, R., Francis, B. and Whittaker, J. (Eds) (1985) *Proceedings of the International Conference on Generalised Linear Models*. Springer-Verlag, New York.

Papers from the second GLIM conference.

GLIM Newsletters (1979–) Numerical Algorithms Group, Oxford.

Twice-yearly newsletters containing articles on the use of GLIM. Many of the articles include useful macros for non-standard analyses.

Green, P.J. (1984) Iteratively reweighted least squares for maximum likelihood estimation and some robust and resistant alternatives (with discussion). *J. R. Statist.Soc. B*, **46**, 149–192.

A detailed discussion of the iterative algorithm at the heart of GLIM.

Hastie, T. and Tibshirani, R.J. (1990) *Generalized Additive Models*. Chapman and Hall, London.

Generalized additive models are an extension of the class of generalized linear models to incorporate non-parametric functions of the explanatory variables. While these models cannot be directly fitted in GLIM a PASS routine has been written, see GLIM Newsletter 22.

Healy, M.J. (1988) *An Introduction to GLIM*. Oxford University Press.

A very elementary introduction to GLIM and generalized linear models. Contains some useful simple examples.

Hinkley, D.V., Reid, N., and Snell, E.J. (Eds) (1990) *Statistical Theory and Modelling*. Chapman and Hall, London.

This collection of review articles in honour of Sir David Cox includes a wide ranging but succinct discussion of generalized linear models, a treatment of general regression diagnostics, and chapters on applied statistics, quasi-likelihood, and lifetable analysis.

Lindsey, J.K. (1989) *Analysis of Categorical Data using GLIM*. Lecture Notes in Statistics **56**. Springer-Verlag, Berlin.

This book is aimed at the social and biomedical sciences and contains extensive examples on contingency table analysis giving both GLIM commands and output. An appendix also lists a number of GLIM macros for particular models.

Lindsey, J.K. (1992) *The analysis of Stochastic Processes using GLIM*. Lecture Notes in Statistics **72**. Springer-Verlag, Berlin.

This book is also aimed at the social and biomedical sciences. It gives many examples of stochastic models fitted using GLIM. These include markov chains, survival curves, growth curves, simple time series and repeated measures. An appendix contains example data sets and GLIM macros for fitting these models.

McCullagh, P., and Nelder, J.A. (1989) *Generalized Linear Models,* (2nd Ed). Chapman and Hall, London.

Definitely the bible on the theory of generalized linear models. This book includes many examples with detailed analyses but no explicit reference to the necessary GLIM commands. Many non-standard problems and extensions are also considered.

Nelder, J.A. and Wedderburn, R.W.M. (1972) Generalized linear models. *J.R. Statist. Soc., A*, **135**, 370–384.

The original paper introducing the concept of generalized linear models.

Pregibon, D. (1981) Logistic regression diagnostics. *Ann. Statist.*, **9**, 705–724.

An introduction to diagnostics for non-normal models.

Wedderburn, R.W.M. (1974) Quasi-likelihood functions, generalized linear models and the Gauss-Newton method. *Biometrika*, **64**, 439–447.

This paper provides the theoretical background for the wide class of models which can be fitted by GLIM through user-defined models, in which the only important elements in the fitting process are the link function and the mean–variance relationship.

Whittaker, J. (1990) *Graphical Models in Applied Multivariate Statistics*. Wiley, Chichester.

Describes the ideas of graphical models as applied to multi-way contingency tables, including a helpful discussion on model selection.

References

Agresti, A. (1984) *Analysis of Ordinal Categorical Data*. Wiley, New York.

Agresti, A. (1990) *Categorical Data Analysis*. Wiley, New York.

Aitkin, M. (1987) Modelling variance heterogeneity in normal regression using GLIM. *Appl. Statist.*, **36,** 332-339.

Aitkin, M., Anderson, D., Francis, B., and Hinde, J. (1989) *Statistical Modelling in GLIM*. Oxford University Press.

Aitkin, M., and Francis, B. (1980) A GLIM macro for fitting the exponential or Weibull distribution to censored survival data. *GLIM Newsletter,* **2**, 19-25.

Anderson, C.W., and Ray, W.D. (1975) Improved maximum likelihood estimators for the gamma distribution. *Communications in Statistics*, **4**, 437-488.

Armitage, P.A., and Berry, G. (1987) *Statistical Methods in Medical Research*. Blackwell, Oxford.

Aston, C.E., and Wilson, S.R. (1984) Comment on M.B. Brown and C. Fuchs, On maximum likelihood estimation in sparse contingency tables. *Computational Statistics and Data Analysis*, **2**, 71-77.

Atkinson, A.C. (1985) *Plots, Transformations and Regression*. Oxford Statistical Science Series, Oxford.

Barndorff-Nielsen, O.E. (1988) *Parametric Statistical Models and Likelihood*. Lecture Notes in Statistics **50**, Springer-Verlag, New York.

Becker, M.P. (1990) Square contingency tables having ordered categories and GLIM. *GLIM Newsletter*, **20**, 22-31.

Bennett, S., and Whitehead, J. (1981) Fitting logistic and log-logistic regression models to censored data using GLIM. *GLIM Newsletter,* **4**, 12-19.

Bishop, Y.M.M., Fienberg S.E., and Holland, P.W. (1975) *Discrete Multivariate Analysis*. M.I.T. Press, Cambridge, Mass.

Bliss, C. (1970), *Statistics in Biology, Vol II*, McGraw-Hill, New York.

Bliss, C.I. (1935) The calculation of the dosage-mortality curve. *Ann. Appl. Biol.*, **22**, 134-67.

Borre, K., and Lauritzen, S.L. (1989) Some geometric aspects of adjustment. In *Festschrift to Torben Karup*. Kejlso, K. et al. (Eds), Geodaetisk Institut, Meddelelse No 58, Copenhagen.

Breslow, N.E., and Day, N.E. (1980) *Statistical Methods in Cancer Research. 1. The Analysis of Case-Control Studies*. I.A.R.C, Lyon.

British National Lymphoma Investigation (1975) Value of prednisone in combination chemotherapy of stage IV Hodgkins disease. *Br. Med. J. iii,* 413.

Brookes, B.C., and Dick, W.F.L. (1951) *Introduction to Statistical Method*. Heinemann, London.

Burn, R. (1982) Log-linear models with composite link functions in genetics. In *GLIM82, Proceedings of the International Conference on Generalised Linear Models*, Lecture Notes in Statistics **14**. Gilchrist, R. (Ed), 144-154, Springer-Verlag, New York.

Clarke, G.M., and Cooke, D. (1983) *A Basic Course in Statistics (2nd edn)*, 166. Edward Arnold, London.

Cook, R.D. (1977) Detection of influential observations in linear regression. *Technometrics*, **19**, 15-18.

Copenhaver, T.W., and Mielke, P.W. (1977) Quality analysis: a quantal assay refinement.

Biometrics, **33**, 175-186.

Cox, D.R. (1967) *Renewal Theory*. Chapman and Hall, London.

Cox, D.R. (1972) Regression models and life tables (with discussion). *J.R.Statist.Soc. B*, **34**, 187-220.

Cox, D.R. (1981) Statistical analysis of time series, some recent developments. *Scand. J. Statist.*, **8**, 93-115.

Cox, B.D., (1987) *The HPRT Health and Lifestyle Survey*. HPRT, Cambridge.

de Falguerolles, A. (1989) Square contingency tables, GLIM and correspondence analysis, *GLIM Newsletter*, **19**, 16-21.

Edwards, D., and Kreiner, S (1983) The analysis of contingency tables by graphical models. *Biometrika*, **70**, 553-565.

Efron, B. (1986) Double exponential families and their use in generalised linear regression. J. *Amer. Statist. Assoc.*, **81**, 709-721.

Finney, D.J. (1971) *Probit Analysis*, Cambridge University Press.

Firth, D. (1987) On the efficiency of quasi-likelihood estimation. *Biometrika*, **74**, 233-45.

Firth, D., and Treat, B.R. (1988) Square contingency tables and GLIM. *GLIM Newsletter*, **16**, 16-20.

Firth, D. (1990) Generalized linear models. In *Statistical Theory and Modelling*. Hinkley, D.V., Reid, N., and Snell, E.J. (Eds). Chapman and Hall, London.

Forcina, A. (1986) Correlated observations with Normal error. *GLIM Newsletter*, **12**, 31-32.

Forni, A. and Sciame, A.(1980), Chromosome and biochemical studies in women occupationally exposed to lead, *Arch. Env. Health*, **35**, 139-145.

Francis, B. (1991) ASR88 - A remark on AS121, the trigamma function. *Appl. Statist.*, **40**, 514.

Gehan, E.A. (1965) A generalized Wilcoxon test for comparing arbitrarily singly censored samples. *Biometrika*, **52**, 203-223.

Gentleman, W.M. (1974) Algorithm AS75: Basic procedure for large sparse or weighted linear least squares problems. *Appl. Statist*, **23**, 448-454.

Gilchrist, R. (1981) Estimation of the parameters of the gamma distribution by means of a maximal invariant. *Communications in Statistics*, **11**, 1095-1110.

Gilchrist, R. (1987) Proposal of the note of thanks to Jørgensen, B. op cit.

Gilchrist, R., and Scallan, A.J. (1984). Parametric link functions in generalised linear models. In *COMPSTAT 84*, 203-8. Physica Verlag, Wein.

Godambe, V.P., and Heyde, C.C. (1988) Quasi-likelihood and optimal estimation. *Int. Statist Review*, **55**, 231-44.

Goldstein, H. (1987) *Multilevel Models in Educational and Social Research*, Griffin, London.

Green, P.J. (1984) Iteratively reweighted least squares for maximum likelihood estimation and some robust and resistant alternatives (with discussion). *J. R. Statist. Soc.*, B, **46**, 149-192.

Green, P.J. and Yandell, B.S. (1985) Semi-parametric generalized linear models. In *Generalized Linear Models*. Lecture Notes in Statistics **32**, 44-55. Gilchrist, R., Francis, B, and Whittaker, J. (Eds). Springer -Verlag, New York.

Hastie, T.J., and Tibshirani, R.J. (1990) *Generalized Additive Models*, Chapman and Hall, London.

Healy, M.J.R. (1972) Statistical analysis of radio-immunoassay data *Biochemical J.* **130**, 207-210.

Hutchinson, D. (1985) Ordinal variable regression using the McCullagh proportional odds model. *GLIM Newsletter*, **9**, 9-17.

John, J.A., and Quenouille, M.H. (1977), *Experiments: Design and Analysis*, p250. Griffin, London.

Jørgensen, B. (1987). Exponential dispersion models. (with discussion). *J. R. Statist. Soc., B,* **49**, 127-162.

Kendall, M., and Stuart, A. (1967) *The Advanced Theory of Statistics*, Vol 2. Griffin, London.

Leimer, H.G., and Rudas, T. (1989) A note on the conversion between GLIM and BMDP type log-linear parameters. *GLIM Newsletter*, **19**, 47.

Liang, K.Y.,and Zeger, S.L. (1986) Longitudinal data analysis using generalized linear models. *Biometrika,* **73**, 13-22.

Lindsey, J.K. (1989) Analysis of Categorical Data using GLIM. Lecture Notes in Statistics **56**, Springer-Verlag, Berlin.

Ludlam, H.A., Nwachukwu, B., Noble, W.C., Swan, A.V., and Phillips, I. (1989) The preservation of micro-organisms in biological specimens stored at -70°C *J. App. Bact.*, **67**, 417-423.

Macleod, A.J. (1989) AS245: A robust and reliable algorithm for the logarithm of the gamma function. *Appl. Statist.,* **38**, 397-401.

McCullagh, P. (1985) Macro to calculate the approximate conditional cumulants of Pearson's goodness-of-fit statistics for binomial and Poisson response models, *GLIM Newsletter*, **10**, 41-42.

McCullagh, P., and Nelder, J.A. (1989) *Generalized Linear Models (2nd Edn).* Chapman and Hall, London.

Nelder, J.A. (1985) Quasi-likelihood and GLIM. In *Generalised Linear Models*. Lecture Notes in Statistics **32**, 120-127. Gilchrist, R., Francis, B., and Whittaker, J. (Eds). Springer-Verlag, Berlin.

Nelder, J.A., and Pregibon, D. (1987) An extended quasi-likelihood function. *Biometrika*, **74**, 221-32.

Nelder, J.A., and Wedderburn, R.W.N. (1972) Generalized linear models. *J. R. Statist. Soc., A,* **135**, 370-384.

Payne, C.D. (1987) *The GLIM System Manual (2nd Edn).* The Numerical Algorithms Group, Oxford.

Pearson, E.S., and Hartley, H.O. (1958) *Biometrika Tables for Statisticians.* Vol 1. Cambridge University Press.

Pregibon, D. (1980) Goodness of link tests for generalized linear models. *Appl. Statist.*, **29**, 15-24.

Pregibon, D. (1981) Logistic regression diagnostics. *Ann.Statist.*, **9**, 705-724.

Pringle, R.M., and Rayner, A.A. (1971) *Generalized Inverse Matrices.* Griffin, London.

Rao, C.R. (1973) *Linear Statistical Inference*, Wiley, New York.

Richards, F.S.G. (1961) A method of maximum likelihood estimation. *J. R. Statist. Soc., B*, **23**, 469-75.

Roger, J.H., and Peacock, S.B. (1982) Fitting the scale as a GLIM parameter for Weibull, extreme value, logistic and log-logistic regression models with censored data. *GLIM Newsletter,* **6**, 30-37.

Roger, J.H. (1985) Using factors when fitting the scale parameter to Weibull and other survival regression models with censored data. *GLIM Newsletter,* **11**, 14-15.

Scallan, A. (1985) Fitting autoregressive processes in GLIM. *GLIM Newsletter,* **9**, 17-22.

Scallan, A.J., and Evans, S.J.W. (1989) Fitting truncated distributions to grouped data using GLIM. *GLIM Newsletter,* **18**, 17-20.

Scallan, A.J., and Evans, S.J. (1989) Application of truncated and mixture distributions to comparisons of birthweight. In *Statistical Modelling.* Lecture Notes in Statistics, **57**, 270-277. Decarli, A., Francis, B.F., Gilchrist, R. and Seeber, G.U.H. (Eds). Springer-Verlag, New York.

Scallan, A. J., and Evans, S. J. (1992) Maximum Likelihood estimation for a Laplace/Normal mixture distribution. *The Statistician*, **3**, to appear.

Scallan, A., Gilchrist, R., and Green, M. (1984) Fitting parametric link functions in generalized linear models. *Comp. Statist. and Data Analysis,* **2**, 37-49.

Schneider, B.E. (1978) AS121: The trigamma function. *Appl. Statist.*, **27**, 97-99.

Searle, S.E. (1971) *Linear Models.* Wiley, New York.

Shea, B.L. (1988) AS239: Chi-squared and incomplete gamma integral. *Appl. Statist.*, **37**, 466-473.

Snedecor, G.W. (1956) *Statistical Methods.* Iowa State University Press, Ames, USA.

Swan, A.V., and Gomes, U. (1989) Regression with a continuous y variable requiring transformation. *The Professional Statistician*, **8**, 6-12.

Swan, A.V. (1985) Statistical methods, Chapter 9 in *Oxford Textbook of Public Health*, Vol III. Holland, W.W., Detels, R., and Knox, G. (Eds). Oxford University Press.

Swan, A.V., Murray, M. and Jarrett, L. (1991) *Smoking Behaviour from Pre-Adolescence to Young Adulthood.* Avebury, Gower Publishing Company, Aldershot and Brookfield, USA.

Thompson, R., and Baker, R.J. (1981) Composite link functions in generalized linear models, *Appl. Stats.*, **30**, 125-131.

Titterington, D.M., Smith, A.F.M., and Makov, U.E. (1985) *Statistical Analysis of Finite Mixture Distributions.* Wiley, Chichestser.

Verbeek, A. (1989) The compactification of generalized linear models. In *Statistical Modelling.* Lecture Notes in Statistics **57**, 314-327. Decarli, A., Francis, B. J., Gilchrist, R. and Seeber, G.U.H. (Eds). Springer-Verlag, New York.

Wardlaw, A.C. (1985) *Practical Statistics for Experimental Biologists.* Wiley, New York.

Wedderburn, R.W.M. (1974a) Generalised linear models specified in terms of constraints. *J. R. Statist. Soc., B,* **36**, 449-454.

Wedderburn, R.W.M. (1974b) Quasi-likelihood functions, generalised linear models and the Guass-Newton method. *Biometrika,* **64**, 439-447.

Wedderburn, R.W.M. (1976) On the existence and uniqueness of the maximum likelihood for certain generalised linear models. *Biometrika*, **63**, 27-32.

Whittaker, J. (1990) *Graphical Models in Applied Multivariate Statistics*, Wiley, New York.

Wiles, C.M., Clarke, C.R.A., Irwin, H.P., Edgar, E.F., and Swan,A.V. (1986) Hyperbaric oxygen in multiple sclerosis: a double blind trial. *B.Med.J* , **292**, 367-371.

Wilkinson, G.N., and Rogers, C.E. (1973) Symbolic description of factorial models for analysis of variance. *Appl. Statist.*, **22**, 392-399.

Williams, D.A. (1987) Generalized linear model diagnostics using the deviance and single case deletions. *Appl. Statist.*, **36**, 181-191.

Zambello de Pinho, S. (1989) The analysis of split plot designs in GLIM. *GLIM Newsletter*, **19**, 39-44.

Zeger, S.L. (1988) A regression model for time series of counts. *Biometrika*, **75**, 621-629.

Zeger, S.L., and Liang, K.Y. (1986) Longitudinal data analysis for discrete and continuous outcomes, *Biometrics,* **42**, 121-30.

Index